T0188824

Lecture Notes in Computer Science 12950

More information about this subseries at http://www.springer.com/series/7407

Osvaldo Gervasi · Beniamino Murgante ·
Sanjay Misra · Chiara Garau ·
Ivan Blečić · David Taniar ·
Bernady O. Apduhan · Ana Maria A. C. Rocha ·
Eufemia Tarantino · Carmelo Maria Torre (Eds.)

Computational Science and Its Applications – ICCSA 2021

21st International Conference
Cagliari, Italy, September 13–16, 2021
Proceedings, Part II

Springer

Editors
Osvaldo Gervasi ⓘ
University of Perugia
Perugia, Italy

Sanjay Misra ⓘ
Covenant University
Ota, Nigeria

Ivan Blečić ⓘ
University of Cagliari
Cagliari, Italy

Bernady O. Apduhan
Kyushu Sangyo University
Fukuoka, Japan

Eufemia Tarantino ⓘ
Polytechnic University of Bari
Bari, Italy

Beniamino Murgante ⓘ
University of Basilicata
Potenza, Potenza, Italy

Chiara Garau ⓘ
University of Cagliari
Cagliari, Italy

David Taniar ⓘ
Monash University
Clayton, VIC, Australia

Ana Maria A. C. Rocha ⓘ
University of Minho
Braga, Portugal

Carmelo Maria Torre ⓘ
Polytechnic University of Bari
Bari, Italy

ISSN 0302-9743   ISSN 1611-3349 (electronic)
Lecture Notes in Computer Science
ISBN 978-3-030-86959-5   ISBN 978-3-030-86960-1   (eBook)
https://doi.org/10.1007/978-3-030-86960-1

LNCS Sublibrary: SL1 – Theoretical Computer Science and General Issues

This Springer imprint is published by the registered company Springer Nature Switzerland AG
The registered company address is: Gewerbestrasse 11, 6330 Cham, Switzerland

Preface

These 10 volumes (LNCS volumes 12949–12958) consist of the peer-reviewed papers from the 21st International Conference on Computational Science and Its Applications (ICCSA 2021) which took place during September 13–16, 2021. By virtue of the vaccination campaign conducted in various countries around the world, we decided to try a hybrid conference, with some of the delegates attending in person at the University of Cagliari and others attending in virtual mode, reproducing the infrastructure established last year.

This year's edition was a successful continuation of the ICCSA conference series, which was also held as a virtual event in 2020, and previously held in Saint Petersburg, Russia (2019), Melbourne, Australia (2018), Trieste, Italy (2017), Beijing. China (2016), Banff, Canada (2015), Guimaraes, Portugal (2014), Ho Chi Minh City, Vietnam (2013), Salvador, Brazil (2012), Santander, Spain (2011), Fukuoka, Japan (2010), Suwon, South Korea (2009), Perugia, Italy (2008), Kuala Lumpur, Malaysia (2007), Glasgow, UK (2006), Singapore (2005), Assisi, Italy (2004), Montreal, Canada (2003), and (as ICCS) Amsterdam, The Netherlands (2002) and San Francisco, USA (2001).

Computational science is the main pillar of most of the present research on understanding and solving complex problems. It plays a unique role in exploiting innovative ICT technologies and in the development of industrial and commercial applications. The ICCSA conference series provides a venue for researchers and industry practitioners to discuss new ideas, to share complex problems and their solutions, and to shape new trends in computational science.

Apart from the six main conference tracks, ICCSA 2021 also included 52 workshops in various areas of computational sciences, ranging from computational science technologies to specific areas of computational sciences, such as software engineering, security, machine learning and artificial intelligence, blockchain technologies, and applications in many fields. In total, we accepted 494 papers, giving an acceptance rate of 30%, of which 18 papers were short papers and 6 were published open access. We would like to express our appreciation for the workshop chairs and co-chairs for their hard work and dedication.

The success of the ICCSA conference series in general, and of ICCSA 2021 in particular, vitally depends on the support of many people: authors, presenters, participants, keynote speakers, workshop chairs, session chairs, organizing committee members, student volunteers, Program Committee members, advisory committee members, international liaison chairs, reviewers, and others in various roles. We take this opportunity to wholehartedly thank them all.

We also wish to thank Springer for publishing the proceedings, for sponsoring some of the best paper awards, and for their kind assistance and cooperation during the editing process.

We cordially invite you to visit the ICCSA website https://iccsa.org where you can find all the relevant information about this interesting and exciting event.

September 2021

Osvaldo Gervasi
Beniamino Murgante
Sanjay Misra

Welcome Message from the Organizers

COVID-19 has continued to alter our plans for organizing the ICCSA 2021 conference, so although vaccination plans are progressing worldwide, the spread of virus variants still forces us into a period of profound uncertainty. Only a very limited number of participants were able to enjoy the beauty of Sardinia and Cagliari in particular, rediscovering the immense pleasure of meeting again, albeit safely spaced out. The social events, in which we rediscovered the ancient values that abound on this wonderful island and in this city, gave us even more strength and hope for the future. For the management of the virtual part of the conference, we consolidated the methods, organization, and infrastructure of ICCSA 2020.

The technological infrastructure was based on open source software, with the addition of the streaming channels on YouTube. In particular, we used Jitsi (jitsi.org) for videoconferencing, Riot (riot.im) together with Matrix (matrix.org) for chat and ansynchronous communication, and Jibri (github.com/jitsi/jibri) for streaming live sessions to YouTube.

Seven Jitsi servers were set up, one for each parallel session. The participants of the sessions were helped and assisted by eight student volunteers (from the universities of Cagliari, Florence, Perugia, and Bari), who provided technical support and ensured smooth running of the conference proceedings.

The implementation of the software infrastructure and the technical coordination of the volunteers were carried out by Damiano Perri and Marco Simonetti.

Our warmest thanks go to all the student volunteers, to the technical coordinators, and to the development communities of Jitsi, Jibri, Riot, and Matrix, who made their terrific platforms available as open source software.

A big thank you goes to all of the 450 speakers, many of whom showed an enormous collaborative spirit, sometimes participating and presenting at almost prohibitive times of the day, given that the participants of this year's conference came from 58 countries scattered over many time zones of the globe.

Finally, we would like to thank Google for letting us stream all the live events via YouTube. In addition to lightening the load of our Jitsi servers, this allowed us to record the event and to be able to review the most exciting moments of the conference.

Ivan Blečić
Chiara Garau

Organization

ICCSA 2021 was organized by the University of Cagliari (Italy), the University of Perugia (Italy), the University of Basilicata (Italy), Monash University (Australia), Kyushu Sangyo University (Japan), and the University of Minho (Portugal).

Honorary General Chairs

Norio Shiratori	Chuo University, Japan
Kenneth C. J. Tan	Sardina Systems, UK
Corrado Zoppi	University of Cagliari, Italy

General Chairs

Osvaldo Gervasi	University of Perugia, Italy
Ivan Blečić	University of Cagliari, Italy
David Taniar	Monash University, Australia

Program Committee Chairs

Beniamino Murgante	University of Basilicata, Italy
Bernady O. Apduhan	Kyushu Sangyo University, Japan
Chiara Garau	University of Cagliari, Italy
Ana Maria A. C. Rocha	University of Minho, Portugal

International Advisory Committee

Jemal Abawajy	Deakin University, Australia
Dharma P. Agarwal	University of Cincinnati, USA
Rajkumar Buyya	University of Melbourne, Australia
Claudia Bauzer Medeiros	University of Campinas, Brazil
Manfred M. Fisher	Vienna University of Economics and Business, Austria
Marina L. Gavrilova	University of Calgary, Canada
Yee Leung	Chinese University of Hong Kong, China

International Liaison Chairs

Giuseppe Borruso	University of Trieste, Italy
Elise De Donker	Western Michigan University, USA
Maria Irene Falcão	University of Minho, Portugal
Robert C. H. Hsu	Chung Hua University, Taiwan
Tai-Hoon Kim	Beijing Jaotong University, China

Vladimir Korkhov	St. Petersburg University, Russia
Sanjay Misra	Covenant University, Nigeria
Takashi Naka	Kyushu Sangyo University, Japan
Rafael D. C. Santos	National Institute for Space Research, Brazil
Maribel Yasmina Santos	University of Minho, Portugal
Elena Stankova	St. Petersburg University, Russia

Workshop and Session Chairs

Beniamino Murgante	University of Basilicata, Italy
Sanjay Misra	Covenant University, Nigeria
Jorge Gustavo Rocha	University of Minho, Portugal

Awards Chair

| Wenny Rahayu | La Trobe University, Australia |

Publicity Committee Chairs

Elmer Dadios	De La Salle University, Philippines
Nataliia Kulabukhova	St. Petersburg University, Russia
Daisuke Takahashi	Tsukuba University, Japan
Shangwang Wang	Beijing University of Posts and Telecommunications, China

Technology Chairs

| Damiano Perri | University of Florence, Italy |
| Marco Simonetti | University of Florence, Italy |

Local Arrangement Chairs

Ivan Blečić	University of Cagliari, Italy
Chiara Garau	University of Cagliari, Italy
Alfonso Annunziata	University of Cagliari, Italy
Ginevra Balletto	University of Cagliari, Italy
Giuseppe Borruso	University of Trieste, Italy
Alessandro Buccini	University of Cagliari, Italy
Michele Campagna	University of Cagliari, Italy
Mauro Coni	University of Cagliari, Italy
Anna Maria Colavitti	University of Cagliari, Italy
Giulia Desogus	University of Cagliari, Italy
Caterina Fenu	University of Cagliari, Italy
Sabrina Lai	University of Cagliari, Italy
Francesca Maltinti	University of Cagliari, Italy
Pasquale Mistretta	University of Cagliari, Italy

Augusto Montisci	University of Cagliari, Italy
Francesco Pinna	University of Cagliari, Italy
Davide Spano	University of Cagliari, Italy
Giuseppe A. Trunfio	University of Sassari, Italy
Corrado Zoppi	University of Cagliari, Italy

Program Committee

Vera Afreixo	University of Aveiro, Portugal
Filipe Alvelos	University of Minho, Portugal
Hartmut Asche	University of Potsdam, Germany
Ginevra Balletto	University of Cagliari, Italy
Michela Bertolotto	University College Dublin, Ireland
Sandro Bimonte	INRAE-TSCF, France
Rod Blais	University of Calgary, Canada
Ivan Blečić	University of Sassari, Italy
Giuseppe Borruso	University of Trieste, Italy
Ana Cristina Braga	University of Minho, Portugal
Massimo Cafaro	University of Salento, Italy
Yves Caniou	University of Lyon, France
José A. Cardoso e Cunha	Universidade Nova de Lisboa, Portugal
Rui Cardoso	University of Beira Interior, Portugal
Leocadio G. Casado	University of Almeria, Spain
Carlo Cattani	University of Salerno, Italy
Mete Celik	Erciyes University, Turkey
Maria Cerreta	University of Naples "Federico II", Italy
Hyunseung Choo	Sungkyunkwan University, South Korea
Chien-Sing Lee	Sunway University, Malaysia
Min Young Chung	Sungkyunkwan University, South Korea
Florbela Maria da Cruz Domingues Correia	Polytechnic Institute of Viana do Castelo, Portugal
Gilberto Corso Pereira	Federal University of Bahia, Brazil
Fernanda Costa	University of Minho, Portugal
Alessandro Costantini	INFN, Italy
Carla Dal Sasso Freitas	Universidade Federal do Rio Grande do Sul, Brazil
Pradesh Debba	The Council for Scientific and Industrial Research (CSIR), South Africa
Hendrik Decker	Instituto Tecnolĕgico de Informática, Spain
Robertas Damaševičius	Kausan University of Technology, Lithuania
Frank Devai	London South Bank University, UK
Rodolphe Devillers	Memorial University of Newfoundland, Canada
Joana Matos Dias	University of Coimbra, Portugal
Paolino Di Felice	University of L'Aquila, Italy
Prabu Dorairaj	NetApp, India/USA
Noelia Faginas Lago	University of Perugia, Italy
M. Irene Falcao	University of Minho, Portugal

Cherry Liu Fang	Ames Laboratory, USA
Florbela P. Fernandes	Polytechnic Institute of Bragança, Portugal
Jose-Jesus Fernandez	National Centre for Biotechnology, Spain
Paula Odete Fernandes	Polytechnic Institute of Bragança, Portugal
Adelaide de Fátima Baptista Valente Freitas	University of Aveiro, Portugal
Manuel Carlos Figueiredo	University of Minho, Portugal
Maria Celia Furtado Rocha	Universidade Federal da Bahia, Brazil
Chiara Garau	University of Cagliari, Italy
Paulino Jose Garcia Nieto	University of Oviedo, Spain
Jerome Gensel	LSR-IMAG, France
Maria Giaoutzi	National Technical University of Athens, Greece
Arminda Manuela Andrade Pereira Gonçalves	University of Minho, Portugal
Andrzej M. Goscinski	Deakin University, Australia
Eduardo Guerra	Free University of Bozen-Bolzano, Italy
Sevin Gümgüm	Izmir University of Economics, Turkey
Alex Hagen-Zanker	University of Cambridge, UK
Shanmugasundaram Hariharan	B.S. Abdur Rahman University, India
Eligius M. T. Hendrix	University of Malaga, Spain/Wageningen University, The Netherlands
Hisamoto Hiyoshi	Gunma University, Japan
Mustafa Inceoglu	EGE University, Turkey
Peter Jimack	University of Leeds, UK
Qun Jin	Waseda University, Japan
Yeliz Karaca	University of Massachusetts Medical School, USA
Farid Karimipour	Vienna University of Technology, Austria
Baris Kazar	Oracle Corp., USA
Maulana Adhinugraha Kiki	Telkom University, Indonesia
DongSeong Kim	University of Canterbury, New Zealand
Taihoon Kim	Hannam University, South Korea
Ivana Kolingerova	University of West Bohemia, Czech Republic
Nataliia Kulabukhova	St. Petersburg University, Russia
Vladimir Korkhov	St. Petersburg University, Russia
Rosa Lasaponara	National Research Council, Italy
Maurizio Lazzari	National Research Council, Italy
Cheng Siong Lee	Monash University, Australia
Sangyoun Lee	Yonsei University, South Korea
Jongchan Lee	Kunsan National University, South Korea
Chendong Li	University of Connecticut, USA
Gang Li	Deakin University, Australia
Fang Liu	Ames Laboratory, USA
Xin Liu	University of Calgary, Canada
Andrea Lombardi	University of Perugia, Italy
Savino Longo	University of Bari, Italy

Tinghuai Ma	Nanjing University of Information Science and Technology, China
Ernesto Marcheggiani	Katholieke Universiteit Leuven, Belgium
Antonino Marvuglia	Research Centre Henri Tudor, Luxembourg
Nicola Masini	National Research Council, Italy
Ilaria Matteucci	National Research Council, Italy
Eric Medvet	University of Trieste, Italy
Nirvana Meratnia	University of Twente, The Netherlands
Giuseppe Modica	University of Reggio Calabria, Italy
Josè Luis Montaña	University of Cantabria, Spain
Maria Filipa Mourão	Instituto Politécnico de Viana do Castelo, Portugal
Louiza de Macedo Mourelle	State University of Rio de Janeiro, Brazil
Nadia Nedjah	State University of Rio de Janeiro, Brazil
Laszlo Neumann	University of Girona, Spain
Kok-Leong Ong	Deakin University, Australia
Belen Palop	Universidad de Valladolid, Spain
Marcin Paprzycki	Polish Academy of Sciences, Poland
Eric Pardede	La Trobe University, Australia
Kwangjin Park	Wonkwang University, South Korea
Ana Isabel Pereira	Polytechnic Institute of Bragança, Portugal
Massimiliano Petri	University of Pisa, Italy
Telmo Pinto	University of Coimbra, Portugal
Maurizio Pollino	Italian National Agency for New Technologies, Energy and Sustainable Economic Development, Italy
Alenka Poplin	University of Hamburg, Germany
Vidyasagar Potdar	Curtin University of Technology, Australia
David C. Prosperi	Florida Atlantic University, USA
Wenny Rahayu	La Trobe University, Australia
Jerzy Respondek	Silesian University of Technology Poland
Humberto Rocha	INESC-Coimbra, Portugal
Jon Rokne	University of Calgary, Canada
Octavio Roncero	CSIC, Spain
Maytham Safar	Kuwait University, Kuwait
Francesco Santini	University of Perugia, Italy
Chiara Saracino	A.O. Ospedale Niguarda Ca' Granda, Italy
Haiduke Sarafian	Pennsylvania State University, USA
Marco Paulo Seabra dos Reis	University of Coimbra, Portugal
Jie Shen	University of Michigan, USA
Qi Shi	Liverpool John Moores University, UK
Dale Shires	U.S. Army Research Laboratory, USA
Inês Soares	University of Coimbra, Portugal
Elena Stankova	St. Petersburg University, Russia
Takuo Suganuma	Tohoku University, Japan
Eufemia Tarantino	Polytechnic University of Bari, Italy
Sergio Tasso	University of Perugia, Italy

Ana Paula Teixeira	University of Trás-os-Montes and Alto Douro, Portugal
Senhorinha Teixeira	University of Minho, Portugal
M. Filomena Teodoro	Portuguese Naval Academy/University of Lisbon, Portugal
Parimala Thulasiraman	University of Manitoba, Canada
Carmelo Torre	Polytechnic University of Bari, Italy
Javier Martinez Torres	Centro Universitario de la Defensa Zaragoza, Spain
Giuseppe A. Trunfio	University of Sassari, Italy
Pablo Vanegas	University of Cuenca, Equador
Marco Vizzari	University of Perugia, Italy
Varun Vohra	Merck Inc., USA
Koichi Wada	University of Tsukuba, Japan
Krzysztof Walkowiak	Wroclaw University of Technology, Poland
Zequn Wang	Intelligent Automation Inc, USA
Robert Weibel	University of Zurich, Switzerland
Frank Westad	Norwegian University of Science and Technology, Norway
Roland Wismüller	Universität Siegen, Germany
Mudasser Wyne	National University, USA
Chung-Huang Yang	National Kaohsiung Normal University, Taiwan
Xin-She Yang	National Physical Laboratory, UK
Salim Zabir	National Institute of Technology, Tsuruoka, Japan
Haifeng Zhao	University of California, Davis, USA
Fabiana Zollo	University of Venice "Cà Foscari", Italy
Albert Y. Zomaya	University of Sydney, Australia

Workshop Organizers

Advanced Transport Tools and Methods (A2TM 2021)

Massimiliano Petri	University of Pisa, Italy
Antonio Pratelli	University of Pisa, Italy

Advances in Artificial Intelligence Learning Technologies: Blended Learning, STEM, Computational Thinking and Coding (AAILT 2021)

Alfredo Milani	University of Perugia, Italy
Giulio Biondi	University of Florence, Italy
Sergio Tasso	University of Perugia, Italy

Workshop on Advancements in Applied Machine Learning and Data Analytics (AAMDA 2021)

Alessandro Costantini	INFN, Italy
Davide Salomoni	INFN, Italy
Doina Cristina Duma	INFN, Italy
Daniele Cesini	INFN, Italy

Automatic Landform Classification: Spatial Methods and Applications (ALCSMA 2021)

Maria Danese	ISPC, National Research Council, Italy
Dario Gioia	ISPC, National Research Council, Italy

Application of Numerical Analysis to Imaging Science (ANAIS 2021)

Caterina Fenu	University of Cagliari, Italy
Alessandro Buccini	University of Cagliari, Italy

Advances in Information Systems and Technologies for Emergency Management, Risk Assessment and Mitigation Based on the Resilience Concepts (ASTER 2021)

Maurizio Pollino	ENEA, Italy
Marco Vona	University of Basilicata, Italy
Amedeo Flora	University of Basilicata, Italy
Chiara Iacovino	University of Basilicata, Italy
Beniamino Murgante	University of Basilicata, Italy

Advances in Web Based Learning (AWBL 2021)

Birol Ciloglugil	Ege University, Turkey
Mustafa Murat Inceoglu	Ege University, Turkey

Blockchain and Distributed Ledgers: Technologies and Applications (BDLTA 2021)

Vladimir Korkhov	St. Petersburg University, Russia
Elena Stankova	St. Petersburg University, Russia
Nataliia Kulabukhova	St. Petersburg University, Russia

Bio and Neuro Inspired Computing and Applications (BIONCA 2021)

Nadia Nedjah	State University of Rio de Janeiro, Brazil
Luiza De Macedo Mourelle	State University of Rio de Janeiro, Brazil

Computational and Applied Mathematics (CAM 2021)

Maria Irene Falcão	University of Minho, Portugal
Fernando Miranda	University of Minho, Portugal

Computational and Applied Statistics (CAS 2021)

Ana Cristina Braga	University of Minho, Portugal

Computerized Evaluation of Economic Activities: Urban Spaces (CEEA 2021)

Diego Altafini	Università di Pisa, Italy
Valerio Cutini	Università di Pisa, Italy

Computational Geometry and Applications (CGA 2021)

Marina Gavrilova   University of Calgary, Canada

Collaborative Intelligence in Multimodal Applications (CIMA 2021)

Robertas Damasevicius   Kaunas University of Technology, Lithuania
Rytis Maskeliunas   Kaunas University of Technology, Lithuania

Computational Optimization and Applications (COA 2021)

Ana Rocha   University of Minho, Portugal
Humberto Rocha   University of Coimbra, Portugal

Computational Astrochemistry (CompAstro 2021)

Marzio Rosi   University of Perugia, Italy
Cecilia Ceccarelli   University of Grenoble, France
Stefano Falcinelli   University of Perugia, Italy
Dimitrios Skouteris   Master-Up, Italy

Computational Science and HPC (CSHPC 2021)

Elise de Doncker   Western Michigan University, USA
Fukuko Yuasa   High Energy Accelerator Research Organization
                               (KEK), Japan
Hideo Matsufuru   High Energy Accelerator Research Organization
                               (KEK), Japan

Cities, Technologies and Planning (CTP 2021)

Malgorzata Hanzl   University of Łódź, Poland
Beniamino Murgante   University of Basilicata, Italy
Ljiljana Zivkovic   Ministry of Construction, Transport and
                               Infrastructure/Institute of Architecture and Urban
                               and Spatial Planning of Serbia, Serbia
Anastasia Stratigea   National Technical University of Athens, Greece
Giuseppe Borruso   University of Trieste, Italy
Ginevra Balletto   University of Cagliari, Italy

Advanced Modeling E-Mobility in Urban Spaces (DEMOS 2021)

Tiziana Campisi   Kore University of Enna, Italy
Socrates Basbas   Aristotle University of Thessaloniki, Greece
Ioannis Politis   Aristotle University of Thessaloniki, Greece
Florin Nemtanu   Polytechnic University of Bucharest, Romania
Giovanna Acampa   Kore University of Enna, Italy
Wolfgang Schulz   Zeppelin University, Germany

Digital Transformation and Smart City (DIGISMART 2021)

Mauro Mazzei National Research Council, Italy

Econometric and Multidimensional Evaluation in Urban Environment (EMEUE 2021)

Carmelo Maria Torre	Polytechnic University of Bari, Italy
Maria Cerreta	University "Federico II" of Naples, Italy
Pierluigi Morano	Polytechnic University of Bari, Italy
Simona Panaro	University of Portsmouth, UK
Francesco Tajani	Sapienza University of Rome, Italy
Marco Locurcio	Polytechnic University of Bari, Italy

The 11th International Workshop on Future Computing System Technologies and Applications (FiSTA 2021)

Bernady Apduhan	Kyushu Sangyo University, Japan
Rafael Santos	Brazilian National Institute for Space Research, Brazil

Transformational Urban Mobility: Challenges and Opportunities During and Post COVID Era (FURTHER 2021)

Tiziana Campisi	Kore University of Enna, Italy
Socrates Basbas	Aristotle University of Thessaloniki, Greece
Dilum Dissanayake	Newcastle University, UK
Kh Md Nahiduzzaman	University of British Columbia, Canada
Nurten Akgün Tanbay	Bursa Technical University, Turkey
Khaled J. Assi	King Fahd University of Petroleum and Minerals, Saudi Arabia
Giovanni Tesoriere	Kore University of Enna, Italy
Motasem Darwish	Middle East University, Jordan

Geodesign in Decision Making: Meta Planning and Collaborative Design for Sustainable and Inclusive Development (GDM 2021)

Francesco Scorza	University of Basilicata, Italy
Michele Campagna	University of Cagliari, Italy
Ana Clara Mourao Moura	Federal University of Minas Gerais, Brazil

Geomatics in Forestry and Agriculture: New Advances and Perspectives (GeoForAgr 2021)

Maurizio Pollino	ENEA, Italy
Giuseppe Modica	University of Reggio Calabria, Italy
Marco Vizzari	University of Perugia, Italy

Geographical Analysis, Urban Modeling, Spatial Statistics (GEOG-AND-MOD 2021)

Beniamino Murgante	University of Basilicata, Italy
Giuseppe Borruso	University of Trieste, Italy
Hartmut Asche	University of Potsdam, Germany

Geomatics for Resource Monitoring and Management (GRMM 2021)

Eufemia Tarantino	Polytechnic University of Bari, Italy
Enrico Borgogno Mondino	University of Turin, Italy
Alessandra Capolupo	Polytechnic University of Bari, Italy
Mirko Saponaro	Polytechnic University of Bari, Italy

12th International Symposium on Software Quality (ISSQ 2021)

Sanjay Misra	Covenant University, Nigeria

10th International Workshop on Collective, Massive and Evolutionary Systems (IWCES 2021)

Alfredo Milani	University of Perugia, Italy
Rajdeep Niyogi	Indian Institute of Technology, Roorkee, India

Land Use Monitoring for Sustainability (LUMS 2021)

Carmelo Maria Torre	Polytechnic University of Bari, Italy
Maria Cerreta	University "Federico II" of Naples, Italy
Massimiliano Bencardino	University of Salerno, Italy
Alessandro Bonifazi	Polytechnic University of Bari, Italy
Pasquale Balena	Polytechnic University of Bari, Italy
Giuliano Poli	University "Federico II" of Naples, Italy

Machine Learning for Space and Earth Observation Data (MALSEOD 2021)

Rafael Santos	Instituto Nacional de Pesquisas Espaciais, Brazil
Karine Ferreira	Instituto Nacional de Pesquisas Espaciais, Brazil

Building Multi-dimensional Models for Assessing Complex Environmental Systems (MES 2021)

Marta Dell'Ovo	Polytechnic University of Milan, Italy
Vanessa Assumma	Polytechnic University of Turin, Italy
Caterina Caprioli	Polytechnic University of Turin, Italy
Giulia Datola	Polytechnic University of Turin, Italy
Federico dell'Anna	Polytechnic University of Turin, Italy

Ecosystem Services: Nature's Contribution to People in Practice. Assessment Frameworks, Models, Mapping, and Implications (NC2P 2021)

Francesco Scorza	University of Basilicata, Italy
Sabrina Lai	University of Cagliari, Italy
Ana Clara Mourao Moura	Federal University of Minas Gerais, Brazil
Corrado Zoppi	University of Cagliari, Italy
Dani Broitman	Technion, Israel Institute of Technology, Israel

Privacy in the Cloud/Edge/IoT World (PCEIoT 2021)

Michele Mastroianni	University of Campania Luigi Vanvitelli, Italy
Lelio Campanile	University of Campania Luigi Vanvitelli, Italy
Mauro Iacono	University of Campania Luigi Vanvitelli, Italy

Processes, Methods and Tools Towards RESilient Cities and Cultural Heritage Prone to SOD and ROD Disasters (RES 2021)

Elena Cantatore	Polytechnic University of Bari, Italy
Alberico Sonnessa	Polytechnic University of Bari, Italy
Dario Esposito	Polytechnic University of Bari, Italy

Risk, Resilience and Sustainability in the Efficient Management of Water Resources: Approaches, Tools, Methodologies and Multidisciplinary Integrated Applications (RRS 2021)

Maria Macchiaroli	University of Salerno, Italy
Chiara D'Alpaos	Università degli Studi di Padova, Italy
Mirka Mobilia	Università degli Studi di Salerno, Italy
Antonia Longobardi	Università degli Studi di Salerno, Italy
Grazia Fattoruso	ENEA Research Center, Italy
Vincenzo Pellecchia	Ente Idrico Campano, Italy

Scientific Computing Infrastructure (SCI 2021)

Elena Stankova	St. Petersburg University, Russia
Vladimir Korkhov	St. Petersburg University, Russia
Natalia Kulabukhova	St. Petersburg University, Russia

Smart Cities and User Data Management (SCIDAM 2021)

Chiara Garau	University of Cagliari, Italy
Luigi Mundula	University of Cagliari, Italy
Gianni Fenu	University of Cagliari, Italy
Paolo Nesi	University of Florence, Italy
Paola Zamperlin	University of Pisa, Italy

13th International Symposium on Software Engineering Processes and Applications (SEPA 2021)

Sanjay Misra   Covenant University, Nigeria

Ports of the Future - Smartness and Sustainability (SmartPorts 2021)

Patrizia Serra	University of Cagliari, Italy
Gianfranco Fancello	University of Cagliari, Italy
Ginevra Balletto	University of Cagliari, Italy
Luigi Mundula	University of Cagliari, Italy
Marco Mazzarino	University of Venice, Italy
Giuseppe Borruso	University of Trieste, Italy
Maria del Mar Munoz Leonisio	Universidad de Cádiz, Spain

Smart Tourism (SmartTourism 2021)

Giuseppe Borruso	University of Trieste, Italy
Silvia Battino	University of Sassari, Italy
Ginevra Balletto	University of Cagliari, Italy
Maria del Mar Munoz Leonisio	Universidad de Cádiz, Spain
Ainhoa Amaro Garcia	Universidad de Alcalà/Universidad de Las Palmas, Spain
Francesca Krasna	University of Trieste, Italy

Sustainability Performance Assessment: Models, Approaches and Applications toward Interdisciplinary and Integrated Solutions (SPA 2021)

Francesco Scorza	University of Basilicata, Italy
Sabrina Lai	University of Cagliari, Italy
Jolanta Dvarioniene	Kaunas University of Technology, Lithuania
Valentin Grecu	Lucian Blaga University, Romania
Corrado Zoppi	University of Cagliari, Italy
Iole Cerminara	University of Basilicata, Italy

Smart and Sustainable Island Communities (SSIC 2021)

Chiara Garau	University of Cagliari, Italy
Anastasia Stratigea	National Technical University of Athens, Greece
Paola Zamperlin	University of Pisa, Italy
Francesco Scorza	University of Basilicata, Italy

Science, Technologies and Policies to Innovate Spatial Planning (STP4P 2021)

Chiara Garau	University of Cagliari, Italy
Daniele La Rosa	University of Catania, Italy
Francesco Scorza	University of Basilicata, Italy

Anna Maria Colavitti   University of Cagliari, Italy
Beniamino Murgante   University of Basilicata, Italy
Paolo La Greca   University of Catania, Italy

Sustainable Urban Energy Systems (SURENSYS 2021)

Luigi Mundula   University of Cagliari, Italy
Emilio Ghiani   University of Cagliari, Italy

Space Syntax for Cities in Theory and Practice (Syntax_City 2021)

Claudia Yamu   University of Groningen, The Netherlands
Akkelies van Nes   Western Norway University of Applied Sciences,
                                 Norway
Chiara Garau   University of Cagliari, Italy

Theoretical and Computational Chemistry and Its Applications (TCCMA 2021)

Noelia Faginas-Lago   University of Perugia, Italy

13th International Workshop on Tools and Techniques in Software Development Process (TTSDP 2021)

Sanjay Misra   Covenant University, Nigeria

Urban Form Studies (UForm 2021)

Malgorzata Hanzl   Łódź University of Technology, Poland
Beniamino Murgante   University of Basilicata, Italy
Eufemia Tarantino   Polytechnic University of Bari, Italy
Irena Itova   University of Westminster, UK

Urban Space Accessibility and Safety (USAS 2021)

Chiara Garau   University of Cagliari, Italy
Francesco Pinna   University of Cagliari, Italy
Claudia Yamu   University of Groningen, The Netherlands
Vincenza Torrisi   University of Catania, Italy
Matteo Ignaccolo   University of Catania, Italy
Michela Tiboni   University of Brescia, Italy
Silvia Rossetti   University of Parma, Italy

Virtual and Augmented Reality and Applications (VRA 2021)

Osvaldo Gervasi   University of Perugia, Italy
Damiano Perri   University of Perugia, Italy
Marco Simonetti   University of Perugia, Italy
Sergio Tasso   University of Perugia, Italy

Workshop on Advanced and Computational Methods for Earth Science Applications (WACM4ES 2021)

Luca Piroddi	University of Cagliari, Italy
Laura Foddis	University of Cagliari, Italy
Augusto Montisci	University of Cagliari, Italy
Sergio Vincenzo Calcina	University of Cagliari, Italy
Sebastiano D'Amico	University of Malta, Malta
Giovanni Martinelli	Istituto Nazionale di Geofisica e Vulcanologia, Italy/Chinese Academy of Sciences, China

Sponsoring Organizations

ICCSA 2021 would not have been possible without the tremendous support of many organizations and institutions, for which all organizers and participants of ICCSA 2021 express their sincere gratitude:

Springer International Publishing AG, Germany (https://www.springer.com)

Computers Open Access Journal (https://www.mdpi.com/journal/computers)

IEEE Italy Section, Italy (https://italy.ieeer8.org/)

Centre-North Italy Chapter IEEE GRSS, Italy (https://cispio.diet.uniroma1.it/marzano/ieee-grs/index.html)

Italy Section of the Computer Society, Italy (https://site.ieee.org/italy-cs/)

University of Perugia, Italy (https://www.unipg.it)

University of Cagliari, Italy (https://unica.it/)

University of Basilicata, Italy
(http://www.unibas.it)

Monash University, Australia
(https://www.monash.edu/)

Kyushu Sangyo University, Japan
(https://www.kyusan-u.ac.jp/)

University of Minho, Portugal
(https://www.uminho.pt/)

Scientific Association Transport Infrastructures,
Italy
(https://www.stradeeautostrade.it/associazioni-e-
organizzazioni/asit-associazione-scientifica-
infrastrutture-trasporto/)

Regione Sardegna, Italy
(https://regione.sardegna.it/)

Comune di Cagliari, Italy
(https://www.comune.cagliari.it/)

Città Metropolitana di Cagliari

Cagliari Accessibility Lab (CAL)
(https://www.unica.it/unica/it/cagliari_
accessibility_lab.page/)

Referees

Nicodemo Abate	IMAA, National Research Council, Italy
Andre Ricardo Abed Grégio	Federal University of Paraná State, Brazil
Nasser Abu Zeid	Università di Ferrara, Italy
Lidia Aceto	Università del Piemonte Orientale, Italy
Nurten Akgün Tanbay	Bursa Technical University, Turkey
Filipe Alvelos	Universidade do Minho, Portugal
Paula Amaral	Universidade Nova de Lisboa, Portugal
Federico Amato	University of Lausanne, Switzerland
Marina Alexandra Pedro Andrade	ISCTE-IUL, Portugal
Debora Anelli	Sapienza University of Rome, Italy
Alfonso Annunziata	University of Cagliari, Italy
Fahim Anzum	University of Calgary, Canada
Tatsumi Aoyama	High Energy Accelerator Research Organization, Japan
Bernady Apduhan	Kyushu Sangyo University, Japan
Jonathan Apeh	Covenant University, Nigeria
Vasilike Argyropoulos	University of West Attica, Greece
Giuseppe Aronica	Università di Messina, Italy
Daniela Ascenzi	Università degli Studi di Trento, Italy
Vanessa Assumma	Politecnico di Torino, Italy
Muhammad Attique Khan	HITEC University Taxila, Pakistan
Vecdi Aytaç	Ege University, Turkey
Alina Elena Baia	University of Perugia, Italy
Ginevra Balletto	University of Cagliari, Italy
Marialaura Bancheri	ISAFOM, National Research Council, Italy
Benedetto Barabino	University of Brescia, Italy
Simona Barbaro	Università degli Studi di Palermo, Italy
Enrico Barbierato	Università Cattolica del Sacro Cuore di Milano, Italy
Jeniffer Barreto	Istituto Superior Técnico, Lisboa, Portugal
Michele Bartalini	TAGES, Italy
Socrates Basbas	Aristotle University of Thessaloniki, Greece
Silvia Battino	University of Sassari, Italy
Marcelo Becerra Rozas	Pontificia Universidad Católica de Valparaíso, Chile
Ranjan Kumar Behera	National Institute of Technology, Rourkela, India
Emanuele Bellini	University of Campania Luigi Vanvitelli, Italy
Massimo Bilancia	University of Bari Aldo Moro, Italy
Giulio Biondi	University of Firenze, Italy
Adriano Bisello	Eurac Research, Italy
Ignacio Blanquer	Universitat Politècnica de València, Spain
Semen Bochkov	Ulyanovsk State Technical University, Russia
Alexander Bogdanov	St. Petersburg University, Russia
Silvia Bonettini	University of Modena and Reggio Emilia, Italy
Enrico Borgogno Mondino	Università di Torino, Italy
Giuseppe Borruso	University of Trieste, Italy

Michele Bottazzi	University of Trento, Italy
Rahma Bouaziz	Taibah University, Saudi Arabia
Ouafik Boulariah	University of Salerno, Italy
Tulin Boyar	Yildiz Technical University, Turkey
Ana Cristina Braga	University of Minho, Portugal
Paolo Bragolusi	University of Padova, Italy
Luca Braidotti	University of Trieste, Italy
Alessandro Buccini	University of Cagliari, Italy
Jorge Buele	Universidad Tecnológica Indoamérica, Ecuador
Andrea Buffoni	TAGES, Italy
Sergio Vincenzo Calcina	University of Cagliari, Italy
Michele Campagna	University of Cagliari, Italy
Lelio Campanile	Università degli Studi della Campania Luigi Vanvitelli, Italy
Tiziana Campisi	Kore University of Enna, Italy
Antonino Canale	Kore University of Enna, Italy
Elena Cantatore	DICATECh, Polytechnic University of Bari, Italy
Pasquale Cantiello	Istituto Nazionale di Geofisica e Vulcanologia, Italy
Alessandra Capolupo	Polytechnic University of Bari, Italy
David Michele Cappelletti	University of Perugia, Italy
Caterina Caprioli	Politecnico di Torino, Italy
Sara Carcangiu	University of Cagliari, Italy
Pedro Carrasqueira	INESC Coimbra, Portugal
Arcangelo Castiglione	University of Salerno, Italy
Giulio Cavana	Politecnico di Torino, Italy
Davide Cerati	Politecnico di Milano, Italy
Maria Cerreta	University of Naples Federico II, Italy
Daniele Cesini	INFN-CNAF, Italy
Jabed Chowdhury	La Trobe University, Australia
Gennaro Ciccarelli	Iuav University of Venice, Italy
Birol Ciloglugil	Ege University, Turkey
Elena Cocuzza	Univesity of Catania, Italy
Anna Maria Colavitt	University of Cagliari, Italy
Cecilia Coletti	Università "G. d'Annunzio" di Chieti-Pescara, Italy
Alberto Collu	Independent Researcher, Italy
Anna Concas	University of Basilicata, Italy
Mauro Coni	University of Cagliari, Italy
Melchiorre Contino	Università di Palermo, Italy
Antonella Cornelio	Università degli Studi di Brescia, Italy
Aldina Correia	Politécnico do Porto, Portugal
Elisete Correia	Universidade de Trás-os-Montes e Alto Douro, Portugal
Florbela Correia	Polytechnic Institute of Viana do Castelo, Portugal
Stefano Corsi	Università degli Studi di Milano, Italy
Alberto Cortez	Polytechnic of University Coimbra, Portugal
Lino Costa	Universidade do Minho, Portugal

Alessandro Costantini — INFN, Italy
Marilena Cozzolino — Università del Molise, Italy
Giulia Crespi — Politecnico di Torino, Italy
Maurizio Crispino — Politecnico di Milano, Italy
Chiara D'Alpaos — University of Padova, Italy
Roberta D'Ambrosio — Università di Salerno, Italy
Sebastiano D'Amico — University of Malta, Malta
Hiroshi Daisaka — Hitotsubashi University, Japan
Gaia Daldanise — Italian National Research Council, Italy
Robertas Damasevicius — Silesian University of Technology, Poland
Maria Danese — ISPC, National Research Council, Italy
Bartoli Daniele — University of Perugia, Italy
Motasem Darwish — Middle East University, Jordan
Giulia Datola — Politecnico di Torino, Italy
Regina de Almeida — UTAD, Portugal
Elise de Doncker — Western Michigan University, USA
Mariella De Fino — Politecnico di Bari, Italy
Giandomenico De Luca — Mediterranean University of Reggio Calabria, Italy
Luiza de Macedo Mourelle — State University of Rio de Janeiro, Brazil
Gianluigi De Mare — University of Salerno, Italy
Itamir de Morais Barroca Filho — Federal University of Rio Grande do Norte, Brazil
Samuele De Petris — Università di Torino, Italy
Marcilio de Souto — LIFO, University of Orléans, France
Alexander Degtyarev — St. Petersburg University, Russia
Federico Dell'Anna — Politecnico di Torino, Italy
Marta Dell'Ovo — Politecnico di Milano, Italy
Fernanda Della Mura — University of Naples "Federico II", Italy
Ahu Dereli Dursun — Istanbul Commerce University, Turkey
Bashir Derradji — University of Sfax, Tunisia
Giulia Desogus — Università degli Studi di Cagliari, Italy
Marco Dettori — Università degli Studi di Sassari, Italy
Frank Devai — London South Bank University, UK
Felicia Di Liddo — Polytechnic University of Bari, Italy
Valerio Di Pinto — University of Naples "Federico II", Italy
Joana Dias — University of Coimbra, Portugal
Luis Dias — University of Minho, Portugal
Patricia Diaz de Alba — Gran Sasso Science Institute, Italy
Isabel Dimas — University of Coimbra, Portugal
Aleksandra Djordjevic — University of Belgrade, Serbia
Luigi Dolores — Università degli Studi di Salerno, Italy
Marco Donatelli — University of Insubria, Italy
Doina Cristina Duma — INFN-CNAF, Italy
Fabio Durastante — University of Pisa, Italy
Aziz Dursun — Virginia Tech University, USA
Juan Enrique-Romero — Université Grenoble Alpes, France

Annunziata Esposito Amideo	University College Dublin, Ireland
Dario Esposito	Polytechnic University of Bari, Italy
Claudio Estatico	University of Genova, Italy
Noelia Faginas-Lago	Università di Perugia, Italy
Maria Irene Falcão	University of Minho, Portugal
Stefano Falcinelli	University of Perugia, Italy
Alessandro Farina	University of Pisa, Italy
Grazia Fattoruso	ENEA, Italy
Caterina Fenu	University of Cagliari, Italy
Luisa Fermo	University of Cagliari, Italy
Florbela Fernandes	Instituto Politecnico de Braganca, Portugal
Rosário Fernandes	University of Minho, Portugal
Luis Fernandez-Sanz	University of Alcala, Spain
Alessia Ferrari	Università di Parma, Italy
Luís Ferrás	University of Minho, Portugal
Ângela Ferreira	Instituto Politécnico de Bragança, Portugal
Flora Ferreira	University of Minho, Portugal
Manuel Carlos Figueiredo	University of Minho, Portugal
Ugo Fiore	University of Naples "Parthenope", Italy
Amedeo Flora	University of Basilicata, Italy
Hector Florez	Universidad Distrital Francisco Jose de Caldas, Colombia
Maria Laura Foddis	University of Cagliari, Italy
Valentina Franzoni	Perugia University, Italy
Adelaide Freitas	University of Aveiro, Portugal
Samuel Frimpong	Durban University of Technology, South Africa
Ioannis Fyrogenis	Aristotle University of Thessaloniki, Greece
Marika Gaballo	Politecnico di Torino, Italy
Laura Gabrielli	Iuav University of Venice, Italy
Ivan Gankevich	St. Petersburg University, Russia
Chiara Garau	University of Cagliari, Italy
Ernesto Garcia Para	Universidad del País Vasco, Spain,
Fernando Garrido	Universidad Técnica del Norte, Ecuador
Marina Gavrilova	University of Calgary, Canada
Silvia Gazzola	University of Bath, UK
Georgios Georgiadis	Aristotle University of Thessaloniki, Greece
Osvaldo Gervasi	University of Perugia, Italy
Andrea Gioia	Polytechnic University of Bari, Italy
Dario Gioia	ISPC-CNT, Italy
Raffaele Giordano	IRSS, National Research Council, Italy
Giacomo Giorgi	University of Perugia, Italy
Eleonora Giovene di Girasole	IRISS, National Research Council, Italy
Salvatore Giuffrida	Università di Catania, Italy
Marco Gola	Politecnico di Milano, Italy

Pavan Kumar	University of Calgary, Canada
Anisha Kumari	National Institute of Technology, Rourkela, India
Ludovica La Rocca	University of Naples "Federico II", Italy
Daniele La Rosa	University of Catania, Italy
Sabrina Lai	University of Cagliari, Italy
Giuseppe Francesco Cesare Lama	University of Naples "Federico II", Italy
Mariusz Lamprecht	University of Lodz, Poland
Vincenzo Laporta	National Research Council, Italy
Chien-Sing Lee	Sunway University, Malaysia
José Isaac Lemus Romani	Pontifical Catholic University of Valparaíso, Chile
Federica Leone	University of Cagliari, Italy
Alexander H. Levis	George Mason University, USA
Carola Lingua	Polytechnic University of Turin, Italy
Marco Locurcio	Polytechnic University of Bari, Italy
Andrea Lombardi	University of Perugia, Italy
Savino Longo	University of Bari, Italy
Fernando Lopez Gayarre	University of Oviedo, Spain
Yan Lu	Western Michigan University, USA
Maria Macchiaroli	University of Salerno, Italy
Helmuth Malonek	University of Aveiro, Portugal
Francesca Maltinti	University of Cagliari, Italy
Luca Mancini	University of Perugia, Italy
Marcos Mandado	University of Vigo, Spain
Ernesto Marcheggiani	Università Politecnica delle Marche, Italy
Krassimir Markov	University of Telecommunications and Post, Bulgaria
Giovanni Martinelli	INGV, Italy
Alessandro Marucci	University of L'Aquila, Italy
Fiammetta Marulli	University of Campania Luigi Vanvitelli, Italy
Gabriella Maselli	University of Salerno, Italy
Rytis Maskeliunas	Kaunas University of Technology, Lithuania
Michele Mastroianni	University of Campania Luigi Vanvitelli, Italy
Cristian Mateos	Universidad Nacional del Centro de la Provincia de Buenos Aires, Argentina
Hideo Matsufuru	High Energy Accelerator Research Organization (KEK), Japan
D'Apuzzo Mauro	University of Cassino and Southern Lazio, Italy
Chiara Mazzarella	University Federico II, Italy
Marco Mazzarino	University of Venice, Italy
Giovanni Mei	University of Cagliari, Italy
Mário Melo	Federal Institute of Rio Grande do Norte, Brazil
Francesco Mercaldo	University of Molise, Italy
Alfredo Milani	University of Perugia, Italy
Alessandra Milesi	University of Cagliari, Italy
Antonio Minervino	ISPC, National Research Council, Italy
Fernando Miranda	Universidade do Minho, Portugal

B. Mishra	University of Szeged, Hungary
Sanjay Misra	Covenant University, Nigeria
Mirka Mobilia	University of Salerno, Italy
Giuseppe Modica	Università degli Studi di Reggio Calabria, Italy
Mohammadsadegh Mohagheghi	Vali-e-Asr University of Rafsanjan, Iran
Mohamad Molaei Qelichi	University of Tehran, Iran
Mario Molinara	University of Cassino and Southern Lazio, Italy
Augusto Montisci	Università degli Studi di Cagliari, Italy
Pierluigi Morano	Polytechnic University of Bari, Italy
Ricardo Moura	Universidade Nova de Lisboa, Portugal
Ana Clara Mourao Moura	Federal University of Minas Gerais, Brazil
Maria Mourao	Polytechnic Institute of Viana do Castelo, Portugal
Daichi Mukunoki	RIKEN Center for Computational Science, Japan
Beniamino Murgante	University of Basilicata, Italy
Naohito Nakasato	University of Aizu, Japan
Grazia Napoli	Università degli Studi di Palermo, Italy
Isabel Cristina Natário	Universidade Nova de Lisboa, Portugal
Nadia Nedjah	State University of Rio de Janeiro, Brazil
Antonio Nesticò	University of Salerno, Italy
Andreas Nikiforiadis	Aristotle University of Thessaloniki, Greece
Keigo Nitadori	RIKEN Center for Computational Science, Japan
Silvio Nocera	Iuav University of Venice, Italy
Giuseppina Oliva	University of Salerno, Italy
Arogundade Oluwasefunmi	Academy of Mathematics and System Science, China
Ken-ichi Oohara	University of Tokyo, Japan
Tommaso Orusa	University of Turin, Italy
M. Fernanda P. Costa	University of Minho, Portugal
Roberta Padulano	Centro Euro-Mediterraneo sui Cambiamenti Climatici, Italy
Maria Panagiotopoulou	National Technical University of Athens, Greece
Jay Pancham	Durban University of Technology, South Africa
Gianni Pantaleo	University of Florence, Italy
Dimos Pantazis	University of West Attica, Greece
Michela Paolucci	University of Florence, Italy
Eric Pardede	La Trobe University, Australia
Olivier Parisot	Luxembourg Institute of Science and Technology, Luxembourg
Vincenzo Pellecchia	Ente Idrico Campano, Italy
Anna Pelosi	University of Salerno, Italy
Edit Pengő	University of Szeged, Hungary
Marco Pepe	University of Salerno, Italy
Paola Perchinunno	University of Cagliari, Italy
Ana Pereira	Polytechnic Institute of Bragança, Portugal
Mariano Pernetti	University of Campania, Italy
Damiano Perri	University of Perugia, Italy

Federica Pes	University of Cagliari, Italy
Marco Petrelli	Roma Tre University, Italy
Massimiliano Petri	University of Pisa, Italy
Khiem Phan	Duy Tan University, Vietnam
Alberto Ferruccio Piccinni	Polytechnic of Bari, Italy
Angela Pilogallo	University of Basilicata, Italy
Francesco Pinna	University of Cagliari, Italy
Telmo Pinto	University of Coimbra, Portugal
Luca Piroddi	University of Cagliari, Italy
Darius Plonis	Vilnius Gediminas Technical University, Lithuania
Giuliano Poli	University of Naples "Federico II", Italy
Maria João Polidoro	Polytecnic Institute of Porto, Portugal
Ioannis Politis	Aristotle University of Thessaloniki, Greece
Maurizio Pollino	ENEA, Italy
Antonio Pratelli	University of Pisa, Italy
Salvatore Praticò	Mediterranean University of Reggio Calabria, Italy
Marco Prato	University of Modena and Reggio Emilia, Italy
Carlotta Quagliolo	Polytechnic University of Turin, Italy
Emanuela Quaquero	Univesity of Cagliari, Italy
Garrisi Raffaele	Polizia postale e delle Comunicazioni, Italy
Nicoletta Rassu	University of Cagliari, Italy
Hafiz Tayyab Rauf	University of Bradford, UK
Michela Ravanelli	Sapienza University of Rome, Italy
Roberta Ravanelli	Sapienza University of Rome, Italy
Alfredo Reder	Centro Euro-Mediterraneo sui Cambiamenti Climatici, Italy
Stefania Regalbuto	University of Naples "Federico II", Italy
Rommel Regis	Saint Joseph's University, USA
Lothar Reichel	Kent State University, USA
Marco Reis	University of Coimbra, Portugal
Maria Reitano	University of Naples "Federico II", Italy
Jerzy Respondek	Silesian University of Technology, Poland
Elisa Riccietti	École Normale Supérieure de Lyon, France
Albert Rimola	Universitat Autònoma de Barcelona, Spain
Angela Rizzo	University of Bari, Italy
Ana Maria A. C. Rocha	University of Minho, Portugal
Fabio Rocha	Institute of Technology and Research, Brazil
Humberto Rocha	University of Coimbra, Portugal
Maria Clara Rocha	Polytechnic Institute of Coimbra, Portugal
Miguel Rocha	University of Minho, Portugal
Giuseppe Rodriguez	University of Cagliari, Italy
Guillermo Rodriguez	UNICEN, Argentina
Elisabetta Ronchieri	INFN, Italy
Marzio Rosi	University of Perugia, Italy
Silvia Rossetti	University of Parma, Italy
Marco Rossitti	Polytechnic University of Milan, Italy

Francesco Rotondo	Marche Polytechnic University, Italy
Irene Rubino	Polytechnic University of Turin, Italy
Agustín Salas	Pontifical Catholic University of Valparaíso, Chile
Juan Pablo Sandoval Alcocer	Universidad Católica Boliviana "San Pablo", Bolivia
Luigi Santopietro	University of Basilicata, Italy
Rafael Santos	National Institute for Space Research, Brazil
Valentino Santucci	Università per Stranieri di Perugia, Italy
Mirko Saponaro	Polytechnic University of Bari, Italy
Filippo Sarvia	University of Turin, Italy
Marco Scaioni	Polytechnic University of Milan, Italy
Rafal Scherer	Częstochowa University of Technology, Poland
Francesco Scorza	University of Basilicata, Italy
Ester Scotto di Perta	University of Napoli "Federico II", Italy
Monica Sebillo	University of Salerno, Italy
Patrizia Serra	University of Cagliari, Italy
Ricardo Severino	University of Minho, Portugal
Jie Shen	University of Michigan, USA
Huahao Shou	Zhejiang University of Technology, China
Miltiadis Siavvas	Centre for Research and Technology Hellas, Greece
Brandon Sieu	University of Calgary, Canada
Ângela Silva	Instituto Politécnico de Viana do Castelo, Portugal
Carina Silva	Polytechic Institute of Lisbon, Portugal
Joao Carlos Silva	Polytechnic Institute of Cavado and Ave, Portugal
Fabio Silveira	Federal University of Sao Paulo, Brazil
Marco Simonetti	University of Florence, Italy
Ana Jacinta Soares	University of Minho, Portugal
Maria Joana Soares	University of Minho, Portugal
Michel Soares	Federal University of Sergipe, Brazil
George Somarakis	Foundation for Research and Technology Hellas, Greece
Maria Somma	University of Naples "Federico II", Italy
Alberico Sonnessa	Polytechnic University of Bari, Italy
Elena Stankova	St. Petersburg University, Russia
Flavio Stochino	University of Cagliari, Italy
Anastasia Stratigea	National Technical University of Athens, Greece
Yasuaki Sumida	Kyushu Sangyo University, Japan
Yue Sun	European X-Ray Free-Electron Laser Facility, Germany
Kirill Sviatov	Ulyanovsk State Technical University, Russia
Daisuke Takahashi	University of Tsukuba, Japan
Aladics Tamás	University of Szeged, Hungary
David Taniar	Monash University, Australia
Rodrigo Tapia McClung	Centro de Investigación en Ciencias de Información Geoespacial, Mexico
Eufemia Tarantino	Polytechnic University of Bari, Italy

Sergio Tasso	University of Perugia, Italy
Ana Paula Teixeira	Universidade de Trás-os-Montes e Alto Douro, Portugal
Senhorinha Teixeira	University of Minho, Portugal
Tengku Adil Tengku Izhar	Universiti Teknologi MARA, Malaysia
Maria Filomena Teodoro	University of Lisbon/Portuguese Naval Academy, Portugal
Giovanni Tesoriere	Kore University of Enna, Italy
Yiota Theodora	National Technical Univeristy of Athens, Greece
Graça Tomaz	Polytechnic Institute of Guarda, Portugal
Carmelo Maria Torre	Polytechnic University of Bari, Italy
Francesca Torrieri	University of Naples "Federico II", Italy
Vincenza Torrisi	University of Catania, Italy
Vincenzo Totaro	Polytechnic University of Bari, Italy
Pham Trung	Ho Chi Minh City University of Technology, Vietnam
Dimitrios Tsoukalas	Centre of Research and Technology Hellas (CERTH), Greece
Sanjida Tumpa	University of Calgary, Canada
Iñaki Tuñon	Universidad de Valencia, Spain
Takahiro Ueda	Seikei University, Japan
Piero Ugliengo	University of Turin, Italy
Abdi Usman	Haramaya University, Ethiopia
Ettore Valente	University of Naples "Federico II", Italy
Jordi Vallverdu	Universitat Autònoma de Barcelona, Spain
Cornelis Van Der Mee	University of Cagliari, Italy
José Varela-Aldás	Universidad Tecnológica Indoamérica, Ecuador
Fanny Vazart	University of Grenoble Alpes, France
Franco Vecchiocattivi	University of Perugia, Italy
Laura Verde	University of Campania Luigi Vanvitelli, Italy
Giulia Vergerio	Polytechnic University of Turin, Italy
Jos Vermaseren	Nikhef, The Netherlands
Giacomo Viccione	University of Salerno, Italy
Marco Vizzari	University of Perugia, Italy
Corrado Vizzarri	Polytechnic University of Bari, Italy
Alexander Vodyaho	St. Petersburg State Electrotechnical University "LETI", Russia
Nikolay N. Voit	Ulyanovsk State Technical University, Russia
Marco Vona	University of Basilicata, Italy
Agustinus Borgy Waluyo	Monash University, Australia
Fernando Wanderley	Catholic University of Pernambuco, Brazil
Chao Wang	University of Science and Technology of China, China
Marcin Wozniak	Silesian University of Technology, Poland
Tiang Xian	Nathong University, China
Rekha Yadav	KL University, India
Claudia Yamu	University of Groningen, The Netherlands
Fenghui Yao	Tennessee State University, USA

Contents – Part II

General Track 3: Geometric Modeling, Graphics and Visualization

General Track 4: Advanced and Emerging Applications

**International Workshop on Advanced Transport Tools
and Methods (A2TM 2021)**

International Workshop on Advances in Artificial Intelligence Learning Technologies: Blended Learning, STEM, Computational Thinking and Coding (AAILT 2021)

International Workshop on Advancements in Applied Machine-learning and Data Analytics (AAMDA 2021)

Short Papers

General Track 3: Geometric Modeling, Graphics and Visualization

Construction of Polyhedra Whose Vertices are Points on Curve Which Lying on Lemniscatic Torus with Mathematica

Ricardo Velezmoro-León[(✉)] [ID], Robert Ipanaqué-Chero[(✉)] [ID], Marcela Velásquez Fernández[(✉)] [ID], and Jorge Jimenez Gomez[(✉)] [ID]

Universidad Nacional de Piura, Urb. Miraflores S/n Castilla, Piura, Peru
{rvelezmorol,ripanaquec,fvelasquezf}@unp.edu.pe

Abstract. Polyhedra are widely used in art, science and technology. Faced with this situation, the following research question is formulated: Can new polyhedral structures be generated from another mathematical object such as a lemniscatic torus? For this question, the results we obtained are two particular cases whose vertices are points that belong to curves that lie on a lemniscatic torus: the first, a new polyhedron that has regular trapezoids in the equatorial zone, and the second, one that has triangles equal to each other. For both polyhedra, there exists an antipodal symmetry in the Arctic and Antarctic zones. Emphasis is placed on the construction of two convex polyhedra above mentioned: a one with 18-faces and other with 36-faces, using the scientific software *Mathematica v.11.2*. We also determine their total areas which respectively approximate 9.51 R^2 and 10.44 R^2. Likewise, the volume of each one is approximately 2.41 R^3 and 4.19 R^3, respectively. Moreover, they being inscribed in a sphere of radius R, and their opposite faces are not parallel.

Keywords: Lemniscatic torus · Polyhedral · Wolfram mathematica

1 Introduction

In general, a torus of revolution is defined considering a circle of radius $b > 0$ in the plane $y = 0$, whose center lie at the point $(a, 0, 0)$ with $a > b$. If a is on the circle, then we rotate it around the z-axis. Next we generate a surface called Torus of Revolution [1]. But according to [2,5,6], lemniscatic tori are generated taking into account the parametric definition of a lemniscate, which lies on a plane parallel to the xz-plane and rotating it in the z-axis.

It is interesting to study polyhedra due to their connection in the field of science, these being the basic structures of microorganisms, forms of tissues and crystals, and architecture [4].

Unlike [8], where a dodecahedron construction was made starting from a regular pentagon, in this paper to construct the 18-faces polyhedron we start from a particular curve that lies on a lemniscate torus which has its equatorial zone in

© Springer Nature Switzerland AG 2021
O. Gervasi et al. (Eds.): ICCSA 2021, LNCS 12950, pp. 3–17, 2021.
https://doi.org/10.1007/978-3-030-86960-1_1

the xy-plane. And taking values for the argument t of the curve we obtain two hexangles, which lie in the positive and negative part of the z-axis. From this a sphere is constructed where the 18-faces polyhedron is inscribed. By means of an analogous process the 36-faces polyhedron is also constructed. In this way, for first one, we obtain a new polyhedra that has regular trapeziums in the equatorial zone, And for second one, we no longer get trapeziums mentioned above, but triangles equal to each other. For both polyhedra, there exists antipodal symmetry in the arctic and antarctic zones.

The lateral area, total area and volume are calculated to these polyhedra, making use of geometric results of areas and volumes of 3D objects already known to the student. The generation of these polyhedra will allow students to acquire spatial and programming skills, as well as a high capacity for deductive reasoning. It will also stimulate interest in seeking the connections of these new structures with the field of science [7].

The structure of this paper is as follows: Sect. 2 introduces the mathematical definitions of the lemniscate, a lemniscatic torus, and two curves that lie on such a lemniscatic torus. In addition, two different partitions are defined on the domain of the curve to produce control points that generate the polyhedra. Then, Sect. 3 introduces the construction of a new 18-faces irregular polyhedron from 12 points which are vertices of two hexangles, and to these points two suitable points are added as poles for the construction of the 18-faces polyhedron inscribed in a R-radius sphere. Here, the lateral areas, total, and the volume of this polyhedron are also calculated. Next, in Sect. 4 an irregular polyhedron with 36 faces is built inscribed in a R-radius sphere. This polyhedron is generated from 20 points of which 18 are vertices of two regular nonagons that are near the equatorial zone, and 2 are the poles. Additionally, lateral and total area, and volume of this polyhedron are calculated. Finally, Sect. 5 closes with the main conclusions of this paper.

2   Mathematical Preliminaries

In general form, we define a lemniscate in the following way:

$$Lemniscate(v, A, B) = \left(B, \frac{A \cos v}{1 + \sin^2 v}, \frac{A \cos v \sin v}{1 + \sin^2 v} \right), \quad 0 < v < 2\pi,$$

which rests on the plane $x = B$ and also this lemniscate is inscribed in a circle that is on the same plane $x = B$ and has a radius A. Keep in mind rotation matrix

$$rot(u) = \begin{pmatrix} \cos u & -\sin u & 0 \\ \sin u & \cos u & 0 \\ 0 & 0 & 1 \end{pmatrix}.$$

The Lemniscatic Torus TL [6] is defined by:

$$TL(u, v, A, B) = Rot(u) \cdot (Lemniscate(v, A, B))^t; 0 < u < 2\pi, \quad -\frac{\pi}{2} < v < \frac{\pi}{2}.$$

If we perform the matrix product, we have the following:

$$TL((u,v),(A,B)) = \left(B\cos u - \frac{A\cos v \sin u}{1+\sin^2 v}, \right.$$
$$\left. B\sin u + \frac{A\cos v \cos u}{1+\sin^2 v}, \frac{A\cos v \sin v}{1+\sin^2 v} \right).$$

In above equation of the TL we have two constants A and B in $\mathbb{R}$ and two parameters u and v. The parameters will be replaced by differentiable ordinary functions. And so we define the curves in the following way:

$$CTL((u,v),(A,B),(\alpha_1,\alpha_2),t) = TL[(u,v),(A,B)] \; /. \; \{u \to \alpha_1, v \to \alpha_2\} \quad (1)$$

Here we have infinite possibilities of obtaining curves that will be resting on the TL. To obtain these possibilities we define the Lemniscatic Torus in Mathematica, using (1):

$$TL[\{u_-,v_-\},\{A_-,B_-\}] := \left\{ BCos[u] - \frac{(ACos[v]Sin[u])}{(1+Sin[v]^2)}, \right.$$
$$\left. BSin[u] + \frac{(ACos[u]Cos[v])}{(1+Sin[v]^2)}, \frac{(ACos[v]Sin[v])}{(1+Sin[v]^2)} \right\}$$

To visualize the lemniscathic torus, we will use:

$$g1 = ParametricPlot3D[TL[\{u,v\},\{2,1\}],\{u,0,2\pi\},\{v,-Pi/2,Pi/2\},$$
$$PlotStyle \to \{Yellow, Opacity[0.75]\}]$$

As a particular case, we have the given definition of the curves in the TL in Mathematica [8]:

$$CTL[t_-,\{A_-,B_-\},\alpha\colon \{_-,_-\}] := TL[\{u,v\},\{A,B\}] \; /. \; \{u \to \alpha[[1]], v \to \alpha[[2]]\}$$

Let $A = 2$ and $B = 1$, we get

$$\left\{ Cos[2t] - \frac{2Cos[3t]Sin[2t]}{1+Sin[3t]^2}, Sin[2t] + \frac{2Cos[2t]Cos[3t]}{1+Sin[3t]^2}, \frac{2Cos[3t]Sin[3t]}{1+Sin[3t]^2} \right\}$$

However, we give the following definition to improve the manipulation of the points that are immersed in the generation of polyhedra.

$$cur[t_-] := CTL[t,\{2,1\},\{2t,3t\}]$$

We can observe a sketch by

$$g2 = ParametricPlot3D[Cur[t],\{t,0,2Pi\}, PlotStyle \to Hue[0.75]]$$

and by $Show[g1,g2]$ we observe the representation of the curve together with the lemniscatic torus (see Fig. 1).

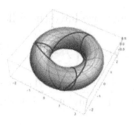

Fig. 1. Lemniscatic torus with $A = 2$, $B = 1$, and the curve $cur[t]$.

3   18-Faces Polyhedron

For domain $t \in [0, 2\pi]$ of the previous curve we will take the following values for the parameter t in the following order $t = \frac{\pi}{4}, \frac{\pi}{12}, \frac{5\pi}{12}, \frac{5\pi}{4}, \frac{19\pi}{12}, \frac{17\pi}{12}, \frac{7\pi}{4}, \frac{7\pi}{12}, \frac{11\pi}{12}, \frac{3\pi}{4}, \frac{13\pi}{12}, \frac{23\pi}{12}$, and then we evaluate the curve at these points in this way:

$$\{p1, p2, p3, p4, p5, p6, p7, p8, p9, p10, p11, p12\} = Cur/@\,\{Pi/4, Pi/12, 5Pi/12,$$
$$5Pi/4, 19Pi/12, 17Pi/12, 7Pi/4, 7Pi/12, 11Pi/12, 3Pi/4, 13Pi/12, 23Pi/12\}$$

The following points are obtained:

$$\left\{ \left\{ \frac{2\sqrt{2}}{3}, 1, -\frac{2}{3} \right\}, \left\{ -\frac{\sqrt{2}}{3} + \frac{\sqrt{3}}{2}, \frac{1}{2} + \sqrt{\frac{2}{3}}, \frac{2}{3} \right\}, \left\{ \frac{\sqrt{2}}{3} - \frac{\sqrt{3}}{2}, \frac{1}{2} + \sqrt{\frac{2}{3}}, \frac{2}{3} \right\}, \right.$$

$$\left\{ -\frac{2\sqrt{2}}{3}, 1, -\frac{2}{3} \right\}, \left\{ -\frac{\sqrt{2}}{3} - \frac{\sqrt{3}}{2}, -\frac{1}{2} + \sqrt{\frac{2}{3}}, -\frac{2}{3} \right\}, \left\{ -\frac{\sqrt{2}}{3} - \frac{\sqrt{3}}{2}, \frac{1}{2} - \sqrt{\frac{2}{3}}, \frac{2}{3} \right\},$$

$$\left\{ -\frac{2\sqrt{2}}{3}, -1, \frac{2}{3} \right\}, \left\{ \frac{\sqrt{2}}{3} - \frac{\sqrt{3}}{2}, -\frac{1}{2} - \sqrt{\frac{2}{3}}, -\frac{2}{3} \right\}, \left\{ -\frac{\sqrt{2}}{3} + \frac{\sqrt{3}}{2}, -\frac{1}{2} - \sqrt{\frac{2}{3}}, -\frac{2}{3} \right\},$$

$$\left. \left\{ \frac{2\sqrt{2}}{3}, -1, \frac{2}{3} \right\}, \left\{ \frac{\sqrt{2}}{3} + \frac{\sqrt{3}}{2}, \frac{1}{2} - \sqrt{\frac{2}{3}}, \frac{2}{3} \right\}, \left\{ \frac{\sqrt{2}}{3} + \frac{\sqrt{3}}{2}, -\frac{1}{2} + \sqrt{\frac{2}{3}}, -\frac{2}{3} \right\} \right\}$$

Taking the length of all these points, using: $Norm/@\{p1, p2, p3, p4, p5, p6, p7, p8, p9, p10, p11, p12\}$, we obtain that the length is constant, i.e., the points lie in the lemniscatic torus and the curve, and on the sphere of radius $\sqrt{\frac{7}{3}}$. We will select the points $\{p2, p3, p6, p7, p10, p11\}$ on the northern hemisphere that form a polygon whose sides measure $-\frac{2\sqrt{2}}{3} + \sqrt{3}$ and $\frac{4\sqrt{2}}{3}$:

$$\left\{ \left\{ -\frac{\sqrt{2}}{3} + \frac{\sqrt{3}}{2}, \frac{1}{2} + \sqrt{\frac{2}{3}}, \frac{2}{3} \right\}, \left\{ \frac{\sqrt{2}}{3} - \frac{\sqrt{3}}{2}, \frac{1}{2} + \sqrt{\frac{2}{3}}, \frac{2}{3} \right\}, \right.$$

$$\left\{ -\frac{\sqrt{2}}{3} - \frac{\sqrt{3}}{2}, \frac{1}{2} - \sqrt{\frac{2}{3}}, \frac{2}{3} \right\}, \left\{ -\frac{2\sqrt{2}}{3}, -1, \frac{2}{3} \right\}, \left\{ \frac{2\sqrt{2}}{3}, -1, \frac{2}{3} \right\},$$

$$\left. \left\{ \frac{\sqrt{2}}{3} + \frac{\sqrt{3}}{2}, \frac{1}{2} - \sqrt{\frac{2}{3}}, \frac{2}{3} \right\} \right\}$$

and from now on we will call it hexangle. It should be understood that a hexangle is a 3-equal-interspersed-sides 6-sides-polygon.

In the same way, for the points $\{p4, p5, p8, p9, p12, p1\}$ on the southern hemisphere, namely:

$$\left\{\left\{-\frac{2\sqrt{2}}{3}, 1, -\frac{2}{3}\right\}, \left\{-\frac{\sqrt{2}}{3} - \frac{\sqrt{3}}{2}, -\frac{1}{2} + \sqrt{\frac{2}{3}}, -\frac{2}{3}\right\},\right.$$

$$\left\{\frac{\sqrt{2}}{3} - \frac{\sqrt{3}}{2}, -\frac{1}{2} - \sqrt{\frac{2}{3}}, -\frac{2}{3}\right\}, \left\{-\frac{\sqrt{2}}{3} + \frac{\sqrt{3}}{2}, -\frac{1}{2} - \sqrt{\frac{2}{3}}, -\frac{2}{3}\right\}$$

$$\left.\left\{\frac{\sqrt{2}}{3} + \frac{\sqrt{3}}{2}, -\frac{1}{2} + \sqrt{\frac{2}{3}}, -\frac{2}{3}\right\}, \left\{\frac{2\sqrt{2}}{3}, 1, -\frac{2}{3}\right\}\right\}$$

it follows that they form the same hexangle as the one on the northern hemisphere. The two hexangles define a zone which we will call: the equatorial zone, which consists of 12 points. We can add the north pole point $pn = \left\{0, 0, \sqrt{\frac{7}{3}}\right\}$ and the south pole point $ps = \left\{0, 0, -\sqrt{\frac{7}{3}}\right\}$ of the sphere centered at the origin. Using Mathematica, we can visualize the northern and southern hemispheres with their respective poles (see Fig. 2a and Fig. 2b).

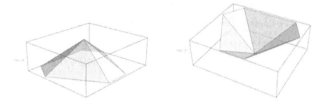

(a) Northern hemisphere, it is a pyramid

(b) Southern hemisphere, it is a pyramid

Fig. 2. Northern and southern Hemisphere

In the equatorial zone we have isosceles trapeziums, see Fig. 3a, it is also verified that the outline of the roof and basis of the figure are hexangles Fig. 3b. When we paste Fig. 2a, Fig. 2b and Fig. 3a we obtain polyhedron 1, see Fig. 3c.

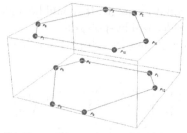

(a) They are six isosceles trape- zoids.

(b) From Fig.3a the roof and basis are hexangle.

(c) Sketch of the obtained polyhe- dron.

Fig. 3. Quadrilaterals of equatorial zone and 18-faces polyhedron.

The points of the lemniscatic torus were normalized to generalize them to polyhedra inscribed in a sphere of radius R.

$$Cur[t\_] := \left\{ Cos[2t] - \frac{2Cos[3t]Sin[2t]}{1 + Sin[3t]^2}, Sin[2t] + \frac{2Cos[2t]Cos[3t]}{1 + Sen[3t]^2}, \right.$$

$$\left. \frac{2Cos[3t]Sin[3t]}{1 + Sin[3t]^2} \right\};$$

$NewPolyhedron[R\_] := Module[\{nn = 4, n, polo1, polo2, p1, p2, p3, p4, p5, p6,$
$p7, p8, p9, p10, p11, p12, lisnNew1, lisNew2, lisnNew3, lisNew4, lisnNew5,$
$lisNew6, lis1, lis2, lis3\}, \{p1, p2, p3, p4, p5, p6, p7, p8, p9, p10, p11, p12\} = R$
$Normalize/@(Cur/@\{Pi/4, Pi/12, 5Pi/12, 5Pi/4, 19Pi/12, 17Pi/12, 7Pi/4,$
$7Pi/12, 11Pi/12, 3Pi/4, 13Pi/12, 23Pi/12\}); lis1 = \{p1, p2, p3, p4, p5, p6, p7,$
$p8, p9, p10, p11, p12\}; lis2 = \{p2, p3, p6, p7, p10, p11\}; lis3 = \{p4, p5, p8, p9, p12,$
$p1\}; listNew1 = Partition[Append[Append[lis1, lis1[[1]]], lis1[[2]]], nn, 2];$
$lisNew2 = Polygon/@Table[Append[lisNew1[[i]], lisNew1[[i]][[1]]], \{i,$
$Length[lisNew1]\}]; n = length[lis2]; polo1 = \{0, 0, R\}; Polo2 = \{0, 0, -R\};$

$lisNew3 = Partition[Append[Append[lis2, lis2[[1]]], lis2[[2]]], 2, 1]; lisNew4 =$
$Polynog/@Table[Append[Append[lisNew3[[i]], polo1], lisNew3[[i]][[1]]], \{i, n\}];$
$lisNew5 = Partition[Append[Append[lis3, lis3[[1]]], lis3[[2]]], 2, 1]; lisNew6 =$
$Polygon/@Table[Append[Append[lisNew5[[i]], polo2], lisNew5[[i]][[1]]], \{i, n\}];$
$Join[lisNew4, lisNew2, LisNew6]]$

Let's visualize Fig. 4 when $R = 1$ and $R = 2$ using above code.

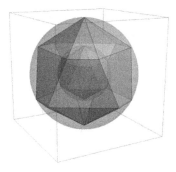

Fig. 4. There are two polyhedra, one is inscribed in a sphere with radius 1 and the other in a sphere of radius 2.

To achieve a generic result in the sense that it is a polygon inscribed in a circle of radius R, we would have to normalize the points and multiply them by R, that is, the distance from each point to the origin is equal to $R > 0$.

$$lis = \{q1, q2, q3, q4, q5, q6, q6, q7, q8, q8, q9, q10, q11, q12\} =$$
$$R * Normalize/@\{p1, p2, p3, p4, p5, p6, p7, p8, p9, p10, p11, p12\}$$

So we have: $Norm/@lis//Simplify = \{R, R, R, R, R, R, R, R, R, R, R, R\}$

It is clear that the north pole is $pn = \{0, 0, R\}$ and the south pole is $ps = \{0, 0, -R\}$. Now, regarding the total area, we have the equatorial zone where there are six isosceles trapeziums. We also have two pyramids, one in the northern hemisphere and the other in the southern hemisphere, both are equal to each other, except for one isometry. This can be seen in Fig. 5.

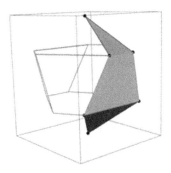

Fig. 5. Types of areas to be calculated in each region.

To find the area we need to automate the calculation of the area of an n-sided convex polygon based on the area of a triangle. For the area of the triangle:

$$AreaTri[P1 : \{\_,\_,\_\}, P2 : \{\_,\_,\_\}, P3 : \{\_,\_,\_\}\}] :=$$

$$\frac{1}{2}Norm[Cross[P2 - P1, P3 - P1]]$$

Area of the convex polygon: $AreaPolyConvex[P : \{\_,\_,\_\}..\}] := Module[\{A\},$ $A = AreaTri/@table[\{P[[1]], P[[i]], P[[i+1]]\}, \{i, 2, Length[P]-1\}]; Suma[A[[i]],$ $\{i, Length[A]\}]$

With respect to the trapeziums observed in Fig. 5, there are 6, and therefore the area would be: $a1 = 6AreaPolyConvex[\{q1, q2, q3, q4\}]//Simplify$

$$= \frac{1}{14} \left(\sqrt{4259 - 1584\sqrt{6}} + 4\sqrt{194 - 24\sqrt{6}} \right) R^2$$

Regarding to the congruent triangles of the northern and southern hemispheres identified by p_2, pn, p_3 there are six, see Fig. 5:

$$a2 = 6AreaTri[\{q2, \{0, 0, R\}, q3\}] = \frac{1}{14}\sqrt{3791 - 1176\sqrt{6} + 576\sqrt{14} - 560\sqrt{21}}R^2$$

Regarding to the congruent triangles of the southern and northern hemispheres identified by p_4, ps, p_5 there are six, see Fig. 5:

$$a3 = 6AreaTri[\{q4, \{0, 0, -R\}, q1\}]//Simplify = \frac{8}{7}\sqrt{17 - 2\sqrt{21}}R^2$$

Therefore, the total area of polyhedron 1 is:

$$a1 + a2 + a3//Simplify \approx 9.50799R^2$$

To calculate the volume we will carry out the following procedure: We first calculate the volume of the pyramid in the northern hemisphere and multiply it by 2, since the pyramid in the southern hemisphere is congruent with the northern one.

Let's calculate the center of gravity of the hexangle in the northern hemisphere:

$$cg = (q2 + q3 + q6 + q7 + q10 + q11)/6//Simplify = \left\{0, 0, \frac{2R}{\sqrt{21}}\right\}$$

Thus, we have that the height of the pyramid is obtained from:

$$\{0, 0, R\} - cg = \left\{0, 0, R - \frac{2R}{\sqrt{21}}\right\}$$

Therefore, the volume of the two pyramids is:

$$Vpira = \frac{2}{3}AreaPolyConvex[\{q2, q3, q7, q10, q11\}]\left(R - \frac{2R}{\sqrt{21}}\right)//Simplify$$

$$-\frac{(-21 + 2\sqrt{21})(19 + 8\sqrt{70 - 24\sqrt{6}} + 4\sqrt{70 + 24\sqrt{6}})R^3}{882\sqrt{3}}$$

Second, we calculate the volume of the equatorial zone, which is inscribed within a prism that has dodecangle bases, and the height is 2 times the distance from the origin to the center of gravity of the northern hemisphere hexangle: $h = \frac{4R}{\sqrt{21}}$. See Fig. 6a.

(a) Polyhedron in equatorial zone inscribed in a polyhedron whose bases are dodecangle.

(b) Polyhedral volume to removen.

Fig. 6. Calculating the volume of the Polyhedron in the equatorial zone inscribed in a polyhedron with dodecangle bases.

In Fig. 6a is shown that the part of the equatorial zone of polyhedron 1 is inscribed in a prism whose basis is a dodecangle. We define a dodecangle as a convex polygon with twelve angles and sides, with the characteristic that its internal angles are equal and its non-adjacent sides are equal.

Now, the volume of polyhedron 1 is the volume of the prism whose basis is the dodecangle, whose vertices are shown by sup, plus that of the two pyramids minus six times the volume of polyhedron, which is shown in Fig. 6b. The area of the basis is calculated based on the points:

$$sup = Table\left[Append\left[Delete\left[lis[[i]], -1\right], \frac{2}{\sqrt{21}}\right], \{i, 12\}\right] //Simplify$$

$$\left\{\left\{2\sqrt{\frac{2}{21}}R, \sqrt{\frac{3}{7}}R, \frac{2R}{\sqrt{21}}\right\}, \left\{\frac{\left(9 - 2\sqrt{6}\right)R}{6\sqrt{7}}, \frac{\left(3 + 2\sqrt{6}\right)R}{2\sqrt{21}}, \frac{2R}{\sqrt{21}}\right\}\right.$$

$$\left\{\frac{\left(-9 + 2\sqrt{6}\right)R}{6\sqrt{7}}, \frac{\left(3 + 2\sqrt{6}\right)R}{2\sqrt{21}}, \frac{2R}{\sqrt{21}}\right\}, \left\{-2\sqrt{\frac{2}{21}}R, \sqrt{\frac{3}{7}}R, \frac{2R}{\sqrt{21}}\right\},$$

$$\left\{-\frac{\left(9 + 2\sqrt{6}\right)R}{6\sqrt{7}}, \frac{\left(-3 + 2\sqrt{6}\right)R}{2\sqrt{21}}, \frac{2R}{\sqrt{21}}\right\}, \left\{-\frac{\left(9 + 2\sqrt{6}\right)R}{6\sqrt{7}}, \frac{\left(3 - 2\sqrt{6}\right)R}{2\sqrt{21}}, \frac{2R}{\sqrt{21}}\right\},$$

$$\left\{-2\sqrt{\frac{2}{21}}R, -\sqrt{\frac{3}{7}}R, \frac{2R}{\sqrt{21}}\right\}, \left\{\frac{\left(-9+2\sqrt{6}\right)R}{6\sqrt{7}}, -\frac{\left(3+2\sqrt{6}\right)R}{2\sqrt{21}}, \frac{2R}{\sqrt{21}}\right\},$$

$$\left\{\frac{\left(9-2\sqrt{6}\right)R}{6\sqrt{7}}, -\frac{\left(3+2\sqrt{6}\right)R}{2\sqrt{21}}, \frac{2R}{\sqrt{21}}\right\}, \left\{2\sqrt{\frac{2}{21}}R, -\sqrt{\frac{3}{7}}R, \frac{2R}{\sqrt{21}}\right\},$$

$$\left\{\frac{\left(9+2\sqrt{6}\right)R}{6\sqrt{7}}, \frac{\left(3-2\sqrt{6}\right)R}{2\sqrt{21}}, \frac{2R}{\sqrt{21}}\right\}, \left\{\frac{\left(9+2\sqrt{6}\right)R}{6\sqrt{7}}, \frac{\left(-3+2\sqrt{6}\right)R}{2\sqrt{21}}, \frac{2R}{\sqrt{21}}\right\}\right\}$$

So, the volume is:

$$Vdodec = (AreaPolyConvex[sup]Norma[P3 - P6])//Simplify = \frac{16}{7}\sqrt{\frac{6}{7}}R^3$$

We calculate the volume of the polyhedron in Fig. 6b, considering 8 labels for the points that will be needed to calculate the volume, where:

$$\{P1, P2, P3, P4, P5, P6\} =$$

$$\left\{\left\{2\sqrt{\frac{2}{21}}R, \sqrt{\frac{3}{7}}R, -\frac{2R}{\sqrt{21}}\right\}, \left\{\frac{\left(9-2\sqrt{6}\right)R}{6\sqrt{7}}, \frac{\left(3+2\sqrt{6}\right)R}{2\sqrt{21}}, \frac{2R}{\sqrt{21}}\right\},\right.$$

$$\left\{\frac{\left(-9+2\sqrt{6}\right)R}{6\sqrt{7}}, \frac{\left(3+2\sqrt{6}\right)R}{2\sqrt{21}}, \frac{2R}{\sqrt{21}}\right\}, \left\{-2\sqrt{\frac{2}{21}}R, \sqrt{\frac{3}{7}}R, -\frac{2R}{\sqrt{21}}\right\},$$

$$\left.\left\{\frac{\left(9-2\sqrt{6}\right)R}{6\sqrt{7}}, \frac{\left(3+2\sqrt{6}\right)R}{2\sqrt{21}}, -\frac{2R}{\sqrt{21}}\right\}, \left\{\frac{\left(-9+2\sqrt{6}\right)R}{6\sqrt{7}}, \frac{\left(3+2\sqrt{6}\right)R}{2\sqrt{21}}, -\frac{2R}{\sqrt{21}}\right\}\right\}.$$

Now, if we consider orthogonal projection of P_5, P_6 on P_4, P_1 we have two points P_8, P_7 which are obtained by $\|P_4 - P_8\| = \|P_7 - P_1\| = x$, $b = \|P_2 - P_3\|$ (minor basis) and $B = \|P_1 - P_4\|$ (major basis) of trapezium $P_1 P_5 P_6 P_4$. $b = Norm[P2 - P3]/Simplify$; $B = Norm[P1 - P4]//Simplify$; $Solve[B == 2x + b, x]$

We take the value of $x = \frac{1}{42}\left(-9\sqrt{7}R + 2\sqrt{42}R + 3\sqrt{77}R\right)$, and we obtain the point $P7 = P1 + x (P4 - P1)/Norm[P4 - P1] = \left\{\frac{(9+2\sqrt{6}-3\sqrt{11})R}{6\sqrt{7}}, \sqrt{\frac{3}{7}}R, -\frac{2R}{\sqrt{21}}\right\}$ Similarly we obtain the point P_8.

$$P8 = P1 + (x + b)(P4 - P1)/Norm[P4 - P1]//Simplify$$

$$\left\{\frac{(-3+2\sqrt{6}-\sqrt{11})R}{2\sqrt{7}}, \sqrt{\frac{3}{7}}R, -\frac{2R}{\sqrt{21}}\right\}$$

Now we calculate the volume of the polyhedron $P_5, P_6, P_8, P_7, P_2, P_3$ which is half of a prism:

$$Vprisma = \frac{1}{2}\left(AreaPolConvex\left[\{P5, P6, P8, P7\}\right] Norm\left[P3 - P6\right]\right) \approx 0.047R^3$$

Because of tetrahedra volumes $P_4P_8P_6 - P_3$ and $P_1P_5P_7 - P_2$ are congruent, we have:

$$V2 = 2\left(\frac{1}{3}AreaPolyConvex\left[\{P4, P8, P6\}\right]Norm\left[P3 - P6\right]\right) \approx 0.0210578R^3$$

Thus, we have the volume of Fig. 6b:

$$Vprisma + V2//Simplify = -\frac{\left(2 - 8\sqrt{6} - 3\sqrt{11} + 2\sqrt{66}\right)}{63\sqrt{7}}R^3$$

Therefore, the volume of polyhedron 1 is:

$$Vpoliedro1 = Vdodec + Vpira - 6\left(Vprisma + V2\right)//FullSiimplify$$

$$\frac{1}{294}\left(252\sqrt{2} + 7\sqrt{3} + 6\sqrt{7} + 40\sqrt{42} + 4\sqrt{77}\left(-3 + 2\sqrt{6}\right)\right)R^3$$

4   36-Faces Polyhedron

We define the curve from the lemniscatic torus when $A = 8$, $B = 5$ and $\alpha_1(t) = 5t$, $\alpha_2(t) = 7$ with Mathematica.

$$Cur1[t] := CTL[t, \{8, 5\}, \{t, 9t\}]$$
$$Cur1[t]$$

$$\left\{5Cos[t] - \frac{8Cos[9t]Sin[t]}{1 + Sin[9t]^2}, 5Sin[t] + \frac{8Cos[t]Cos[9t]}{1 + Sin[9t]^2}, \frac{8Cos[9t]Sin[9t]}{1 + Sin[9t]^2}\right\}$$

Now, we vary the parameter t in the order given:

$$t = \frac{\pi}{4}, \frac{3\pi}{4}, \frac{17\pi}{36}, \frac{35\pi}{36}, \frac{25\pi}{36}, \frac{43\pi}{36}, \frac{11\pi}{12}, \frac{17\pi}{12}, \frac{41\pi}{36}, \frac{59\pi}{36}, \frac{49\pi}{36}, \frac{67\pi}{36}, \frac{19\pi}{12},$$
$$\frac{\pi}{12}, \frac{65\pi}{36}, \frac{11\pi}{36}, \frac{\pi}{36}, \frac{19\pi}{36}$$

$m1 = \{r1, r2, r3, r4, r5, r6, r7, r8, r9, r10, r11, r12, r13, r14, r15, r16, r17, r18\} = $
$Normalize/@Cur1/@\{\pi/4, 3\pi/4, 17\pi/36, 35\pi/36, 25\pi/36, 43\pi/36, 11\pi/12, 17\pi/12,$
$41\pi/36, 59\pi/36, 49\pi/36, 67\pi/36, 19\pi/12, \pi/12, 65\pi/36, 11\pi/36,$
$\pi/36, 19\pi/36\}//Simplify;$

Taking the lists

$r2, r4, r6, r8, r10, r12, r14, r16, r18$ and $r1, r3, r5, r7, r9, r11, r13, r15, r17$

two regular nonagons are obtained which will define the equatorial zone together with 18 congruent triangles, as shown in the Fig. 7.

Fig. 7. Regular nonagons in the equatorial zone.

To form the zone of the southern hemisphere we will have to add $\{0, 0, -1\}$:

$$ps = \{\{r2, r4, r6, r8, r10, r12, r14, r16, r18\}, \{0, 0 - 1\}\}$$

To form the zone of the northern hemisphere we will have to add $\{0, 0, 1\}$:

$$pn = \{\{r1, r3, r5, r7, r9, r11, r13, r15, r17\}, \{0, 0, 1\}\}$$

Fig. 8. Constructions in the north pole and south pole area.

Uniting the equatorial zone together with the zone of the southern and northern hemisphere (see Fig. 8) we have the polyhedron and the polyhedron inscribed in a unit sphere with three points $r2, r1, r3$, as shown in Fig. 9. The three points are used to find the area of the equatorial zone.

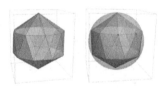

Fig. 9. Polyhedron 2 and polyhedron 2 with 3 points to determine the area located in the equatorial zone.

The one with 18 similar triangles between the northern and southern hemispheres also has 18 congruent triangles in the equatorial zone, but they are different from those in the southern and northern hemispheres, so this polyhedron 2 has 36 faces, 54 edges and 20 vertices of which 18 are on the lemniscatic torus and 2 in the sphere of unit radius. In addition Euler's formula is satisfaced: $36 - 54 + 20 = 2$. Now, let's calculate the total area of polyhedron 2: First, we calculate the area of a triangle in the southern hemisphere and multiply it by 18. To do this we take $\{r2, r4, \{0, 0, -1\}\}$ and the south pole.

$$azhnzhs = 18 * AreaPolyConvex[\{r2, r4, \{0, 0, -1\}\}]//N \approx 5.99$$

Equatorial zone area: $aze = 18 * AreaPolyConvex[\{r2, r1, r3\}]//N \approx 4.45$

So the total area is: $azhnzhs + aze//N \approx 10.44$

To calculate the volume of the polyhedron 2: First we calculate the volume of the pyramid of the southern hemisphere, whose vertices are: $\{r2, r4, r6, r8, r10, r12, r14, r16, r18\}$, and the height $h1$ is: $h1 = 1 - \frac{h}{2} = 1 - \frac{8}{\sqrt{417}}$. So the volume of the pyramid in the southern hemisphere is:

$$Vpisur = \frac{1}{3} AreaPolyConvex\left[\{r2, r4, r6, r8, r10, r12, r14, r16, r18\}\right]$$

$$h1//Simplify \approx 0.361364$$

The pyramid in the northern hemisphere has the same value as the volume of the pyramid in the southern hemisphere.

Second, we construct two 18-sided polygons, at the top and bottom of the equatorial zone, with the following lists:

$$polsup = Table\left[Append\left[Delete\left[m1\left[[i]\right], -1\right], \frac{8}{\sqrt{417}}\right], \{i, 1, 18\}\right];$$

$$polinf = Table\left[Append\left[Delete\left[m1\left[[i]\right], -1\right], -\frac{8}{\sqrt{417}}\right], \{i, 1, 18\}\right];$$

Fig. 10. 18-sided polygons, top to bottom of the equatorial zone, with the *polsup* and *polinf* lists.

In Fig. 10 it is shown that the part of the equatorial zone of polyhedron 2 is inscribed in a prism whose bases are regular nonagon.

Furthermore, we project rays, all parallel to z-axis and we will choose a tetra-hedron as shown in the Fig. 10 (right). The four points involved are $polinf[[1]]$,

$polinf[[2]]$, $polinf[[18]]$ for the basis and $polsup[[1]]$ for the top vertex, the height of the tetrahedron is:

$$h = Norm\left[polinf\left[[2]\right] - polsup\left[[2]\right]\right] //Simplify$$

Thus, we have that the volume of the tetrahedron in question is:

$$Vt = \left(\frac{1}{3}\right) AreaPolyConvex\left[\{polinf\left[[1]\right], polinf\left[[2]\right], polinf\left[[18]\right]\}\right]$$

$h//Simplify \approx 0.00437139$

Now, we calculate the volume of the prism, whose basis is $polinf$ or $polsup$ and the height is the same as that of the tetrahedron, said volume is:

$$Vpri = AreaPolyConvex[polinf]h//N \approx 2.04$$

Therefore, the volume of the equatorial zone is:

$$Vzecua = Vpri - 18Vt//N \approx 1.96$$

Consequently, we get that the volume of polyhedron 2 is:

$$Vpoliedro2 = 2 * Vpisur + Vzecua - 18Vt//FullSimplify = \frac{4}{3}\pi \approx 4.19$$

5   Conclusions

This article illustrates an approach based on the use of numerical and symbolic programming of the scientific software Mathematica v.11.2 [1]. This approach is used for the construction and analysis of the particular characteristics of the two convex polyhedra, the 18-faces one and the 36-faces one. We started from the equation of the parametrically defined lemniscatic torus. If we substitute the parameters for differentiable ordinary functions, there are infinite possibilities of curves that lie on the lemniscatic torus. Particular cases were taken in such a way that when evaluating the curves, we obtained points that were taken as the vertices of such polyhedra, except for the vertices that coincide with the poles of the sphere that inscribes them.

The construction of the polyhedra under study exhibits interesting characteristics in its construction. In the 18-faces polyhedron, on the faces of the equatorial zone, two roof and basis hexangles were found, whose internal angles are equal and their non-adjacent sides are equal. Likewise, it was observed that the same region is inscribed in a prism whose basis is a dodecangle with equal internal angles and equal non-adjacent sides. Similarly, the 36-faces polyhedron was studied. On the sides of the equatorial zone, a regular nonagon was found as a roof and basis. The interesting thing about this region is that it is inscribed in a prism whose basis is an octadecangle with the characteristic that its internal angles are equal and its non-adjacent sides are equal.

The total area of the polyhedron of 18 and 36 faces was calculated, which is approximately 9.51 R^2 and 10.44 R^2, respectively. Also, the volume is 2.41 R^3 and 4.19 R^3, respectively. In addition, it was found that they satisfy Leonhard Euler's theorem, which indicates the relationship between the number of faces, edges and vertices of a convex polyhedron: $C - R + V = 2$.

The generation of these polyhedra will allow students to acquire spatial and programming skills, as well as a high capacity for deductive reasoning. It will also stimulate interest in seeking the connections of these new structures with the field of science.

Future research includes the extension of the study of the isometries that leave invariant convex irregular polyhedra with 18 and 32 faces. Likewise, the construction of new polyhedra generated from the lemniscatic torus and possible ranges of application in different fields such as morphogeometry can also be carried out.

References

1. Hasser, N., LaSalle, J., Sullivan, J.: A Course in Mathematical Analysis, vol. 2. Blaisdell Publishing Company, New York (1964)
2. Vega, S., Silupu, C.: Construcción de toros de revolución, apartir de curvas planas y espaciales con curvatura no constante o torsión no nula, utilizando el Mathematica. Universidad Nacional de Piura, Piura-Perú (2018)
3. Torrence, F., Torrence, A.: The Student's Introduction to Mathematica and the Wolfram Language. Cambridge University Press, United Kingdom (2019)
4. Briz, A., Serrano, A.: Learning Mathematics through the R Programming Language in Secondary Educ. Mat. **30**(1), 133-162 (2018) https://doi.org/10.24844/em3001. 05
5. Ipanaque, R., Iglesias, A., Velezmoro, R.: Symbolic computational approach to construct a 3D torus via curvature. In: Proceedings of Fourth International Conference Ubiquitous Computing and Multimedia Applications, UCMA Lecture Notes in Computer Science, vol. 22 (2013)
6. Ipanaque, R., Iglesias, A., Velezmoro, R.: Parameterization of some surfaces of revolution through curvature-varying curves: a computational analysis. Int. J. Hybrid Inf. Technol. **6** (2013)
7. Sinclair, N., et al.: Recent research on geometry education: an ICME-13 survey team report. ZDM **48**(5), 691–719 (2016). https://doi.org/10.1007/s11858-016-0796-6
8. Velezmoro León, R., Velásquez Fernández, M., Jimenez Gomez, J.: Construction of the Regular Dodecahedron with the MATHEMATICA. In: Gervasi, O., et al. (eds.) ICCSA 2020. LNCS, vol. 12249, pp. 360–375. Springer, Cham (2020). https://doi. org/10.1007/978-3-030-58799-4_27

Assessment of Linear and Non-linear Feature Projections for the Classification of 3-D MR Images on Cognitively Normal, Mild Cognitive Impairment and Alzheimer's Disease

Marcelo R. Moura Araújo⬤, Katia M. Poloni⬤, and Ricardo J. Ferrari$^{(\boxtimes)}$⬤

Department of Computing, Federal University of São Carlos, São Carlos,
SP 13565-905, Brazil
rferrari@ufscar.br
https://www.bipgroup.dc.ufscar.br

Abstract. Alzheimer's disease (AD) is an age-related neurodegenerative disease and the most common form of dementia. It is a brain disorder that impacts the daily life of the patient due to memory loss and cognitive changes. Due to population aging and the fact that dementia incidence rising sharply at ages greater than 75, AD has become a major public health problem. Currently, hippocampal atrophy assessed on structural magnetic resonance (MR) images is the most used imaging biomarker of AD. Among many methods applied for automated classification of cognitively normal (CN), mild cognitive impairment (MCI), and AD, the linear PCA projection method, also known as eigenbrain, has shown effectiveness in AD subject prediction even with its restriction of using only linear projections. This study investigates both linear (PCA) and non-linear (kernel PCA) projection method performances to classify 3-D structural MR images on the CN, MCI, and AD classes. Support vector machines (SVMs) were trained using feature vectors with different space dimensions obtained by projecting MR study images into the previously created "eigenbrain spaces." We tested the method using different kernel functions for both cases, the eigenbrain projections and SVM classifiers. We also conducted our analyses separately using the whole-brain and the gray-matter (GM) regions. Comparison results between PCA and kernel PCA methods showed that non-linear projections improved the classification of MR images in both class groups, particularly when processing the GM region - CN × MCI (PCA: AUC = 0.63 versus KPCA: AUC = 0.76), MCI × DA (PCA: AUC = 0.67 versus KPCA: AUC = 0.74), and CN × AD (PCA: AUC = 0.86 versus KPCA: AUC = 0.89).

Keywords: Alzheimer's disease · Eigenbrain · kernel PCA · Support vector machines · Magnetic resonance imaging

© Springer Nature Switzerland AG 2021
O. Gervasi et al. (Eds.): ICCSA 2021, LNCS 12950, pp. 18–33, 2021.
https://doi.org/10.1007/978-3-030-86960-1_2

1   Introduction

Alzheimer's disease (AD) is the most devastating brain disorder of elderly humans and the most common cause of dementia, accounting for 70 to 80% of all cases. There is currently no cure or even an effective treatment available that will alter the progressive course of AD, which can eventually lead to death. In 2019, nearly 50 million patients were diagnosed with AD globally, with a prediction of 115.5 million by 2050. As the world's population is rapidly aging, AD has become one of the most critical public health problems [35].

Histologically, AD is caused by neuronal death, caused mainly by two factors: the excessive accumulation of amyloid-β protein extracellular plaques that form senile plaques [11], which have a toxic effect on neurons and blocks the synaptic connections; and the deposition of hyperphosphorylated tau protein in conjunction with the neuronal microtubules disassociation, that creates intracellular neurofibrillary tangles [6], which causes a fail on transportation of nutrients and other essential elements between cells, causing neuronal death. While many neurotransmitters are affected, there is widespread loss of neurons containing acetylcholine, which correlates well with the degree of cognitive impairment observed clinically. Both neurofibrillary tangles and senile plaques occur in normal aging but are more numerous and widespread in AD [9]. This entire process is responsible for the inevitable downstream consequence of neurodegeneration that is brain atrophy.

Currently, the diagnosis of AD is based on clinical criteria in which doctors conduct tests to assess memory impairment and other thinking skills of a patient. They also perform a series of tests to rule out other possible causes of impairment. In addition to the clinical assessment, magnetic resonance imaging (MRI) has been playing an increasingly important role in helping doctors to identify structural brain changes and, consequently, to support their decision-making process [10,13], since this imaging technique allows to obtain high-quality images to observe subtle changes in anatomical brain structures [13]. Such information has also been used to build imaging biomarkers for AD.

Many computerized image analysis methods have been published in the last decade to help radiologists diagnose AD more accurately [2,4,5,8,25,36]. These methods use image information extracted either from magnetic resonance (MR), positron emission tomography (PET), or single-photon emission computed tomography (SPECT), with some methods concentrating the analysis on the whole brain [4,23] while others on specific brain regions - region of interest (ROI) - such as the hippocampus [5,19].

Among the methods, the eigenbrain [3,4,22,29,30,38–40], which is a relatively simple multivariate approach based on the Principal Component Analysis (PCA) technique [17], has shown a competitive performance compared to other more complex approaches, such as atlas-based [24] and deep learning [16,20]. Zhang et al. [39], for instance, selected 2D coronal slices with inter-class variance and created an eigenbrain set with PCA. They selected the first projection weights (PWs) for each 2D slice and used them as feature inputs to an SVM classifier to aid AD diagnosis. Moreover, Kheder et al. [22] exploited the eigenbrain construction using partial least squares (PLS) since the PLS considers the samples' variance and the class labels and compared them with the PCA. They used only GM voxel intensities and inserted the most discriminant PWs as features into an SVM to distinguish patients with MCI compared to CN and AD. In the same year, Salvatore et al. [29] has also created eigenbrain sets. The authors have assessed the method with GM and WM voxel intensities and have ordered the PWs by the Fisher Discriminant Ratio to include the class discriminative power and used them to predict the conversion from MCI to AD.

Despite its performance, one limitation of the original eigenbrain method is the linear orthogonal projections that may fail to capture the underlying structure of the data. Aiming to overcome this limitation, Alam et al. [3] have created eigenbrains with a slightly modified version of this method that uses a non-linear generalization of PCA, the kernel PCA (KPCA). The authors have created the eigenbrain set from GM voxel intensities and used the LDA for feature selection and inclusion of the class as a criterion; they compared the PCA with the Gaussian kernel PCA. In this study, we conducted a comparative assessment of the original eigenbrain (standard PCA) with four different PCA kernels [31] using the GM and the whole-brain voxel intensities. We compared the sigmoid, radial basis function (RBF), cosine, and Polynomial PCA kernels with the (linear) PCA and evaluated their predictive power using only the PWs as features, without including any class information. The methods were evaluated using image information extracted from the whole-brain image and the gray matter (GM) tissue regions.

Using feature vectors obtained by projecting each ADNI training image from the CN, MCI, and AD groups into a previously defined PCA and KPCA "eigenbrain spaces," we designed SVM classifiers with three different kernels to classify unseen MR images in the CN × MCI, MCI × AD, and CN × AD cases. The classification results of our experiments were assessed using different quantitative metrics.

The rest of this paper is structured as follows: Sect. 2 presents the image datasets used in this study. Section 3 describes the developed methods, providing details on how the feature extraction is performed. Section 4 present the results and discussions of our experiments, and, finally, Sect. 5 concludes the paper.

2   Image Datasets

This study uses two image datasets: the Alzheimer's Disease Neuroimaging Initiative[1] (ADNI) and the Neuroimage Analysis Center[2] (NAC).

The ADNI dataset includes T1-weighted (T1-w) images acquired using a wide variety of 1.5T and 3T scanners and protocols from the three leading manufacturers (Philips, General Electric, and Siemens), so the image specifications (size, resolution, etc.) are diverse and will be omitted here. For this reason, as we will discuss further, the images were standardized to a reference image previously to

---

[1] http://adni.loni.usc.edu/.
[2] http://projects.iw.harvard.edu/nac.

processing. A total of 774 T1-w images, 258 for each group (CN, MCI, and AD), were used in this study. According to the ADNI documentation, CN patients showed no signs of depression, mild cognitive impairment, or dementia. Also, mild-AD patients were evaluated and met established criteria for probable AD. Table 1 summarizes relevant patient information from this dataset, including sex, age, and clinical information.

Table 1. Clinical and demographic data for CN, MCI and AD patients.

Group	#Subjects	Age $(\mu \pm \sigma)$ [range]	Sex (M/F)	MMSE score $(\mu \pm \sigma)$ [range]
CN	258	74.0 ± 6.3 [58–90]	114/144	29.1 ± 1.1 [25–30]
MCI	258	74.7 ± 8.0 [54–97]	164/94	26.8 ± 1.9 [20–30]
AD	258	75.8 ± 8.1 [55–91]	140/1118	21.8 ± 4.4 [8–28]

MMSE stands for mini-mental state exam.

The NAC dataset provides a T1-w reference image with an isotropic spatial resolution of 1 mm and a matrix size of $256 \times 256 \times 256$ voxels used as a reference image for the standardization of all ADNI images.

3   Methods

In our proposed methodology, first, we preprocessed all ADNI images and extracted the GM tissue to delimited the two regions of analysis, the whole-brain and GM voxel intensities. Next, we applied the eigenbrain technique using two distinct projections, the linear PCA and the kernel PCA, with four different kernel functions. Finally, we trained SVM classifiers with three kernels using 10-fold stratified cross-validation to perform the brain MR image classification. All steps are described as follows.

3.1   Preprocessing

Several preprocessing steps were applied to the MR images to improve the image quality and allow proper analysis. First, the images were denoised using the Non-Local Means (NLM) method proposed by Buades, Coll and Morel [7]. Then, we used the N4-ITK method [34] to mitigate the bias field effect [18]. Next, using the T1-w template image from the NAC dataset (as indicated in Sect. 2) as a reference, we applied the image standardization method proposed by Nyul, Udupa, and Zhang [27] and aligned all ADNI images to the same spatial reference with affine transformations using the Nifty-Reg tool[3]. Then, we resampled all images to have an isotropic voxel size of 1mm and a matrix of dimension 256 × 256 × 256 voxels. Finally, we applied a skull stripping technique [15] to remove unnecessary information from the analysis and reduce the computational time.

---

[3] http://cmictig.cs.ucl.ac.uk/wiki/index.php/NiftyReg.

3.2   Gray Matter Segmentation

Besides the whole-brain analysis, in this study, we also limited our analysis to the gray matter tissue region, which is the most affected tissue type in AD. For that, we segmented the three main tissues (GM, WM, and CSF) in all MR images using the fully automated brain tissue segmentation software FAST [37] (version 4.1), which is considered state of the art on tissue segmentation [28] and is available on the FSL package[4]. The FAST implementation is based on a Hidden Markov Random Field Model, optimized using the Expectation-Maximization (EM) algorithm. A detailed description of the algorithm is available in [37]. The input image required by the FAST is a skull stripped MR image, which we obtained as a result of the preprocessing step using the ROBEX [15] technique. The outputs are an intensity non-uniformity corrected version of the input image and the segmented GM, WM, and CSF brain tissues.

3.3   Eigenbrain Technique

Eigenbrain is a 3-D extension of the 2-D eigenface technique [33] applied to identify MR brain images of subjects with Alzheimer's disease [39]. The technique captures different characteristic variances of anatomical brain structures from a collection of 3-D brain images and uses them to encode and decode a brain, within a machine learning context, without considering the full image information. This process reduces computation and space complexity.

Mathematically, the original eigenbrain technique uses the PCA technique [17], which is a linear transform commonly used for dimensionality reduction, to calculate the linear brain space bases, represented by unidimensional vectors reformatted from a set of training 3D images. The base vectors correspond to the eigenvectors of the covariance matrix constructed from the training dataset and provide the directions of the greatest variances of the data, under the constraint that the base vectors are orthogonal to each other [1]. These eigenvectors, defined in the image space, look like ghost brain images after their reshaping to three-dimensional arrays and are usually referred to as eigenbrains. The first eigenbrain is the average brain, while the rest of the eigenbrains represent variations from the average brain. Figure 1 shows the formulation of the technique and illustrates the eigenbrains reshaped to the three-dimensional array using 25, 50, and 100 projection weights, respectively.

Eigenbrains
To compute the eigenbrains, we used a group of N_e MR brain images from the ADNI dataset. Then, to avoid unnecessary computation we used a brain mask from NAC to ignore background voxels containing no information. The resulting voxels were then transformed to unidimensional column vectors of size $M \times 1$, with $M = 1642084$ elements, corresponding to the number of useful brain voxels

---

[4] http://fsl.fmrib.ox.ac.uk/fsl/fslwiki.

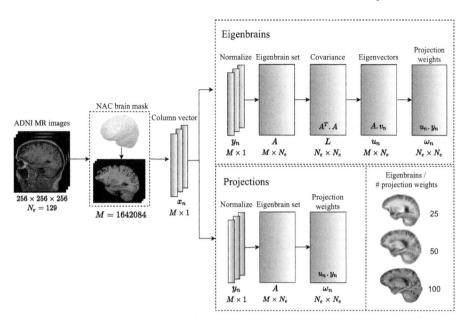

Fig. 1. Flowchart for the eigenbrain technique.

in the MR images. Next, as required by the PCA technique, we normalized the vector images by shifting them to have zero-mean, as

$$y_n = x_n - \bar{x}, \tag{1}$$

where x_n is the nth vector image, and

$$\bar{x} = \frac{1}{N_e} \sum_{n=1}^{N_e} x_n, \tag{2}$$

is the average brain image. The normalized vector brain images, forming the eigenbrain set $\Delta = \{y_n\}_{n=1}^{N_e}$, are organized as a matrix $A = \begin{bmatrix} y_1 & y_2 & \cdots & y_{N_e} \end{bmatrix}$ of size $M \times N_e$.

Now, if we consider finding the eigenvalues λ_n and eigenvector u_n for the covariance matrix C

$$C = \frac{1}{N_e} \sum_{n=1}^{N_e} y_n y_n^T = AA^T, \tag{3}$$

this would imply to solve

$$Cu_n = AA^T u_n = \lambda_n u_n, \tag{4}$$

which would be a prohibitive task due to the large size ($M \times M$ elements) of matrix C.

To overcome this issue, instead of solving Eq. 4 as it is, we replace the covariance matrix $\boldsymbol{C}$ by matrix

$$L = \frac{1}{N_e} \sum_{n=1}^{N_e} \boldsymbol{y}_n^T \boldsymbol{y}_n = \boldsymbol{A}^T \boldsymbol{A}, \tag{5}$$

which has a size of $N_e \times N_e$ ($N_e \ll M$) and whose eigenvalues/vectors hold a close relationship with the ones of matrix $\boldsymbol{C}$. In fact, the eigenvectors $\boldsymbol{v}_n$ of matrix $\boldsymbol{L}$ can be determined by solving

$$\boldsymbol{L}\boldsymbol{v}_n = \boldsymbol{A}^T \boldsymbol{A}\boldsymbol{v}_n = \lambda_n \boldsymbol{v}_n, \tag{6}$$

which is equivalent to

$$\begin{aligned} \boldsymbol{A}\boldsymbol{A}^T \boldsymbol{A}\boldsymbol{v}_n &= \lambda_n \boldsymbol{A}\boldsymbol{v}_n, \\ \boldsymbol{C}\boldsymbol{A}\boldsymbol{v}_n &= \lambda_n \boldsymbol{A}\boldsymbol{v}_n, \\ \boldsymbol{C}\boldsymbol{u}_n &= \lambda_n \boldsymbol{u}_n, \end{aligned} \tag{7}$$

where $\boldsymbol{u}_n = \boldsymbol{A}\boldsymbol{v}_n$. Thus, the matrices $\boldsymbol{C}$ and $\boldsymbol{L}$ have the same eigenvalues λ_n, and $\boldsymbol{u}_n = \boldsymbol{A}\boldsymbol{v}_n$ relates their eigenvectors. The eigenvalues/vectors of matrix $\boldsymbol{L} = \boldsymbol{A}^T \boldsymbol{A}$ represent the N_e largest eigenvalues/vectors of matrix $\boldsymbol{C} = \boldsymbol{A}\boldsymbol{A}^T$, with the remaining eigenvectors being associated with eigenvalues of zero value [33].

Using the correspondence between matrices $\boldsymbol{C}$ and $\boldsymbol{L}$ as described above, all calculations are greatly reduced, from the order of the number of image voxels, $\mathcal{O}(M^3)$, to the number of images in the normalized set, $\mathcal{O}(N_e)$.

After normalizing the eigenvectors $\boldsymbol{u}_n$ to have a unit norm, i.e., $\|\boldsymbol{u}_n\| = 1$, we keep only the K eigenvectors ($K \leq N_e$) corresponding to the K largest eigenvalues to represent the images with minimal inaccuracy. By using all K bases, each brain image from our eigenbrain set Δ can be represented as a linear combination of the best K eigenvectors as

$$\hat{\boldsymbol{y}}_n = \sum_{j=1}^{K} \omega_j \boldsymbol{u}_j, \tag{8}$$

where $\omega_j = \boldsymbol{u}_j^T \boldsymbol{y}_n$ represents the PWs, and the $\boldsymbol{u}_j$ eigenvectors are referred to as the eigenbrains. Finally, with the aim of pattern classification, each normalized vector image $\boldsymbol{y}_n$ can be represented in this K bases by the vector $\Omega = [\omega_1, \omega_2, \ldots, \omega_K]$

Kernel PCA

In eigenbrain analysis, both PCA and Kernel PCA [31,32] have a similar task, which is to extract relevant features from the brain images. However, while PCA extracts features via orthogonal projections, Kernel PCA uses a kernel function $\varphi(\cdot)$ to implicitly map an image from the input space (image space $\boldsymbol{R}^M$) to a higher dimensional feature space (kernel space $\boldsymbol{F}$), possibly in a non-linear way, as

$$\varphi : \boldsymbol{R}^M \to \boldsymbol{F}, \tag{9}$$

and there it performs PCA on the images $\boldsymbol{y}_n$.

To avoid explicit calculations in the kernel space, a $N_e \times N_e$ kernel matrix $\boldsymbol{G}$, also known as Gram matrix [26], is defined, whose elements are

$$g_{i,j} = g(\boldsymbol{y}_i, \boldsymbol{y}_j) = \langle \varphi(\boldsymbol{y}_i), \varphi(\boldsymbol{y}_j) \rangle, \tag{10}$$

where $i, j = 1, 2, \ldots, N_e$ and $g(\cdot, \cdot)$ is a kernel function that allows computing inner products, denoted by $\langle \cdot, \cdot \rangle$, in $\boldsymbol{F}$ [31,32].

To obtain the eigenvectors in this new space, the covariance matrix is defined as

$$\boldsymbol{C}_\varphi = \frac{1}{N_e} \sum_{n=1}^{N_e} \varphi(\boldsymbol{y}_n) \varphi(\boldsymbol{y}_n)^T = \frac{1}{N_e} \boldsymbol{A}_\varphi \boldsymbol{A}_\varphi^T, \tag{11}$$

where $\boldsymbol{A}_\varphi = [\varphi(\boldsymbol{y}_1)\ \varphi(\boldsymbol{y}_2)\ \cdots\ \varphi(\boldsymbol{y}_{N_e})]$.

Similarly to PCA, the eigenvalues λ_i and eigenvectors $\boldsymbol{w}_i$ of matrix $\boldsymbol{C}_\varphi$ are computed by solving

$$\boldsymbol{C}_\varphi \boldsymbol{w}_i = \lambda_i \boldsymbol{w}_i. \tag{12}$$

By replacing $\boldsymbol{C}_\varphi$ in the above equation with its definition in Eq. 11 and rearranging the terms we have

$$\boldsymbol{w}_i = \sum_{n=1}^{N_e} \{ \frac{\varphi(\boldsymbol{y}_n)^T \boldsymbol{w}_i}{\lambda_i N_e} \} \varphi(\boldsymbol{y}_n)$$

$$= \sum_{n=1}^{N_e} \alpha_{i,n} \varphi(\boldsymbol{y}_n) = \boldsymbol{A}_\varphi \boldsymbol{\alpha}_i, \tag{13}$$

where $\boldsymbol{\alpha}_i = [\alpha_{i,1}, \alpha_{i,2}, \ldots, \alpha_{i,M}]^T$. Therefore, since $\varphi(\cdot)$ is a predetermined kernel function, to find the eigenvectors $\boldsymbol{w}_i$ is equivalent to find the coefficients $\alpha_{i,m}$.

In order to compute the $\alpha_{i,m}$, the eigenvalue problem from Eq. 12 is projected on each column of matrix $\boldsymbol{A}_\varphi$ as

$$\langle \varphi_n(\boldsymbol{y}_n), \boldsymbol{C}_\varphi \boldsymbol{w}_i \rangle = \lambda_i \langle \varphi_n(\boldsymbol{y}_n), \boldsymbol{w}_i \rangle, \quad n = 1, \ldots, N_e. \tag{14}$$

Now, substituting Eq. 13 into Eq. 14 and using the definition of the Gram matrix $\boldsymbol{G}$, all $\boldsymbol{\alpha}_i$'s can be determined by solving

$$\frac{1}{N_e} \boldsymbol{G} \boldsymbol{\alpha} = \lambda \boldsymbol{\alpha}. \tag{15}$$

As the last step, the eigenvectors $\boldsymbol{w}_i$ need to be normalized in the feature space to ensure the unit norm, which is done by imposing the following condition $\langle \boldsymbol{w}_i, \boldsymbol{w}_i \rangle = 1$ and leads to

$$
\begin{aligned}
1 &= \sum_{i,j=1}^{N_e} \alpha_i \alpha_j \langle \varphi_i(\boldsymbol{y}_i), \varphi_j(\boldsymbol{y}_j) \rangle, \\
&= \sum_{i,j=1}^{N_e} \alpha_i \alpha_j \boldsymbol{G}_{ij}, \\
&= \lambda \langle \boldsymbol{\alpha}, \boldsymbol{G}\boldsymbol{\alpha} \rangle,
\end{aligned}
\tag{16}
$$

where the sum is taken over all values of i, j. The normalization implies that each α_i be divided by the square root of the corresponding eigenvalue ($\sqrt{\lambda_i}$).

Now, assuming that $\boldsymbol{y}$ and $\varphi(\boldsymbol{y})$ are, respectively, a normalized vector brain image and its mapping in the feature space, then the projection of $\varphi(\boldsymbol{y})$ on the eigenvectors $\boldsymbol{w}_i$ is computed as

$$
\langle \boldsymbol{w}_i, \varphi(\boldsymbol{y}) \rangle = \sum_{n=1}^{N_e} \alpha_n \langle \varphi(\boldsymbol{y}_n), \varphi(\boldsymbol{y}) \rangle = \sum_{n=1}^{N_e} \alpha_n g(\boldsymbol{y}_n, \boldsymbol{y}),
\tag{17}
$$

and provides non-linear principal components.

Therefore, we can extract the K non-linear principal components by working implicitly in the feature space.

3.4   Experiments Setup

In this study, we conducted six different classification experiments, i.e., CN × MCI, MCI × AD, and CN × AD, for features extracted from the whole-brain (whole-brain analysis - WBA) and for the GM tissue only (gray-matter analysis - GMA). We randomly split the entire ADNI dataset into two sets of equal sizes, with the number of class images for each experiment equal to 129 CN, 129 MCI, and 129 DA subjects for each split. Images from the first split were used to create the eigenspace. Images from the second split were projected into the eigenspace to produce the feature vectors for the training, validation, and test of the SVM classifiers. We applied this strategy to all experiments, i.e., whole-brain and GM analysis using both the PCA and KPCA techniques.

3.5   Pattern Classification Design

In this study, SVM classifiers were designed using the projection weights obtained from either the PCA or KPCA techniques using a training dataset as feature vectors. The SVM kernels used in the analyses were the Polynomial (degree 1), RBF, and the sigmoid kernels with $\gamma = \frac{1}{n_{PWs} * var(\omega_{ij})}$. The regularization parameter C, which controls the influence of the misclassification on the objective function in SVM, was defined for each kernel using a 10-fold stratified cross-validation evaluation scheme [14], with each training set varying C penalty using the grid search algorithm from 0,01 to 100. The balanced accuracy metric [12] was used as the figure of merit to be maximized. We used this approach by increasing the number of PWs, i.e., we applied to the first PC, then with first and second, so forth.

The metrics used to assess our classification performance were sensitivity, specificity, balanced accuracy, which correspond to the sensitivity and specificity average, and the area under the roc curve (AUC). A detailed description of these metrics can be found in [21].

4   Results and Discussion

As indicated in Sect. 3.4, our experiments were organized into two groups: (a) whole-brain analysis and (b) gray-matter analysis (GMA). Tables 2a to 3c are organized to show the obtained results using the combination of the feature extraction technique (PCA or KPCA) and the SVM classifier (kernels RBF or sigmoid or Polynomial). Also, for each combination of techniques, the number of projection weights retained for the analysis is indicated.

4.1   Results of the Whole-Brain Analysis

Table 2a shows the results for the CN × MCI experiments with an indication in bold for the best result obtained for each combination of feature extraction technique and SVM-kernel classifier. As can be noticed, the kernel PCA technique achieved slightly better results (AUC values of 0.74 ± 0.12 for the Polynomial-Sigmoid and Cosine-Polynomial combinations) compared to the traditional PCA (AUC value of 0.73 ± 0.11 for using the SVM-sigmoid classifier). The highest accuracy value (73.17 ± 9.12 %) for this experiment was obtained using the Cosine-RBF combination. However, it should be noted that the cross-validation standard deviation values for all metrics were relatively high. The best numbers of projection weights for most of the experiments were considerately small, especially for the two ones with the highest AUC values (13 out of 256).

The results for the MCI × AD experiments are shown in Table 2b. Similar to the CN × MCI experiments, the AUC values for the PCA and kernel PCA techniques were close, with a slightly higher average value for the kernel-Cosine/SVM-Polynomial case (AUC value of 0.72 ± 0.07). The numbers of projection weights were significantly higher than the CN × MCI experiments, which can be explained by the difficulty in differentiating the MCI and AD brains. The standard deviation values for all metrics were also considerably high.

Table 2c presents the results for the CN × AD experiments. Following the same trend seen in the two previous cases (CN × MCI and MCI × AD), it can be noticed that the results using the kernel PCA technique were slightly better than the PCA. Also, the numbers of projection weights were relatively higher when compared to the CN × MCI experiments.

As expected, the results have shown higher AUC values for the CN × AD experiments when compared to the CN × MCI and MCI × AD, which can be explained by the higher structural brain differences between these two extreme groups. That is corroborated by noticing that the CN × MCI case results are slightly better than the MCI × AD, for which the brains may show similar structural changes.

4.2   Results of the Gray-Matter Analysis

Considering the dynamically spreading wave of gray matter loss in the brains of patients with AD, which results in brain atrophy as a consequence, in this study, we investigate the PCA and kernel PCA techniques applied only to the gray-matter region of the brain. For that, as mentioned in Sect. 3.2, the three main brain tissues were segmented, and the GM was isolated for further analysis. Table 3a shows the results for the CN × MCI experiments. Different from the experiments using the whole-brain analysis, in this case, we can notice a significant average difference between the PCA and kernel PCA techniques, i.e., AUC value of 0.63 ± 0.08 for the PCA/SVM-polynomial classifier versus 0.76 ± 0.13 for the polynomial/SVM-polynomial classifier. The standard deviation values for these experiments have also shown to be high. Also, with a few exceptions, the numbers of projection weights were relatively small (35 out of 256, for the higher AUC value).

For the MCI × AD experiments (Table 3b), the AUC average values were marginally better than the CN × MCI case for the traditional PCA technique, with the highest AUC value of 0.67 ± 0.11, and slightly worse for the kernel PCA techniques, with the highest AUC value of 0.74 ± 0.11, achieved using the kernel PCA RBF and the SVM-polynomial classifier. The standard deviation values were similar or even smaller than all other experiments, while the number of projection weights for the best result was higher (145 out of 256).

Finally, Table 3c shows the results for the CN × AD experiments. In this case, we can see a significant improvement in the results, with the highest average AUC value of 0.89 achieved by using two different PCA kernels (RBF and Polynomial). The numbers of projection weights were relatively small (43 and 30) in these two cases. The best average AUC value obtained by the traditional PCA was 0.86 using the SVM-RBF classifier, which is higher than any other experiment.

Table 2. Whole-brain analysis

	Feature extraction	kernel SVM	#PWs	Bal.Accuracy (%)	Sensitivity (%)	Specificity (%)	AUC
(a) CN × MCI							
kernel PCA	PCA	RBF	20	69.32 ± 9.57	71.41 ± 17.12	67.24 ± 17.49	0.71 ± 0.13
		Sigmoid	**8**	**70.42 ± 8.80**	**64.42 ± 18.47**	**76.41 ± 14.50**	**0.73 ± 0.11**
		Polynomial	10	70.38 ± 10.85	68.14 ± 18.17	72.63 ± 13.50	0.71 ± 0.13
	RBF	**RBF**	**12**	**71.54 ± 5.78**	**71.28 ± 17.94**	**71.79 ± 10.63**	**0.73 ± 0.07**
		Sigmoid	11	68.56 ± 6.79	65.19 ± 17.84	71.92 ± 13.38	0.71 ± 0.10
		Polynomial	18	71.22 ± 6.99	71.99 ± 18.16	70.45 ± 16.74	0.73 ± 0.11
	Sigmoid	RBF	20	65.29 ± 8.09	62.63 ± 15.48	67.95 ± 13.09	0.66 ± 0.11
		Sigmoid	42	68.11 ± 8.17	69.92 ± 15.23	66.60 ± 13.20	0.72 ± 0.11
		Polynomial	**50**	**71.19 ± 7.44**	**69.55 ± 15.92**	**72.82 ± 10.85**	**0.72 ± 0.10**
	Polynomial	RBF	20	68.56 ± 10.29	70.64 ± 18.69	66.47 ± 18.86	0.71 ± 0.13
		Sigmoid	**13**	**72.85 ± 7.71**	**72.12 ± 15.41**	**73.59 ± 15.74**	**0.74 ± 0.12**
		Polynomial	10	70.38 ± 10.85	68.14 ± 18.17	72.63 ± 13.50	0.73 ± 0.11
	Cosine	RBF	21	73.17 ± 9.12	75.26 ± 19.94	71.09 ± 13.78	0.74 ± 0.13
		Sigmoid	69	69.07 ± 12.22	70.58 ± 14.85	67.56 ± 17.69	0.71 ± 0.13
		Polynomial	**13**	**71.99 ± 8.53**	**70.45 ± 19.27**	**73.53 ± 16.48**	**0.74 ± 0.12**
(b) MCI × AD							
kernel PCA	PCA	RBF	98	64.17 ± 9.76	61.54 ± 18.76	66.79 ± 20.83	0.63 ± 0.09
		Sigmoid	77	65.61 ± 7.07	64.62 ± 14.65	66.60 ± 13.94	0.69 ± 0.08
		Polynomial	**31**	**65.42 ± 8.29**	**65.04 ± 15.00**	**65.19 ± 17.51**	**0.71 ± 0.07**
	RBF	RBF	51	63.53 ± 8.22	63.40 ± 14.38	63.65 ± 12.24	0.67 ± 0.11
		Sigmoid	**19**	**67.05 ± 9.59**	**61.15 ± 16.72**	**72.95 ± 14.19**	**0.71 ± 0.07**
		Polynomial	10	63.94 ± 10.51	60.38 ± 19.04	67.50 ± 11.68	0.64 ± 0.11
	Sigmoid	**RBF**	**53**	**64.10 ± 10.06**	**62.24 ± 18.29**	**65.96 ± 13.27**	**0.66 ± 0.07**
		Sigmoid	68	57.63 ± 8.56	57.88 ± 13.62	57.37 ± 15.04	0.57 ± 0.09
		Polynomial	68	62.40 ± 7.29	68.01 ± 14.27	56.79 ± 19.03	0.58 ± 0.09
	Polynomial	RBF	31	65.53 ± 4.25	69.55 ± 16.62	57.50 ± 16.73	0.65 ± 0.07
		Sigmoid	125	66.12 ± 6.56	67.82 ± 13.03	64.42 ± 11.32	0.67 ± 0.09
		Polynomial	**120**	**63.97 ± 7.07**	**68.14 ± 15.84**	**59.81 ± 20.43**	**0.68 ± 0.08**
	Cosine	RBF	34	67.34 ± 10.15	67.18 ± 14.97	67.50 ± 17.72	0.70 ± 0.13
		Sigmoid	28	64.58 ± 10.31	64.87 ± 15.00	64.29 ± 16.66	0.68 ± 0.12
		Poly	**85**	**64.13 ± 6.14**	**63.27 ± 7.74**	**65.00 ± 16.18**	**0.72 ± 0.07**
(c) CN ×AD							
kernel PCA	PCA	RBF	89	72.33 ± 8.04	76.47 ± 11.27	68.20 ± 13.95	0.734 ± 0.143
		Sigmoid	**55**	**75.06 ± 5.55**	**74.23 ± 13.60**	**75.89 ± 11.30**	**0.801 ± 0.092**
		Polynomial	92	73.87 ± 9.36	72.56 ± 12.57	75.19 ± 18.13	0.747 ± 0.117
	RBF	**RBF**	**62**	**72.76 ± 4.31**	**73.52 ± 7.33**	**71.41 ± 9.58**	**0.768 ± 0.054**
		Sigmoid	67	72.17 ± 11.44	71.85 ± 11.37	72.50 ± 17.23	0.759 ± 0.067
		Polynomial	28	72.91 ± 8.46	73.58 ± 16,91	72.24 ± 6.96	0.758 ± 0.077
	Sigmoid	**RBF**	**43**	**71.25 ± 8.34**	**68.58 ± 14.80**	**73.91 ± 13.47**	**0.733 ± 0.111**
		Sigmoid	12	69.13 ± 6.05	66.08 ± 9.51	72.17 ± 11.64	0.652 ± 0.109
		Polynomial	38	71.31 ± 10.73	66.08 ± 10.22	76.53 ± 17.66	0.723 ± 0.104
	Polynomial	RBF	59	75.83 ± 6.47	76.02 ± 9.46	75.64 ± 12.60	0.798 ± 0.053
		Sigmoid	162	75.80 ± 7.79	74.29 ± 10.94	77.30 ± 13.76	0.758 ± 0.092
		Polynomial	**56**	**75.44 ± 6.12**	**73.65 ± 11.88**	**77.24 ± 10.98**	**0.804 ± 0.042**
	Cosine	**RBF**	**27**	**74.93 ± 6.43**	**74.42 ± 10.68**	**75.44 ± 12.61**	**0.793 ± 0.075**
		Sigmoid	28	75.44 ± 6.73	76.92 ± 7.88	73.97 ± 11.14	0.790 ± 0.093
		Polynomial	58	75.06 ± 8.2	72.05 ± 10.60	78.07 ± 11.09	0.763 ± 0.058

Table 3. Gray-matter analysis

	Feature extraction	kernel SVM	#PWs	Bal.Accuracy (%)	Sensitivity (%)	Specificity (%)	AUC
(a) CN × MCI							
kernel PCA	PCA	RBF	25	63.69 ± 10.49	64.23 ± 10.33	63.14 ± 13.47	0.63 ± 0.13
		Sigmoid	77	62.82 ± 10.50	68.21 ± 10.58	57.44 ± 20.67	0.60 ± 0.11
		Polynomial	30	**63.33 ± 8.58**	68.08 ± 11.67	58.59 ± 10.94	0.63 ± 0.08
	RBF	RBF	33	67.66 ± 10.83	69.74 ± 19.92	65.58 ± 15.16	0.69 ± 0.13
		Sigmoid	11	68.30 ± 10.35	72.05 ± 8.69	64.55 ± 16.40	0.71 ± 0.15
		Polynomial	36	**71.60 ± 10.84**	**71.99 ± 14.94**	**71.22 ± 22.72**	**0.74 ± 0.11**
	Sigmoid	RBF	118	66.92 ± 11.47	73.59 ± 13.95	60.26 ± 22.16	0.66 ± 0.10
		Sigmoid	32	70.03 ± 9.97	70.58 ± 11.73	69.49 ± 13.23	0.65 ± 0.14
		Polynomial	12	**68.69 ± 11.15**	**71.28 ± 10.99**	**66.09 ± 13.19**	**0.72 ± 0.16**
	Polynomial	RBF	31	69.62 ± 8.69	69.62 ± 16.36	69.62 ± 15.01	0.71 ± 0.14
		Sigmoid	34	70.80 ± 9.03	71.28 ± 17.27	70.32 ± 20.66	0.73 ± 0.12
		Polynomial	35	**75.10 ± 8.82**	**77.44 ± 14.13**	**72.76 ± 16.23**	**0.76 ± 0.13**
	Cosine	RBF	34	68.43 ± 11.73	73.59 ± 13.95	63.27 ± 23.93	0.67 ± 0.16
		Sigmoid	4	69.13 ± 8.65	65.06 ± 9.44	73.21 ± 12.25	0.71 ± 0.13
		Polynomial	37	**70.71 ± 11.62**	**75.00 ± 16.96**	**66.41 ± 24.48**	**0.73 ± 0.13**
(b) MCI × AD							
kernel PCA	PCA	RBF	17	67.02 ± 10.94	64.23 ± 17.41	69.81 ± 13.87	0.66 ± 0.15
		Sigmoid	18	**66.67 ± 11.30**	**63.46 ± 17.96**	**69.87 ± 15.05**	**0.67 ± 0.11**
		Polynomial	15	67.34 ± 9.19	64.87 ± 20.11	69.81 ± 11.54	0.65 ± 0.10
	RBF	RBF	72	65.87 ± 9.11	61.79 ± 14.97	69.94 ± 18.59	0.68 ± 0.14
		Sigmoid	146	64.97 ± 4.74	60.96 ± 12.77	68.97 ± 9.76	0.59 ± 0.04
		Polynomial	145	**66.57 ± 5.42**	**65.64 ± 14.60**	**67.50 ± 13.11**	**0.74 ± 0.11**
	Sigmoid	**RBF**	83	**68.59 ± 8.11**	**65.71 ± 11.81**	**71.47 ± 15.70**	**0.72 ± 0.08**
		Sigmoid	88	65.42 ± 9.51	65.58 ± 9.38	65.26 ± 18.79	0.65 ± 0.13
		Polynomial	89	65.35 ± 8.89	66.28 ± 11.29	64.42 ± 17.48	0.67 ± 0.07
	Polynomial	RBF	74	63.43 ± 10.91	59.36 ± 16.66	67.50 ± 13.99	0.64 ± 0.11
		Sigmoid	64	64.97 ± 7.75	63.97 ± 11.84	65.96 ± 14.55	0.61 ± 0.11
		Polynomial	55	**63.85 ± 6.48**	**63.27 ± 10.91**	**64.42 ± 11.83**	**0.65 ± 0.06**
	Cosine	**RBF**	141	**64.90 ± 7.46**	**60.583 ± 17.43**	**68.97 ± 15.78**	**0.68 ± 0.08**
		Sigmoid	136	63.40 ± 6.74	59.36 ± 16.83	67.44 ± 18.13	0.64 ± 0.09
		Polynomial	138	65.29 ± 8.03	60.77 ± 20.46	69.81 ± 15.86	0.63 ± 0.07
(c) CN × AD							
kernel PCA	PCA	**RBF**	34	**80.74 ± 9.25**	**81.22 ± 11.07**	**80.26 ± 19.80**	**0.86 ± 0.13**
		Sigmoid	78	77.24 ± 10.34	70.19 ± 16.07	84.29 ± 11.58	0.80 ± 0.12
		Polynomial	69	77.56 ± 11.17	69.42 ± 14.35	85.71 ± 17.20	0.81 ± 0.12
	RBF	**RBF**	43	**82.72 ± 8.33**	**84.42 ± 11.03**	**81.03 ± 18.88**	**0.89 ± 0.07**
		Sigmoid	139	83.46 ± 10.41	83.65 ± 10.15	83.27 ± 22.53	0.85 ± 0.14
		Polynomial	137	82.69 ± 9.85	82.12 ± 9.88	83.27 ± 22.53	0.86 ± 0.10
	Sigmoid	**RBF**	146	**82.31 ± 11.11**	**88.33 ± 7.17**	**76.28 ± 23.62**	**0.88 ± 0.13**
		Sigmoid	149	81.15 ± 12.64	83.72 ± 12.61	78.59 ± 24.67	0.87 ± 0.10
		Polynomial	148	81.89 ± 13.11	84.42 ± 12.05	79.36 ± 24.97	0.88 ± 0.10
	Polynomial	**RBF**	30	**83.11 ± 9.37**	**84.36 ± 7.84**	**81.86 ± 17.52**	**0.89 ± 0.11**
		Sigmoid	146	78.40 ± 10.95	73.53 ± 12.99	83.27 ± 22.79	0.74 ± 0.12
		Poly	144	78.43 ± 9.31	71.99 ± 11.50	84.87 ± 20.56	0.82 ± 0.13
	Cosine	**RBF**	39	**82.82 ± 6.09**	**85.26 ± 8.05**	**80.38 ± 15.04**	**0.88 ± 0.09**
		Sigmoid	134	78.46 ± 8.86	78.97 ± 10.89	77.95 ± 19.60	0.84 ± 0.07
		Polynomial	103	80.03 ± 9.99	84.42 ± 10.28	75.64 ± 17.51	0.87 ± 0.07

5 Conclusions

The eigenbrain method was initially proposed as a simple extension of the eigen-faces for the Alzheimer's disease classification problem, and, consequently, both techniques work by projecting the original data (3-D MR images) into PCA orthogonal bases.

In this study, we investigated the application of linear (PCA) and non-linear (kernel PCA) feature projections for the classification of 3-D MR images on CN, MCI, and AD, with the goal of contrast the results obtained from these two types of projections. In addition, our analyzes were performed using information extracted from the whole-brain MR images and from the gray-matter tissues, segmented from the T1-w MR images using the FAST segmentation technique.

With the use of the KPCA, compared with the standard PCA, we included information about the non-linearity of the features. These non-linear features were capable of outperforming the standard PCA in terms of classification accuracy when predicting CN × MCI, MCI × AD, and CN × AD, with a more advantage difference from the experiments performed for the GM analysis. In this case, the highest average AUC values (0.89 ± 0.07 and 0.89 ± 0.11) were obtained for the CN × AD experiment using two different combinations of the PCA kernels and SVM-RBF classifiers.

Acknowledgment. Funding for ADNI can be found at http://adni.loni.usc.edu/about/#fund-container.

Funding Statement. This study was financed by the Fundação de Amparo à Pesquisa do Estado de São Paulo (FAPESP) (grant numbers 2018/08826-9 and 2018/06049-5) and the Coordenação de Aperfeiçoamento de Pessoal de Nível Superior (CAPES) - Finance Code 001.

References

1. Abdi, H., Williams, L.J.: Principal Component Analysis. Wiley Interdisc. Rev. Comput. Stat. **2**(4), 433–459 (2010)
2. Ahmed, M.R., Zhang, Y., Feng, Z., Lo, B., Inan, O.T., Liao, H.: Neuroimaging and machine learning for dementia diagnosis: recent advancements and future prospects. IEEE Rev. Biomed. Eng. **12**, 19–33 (2018)
3. Alam, S., Kwon, G., Initiative, A.D.N.: Alzheimer disease classification using kpca, lda, and multi-kernel learning svm. Int. J. Imaging Syst. Technol. **27**(2), 133–143 (2017)
4. Álvarez, I., et al.: Alzheimer's diagnosis using eigenbrains and support vector machines. Electron. Lett. **45**(7), 342–343 (2009)
5. Amoroso, N., et al.: Alzheimer's disease diagnosis based on the hippocampal unified multi-atlas network (human) algorithm. Biomed. Eng. Online **17**(1), 1–16 (2018)
6. Bassiony, H.S., Zickri, M.B., Metwally, H.G., Elsherif, H.A., Alghandour, S.M., Sakr, W.: Comparative histological study on the therapeutic effect of green tea and stem cells in Alzheimer's disease complicating experimentally induced diabetes. Int. J. Stem Cells **8**(2), 181–190 (2015)
7. Buades, A., Coll, B., Morel, J.M.: A review of image denoising algorithms, with a new one. Multiscale Model. Simul. **4**(2), 490–530 (2005)
8. Cao, P., et al.: Nonlinearity-aware based dimensionality reduction and over-sampling for AD/MCI classification from MRI measures. Comput. Biol. Med. **91**, 21–37 (2017)

9. Deture, M., Dickson, D.: The neuropathological diagnosis of Alzheimer's disease. Mol. Neurodegeneration **14**(32), 1–18 (2019)
10. Devanand, D.P., Bansal, R., Liu, J., Hao, X., Pradhaban, G., Peterson, B.S.: MRI hippocampal and entorhinal cortex mapping in predicting conversion to Alzheimer's disease. Neuroimage **60**(3), 1622–1629 (2012)
11. Elahi, F.M., Miller, B.L.: A clinicopathological approach to the diagnosis of dementia. Nat. Rev. Neurol. **13**(8), 457–476 (2017)
12. Fawcett, T.: An introduction of ROC analysis. Pattern Recogn. Lett. **27**(8), 861–874 (2006)
13. Frisoni, G.B., Fox, N.C., Jack-Jr, C.R., Scheltens, P., Thompson, P.M.: The clinical use of structural MRI in Alzheimer disease. Nat. Rev. Neurol. **6**(2), 67–77 (2010)
14. Hsu, C.W., Chang, C.C., Lin, C.J.: A practical guide to support vector classification. Technical Report, Department of Computer Science, National Taiwan University, Taiwan (May 2016). https://www.csie.ntu.edu.tw/~cjlin/papers/guide/guide.pdf
15. Iglesias, J.E., Liu, C.Y., Thompson, P.M., Tu, Z.: Robust brain extraction across datasets and comparison with publicly available methods. IEEE Trans. Med. Imaging **30**(9), 1617–1634 (2011)
16. Jo, T., Nho, K., Saykin, A.: Deep learning in Alzheimer's disease: diagnostic classification and prognostic prediction using neuroimaging data. Front. Aging Neurosci. **11**(220), 1–14 (2019)
17. Jolliffe, I.T., Cadima, J.: Principal component analysis: a review and recent developments. Phil. Trans. Royal Soc. Math. Phys. Eng. Sci. **374**(2065), 1–16 (2016)
18. Juntu, J., Sijbers, J., Van Dyck, D., Gielen, J.: Bias Field Correction for MRI Images. In: Kurzy?ski, M., Puchała, E., Woźniak, M., żołnierek, A. (eds) Computer Recognition Systems. Advances in Soft Computing, vol. 30, pp. 543-551. Springer, Berlin (2005) https://doi.org/10.1007/3-540-32390-2_64
19. Hett, K., Ta, V.-T., Manjón, J.V., Coupé, P.: Graph of hippocampal subfields grading for Alzheimer's disease prediction. In: Shi, Y., Suk, H.-I., Liu, M. (eds.) MLMI 2018. LNCS, vol. 11046, pp. 259–266. Springer, Cham (2018). https://doi.org/10.1007/978-3-030-00919-9_30
20. Kanghan, O., Young-Chul, C., Ko, W., Woo-Sung, K., Il-Seok, O.: Classifcation and visualization of Alzheimer's disease using volumetric convolutional neural network and transfer learning. Sci. Reports **9**(18150), 1–16 (2019)
21. Kelleher, J.D., Mac Namee, B., D'arcy, A.: Fundamentals of Machine Learning for Predictive Data Analytics: Algorithms, Worked Examples, and Case Studies. 1 edn. MIT Press, Cambridge (2015)
22. Khedher, L., Ramirez, J., Gorriz, J.M., Brahim, A., Segovia, F., Initiative, A.D.N., et al.: Early diagnosis of alzheimer's disease based on partial least squares, principal component analysis and support vector machine using segmented mri images. Neurocomputing **151**, 139–150 (2015)
23. Liu, J., Li, M., Lan, W., Wu, F., Pan, Y., Wang, J.: Classification of alzheimer's disease using whole brain hierarchical network. IEEE Trans. Comput. Biol. Bioinf. **15**(2), 624–632 (2018)
24. Liu, M., Zhang, D., Shen, D.: View-centralized multi-atlas classification for Alzheimer's disease diagnosis. Hum. Brain Mapp. **36**(5), 1847–1865 (2015)
25. Liu, M., Zhang, J., Nie, D., Yap, P.T., Shen, D.: Anatomical landmark based deep feature representation for MR images in brain disease diagnosis. IEEE J. Biomed. Health Inf. **22**(5), 1476–1485 (2018)
26. Nell, C., Shawe-Taylor, J.: An Introduction to Support Vector Machines and Other Kernel-Based Learning Methods. Cambridge University Press, Cambridge (2000)

27. Nyúl, L.G., Udupa, J.K., Zhang, X.: New variants of a method of MRI scale standardization. IEEE Trans. Med. Imaging **19**(2), 143–150 (2000)
28. Rehman, H.Z.U., Hwang, H., Lee, S.: Conventional and deep learning methods for skull stripping in brain MRI. Appl. Sci. **10**(5), 1773 (2020)
29. Salvatore, C., Cerasa, A., Battista, P., Gilardi, M.C., Quattrone, A., Castiglioni, I.: Magnetic resonance imaging biomarkers for the early diagnosis of Alzheimer's disease: a machine learning approach. Front. Neurosci. **9**, 307 (2015)
30. Sarwinda, D., Arymurthy, A.M.: Feature selection using kernel PCA for alzheimer's disease detection with 3D MR images of brain. In: 2013 International Conference on Advanced Computer Science and Information Systems (ICACSIS), pp. 329–333. IEEE, Bali, IDN (September 2013)
31. Schölkopf, B.: Nonlinear component analysis as a kernel eigenvalue problem. Neural Comput. **10**(5), 1299–1319 (1998)
32. Schölkopf, B., Smola, A., Müller, K.-R.: Kernel principal component analysis. In: Gerstner, W., Germond, A., Hasler, M., Nicoud, J.-D. (eds.) ICANN 1997. LNCS, vol. 1327, pp. 583–588. Springer, Heidelberg (1997). https://doi.org/10.1007/BFb0020217
33. Turk, M., Pentland, A.: Eigenfaces for recognition. J. Cogn. Neurosci. **3**(1), 71–86 (1991)
34. Tustison, N.J., et al.: N4itk: improved N3 bias correction. IEEE Trans. Med. Imaging **29**(6), 1310–1320 (2010)
35. World Health Organization: Dementia: A public health priority. Technical Report, World Health Organization, Geneva, Switzerland (2019)
36. Zhang, J., Liu, M., An, L., Gao, Y., Shen, D.: Alzheimer's disease diagnosis using landmark-based features from longitudinal structural MR images. IEEE J. Biomed. Health Inf. **21**(5), 1607–1616 (2017)
37. Zhang, Y., Brady, M., Smith, S.: Segmentation of brain MR images through a hidden markov random field model and the expectation-maximization algorithm. IEEE Trans. Med. Imaging **20**(1), 45–57 (2001)
38. Zhang, Y.D., Wang, S., Dong, Z.: Classification of Alzheimer disease based on structural magnetic resonance imaging by kernel support vector machine decision tree. Prog. Electromagnet. Res. **144**, 171–184 (2014)
39. Zhang, Y., et al.: Detection of subjects and brain regions related to Alzheimer's disease using 3D MRI scans based on eigenbrain and machine learning. Front. Comput. Neurosci. **9**, 66 (2015)
40. Zhang, Y., Wang, S., Phillips, P., Yang, J., Yuan, T.F.: Three-dimensional eigenbrain for the detection of subjects and brain regions related with Alzheimer's disease. J. Alzheimer's Dis. **50**(4), 1163–1179 (2016)

Turning Meshes into B-reps with T-splines

Luigi Barazzetti[✉]

ABC Department, Politecnico di Milano, via Ponzio 31, 20133 Milano, Italy
`luigi.barazzetti@polimi.it`

Abstract. This paper presents the first results of a semi-automatic procedure able to turn tetra-meshes into solid models using T-splines. The paper deals with free-form objects where manual modelling based on NURBS curves and surfaces is not a simple process or could result in approximate reconstructions with significant geometric simplification.

The proposed solution relies on a multi-step processing workflow with several conversions and metric accuracy evaluation after each step. First, a closed triangular mesh is turned into a quad mesh, which is also made lighter and cleaner without losing significant level of detail. The quad mesh is then converted into a surface based on T-splines, which are powerful tools to model organic shapes. The final step of the method is the conversion of the closed surface into a novel B-rep based on watertight NURBS surfaces.

Keywords: Accuracy · B-rep · Mesh · Point clouds · Quad-mesh · T-splines

1 Introduction

Boundary representation (B-rep) [1, 2] models are used in several applications such as manufacturing products, prototyping, numerical simulation, and applications in the medical or entertainment industries. B-rep models are composed of faces (i.e. surfaces all connected without gaps) and topological relationships that allow defining the boundary of the object.

B-rep models are common solutions for representing solid bodies composed of simple and more complex NURBS(non-uniform rational B-splines)-based models. Indeed, CSG (constructive solid geometry [3]) based on geometric primitives (cubes, cylinders, cones, etc.) processed with Boolean operations (union, difference, intersection) do not give a sufficient flexibility in modeling projects requiring advanced geometric shapes, especially when objects cannot be represented by single or a combination of multiple primitives processed using Boolean operations.

B-rep is therefore the current standard in several engineering applications that require numeric simulation. The opportunity to model complex surfaces with NURBS provides advanced modeling tools for more complex shapes [4, 5]. On the other hand, the flexibility of such modeling tools has limitations when irregular surfaces need to be reconstructed. Modeling tools based on NURBS surfaces are not the best solution to reconstruct free-form objects like toys, sculptures, jewels, sporting goods, etc. For instance, the famous Stanford bunny (Fig. 1, [6]) created via laser scanning technology and made available

© Springer Nature Switzerland AG 2021
O. Gervasi et al. (Eds.): ICCSA 2021, LNCS 12950, pp. 34–44, 2021.
https://doi.org/10.1007/978-3-030-86960-1_3

with a mesh form could be a rather complicated objects to model using manual modeling based on B-rep.

The aim of this paper is to present a workflow able to answer the following question: can we get a B-rep model of a complex free-form objects reconstructed using point clouds or meshes? The idea is shown in Fig. 1, where the B-rep model of the Stanford bunny (b) was generated with the proposed workflow.

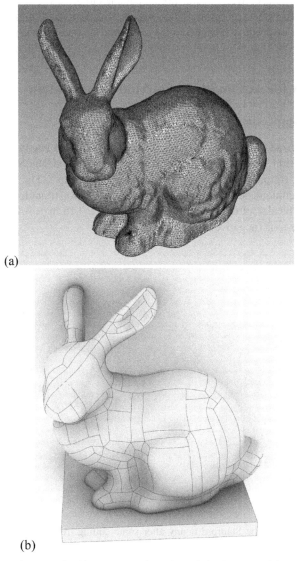

(a)

(b)

Fig. 1. (a) The original Stanford bunny as a tri-mesh, and (b) B-rep model generated using the initial mesh. The only part manually added was the rectangular base, whereas the conversion of the initial mesh into a solid B-rep model was fully automatic.

The model was generated using a semi-automatic approach that uses the available mesh of the bunny, which is converted into a B-rep model and saved using the STEP format [7]. Manual modeling is not necessary with the proposed workflow, that is based on a multi-step workflow in which the initial mesh is converted into a sequence a products, whose metric accuracy (i.e., the discrepancy between the original mesh and all the intermediate products generated during the conversion pipeline) is checked to verify the magnitude of the error. The only "manually modeled" element is the base of the bunny.

Automation in the conversion was possible thanks to the use T-splines [8, 9], which are better than NURBS to reconstruct the geometry of irregular shapes. T-splines are an extension of tensor-product NURBS and are defined on a control grid called T-mesh, which allows a row of control points to terminate, obtaining a T-junction.

2   Workflow for Automatic B-rep Model Generation

2.1   Mesh Pre-processing

The case study used in this paper is the well-known Stanford bunny, as mentioned in the previous section. The model is available on the internet on The Stanford 3D Scanning Repository (http://graphics.stanford.edu/data/3Dscanrep/). The mesh was generated using the technique described in [6] and has about 69,400 triangles with some holes in the bottom. The size of the model is about 0,15 m × 0,15 m × 0.11 m. The gravity direction is the y-axis (negative).

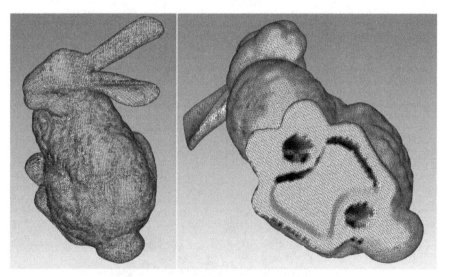

Fig. 2. The tri-mesh of the Stanford bunny after some editing: removal of holes and smooth filtering to reduce spikes.

One of the requirements to produce the model with the procedure proposed in this paper is the availability of a closed mesh. This means that holes must be removed. Other important requirements consist of the absence of intersecting triangles and non-manifold edges. The mesh should also have a rather uniform size of triangles (although this is not always possible, e.g., for small parts), which could be obtained with remeshing techniques. The preliminary pre-processing consists of a set of (semi) automatic operations aimed at correcting such effects and can be done in different commercial software able to handle meshes generated from point clouds.

Some images of the mesh are shown in Fig. 2. Holes in the bottom part were fixed using Geomagic Wrap obtaining a total number of 70,458 triangles. A smoothing filter was also applied to reduce spikes, slightly changing the initial shape of the bunny. As the goal of the procedure described in this paper is a good geometric consistency between the final model and the original mesh, an accuracy analysis was carried out using mesh-to-mesh comparison algorithms. The discrepancy error was about 0.0001 m, which is about 0.04% in terms of relative error (i.e., compared to the diagonal of the model bounding box).

2.2 Tri- to Quad-Mesh Conversion

Most mesh-based models produced with photogrammetric or laser scanning point clouds are provided as triangular (tri-) meshes. The output of the previous phase (Sect. 2.1) is a tri-mesh formed by adjacent triangles and saved using the obj format. Such type of mesh is very popular in the field of reality-based modeling, where a set of 3 control points defines the basic triangle of the mesh. 3D applications with huge meshes (million triangles) are reported in technical literature because generation and processing of tri-meshes is easier than quadrilateral (quad-)meshes.

Quad meshes are mathematically more complex, notwithstanding they are easier to manipulate and have significant use in graphics applications [10]. The big advantage of a quad-meshes exploited in this work is their close relationship with T-meshes, and therefore a reconstruction based on T-splines. The initial conversion from the tri- to a novel quad-mesh is a quite common operations in different applications, and different procedures and algorithms were developed to automatically get lighter and more efficient representation without losing level of detail [11]. Two quad meshes of the bunny are shown in Fig. 3. They feature 1,105 (left) and 9,997 (right) faces, respectively. Both where created with Recap Photo. The possibility to work with a model with less faces is of primary interest in the proposed approach, so the "coarse" quad-mesh was compared to the original tri-mesh, obtaining a discrepancy of about 0.0005 m, which was considered sufficient. The complete error map is shown in Fig. 4.

The selection of the optimal number of faces in a project requires repeated conversions by changing the resolution and comparisons to check if sufficient metric accuracy of the quad-mesh has been achieved. It is not simple to define a priori the optimal number of faces because the intrinsic geometry of the considered object determines the result.

2.3 Generation of a Model Based on T-splines and B-rep Conversion

Quad-meshes are particularly suitable to generate a novel surface based on T-splines. The quad-mesh can be used as T-mesh, which is the control grid of a T-splines [12]. Such procedure consists in the production of a novel surface that can be interactively edited.

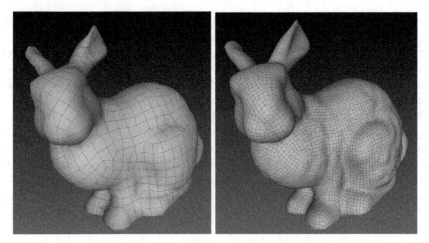

Fig. 3. Examples of quad-meshes with different number of faces generated from the tri-mesh. The solution on the left has 1105 faces and was chosen for further processing.

Fig. 4. Comparison between the quad-mesh with 1105 faces and the original tri-mesh available on the Stanford repository. The overall discrepancy is 0,0005 m.

The modeling approach is also very suitable for organic shapes, i.e., those irregular bodies in which direct modeling with NURBS could be rather difficult.

T-splines can be defined as NURBS with T-junctions. They allow a row of control points to terminate and offer a more flexible way for local refinements. T-splines significantly reduce the number of control points necessary to satisfy topological constraints in NURBS-based modeling, in which all points must lie in a rectangular grid.

T-splines have become popular with the development of the T-splines plugin for Rhino, which however is no longer supported by Autodesk. The workflow proposed in this paper was carried out with Autodesk Fusion 360, in which T-splines tools have been integrated.

The quad-mesh was imported and converted into a T-spline, as shown in Fig. 5 (left). The surface can now be manipulated with different editing tools that can work on points, edges, or faces. Figure 5 (right) shows an example of possible modifications, in which the head of the bunny has been rotated, and the ears are now lying down. Although such simple modifications are not necessary in this work, they prove the advantage of T-splines as sculpting tools.

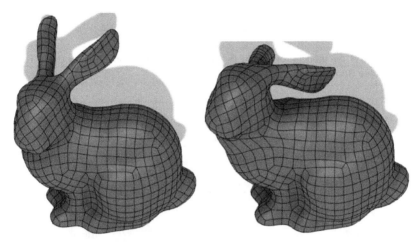

Fig. 5. The T-splines model from the quad-mesh (left), and an example of editing on the head of the bunny (right).

Finally, the T-splines can be converted into NURBS surfaces using knot insertion, and the method can provide a complete B-rep model if the initial surface is closed. The operation is still carried out in Fusion 360, and the obtained solid model was saved as STEP. Such operation provides the model shown in Fig. 6 (left).

The model is now a solid and Boolean operations are now possible, like in the case of Fig. 6 (right) where the difference with a sphere and a transversal cylinder was computed for the head and the body, respectively. A base was also added to the model. It was decided to add the base to show that Boolean operations are feasible. A comparison between the initial tri-mesh available in the Stanford repository and the B-rep model shows an overall discrepancy of 0,0007 m (Fig. 7).

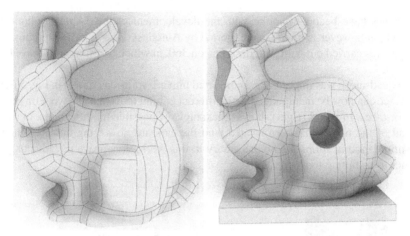

Fig. 6. The final B-rep model provided by the procedure described in the paper (left). The same model after some Boolean operations (right).

Fig. 7. Comparison between the final B-rep and the initial tri-mesh.

3   Testing the Influence of Multiple Model Conversions

The used workflow includes several conversions of the initial mesh into multiple products to generate a final B-rep model. A second test based on a relatively simple object is presented to check the influence of the different operations. The idea of this test is the comparison between results achieved with a body directly modeled using B-rep, and the same model initially converted into a closed tri-mesh. The workflow presented in the previous sections should be able to provide a novel B-rep model, which can be compared to the initial one.

The solid model of a beam was generated by extruding a planar curve representing the transversal section (Fig. 8). The beam is $L = 3$ m long, whereas the section is visible in the figure and has overall dimensions of 1.2 m × 0.4 m. The edges where smoothed using circumferences of 0.1 m radius to simplify the generation of the quad-mesh. Although T-splines can also handle creased edges, the use of a smooth transition in the section was preferred.

The solid model was preliminary converted into a surface tri-mesh denser close to discontinuities.

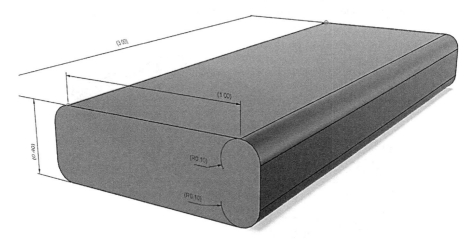

Fig. 8. The simulated beam and its dimensions in meters.

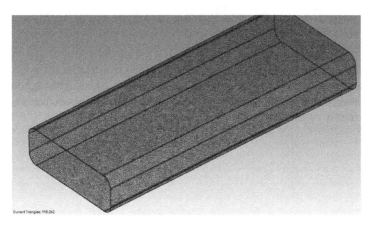

Fig. 9. The initial tri-mesh of the beam

The initial tri-mesh is shown in Fig. 9 and features more than 118,000 triangles. The model was converted into a quad-mesh using Instant Meshes, which is available for Windows, OS X, and Linux, and it can retopologize complex meshes.

The quad-mesh was then imported into Fusion 360 and converted into a T-spline surface. The result is the surface shown in Fig. 10. It features 6217 faces and is smoother, especially the transitions which were effective discontinuities in the initial model. The B-rep model appears as split into different surfaces. Indeed, such conditions are usually fixed working on edge or surfaces, which should follow the logic of construction of the element. This could be a less significant issue for free-form organic objects, in which most surfaces are irregular since the beginning. In the case of regular elements, the automatic generation of a B-rep model could result in a solid with overall good geometry, but the boundary could be made up of multiple surfaces with a quite irregular pattern (Fig. 11).

Fig. 10. The T-spline model of the beam created from a quad-mesh.

The overall discrepancy between the initial solid model and the final one recon-structed with the proposed solution is 0.0019 m. The error map is shown in Fig. 12, where errors are clearly more visible close to the traversal edges. Although the overall error is relatively good, the problem related to the irregular subdivision pattern obtained must be carefully considered. This will be done in future work.

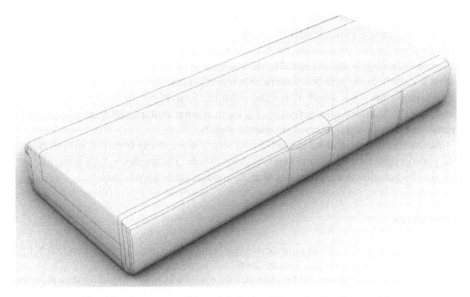

Fig. 11. The B-rep solid model obtained from the T-spline model.

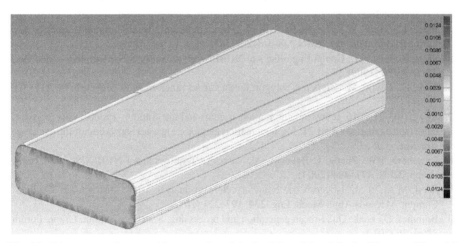

Fig. 12. Discrepancy (in meters) between the original solid model and the final B-rep solid model regenerated with the proposed workflow.

4 Conclusions

The paper presented the first results of a workflow for solid modeling of irregular bodies digitized with laser scanning. Starting from a tri-mesh of an object a B-rep model is created using consecutive conversions based on quad-meshes and T-splines, which are suitable functions for free-form objects.

Accuracy evaluation is carried out after each conversion to verify metric quality. The aim is to generate a B-rep model as similar as possible to the initial tri-mesh. Metric evaluation performed on a simple model of a beam have revealed good metric accuracy. The main limitation found is related to the connection of trimmed surfaces, which could follow a quite inhomogeneous pattern. Such result becomes evident when the object features discontinuities, which is not the considered target for such workflow. Indeed, results on the Stanford bunny provided a more regular pattern since the object itself cannot be decomposed into basic geometric shapes.

More experiments will be carried out in future work, using both models generated with photogrammetric and laser scanning point clouds. The method seems attractive not only for the generation of solids, but also 3D models made up of surfaces.

References

1. Stroud, I.: Boundary Representation Modelling Techniques, p. 808, Springer, London (2010). Softcover reprint of hardcover 1st edn. 2006
2. Kaufmann, M.: Geometric and Solid Modeling: An Introduction. Morgan Kaufmann Series in Computer Graphics and Geometric Modeling, p. 338 (1989). ISSN 1046-235X
3. Requicha, A.A.G.: Solid modeling: a 1988 update. In: Ravani, B. (ed.) CAD Based Programming Sensory Robots, pp. 3–22. Springer Verlag, New York (1988). https://doi.org/10.1007/978-3-642-83625-1_1
4. Piegl, L., Tiller, W.: The NURBS Book. Monographs in Visual Communication, p. 646 Springer, Berlin (1995). ISBN 978-3-642-97385-7
5. Piegl, L., Tiller, W.: Computing offsets of NURBS curves and surfaces. Comput. Aided Des. **31**, 147–156 (1999)
6. Turk, G., Levoy, M.: Zippered polygon meshes from range images. Siggraph **94**, 311–318 (1994)
7. ISO 10303-21:2016. Industrial automation systems and integration – Product data representation and exchange – Part 21: Implementation methods: clear text encoding of the exchange structure
8. Sederberg, T.W., Zheng, J., Bakenov, A., Nasri, A.: T-splines and T-NURCCS. ACM Trans. Graph. **22**(3), 477–484 (2003)
9. Scott, M.A., et al.: Isogeometric boundary element analysis using unstructured T-splines. Comput. Methods Appl. Mech. Eng. **254**, 197–221 (2013)
10. Bommes, D., et al.: Quad-mesh generation and processing: a survey. Comput. Graph. Forum **32**(6), 26 (2013)
11. Jakob, W., Tarini, M., Panozzo, D., Sorkine-Hornung, O.: Instant field-aligned meshes. In: ACM SIGGRAPH ASIA 2015, p. 15 (2015)
12. Wang, W., Zhang, Y., Scott, M.A., Hughes, T.J.R.: Converting an unstructured quadrilateral mesh to a standard T-spline Surface. ICES REPORT 10–50, p. 23. The Institute for Computational Engineering and Sciences (2010)

Deep Learning for Accurate Corner Detection in Computer Vision-Based Inspection

M. Fikret Ercan$^{(\boxtimes)}$ (iD) and Ricky Ben Wang

School of Electrical and Electronic Engineering, Singapore Polytechnic, 500 Dover Rd,
Singapore, Singapore
M_Fikret_Ercan@sp.edu.sg

Abstract. This paper describes application of deep learning for accurate detection of corner points in images and its application for an inspection system developed for the worker training and assessment. In our local built and construction industry, workers need to be certified for their technical skills through a training and assessment process. Assessment involves trainees to understand a task given with a technical drawing, e.g. electrical wiring and trunking wall assembly, and implement it accurately in a given period of time. Typically experts manually/visually evaluate the finished assembly and decide if it's done correctly. In this study, we employed computer vision techniques for the assessment process in order to reduce significant man hour. Computer vision based system measures dimensions, orientation and position of the wall assembly and produces a report accordingly. However, analysis depends on accurate detection of the objects and their corner points which are used as reference points for measurements. Corner detection is widely used in image processing and there are numerous algorithms available in the literature. Conventional algorithms are founded upon pixel based operations and they return many redundant or false corner points. In this study, we employed a hybrid approach using deep learning and Minimum Eigen value corner detection for this purpose and achieved highly accurate corner detection. This subsequently improved the reliability of the inspection system.

Keywords: Computer vision · Deep learning · Corner detection · Quality construction

1 Introduction

The application of computer vision for inspecting and quality control has been widely used in industry. It has wide and diverse range of applications extending from manufacturing [1] to farming [2], civil engineering [3] to biomedical engineering [4]. Most of these applications concerned with detecting patterns, irregularities or deriving information from the objects analyzed which can be broadly defined as recognition tasks. Computer vision is also widely used in automated assembly [5] and vision based control of mobile robots [6]. Among these, manufacturing is a typical application where geometric measurement tests for parts and objects are performed. In such computer vision application, inspection of parts or an assembly achieved by automatic recognition of the

© Springer Nature Switzerland AG 2021
O. Gervasi et al. (Eds.): ICCSA 2021, LNCS 12950, pp. 45–54, 2021.
https://doi.org/10.1007/978-3-030-86960-1_4

object and then analyzing geometric measurements. Using computer vision in manufacturing for this purpose saves time, minimize errors and improve the automation process. There is a vast number of application examples presented in literature [7, 8].

The application concerned in this paper is an automated inspection and assessment system developed to be used in construction industry. In the local construction industry, workers are required to go through a training and certification process in order to ensure quality of building and construction standards. After the training period, workers go through an assessment process for their certification in which they are expected to deliver an assembly from its technical drawing in a given time period. The quality of the work produced is examined by a team of experts manually and visually. The assessment involves checking measurements, alignments, and positions of the assembly and rating is based on its compliance with the technical drawing given. Naturally, it is an arduous and time consuming process for the training providers. Therefore, a computer vision solution is developed to automate the assessment process. Computer vision and image processing techniques in built and construction industry has long been utilized typically for construction safety, resource tracking and activity monitoring, surveying, inspection and condition monitoring. For a comprehensive survey on computer vision application in built and construction industry refer to Martinez et al. [3].

In our application, vision system inspects the work produced by a trainee in terms of the geometric measures, orientation and position. It produces a report highlighting discrepancies from the specifications. A fundamental object feature used in the process is corner detection. Performance of the system is therefore highly dependent on accurate detection of the corner points of the objects. Corner detection is widely used in image processing and there are numerous algorithms available in the literature most popular ones being FAST [9], SUSAN [10] and Harris corner detection algorithms [11]. These algorithms are pixel based operations and they often return redundant or false corner points. In this study, we experimented with FAST, Harris and Minimum Eigen value corner detection and found that Minimum Eigen value performed well. However, it is computationally costly to filter out noisy detections in order to identify corners that are belong to the objects. On the other hand, corner detectors utilizing neural network and evolutionary techniques present competitive results especially with noisy images [12]. Recently, deep learning algorithms become very popular in computer vision due to their remarkable performance. Our experiments showed that deep learning can identify object corners effectively however it is within a bounded box. Using a hybrid approach that is filtering Minimum Eigen value result with deep learning output, we are able to detect object corners very accurately which makes the geometric measurement inspection robust.

In the following an overview of the system, together with hybrid solution and performance results will be presented.

2   System Overview

Computer vision hardware used in this study is made of off –the-shelve components and the hardware setup is rather straightforward as shown in Fig. 1(a). The camera is a Nikon D800 DSLR with an AF-S Micro NIKKOR 60 mm f/2.8G ED Lens. Optoma

LED projector serves as an illuminator, providing a constant lighting condition which improves image processing accuracy. A set of wall assembly developed by trainee is placed 211 cm from the camera lens. The camera is set at the same height as the midline of the assembly. Matlab is used for the software development.

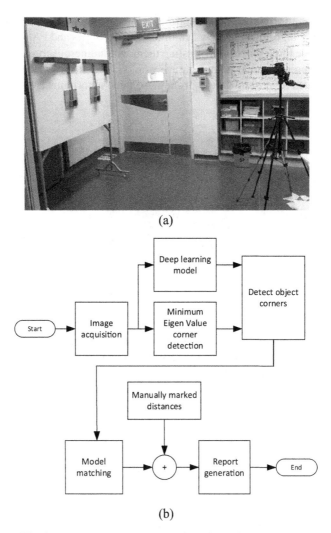

(a)

(b)

Fig. 1. (a) Hardware setup. (b) A flow chart of the operations.

Figure 1 (b) shows a flow of operations. After image acquisition, the first step of image processing algorithm is to detect corner points using hybrid method. These corner points will be used as reference points in measuring objects dimensions, alignments, angles, and so on. A model of the objects in the wall assembly is available as CAD drawing. The next step of the process is to match detected features with the model and determine discrepancies. Furthermore, a user can also manually mark any points of interest in the image and get the measurements for them. Finally, a report is generated providing assessment results.

2.1 Detecting Corner Points

There is a vast literature on corner detection algorithms. Amongst those, we experimented with three popular algorithms namely Minimum Eigen, Harris and FAST. After experimenting with these three functions by testing them with numerous trunking images, we observed that reference corner points that we would like to detect was mostly captured by Minimum Eigen algorithm. Output from these three algorithms are shown in Fig. 2 for comparison.

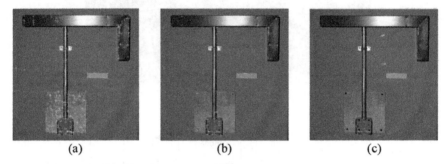

(a)   (b)   (c)

Fig. 2. Corner detection algorithms. Results for (a) MinEigen (b) Harris (c) Fast

Although the Minimum Eigen algorithm is capable of detecting the corner points of the objects that we are interested, there are still redundancy and errors present as seen from Fig. 3(a). The quality of the image including its resolutions, lighting conditions also contribute to these errors. On the other hand, recent developments in computer vision demonstrate that deep learning algorithms are very effective in object detection, particularly in natural environment with many background clutter. In the next phase of corner detection, a trained deep learning network is used for detecting key reference points in the image. The bounding boxes generated by the trained network as the region of interest (ROI) are not as accurate as the pixel level detection of corner points though using the ROI we are able to filter out erroneous corner points as seen on Fig. 3(b).

2.2 Deep Learning for Detecting Trunking Corners

Deep Learning network architecture used in this study is YOLO version 2. Other meta architectures that are light weight and fast such as SSD (Single Shot Multibox Detector), Faster R-CNN were also experimented but not considered in this application since accuracy is more imperative than the real-time performance. YOLO v2 is trained to do classification and bounding box regression at the same time. Additionally, YOLO v2 learns generalizable representations of objects therefore it can perform better when applied to new domains or unexpected inputs. For the feature extraction network, we

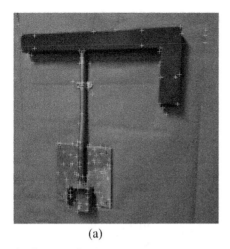 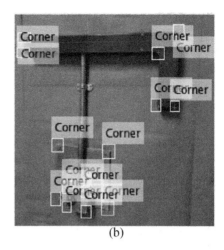

(a)   (b)

Fig. 3. Error filtering using trained deep learning networks. (a) Detected corner points before filter (b) Corners detected after error filtering

employed Resnet-50 based on our earlier empirical study in comparing network and feature extractor architectures [13]. Resnet-50 consists of 50 layers and the pre-trained network can classify images in up to a thousand object categories. The detection sub-network is a small CNN compared to the feature extraction network and is composed of a few convolutional layers and layers specific for YOLO v2.

For the training, we have used 500 images of trunking wall assembly as shown in Fig. 1 under different light conditions and angles. After multiple iterations of training, a model with a precision of 92% achieved. To test the accuracy and consistency of the trained network, we used 20 test images with different light and camera angles. The accuracy is defined as the ratio of accurately detected corner points over actual number of corner points in the image. Table 1 tabulates these test results. Majority of test results lie above 93% accuracy, however there were also some outliers with accuracies as low as 75% due to poor image quality. Overall a median of 93% with a confidence interval of 95% was satisfactory to use such model in the corner detection problem.

Table 1. Network performance on test images

Image	Corners expected	Corners detected	Corners missed	Accuracy	Wrong detection
1	12	12	0	100%	1
2	12	12	0	100%	1
3	12	12	0	100%	1
4	12	12	0	100%	3

(continued)

Table 1. (*continued*)

Image	Corners expected	Corners detected	Corners missed	Accuracy	Wrong detection
5	12	12	0	100%	1
6	12	12	0	100%	1
7	12	12	0	100%	1
8	12	12	0	100%	0
9	12	12	0	100%	1
10	12	12	0	100%	2
11	12	11	1	91.7%	0
12	12	10	2	83.3%	0
13	12	11	1	91.7%	0
14	12	10	2	83.3%	1
15	12	9	3	75%	1
16	12	8	4	66.7%	0
17	12	10	2	83.3%	0
18	12	10	2	83.3%	0
19	12	12	0	100%	0
20	12	12	0	100%	0
Average				92.2%	

2.3   Template Matching and Operation of the System

Hybrid approach provided a robust detection of corner points that are needed for further analysis. However, there may still be a few remnant stray corner points. As the technical drawing of the wall assembly is known prior, in the final step of analysis a template matching method is used to establish the most suitable corner points to represent the corners of the trunking object. Template is constructed starting from the top left corner point using geometric methods. The selection of valid corner points is done by accepting the corner points that are closest to the template's corner points, thereby completing the corner point selection process. Figure 4(a) shows the template and Fig. 4(b) the final reference corner points extracted. Once these reference points obtained measurements for assessment are done using simple geometric techniques. Finally, a report as an excel spreadsheet is created listing the result of analysis as shown in Fig. 4(c).

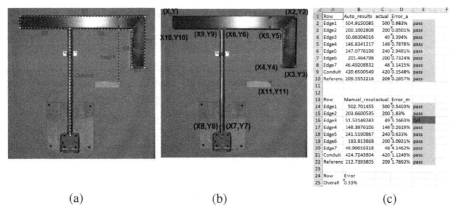

Fig. 4. Model matching with template. (a) Template of the object. Edges labeled 1 to 6 and conduit length are some of the key measurements (b) Final reference corner points marked. (c) Final report generated for the trainee assessment.

3   Performance of the System

In order to evaluate the performance of our system, we used a test setup which is fabricated accurately as indicated in the technical drawings given to trainees. We have taken images from the fixed point under various lighting conditions as lighting will immediately affect the image processing performance. Table 2 shows measurement results for the selected assessment points using five images taken under good lighting conditions. These results compared with the actual dimensions of the test setup. Difference between them implies the measurement error of the vision system.

Table 2 shows all the dimensions detected by the program for each of the test images. As seen on the Table 2, most of the average percentage of error lies below 1%, which implies a negligible measurement error. However, it is also noticeable that the average percentage of error for "Edge 4"and "Edge 5" is relatively higher than the ones of the other edges. These larger errors are due to the inaccuracies generated during template matching particularly caused by shadows of the trunking.

The template matching method helped to find the most suitable/correct corner points to be used for the calculation of the trunking's dimensions, however if there are errors present, such as an inaccurate camera calibration, it will affect the selection of the most suitable corner points, hence causing the chosen corner points to be inaccurate and resulting a miscalculation of the trunking's dimensions.

A function namely "display template" added to the GUI of the system. It enables user to have a clear visual of the areas of error. And displays the overall error at the bottom left side for reference as shown on Fig. 5(a). Standard trunking's area displayed in white pixels based on the dimensions of the template. While the display of the test trunking's area (in red) is based on the detected corner points of the test trunking. Hence, the differences in area (the error) can be spotted easily by overlapping these two areas together.

Table 2. Average measurement (in millimeters) error for test images

Feature	Image 1	Image 2	Image 3	Image 4	Image 5	Actual	Average measurement error (%)
Edge 1	504.6	504.3	502.7	504.2	502.9	500	0.75
Edge 2	199.8	199.6	199.5	199.8	199.3	200	0.20
Edge 3	50.1	50.7	50.5	50.4	50.4	50	0.84
Edge 4	145.7	146.6	146.4	146.6	146.9	150	2.37
Edge 5	231.7	246.4	231.9	231.8	231.9	240	3.26
Edge 6	201.1	201.2	200.6	200.5	200.6	200	0.40
Conduit	419.9	410.3	408.5	421.0	420.3	420	1.08
Angle	130.7°	130.5°	130.4°	131.3°	130.8°	131°	0.29

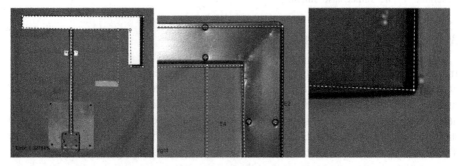

Fig. 5. Example of result (a) with error (percentage error calculated and indicated at the bottom left) (b) error due to image quality (c) due to shadows. (Color figure online)

The image in Fig. 5 (b) shows an example of inaccuracy when detecting suitable corner points. The picture of the trunking shown in this figure is slightly tilted by 1° anti-clockwise. Since the program is unable to detect such shortcomings, the template is not tilted to fit the image. Hence, the selection for the correct corner points will be based on the inaccurate template match therefore producing errors in its dimensions. Based on our experiments, we note that this error in inaccuracy usually within 5%. If the error in accuracy is more than 5%, we observed other external factors are contributing to the error in measurement accuracy. Figure 5(c) shows the error in corner detection due to the presence of shadows. Shadows appear when there isn't an even distribution of light source on the trunking when taking the image using a camera. Therefore, if the difference in shade between the shadow and the trunking itself isn't stark enough, the corner detector algorithm will detect the shadow's corner point, instead of the trunking's. Hence, if the process of template matching will not be unable to correct this error, the shadow's corner will be used to calculate the trunking's dimensions, producing an inaccurate result. This can error can easily be prevented with good lighting arrangement in hardware.

In conclusion, test results presented in Table 2 imply that the computer vision based system delivers accurate measurement results. Minor inaccuracies are mainly due to environmental factors. Current set-up utilizes a commercial camera and light system. The performance can be further improved if the vision set up is replaced with an industrial grade lighting and camera hardware.

4 Conclusions

This paper described a computer vision based inspection system for automated assessment of work done by trainees in construction industry for their certification. A key feature used in the geometric measurement of the assembly was corner points of the objects. Accurate detection of them significantly contributes to the system performance. Minimum Eigen Value corner detection algorithm combined with a deep learning model for the object corners provided accurate corner detection. Using template matching, detected objects dimensions are compared with the model obtained from technical drawings. Our experimental results revealed that vision system has minor measurement error under good lighting conditions. Our future study involves developing a generic system where various assessment configurations can be programmed by the instructors creating a wide range of scenarios that can be used for examining trainees work and grading them.

Acknowledgement. This project is sponsored by Singapore Polytechnic on under the grant number 03-11000-36-J723 and cosponsored by Fonda Global Engineering Pte Ltd. Authors acknowledge the contributions of interns Fong Kah Kian and Jacky Wijaya.

References

1. Batchelor, B.G.: Machine vision for industrial applications. In: Batchelor, B.G. (eds.) Machine Vision Handbook. Springer, London (2012). https://doi.org/10.1007/978-1-84996-169-1
2. Mavridou, E., Vrochidou, E., Papakostas, G.A., Pachidis, T., Kaburlasos, V.G.: Machine vision systems in precision agriculture for crop farming. J. Imag. **5**(89), 1–32 (2019)
3. Martinez, P., Al-Hussein, M., Ahmad, R.: A scientometric analysis and critical review of computer vision applications for construction. Autom. Constr. **107**, 102947 (2019)
4. Park, C., Took, C.C., Seong, J.-K.: Machine learning in biomedical engineering. Biomed. Eng. Lett. **8**(1), 1–3 (2018)
5. Pierleoni, P., Belli, A., Palma, L., Palmucci, M., Sabbatini, L.: A machine vision system for manual assembly line monitoring. In: International Conference on Intelligent Engineering and Management (ICIEM), London, UK, pp. 33–38 (2020)
6. Sankowski, D., Nowakowski, J.: Computer Vision in Robotics and Industrial Applications. World Scientific Publishing, New Jersey (2014)
7. Okarma, K., Fastowicz, J.: Computer vision methods for non-destructive quality assessment in additive manufacturing. In: Burduk, R., Kurzynski, M., Wozniak, M. (eds.) CORES 2019. AISC, vol. 977, pp. 11–20. Springer, Cham (2020). https://doi.org/10.1007/978-3-030-197 38-4_2
8. Rajan, A.J., Jayakrishna, K., Vignesh, T., Chandradass, J., Kannan, T.T.M.: Development of computer vision for inspection of bolt using convolutional neural network. Mater. Today Proc. **45**(7), 6931–6935 (2021)

9. Trajković, M., Hedley, M.: Fast corner detection. Image Vis. Comput. **16**(2), 75–87 (1998)
10. Smith, S., Brady, J.: SUSAN-A new approach to low level image processing. Int. J. Comput. Vis. **23**(1), 45–48 (1997)
11. Harris, C., Stephens, M.: A combined corner and edge detector. In: Proceedings of the 4th Alvey Vision Conference, pp. 147–151 (1988)
12. Cuevas, E., Rodríguez, A., Alejo-Reyes, A., Del-Valle-Soto, C.: Corner detection algorithm based on Cellular Neural Networks (CNN) and Differential Evolution (DE). In: Recent Meta-heuristic Computation Schemes in Engineering. Studies in Computational Intelligence, vol. 948. Springer, Cham (2021). https://doi.org/10.1007/978-3-030-66007-9_4
13. Ercan, M.F., Qiankun, A.L., Sakai, S.S, Miyazaki, T.: Circle detection in images: a deep learning approach. In: Proceedings of IEEE OCEANS Conference (2020)

Improving Parametric PCA Using KL-divergence Between Gaussian-Markov Random Field Models

Alexandre L. M. Levada$^{(\boxtimes)}$ [iD]

Computing Department, Federal University of São Carlos, São Carlos, SP, Brazil
`alexandre.levada@ufscar.br`

Abstract. Parametric PCA is a dimensionality reduction based metric learning method that uses the Bhattacharrya and Hellinger distances between Gaussian distributions estimated from local patches of the KNN graph to build the parametric covariance matrix. Later on, PCA-KL, an entropic PCA version using the symmetrized KL-divergence (relative entropy) was proposed. In this paper, we extend this method by replacing the Gaussian distribution by a Gaussian-Markov random field model. The main advantage is the incorporation of the spatial dependence by means of the inverse temperature parameter. A closed form expression for the KL-divergence is derived, allowing fast computation and avoiding numerical simulations. Results with several real datasets show that the proposed method is capable of improving the average classification performance in comparison to PCA-KL and some state-of-the-art manifold learning algorithms, such as t-SNE and UMAP.

Keywords: Dimensionality reduction · Metric learning · PCA · KL-divergence · Markov random fields

1 Introduction

Metric learning is concerned with the process of finding suitable and adaptive distance functions prior to pattern classification problems. [9]. A powerful strategy for achieving this goal consists in finding a more compact and meaningful representation for the observed data, which is the mathematical framework of dimensionality reduction methods [19,20].

Linear dimensionality reduction methods have been developed throughout many areas of science such as statistics, optimization, machine learning, and other applied fields for over a century, and these methods have become powerful mathematical tools for analyzing high dimensional and noisy data. Part of the success of the linear methods comes from the fact that they have simple geometric interpretations and typically attractive computational properties [3].

Nonlinear dimensionality reduction involves finding low-dimensional representations in high-dimensional space. This problem arises naturally when analyzing data like hyperspectral imagens, human faces, speech waveforms and handwritten characters. Linear algorithms often fail to capture the hidden non-linear

© Springer Nature Switzerland AG 2021
O. Gervasi et al. (Eds.): ICCSA 2021, LNCS 12950, pp. 55–69, 2021.
https://doi.org/10.1007/978-3-030-86960-1_5

representation of the data in the presence of non-Euclidean geometry. It has been found that manifolds are used to explain how our visual perception and learning work by recognizing the variability of perceptual stimuli and other types of high-dimensional data [10, 12, 15].

Manifold learning algorithms explore intrinsic geometric properties of the datasets to build a low dimensional representation using non-Euclidean distances [16]. However, one limitation with these methods is that it is not clear how to deal with out-of-sample problems, that is, how to find the low dimensional representation of samples that do not belong to the training set.

Recently, Parametric PCA has been proposed as an alternative to dimensionality reduction based metric learning algorithms. The basic idea consists in using the Bhattacharyya and Hellinger distances between univariate Gaussian distributions learned from the local patches along the KNN graph to compute a surrogate for the covariance matrix [6]. Later on, PCA-KL has been proposed as a generalization of Parametric PCA, in which the symmetrized KL-divergence (relative entropy) is employed to build the entropic covariance matrix [8].

In this paper, we propose a contextual version of PCA-KL that has two main contributions in comparison with regular PCA-KL: 1) to capture the dependencies between pieces of data inside a local patch, instead of using a Gaussian distribution, we use a Gaussian-Markov random field to model the intra-patch correlations; 2) to avoid k-regular graphs, instead of using the KNN rule, we use the ϵ-neighborhood rule, allowing patches of variable sizes along the graph. Another positive aspect of the proposed method is that by replacing a pointwise similarity measure (Euclidean distance) by a path-based one (symmetric KL-divergence), we can make the whole process more robust to the presence of noise and outliers in the observations.

The remaining of the paper is organized as follows: Sect. 2 and its subsections describe the proposed method in details, from the random field model and the computation of the KL-divergence to the contextual PCA-KL algorithm. Section 3 shows the experiments and the obtained results from an objective comparison of the proposed method with several dimensionality reduction methods in different real world datasets. Finally, Sect. 4 brings our conclusions, final remarks and future directions for research.

2   The Proposed Method

In this section, we describe the proposed method in details, from the definition of the statistical model to the derivation of a closed-form expression for the KL-divergence between pairwise GMRF's and the contextual PCA-KL algorithm.

2.1   The Random Field Model

Pairwise Gaussian-Markov random fields (GMRF's) are mathematical structures particularly suitable to study dependent continuous random variables by means of non-linear interactions between neighboring variables that are part of a grid.

The main advantage of this model in comparison to other random field models is related to mathematical tractability. Often, we cannot derive closed-form expressions for statistical divergences, such as the Kullback-Leibler divergence, due to the presence of expected values. In pairwise GMRF's, it is possible to compute these measures analytically, which reduces the computational burden. Moreover, by invoking the Hammersley-Clifford theorem [4], which states the equivalence between Gibbs random fields (global models) and Markov random fields (local models), we can characterize an isotropic pairwise Gaussian-Markov random field by the set of local conditional density functions, given by:

$$p\left(x_i | \eta_i, \vec{\theta}\right) = \frac{1}{\sqrt{2\pi\sigma^2}} exp \left\{ -\frac{1}{2\sigma^2} \left[x_i - \mu - \beta \sum_{j \in \eta_i} (x_j - \mu) \right]^2 \right\} \qquad (1)$$

where η_i denotes the neighborhood system comprised by the nearest neighbors of x_i, $\vec{\theta} = (\mu, \sigma^2, \beta)$ denotes the vector of model parameters, with μ and σ^2 being the expected value (mean) and the variance of the random variables in the grid, and β representing the inverse temperature, which encodes the correlation between the variables in η_i. Note that when $\beta = 0$, the model degenerates a the simple Gaussian distribution. The main advantage of using the local model is that we avoid the joint Gibbs distribution and the need of intense Markov Chain Monte Carlo (MCMC) simulations. In this work, we use a pairwise GMRF to model the dependency between Gaussian random variables in a 1-D grid. Figure 1 illustrates different neighborhood systems, from first to third order. Note that the number of neighbors is twice the order of the neighborhood system.

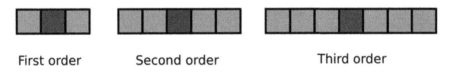

First order   Second order   Third order

Fig. 1. Several neighborhood systems defined in a 1-D grid: from left to right, we have first, second and third order systems.

2.2   The KL-divergence Between GMRF Models

The entropy of a random variable x is the expected value of the self-information:

$$H(p) = - \int p(x)[log\ p(x)]dx = -E\left[log\ p(x)\right] \qquad (2)$$

where $p(x)$ is the probability density function (pdf) of x. For a Gaussian random variable x, distributed as $N(\mu, \sigma^2)$, the entropy can be computed as:

$$H(p) = \frac{1}{2}log\ (2\pi\sigma^2) + \frac{1}{2\sigma^2}E[(x-\mu)^2] = \frac{1}{2}\left(1 + log\ (2\pi\sigma^2)\right) \qquad (3)$$

The cross-entropy between two random variables with pdf's $p(x)$ and $q(x)$ is defined by:

$$H(p, q) = -\int p(x)[log\ q(x)]dx \tag{4}$$

The Kullback-Leibler divergence is the relative entropy, that is, the difference between the cross-entropy of $p(x)$ and $q(x)$ and the entropy of $p(x)$:

$$D_{KL}(p, q) = H(p, q) - H(p) = -\int p(x)[log\ q(x)]dx + \int p(x)[log\ p(x)]dx$$

$$= \int p(x)log\left(\frac{p(x)}{q(x)}\right)dx = E_p\left[log\left(\frac{p(x)}{q(x)}\right)\right] \tag{5}$$

It should be mentioned that the relative entropy is always non-negative, that is, $D_{KL}(p, q) \geq 0$, being equal to zero if, and only if, $p(x) = q(x)$. Let $p(x)$ and $q(x)$ be univariate Gaussian densities, $N(\mu_1, \sigma_1^2)$ and $N(\mu_2, \sigma_2^2)$. Then, the KL-divergence between these two pdf's is:

$$D_{KL}(p, q) = E_p\left[-log\ \sigma_1 - \frac{1}{2\sigma_1^2}(x - \mu_1)^2 + log\ \sigma_2 - \frac{1}{2\sigma_2^2}(x - \mu_2)^2\right]$$

$$= log\left(\frac{\sigma_2}{\sigma_1}\right) + \frac{1}{2\sigma_2^2}E_p[(x - \mu_2)^2] - \frac{1}{2\sigma_1^2}E_p[(x - \mu_1)^2] \tag{6}$$

From basic statistics, it is known that:

$$E_p[(x - \mu_1)^2] = \sigma_1^2 \tag{7}$$

$$E_p[(x - \mu_2)^2] = E[x^2] - 2E[x]\mu_2 + \mu_2^2 \tag{8}$$

$$E[x^2] = Var[x] + E^2[x] = \sigma_1^2 + \mu_1^2 \tag{9}$$

which leads to:

$$D_{KL}(p, q) = log\left(\frac{\sigma_2}{\sigma_1}\right) + \frac{1}{2\sigma_2^2}(\sigma_1^2 + \mu_1^2 - 2\mu_1\mu_2 + \mu_2^2) - \frac{1}{2}$$

$$= log\left(\frac{\sigma_2}{\sigma_1}\right) + \frac{\sigma_1^2 + (\mu_1 - \mu_2)^2}{2\sigma_2^2} - \frac{1}{2} \tag{10}$$

Note that $D_{KL}(p, q) \neq D_{KL}(q, p)$, that is, the relative entropy is not symmetric. The symmetrized KL-divergence between $p(x)$ and $q(x)$ is:

$$D_{KL}^{sym}(p, q) = \frac{1}{2}[D_{KL}(p, q) + D_{KL}(q, p)]$$

$$= \frac{1}{4}\left[\frac{\sigma_1^2 + (\mu_1 - \mu_2)^2}{\sigma_2^2} + \frac{\sigma_2^2 + (\mu_1 - \mu_2)^2}{\sigma_1^2} - 2\right]$$

$$= \frac{1}{4\sigma_1^2\sigma_2^2}\left[(\sigma_1^2 - \sigma_2^2)^2 + (\mu_1 - \mu_2)^2(\sigma_1^2 + \sigma_2^2)\right] \tag{11}$$

Similarly, the KL-divergence between two GMRF's with parameter vectors $\vec{\theta}_1 = (\mu_1, \sigma_1^2, \beta_1)$ and $\vec{\theta}_2 = (\mu_2, \sigma_2^2, \beta_2)$ is given by:

$$D_{KL}(p,q) = log\left(\frac{\sigma_2}{\sigma_1}\right) + \frac{1}{2\sigma_2^2} E_p\left\{\left[x - \mu_2 - \beta_2 \sum_{j \in \eta_i}(x_j - \mu_2)\right]^2\right\} \qquad (12)$$

$$- \frac{1}{2\sigma_1^2} E_p\left\{\left[x - \mu_1 - \beta_1 \sum_{j \in \eta_i}(x_j - \mu_1)\right]^2\right\}$$

It is possible to expand the second expected value in Eq. (12) as:

$$E_p\left\{\left[x - \mu_1 - \beta_1 \sum_{j \in \eta_i}(x_j - \mu_1)\right]^2\right\} \qquad (13)$$

$$= E_p\left\{(x - \mu_1)^2 - 2\beta_1 \sum_{j \in \eta_i}(x_i - \mu_1)(x_j - \mu_1) + \beta_1^2 \sum_{j \in \eta_i}\sum_{k \in \eta_i}(x_j - \mu_1)(x_k - \mu_1)\right\}$$

$$= \sigma_1^2 - 2\beta_1 \sum_{j \in \eta_i}\sigma_{ij} + \beta_1^2 \sum_{j \in \eta_i}\sum_{k \in \eta_i}\sigma_{jk}$$

where σ_{ij} denotes the covariance between the central variable x_i and a neighboring variable x_j and σ_{jk} denotes the covariance between two different variables belonging to the same neighborhood system η_i. Now, we expand the first expected value as:

$$E_p\left\{\left[x - \mu_2 - \beta_2 \sum_{j \in \eta_i}(x_j - \mu_2)\right]^2\right\} \qquad (14)$$

$$= E_p\left\{(x - \mu_2)^2 - 2\beta_2 \sum_{j \in \eta_i}(x_i - \mu_2)(x_j - \mu_2) + \beta_2^2 \sum_{j \in \eta_i}\sum_{k \in \eta_i}(x_j - \mu_2)(x_k - \mu_2)\right\}$$

We need to break Eq. (14) in three distinct expected values in order to simplify it. The first one remains identical to the corresponding term in the Gaussian case:

$$E_p\left\{(x - \mu_2)^2\right\} = \sigma_1^2 + (\mu_1 - \mu_2)^2 \qquad (15)$$

The second expected value is given by:

$$-2\beta_2 \sum_{j \in \eta_i} E_p\{(x_i - \mu_2)(x_j - \mu_2)\} = -2\beta_2 \sum_{j \in \eta_i} E_p\left\{x_i x_j - \mu_2 x_i - \mu_2 x_j + \mu_2^2\right\}$$

$$= -2\beta_2 \sum_{j \in \eta_i}\left[\sigma_{ij}^{(p)} + (\mu_1 - \mu_2)^2\right] \qquad (16)$$

where $\sigma_{ij}^{(p)}$ denotes the covariance between x_i and x_j from the covariance matrix Σ_p. In a similar way, the third expected value leads to:

$$\beta_2^2 \sum_{j\in\eta_i} \sum_{k\in\eta_i} E_p \{(x_j - \mu_2)(x_k - \mu_2)\} = \beta_2^2 \sum_{j\in\eta_i} \sum_{k\in\eta_i} \left[\sigma_{jk}^{(p)} + (\mu_1 - \mu_2)^2\right] \quad (17)$$

Thus, the KL-divergence can be expressed as:

$$D_{KL}(p,q) = log\left(\frac{\sigma_2}{\sigma_1}\right) - \frac{1}{2\sigma_1^2}\left[\sigma_1^2 - 2\beta_1 \sum_{j\in\eta_i} \sigma_{ij} + \beta_1^2 \sum_{j\in\eta_i}\sum_{k\in\eta_i} \sigma_{jk}\right] \quad (18)$$

$$+ \frac{1}{2\sigma_2^2}\left\{\sigma_1^2 + (\mu_1 - \mu_2)^2 - 2\beta_2 \sum_{j\in\eta_i}\left[\sigma_{ij}^{(p)} + (\mu_1 - \mu_2)^2\right]\right.$$

$$\left. + \beta_2^2 \sum_{j\in\eta_i}\sum_{k\in\eta_i}\left[\sigma_{jk}^{(p)} + (\mu_1 - \mu_2)^2\right]\right\}$$

By simplifying the last term of the previous equation, we have the final expression for the KL-divergence as:

$$D_{KL}(p,q) = log\left(\frac{\sigma_2}{\sigma_1}\right) - \frac{1}{2\sigma_1^2}\left[\sigma_1^2 - 2\beta_1 \sum_{j\in\eta_i} \sigma_{ij}^{(p)} + \beta_1^2 \sum_{j\in\eta_i}\sum_{k\in\eta_i} \sigma_{jk}^{(p)}\right] \quad (19)$$

$$+ \frac{1}{2\sigma_2^2}\left\{\left[\sigma_1^2 - 2\beta_2 \sum_{j\in\eta_i} \sigma_{ij}^{(p)} + \beta_2^2 \sum_{j\in\eta_i}\sum_{k\in\eta_i} \sigma_{jk}^{(p)}\right] + (\mu_1 - \mu_2)^2(1 - \Delta\beta_2)^2\right\}$$

where Δ denotes the number of neighbors in the neighborhood system η_i. Note that, when the inverse temperatures are zero, that is, $\beta_1 = \beta_2 = 0$, $D_{KL}(p,q)$ becomes Eq. (10), which is the KL-divergence between two Gaussian pdf's.

By analogy, the expression for $D_{KL}(q,p)$ can be written as:

$$D_{KL}(q,p) = log\left(\frac{\sigma_1}{\sigma_2}\right) - \frac{1}{2\sigma_2^2}\left[\sigma_2^2 - 2\beta_2 \sum_{j\in\eta_i} \sigma_{ij}^{(q)} + \beta_2^2 \sum_{j\in\eta_i}\sum_{k\in\eta_i} \sigma_{jk}^{(q)}\right] \quad (20)$$

$$+ \frac{1}{2\sigma_1^2}\left\{\left[\sigma_2^2 - 2\beta_1 \sum_{j\in\eta_i} \sigma_{ij}^{(q)} + \beta_1^2 \sum_{j\in\eta_i}\sum_{k\in\eta_i} \sigma_{jk}^{(q)}\right] + (\mu_1 - \mu_2)^2(1 - \Delta\beta_1)^2\right\}$$

Finally, after some algebraic manipulations, the symmetrized KL-divergence between two pairwise Gaussian-Markov random field models is given by:

$$D_{KL}^{sym}(p,q) = \frac{1}{2}[D_{KL}(p,q) + D_{KL}(q,p)]$$

$$= \frac{1}{4\sigma_1^2\sigma_2^2} \left\{ \left(\sigma_1^2 - \sigma_2^2\right)^2 - 2(\beta_2\sigma_1^2 - \beta_1\sigma_2^2) \sum_{j\in\eta_i} \left(\sigma_{ij}^{(p)} - \sigma_{ij}^{(q)}\right) \right.$$

$$+ (\beta_2^2\sigma_1^2 - \beta_1^2\sigma_2^2) \sum_{j\in\eta_i} \sum_{k\in\eta_i} \left(\sigma_{jk}^{(p)} - \sigma_{jk}^{(q)}\right) \tag{21}$$

$$\left. + (\mu_1 - \mu_2)^2 \left[\sigma_1^2(1 - \Delta\beta_2)^2 + \sigma_2^2(1 - \Delta\beta_1)^2\right] \right\}$$

Once again, note that when $\beta_1 = \beta_2 = 0$, the expression becomes the symmetrized KL-divergence between two Gaussian random variables, as indicated by Eq. (11).

2.3   Maximum Pseudo-likelihood Estimation

The KL-divergence between two GMRF's is a function of the model parameters: mean, variances, covariances and the inverse temperature. In our experiments, the GMRF's parameters μ and Σ, the covariance matrix of the patches, are both estimated via maximum likelihood. However, maximum likelihood estimation of the inverse temperature parameter is intractable, because we need to compute the partition function in the joint Gibbs distribution. To overcome this limitation, we perform maximum pseudo-likelihood estimation, which is based on the conditional independence principle [2]. In the pairwise Gaussian-Markov random field model, we have to maximize the logarithm of the product of all the local conditional density functions:

$$log \, L\left(\vec{\theta}; \mathbf{X}\right) = -\frac{n}{2}log\left(2\pi\sigma^2\right) - \frac{1}{2\sigma^2} \sum_{i=1}^{n} \left[x_i - \mu - \beta \sum_{j\in\eta_i} (x_j - \mu)\right]^2 \tag{22}$$

If we differentiate Eq. (22) with respect to β and properly solve the pseudo-likelihood equation, we have [7]:

$$\hat{\beta}_{MPL} = \frac{\sum_{i=1}^{n} \left[\sum_{j\in\eta_i} (x_i - \mu)(x_j - \mu)\right]}{\sum_{i=1}^{n} \left[\sum_{j\in\eta_i} (x_j - \mu)\right]^2} \tag{23}$$

Another simpifying assumption is that in our case, the random field is defined on a 1-D grid, where the size of the neighborhood system is fixed ($\Delta = 2, 4, 6, ...$). Hence, the maximum pseudo-likelihood estimator for the inverse temperature can be expressed in terms of the covariances:

$$\hat{\beta}_{MPL} = \frac{\sum\limits_{j\in\eta_i} \sigma_{ij}}{\sum\limits_{j\in\eta_i}\sum\limits_{k\in\eta_i} \sigma_{jk}} \tag{24}$$

which means that we can use the covariance matrix of the patches extracted from the random sample.

2.4   Numerical Computations

In order to speed up the numerical computations, we propose to express the equations for the maximum pseudo-likelihood estimator of the inverse temperature parameter and the KL-divergence between two GMRF's using tensor notation. First, note that given a random sample of a Gaussian variable with size n, denoted by $X = \{x_1, x_2, ..., x_n\}$, we can build a $n \times (\varDelta+1)$ dataset in which the i-th row represents the i-th patch. For instance, considering a first order neighborhood system, that is, $\varDelta = 2$, we have $p_i = [x_{i-1}, x_i, x_{i+1}]$. Then, we compute the covariance matrix of these patches, denoted by $\varSigma$. From this covariance matrix, we extract two main components: 1) a vector $\vec{\rho}$, composed by the elements of the central row of $\varSigma$, excluding the middle one, which denotes the variance of x_i (we want only the covariances between x_i and x_j, for $j \neq i$; and 2) a sub-matrix $\varSigma^-$, obtained by removing the central row and central column from $\varSigma$ (we want only the covariances between $x_j \in \eta_i$ and $x_k \in \eta_i$). Figure 2 shows the decomposition of the covariance matrix $\varSigma$ into the sub-matrix $\varSigma^-$ and the vector $\vec{\rho}$ in a fourth order neighborhood system ($\varDelta = 8$).

We can now use Kronecker products to rewrite the Fisher information matrices in a tensor notation, providing a computationally efficient way to compute the maximum pseudo-likelihood estimator of the inverse temperature parameter as:

$$\hat{\beta}_{MPL} = \frac{\|\vec{\rho}\|_+}{\|\varSigma^-\|_+} \tag{25}$$

where $\|A\|_+$ denotes the summation of the elements of the vector/matrix A. Using the same logic, the expression for the symmetrized KL-divergence between two GMRF's can be expressed as:

$$D_{KL}^{sym}(p, q) = \frac{1}{4\sigma_1^2\sigma_2^2}\left\{ (\sigma_1^2 - \sigma_2^2)^2 - 2(\beta_2\sigma_1^2 - \beta_1\sigma_2^2)\left(\left\|\vec{\rho}^{(p)}\right\|_+ - \left\|\vec{\rho}^{(q)}\right\|_+\right)\right. \tag{26}$$

$$+ (\beta_2^2\sigma_1^2 - \beta_1^2\sigma_2^2)\left(\left\|\varSigma_{(p)}^-\right\|_+ - \left\|\varSigma_{(q)}^-\right\|_+\right)$$

$$\left. + (\mu_1 - \mu_2)^2\left[\sigma_1^2(1 - \varDelta\beta_2)^2 + \sigma_2^2(1 - \varDelta\beta_1)^2\right]\right\}$$

where $\vec{\rho}^{(p)}$ and $\varSigma_{(p)}^-$ are extracted from the covariance matrix of the first GMRF and $\vec{\rho}^{(q)}$ and $\varSigma_{(q)}^-$ are extracted from the covariance matrix of the second GMRF.

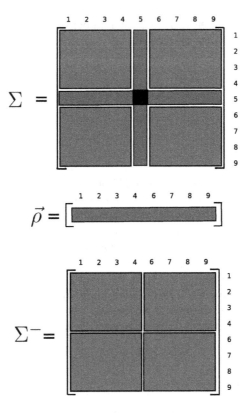

Fig. 2. Decomposition of Σ into Σ^- and $\vec{\rho}$ on a **fourth order neighborhood system** ($\Delta = 8$). By rewriting maximum pseudo-likelihood estimator of the inverse temperature parameter and the KL-divergence between two GMRF's in terms of Kronocker products, we speed up the numerical computations.

2.5   PCA-KL with Pairwise GMRF's

The contextual PCA-KL using pairwise GMRF's proposed here is inspired by Parametric PCA, a recent dimensionality reduction based metric learning algorithm that computes the entropic covariance matrix of the data through information-theoretic divergences estimated in local patches along the KNN graph [6]. Previous works proposed extensions to Parametric PCA by replacing the Bhatthacharyya and Hellinger distances by the symmetrized KL-divergence [8]. The main difference between the proposed method and PCA-KL concerns the statistical model employed to fit the data.

Let $X = \{\vec{x}_1, \vec{x}_2, ..., \vec{x}_n\}$, where $\vec{x}_i \in R^m$, represent our input data matrix and $G = (V, E)$ be the ϵ-neighborhood graph induced from X by linking each pair of samples $\vec{x}_i$ and $\vec{x}_j$ if $d_E(\vec{x}_i, \vec{x}_j) < \epsilon$, where $d_E(.,.)$ is the Euclidean distance. Hence, we can define a patch P_i as the set composed by a sample $\vec{x}_i$ and its neighbors in the graph. Let k_i denote the number of neighbors of $\vec{x}_i$ in G. Then, a patch P_i is a $(k+1) \times m$ matrix:

$$P_i = \begin{bmatrix} \vec{x}_i \\ \vec{x}_{i1} \\ \vec{x}_{i2} \\ \vec{x}_{i3} \\ \vdots \\ \vec{x}_{ik_i} \end{bmatrix} = \begin{bmatrix} \vec{x}_i(1) & \vec{x}_i(2) & \cdots & \vec{x}_i(m) \\ \vec{x}_{i1}(1) & \vec{x}_{i1}(2) & \cdots & \vec{x}_{i1}(m) \\ \vec{x}_{i2}(1) & \vec{x}_{i2}(2) & \cdots & \vec{x}_{i2}(m) \\ \vec{x}_{i3}(1) & \vec{x}_{i3}(2) & \cdots & \vec{x}_{i3}(m) \\ \vdots & \vdots & \ddots & \vdots \\ \vec{x}_{ik_i}(1) & \vec{x}_{ik_i}(2) & \cdots & \vec{x}_{ik_i}(m) \end{bmatrix}$$

where $\vec{x}_{i1}, \vec{x}_{i2}, \ldots, \vec{x}_{ik_i}$ denote the k_i neighbors of $\vec{x}_i$ in the ϵ-neighborhood graph. The idea is to consider that each column of the matrix P_i is a sample of a pairwise Gaussian-Markov random field, characterized a set of local conditional density functions. In the proposed method, we estimate the parameter vetors $\vec{\theta}_k = (\mu_k, \Sigma_k, \beta_k)$ for $k = 1, 2, \ldots, \ldots, m$ for each patch of the ϵ-neighborhood graph.

The proposed contextual PCA-KL method maps each patch P_i to a m-dimensional vector composed by 3-dimensional tuples, where each tuple j contains the estimatives of the model parameters. As we have m distinct features, we will have exactly m tuples. The parametric feature vector $\vec{p}_i$ for the patch P_i is defined as:

$$\vec{p}_i = \left[\vec{\theta}_1^{(i)}, \vec{\theta}_2^{(i)}, \ldots, \vec{\theta}_m^{(i)} \right] \tag{27}$$

where each component is a tuple of parameters:

$$\vec{\theta}_j^{(i)} = \left(\mu_j^{(i)}, \Sigma_j^{(i)}, \beta_j^{(i)} \right) \tag{28}$$

The space spanned by $\vec{p}_i$, for $i = 1, 2, \ldots, n$, is the parametric feature space. The sample average of all $\vec{p}_i$'s represents the mean distribution and defines the centroid of this space:

$$\tilde{p} = \frac{1}{n} \sum_{i=1}^{n} \vec{p}_i \tag{29}$$

In PCA-KL, the difference between two parametric vectors is computed by the symmetrized KL-divergence between each one of the tuples, that is:

$$\vec{p}_i - \vec{p}_j = \left[d_{KL}(\vec{\theta}_1^{(i)}, \vec{\theta}_1^{(j)}), \ldots, d_{KL}(\vec{\theta}_m^{(i)}, \vec{\theta}_m^{(j)}) \right] = \vec{d}_{KL}(\vec{p}_i, \vec{p}_j) \tag{30}$$

For pairwise Gaussian-Markov random field models, a closed-form expression for the symmetrized KL-divergence is given by Eq. (21). Then, the entropic covariance matrix C is defined by:

$$C = E \left[\vec{d}_{KL}(\vec{p}_i, \tilde{p}) \, \vec{d}_{KL}(\vec{p}_i, \tilde{p})^T \right] = \frac{1}{n-1} \sum_{i=1}^{n} \vec{d}_{KL}(\vec{p}_i, \tilde{p}) \, \vec{d}_{KL}(\vec{p}_i, \tilde{p})^T \tag{31}$$

Note that with the proposed method it is really straightforward to deal with the out-of-sample problem, since instances that do not belong to the training set, can be mapped to their new representation with the projection matrix. This is an advantage of PCA-KL in comparison to manifold learning algorithms, such as Laplacian Eigenmaps [1], t-SNE [18] and UMAP [11].

In the following, we present the an algorithm for unsupervised metric learning with contextual PCA-KL. The input is a data matrix $X_{m \times n}$, in which each column $\vec{x}_j \in R^m$ is a sample.

1. From the input data $\vec{x}_1, \vec{x}_2, ..., \vec{x}_n \in R^m$ build an undirected proximity graph using the ϵ-neighborhood rule;
2. For each patch, that is, $\vec{x}_i \cup \eta_i$, where η_i denotes the local neighborhood around $\vec{x}_i$, compute the sample mean of each feature. Additionally, estimate the covariance matrix of the patches and the inverse temperature for each feature ($k = 1, 2, ..., m$). At the end, this step generates for each patch, the following parametric vector:

$$\vec{p}_i = [(\mu_1, \Sigma_1, \beta_1), (\mu_2, \Sigma_2, \beta_2), ...(\mu_m, \Sigma_m, \beta_m)] \tag{32}$$

3. Compute the mean parametric vector for all patches, $\tilde{p}$, which represents the average distribution, given all the dataset:

$$\tilde{p} = \left[(\tilde{\mu}_1, \tilde{\Sigma}_1, \tilde{\beta}_1), (\tilde{\mu}_2, \tilde{\Sigma}_2, \tilde{\beta}_2), ..., (\tilde{\mu}_m, \tilde{\Sigma}_m, \tilde{\beta}_m) \right] \tag{33}$$

4. Compute the entropic covariance matrix C, a surrogate for the usual covariance matrix based in the symmetrized KL-divergence between each parametric vector $\vec{p}_i$ and the average distribution $\tilde{p}$ as:

$$C = \frac{1}{n-1} \sum_{i=1}^{n} \vec{d}_{KL}(\vec{p}_i, \tilde{p}) \; \vec{d}_{KL}(\vec{p}_i, \tilde{p})^T \tag{34}$$

where $\vec{d}_{KL}(\vec{p}_i, \tilde{p})$ is a column vector of KL-divergences:

$$\vec{d}_{KL}(\vec{p}_i, \tilde{p}) = \left[D_{KL}^{sym} \left(\vec{\theta}_1, \tilde{\vec{\theta}}_1 \right), ..., D_{KL}^{sym} \left(\vec{\theta}_m, \tilde{\vec{\theta}}_m \right) \right]^T \tag{35}$$

with $\vec{\theta}_i = (\mu_i, \Sigma_i, \beta_i)$ and $\tilde{\vec{\theta}}_i = (\tilde{\mu}_i, \tilde{\Sigma}_i, \tilde{\beta}_i)$.

5. Select the $d < m$ eigenvectors associated to the d largest eigenvectors of the matrix C to compose the projection matrix W.
6. Project the data into the subspace spanned by W.

3   Experiments and Results

To evaluate the performance of the proposed contextual PCA-KL for dimensionality reduction based metric learning, we conducted a set of computational experiments to compare the average classification accuracies of eight different supervised classifiers (KNN, Naive Bayes, SVM, Decision Tree, Bayesian classifier under Gaussian hypothesis, Multilayer Perceptron, Gaussian Process Classifier and Random Forest) after dimensionality reduction to 2-D spaces. Our hypothesis is that if the underlying distance functions are correctly learned during the feature extraction stage, we should observe a significant increase in the average scores when considering several real world datasets. We compared our method against PCA [5], Kernel PCA [14], PCA-KL (under Gaussian hypothesis) [8], ISOMAP [17], LLE [13], Laplacian Eigenmaps [1], t-SNE [18] and UMAP [11]. In all experiments, we selected 50% of the samples for training and 50% of the samples for testing. In the proposed method, we use a first order neighborhood system. Table 1 shows the average classification accuracies obtained from eight classifiers for several datasets after dimensionality reduction. The best result in a line is boldfaced. All 32 datasets, as well as detailed information regarding the number of instances, features and classes for each one of them, are freely available at openML.org.

To test if the results obtained by the proposed method are statistically superior, we performed a non-parametric Friedman test. There are strong evidences against the null hypothesis that all methods are identical ($p = 1.11 \times 10^{-16}$) for a significance level $\alpha = 0.01$. To check which groups are significantly different, we performed a Nemenyi post-hoc test. According to the test, for the same significance level, $\alpha = 0.01$, there are strong evidences that PCA-KL with GMRF's produced significantly higher accuracies than PCA ($p = 2.09 \times 10^{-11}$), Kernel PCA ($p = 3.11 \times 10^{-14}$), ISOMAP ($p = 1.80 \times 10^{-11}$), LLE ($p < 10^{-17}$), Laplacian Eigenmaps ($p < 10^{-17}$), t-SNE ($p = 3.28 \times 10^{-7}$) and UMAP ($p = 9.95 \times 10^{-10}$). Furthermore, there are evidences that PCA-KL with GMRF's produced significantly higher accuracies than traditional PCAKL ($p = 0.002$), which suggests that the Gaussian-Markov random field model is a better option in terms of dimensionality reduction based metric learning in these datasets. With the incorporation of the inverse temperature parameter in the model, we can capture relevant contextual information for data discrimination, something that is ignored by the usual Gaussian model, which assumes independence between the observations within a patch from the ϵ-neighborhood graph.

One possible limitation with the proposed method concerns the definition of the parameter ϵ (radius) that controls the patch size (number of neighbors of a given sample in the ϵ-neighborhood graph). Our experiments revealed that the classification accuracies are quite sensitive to changes in this parameter. In this study, we employed the following strategy: for each dataset, we build the complete graph by linking a sample to every other sample. Then, for each sample $\vec{x}_i$, we compute the approximate distribution of the distances from $\vec{x}_i$ to any other sample $\vec{x}_j$. According to our computational experiments, the most interesting values of ϵ (radius) are obtained by the percentiles p of the distribution belonging to the interval $P = [1, 50]$, but we choose the best model as the one that maximizes the classification accuracy among all values of $p \in [1, 99]$, a process that require the construction of several different graphs for each dataset. Note that unlike the KNN graph, our graph is not regular, in the sense that we allow the degree of the vertices to be quite different. It is worth mentioning that we are using the class labels to choose the best model (model selection), however the feature extraction stage is not supervised, in the sense that the proposed method performs unsupervised metric learning.

Table 1. Average accuracies among KNN, NB, SVM, DT, QDA, MPL, GPC and RFC after dimensionality reduction with PCA, Kernel PCA, PCAKL, GMRF (proposed method), ISOMAP, LLE, Laplacian Eigenmaps, t-SNE and UMAP for several openML.org datasets (2-D case).

	PCA	PCAKL	GMRF	KPCA	ISO	LLE	LAP	tSNE	UMAP
iris	0.941	0.939	**0.978**	0.858	0.906	0.833	0.650	0.957	0.96
prnn_crabs	0.566	0.539	**0.869**	0.578	0.588	0.603	0.558	0.740	0.786
hapiness	0.219	0.494	**0.521**	0.198	0.245	0.182	0.188	0.313	0.245
mux6	0.615	0.591	**0.820**	0.695	0.526	0.630	0.457	0.785	0.636
threeOf9	0.589	0.795	**0.833**	0.532	0.639	0.678	0.588	0.725	0.753
diggle	0.86	0.962	**0.989**	0.897	0.879	0.856	0.784	0.886	0.886
Breast-tissue	0.422	0.610	**0.620**	0.469	0.448	0.492	0.476	0.502	0.469
Hayes-roth	0.581	0.751	**0.772**	0.634	0.638	0.623	0.611	0.666	0.56
AIDS	0.375	0.54	**0.64**	0.35	0.365	0.36	0.37	0.54	0.52
lupus	0.786	0.801	**0.813**	0.758	0.795	0.71	0.704	0.758	0.744
fri_c3_100_5	0.667	0.697	**0.77**	0.667	0.687	0.625	0.668	0.682	0.66
creditscore	0.795	0.843	**0.89**	0.72	0.795	0.76	0.765	0.735	0.78
blogger	0.675	0.75	**0.758**	0.655	0.712	0.712	0.635	0.735	0.645
vineyard	0.793	0.81	**0.817**	0.778	0.793	0.759	0.735	0.788	0.798
veteran	0.668	0.701	**0.741**	0.692	0.643	0.657	0.648	0.708	0.713
Monks	0.587	0.772	**0.880**	0.535	0.557	0.568	0.549	0.649	0.6
xd6	0.682	0.847	**0.858**	0.678	0.728	0.658	0.686	0.78	0.726
corral	0.831	0.909	**1.000**	0.817	0.825	0.675	0.754	0.865	0.796
boxing	0.691	0.695	**0.731**	0.672	0.65	0.654	0.639	0.664	0.645
baskball	0.645	0.609	**0.703**	0.645	0.578	0.604	0.546	0.643	0.614
bolts	0.781	0.917	**0.925**	0.843	0.843	0.675	0.743	0.837	0.818
newton	0.681	0.685	**0.732**	0.690	0.699	0.589	0.596	0.694	0.728
sleuth	0.556	0.565	**0.742**	0.550	0.572	0.604	0.508	0.610	0.584
parity5	0.445	0.546	**0.586**	0.414	0.367	0.351	0.398	0.461	0.367
confidence	0.843	0.865	**0.885**	0.84	0.84	0.802	0.798	0.829	0.85
census6	0.556	0.568	**0.591**	0.552	0.550	0.546	0.550	0.550	0.543
diabetes	0.706	0.715	**0.740**	0.687	0.693	0.664	0.664	0.699	0.689
bodyfat	0.783	0.761	**0.823**	0.702	0.757	0.720	0.637	0.781	0.766
machine_cpu	0.945	0.948	**0.95**	0.935	0.941	0.866	0.844	0.927	0.904
pm10	0.507	0.57	**0.579**	0.529	0.515	0.497	0.508	0.532	0.515
KnuggetChase3	0.764	0.777	**0.808**	0.762	0.773	0.771	0.770	0.768	0.778
fri_c2_100_10	0.697	0.747	**0.787**	0.587	0.650	0.487	0.612	0.650	0.672
Mean	0.664	0.729	**0.788**	0.654	0.662	0.632	0.614	0.702	0.680
Median	0.678	0.749	**0.798**	0.675	0.669	0.656	0.636	0.717	0.701
Minimum	0.219	0.494	**0.558**	0.198	0.245	0.182	0.188	0.313	0.245
Maximum	0.945	0.962	**1.000**	0.935	0.941	0.866	0.844	0.957	0.960
Std. Dev.	0.162	0.136	**0.121**	0.160	0.165	0.149	0.140	0.139	0.155

4   Conclusions and Final Remarks

Unsupervised metric learning and manifold learning are intrinsically related, because, in general, to build robust non-linear features from the observed data, we have to overcome the limitations of the Euclidean metric. In this paper, a generalization of Parametric PCA was proposed to replace the Euclidean distance by the symmetrized KL-divergence between Gaussian-Markov random field models estimated from patches along the ϵ-neighborhood graph. Our claim is that the proposed method is a promising alternative to several dimensionality reduction methods, since the computational experiments supported one main point: the obtained non-linear features can be more discriminative in supervised classification than features obtained by state-of-the-art manifold learning algorithms.

Basically, the good performance of the proposed method can be explained by the incorporation of a patch-based distance function to measure the similarity between the samples, which can be more robust than the pointwise Euclidean distance to deal with the presence of noise and outliers in data. The GMRF model captures the dependence structure of the elements inside a neighborhood system in the graph by means of the inverse temperature parameter and the covariance matrix of the patches. Additionally, unlike autoencoders and other deep-learning based approaches, the proposed method can be used in small sample size problems, without data augmentation techniques, since it does not require large amount of data for convergence.

Future works may include the use of other information-theoretic measures, such as the Fisher information matrix, which is the metric tensor of the parametric space, used to approximate geodesic distances between different statistical models. Different probability density functions can be used to model the features. Many datasets have multi-modal features, that is, their distribution is highly non-Gaussian. Gaussian Mixture Models can be employed to capture this kind of behavior. Generalized Gaussians are also versatile and tractable models for non-Gaussian data. Furthermore, a supervised version of the proposed method can be designed in a way that only neighboring samples of the same class are considered in a patch. Finally, better ways to automatic select the radius parameter in the ϵ-neighborhood graph can be explored, such as the use of different radiuses for each vertex of the graph.

Acknowledgments. This study was financed in part by the Coordenação de Aperfeiçoamento de Pessoal de Nível Superior - Brasil (CAPES) - Finance Code 001.

References

1. Belkin, M., Niyogi, P.: Laplacian eigenmaps for dimensionality reduction and data representation. Neural Comput. **15**(6), 1373–1396 (2003)
2. Besag, J.: Spatial interaction and the statistical analysis of lattice systems. J. Royal Stat. Soc. Ser. B (Methodological) **36**(2), 192–236 (1974)
3. Cunningham, J.P., Ghahramani, Z.: Linear dimensionality reduction: survey, insights, and generalizations. J. Mach. Learn. Res. **16**, 2859–2900 (2015)

4. Hammersley, J.M., Clifford, P.: Markov field on finite graphs and lattices (preprint) (1971). www.statslab.cam.ac.uk/~grg/books/hammfest/hamm-cliff.pdf
5. Jolliffe, I.T.: Principal Component Analysis. 2 edn. Springer, New York (2002). https://doi.org/10.1007/b98835
6. Levada, A.L.M.: Parametric PCA for unsupervised metric learning. Pattern Recogn. Lett. **135**, 425–430 (2020)
7. Levada, A.L.M.: Information geometry, simulation and complexity in gaussian random fields. Monte Carlo Methods Appl. **22**, 81–107 (2016)
8. Levada, A.L.M.: PCA-KL: a parametric dimensionality reduction approach for unsupervised metric learning. Adv. Data Anal. Classif. 1–40 (2021). https://doi.org/10.1007/s11634-020-00434-3
9. Li, D., Tian, Y.: Survey and experimental study on metric learning methods. Neural Netw. **105**, 447–462 (2018)
10. McClurkin, J., L.M. Optican, B., Gawne, T.: Concurrent processing and complexity of temporally encoded neuronal messages in visual perception. Science **253**, 675–677 (1991)
11. McInnes, L., Healy, J., Melville, J.: UMAP: Uniform manifold approximation and projection for dimension reduction. arXiv arXiv:1802.03426 (2020)
12. Murase, H., Nayar, S.: Visual learning and recognition of 3d objects from appearance. Int. J. Comput. Vis. **14**, 5–24 (1995)
13. Roweis, S., Saul, L.: Nonlinear dimensionality reduction by locally linear embedding. Science **290**, 2323–2326 (2000)
14. Schölkopf, B., Smola, A., Müller, K.-R.: Kernel principal component analysis. In: Gerstner, W., Germond, A., Hasler, M., Nicoud, J.-D. (eds.) ICANN 1997. LNCS, vol. 1327, pp. 583–588. Springer, Heidelberg (1997). https://doi.org/10.1007/BFb0020217
15. Seung, H.S., Lee, D.D.: The manifold ways of perception. Science **290**, 2268–2269 (2000)
16. Tasoulis, S., Pavlidis, N.G., Roos, T.: Nonlinear dimensionality reduction for clustering. Pattern Recog. **107**, 107508 (2020)
17. Tenenbaum, J.B., de Silva, V., Langford, J.C.: A global geometric framework for nonlinear dimensionality reduction. Science **290**, 2319–2323 (2000)
18. Van Der Maaten, L., Hinton, G.E.: Visualizing high-dimensional data using t-sne. J. Mach. Learn. Res. **9**, 2579–2605 (2008)
19. Van Der Maaten, L., Postma, E., Van den Herik, J.: Dimensionality reduction: a comparative review. J. Mach. Learn. Res. **10**, 66–71 (2009)
20. Wang, F., Sun, J.: Survey on distance metric learning and dimensionality reduction in data mining. Data Min. Knowl. Disc. **29**(2), 534–564 (2014). https://doi.org/10.1007/s10618-014-0356-z

The Importance of Color Spaces for Image Classification Using Artificial Neural Networks: A Review

Ronny Velastegui$^{(\boxtimes)}$ (iD), Linna Yang(iD), and Dong Han(iD)

Norwegian University of Science and Technology, Gjøvik, Norway
ronnyxv@stud.ntnu.no

Abstract. Image classification is one of the most important applications of artificial neural networks in the field of industry and research. In most research work, when implementing ANN-based image classification models, the images used for training and testing are always represented in the RGB color space. But recent articles show that the use of certain color spaces, other than RGB, can lead to better precision in ANN-based image classification models. Thus, in this work, we present an analysis of several relevant research articles about the importance of color spaces for image classification using artificial neural networks. Thus, through the review of these articles, we will evaluate the behavior and efficiency of several ANN architectures, in different image classification contexts, and using images data sets represented in several color spaces. In the end, we not only found that there is a clear influence of the color spaces in the final accuracy of this type of tasks. But, we also found that both the creation of new special ANN architectures, and the creation of new color spaces formed from the combination of others, can lead to an increase in the performance of ANN-based image classification models.

Keywords: Image classification · Color spaces · Artificial neural networks

1 Introduction

Image Classification is an important task nowadays for many applications in both the industrial and research field [20]. For example, in the field of medical images analysis, we could detect if a patient has a tumor or not, based on classifying tissue images [15]. Also, in the food industry, we could classify fruits according to their ripeness level [7]. Moreover, in a social media context, we could apply image classification to automatically detect inappropriate or offensive images, for later censorship [32].

As the applications of image classification have evolved, so have the methods of implementing it. In practice, image classification implementations are based on Machine Learning techniques [16]. The most relevant machine learning techniques to do this task include Decision Trees, Support Vector Machines,

© Springer Nature Switzerland AG 2021
O. Gervasi et al. (Eds.): ICCSA 2021, LNCS 12950, pp. 70–83, 2021.
https://doi.org/10.1007/978-3-030-86960-1_6

K-Nearest Neighbor, and Artificial Neural Networks. Each one of them has its advantages and disadvantages when classifying images, in this way we cannot define for sure, which of the methods is better in all cases [14]. But, without any doubt, the more promising techniques in the previous list are which imitate the functioning of biological neural, these are the Artificial Neural Networks [24].

Now, there are different color spaces through which an image can be represented, such as RGB, XYZ, CIELAB, HSV, CMYK, YIQ, etc. Each of these spaces has its way of representing images, and for this reason, they have particular uses [19]. For example, CMYK space is preferred for printing devices because it creates colors in a "subtractive" way. While the YIQ space was preferred in the first color televisions because it allowed backward compatibility with B/W television signals [31].

Although there are several numbers of different color spaces that have their particular advantages and disadvantages when representing images. Most of the time, the images used to perform Image Classification using Artificial Neural Networks are represented in the RGB Color Space [8]. Perhaps, this is because RGB is the most widely used intuitive color space for rendering images on devices [19]. But, is RGB the most recommended space to perform image classification using artificial neural networks? If not, which is the best color space to perform this type of task?

In this work, we will try to answer these and other questions by reviewing some relevant articles about it. In this way, we will evaluate the behavior and efficiency of several ANN architectures, in different image classification contexts, and using images data sets represented in several color spaces.

1.1   Structure of the Work

The general structure of this work is as follow:

In the Theoretical Framework, we will briefly cover the definitions of Image Classification, Artificial Neural Networks, and Color Spaces, which are very important concepts to clearly understand this literature review.

In the Review Section, we will analyze seven research articles, which contains very relevant information and results about the current state of the art related to the Importance of Color Spaces for Image Classification using ANN. At the final of this section, we will include a table defining the key details of each one of these research articles. So, this table will allow us to appreciate and compare the relationship between them.

In the Conclusions Section, we will clearly define the existing relationship between the Color Spaces and Image Classification using ANN, also how much the color spaces influence the overall performance of this type of task. Finally, we will summarize and propose some ideas for future research projects, which can help both fill research gaps, as well as expand the practical applications of this research area.

2   Theoretical Framework

2.1   Image Classification

Image Classification is an important task of Image Recognition. That means that it incorporates aspects from Computer Vision and Artificial Intelligence for detecting and analyzing images [22]. Because Image Detection and Segmentation are also subfields of Image Recognition, usually, the definitions between them are often confused. The key difference is that, in the Image Classification, each image is considered as a whole and is assigned to a particular class. While in the Image Detection and Segmentation, each image can contain several objects of interest, and the objective is to locate the exact position of each of those objects within the image [16]. Thus, although Image Classification, Detection, and Segmentation are similar, the practical applications of them are very different.

In that way, the specific objective of Image Classification is to separate images into "classes" or "labels". For example, let's suppose that you have a data set with different images of animals. You could classify them into two classes: wild or domestic animals, this problem is called Binary Classification. Also, you could classify those images in more than two classes (for example: cat, dog, rabbit, horse, bird, and so on) this problem is called Multiclass Classification [18].

But the applications of Image Classification extend beyond just classifying images of animals. Today, image classification applications are present in many aspects of industrial and research fields [20]. So, image classification is present in facial recognition, fruit image classification, satellite image analysis, mineral image classification, medical image analysis, scene classification, and more. Its importance is so high that even world events, such as ILSVRC, are held to compare new image classification methods in terms of accuracy and efficiency [3].

So, as we mentioned before, the image classification tasks can be implemented in several ways. The most effective methods belong to the area of Machine Learning. Those methods include: Decision Trees, Support Vector Machines, K-Nearest Neighbor, and Artificial Neural Networks [24]. Although the use of each of them depends on the specific application, it must be recognized that the most efficient image classification methods today are based on Artificial Neural Networks [20].

Now, before implementing an image classification method, those images must go through a pre-processing stage. Thus, image pre-processing is a crucial stage to achieve good classification results [16]. Some pre-processing techniques consist of reducing noise, improving edges, improving contrast, or even transforming the color space in which the images are being represented. Each of these techniques, no matter how simple it may seem, can lead to a great improvement in any image classification method to be implemented [22].

2.2   Artificial Neural Networks

Artificial Neural Networks (ANN) are powerful computational models that belong to the field of Artificial Intelligence [29]. Unlike other Machine Learning techniques, these models are inspired by the biological neural networks of animals brains. In that way, the Artificial Neural Networks are composed of basic nodes called Neurons, which are equivalent to Biological Neurons. These nodes communicate between them through connections called Weights, which are equivalent to the Synaptic Process [10].

In general terms, the functioning of an Artificial Neural Network is as follows. First, the ANN takes the input information through Input Neurons. Then, the input information is transferred to other neurons, usually called Hidden Neurons, which are in charge of processing this information, step by step. Finally, once the input information was processed, the ANN gives us the final result through Output Neurons [12].

Now, before a neural network can correctly process the information and gives us the desired output, like other Machine Learning methods, it must first go through a training stage [2]. So, during the training phase, the network architecture is configured for solving a specific problem. This is done through automatic adjustment of the weights. So, when the training phase ends, all the learning has been stored in the weights. And now, the network is ready to complete the task for which it has been trained [9].

Nowadays, Artificial Neural Networks are present in many fields of research and industry, so the number of applications is very huge. In the field of economy, several ANN architectures have been implemented to perform sales forecasting with high accuracy [30]. In the field of information security, some ANN architectures have been implemented to predict credit card frauds with goods results [26]. In the field of Language, ANNs are used to accurately translate both text and voice in real-time between different languages [1]. Also, in the field of image recognition and analysis, several types of neural networks have been applied to perform different tasks, either: Classification, Tagging, Detection, and Segmentation of Images [23].

2.3   Color Spaces

The Color Spaces, also known as Color Models, are essentially systems to organize colors [8]. It also can be defined as a mathematical model to describe colors. So, those colors are usually represented using tuples of three or four numbers. Some of the most common color spaces used nowadays are: RGB, CMYK, CIELAB, XYZ, HSV, YIQ, etc. [19].

The Color Spaces can be classified into two different groups: Device-dependent color spaces and Device-independent color spaces. Some Device-dependent color spaces are: RGB, CMYK, HSV, HSL, HSI, YUV, YIQ, YCbCr, and YPbPr. And, in the case of Device-independent color spaces, the most used are XYZ, and CIELAB [31].

On the one hand, a Device-dependent color space is a color space in which the final color representation depends on the physical device configuration and quality used to produce it [17]. For example, if we choose the same CMYK color value and print it on two different types of paper, the printed colors will look different from each other. On the other hand, a Device-independent color space is a color space in which the numerical color values used to specify a color will produce the same color no matter the device. Due to the ability to represent colors in an "absolute way", these color spaces are usually used as Profile Connection Space (PCS) [21]. In Color Management, the Profile Connection Spaces are very important because they allow us to convert between color spaces [11].

The RGB Color Space is one of the most popular Device-dependent color spaces. In this system, each color is formed by the addition of three primary colors, which are: red, green, and blue. So, in this space, a color is represented using a tuple of 3 values, usually between the range of 0 and 255, which represents the amount of each primary color. This color space is widely used on multimedia devices such as: cameras, displays, scanners, etc. [27]

CIE XYZ Color Space is a Device-independent color space derived from the curves of LMS Color Space which represent the response of the three types of cones in the human visual system. Because of this, CIE XYZ Color Space allows us to represent all possible colors that the average human eye is capable of seeing [21].

CIE L*a*b* Color Space, usually called CIELAB, is a Device-independent color space defined by the International Commission on Illumination (CIE) in 1976. This color space was derived from the previous CIE XYZ Color Space. In the CIELAB color space, the colors are represented using three real values for L*, a* and b*, which represent the lightness, the green-red component, and the blue-yellow component respectively [11].

3   Literature Review

In this section, we will analyze seven research articles, which contains very relevant information and results about the current state of the art related to the Importance of Color Spaces for Image Classification using ANN. At the final of this section, we will include a table defining the key details of each one of these research articles. So, this table will allow us to appreciate and compare the relationship between them.

3.1   Javier Diaz Cely, et al. (2019)

In this research [6], the authors explore the impact of the color spaces in the transfer learning of Convolutional Neural Networks. So, the authors analyze the behavior of some pre-trained convolutional neural networks during the classification of a new data set. This new data set has been transformed into other color spaces, and not only in the default RGB color space.

The authors use three popular convolutional neural architectures: ResNet, MobileNet, and Inception-V3. All three were originally pre-trained using the ImageNet data set. This data set is one of the most used data set to pre-train convolutional neural networks for image classification tasks, and it contains 1.2 million RGB images divided into 1000 classes.

Once the pre-trained CNN architectures are established, the authors define the new image classification problem to perform. So, this problem is basically to classify, in two possible classes, a data set of images of cats and dogs. This data set contains 25000 images represented in the RGB color space. So, for the experiment, the authors perform this classification task using this data set in the original RGB color space, and after that, the authors perform the same task again for other color spaces.

So, the color spaces used to represent the data set are RGB and LAB, which are examples of Device-dependent and Device-independent color spaces respectively. Also, in order to explore how much each component of LAB Color Space affects the learning transfer, the authors simulate three other color spaces. Those color spaces are: LLL, AAA, and BBB, which are the three repeated components of the LAB Color Space.

In the results, the authors found that each one of the three CNN architectures obtains different accuracy levels in different color spaces. For example, in the case of ResNet architecture, the model obtains a slightly better performance in the RGB color space, compared with the LAB, LLL, AAA, and BBB color spaces. While, in the case of MobileNet, the same level of accuracy is reached using the five color spaces. This is interesting because even using just the LLL, AAA, or BBB color spaces (each one representing one LAB component) the same accuracy as using the RGB or LAB color spaces is reached. Finally, the Inception-V3 architecture obtained the most interesting result because, in this case, the RGB color space reaches a low accuracy compared to all the other color spaces. That means that even the LLL, AAA, or BBB color spaces obtain better results than the original RGB color space.

The authors conclude that these accuracy differences could be due to the differences in the internal architecture and functioning of each convolutional neural network. Because, for example, some architectures are much deeper than others and therefore they become more specialized in handling the color space in which they were originally pre-trained. Also, they assume that some CNN architectures perform the classification based mostly on spatial characteristics, rather than fine color details. Finally, as future works, they propose that it would be very interesting to perform this type of comparison using other architectures, data sets, and color spaces.

3.2   Shreyank Gowda, and Chun Yuan (2019)

In this research [8], the authors explore the influence of different color spaces on convolutional neural networks to perform image multi-class classification. Also, they propose a novel CNN architecture for image classification which processes images represented in seven different color spaces at the same time.

In order to do so, first, the authors implement a simple CNN architecture and perform multiclass classification using CIFAR-10 Data Set (In RGB Color Space). Then, this is repeated using the other 6 color spaces, which are: YIQ, LAB, HSV, YUV, YCbCr, and HED. After that, they compared the classification accuracy obtained using each color space. So, through confusion matrixes, they found that some color spaces improve the classification accuracy of some classes in particular.

So, based on that, they propose a novel CNN architecture for image classification that uses seven different color spaces, at the same time, to classify a particular image. This new architecture, named ColorNet, is composed of seven independent DenseNets that process the images dataset in different color spaces. Then, the output of those seven DenseNets is combined using a final dense layer which gives the final classification result. The authors also mention that due to the redundancy produced by introducing the same image in seven color spaces, the original DenseNet architecture can be modified to reduce the number of parameters and thus to avoid a computational overhead.

In the results, the architecture proposed by the authors obtains slightly better accuracy than the traditional CNN architectures that use images represented just in the RGB color space. Also, due to the reduced number of parameters to adjust during training, they mention that the efficiency of this architecture is slightly better than other complex CNN architectures with many hidden layers and parameters.

3.3   Wilson Castro, et al. (2019)

In this work [5], the authors implement four machine learning techniques, combined with three different color spaces, in order to compare and determine which machine learning technique and color space are the best to classify fruit images according to their ripeness level.

The data set used in this work contains 925 images of Cape Gooseberry Fruit. This data set was labeled into 7 different ripeness levels by human experts. Also, some image pre-processing techniques, such as image segmentation and enhancement, were applied to the original data set. Finally, the resultant enhanced image data set, which is represented in the RGB color space, was transformed to HSV and LAB color spaces. So, these three image data sets will be used to train and test four image classification approaches.

So, the machine learning approaches that have been implemented and compared in this work are Artificial Neural Networks (ANN), K-nearest Neighbors (KNN), Decision Trees (DT), and Support Vector Machines (SVM). With respect to the ANN Image Classification Method, which is relevant to this literature review work, the specific architecture used is the Radial Basis Function Artificial Neural Network (RBF-ANN). This type of ANN architecture is an improvement of the standard ANN because it allows for faster learning and convergence. So, all these machine learning approaches were individually trained using the three-color spaces mentioned before, and the performances were analyzed.

In the results, the authors found out that the models that reached the best accuracy were: ANN and SVM. Also, they noticed that the precision of both models is very sensitive to the color space used during the training phase. For this reason, both models reached a very high accuracy in the CIELAB color space, but a very low accuracy in the other color spaces.

3.4   Vanessa Buhrmester, et al. (2019)

In this work [4], the authors explore two principal ideas about image classification using CNN. First, they try to understand the behavior of CNN architectures when there is a color space transformation in the input images data set. Second, they explored which are the color spaces more prone (and less prone) against image disturbances, for example, noise or blur. In both cases, they try to calculate the accuracy difference performing image classification, using different types of data set. In this way, they try to find the color space that leads to better classification accuracy and that has more robustness against disturbances.

So, they use four different image data sets: PersonFinder, FlickrScene, CIFAR-10, and CIFAR-100. The first data set contains 15876 RGB images labeled in two classes, person and background. The second data set contains 10000 RGB landscape images labeled in four classes: desert, forest, snow, and urban. Finally, the third and fourth are both well-known data sets for image classification, each one contains 60000 RGB images of different objects divided into 10 and 100 classes respectively.

Now, for the first part of this project, they perform color space transformation in the four original image data sets. So, they obtain five different versions: RGB, HSV, HSI, YUV, and YIQ. After that, for the second part, they create the disturbed versions of those data sets through the addition of Gaussian Noise and Gaussian Blur. Then, with each version of those data sets, they train a simple CNN architecture, which is formed by two convolutional-pooling layers pairs. Finally, they analyzed and compared the accuracy obtained in each case.

So, in the results, they conclude that the color space information is very important in image classification using CNN. In fact, they found that in some data sets, especially with a large number of classes, the accuracy of classification could be very different if we change the original color space. For example, in the CIFAR-100 data set, the model with the best accuracy was trained with RGB Color Space. Instead, in the FlickrScene data set, the YIQ color space outperformed the other implemented spaces. They also conclude that HSV is the less robust color space against image disturbances, and they also mention that the Luminance component itself, present in some color spaces, is very robust against this type of disturbances.

3.5   Parham Khojasteh, et al. (2018)

In this work [13], the authors compare the performance of different color spaces during the detection of exudates through retinal image classification using CNN. Also, they perform a Principal Component Analysis (PCA) to generate the eigenchannels of each color space used, and they measure the accuracy with those color spaces versions. Finally, they propose a novel color space, formed by the combination of different components of different color spaces, which leads to high accuracy in this retinal image analysis.

The data set used in this work is a fusion of two publicly available image data set: DIARETDB1 and e-Ophtha. They contain RGB images of the retina which were manually labeled in two independent classes: Exudate and Non-Exudate. After that, this image data set was converted into two other color spaces: LUV and HSI. Finally, a PCA Analysis was applied to those color spaces obtaining the PCA versions: PCA-RGB, PCA-LUV, and PCA-HSI.

After training a simple convolutional neural network using six color spaces versions, the authors obtained that the HSI space reaches the highest accuracy level with 97.62%. And the second-best color space was PCA-RGB with a 96% accuracy level. Also, they found out that the second component of each PCA color space version shows a higher contrast between exudate and non-exudate image regions. With these results, the author proposes a new color space formed by the combination of HSI and PCA-RGB color spaces. This new three-channel color space, called "PHS", combines the second channel of PCA-RGB with the Hue and Saturation channels of the HSI Color Space. So, using this new color space, the classification model achieves higher accuracy than all previously tested models.

In the end, the authors conclude that the selection of a specific color space is a very important factor for image classification using CNN, especially in the retinal image analysis. They also demonstrate that it is possible to form a new three-channel color space, from the combination of individual components from other spaces, which can lead to better accuracy in image classification. Finally, as future works, they propose to carry out these experiments on other types of medical images, in order to improve the precision of current image classification systems based on convolutional neural networks.

3.6   Sachin Rajan, et al. (2018)

In this research [25], a pre-trained convolutional neural network is implemented to classify a scene image data set. This data set is converted to different color spaces and intensity plane representations. The objective of the authors is to find out which color space or intensity plane leads to obtain the highest accuracy in performing this particular image classification task.

With respect to the color spaces used, the authors analyze the effect of 4 different color spaces: RGB, HSV, CIELAB, and YCbCr. Also, they analyze 4 intensity planes obtained from the color spaces mentioned before. These intensity planes are V Plane, L Plane, Y Plane, and RGB2Gray obtained from HSV, CIELAB, YCbCr, and RGB color spaces respectively.

The experiment was implemented using the Oliva Torralba (OT) data set which contains 2688 RGB Images (256 × 256 pixels in size) of outside scenes. This data set is divided into 8 different classes: Open Country, Coast, Forest, Highway, Inside City, Street, Mountain, and Tall Building. For this experiment, the image data set is divided into 1888 images for training, and 800 images for testing. Finally, all these images are converted to obtain the versions in 4 different color spaces and 4 intensity planes mentioned before.

So, the convolutional neural network architecture used in this work is called Places205-CNN. This is a pre-trained CNN used to classify scene images. In this work, the CNN does not perform the classification by itself, but this is used just as a feature extractor. In this way, the CNN process the input images obtaining a feature map which will be processed later by another simple classification method. So, for this part, the authors implement two final classification methods, which are: Random Forest, and Extra Tree Classifiers.

In the results, the authors realized that each of the eight classes is classified better or worse than others in a specific color space or intensity plane. In this way, they conclude that there is no unique color space (or intensity plane) that obtains the perfect accuracy for all the classes at the same time. Finally, they conclude that choosing a determined color space is very important for image classification since if we ignore this, it can have a negative impact on the general accuracy of this type of task.

3.7 Jiasong Wu, et al. (2017)

In this work [28], the authors implement a PCANet-SVM to perform image classification using three different data sets and using different color spaces. So, the main objective is to find the color space which leads to the best accuracy performing these image classification tasks.

The three data sets that the authors use are CURet Texture Dataset, UC Merced Land Use Dataset, and Georgia Tech Face Dataset. So, the first data set contains images of surfaces of 61 different materials classes, where each class contains 92 images in RGB color space. The second data set contains satellite image sections of an urban map, which is divided into 21 classes with 100 images per class. The last data set, Georgia Tech Face Dataset, contains images of faces of 50 people, with 15 images for each person.

So, after defining and converting those three data sets into different color spaces (such as CIELAB, HSV, HSI, YUV, etc.), the proposed PCANet-SVM architecture was implemented. This architecture consists of two fundamental parts. The first part is the PCANet Convolutional Neural Network which works as a feature extractor of input images. The second part is the Support Vector Machine (SVM) which receives the features extracted in the first part and gives us the final classification result.

At the final, the authors conclude that, in most cases, the best color spaces to perform image classification using PCANet Convolutional Neural Network are: YUV and YIQ color spaces. Also, they found that the use of Hue-Saturation Color Spaces, which are HSV, HSL, and HSI, produces very poor image classification performance compared with other color spaces.

3.8 Comparison Table

In Table 1, a comparison of the main aspects of the seven scientific articles reviewed is presented. This table describes the specific image classification task, including the number of classes. Also, the type of ANN architecture is mentioned. And finally, the color spaces used in each article are detailed.

Table 1. Comparison table of the reviewed scientific articles

Article	Author	Image Classification Task	Number of Classes	ANN Architecture	Color Space Used
The Effect of Color Channel Representations on the Transferability of Convolutional Neural Networks. (2019)	Javier Diaz-Cely, et al.	Cat/Dog Image Classification.	2	1)InceptionV3 + FC-ANN 2)ResNet + FC-ANN 3)MobileNet + FC-ANN	RGB, LAB, LLL, AAA, BBB.
ColorNet: Investigating the importance of color spaces for image classification. (2019)	Shreyank Gowda, and Chun Yuan.	Objects and Animals Images Classification. (CIFAR-10)	10	ColorNet, (7 DenseNets Fusion)	RGB, YIQ, LAB, HSV, YUV, YCbCr, HED
Using machine learning techniques and different color spaces for the classification of Cape gooseberry (Physalis peruviana L.) fruits according to ripeness level. (2019)	Wilson Castro, et al.	Fruit Images Classification According Ripeness Level	7	RBF-ANN	RGB, HSV, LAB.
Evaluating the Impact of Color Information in Deep Neural Networks. (2019)	Vanessa Buhrmester, et al.	1) Person/Background Image Classification 2) Landscape Image Classification 3) CIFAR-10 4) CIFAR-100.	2, 4, 10, 100	Simple CNN	RGB, HSV, HSI, YUV, YIQ
A novel color space of fundus images for automatic exudates detection. (2018)	Parham Khojasteh, et al.	Exudate/Non-Exudates Retinal Images Classification	2	Simple CNN	Color Spaces: RGB, LUV, HSI. PCA Transformations: PCA-RGB, PCA-LUV, PCA-HSI
Dependency of Various Color and Intensity Planes on CNN Based Image Classification. (2018)	Sachin Rajan, et al.	Outside Scene Images Classification.	8	Places205-CNN + RF/ET.	Color Spaces: RGB, HSV, LAB, YCbCr. Intensity Planes: RGB2Gray, L Plane, V Plane, Y Plane
PCANet for Color Image Classification in Various Color Spaces. (2017)	Jiasong Wu, et al.	1) Texture Image Classification 2) Urban Areas Image Classification 3) Face Image Classification	61,21,50	PCANet + SVM, (PCANet is used as a feature extractor)	RGB, YUV, YIQ, YPbPr, YCbCr, YDbDr, HSI, HSV, HSL, CIEXYZ, CIELCH, CIELAB.

4   Conclusions

In this work, we have reviewed some relevant research articles about the importance of color spaces for image classification using ANN. So, we have explored the behavior of different ANN-based image classification architectures when different data sets are represented in different color spaces.

In this way, through this literature review, we have explored the use and effectiveness of different color spaces when performing image classification tasks. Although the RGB color space is the most used space to perform these types of tasks, we were able to verify that the use of other color spaces, such as HSV or YUV, can lead to better classification accuracy. Furthermore, the effectiveness of new proposed color spaces, such as "PHS" which is a combination of different color spaces, was reviewed in this work.

Also, through the scientific articles reviewed in this paper, we have explored different ANN-based image classification architectures. So, we have reviewed from relatively simple ANNs, such as an RBF-ANN, to more complex ANNs architectures, such as DenseNet, PCANet, InceptionV3, among others. Also, in some articles, we were able to review new architectures such as ColorNet, which exploits the potential of combining different color spaces to improve the general image classification accuracy.

So, we not only reviewed different methods of classifying images in different color spaces, but we also reviewed their performance in different contexts of classifying images. Since, in the reviewed articles, the authors used different data sets to carry out different image classification applications. Applications such as face classification, medical image classification, fruit classification, among others. Each one of these applications being of great importance today for the field of industry and research.

Thus, after evaluating these articles, we have been able to confirm that the use of different color spaces can lead to an increase, or decrease, in the overall ANN-based image classification accuracy. Thus, certain architectures show great accuracy when trained with images in a given color space, while the same architectures show much lower accuracy when using other color spaces. Finally, we were able to observe that some classes, within the same dataset, are likely to be classified more accurately than the rest of the classes when using a specific color space.

For future works, all the authors agree that it would be good to implement these types of comparisons using other image data sets and applications. Also, the authors agree that it would be good to test the influence of color spaces in new types of ANN architectures, in order to improve the general accuracy when performing image classification tasks.

References

1. Akmeliawati, R., Ooi, M.P.L., Kuang, Y.C.: Real-time Malaysian sign language translation using colour segmentation and neural network. In: 2007 IEEE Instrumentation & Measurement Technology Conference IMTC 2007, pp. 1–6. IEEE (2007)
2. Basheer, I.A., Hajmeer, M.: Artificial neural networks: fundamentals, computing, design, and application. J. Microbiol. Methods **43**(1), 3–31 (2000)

3. Berg, A., Deng, J., Fei-Fei, L.: Large scale visual recognition challenge (ILSVRC) 2010. http://www.image-net.org/challenges/LSVRC 3 (2010)
4. Buhrmester, V., Münch, D., Bulatov, D., Arens, M.: Evaluating the impact of color information in deep neural networks. In: Morales, A., Fierrez, J., Sánchez, J.S., Ribeiro, B. (eds.) IbPRIA 2019. LNCS, vol. 11867, pp. 302–316. Springer, Cham (2019). https://doi.org/10.1007/978-3-030-31332-6_27
5. Castro, W., Oblitas, J., De-La-Torre, M., Cotrina, C., Bazán, K., Avila-George, H.: Classification of cape gooseberry fruit according to its level of ripeness using machine learning techniques and different color spaces. IEEE Access 7, 27389–27400 (2019)
6. Diaz-Cely, J., Arce-Lopera, C., Mena, J.C., Quintero, L.: The effect of color channel representations on the transferability of convolutional neural networks. In: Arai, K., Kapoor, S. (eds.) CVC 2019. AISC, vol. 943, pp. 27–38. Springer, Cham (2020). https://doi.org/10.1007/978-3-030-17795-9_3
7. El-Bendary, N., El Hariri, E., Hassanien, A.E., Badr, A.: Using machine learning techniques for evaluating tomato ripeness. Expert Syst. Appl. 42(4), 1892–1905 (2015)
8. Gowda, S.N., Yuan, C.: ColorNet: investigating the importance of color spaces for image classification. In: Jawahar, C.V., Li, H., Mori, G., Schindler, K. (eds.) ACCV 2018. LNCS, vol. 11364, pp. 581–596. Springer, Cham (2019). https://doi.org/10.1007/978-3-030-20870-7_36
9. Graupe, D.: Principles of artificial neural networks, vol. 7. World Scientific (2013)
10. Hassoun, M.H., et al.: Fundamentals of Artificial Neural Networks. MIT Press, Cambridge (1995)
11. Hill, B., Roger, T., Vorhagen, F.W.: Comparative analysis of the quantization of color spaces on the basis of the CIELAB color-difference formula. ACM Trans. Graph. (TOG) 16(2), 109–154 (1997)
12. Jain, A.K., Mao, J., Mohiuddin, K.M.: Artificial neural networks: a tutorial. Computer 29(3), 31–44 (1996)
13. Khojasteh, P., Aliahmad, B., Kumar, D.K.: A novel color space of fundus images for automatic exudates detection. Biomed. Signal Process. Control 49, 240–249 (2019)
14. Kim, J., Kim, B., Savarese, S.: Comparing image classification methods: K-nearest-neighbor and support-vector-machines. In: Proceedings of the 6th WSEAS International Conference on Computer Engineering and Applications, and Proceedings of the 2012 American Conference on Applied Mathematics, vol. 1001, pp. 48109–2122 (2012)
15. Kumar, A., et al.: Deep feature learning for histopathological image classification of canine mammary tumors and human breast cancer. Inf. Sci. 508, 405–421 (2020)
16. Lu, D., Weng, Q.: A survey of image classification methods and techniques for improving classification performance. Int. J. Remote Sens. 28(5), 823–870 (2007)
17. Mahy, M., Van Eycken, L., Oosterlinck, A.: Evaluation of uniform color spaces developed after the adoption of CIELAB and CIELUV. Color Res. Appl. 19(2), 105–121 (1994)
18. Matuska, S., Hudec, R., Kamencay, P., Benco, M., Zachariasova, M.: Classification of wild animals based on SVM and local descriptors. AASRI Procedia 9, 25–30 (2014)
19. McLaughlin, B.P.: Colors and color spaces. Proc. Twentieth World Congress Philos. 5, 83–89 (2000)

20. Nath, S.S., Mishra, G., Kar, J., Chakraborty, S., Dey, N.: A survey of image classification methods and techniques. In: 2014 International Conference on Control, Instrumentation, Communication and Computational Technologies (ICCICCT), pp. 554–557. IEEE (2014)
21. Paschos, G.: Perceptually uniform color spaces for color texture analysis: an empirical evaluation. IEEE Trans. Image Process. **10**(6), 932–937 (2001)
22. Perez, L., Wang, J.: The effectiveness of data augmentation in image classification using deep learning. arXiv preprint arXiv:1712.04621 (2017)
23. Quraishi, M.I., Choudhury, J.P., De, M.: Image recognition and processing using artificial neural network. In: 2012 1st International Conference on Recent Advances in Information Technology (RAIT), pp. 95–100. IEEE (2012)
24. Rawat, W., Wang, Z.: Deep convolutional neural networks for image classification: a comprehensive review. Neural Comput. **29**(9), 2352–2449 (2017)
25. Sachin, R., Sowmya, V., Govind, D., Soman, K.P.: Dependency of various color and intensity planes on CNN based image classification. In: Thampi, S.M., Krishnan, S., Corchado Rodriguez, J.M., Das, S., Wozniak, M., Al-Jumeily, D. (eds.) SIRS 2017. AISC, vol. 678, pp. 167–177. Springer, Cham (2018). https://doi.org/10.1007/978-3-319-67934-1_15
26. Sahin, Y., Duman, E.: Detecting credit card fraud by ANN and logistic regression. In: 2011 International Symposium on Innovations in Intelligent Systems and Applications, pp. 315–319. IEEE (2011)
27. Süsstrunk, S., Buckley, R., Swen, S.: Standard RGB color spaces. In: Color and Imaging Conference, vol. 1999, pp. 127–134. Society for Imaging Science and Technology (1999)
28. Wu, J., Qiu, S., Zeng, R., Senhadji, L., Shu, H.: PCANet for color image classification in various color spaces. In: Sun, X., Chao, H.-C., You, X., Bertino, E. (eds.) ICCCS 2017. LNCS, vol. 10603, pp. 494–505. Springer, Cham (2017). https://doi.org/10.1007/978-3-319-68542-7_42
29. Yegnanarayana, B.: Artificial neural networks. PHI Learning Pvt, Ltd (2009)
30. Yu, Y., Choi, T.M., Hui, C.L.: An intelligent fast sales forecasting model for fashion products. Expert Syst. Appl. **38**(6), 7373–7379 (2011)
31. Zeng, H., Hudson, K.R.: Method and system for management of color through conversion between color spaces. US Patent 7,054,035, 30 May 2006
32. Zuo, H., Hu, W., Wu, O.: Patch-based skin color detection and its application to pornography image filtering. In: Proceedings of the 19th International Conference on World Wide Web, pp. 1227–1228 (2010)

Classifying Historical Azulejos from Belém, Pará, Using Convolutional Neural Networks

Wanderlany Fialho Abreu, Rafael Lima Rocha,
Rafael Nascimento Sousa, Tiago Davi Oliveira Araújo,
Bianchi Serique Meiguins, and Carlos Gustavo Resque Santos[✉]

Universidade Federal do Pará, Belém, Brazil
{wanderlany.abreu,rafael.lima,rafael.nascimento.sousa}@icen.ufpa.br,
{tiagoaraujo,bianchi,carlosresque}@ufpa.br

Abstract. The cultural heritage of a city is of great importance for the maintenance and enhancement of its history. Users can use innovative technologies such as Computer Vision to emphasize the city's treasures attractively and playfully. In Belém, Pará, azulejo is a meaningful cultural heritage case, which goes back to its foundation. The image recognition can facilitate and speed up the search for an azulejo and its historical information since, although cataloged, its visual appearance identifies it better than a name, which implies that the image-based search is much more natural than a text-based search. In this way, this work presents a prototype that uses Convolutional Neural Networks (CNN) to classify the azulejos from Belém by an image-based search. CNN's training used two image datasets. The first contains images that show azulejos and other environmental elements (for instance, walls, doors, streets, and people). The second dataset contains images that show only azulejos, and in both datasets, they only have one type of azulejo per image. The trained model consists of twelve different types of azulejos, representing the recognizable classes. Thus, after training, the tflite (Tensorflow Lite) model is generated with azulejos classes to be used in the mobile device image classification task. Finally, we developed an application in which the user takes a photo, and the system sends it to the classification module that contains the trained CNN model. After the image classification process, the module returns the five classes' values with the best accuracy and historical details about the azulejos.

Keywords: Image recognition · Cultural heritage · Azulejo · Convolutional neural networks

1 Introduction

Tourism is a fundamental activity for city economics and even cultural identity [4]. The visitation and presentation of a city's cultural heritage are among the tourism modality that most express the city's identity. The heritage can be

O. Gervasi et al. (Eds.): ICCSA 2021, LNCS 12950, pp. 84–98, 2021.
https://doi.org/10.1007/978-3-030-86960-1_7

physical (such as buildings, avenues, monuments) or exhibitions of their histories and traditions [29].

The Belém, Pará, Brazil, city has a vast cultural heritage, including physical heritage built at different times in the city's history. The collection of Azulejos is a prominent case of physical heritage. They were manufactured in Europe and imported to decorate the home of influential people between the 18th and 20th centuries [3]. The expressive amount of Azulejos in the heritage of Belém increased mainly in the period of economic growth of the city [1] when it ceases to be a subsistence economy and starts to be market-based.

Belém has an expressive diversity of different Azulejos patterns. Alcântara et al. [1] book, published by IPHAN (National Historical and Artistic Heritage Institute), is the most extensive catalog of Belém's Azulejos until now, containing various types, such as patterns, panels, and devotional art, along with its historical information.

This cultural treasure is already presented to the tourist in a narrated way or through books. However, new technologies could transmit this cultural heritage more interactively and playfully. The tourist activity itself is changing more and more by the visitor's independence with the help of technologies, such as mobile applications that assist the visitation and exploration of points of interest and associated content [18].

Mobile applications, smart cities, virtual and augmented reality, computer networks, image processing, and computer vision are among the areas of information technology that can assist visitors. This work focuses on applying computer vision techniques to facilitate the search for content of a particular Azulejo only from a photo of it obtained through a cell phone camera.

Therefore, this work uses one of the leading computer vision techniques, the Convolutional Neural Network (CNN). CNN combines parts of the images provided, making comparisons between these small parts called masks, which are patterns used to verify the new images. The set of masks forms a filter, and a convolution operation is performed on each layer where there are filters to reveal new patterns. The values for the masks in the filters are learned during the CNN training stage.

We used two datasets of Azulejos [28] to train our classifier. The first dataset contains images showing only Azulejos, and the second contains images showing Azulejos and adjacent environmental elements (such as streets, walls, and vegetation) in the image. Thus, the CNN model's training has used these datasets to classify Azulejos from new photos obtained by the user of the application. Additionally, we have implemented a prototype application to test the model's functioning on a mobile device.

The rest of the article is divided as follows: Sect. 2 describes the related works, Sect. 3 presents the datasets used for training the model and the dataset used to show historical content through the prototype application, Sect. 4 explains the training of the CNN model, Sect. 5 shows the mobile application developed to test the image-based search, and Sect. 6 concludes this work and draws directions for future work.

2   Related Works

This section presents works related to the main subjects addressed in this research, showing their objectives, techniques, and conclusions.

Table 1 presents an overview of the use of technologies concerning the related works, emphasizing Augmented Reality (AR), Computer Vision (CV) techniques, and the area of application.

Table 1. The overview of the related works.

Reference	Year	AR technique	CV technique	Area
[11]	2014	GPS-based	Image-recognition AR	Tourism
[20]	2015	GPS-based	–	Education
[24]	2015	GPS-based	–	Tourism
[27]	2015	QRcode	–	Journalism
[15]	2018	Augmenting immersive	–	Museum and art
[10]	2018	GPS-based	IFT, SURF	Tourism
[23]	2018	GPS-based	–	Tourism
[13]	2019	GPS-based	R-CNN	Tourism
[8]	2019	GPS-based	–	Tourism
[21]	2019	–	Machine learning, CNNs	Archaeology
[26]	2020	–	Point clouds, 3D reconstruction	Heritage
[12]	2020	Image-based	SURF, CNN model, SVM	Archaeology
[14]	2020	–	TU-DJ-Cluster and deep learning	Tourism
[16]	2020	–	Deep feature, image stitching	Tourism
[9]	2020	Image-based	–	Heritage
[22]	2020	–	VGG	Museum and Art
[17]	2021	Image-based	–	Tourism
[30]	2021	–	R-CNN	Archaeology
[25]	2021	Multisensory	–	Heritage
[2]	2021	3D Model	–	Tourism
[7]	2021	–	R-CNN	Archaeology

Rocha [27] analyzes augmented reality, aiming to improve journalism by innovating the presentation of news and proposing new ways of interacting with the reader. His work presents the chronological evolution of augmented reality technology and the uses in the journalistic medium, using mainly QRcode, for example, on magazine covers. As a result, the adherence to technology depends on the user's knowledge of how to use it, the different possible uses, and changes in society in the future.

Godoy et al. [13] present a mobile application focused on tourism, which allows the exploration and access to historical and cultural information in different places in Santiago, RS, Brazil. They used JavaScript, HTML, and CSS to create the application screens, in addition to the JavaScript framework Vue.Js. They used an API provided by Cordova (library) to implement the geolocation, while OpenCV was responsible for the computer vision techniques. The possibility of obtaining cultural and historical knowledge about the location during visitation is their main contribuiton.

Martins et al. [24] present a study on augmented reality and its use for tourism, which according to the authors, presents a significant growth in investments in technologies. Besides, they presented the project "Viseu na Palma da Mão" to better know the city by cell phone. In addition to visualizing, the tool allows manipulating virtual objects just by pointing the device at the real objects.

Various authors [15, 22] present the use of augmented reality by immersion and image processing, using museum and art images as a database, showing the potential of augmented reality in creating innovative experiences in various sectors of museums and arts. The visitors' evaluation states that the more real the museum experience and visualization of the arts, the greater the interest in visiting and buying art.

Various authors [9, 25, 26] present immersive virtual reality, deep learning, and neural networks concerning images of cultural heritage sites to provide visitor experiences in various external cultural heritage environments.

Various authors [7, 12, 21, 30] use new methods of deep neural networks, image processing, and augmented reality in archaeological images to detect unknown archaeological objects, from different places, in an innovative, accurate, and efficient way.

Kysela and Štorková [20] apply augmented reality technology to teaching history and tourism through a mobile application based on geolocation to offer innovative teaching methods attractively and effectively.

Various authors [2, 8, 10, 11, 14, 17, 23] use augmented reality resources, deep learning, and image recognition through images, geolocation, and immersion to provide sightseeing tours through various tourist spots.

3   Datasets

This section presents the datasets used in this work and their characteristics, structure, and purpose for their use. We have used two types of datasets to provide historical/cultural content about the Azulejos through the prototype application and another to train the CNN model.

3.1   Historical Content Dataset

Alcântara et al. book [1] provided the Belém, Pará, Azulejo's inventory that contains information on the cultural heritage of the Azulejos from Belém.

We transcribed the tables in the book (which are in printed format) to the digital format using a spreadsheet editor.

We created a spreadsheet for each table present in the book, containing a table with the information separated into the following columns: identification code, address, neighborhood, and information regarding the manufacture, characteristics, and application method. After registering all information in the spreadsheet editor, we have exported the datasets to the CSV (Comma-Separated Values) format to be easily imported into other applications. The aim of transforming this information into the digital format is to make it available for use by computer systems, thus facilitating further applications with this theme.

The historical/cultural dataset presents a total of 274 instances of Azulejos of pattern type, with eight attributes to describe each of them. The first column of Table 2 shows the attributes' name, the second has a brief description of that attribute, and the last column shows a sample of this dataset.

Table 2. Historical/Cultural dataset attributes on the Azulejos of pattern type

Attribute's name	Description	Sample
Azulejo-code	ID assigned for unique identification of records	Azulejo-PE1-1-1
Origin	Original manufacturing site of an Azulejo	Lisboa/Portugal
Factory	Original factory name	Viúva Lamego-Constância
Dimension-format	Size of individual tile piece	13.4 × 13.4 cm
Manufacturing technique	Material and techniques used in the manufacture of the Azulejo	Ceramic and glazed print
Address	Places in Belém where it is possible to find such Azulejos	Trav. Frutuso Guimarães n° 18/30/36, 257/259 ...
District	District containing the addresses	Cidade Velha, Campina
Use-application	Position and technique of tile application	Facade cladding, laid by compression against fresh mortar

A part of the information available in the historical/cultural dataset is used as visualized content in the prototype application of this work.

3.2   Azulejos Image Dataset for CNN Training

Despite the data and images obtained from the Azulejo's inventory, a larger amount of images with a variety of sizes, lighting, deformation, and distance is necessary to create a suitable dataset for training the CNN so that the final result is reliable.

Considering the purpose of using CNNs to classify Azulejos, the selected dataset [28] has pictures of Azulejos in the city taken from three different cell phones. Being divided into two datasets, one containing 191 pictures showing Azulejos and environmental elements and the other 865 images showing only the Azulejos. Further, the current situation and the impossibility of visiting some places' interior have created an obstacle in obtaining new images.

The database has images of historic mansions in Belém, with Azulejos decorating the interiors and facades. Figure 1 shows two samples of images in the two datasets, which are used for training the CNN model.

Fig. 1. Sample images present in the 1st database, containing Azulejos and other elements (left), and in the 2nd database, containing only tiles (right). [28]

The 1st dataset contains 191 images collected using three different cell phones, and the images have different types of quality, angles, and shadows. Additionally, other environmental elements are shown, such as doors, windows, and elements of buildings' facades, making it challenging to identify the Azulejo shown in the image. In short, it is the one that best represents the use of the application by the final user.

The 2nd database contains 865 images, consisting of cropped images created from the 191 images in the 1st database. It aims to filter the images focusing on the Azulejos and discard environmental elements present in the first dataset.

Moreover, the dataset generated by the training presents twelve classes of different Azulejos. Figure 2 shows the twelve classes of Azulejos present in the dataset, presented in a 2×2 grid (4 tiles) to illustrate the training classes.

4   Model Training

This section presents the phases of model training, the image preprocessing, the selection of the convolutional neural network architecture, and the training itself.

Fig. 2. 2×2 (4 tiles) photo of the twelve classes of Azulejos present in the dataset. [28]

4.1 Preprocessing

This phase uses preprocessing to standardize the dimensions of the images in the dataset. The necessary data is loaded in the disk and is performed the standardization of dimensions, adding vertical or horizontal black padding to images at 256×256 pixels without stretching the tiles. Figure 3 presents a demonstrative compilation of four images of one of twelve tiles classes to show the images after the standardization process.

Fig. 3. Demonstrative compile of images from same tile class after the standardization of dimensions.

Besides, we used the Samplewise normalization of mean and standard deviation to normalize the network's input images. Each sample or image is normalized by subtraction of mean of image pixels values and then divided by the standard deviation of image pixels values. Finally, each input image will tend to zero mean and unit standard deviation of pixels values.

The Eq. 1 presents the normalization performed per image. The image before the normalization is represent by I, and μ and σ are the mean and standard deviation of image I, while I_n represent the normalized image witch the pixels values has zero mean and unit standard deviation.

$$I_n = \frac{I - \mu}{\sigma} \tag{1}$$

4.2 CNN Selection

We have performed a random search to find the best architecture and parameters to classify the Azulejos, where given a finite set o parameters, is investigated a random combinations of parameters to find the ones that best fit the proposed task in this work [5].

It was investigated the number of layers of the CNN and dense layers, besides the number of filters and neurons as shown in Table 3. For example, the Conv layers block 1 (Table 3) represents the first block of convolutional layers, varying between one to three layers, with steps of one, while Neurons represent the number of neurons in dense layers, starting with 128 until 1024, with increments of 128.

The predefined architectures ResNet50, VGG-19, and InceptionV3 also are evaluated to perform Azulejos classification.

Table 3. Set of parameters used in random search.

Search	Min	Max	Step
Conv layers block 1	1	3	1
Conv layers block 2	1	3	1
Conv layers block 3	1	3	1
Dense layers	1	3	1
Neurons	128	1024	128
Filters	16	128	16

The evaluation of predefined architectures suggests a poor classification performance of Azulejos. Nevertheless, the random search to find our architecture is successful. The obtained architecture has 18 layers, as shown in Table 4.

The selected architecture has seven convolutional layers, seven pooling layers (max pooling), and four dense layers, as shown in Table 4. The convolutional layers have filters of size 3×3, and have the number of filters present in the column filters/neurons in Table 4. The first three dense layers have 1024, 512, and 128 neurons, respectively. While the fourth and last dense layer present 12 neurons, expressing the twelve classes of Azulejos.

4.3 Training

We have added Azulejos' images taken from photos of the computer screen due to the pandemic situation that the world is currently experiencing, making it unfeasible to test the prototype in person. These images are sensitive to illumination interference, distortions, and others. Therefore, the model can learn to recognize images captured from a computer screen and in-loco visitation.

Table 4. Summary of selected architecture.

Layer	Type	Filters/neurons	Number of parameters
1	Convolutional	64	1792
2	Pooling		
3	Convolutional	64	36928
4	Pooling		
5	Convolutional	32	18464
6	Pooling		
7	Convolutional	32	9248
8	Pooling		
9	Convolutional	16	4624
10	Pooling		
11	Convolutional	16	2320
12	Pooling		
13	Convolutional	16	2320
14	Pooling		
15	Dense	1024	66560
16	Dense	512	524800
17	Dense	128	65664
18	Dense	12	1548
Total of parameters			734.268
Trainable parameters			734.268

The final dataset has 1164 images distributed in twelve classes of Azulejos. We divided the dataset into 80% of images for training and 20% of images for test the generated model, 931 and 233 images, respectively.

We used the RMSprop optimizer to minimize the cost function (known as loss function) to reduce the difference between the expected output and the network output [6]. The learning rate used is 0.001. We used 100 epochs and 16 batches of images.

The training image set is small to train a CNN accurately. Thus, we generated batches of synthetic images in real-time to the model training through the data augmentation approach [19]. The transformations applied were vertical and horizontal flips, translations of −30 to 30 pixels per axis, besides the scale transformation of 70–130% of image size per axis.

Figure 4 presents some transformations performed by data augmentation. The left image presents the original image, without transformations, of the Azulejo of the "soberano" class, and the subsequent images represent the transformations applied to the original image.

Table 5 presents the obtained results in terms of accuracy and loss after the Azulejos classification model generation. The values of training classification

Fig. 4. Transformations applied by data augmentation. The left image shows the original, and the three right images show sample transformations applied by data augmentation.

results show that data is adjusted to model learning. The training set reaches an accuracy of 98.60%, indicating that almost all images are correctly classified. Furthermore, the loss value (0.0622) of Table 5 shows the efficiency of optimization performed by RMSprop with training data.

Table 5. Results of train and test sets.

	Accuracy (%)	Loss
Training	98.60	0.0622
Test	92.70	0.2875

The test results show the quality of both accuracy and loss results. The accuracy of 92.70%, shown in Table 5, indicates that 219 images are correctly classified of 233 available images. Besides, both training and test accuracies indicate no occurrence of model overfitting, i.e., the classification model of Azulejos showed up the efficiency to generalize and classify new images, which were not used in training. Finally, the loss value of the test set shows the optimization quality performed by RMSprop.

The database shows a certain level of imbalance between the number of Azulejo images per class. Therefore, to better elucidate the test images' results, the confusion matrix is generated, as shown in Fig. 5. The 'true class' shows the number of images that belong to a specific Azulejo class, while the 'predicted class' column represents the generated model's predictions.

Seventeen images are classified incorrectly, as shown in Fig. 5. The "pinho" class stands out, with five of twenty-five incorrectly classified. Two are classified as "particular2" and "sesc", while one is classified as "ihgp".

5 Prototype

In this section, we present the main activities of the azulejo recognition prototype. The application's main interactions are: take a picture of an azulejo,

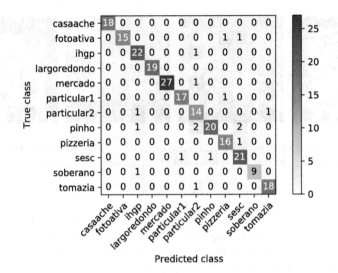

Fig. 5. Classification's confusion matrix of validation Azulejos images.

select the correct azulejo from the list of possibilities, and view content about the selected azulejo. Figure 6a presents the prototype's initial activity, showing the camera pointing to the image containing the azulejo to be recognized. After capturing the image, using the "FOTO" button, the smartphone screen displays the captured image.

Figure 6b shows the prototype's classification screen. It shows the four classes with the highest chance of being correct and their respective probability. The items on the screen are sorted from top to bottom in descending order of probability. In the example of Fig. 6b, the classes list shows "mercado" with ∼99% of probability, "casaache" with ∼0%, "pinho" with ∼0%, and "largoredondo" with ∼0%.

The activity of Fig. 6c presents the selected Azulejo's content of the classes list. This example shows the "mercado" class. This screen shows an Azulejo image alongside a textual description of its history, fabrication date, and location.

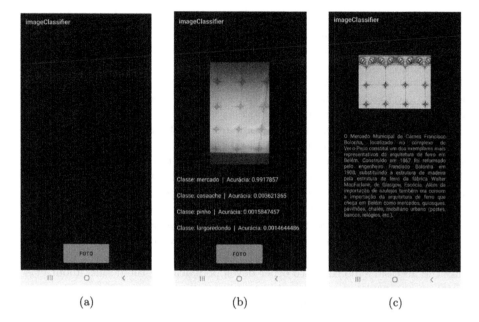

(a)   (b)   (c)

Fig. 6. Activities of the application in order of interaction. (A) The home screen has a button that opens the device's camera to take the picture. (B) Activity containing the taken photo and the name of the classes with the accuracy of the classification task. (C) Screen with details about the azulejos, accessed when the user clicks on an item listed in screen (B).

6   Conclusion

This work presents the application of a convolutional neural network in the city of Belém, Pará. We developed an image classification module using the Tensorflow environment libraries to reuse the created module in different projects that need the classification task. Besides, a table containing an inventory of the Azulejos from Belém, Pará, was transcribed into the digital format.

We researched several articles about other applications developed with similar themes aiming to acquire the necessary knowledge for the execution of the work. Besides, we detailed the technology they have used and the area of the application. to have a factual vision of what could be developed.

Regarding the use of a CNN for an image classification task, was used an eighteen layers architecture obtained through an exploratory search. The selected dataset for the training has twelve Azulejos classes, 1164 images taken from different cell phones, showing or not environmental elements, and having different angles, lighting, and shadow conditions.

The dataset was divided between images for the training itself, with 80%, and images for the model test, with 20%. Due to the relatively small number of images, we used the "Data Augmentation" technique to generate synthetic images (original images with subtle transformations).

Due to an in-loco test's unavailability, images of Azulejos taken from the computer monitor were inserted, making it necessary to use transformations related to lighting interference. During the training, the accuracy value was 98.60%, indicating that almost all training images were classified correctly, and the loss value was 0.0622%. During the test phase with the application, the accuracy and loss values marked the values of 92.70% and 0.2875% respectively, attesting to the training quality.

A confusion matrix was used to compare the obtained classification results with the true class. Additionally, we used the data augmentation technique to reduce repetition in deeper layers of training. As for the result obtained, among the exploratory search, our 18 layers architecture did better with the False Negative errors and the cost function and opened the possibility of future studies involving this architecture, using a larger dataset.

Regarding the coding contributions of this work, we decided to create a classification module, which brought together the necessary functions and interactions to perform the image classification task. The developed module aims to present only the necessary interfaces to other developers, leaving the details about classification encapsulated.

Besides, it is important to mention the transcription of the table present in the Azulejos' inventory from Belém, Pará, to the digital format, with 479 different specimens and various information about each of them. Thus, the table created can be used both concerning the information contained, offer a good source of data for building larger models, and the availability of information, which can be used in other projects for other purposes.

All used prototype code, trained CNN, and recognition module are available on the following link[1].

The future works are the integration of the prototype or the recognition module in other applications for tourism purposes, the integration with the location technology (GPS), the integration of the AR technology. Also, we will re-train the model with more classes when we could visit the local to take pictures.

References

1. Alcântara, D.M.E.S., Brito, S.R.S., Sanjad, T.A.B.C.: Azulejaria em Belém do Pará: Inventário - Arquitetura civil e religiosa - Século XVIII ao XX. Biblioteca Aloísio Magalhães IPHAN, Bras'ilia, DF (2016)
2. Marino Alfonso, J.L., Poblete Piedrabuena, M.Á., Beato Bergua, S., Herrera Arenas, D.: Geotourism Itineraries and Augmented Reality in the Geomorphosites of the Arribes del Duero Natural Park (Zamora Sector, Spain). Geoheritage **13**(1), 1–17 (2021). https://doi.org/10.1007/s12371-021-00539-x
3. Arruda, T.C., Sanjad, T.A.B.C.: Ornamentos de platibanda em edificações de belém entre os séculos XIX e XX: inventário e conservação. Anais do Museu Paulista: História e Cultura Material **25**(3), 341–388 (2017). https://doi.org/10. 1590/1982-02672017v25n0310

---

[1] http://labvis.ufpa.br/AzulejAR/.

4. Awang, K.W., Hassan, W.M.W., Zahari, M.S.M.: Tourism development: a geographical perspective. Asian Soc. Sci. 5(5) (2009). https://doi.org/10.5539/ass.v5n5p67
5. Bergstra, J., Bengio, Y.: Random search for hyper-parameter optimization. J. Mach. Learn. Res. **13**(2), 281–305 (2012)
6. Bishop, C.M.: Pattern Recognition and Machine Learning, 5th edn. Information Science and Statistics. Springer, Heidelberg (2006)
7. Bonhage, A., Eltaher, M., Raab, T., Breuß, M., Raab, A., Schneider, A.: A modified mask region-based convolutional neural network approach for the automated detection of archaeological sites on high-resolution light detection and ranging-derived digital elevation models in the north German lowland. Archaeological Prospection, February 2021). https://doi.org/10.1002/arp.1806
8. Cauchi, M., Scerri, D.: Enriching tourist UX via a location based AR treasure hunt game. In: 2019 IEEE 9th International Conference on Consumer Electronics (ICCE-Berlin). IEEE, September 2019. https://doi.org/10.1109/icce-berlin47944.2019.8966141
9. Conde, M.B.: Nuevas tecnologías y difusión del turismo cultural: descubriendo a goya con realidad aumentada. ROTUR. Revista de Ocio y Turismo **14**(1), 81–93 (2020). https://doi.org/10.17979/rotur.2020.14.1.5945
10. Demir, O.F., Karaarslan, E.: Augmented reality application for smart tourism: GökovAR. In: 2018 6th International Istanbul Smart Grids and Cities Congress and Fair (ICSG). IEEE, April 2018. https://doi.org/10.1109/sgcf.2018.8408965
11. Fukada, H., Kasai, K., Ohtsu, S.: A field experiment of system to provide tourism information using image recognition type AR technology. In: Lecture Notes in Electrical Engineering, pp. 381–387. Springer International Publishing, November 2014. https://doi.org/10.1007/978-3-319-06764-3_47
12. Godewithana, N., Jayasena, K., Nagarawaththa, C., Croos, P., Harshanath, B., Alosius, J.: Historical places & monuments identification system. In: 2020 IEEE Region 10 Conference (TENCON). IEEE, November 2020. https://doi.org/10.1109/tencon50793.2020.9293882
13. Godoy, R.C., Cavalheiro, B.O., de Oliveira Castanho, C.L., da Silva, E.F., Spies, E.H., Alves, V.M.: Turismo e realidade aumentada: Desenvolvimento de um aplicativo para a cidade de santiago/rs. Anais da X edição do Encontro Anual de Tecnologia da Informação - EATI. **9**(1), 110–117 (2019). http://anais.eati.info:8080/index.php/2019/article/view/20/17
14. Han, S., Ren, F., Du, Q., Gui, D.: Extracting representative images of tourist attractions from Flickr by combining an improved cluster method and multiple deep learning models. ISPRS Int. J. Geo-Inf. **9**(2), 81 (2020). https://doi.org/10.3390/ijgi9020081
15. He, Z., Wu, L., Li, X.R.: When art meets tech: the role of augmented reality in enhancing museum experiences and purchase intentions. Tourism Manage. **68**, 127–139 (2018). https://doi.org/10.1016/j.tourman.2018.03.003
16. Hoang, V.D., Tran, D.P., Nhu, N.G., Pham, T.A., Pham, V.H.: Deep feature extraction for panoramic image stitching. In: Intelligent Information and Database Systems, pp. 141–151. Springer International Publishing (2020). https://doi.org/10.1007/978-3-030-42058-1_12
17. Huang, T.L.: Restorative experiences and online tourists' willingness to pay a price premium in an augmented reality environment. J. Retail. Consum. Serv. **58**, 102256 (2021). https://doi.org/10.1016/j.jretconser.2020.102256

18. Kounavis, C.D., Kasimati, A.E., Zamani, E.D.: Enhancing the tourism experience through mobile augmented reality: Challenges and prospects. Int. J. Eng. Bus. Manage. **4**, 10 (2012). https://doi.org/10.5772/51644

19. Krizhevsky, A., Sutskever, I., Hinton, G.E.: ImageNet classification with deep convolutional neural networks. In: Advances in Neural Information Processing Systems, pp. 1097–1105 (2012)

20. Kysela, J., Štorková, P.: Using augmented reality as a medium for teaching history and tourism. Procedia - Soc. Behav. Sci. **174**, 926–931 (2015). https://doi.org/10.1016/j.sbspro.2015.01.713

21. Lambers, K., van der Vaart, W.V., Bourgeois, Q.: Integrating remote sensing, machine learning, and citizen science in Dutch archaeological prospection. Remote Sens. **11**(7), 794 (2019). https://doi.org/10.3390/rs11070794

22. Lei, Q., Wen, B., Ouyang, Z., Gan, J., Wei, K.: Research on image recognition method of ethnic costume based on VGG. In: Machine Learning for Cyber Security, pp. 312–325. Springer International Publishing (2020). https://doi.org/10.1007/978-3-030-62463-7_29

23. Llerena, J., Andina, M., Grijalva, J.: Mobile application to promote the Malecón 2000 tourism using augmented reality and geolocation. In: 2018 International Conference on Information Systems and Computer Science (INCISCOS). IEEE, November 2018. https://doi.org/10.1109/inciscos.2018.00038

24. Martins, M., Malta, C., Costa, V.: Viseu mobile: a location based augmented reality tour guide for mobile devices. Dos Algarves: a Multidisciplinary e-journal **26**(1), 8–26 (2015). DOi: https://doi.org/10.18089/damej.2015.26.1.1

25. Marto, A., Melo, M., Goncalves, A., Bessa, M.: Development and evaluation of an outdoor multisensory AR system for cultural heritage. IEEE Access **9**, 16419–16434 (2021). https://doi.org/10.1109/access.2021.3050974

26. Poux, F., Valembois, Q., Mattes, C., Kobbelt, L., Billen, R.: Initial user-centered design of a virtual reality heritage system: applications for digital tourism. Remote Sens. **12**(16), 2583 (2020). https://doi.org/10.3390/rs12162583

27. Rocha, P.M.A.: A exploração da realidade aumentada pelo jornalismo: a exposição da informação dos média num espaço aumentado. CECS - Centro de Estudos de Comunicação e Sociedade Universidade do MinhoBraga. Portugal. Literacia, Media e Cidadania - Livro de Atas do 4.° Congresso, pp. 475–491, November 2017

28. Santos, C., Junior, P.C., Araújo, T., Neto, N., Meiguins, B.: Recognizing and exploring azulejos on historic buildings facades by combining computer vision and geolocation in mobile augmented reality applications. J. Mob. Multimedia **13** (2017). https://dl.acm.org/doi/abs/10.5555/3177197.3177201

29. Williams, S., Lew, A.A.: Tourism Geography: Critical Understandings of Place, Space and Experience. Taylor and Francis Ltd. Abingdon (2014)

30. Trier, Ø.D., Reksten, J.H., Løseth, K.: Automated mapping of cultural heritage in Norway from airborne lidar data using faster r-CNN. Int. J. Appl. Earth Obs. Geoinf. **95**, 102241 (2021). https://doi.org/10.1016/j.jag.2020.102241

The Concatenated Dynamic Convolutional and Sparse Coding on Image Artifacts Reduction

Linna Yang$^{(\boxtimes)}$ ⓘ and Ronny Velastegui ⓘ

Norwegian University of Science and Technology, Gjøvik, Norway
`linnay@stud.ntnu.no`

Abstract. In order to enhance compressed JPEG image, a deep convolutional sparse coding network is proposed in this article. The network integrates state-of-the-art dynamic convolution to extract multi-scale image features, and uses convolutional sparse coding to separate image artifacts to generate coded feature for the final image reconstruction. Since this architecture consolidates model-based convolutional sparse coding with deep neural network, that allow this method has more interpretability. Also, compared with the existing network, which uses a dilated convolution as a feature extraction approach, this proposed concatenated dynamic method has improved de-blocking result in both numerical experiments and visual effect. Besides, in the higher compressed quality task, the proposed model has more pronounced improvement in reconstructed image quality evaluations.

Keywords: Deep learning · Image reconstruction · Sparse coding · Dynamic convolution

1 Introduction

Generally, there are two main image compression methods used recently. One of them is lossless compression, which usually exploit statistical redundancy in such a way as to represent the sender's data more concisely, but nevertheless perfectly [18]. For example, the run-length encoding [17], is often used in medical, high-tech and comics fields. And the variable-length coding (VLC) [29], whose the most significant property is more frequent symbols receive shorter codes. Several famous algorithms like Shannon-Fano Algorithm [32], Huffman coding [23] are all VLC.

Another one is lossy compression, this compression algorithm does not deliver high enough compression ratios and are widely used in the web and other areas that do not mind the loss of fidelity but require drastically reduction of bit rate [34]. Hence, most multimedia compression algorithms are lossy and commonly apply concept of perceptually lossless compression where the perceptual distortion metrics are needed. Although lossy compression methods can give acceptable

© Springer Nature Switzerland AG 2021
O. Gervasi et al. (Eds.): ICCSA 2021, LNCS 12950, pp. 99–114, 2021.
https://doi.org/10.1007/978-3-030-86960-1_8

result in most cases, there is always a trade-off between the compressed rate and distortion. They will still introduce compression artifacts. These artifacts and blocking might severely reduce the visually perceived quality and subsequent computer vision systems [31].

JPEG is one of the most commonly used lossy image compression standard, primarily used for natural images, was developed by the Joint Photographic Experts Group and was formally accepted as an international standard in 1992 [21]. JPEG compression is achieved by implementing discrete cosine transform (DCT) [1], a type of Fourier-related transform, that also used in more recent high-efficiency image file format (HEIF) [15]. The role of DCT is to decompose the original signal into its constant magnitude and periodic variations components. And then, the inverse discrete cosine transform (IDCT) is used to reconstruct the signal. The effectiveness of the DCT transforms coding method in JPEG relies on three major observations: Spatial redundancy; Lower sensitivity to loss in higher spatial frequencies in human eyes; Less visual acuity for color than gray [22].

There are several steps in JPEG Image Compression. After color conversion and chroma sampling, DCT is implied on image blocks. In this step, each image is divided into 8×8 block. Due to the DCT has a strong "energy concentration" characteristic: most of the natural signal (including sound and image) energy is concentrated in the additional part after the discrete cosine transform. It can compress the size of pictures to a pretty small level [22]. By using block, however, it has the effect of isolating each block from its neighboring context, sometimes it has visible change from block to block, that is why JPEG images look choppy ("blocky") when the user specifies a high compression ratio [5]. This discontinuity at the boundaries of block could sharply degrade the visual perception of image. Because of that, artifacts reduction of lossy compression is a necessary task in computer vision.

This article describes a JPEG image reconstruction method that combines the advantages of two popular image enhancement types in deep learning and utilizes state-of-the-art dynamic convolution. Compared with the previous counterpart with dilation convolution, the complement of dynamic convolution of this proposed model achieves better restoration performance in both numerical result and visual perception. Furthermore, with the improvement of picture quality, the role of new loss function becomes more notable and positively impacts the de-blocking performance.

2   Background

In order to enhance the compressed images quality, various strategies have been proposed. Generally, these image reconstruction methods can be roughly divided into two types. One is model-based [10,25–27], and the other one is learning-based [4,12,18,36,39]. The former is usually modeled using domain knowledge, especially some specific physical meaning, but the cost is time-consuming optimization. The latter focus on learn non-linear mapping function from training dataset directly, that allows this kind of methods have faster speed, while have less sufficient interpretation. These two classes of strategy have complementary advantages, although learning-based methods usually have more agreeable performance compared with another one.

Early JPEG image reconstruction methods depended on design heavily. For example, the filtering in the image domain or the transform domain, like using joint image domain filtering method [14]. Another way is regarding the artifacts reduction as the ill-posed inverse problem through optimization, like non-local self-similarity property. In this direction, sparsity as an effective technique to solve this ill-posed problem has been fully explored.

Moreover, deep convolutional neural networks (CNNs) made substantial progress in the past decade. In the image compressed and artifacts reduction task, this learning-based model has an agreeable ability in learning a nonlinear mapping from image pairs, which consist of the uncompressed original image and its compressed JPEG counterpart. [18] first proposed the deep convolutional neural networks method for JPEG artifacts reduction by utilizing the super-resolution network [8] as its cornerstone. Because of the inspiration from dense connection and residual learning, some deep CNNs approaches were introduced for image reconstruction and denoising [32,38].

Recent work gives some combinable solutions to keep the merits of both model-based and learning-based techniques, using a competitive deep learning model with model-based sparse coding (SC) [35]. Furthermore, Xueyang Fu et al. [11] adopt dilated convolutions to address the different image qualities problem. They propose a method that implements convolutional sparse coding (CSC) on fully connected layers. Due to the space-invariant characteristics of CSC, it can process the entire image, typically suitable for some vision tasks without high-level requests. This method is based on the work of Xueyang Fu et al., but using a different way to solve the multi-scale feature extraction issue and made some improvements in the loss function.

This proposed deep learning model has better explainability that benefits from the idea of utilization of CSC [16], which learned iterative shrinkage threshold algorithm (LISTA) can be applied to separate artifacts. And since the deep CSC part is built from the idea of the classical optimization algorithm, that making this model more structured and clear. Additionally, dynamic convolution makes it possible to process multiple image qualities. This high flexibility allows the proposed method to have a broader range of applications.

3   Methodology

This proposed network contains three main integrants designed for each specific task: multiscale feature extraction (using dynamic convolution), followed by a classic CSC approach – LISTA, and the final image reconstruction. Figure 1 shows the workflow of this model, which is quite composable, which using compressed JPEG image $\mathbf{J}$ as input and achieved de-blocking image $\mathbf{O}$ as output. In this networks, firstly, dynamic convolutions are designed for compressed image $\mathbf{J}$

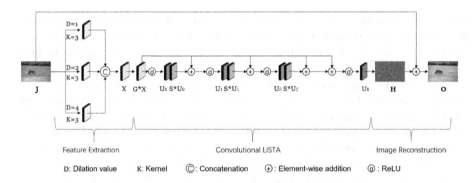

Fig. 1. The workflow of proposed network for image artifacts reduction

feature extraction in three different clarities. Secondly, a convolutional LISTA is used to build sparse code **X** for artifacts identification and separation. Thirdly, to predict the final reconstructed image **O**, the trained sparse codes $\mathbf{U_R}$ go to generate the residual **H**.

3.1   Network Architecture

Dynamic Convolution. Due to the compression quality of JPEG could be varied, only using one fixed model cannot meet the demand. In order to solve this problem, some learning-based strategies build several models based on different image qualities [8], some of them at the cost of greater parameter burden, train networks for the increased receiving field [38], some of them apply dilated convolutions instead [11]. In this network, dynamic convolution is adopted to handle the multi-scale feature extraction procedure.

The propose of dynamic convolution is mainly from the need for light-weight CNNs, typically for mobile devices that require to enable many functions and generate output in real-time. And in CNN architecture, it usually faces the problem that when computational constraint goes lower, the performance shows obviously degradation, since the extremely low computational constraint not only influences the depth of network but also the number of channels, which are pivotal for network performance [3].

Chen et al. [3] come up with a new operator design that presents aggregated multiple convolution kernels with constant network both in depth and width, which has remarkably improvement compared with its single kernel counterpart. The goal of this dynamic convolutional neural networks (DY-CNNs) is to figure out the dilemma between network efficiency and acceptable computational cost. DY-CNN does not increase either the layers or the channels of network, but the introduced K kernels, which determined by different input, play an essential role in enhancing the model capability.

Firstly, this DY-CNN aggregates K linear functions, the weight and bias are defined as Eq. (1), where the π_k is between 0 and 1, representing the attention weight of various linear function, that differ from every input images. Thus, the dynamic perception model is defined in Eq. (2), a non-linear function with a higher representation ability.

$$\tilde{W}(x) = \sum_{k=1}^{K} \pi_k(x)\tilde{W}_k, \tilde{b}(x) = \sum_{k=1}^{K} \pi_k(x)\tilde{b}_k \tag{1}$$

$$y = g(\tilde{W}^T(x)x + \tilde{b}(x)) \tag{2}$$

In addition, assembled parallel convolution kernels share the same input and output channels. Hence the network width or depth is not changed, which gives dynamic convolution great compatibility. And although the introduction of attention weights and K kernels brings some extra computational cost, this induced cost is still negligible compared with the convolution operation.

In this networks, a dynamic convolution using classic CNN design [3] (Fig. 1) is used in the first step. Before the beginning, by using the squeeze-and-excitation method [19], the kernel attentions can be computed. At first, the global average pooling layers squeezes the spatial information. It further goes through two fully connected (FC) layers with one ReLU between them and one softmax after them to create normalized attention weights. After the attention is all computed, the aggregated convolution output is passed a batch normalization (BN) layer and goes to the last activation in this part.

Specifically, to get a more adaptable solution of this model, using three different dilation values (In this work, 1,2,4 are used), a concatenated feature extraction layer is generated. In Eq. (3), X_D is the output of the dynamic convolutions using different dilation values, *concat* indicates the concatenation.

$$X = concat(X_D) \tag{3}$$

Then, this concatenated X after extracted image features goes into the following procedure.

Convolutional LISTA. In order to obtain the complementary advantages of learning-based and model-based methods, the middle part of this network is designed to use LISTA [38], which is a classic CSC method.

Sparse coding is to use a set of over-complete bases to represent a vector, and the obtained vector has a certain sparseness. At the same time, the input vector is a linear combination of these bases. And the issue it originally handled is to find a suitable sparse code that can minimize function 4 with L1 Regularization. In this function, x is the input, and the u is the sparse code correspondingly, with Φ as an over-complete dictionary. And λ is positive a parameter, F represents Frobenius norm.

$$\arg\min_{u} \|x - \Phi u\|_F + \lambda \|u\|_1 \tag{4}$$

To find a way to solve the problem in function 4, the following iterative Eq. (5) that to obtain optimized result was introduced in iterative shrinkage threshold algorithm (ISTA) originally, where the σ_θ is the shrinkage function that with a θ as threshold and r is the iteration parameter. L is a constant and must be the upper limit of the maximum eigenvalue of $\mathbf{\Phi}^T\mathbf{\Phi}$. In addition, note that $\mathbf{G}$ and $\mathbf{S}$ have a coupling relationship, that will cause the degradation of flexibility and capacity in the model. Using independent kernels to $\mathbf{G}$ and $\mathbf{S}$ respectively helps taking full advantage of deep learning. LISTA learn parameters from data, which approximate the SC of ISTA, that can give it faster speed, especially in real-time implements.

$$
\begin{aligned}
\mathbf{u}_r &= \sigma_\theta(\mathbf{u}_{r-1} + \frac{1}{L}\mathbf{\Phi}^T(\mathbf{x} - \mathbf{\Phi}\mathbf{u}_{r-1})) \\
&= \sigma_\theta(\frac{1}{L}\mathbf{\Phi}^T\mathbf{x} + (\boldsymbol{I} - \frac{1}{L}\mathbf{\Phi}^T\mathbf{\Phi})\mathbf{u}_{r-1}) \\
&= \sigma_\theta(\mathbf{G}\mathbf{x} + \mathbf{S}\mathbf{u}_{r-1}) \\
\mathbf{u}_0 &= \sigma_\theta(\mathbf{G}\mathbf{x})
\end{aligned}
\tag{5}
$$

Although CSC has been applied to several image reconstruction tasks, the multiple features extracted through these methods are actually shifted versions of the same one. To solve this problem, CSC methods have been introduced to construct the objective function in a shift-invariant manner, which can be represented in the following ways:

$$
\arg\min_{\mathbf{w},\mathbf{U}}\|\mathbf{X} - \sum_{m=1}^{M}\mathbf{w}(m) * \mathbf{U}(m)\|_F + \lambda\sum_{m=1}^{M}\|\mathbf{U}(m)\|_1
\tag{6}
$$

The approximated result of input image $\mathbf{X}$ can be achieved when $\mathbf{w} * \mathbf{U}$, where $\mathbf{w}$ is convolutional dictionaries and U is sparse coefficient and the number of both of them are M. By converting the kernel into a circulant matrix, the convolution operation can be executed as a matrix multiplication [28]. Because of that, the convolutional sparse coding model could be regarded as a special case of the normal sparse coding model. Thus, compared with function 4, function 6 uses convolutional operation instead to address this optimization issue.

After this, the feature map $\mathbf{U}_r$ can obtained from sparse feature coefficients by embedding the convolutional dictionaries to kernel $\mathbf{G}$ and $\mathbf{S}$, which are learnt from last stage. Rectified Linear Unit (ReLU) [24] is used as the non-linear activation function, because sparsity can be introduced by ReLU.

Image Restoration. At this time, the final feature maps $\mathbf{U}_r$ are achieved by the R iterations in the last step, and can be used in generating output image. In this method, the residual image $\mathbf{H}$ mapped by $\mathbf{U}_r$ (7), that formed residuals are used to simplify learning problem [38]. $\mathbf{W}_R$ and $\mathbf{b}_R$ are the convolutional weights and bias parameters respectively.

$$
\mathbf{H} = \mathbf{W}_R * \mathbf{U}_R + \mathbf{b}_R
\tag{7}
$$

Because of preservation of estimated residual in $\mathbf{H}$, the output image $\mathbf{O}$ which removed blocking artifacts can be calculated as following (8):

$$
\mathbf{O} = \mathbf{J} + \mathbf{H}
\tag{8}
$$

3.2   Loss Function

Although mean squared error (MSE) is the widely used in image reconstruction tasks, it usually produces results that are too smooth because the square penalty does not work well at the edges of the image. Since mean absolute error (MAE) can keep edge information better due to its processing method with large errors [33]. For given N input image sets $\{\mathbf{O}^i, \mathbf{J}^i\}_{i=1}^N$, function 9 need to be minimized as its goal, where $\mathbf{\Theta}$ means all training parameters and $f()$ represents this entire network.

$$\mathcal{L}(\mathbf{\Theta}) = \frac{1}{N} \sum_{i=1}^N \|f(\mathbf{J}^i; \mathbf{\Theta}) - \mathbf{O}^i\|_1 \qquad (9)$$

To avoid amplify noise, some regular items need to be added of this optimization problem model to maintain the smoothness of the image. Total Variation loss (TV loss) is a commonly used regular item, inspired by the methods of [13], TV Loss is attached into the total loss function to constrain noise as well. For 2D images, The total-variation loss (10) proposed by Rudin at el. [30] is

$$\mathcal{R}_{V^\beta}(\mathbf{x}) = \sum_{i,j} \left((x_{i+1,j} - x_{i,j})^2 + (x_{i,j+1} - x_{i,j})^2 \right)^{\frac{\beta}{2}} \qquad (10)$$

where $\mathbf{x}$ is the input image, i and j shows the coordinates of one pixel, and in this case β is always equal to 2. By introducing TV loss, spatial smoothness in the generated image can be enhanced. Additionally, when β is larger than one, it will lose some clarity as sacrifice, and since TV loss is a regularization part, the influence coefficient of TV loss on the total loss function is adjusted to 0.1.

3.3   Method Details

Parameters Setting. All convolutional kernels are in 3 by 3 size, and iteration number is 75. Noting that in dynamic convolution, there are three different dilation values: 1,2 and 4, in order to concatenate them in the following step without changing the size of feature maps, the padding values are set to 1, 2 and 4 correspondingly, and the kernel number K set as 3. For training batch, the size is 8, and the testing batch size is 4. The learning rate is fixed to 10^{-4}.

Training and Testing Data-Set. There are 200 different JPEG images in both training and testing data set. All the images are from BSD 500 [2], and the training and testing images are non-associated with the resolution of 481×321. To have the multi-quality images, every image is compressed to the 10, 20 and 30 as quality values. And it is worth to emphasize that the training process is conducted toward these three different compressed images in the same model. So in total, there are 600 pairs of image set in each epoch, that are from 200 raw images and its three compressed version counterpart.

Evaluation Setting. After the testing stage, peak signal-to-noise ratio (PSNR) and structural similarity (SSIM) are used to evaluate the results as two important indicators, which are commonly used objective measurement methods for evaluating image quality.

Given a grayscale image I with a size of $m \times n$ and a noisy image K, the mean square error (MSE) is defined as (11):

$$MSE = \frac{1}{mn} \sum_{i=0}^{m-1} \sum_{j=0}^{n-1} \left[I(i,j) - K(i,j) \right]^2 \tag{11}$$

Based on MSE, PSNR (in decibel) defined as (12),

$$PSNR = 10 \cdot \log_{10}(\frac{MAX_I^2}{MSE}) \tag{12}$$

where the MAX_I^2 represents the maximum possible pixel value. For color image with three RGB channels, like in this method, the only different with the monochrome image is while calculating the MSE, should sum all squared value differences.

One drawback of PSNR is that it cannot be precisely the same as the visual quality seen by the human eye because the sensitivity of human vision to errors is not absolute. Its perception results will be affected by many factors [20]. For example, human eye is more sensitive to contrast differences with lower spatial frequencies, and the human eye is more sensitive to brightness contrast differences than chroma. But although PSNR has some uncertainty between human vision and its results, it is still worth using as a picture quality quantification index.

Another measurement is SSIM, which denote the similarity of two images. The definition of SSIM is (13),

$$SSIM(x,y) = \frac{(2\mu_x \mu_y + c_1)(2\sigma_{xy} + c_2)}{(\mu_x^2 + \mu_y^2 + c_1)(\sigma_x^2 + \sigma_y^2 + c_2)} \tag{13}$$

in which μ_x and μ_y are the average of x and y respectively. σ_x^2 and σ_y^2 are the corresponding variance value, and σ_{xy} means the covariance of x and y. $c_1 = (k_1 L)^2$, $c_2 = (k_2 L)^2$ are the two are constants used to maintain stability with L as its pixel values range. By default, $k_1 = 0.01$ and $k_2 = 0.03$. The range of SSIM is from –1 to 1, when the two images are exactly the same, the value of SSIM is equal to 1.

As the realization of the structural similarity theory, the SSIM defines structural information from the perspective of image composition as being independent of brightness and contrast, reflecting the properties of the object structure in the scene, and modelling distortion as a combination of three aspects: brightness, contrast and structure.

Both PSNR and SSIM can reflect the quality of the reconstructed pictures to a certain extent and provide a reference for the generated model evaluation.

4   Experiments

4.1   Ablation Comparisons

Compared with Fu. Xueyang et al. work [11], in this proposed network architecture, two different features have been modified.

The first one is replacing dilation convolution with dynamic convolution in the feature extraction part. For the dilation convolution, it uses the same filter in different scales, which can enlarge the contextual area without adding extra parameters [37]. In [11], they adopt this method to address various receptive field problem, that share the same propose with this approach. Also, it concatenate these three dilation layers at the end of getting feature maps from different image qualities.

Another one is adding TV loss in the total loss function in order to restrain the noise further. And as mentioned in [11], compared to MAE, MSE is not struggling to preserve image structures and more tolerant with minor errors, so in this discussed networks, MSE loss is also used as the based loss function.

To distinguish their independent influence on the original network by Fu. Xueyang et al., four experiments were performed: the initial network proposed by Fu. Xueyang et al. in [11], the modified network with only adding TV loss function, the adjusted network with only replacing dilation convolution with dynamic convolution, and the combined network with both of attaching TV loss and switching dynamic convolution. To facilitate the following description, unless otherwise specified, the network **A**, **B**, **C** and **D** represent those four networks described above. Noting that although the network **A** is based on the work of Fu. Xueyang et al., it using different testing image set and training epoch number, so all the comparisons are all among these mentioned networks in the current setting, the result of network **A** could be different with [11].

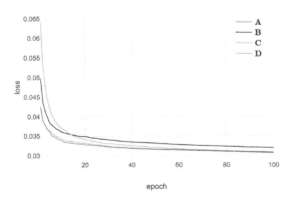

Fig. 2. Traing loss of four networks

Table 1. Comparison of four networks testing result

Network	Compressed JPEG	A	B	C	D
PSNR	Quality = 10	27.20195220	<u>27.15529762</u>	**27.49716605**	27.35324233
	Quality = 20	29.51739216	<u>29.49505357</u>	**29.92748446**	29.81997779
	Quality = 30	<u>30.59214258</u>	30.60956523	**31.06809419**	31.00578163
Overall	27.511688	28.710313	<u>28.704696</u>	**28.936940**	28.918165
SSIM	Quality = 10	0.866696935	<u>0.866316018</u>	**0.871134491**	0.869668812
	Quality = 20	<u>0.912456598</u>	0.912632115	0.916135771	**0.916801777**
	Quality = 30	<u>0.928541638</u>	0.928925026	0.931885152	**0.933535218**
Overall	0.881532	<u>0.902565</u>	0.902624	0.906385	**0.906669**

Training. Firstly, for the training part, all the networks are trained for the one hundred epochs. Figure 2 indicates the different networks training loss. In this chart, all networks tend to convergence after the 40 epoch, but network **D** has a slower convergence speed compared with the other three. And for the network **B** and network **D**, both of them have added the TV loss, show the higher loss at the first several epoch, because at the original few epochs, the TV loss would give additional errors for the total loss.

Figure 3 and 4 indicate the PSNR and SSIM respectively. From these two charts, there is not obviously difference between these four network only from the training progress. They all have the comparable results with each other.

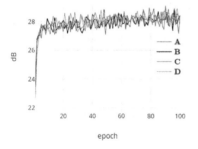

Fig. 3. Training PSNR of four networks

Fig. 4. Training SSIM of four networks

Testing Result. In testing part, there are 150 images are tested by each network, and for three different image qualities, 50 images are tested under each quality. Also, the PSNR and SSIM are calculated for the quantitative evaluation of each network as well as the compressed JPEG image, which are shown in Table 1. The best result in the certain quality are all in bold (including the overall results), while the lowest values have underlines.

By comparing the testing result among these four networks, each of them gives a better result than the original compressed JPEG image. For PSNR, the network **C** has the best results in all three image qualities, as well as the overall result. Network **D** has a comparable testing PSNR, and with the quality increasing, the gap between those two networks is narrow down. In SSIM assessment, the network **D** has a higher structural similarity with the ground truth than the others except when the image quality is 10, in which the network **C** has a little better than the others.

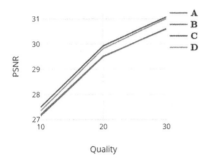
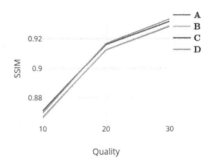

Fig. 5. Testing PSNR of four networks   **Fig. 6.** Testing SSIM of four networks

Among these four networks, network **B** has comparable values in both PSNR and SSIM with network **A**. In the comparisons of different image qualities, when the quality is 10, network **B** does not have better PSNR than network **A**, when the compressed image quality increase, especially when quality is 30, the network **B** overtakes **A** in both PSNR and SSIM. Network **C** and **D** have better performance than those previous two models, while the former has the best overall PSNR result and the latter have the most competitive result in SSIM. The same situation also appeared in these two networks. With the higher image quality, the network **D** is prone to have an accelerated better result than the network **C**, especially in the SSIM. Figure 5 and 6 present the testing result in terms of different image qualities of these four networks.

4.2 Reconstructed JPEG Image

Figure 7, 8 and 9 show the restored image results from BSD500 during the testing session of four networks, and their corresponding PSNR and SSIM are shown below them. The values in bold indicate the highest value among its counterparts. In the low compressed level, the network **C** usually gives better de-blocking result, while the reconstructed images that from high quality dataset, have the better visualization using network **D**.

(a) **A**: 26.145846—0.851684

(b) **B**: 26.179224—0.851521

(c) **C**: **28.583516**—**0.880268**

(d) **D**: 28.432769—0.879082

Fig. 7. Visual comparison of reconstructed images (quality = 10). The image of network **C** has the straighter and smoother edges.

(a) **A**: 29.434875—0.885535

(b) **B**: 29.551224—0.886465

(c) **C**: **31.263746**—0.927009

(d) **D**: 31.222759—**0.928812**

Fig. 8. Visual comparison of reconstructed images (quality = 20). The colors in images of network **C** and network **D** are more uniform.

(a) **A**: 29.100215—0.910681   (b) **B**: 29.086936—0.911006

(c) **C**: 33.481524—0.945058   (d) **D**: **33.565527—0.947649**

Fig. 9. Visual comparison of reconstructed images (quality = 30). There are less artifacts around the grass in the image of network **D**.

5   Analysis

From the previous experiments, the network **C** and **D** slightly overperform the other two networks, which are using dilation convolution in the feature extraction part. The utilization of dynamic convolution expresses its advantages comprehensively that it has a good capability in solving the multi-task problem in shallow neural networks. Compared with the TV loss, the dynamic convolution contributes more to better performance, while the former modification in this network does not have a distinct impact. However, in the intra-comparison, the TV loss has a greater influence on the high-level compressed images and has the tendency to give a better result when the quality keeps increasing. That is one of the properties of TV loss, which has a trade-off between the artifact reduction and image smoothness.

6   Conclusion

This proposed network, which adopts dynamic convolution, benefited by the combination of sparse coding and the CNN, achieves competitive anti-artifact performance. The multi-scale feature extraction method with DCSC approach is straightforward and well-structured. The introducing of concatenated dynamic convolution allows the single lightweight model to handle the different compression qualities at the same time. Typically, in the high compressed level, this network gives a more noticeable result due to the added TV loss part. Overall, this network has a good explainability and adaptability on image artifacts reduction.

For the future work, the consolidation of DCSC and deep learning with dynamic convolution has the potential ability to deal with the other restoration tasks. For instance, the DCT domains can be further explored using this model.

References

1. Ahmed, N., Natarajan, T., Rao, K.: Discrete cosine transform. IEEE Trans. Comput. **C-23**, 90–93 (1974)
2. Arbeláez, P., Maire, M., Fowlkes, C., Malik, J.: Contour detection and hierarchical image segmentation. IEEE Trans. Pattern Anal. Mach. Intell. **33**, 898–916 (2011)
3. Chen, Y., Dai, X., Liu, M., Chen, D., Yuan, L., Liu, Z.: Dynamic convolution: attention over convolution kernels. In: IEEE/CVF Conference on Computer Vision and Pattern Recognition on Proceedings, pp. 11030–11039 (2020)
4. Chen, Y., Pock, T.: Trainable nonlinear reaction diffusion: a flexible framework for fast and effective image restoration. IEEE Trans. Pattern Anal. Mach. Intell. **39**, 1256–1272 (2017)
5. Chung-Bin, W., Bin-Da, L., Jar-Ferr, Y.: Adaptive postprocessors with DCT-based block classifications. IEEE Trans. Circuits Syst. Video Technol. **13**, 365–375 (2003)
6. Connell, J.: A Huffman-Shannon-Fano code. Proc. IEEE **61**, 1046–1047 (1973)
7. Dong, C., Deng, Y., Loy, C. C., Tang, X.: Compression artifacts reduction by a deep convolutional network. In: IEEE International Conference on Computer Vision on Proceedings, pp. 576–584 (2015)
8. Dong, C., Loy, C.C., He, K., Tang, X.: Learning a deep convolutional network for image super-resolution. In: Fleet, D., Pajdla, T., Schiele, B., Tuytelaars, T. (eds.) ECCV 2014. LNCS, vol. 8692, pp. 184–199. Springer, Cham (2014). https://doi.org/10.1007/978-3-319-10593-2_13
9. Dyer, E.L., Johnson, D.H., Baraniuk, R.G.: Learning modular representations from global sparse coding networks. BMC Neurosci. **11**, P131 (2010). https://doi.org/10.1186/1471-2202-11-S1-P131
10. Foi, A., Katkovnik, V., Egiazarian, K.: Pointwise shape-adaptive DCT for high-quality denoising and deblocking of grayscale and color images. IEEE Trans. Image Process. **16**, 1395–1411 (2007)
11. Fu, X., Zha, Z.J., Wu, F., Ding, X., Paisley, J.: Jpeg artifacts reduction via deep convolutional sparse coding. In: IEEE/CVF International Conference on Computer Vision on Proceedings, pp. 2501–2510 (2019)
12. Galteri, L., Seidenari, L., Bertini, M., Bimbo, A.: Deep universal generative adversarial compression artifact removal. IEEE Trans. Multimedia **21**, 2131–2145 (2019)
13. Gatys, L.A., Ecker, A.S., Bethge, M.: Image style transfer using convolutional neural networks. In: IEEE Conference on Computer Vision and Pattern Recognition on Proceedings, pp. 2414–2423 (2016)
14. Guangtao, Z., Wenjun, Z., Xiaokang, Y., Weisi, L., Yi, X.: Efficient deblocking with coefficient regularization, shape-adaptive filtering, and quantization constraint. IEEE Trans. Multimedia **10**, 735–745 (2008)

15. Hannuksela, M., Lainema, J., Malamal Vadakital, V.: The high efficiency image file format standard [Standards in a Nutshell]. IEEE Signal Process. Mag. **32**, 150–156 (2015)

16. Heide, F., Heidrich, W., Wetzstein, G.: Fast and flexible convolutional sparse coding. In: Proceedings of the IEEE Conference on Computer Vision and Pattern Recognition on Proceedings, pp. 5135–5143 (2015)

17. Hinds, S., Fisher, J., D'Amato, D.: A document skew detection method using run-length encoding and the Hough transform. In: Proceedings. 10th International Conference on Pattern Recognition (1990)

18. Howard, P., Vitter, J.: Fast and efficient lossless image compression. In: Proceedings DCC 1993: Data Compression Conference (1993)

19. Hu, J., Shen, L., Albanie, S., Sun, G., Wu, E.: Squeeze-and-excitation networks. IEEE Trans. Pattern Anal. Mach. Intell. **42**, 2011–2023 (2020)

20. Huynh-Thu, Q., Ghanbari, M.: Scope of validity of PSNR in image/video quality assessment. Electron. Lett. **44**, 800 (2008)

21. JPEG - JPEG. https://jpeg.org/jpeg/. Accessed 26 Nov 2020

22. Khayam, S.A.: The discrete cosine transform (DCT): theory and application. Mich. State Univ. **114**, 1–31 (2003)

23. Knuth, D.: Dynamic huffman coding. J. Algorithms **6**, 163–180 (1985)

24. Krizhevsky, A., Sutskever, I., Hinton, G.: ImageNet classification with deep convolutional neural networks. Commun. ACM **60**, 84–90 (2017)

25. Li, Yu., Guo, F., Tan, R.T., Brown, M.S.: A contrast enhancement framework with JPEG artifacts suppression. In: Fleet, D., Pajdla, T., Schiele, B., Tuytelaars, T. (eds.) ECCV 2014. LNCS, vol. 8690, pp. 174–188. Springer, Cham (2014). https://doi.org/10.1007/978-3-319-10605-2_12

26. List, P., Joch, A., Lainema, J., Bjontegaard, G., Karczewicz, M.: Adaptive deblocking filter. IEEE Trans. Circuits Syst. Video Technol. **13**, 614–619 (2003)

27. Liu, X., Wu, X., Zhou, J., Zhao, D.: Data-driven soft decoding of compressed images in dual transform-pixel domain. IEEE Trans. Image Process. **25**, 1649–1659 (2016)

28. Nagy, J., O'Leary, D.: Restoring images degraded by spatially variant blur. SIAM J. Sci. Comput. **19**, 1063–1082 (1998)

29. Redmill, D., Kingsbury, N.: The EREC: an error-resilient technique for coding variable-length blocks of data. IEEE Trans. Image Process. **5**, 565–574 (1996)

30. Rudin, L., Osher, S., Fatemi, E.: Nonlinear total variation based noise removal algorithms. Phys. D: Nonlinear Phenomena **60**, 259–268 (1992)

31. Rothe, R., Timofte, R., Van Gool, L.: Efficient regression priors for reducing image compression artifacts. In: 2015 IEEE International Conference on Image Processing (ICIP) (2015)

32. Tai, Y., Yang, J., Liu, X., Xu, C.: Memnet: a persistent memory network for image restoration. In: IEEE International Conference on Computer Vision on Proceedings, pp. 4539–4547 (2017)

33. Tuchler, M., Singer, A., Koetter, R.: Minimum mean squared error equalization using a priori information. IEEE Trans. Signal Process. **50**, 673–683 (2002)

34. Usevitch, B.: A tutorial on modern lossy wavelet image compression: foundations of JPEG 2000. IEEE Signal Process. Mag. **18**, 22–35 (2001)

35. Wang, Z., Bovik, A., Sheikh, H., Simoncelli, E.: Image quality assessment: from error visibility to structural similarity. IEEE Trans. Image Process. **13**, 600–612 (2004)

36. Wang, Z., Liu, D., Chang, S., Ling, Q., Yang, Y., Huang, T.S.: D3: deep dual-domain based fast restoration of JPEG-compressed images. In: IEEE Conference on Computer Vision and Pattern Recognition on Proceedings, pp. 2764–2772 (2016)
37. Yu, F. and Koltun, V.: Multi-scale context aggregation by dilated convolutions. arXiv preprint arXiv:1511.07122 (2015)
38. Zhang, K., Zuo, W., Chen, Y., Meng, D., Zhang, L.: Beyond a gaussian denoiser: residual learning of deep CNN for image denoising. IEEE Trans. Image Process. **26**, 3142–3155 (2017)
39. Zhang, X., Yang, W., Hu, Y., Liu, J.: DMCNN: dual-domain multi-scale convolutional neural network for compression artifacts removal. In: 2018 25th IEEE International Conference on Image Processing (ICIP), pp. 390–394. IEEE (2018)

Medical Image Denoising in MRI Reconstruction Procedure

Dong Han$^{(\boxtimes)}$ and Ronny Velastegui

Norwegian University of Science and Technology, Gjøvik, Norway
dongha@stud.ntnu.no

Abstract. With the rapid development of computer technology, deep learning can be used in nearly every field, and it always has the potential to achieve a high efficiency performance. Specifically, in the field of medical images, it makes doctors possible to distinguish and diagnose diseases in a more accurate way. Medical images like any other form of imaging techniques are affected by noise and artifacts. There are many types of noise, such as quantum, random, electric, and gaussian noise, etc. The presence of noise affects image clarity and may obstruct the recognition and analysis of diseases. The traditional image denoising method has much more limitation when came to medical images, and the results cannot meet some specific medical image standards. Hence, denoising of medical images in deep learning can be an important technique for further medical image processing. In this work, we conducted a deep learning method, which is a sparse dilated convolution neural network based on compressed sensing, for medical image denoising.

Keywords: Medical image denoising · Medical resonance imaging · Compressed sensing · Dilated convolution · Deep learning

1 Introduction

The digital images are essential in the medical field in which it has been employed to anatomize human body. These medical images can be utilized to recognize different diseases. Unfortunately, medical images require high resolution, and the presence of noise is not beneficial to disease diagnosis and sometimes even leads to the possibility of misdiagnosis [2]. Hence, the need for effective image noise reduction technology is urgent. Researchers have noticed this problem and provided many examples and principles used in the process of noise reduction in medical images [26].

Computed tomography (CT) is to use X-rays to scan the human body layer by layer, and then according to the difference in absorption of X-rays by various tissues in the human body. The attenuation coefficient of X-rays in the human body is determined by using a certain mathematical method. After a specific computer processing, the two-dimensional distribution matrix of the attenuation coefficient values in the human body layer is obtained and transformed into the

© Springer Nature Switzerland AG 2021
O. Gervasi et al. (Eds.): ICCSA 2021, LNCS 12950, pp. 115–130, 2021.
https://doi.org/10.1007/978-3-030-86960-1_9

gray-scale distribution of the image. This is a procedure to realize the modern medical imaging technology of building tomographic images [1]. There are mainly two types of noise in the CT image, quantum noise, and electronic noise [4].

Medical resonance imaging (MRI) is an imaging technology based on a nuclear physical phenomenon, which uses the signal generated by the resonance of the nucleus in a strong magnetic field to be reconstructed from the image [32]. Medical resonance (MR) diagnosis has been widely applied in clinics due to its advantages of high resolution, non-invasiveness, and non-radiation [8]. With the help of MRI, doctors more easily specify the difference between various types of tissues judged by these magnetic properties. Impulse and gaussian noise are the two main noise types in MRI. Most noise in MRI is introduced from the electronic circuits or coils and radio frequency according to the movement of the ions inside the human body [25]. Noise in MRI can cause random fluctuations and result in a low level of contrast because of the inclination in signals. This interferes with accurate subjective and numerical assessment and feature detection of MR images. In most medical application scenarios, MRI has a high signal-to-noise ratio, leading to gaussian noise [10].

Compressed sensing (CS), is a very efficient way to reconstruct the fast MRI and positively affects MRI denoising [18, 19, 28]. For the most traditional CS-MRI approaches, the CS-MRI model normally has two elements: the data fidelity part and the regularization part. The first one is always in the k-space form and the second is associated with sparsifying operation to avoid the overfitting. However, this conventional model can deal with the problem properly but there are some missing detailed information and noise-like artifact occurs during in the reconstruction procedure, especially when it comes up to the high-rate undersampled measurements [24]. Many related research works have been managed in order to alleviate this kind of issue and most studies can be classified as two parts. The one focus on introducing more accurate spasifying transforms or use the nonlocal means. For instance, the singular value decomposition (SVD) [12] and blind compressive sensing (BCS) [21], which are both based on basic discrete cosine transform (DCT), patch-based directional wavelets based on bandelet transform [29], L1 norm method in data fidelity constraints [23], sparsifying dictionary for removing the aliasing [31], and non-local total variation regularization for noise elimination [20]. Nonetheless, the approaches mentioned above have good performance in improving the accuracy of the reconstructed MR image but with high computational cost or requirement for stacking considerable non-local operators [30]. Another part is about the feature restoration and refinement which uses the augmented lagrangian (AL) based on SENSitivity Encoding (SENSE) [31]. However, although this strategy reconstructs the most detailed fine structure, it still brings some noise to the output MR image.

Deep neural networks evolve very fast and have achieved pleasant performance in many research areas because they have efficient learning abilities for different datasets. Analyzing unstructured data and the ability to convey high-quality results are the principal superiorities of deep neural networks. Nevertheless, the normal data-driven network requires huge datasets, and it is difficult to acquire huge medical data because of the sensitivity of privacy. The deep learning method is one of the approaches to alleviate this issue by training the network with natural images to reconstruct the medical image [39]. Another method to mitigate this difficulty is the restoration of medical images through the model-based with a combination of deep learning networks [17]. As the deep learning method is developing in various areas, its utility in medical imaging denoising is confirmed, and their neural network has many varieties to choose from. Image denoising using deep learning techniques plays a vital part in many application areas of medical imaging such as CT and MRI images.

A convolution neural network (CNN) is a special artificial neural network, which is designed to preserve the spatial relationship between data, with few connections between each layer. The traditional CNN includes four main parts: convolutional layer, activation layer, pooling layer, and fully connected layer. CNN can form an efficient representation of input data. The input data in CNN are stack in an array and then flowed through layers that retain these correlations. Different parameters in each layer can achieve optimization in the end. Like a standard artificial neural network, it uses backpropagation and gradient descent for training [22].

In specific, deep learning has good performance in solving the noise artifact that cannot be mitigated purely in the CS model as mentioned previously. In this paper inspired by classical CNN, we mainly introduce a dilated convolution network with batch normalization based on variable splitting for medical MR image denoising. Two metrics are used for evaluating image quality: the peak signal-to-noise ratio (PSNR) and the structural similarity index (SSIM). We combine the model-based approach with the deep learning method, and the sparse dilated network (SD-Net) is introduced to improve denoising performance during the training procedure of the variable splitting-based CS model. The inconsistency of network between training and testing performance is analyzed, the potential reasons are given. Finally, the SD-Net-H network based on hybrid dilated convolution (HDC) is proposed to solve the grid problem in SD-Net and better denoising performance is achieved.

2   Preliminaries

2.1   Compressed Sensing

In CS, sensing describes that in order to express and restore a certain signal or object, a certain sensing method is adopted for it (the sensing here includes the imaging, the sampling of the continuous signal, etc.) to obtain this measurement under this sensing modality so that the subsequent signal can be reconstructed and analyzed or for other applications [7]. There are mainly three types of sensing, full-sampling, over-sampling, and under-sampling. Without signal priors, we record the minimum required sensing sample dimension as the critical sampling rate. Then you can also choose a measurement with less than the critical rate. In this case, we call it under-sampling. This will compress the full sensing, that is, compressed [5]. According to the classic signal processing theory, the target image cannot be restored uniquely and perfectly. This type of problem is called the ill-posed problem, and CS is used to solve ill-posed inverse problems. CS is able to recover the target image perfectly in the case of under-sampling with an effective signal prior [23].

2.2 Batch Normalization

Batch normalization (BN) was suggested by Google researchers [13] and was also implemented in GoogLeNet [33]. The BN algorithm accelerates the training process to a large extent and relaxes the conditions for network initialization. The BN can be used without Dropout, and in the meantime, the recognition effect of the network can be improved to a certain extent. There are widespread applications in new networks ResNet [3,27].

In the neural network training process, input sample features are generally normalized so that the data becomes a distribution with a mean of 0 and a standard deviation of 1, or a distribution with a range of 0 to 1. The sample feature distribution is scattered, if data is not normalized, it may cause the neural network to learn slowly or even difficult to learn [9]. BN can be seen as a component in the neural network that normally functions with an activation function [15]. The algorithm of BN is as follows:

The range of input values x in a mini-batch is $B = \{x_i...x_m\}$; the parameters to be learned are γ, β and final output value is y_i.

$$\mu_B = \frac{1}{m}\sum_{i=1}^{m} x_i \tag{1}$$

$$\sigma^2{}_B = \frac{1}{m}\sum_{i=1}^{m}(x_i - \mu_B)^2 \tag{2}$$

$$\hat{x}_i = \frac{x_i - \mu_B}{\sqrt{\sigma^2{}_B + \epsilon}} \tag{3}$$

$$y_i = \gamma\hat{x}_i + \beta \tag{4}$$

First, the mean μ_B and the variance $\sigma^2{}_B$ of the small size of training data are calculated, then used to normalize the batch of training data to obtain a 0–1 distribution. The ϵ is a tiny positive number in the normalization step $\hat{x}_i$ to exclude the situation when the number is divided by zero. Finally, do the scale transformation and offset by y_i: adjust x_i by multiply by γ, and add β to increase the offset to get y_i, where γ is the scale factor and β is the translation factor. This step is the essence of BN, the normalized x_i will be restricted to a normal distribution, which will reduce the learning ability of the network. To solve this problem, two new parameters y_i, β are introduced and learned by the network during training. The role of BN in deep neural networks is undeniable: if neural network training encounters slow convergence, or "gradient explosion" and other untrainable situations occur, BN is a good way to solve those problems. Simultaneously, BN can also be added to accelerate model training and even improve model accuracy [14].

2.3   Dilated Convolution

The receptive field indicates the input area which can be "sensed" by the neuron in the neural network. In CNN, the calculation of an element on the feature map is affected by a certain area on the input image. This area is the receptive field of the element. The neurons in deeper layers can see the larger input area. The larger receptive field can contain more detailed information of the input image which is detected. Normally, increasing the size of the receptive field can collect much more detailed information. There mainly two ways to get the wide receptive field. One is increasing depth, and another is increasing width of CNN. In the first method, it is time-consuming and difficult to learn for networks. The second approach might introduce a number of parameters so that the network structure becomes more complex.

Adopting larger filters or increasing the number of layers will greatly increase the number of parameters to be calculated, which means that more memory resources will be required. Some researchers came up with dilated convolution for CNN in 2015 [38], so that only a small part of the weight can increase the receiving field. Dilated Convolution is to inject holes into the classical CNN to enlarge the receptive field of the network. Compared with the original normal convolution operation, dilation convolution has one more parameter: dilation rate, which refers to the total intervals between the convolution kernel points. For instance, the dilated rate of the conventional convolution operation is 1.

Receptive field size can be calculated by $2(dilated\ rate - 1) * (kernel\ size - 1) + kernel\ size$. According to this calculation, when the kernel size is 3×3, the receptive field size of standard convolution is 3, and the dilation convolution with dilated rate 2 has a larger receptive field size of 7. In other words, it is a 3×3 size kernel with 9 weights, and other elements are all set to zero. Therefore, it still has the same parameter as standard convolution but with the increased receptive field size 7.

3   Proposed Network

In MRI, CS is defined to realize authentic restoration based on a small set of k-space data instead of using the fully k-space data. The traditional CS-MRI model has been introduced and there are mainly two parts of this model. One is the data fidelity term and another is the regularization term. The first term guarantees the solution accords with the degradation process. The second term enforces desired property of the output.

$$\min_x \frac{\gamma}{2} \sum_{i=1}^{n_c} ||DFS_i m - y_i||_2^2 + R(m) \qquad (5)$$

This is one of the general CS-MRI equations proposed by Jinming Duan et al. [6]. As stated by CS concepts, the image m which is reconstructed can be estimated by solving the unconstrained optimization problem shown above. The first term is the data fidelity term. m is MR image needed to be reconstructed,

y_i is the undersampled k-space data, F is the undersampled Fourier encoding matrix. D is the matrix that some positions are filled with zeros which can be seen as a sampling function. S_i is the ith coil sensitivity which is a diagonal matrix, and λ is a weight to make a trade-off between the two terms. The second term is a common sparse regularization term and it has many variants.

3.1   Variable Splitting

With intention of solving this optimisation issue, the variable splitting method can be apply to break down the relation between variable and other functions. In specific, the auxiliary variables u is introduced and it is equal to m and also enable $S_i m$ equal to x_i. According to the penalty function concept, these constraints are fed back into the model and the problem is converted to multi-optimisation problem:

$$\min_{m,u,x_i} \frac{\gamma}{2}\sum_{i=1}^{n_c}||DFx_i - y_i||_2^2 + R(u) + \frac{\alpha}{2}\sum_{i=1}^{n_c}||x_i - S_i m||_2^2 + \frac{\beta}{2}||u - m||_2^2 \qquad (6)$$

here α and β are added into formula as penalty weights. In order to optimise (6), we can split it into three independent optimisation problems:

$$u^{k+1} = \arg\min_u \frac{\beta}{2}||u - m^k||_2^2 + R(u)$$

$$x_i^{k+1} = \arg\min_{x_i} \lambda\sum_{i=1}^{n_c}||DFx_i - y_i||_2^2 + \frac{\alpha}{2}\sum_{i=1}^{n_c}||x_i - S_i m^k||_2^2 \qquad (7)$$

$$m^{k+1} = \arg\min_m \frac{\alpha}{2}\sum_{i=1}^{n_c}||x^{k+1} - S_i m||_2^2 + \frac{\beta}{2}||u^{k+1} - m||_2^2$$

where k is the index number of kth iteration. Employing gradient descent and least-square (LS) to the middle and last equations of (7), the (8) could be computed out.

$$u^{k+1} = denoiser(m^k)$$
$$x_i^{k+1} = F^{-1}((\lambda D^T D + \alpha I)^{-1}(\alpha F S_i m^k + \lambda D^T y_i)) \qquad (8)$$
$$m^{k+1} = (\beta I + \alpha\sum_{i=1}^{n_c}S_i^H S_i)^{-1}(\beta u^{k+1} + \alpha\sum_{i=1}^{n_c}S_i^H x_i^{k+1})$$

Now, the conventional CS problem (5) is converted into a denoising operation and two mathematical optimization questions. The middle equation is called data consistency because it intuitively shows the consistency between frequency data and spatial data. The bottom equation combines the weight acquired from the previous two equations.

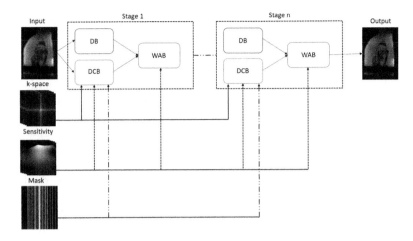

Fig. 1. The framework of VS-Net

In the end, the proposed framework based on Eq. (8) is constructed. The architecture of the network is shown in Fig. 1.

This network consists of three parts: denoiser block (DB), data consistency block (DCB), and weighted average block (WAB), they are coincided with Eq. (8) from upper to bottom respectively.

The basic CNN in the denoiser block is used [6]. Due to the simple architecture, it has limited performance in the denoising procedures. For the denoising block, there many the state of art denoising CNN can be implemented. In this research, inspired by sparse representation (SR) for image denoising in [34], we proposed a sparse dilated CNN denoising architecture to make an improvement on denoising outcomes. In specific, it could increase the depth of the denoising network by enlarging the receptive field without greatly increasing the running time and memory usage.

The presented architecture of sparse dilated network (SD-Net) is shown in Fig. 2.

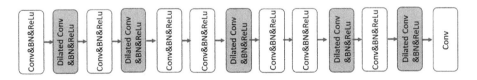

Fig. 2. The structure of the sparse dilated network (SD-Net)

In specific, the defined SD-Net based on dilated convolution, batch normalization, and standard convolution is unlike the normal denoising block as it incorporates a sparse mechanism. The 13 layers SD-Net is made of two main parts: the Dilated block and Conv block.

The Dilated block stands for the dilation convolution with the dilated rate of 2, BN, and the rectified linear function, ReLU [16] are linked. The Conv block is made of classical convolution, BN, and ReLU. The Dilated block is set at the second, fourth, seventh, tenth, and twelve layers in SD-Net.

The dilation convolutions are able to contain more detailed information [37]. According to this principle, the Dilated block can be addressed as high-powered points. The Conv block is set at the first, third, fifth, sixth, eighth, ninth, and eleventh layers in SD-Net, which can be regarded as low-powered points.

In particular, the kernel size applied in convolution layers is 3×3 which is a typical size used broadly. The input of the first layer is m (sensitivity-weighted under-sampled image) and it is actually a complex-valued image, in order to deal with complex values, we convert the complex value of the under-sampled image input into a common two-channel image. Therefore, in this case, the number of channels in the input is 2. In layers from 2 to 12, the input channels and output channels of filters are all 64. The dilated convolution layers are not placed in a contiguous sequence, which introduces the sparsity mechanism into the network by stacking some high-powered points and low-powered points. Furthermore, the SD-Net implements fewer high-powered points rather than many high-powered points to obtain a more comprehensive context. With this setting, SD-Net can not merely boost the denoising performance and ensure effective learning outcomes but also avoid the redundancy structure in the network.

3.2   Loss Function

The loss function plays an important role in the parameterization procedure and it could purify all parameters of the model as a single number that the improvements in this number indicate the quality of the network model. For MR reconstruction and denoising, the loss function is usually associated with the sameness relationship between the image which is reconstructed from the under-sampled image and an original image. For instance, the mean squared error (MSE) is a popular loss function that is also implemented in our network.

$$L(\Theta) = \min_{\Theta} \frac{1}{2n_i} \sum_{i=1}^{n_i} ||m_i^{n_{it}}(\Theta) - g_i||_2^2 \qquad (9)$$

The n_i is the index number of training images, and g is the ground truth image, which is computed by fully sampling through $\sum_j^{n_c} S_j^H F^{-1} f_j$. The n_{it} is the stage number that indicates the iteration times going through the three network blocks DB, DCB and WAB. The f_j denotes the fully sampled raw data of the corresponded coil. Θ can be learned as the dilated parameters during the network training procedure. In specific, the Θ includes weight parameter λ in denoiser network and penalty weights α, β in DCB and WAB. They are all learnable during the training.

4   Experiments

4.1   Dataset

Based on the framework of VS-Net, we incorporated the SD-Net into the denoiser block instead of the old one. For the dataset, the publicly clinical knee dataset is available in [11].

There are mainly 5 types of image acquisition protocols, we picked up coronal proton-density (PD) as the data of our experiments. For PD, the same 20 contents were scanned and the scan of each content contain around 40 slices with each slice has 15 channels. Coil sensitivity maps are also presented in the dataset were computed by BART [35]. In our case, we set the acceleration factor (AF) as 4-fold and use half of the subjects for training, the other half for testing. The trained SD-Net is to denoise the image during the reconstruction procedure of each 2D slice. Same as [6], the epoch number was set as 200 where the parameters of the network were normally optimized, with the learning rate equal to 10^{-3} and batch size 1.

4.2   Performance

With the purpose of investigating the influence of the stage number n_{it}, the n_{it} is set from 1–6 and visualized the training and testing numerical curves with epoch numbers in the first and last row of Fig. 3. The network has better performance when the number of stages is increased. This situation can be attributed to two possible reasons: the first reason is that the learning ability of the network is improved due to the increased weight; the second reason is that the optimization

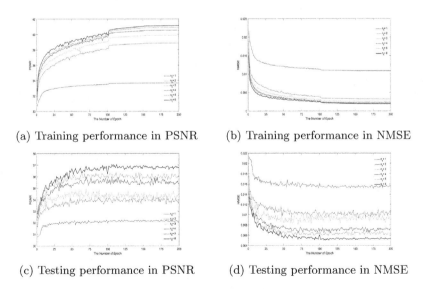

(a) Training performance in PSNR   (b) Training performance in NMSE

(c) Testing performance in PSNR   (d) Testing performance in NMSE

Fig. 3. Performance of SD-Net from stage 1–6

can be computed more precisely when the iterations are increased with larger
stage numbers. Due to the limitation of GPU in the lab, we only increase the
stage number to 6. Judging by the tendency of PSNR and NMSE in training
and testing, it is possible to get better results by increasing the stage number to
a larger number. However, it will introduce a high computational burden and is
time-consuming.

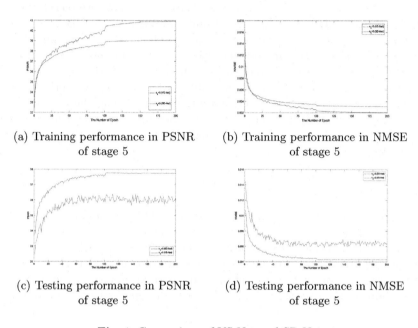

(a) Training performance in PSNR   (b) Training performance in NMSE
        of stage 5   of stage 5

(c) Testing performance in PSNR   (d) Testing performance in NMSE
        of stage 5   of stage 5

Fig. 4. Comparison of VS-Net and SD-Net

For the sake of exploring the effectiveness of denoising performance, the performance in stage number 5 is chosen to make a comparison between VS-Net and
SD-Net. According to Fig. 4, in the same stage, it is obvious that our SD-Net has
high performance either in PSNR or reconstruction error (NMSE) in training
performance. However, we got slightly decreased performance for testing results,
and the fluctuations in PSNR and NMSE lines of SD-Net are observed. Couples
of reasons can lead to this situation, such as high learning rate, unsuitable batch
size, and overfitting.

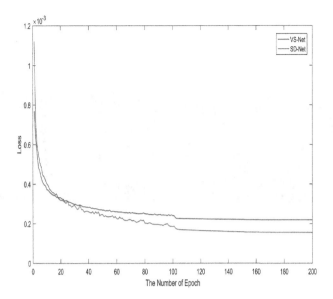

Fig. 5. Training loss in stage 5

According to the training loss performance in Fig. 5, we can notice that our SD-Net has a lower training loss than VS-Net. This effect might be brought with batch normalization operation in SD-Net. The batch normalization prevents the network from getting stuck in local minima and it can give a better error surface, the loss is low and converges to low training error. Additionally, batch normalization accelerates the training procedure and enables the network to converge in very few epochs.

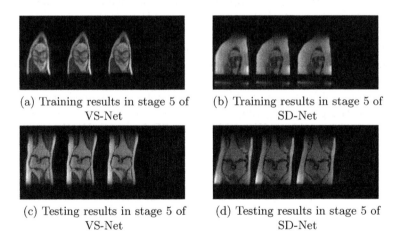

(a) Training results in stage 5 of VS-Net

(b) Training results in stage 5 of SD-Net

(c) Testing results in stage 5 of VS-Net

(d) Testing results in stage 5 of SD-Net

Fig. 6. Visual Comparison of VS-Net and SD-Net, From left to right: under-sampled image, reconstructed image, ground truth and difference between reconstructed and ground truth image.

In Fig. 6, the different images are collected to show the accuracy consistency of various images. We can see the images reconstructed from corresponded under-sampled images are very similar to their ground truth images, respectively.

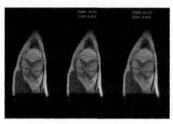 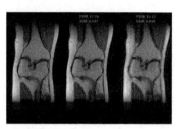

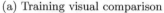

(a) Training visual comparison   (b) Testing visual comparison

Fig. 7. Visual comparison of VS-Net and SD-Net, From left to right: ground truth, reconstructed image from VS-Net, reconstructed image from SD-Net

In this comparison, both VS-Net and SD-Net were trained and tested in stage 5 in Fig. 7. For training visual comparison, the SD-Net has improvements in PSNR and SSIM. For testing visual comparison, the SD-Net has lightly unexpected performance both in PSNR and SSIM. We can see some local information is lost from the testing result of SD-Net. It can be caused by a gridding problem [36] in dilated convolution network. For the output of a certain layer in dilated convolution, the adjacent pixels are obtained from the convolution of independent subsets, and there is no correlation between each other. The gridding problem has two main features. One is the loss of local information which means that since the calculation method of the dilated convolution is similar to the checkerboard format, the convolution results obtained in a certain layer come from the independent set of the previous layer. Hence, there is no correlation among the convolution outputs of this layer. Another main feature is irrelevant information gathered from a distance. Due to the sparsely sampled input signal of the dilated convolution, there is no correlation between the information obtained by the long-distance convolution, which affects the result.

In order to mitigate the gridding effect in testing procedure in SD-Net. According to the paper [36], the hybrid dilated convolution (HDC) method is used to solve this problem. We combine the SD-Net with HDC to generate a new network SD-Net-H showing in Fig. 8 to make an improvement of SD-Net.

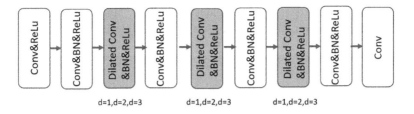

Fig. 8. SD-Net-H network structure

This solution forms a certain number of layers into a group, and then each group uses a continuously increasing dilated rate with repeating the same structure. Same as SD-Net, the low-powered points are introduced in the SD-Net-H to reduce the computational cost. In SD-Net-H, we use a combination of three different dilated rates (1, 2, 3). Moreover, the SD-Net-H network can obtain information from a wider range of pixels, avoiding the grid problem. At the same time, we can also adjust the receptive field arbitrarily by modifying the rate.

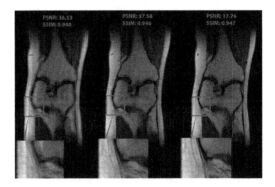

Fig. 9. Visual comparison of the reconstructed image during testing at stage 5, From left to right: SD-Net, SD-Net-H, VS-Net

As the testing results shown in Fig. 9, we can notice that there are some black and white stripes existing in reconstructed image from SD-Net while the SD-Net-H has alleviated the gridding effect and reduced noise artifacts. Besides, the local detailed information can be reconstructed more accurately than SD-Net by checking the same block areas of those three reconstructed images in terms of microstructure distribution. The global sparsity mechanism and local contiguous dilation are integrated together in the SD-Net-H to achieve a comparable performance with VS-Net. The PSNR and SSIM values computed from SD-Net-H and VS-Net are quite close to each other, which gives possibility to outperform the VS-Net by adjusting batch size and epoch number because SD-Net-H has more complex structure in denoiser block. Moreover, since we introduced the local contiguous dilation, the SD-Net-H is more flexible in receptive field than VS-Net.

5   Conclusions

In this paper, we proposed the sparse dilated network (SD-Net) based on variable splitting (VS) for MRI denoising during an accelerated reconstruction procedure. We have introduced the SD-Net architecture and its excellent advantages for medical image denoising in a deep learning framework and every training stage greatly associated with the epoch number during the reconstruction process. For experimentation, we have demonstrated that the accuracy of SD-Net moderately improves with increasing stage number of network, and the parameters in each stage can be effectively learned due to the VS structure. Finally, we made comparisons and analyzed SD-Net with the empirical network VS-Net on an under-sampled Coronal PD dataset for 4-fold factors, and indicated its robustness. For training performance, we got better results but the testing performance is a little bit unexpected with some fluctuation. To mitigate this problem, we introduced HDC features into SD-Net to generated a new network. For future work, we will try to find specific reasons and propose an efficient solution for inconsistency between training and testing, and also achieve more higher performances in reconstruction with pleasant denoising results. The other non-medical state-of-the-art noisy images database can also be tested to see the benefit of the approach.

References

1. Buzug T.M.: Computed tomography. In: Kramme, R., Hoffmann, K.P., Pozos, R.S. (eds.) Springer Handbook of Medical Technology. Springer Handbooks, pp. 311–342. Springer, Berlin (2011). https://doi.org/10.1007/978-3-540-74658-4_16
2. Chow, L.S., Paramesran, R.: Review of medical image quality assessment. Biomed. Signal Process. Control **27**, 145–154 (2016)
3. Chung, Y.A., Weng, W.H.: Learning deep representations of medical images using siamese cnns with application to content-based image retrieval. arXiv preprint arXiv:1711.08490 (2017)
4. Diwakar, M., Kumar, M.: A review on CT image noise and its denoising. Biomed. Signal Process. Control **42**, 73–88 (2018)
5. Donoho, D.L.: Compressed sensing. IEEE Trans. Inf. Theor. **52**(4), 1289–1306 (2006)
6. Duan, J., et al.: VS-net: variable splitting network for accelerated parallel MRI reconstruction. In: Shen, D., et al. (eds.) MICCAI 2019. LNCS, vol. 11767, pp. 713–722. Springer, Cham (2019). https://doi.org/10.1007/978-3-030-32251-9_78
7. Eldar, Y.C., Kutyniok, G.: Compressed Sensing: Theory and Applications. Cambridge University Press, Cambridge (2012)
8. Fatahi, M., Speck, O., et al.: Magnetic resonance imaging (MRI): a review of genetic damage investigations. Mutat. Res./Rev. Mutat. Res. **764**, 51–63 (2015)

9. Garbin, C., Zhu, X., Marques, O.: Dropout vs. batch normalization: an empirical study of their impact to deep learning. Multimedia Tools Appl. **79**, 1–39 (2020). ISSN: 1380-7501
10. Goyal, B., Agrawal, S., Sohi, B.: Noise issues prevailing in various types of medical images. Biomed. Pharmacol. J. **11**(3), 1227 (2018)
11. Hammernik, K., et al.: Learning a variational network for reconstruction of accelerated MRI data. Magn. Reson. Med. **79**(6), 3055–3071 (2018)
12. Hong, M., Yu, Y., Wang, H., Liu, F., Crozier, S.: Compressed sensing MRI with singular value decomposition-based sparsity basis. Phys. Med. Biol. **56**(19), 6311 (2011)
13. Ioffe, S., Szegedy, C.: Batch normalization: Accelerating deep network training by reducing internal covariate shift. arXiv preprint arXiv:1502.03167 (2015)
14. Ismail, A., Ahmad, S.A., Soh, A.C., Hassan, K., Harith, H.H.: Improving convolutional neural network (CNN) architecture (miniVGGNet) with batch normalization and learning rate decay factor for image classification. Int. J. Integr. Eng. **11**(4) (2019). ISSN: 2229-838X
15. Jung, W., Jung, D., Lee, S., Rhee, W., Ahn, J.H., et al.: Restructuring batch normalization to accelerate CNN training. arXiv preprint arXiv:1807.01702 (2018)
16. Krizhevsky, A., Sutskever, I., Hinton, G.E.: Imagenet classification with deep convolutional neural networks. Commun. ACM **60**(6), 84–90 (2017)
17. Lee, D., Yoo, J., Ye, J.C.: Deep artifact learning for compressed sensing and parallel MRI. arXiv preprint arXiv:1703.01120 (2017)
18. Liang, D., DiBella, E.V., Chen, R.R., Ying, L.: k-t ISD: dynamic cardiac MR imaging using compressed sensing with iterative support detection. Magn. Reson. Med. **68**(1), 41–53 (2012)
19. Liang, D., Liu, B., Wang, J., Ying, L.: Accelerating sense using compressed sensing. Magn. Reson. Med. Official J. Int. Soc. Magn. Reson. Med. **62**(6), 1574–1584 (2009)
20. Liang, D., Wang, H., Chang, Y., Ying, L.: Sensitivity encoding reconstruction with nonlocal total variation regularization. Magn. Reson. Med. **65**(5), 1384–1392 (2011)
21. Lingala, S.G., Jacob, M.: Blind compressive sensing dynamic MRI. IEEE Trans. Med. Imaging **32**(6), 1132–1145 (2013)
22. Lundervold, A.S., Lundervold, A.: An overview of deep learning in medical imaging focusing on MRI. Zeitschrift für Medizinische Physik **29**(2), 102–127 (2019)
23. Lustig, M., Donoho, D., Pauly, J.M.: Sparse MRI: the application of compressed sensing for rapid MR imaging. Magn. Reson. Medi. Official J. Int. Soc. Magn. Reson. Med **58**(6), 1182–1195 (2007)
24. Lustig, M., Donoho, D.L., Santos, J.M., Pauly, J.M.: Compressed sensing MRI. IEEE Signal Process. Mag. **25**(2), 72–82 (2008)
25. Macovski, A.: Noise in MRI. Magn. Reson. Med. **36**(3), 494–497 (1996)
26. Mredhula, L., Dorairangasamy, M.: An extensive review of significant researches on medical image denoising techniques. Int. J. Comput. Appl. **64**(14) (2013). ISSN: 0975-8887
27. Näppi, J.J., Hironaka, T., Yoshida, H.: Detection of colorectal masses in CT colonography: application of deep residual networks for differentiating masses from normal colon anatomy. In: Medical Imaging 2018: Computer-Aided Diagnosis. vol. 10575, p. 1057518. International Society for Optics and Photonics (2018)
28. Otazo, R., Kim, D., Axel, L., Sodickson, D.K.: Combination of compressed sensing and parallel imaging for highly accelerated first-pass cardiac perfusion MRI. Magn. Reson. Med. **64**(3), 767–776 (2010)

29. Qu, X., et al.: Undersampled MRI reconstruction with patch-based directional wavelets. Top. Magn. Reson. Imaging **30**(7), 964–977 (2012)
30. Qu, X., Hou, Y., Lam, F., Guo, D., Zhong, J., Chen, Z.: Magnetic resonance image reconstruction from undersampled measurements using a patch-based non-local operator. Med. Image Anal. **18**(6), 843–856 (2014)
31. Ravishankar, S., Bresler, Y.: MR image reconstruction from highly undersampled k-space data by dictionary learning. IEEE Trans. Med. Imaging **30**(5), 1028–1041 (2010)
32. Reimer, P., Parizel, P.M., Meaney, J.F., Stichnoth, F.A.: Clinical MR imaging. Springer, Berlin (2010) https://doi.org/10.1007/978-3-540-74504-4
33. Szegedy, C., et al.: Going deeper with convolutions. In: Proceedings of the IEEE Conference on Computer Vision and Pattern Recognition, pp. 1–9 (2015)
34. Tian, C., Zhang, Q., Sun, G., Song, Z., Li, S.: FFT consolidated sparse and collaborative representation for image classification. Arab. J. Sci. Eng. **43**(2), 741–758 (2018)
35. Uecker, M., et al.: Software toolbox and programming library for compressed sensing and parallel imaging. In: ISMRM Workshop on Data Sampling and Image Reconstruction, p. 41. Citeseer (2013)
36. Wang, P., et al.: Understanding convolution for semantic segmentation. In: 2018 IEEE Winter Conference on Applications of Computer Vision (WACV), pp. 1451–1460. IEEE (2018)
37. Wang, Y., Wang, G., Chen, C., Pan, Z.: Multi-scale dilated convolution of convolutional neural network for image denoising. Multimedia Tools Appl. **78**(14), 19945–19960 (2019). https://doi.org/10.1007/s11042-019-7377-y
38. Yu, F., Koltun, V.: Multi-scale context aggregation by dilated convolutions. arXiv preprint arXiv:1511.07122 (2015)
39. Zhu, B., Liu, J.Z., Cauley, S.F., Rosen, B.R., Rosen, M.S.: Image reconstruction by domain-transform manifold learning. Nature **555**(7697), 487–492 (2018)

Automatic Extraction of the Midsagittal Surface from T1-Weighted MR Brain Images Using a Multiscale Filtering Approach

Fernando N. Frascá, Katia M. Poloni⬤, and Ricardo J. Ferrari$^{(\boxtimes)}$⬤

Department of Computing, Federal University of São Carlos,
São Carlos, SP 13565-905, Brazil
`rferrari@ufscar.br`
`https://www.bipgroup.dc.ufscar.br`

Abstract. The left and right hemispheres of the human brain are separated by a fissure called interhemispheric (IHF), which is commonly shaped by a geometric plane known as the midsagittal plane (MSP). However, despite the name, the MSP does not always resemble a plan since the brain is not perfectly symmetric. The detection of the MSP in human brain images is an essential task to segment the brain hemispheres. Studies suggest that abnormal values of brain asymmetry may be related to traumas and neurological diseases. Through the years, several computer methods were proposed to detect the MSP automatically. Nonetheless, they constrain the detection to a plane without considering brain asymmetries. In this study, an automatic computer technique is developed to detect a midsagittal surface (MSS) in Magnetic Resonance (MR) images using an MSP as a reference, following by multiscale analysis. Different from the MSP, the MSS follows the natural form of the IHF. The proposed method uses a reference MSP to guide a region of interest that potentially contains the IHF. The 3D MR image is decomposed in 2D slices that are processed, filtered, and piled to form an MSS. The proposed method results have shown an accurate detection in all metrics assessed, i.e., DSC of 0.99 and x-distance of 2.13, and a significant improvement of 2.953 for the x-distance compared with the MSP method.

Keywords: Brain asymmetry · Midsagittal surface · Interhemispheric fissure · Multiscale filter

1 Introduction

Based on the assumption that the human brain exhibits a high level of bi-fold symmetry, one common approach to detect brain pathologies and structural changes in Magnetic Resonance (MR) images is to compare the left and right brain hemispheres in the search for violations of a devised symmetric measure.

ⓒ Springer Nature Switzerland AG 2021
O. Gervasi et al. (Eds.): ICCSA 2021, LNCS 12950, pp. 131–146, 2021.
https://doi.org/10.1007/978-3-030-86960-1_10

Several studies have assessed brain asymmetry seeking abnormal asymmetry values due to gender [2,14,23,27,28] and handedness [14,18] differences. These values can also be associated with traumas and neurological diseases, such as schizophrenia [6,9,33], dyslexia [19], autism and developmental language disorder [17], epilepsy [22,30,41], Alzheimer [1,4,5,35,38,39], and Parkinson [31].

In the human brain, the asymmetries are defined based on proportion measures between the brain hemispheres, given that the boundary between each hemisphere is delimited by the interhemispheric fissure (IHF) [15], a deep groove that divides the brain hemispheres in left and right. For quantitative studies, the IHF is represented as a surface that cuts the sagittal axis of the brain, commonly called midsagittal surface (MSS). The precise location of the MSS is an important step in the analysis of neuroimages, and it is of high interest in various medical applications and researches, particularly in studies involving brain pathologies associated with asymmetries. Liu et al. [26] reviewed brain asymmetry analysis methods and their implications to computer-aided diagnosis and reported the importance of the precise identification of the symmetry plane to perform a correct evaluation of hemisphere-wise asymmetries.

Over the last decade, many methods were proposed for brain asymmetries analysis. However, due to the inherent difficulty to precisely detect the IHF, most of them end up handling the MSS as a plane (MSP), which is often considered as a first-order approximation of the MSS [24]. However, the MSP may bias the asymmetry analysis since it may fail to provide a precise representation of the IHF in many cases. Therefore, in this study, we propose a new automatic method to assess brain asymmetries using the MSS detection in T1 weighted (T1-w) MR images of the human brain. The method uses the automatic detected MSP using a previous work developed by our group [11,13] as a starting point for our MSS searching and a 2D multiscale filtering analysis. The aim is to obtain a surface that represents more precisely possible brain asymmetries, overcoming the constraint imposed by other methods presented in the literature, which approximates the IHF as a plane [25].

Our method was evaluated using a set of fifty 2D slices of the axial view of the human brain, extracted from images of three distinct datasets. The automatic MSS projections were compared to manual annotations made by two specialists from the medical field.

The rest of this paper is structured as follows: Sect. 2 provides information about the image dataset used in this study, Sect. 3 present the proposed method and its concepts, Sect. 4 presents the validation metrics, Sect. 5 present the obtained results and Sect. 6 provide the conclusions of this work.

2   Image Datasets

This section presents the datasets used in this study and the set of images used to validate the proposed method. We manually selected thirty-eight MR images from three distinct neuroimaging datasets to include the diversity of the populations on which the method might potentially be deployed, and be certain subjects with brain torque were selected.

We validate our method using manual annotations from two medical doctors (MDs), herein referred to as specialists. We randomly selected fifty 2D slices to validate our experiment results due to manual annotation being an expensive and exhaustive task. The 2D processing occurs after we preprocess the 3D images. Due to the variety of scanners used to acquire the images among databases, all images were resampled to an isotropic resolution of 1 mm.

Neuroimage Analysis Center (NAC): The NAC [16] is a research center affiliated with the Surgical Planning Laboratory and Harvard University. The dataset consists of 149 3-D triangular meshes of distinct brain structures spatially aligned to a T1-weighted (T1-w) image of a healthy 42-year-old male subject and had 1mm isotropic resolution and a matrix size of 256 × 256 × 256 voxels. For this study, we only used the T1-w image as a reference for the preprocessing steps to mitigate problems inherent to image acquisitions and standardize all study images for processing.

Alzheimer's Disease Neuroimaging Initiative (ADNI): The ADNI dataset [21] is part of a project released at the beginning of 2003, idealized by a group of research institutions, private pharmaceutical companies, and non-governmental organizations. The initiative has been looking for ways to determine the progression of Alzheimer's by developing clinical biomarkers for detecting the disease in its early stages. For this study, we used nineteen T1-w MR images of individuals with Alzheimer's disease, from which we randomly selected twenty-two axial slices.

Information eXtraction from Images (IXI): The IXI dataset[1] is developed and maintained by the Neuroimage Analysis Center, from Imperial College London, in the United Kingdom, and it is part of the project Brain Development. Its data are composed of approximately 600 MR brain images acquired with T1-, T2-, and PD-weighting. We used eighteen T1-w RM images of healthy individuals from this dataset, from which we randomly selected twenty-three axial slices.

Simulated Brain Database (BrainWeb): The BrainWeb [3] is a dataset composed of synthetic images of the human brain, provided by the McConnel Brain Imaging Centre, from McGill University, Canada, to help validate research experiments involving MR images of the human brain. We selected only one T1-w MR image of a healthy individual from this dataset, which we randomly selected five axial slices.

3   Methods

This section presents the proposed method for the extraction of the midsagittal surface from T1–w MR brain images.

---

[1] https://brain-development.org/ixi-dataset/.

3.1   Overview of the Proposed Method

The main idea of the proposed method is to develop a method to detect the MSS
that can allow studies to perform a more accurate asymmetry assessment when
comparing the brain hemispheres. Our method is composed of three main steps:
(a) image preprocessing, (b) IFH sampling, and (c) MSS extraction. Figure 1
presents the general framework of the proposed method.

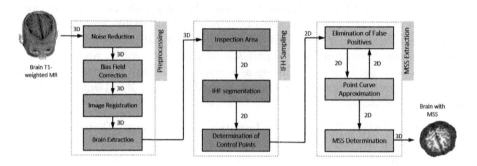

Fig. 1. A schematic block diagram of the proposed method

3.2   Preprocessing

The MR images were first preprocessed for noise reduction using the Non-Local
Means algorithm [8] and bias field correction using the N4-ITK technique [40].
This task was performed to minimize the negative effects that noise and bias field
may have. Next, image registration was conducted using the Nifty-Reg tool [32]
to align all images to the NAC reference image. Finally, brain region extraction
was achieved using the Robust Brain Extraction (ROBEX) technique [20] to
avoid unnecessary processing.

3.3   IFH Sampling

Following the preprocessing steps, we detected the MSP using the technique
proposed in [11,13] and used it as a reference to define the region of interest
(ROI) to search for the MSS. Next, we extracted only the even axial slices from
the 3D MR images, leading to a set of 128 axial slices. Then, we used the MSP
projection for each slice (a straight line) as a reference, herein called midsagittal
line (MSL), to define a limiting box (ROI) to search for the MSS, thus avoiding
unnecessary processing and minimizing the interference of nearby structures to
the MSS. After that, we positioned the limiting box in the longitudinal direction
with its sides positioned paralleling and equidistant from the MSL marked on
the image slice. Finally, we calculated the length of the limiting box based on the
most extreme pixels of the longitudinal axis that presented values of intensity
different from zero. Figure 2(a) shows an example of the ROI (rectangle in red
colour) overlaid on an axial image slice.

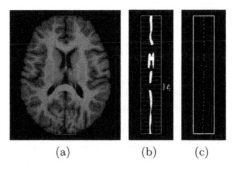

(a)   (b)   (c)

Fig. 2. Example of the ROI (a) delimiting the axial slice, (b) showing the fragmentation lines, and (c) the control points detected

After the ROI demarcation, we enhanced the IHF in each slice image using the multiscale Hessian filter-based technique proposed in [12] for vessel image enhancement. The reason for using this approach is based on the observation that the cerebral sulci, including the IHF, look like a net of curvaceous tubular structures when projected to an axial slice, which may help to identify the MSS 2D projection. The multiscale Hessian filter-based technique works by assessing the relationships between the eigenvalues (λ_1 and λ_2) of a 2×2 Hessian matrix locally created from the neighborhood of each pixel. When visualized on the T1-w MR images, the IHF appears similar to a dark tubular structure (or vessel) on the light background and, according to Table 1 from [12], we have to search for regions in the image where the eigenvalues, λ_1 and λ_2, of the Hessian matrix are low and positive high, respectively to detect tubular structures.

Considering the IHF is naturally connected by other peripheral vessel-like structures, the multiscale Hessian filter-based filtering technique should highlight the IHF sulcus and other undesirable peripheral sulci. To this end, we used morphological operators (erosion and dilation) defined by a thick stick structuring element oriented with its length aligned to the ROI length to eliminate the unwanted structures that mostly manifest in a perpendicular direction to the IHF and are proportionally smaller than it.

Finally, after processing the ROI on the 2D slices to enhance the IHF, we divided the region into equally broad stripes of e_c pixels along the longitudinal axis, as illustrated in Fig. 2(b). Then, we defined the control points (CP) as the centroid point of the IHF fragment present in each stripe, as shown in Fig. 2(c). The stripe widths, e_c, were experimentally defined as 2% of the longitudinal size of the ROI to allow a significant number of CPs to represent the IHF.

Table 1. Possible local patterns that may be detected in an image through analysis of the self-values of the Hessian matrix.

Eigenvalues		Orientation pattern
λ_1	λ_2	
Noise	Noise	Noise, no preferred direction
Low	High−	Tubular structure (bright)
Low	High+	Tubular structure (dark)
High−	High−	Blob-like structure (bright)
High+	High+	Blob-like structure (dark)

3.4   MSS Extraction

After the IHF sampling, we performed a two-step postprocessing. First, we determined the most suitable CPs to build the curve better representing the IHF in each slice and, further, used all IHF curves detected in the 2D planes to create a 3D structure representing the MSS.

The first step was used to eliminate outliers, i.e., control points whose horizontal spatial positions were offset by a certain distance from the reference position defined by the central vertical line of the ROI. To this end, we applied the statistical Z-score test [7] to each test point, with mean and standard deviation parameters estimated from the CP spatial positions. Considering the maximum number of CPs per ROI is lower than 50, we set the Z-score threshold as 2.5 and used it as a criterion for a potential outlier in our experiments , i.e., all points with Z-score value outside the -2.5 and 2.5 range were considered outliers.

The second step used an iterative procedure to remove CP points that may cause disruption in the smoothness of the final reconstructed MSS. This procedure verified if any CP position on the analyzed slice (S_a) was shifted outside a tolerance range (β) when compared to the corresponding point position in a reference slice (S_i). If this happened, then the point was not considered for the construction of the IHF curve. The initial reference slice (S_{i_o}) was defined as the one at the position comprising 40% of the slice sequence, measured from the top to the bottom of the brain. This procedure was based on the observation that the S_{i_o} position in the brain corresponds to the most prominent view of the IHF. The iterative process was performed considering that corresponding points (equal spatial positions) on consecutive slices should not deviate above a tolerance margin; otherwise, they would represent structures with an acute shape, which are unlikely to occur in IHF. At the end of each iteration, the analyzed slice (S_a), with all control points outside the distance tolerance removed, becomes the reference slice (S_i) of the subsequent slice. The value for the tolerance range, β, was experimentally set to 8 pixels.

Finally, we resampled all IHF curves created on each image slice using the B-Spline interpolation [37] to create a new set of control points, each with their respective (x, y) coordinates in the Cartesian plane, and build the 3D structure of the MSS. We stacked the points so that each point, previously sampled in the (x, y) space, received a new z coordinate, which represented the index of the slice to which that point was associated to build the MSS. We performed this process covering each slice of the image in sequential order and converting the respective points associated with the slice to the 3D plane. At the end of the process, we have a grid of points on the 3D plane representing the MSS.

4   Validation Metrics

We assessed the accuracy of the proposed approach by using empirical discrepancy evaluation methods based on region overlap and average x-distance (in pixels) [34]. The discrepancy was computed between our automatic MSS detection and the two specialists' manual annotations. As mentioned in Sect. 2, we randomly selected fifty 2D slices to validate our experiment results, from which the two MD specialists, abbreviated as specialist 1 (S1) and specialist 2 (S2), have annotated the MSS. They used the Fiji tool [36] with the freehand spline drawing technique to manually annotate the curve that divides the cerebral hemispheres.

Dice Similarity Coefficient (DSC): We used the DSC [10] to evaluate similarity based on the overlap between the masks generated by our detection and the ones of the specialists. The mask images were obtained using the IHF manual annotation and the MSS to define which pixels are belonging to the left and right hemispheres on the ROBEX output mask (i.e., the brain extraction mask). The DSC is calculated as

$$\mathrm{DSC} = \frac{2\,S \cap A|}{|S| + |A|},\tag{1}$$

where S and A represent the cerebral hemispheres delimiting the masks obtained by the specialist's annotation and by the automatic method, respectively. The DSC values range from 0 to 1, being 1 when there is a complete agreement between the two masks and 0 when there is a complete lack of agreement.

Average x-distance: The x-distance value represents the distance between the curves marked by the automatic method and by a specialist. The measure is an adaptation of the method proposed in [34] and is calculated as

$$x\text{-distance} = \frac{\sum_y(|x^S_{coord} - x^A_{coord}|)}{N},\tag{2}$$

where x^S_{coord} and x^A_{coord} represent the x coordinate of IHF curve points marked by the specialist and automatic method, respectively, and N is the number of curve points compared along the y axis. Smaller values indicate more proximity between the two curves and larger values otherwise.

Statistical Analysis: We used the Mann-Whitney U test [29] with the two-tailed hypothesis with a significance level $\alpha = 0.01$ to verify if there were statistically significant differences between the results obtained for the DSC values when comparing the left and right hemispheres and each specialist annotation with our MSS method. Further, we also verify the average x-distance obtained between each specialist annotation with our MSS method and with a method used as a comparison, the MSP [11, 13].

5 Results

The results of the proposed technique for the MSS detection in brain MR images are presented as follows.

DSC Results
Figure 3 shows a visual example of three axial image slices with the MSS delimitation applied to segment the hemispheres. The overlaid red region indicates the left hemisphere and the other the right. From these slices, we can visually notice that the proposed MSS method accurately captured the surface of the hemisphere.

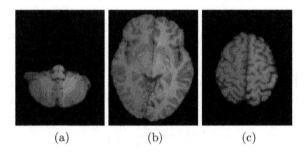

     (a)   (b)   (c)

Fig. 3. Visual example of segmentation of the cerebral hemispheres from automatic delineation.

To ensure we captured the surface accurately, we quantitatively assessed the region overlap results using the DSC. The boxplots in Fig. 4 show the DSC values calculated between the MSS segmentation mask and the two specialists masks, S1 and S2 (considered ground-truths), using three different databases: ADNI, IXI, and BrainWeb. Since we need to separate the left and right hemispheres to generate their masks, we show plots for both hemispheres separately.

For the left hemisphere, we obtained the highest mean DSCs of 0.989 ± 0.005 and 0.989 ± 0.005 for the ADNI dataset when compared with the S1 and S2 specialists, respectively, and for the BrainWeb the corresponding values were 0.989 ± 0.003, and 0.989 ± 0.005. We obtained the smallest interquartile range (IQR) for the ADNI images and the smallest whiskers and number of outliers for the BrainWeb dataset, both compared to the specialist S1. For the IXI dataset, we obtained the lowest mean DSC and the highest IQR, with values of 0.986 ± 0.005 compared with S1 and 0.987 ± 0.003 compared with S2. When tested using

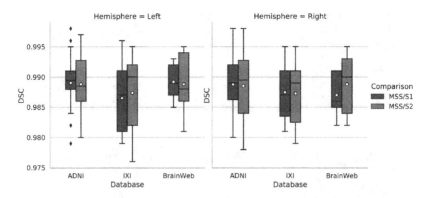

Fig. 4. DSC values for the left and right hemispheres using three different databases: ADNI, IXI, and BrainWeb. S1 and S2 refer to the specialist 1 and 2.

all dataset images, we obtained a mean DSC of 0.988 ± 0.005 and 0.988 ± 0.006, compared with S1 and S2.

For the right hemisphere, we obtained the highest mean DSC of 0.989 ± 0.005 and 0.989 ± 0.006 for the ADNI dataset, a mean DSC of 0.987 ± 0.004 and 0.987 ± 0.004 for the IXI dataset, and 0.987 ± 0.004 and 0.989 ± 0.004 for the BrainWeb, all values compared with S1 and S2, respectively. Using all dataset images, we obtained a mean DSC of 0.988 ± 0.004 compared with S1 and 0.988 ± 0.006 compared with S2.

To assess if the DSC results presented statically significant differences, we performed the Mann-Whitney U test [29]. We compared the left and right hemispheres' DSC results joining and grouping by specialists. Comparing the left and right DSC values, we obtained a p-value of 0.912; while comparing the left and right DSC grouped by specialists, the p-values were 0.438 for the S1 and 0.497 for S2.

We also compared the DSC values for hemisphere and specialists. We obtained a p-value of 0.367 for the comparison between S1 and S2 for the left hemisphere and 0.471 for the right. As it can be noticed, all p-values were bigger than the significance level ($\alpha = 0.01$), meaning that the null test hypothesis was accepted, claiming that both data came from the same distribution. Therefore, we can conclude that despite some minor differences between values, our MSS segmentation obtained similar DSC values for both hemispheres and, more importantly, for the specialist annotations.

Average x-distance
Figure 5 shows a visual example of four distinct axial image slices comparing the MSS application with the two specialists annotations and the MSP proposed in [11, 13] as a qualitative assessment.

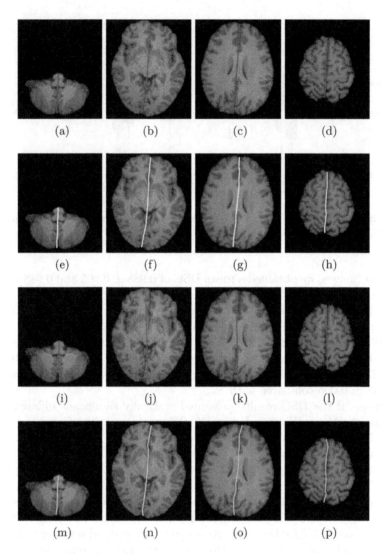

Fig. 5. Example showing four distinct axial slices with (a–d) annotations of the specialist 1 (e–h) annotations of the specialist 2, (i–l) results of the MSP, and (m–p) results of the proposed MSS method.

As it can be noticed, since the MSP assumes that the IHF (midsagittal separation) can be approximated by a plane, it failed the detection in brains with torque, whereas our proposed MSS method successfully captured these changes. We can also notice a very small disagreement between the two specialist's annotations and the MSS method.

The box plots in Fig. 6 and 7 show a comparison of the x-distance between the specialists 1 and 2 (ground-truths) and the MSS detection using three different databases: ADNI, IXI, and BrainWeb. Further, a comparison was performed between the two specialists' annotations to provide an intra-observer measure, and the MSP detection results were compared with the same ground-truths.

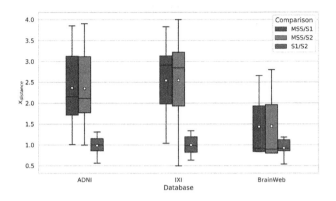

Fig. 6. Boxplot of the x-distance values obtained from the MSS detection compared with specialist 1 (MSS/S1) and specialist 2 (MSS/S2) and the two specialists annotations (S1/S2) compared using three different databases: ADNI, IXI, and BrainWeb.

To better understand the results and show that our results were compared with consistent annotations, Fig. 6 provides a quantitative assessment of the specialists' annotations in the boxplots of green color (S1/S2). Based on the small IQR values between the annotations, we can state that the specialists have a high degree of agreement, with a mean x-distance of 0.985 ± 0.207 for the ADNI slices, 0.998 ± 0.21 for the IXI slices, and 0.922 ± 0.256 for the BrainWeb slices, allowing an overall mean value of 0.968 ± 0.224.

When assessing our MSS detection, by comparing it with S1 and S2 annotations, we achieved mean x-distance results of 2.364 ± 0.897 and 2.347 ± 0.908 for the ADNI, with the highest IQR values in both comparisons. For the IXI, we achieved the highest mean x-distance values, with 2.54 ± 0.853 and 2.544 ± 0.93, and also the minimum and maximum distances. Finally, for the BrainWeb, we obtained the lowest value of mean x-distance values, with 1.434 ± 0.862 and 1.449 ± 0.9 and with the smallest IQR values. The assessment using all image datasets showed mean x-distance values of 2.113 ± 0.913 and 2.113 ± 0.913 compared with the specialists 1 with 2, respectively.

As shown in the box plots in Fig. 7, comparison between the MSP method with S1 and S2 using the ADNI images showed a mean x-distance result of 6.144 ± 1.779 for the specialist S1 and 6.133 ± 1.796 for S2. The same experiment with the IXI images resulted in mean x-distance results of 6.429 ± 2.815 and 6.431 ± 2.766 for the specialists S1 and S2. The same analysis using images from the Brain Web dataset showed mean x-distance values of 2.626 ± 1.225 and 2.604 ± 1.226 for the specialists 1 and 2, respectively. The highest whisker

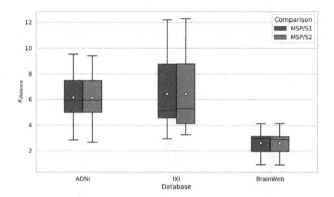

Fig. 7. Boxplot of the x-distance values obtained from the MSP detection compared with specialist 1 (MSP/S1) and specialist 2 (MSP/S2) compared using three different databases: ADNI, IXI, and BrainWeb.

values were noticed for the IXI dataset. The assessment using all image datasets showed a mean with values of 5.066 ± 1.94 for S1 and 5.056 ± 1.929 for S2.

We also used the Mann-Whitney U test [29] to check if there are statically significant differences for the x-distance results obtained by comparing the MSS and MSP methods with S1 and S2. First, we compared the x-distance results for the MSS method with S1 and S2 specialists and obtained a p-value of 0.466, meaning that the test failed to reject H0 and there are no significant differences between the two distributions. Therefore, despite some differences between the specialist's annotations, our method has shown consistent results with both. Next, joining the specialists, we compared the results of our method with the MSP and obtained a p-value of 3.117e–16. After that, grouping by specialists, we repeated the previous experiment and obtained p-values of 1.22e–13 for S1 and 2.937e–14 for S2. Considering the p-value results and the significance level of 0.01, we can discard the null test hypothesis (H0) and conclude that the results compared with the MSP are statistically distinct and do not come from the same population.

For the final analysis, we evaluated the MSS performance using images with curved FIH, i.e., high brain torque. Both specialists manually labeled the images according to two categories: visible torque and non-visible torque. The term brain torque refers to the opposite asymmetry of brain regions. The results are shown in Table 2. For comparison purposes, the results of applying the MSP method [11,13] are also shown. With these results, we noticed that our method is more robust in capturing correctly different brain torques. We obtained similar values for the two categories, i.e., visible torque and non-visible torque, compared to the two specialists. In contrast, when analyzing the MSP results, we noticed that the method was not able to capture the brain torques, which resulted in differences of about 3.5 in the x-distance between the two categories. Furthermore, our results were more accurate than the MSP even when comparing brains with visible torque.

Table 2. Result of the x-distance grouped by category

	x-distance			
	MSS/S1	MSS/S2	MSP/S1	MSP/S2
Visible torque	2.36 ± 0.70	2.37 ± 0.73	7.48 ± 2.09	7.48 ± 2.0
Non-visible torque	2.34 ± 1.103	2.32 ± 1.16	3.94 ± 1.18	3.93 ± 1.15

6   Conclusions

Segmentation of brain hemispheres is widely used to assess brain asymmetries and their implications to computer-aided diagnosis. Most studies have focused on identifying a plane to represent the interhemispheric fissure that divides both hemispheres. However, a plane cannot truly represent the fissure when working with brains presenting torque. According to Toga et al. [39], brain asymmetry levels vary with aging, gender and handedness differences, and the development of diseases. Some diseases, such as Alzheimer's, progress asymmetrically, increasing the brain torque. Therefore, to overcome this limitation, we developed a method to obtain the surface that can precisely represent the interhemispheric fissure, dealing with torque and, consequently, capturing brain asymmetries.

We assess our results using fifty axial slices from T1–w MR brain images, in which two domain specialists manually annotated to use as ground-truths. Our method achieved an accurate detection in quantitative metrics, with a DSC of 0.99 and x-distance of 2.13. We also compared the specialists' annotations distances and verified that the ground-truths have a high degree of agreement. Furthermore, we compared our results with the MSP method [11,13] and achieved x-distance values with a difference of 2.95 between the methods. Lastly, we contrast the method results in brains with different torque degrees and verify the stability of our method to torque.

Acknowledgment. Funding for ADNI can be found at http://adni.loni.usc.edu/about/#fund-container.

Funding Statement. This study was financed by the Fundação de Amparo à Pesquisa do Estado de São Paulo (FAPESP) (grant numbers 2018/08826-9 and 2018/06049-5).

References

1. Ardekani, B.A., Hadid, S.A., Blessing, E., Bachman, A.H.: Sexual dimorphism and hemispheric asymmetry of hippocampal volumetric integrity in normal aging and Alzheimer disease. Am. J. Neuroradiol. **40**(2), 276–282 (2019)
2. Ashburner, J., Hutton, C., Frackowiak, R., Johnsrude, I., Price, C., Friston, K.: Identifying global anatomical differences: deformation-based morphometry. Hum. Brain Mapp. **6**(5–6), 348–357 (1998)

3. Aubert-Broche, B., Griffin, M., Pike, G.B., Evans, A.C., Collins, D.L.: Twenty new digital brain phantoms for creation of validation image data bases. IEEE Trans. Med. Imaging **25**(11), 1410–1416 (2006)
4. Barbará-Morales, E., Pérez-González, J., Rojas-Saavedra, K.C., Medina-Bañuelos, V.: Evaluation of brain tortuosity measurement for the automatic multimodal classification of subjects with Alzheimer's disease. Comput. Intell. Neurosci. **2020**, 11 (2020)
5. Barnes, J., Scahill, R.I., Schott, J.M., Frost, C., Rossor, M.N., Fox, N.C.: Does Alzheimer's disease affect hippocampal asymmetry? Evidence from a cross-sectional and longitudinal volumetric MRI study. Dement. Geriatr. Cogn. Disord. **19**(5–6), 338–344 (2005)
6. Bartko, J.J.: Measurement and reliability: statistical thinking considerations. Schizophr. Bull. **17**(3), 483–489 (1991)
7. Brase, C.H., Brase, C.P.: Understanding Basic Statistics. Nelson Education, Toronto (2013)
8. Buades, A., Coll, B., Morel, J.M.: A review of image denoising algorithms, with a new one. Multiscale Model. Simul. **4**(2), 490–530 (2005)
9. Crow, T.J.: Schizophrenia as an anomaly of cerebral asymmetry. In: Imaging of the Brain in Psychiatry and Related Fields, pp. 3–17. Springer International Publishing, Heidelberg (1993). https://doi.org/10.1007/978-3-642-77087-6_1
10. Dice, L.R.: Measures of the amount of ecologic association between species. Ecology **26**(3), 297–302 (1945)
11. Ferrari, R.J., Villa-Pinto, C.H., Moreira, C.A.F.: Detection of the midsagittal plane in MR images using a sheetness measure from Eigen analysis of local 3D phase congruency responses. In: International Conference on Image Processing (ICIP), pp. 2335–2339. IEEE, Phoeniz, Arizona (2016)
12. Frangi, A.F., Niessen, W.J., Vincken, K.L., Viergever, M.A.: Multiscale vessel enhancement filtering. In: Wells, W.M., Colchester, A., Delp, S. (eds.) MICCAI 1998. LNCS, vol. 1496, pp. 130–137. Springer, Heidelberg (1998). https://doi.org/10.1007/BFb0056195
13. de Lima Freire, P.G., Gregório da Silva, B.C., Villa Pinto, C.H., Ferri Moreira, C.A., Ferrari, R.J.: Midsaggital plane detection in magnetic resonance images using phase congruency, hessian matrix and symmetry information: a comparative study. In: Gervasi, O., et al. (eds.) ICCSA 2018. LNCS, vol. 10960, pp. 245–260. Springer, Cham (2018). https://doi.org/10.1007/978-3-319-95162-1_17
14. Good, C.D., Johnsrude, I., Ashburner, J., Henson, R.N., Friston, K.J., Frackowiak, R.S.: Cerebral asymmetry and the effects of sex and handedness on brain structure: a voxel-based morphometric analysis of 465 normal adult human brains. Neuroimage **14**(3), 685–700 (2001)
15. Guillemaud, R., Marais, P., Zisserman, A., McDonald, B., Crow, T.J., Brady, M.: A three dimensional mid sagittal plane for brain asymmetry measurement. Schizophr. Res. **2**(18), 183–184 (1996)
16. Halle, M., et al.: Multi-modality MRI-based atlas of the brain. http://www.spl.harvard.edu/publications/item/view/2037 (2017)
17. Herbert, M.R., et al.: Brain asymmetries in autism and developmental language disorder: a nested whole-brain analysis. Brain **128**(1), 213–226 (2005)
18. Hervé, P.Y., Crivello, F., Perchey, G., Mazoyer, B., Tzourio-Mazoyer, N.: Handedness and cerebral anatomical asymmetries in young adult males. Neuroimage **29**(4), 1066–1079 (2006)

19. Hynd, G.W., Semrud-Clikeman, M., Lorys, A.R., Novey, E.S., Eliopulos, D.: Brain morphology in developmental dyslexia and attention deficit disorder/hyperactivity. Arch. Neurol. **47**(8), 919–926 (1990)
20. Iglesias, J.E., Liu, C.Y., Thompson, P.M., Tu, Z.: Robust brain extraction across datasets and comparison with publicly available methods. IEEE Trans. Med. Imaging **30**(9), 1617–1634 (2011)
21. Jack, C.R.J., et al.: The Alzheimer's disease neuroimaging initiative (ADNI): MRI methods. J. Magn. Reson. Imaging **27**(4), 685–691 (2017)
22. Kim, J.S., Koo, D.L., Joo, E.Y., Kim, S.T., Seo, D.W., Hong, S.B.: Asymmetric gray matter volume changes associated with epilepsy duration and seizure frequency in temporal-lobe-epilepsy patients with favorable surgical outcome. J. Clin. Neurol. (Seoul, Korea) **12**(3), 323 (2016)
23. Kovalev, V.A., Kruggel, F., von Cramon, D.Y.: Structural brain asymmetry as revealed by 3D texture analysis of anatomical MR images. In: Object Recognition Supported by User Interaction for Service Robots, pp. 808–811. IEEE, Quebec, Canada (2002)
24. Kruggel, F., von Cramon, D.Y.: Alignment of magnetic-resonance brain datasets with the stereotactical coordinate system. Med. Image Anal. **3**(2), 175–185 (1999)
25. Kuijf, H.J., Van Veluw, S.J., Geerlings, M.I., Viergever, M.A., Biessels, G.J., Vincken, K.L.: Automatic extraction of the midsagittal surface from brain MR images using the Kullback-Leibler measure. Neuroinformatics **12**(3), 395–403 (2014)
26. Liu, S.X.: Symmetry and asymmetry analysis and its implications to computer-aided diagnosis: a review of the literature. J. Biomed. Inform. **42**(6), 1056–1064 (2009)
27. Lucarelli, R.T., et al.: MR imaging of hippocampal asymmetry at 3T in a multiethnic, population-based sample: results from the Dallas Heart Study. Am. J. Neuroradiol. **34**(4), 752–757 (2013)
28. Luders, E., Gaser, C., Jancke, L., Schlaug, G.: A voxel-based approach to gray matter asymmetries. Neuroimage **22**(2), 656–664 (2004)
29. Mann, H.B., Whitney, D.R.: On a test of whether one of two random variables is stochastically larger than the other. Ann. Math. Statist. **18**(1), 50–60 (1947)
30. Martins, S.B., Benato, B.C., Silva, B.F., Yasuda, C.L., Falcão, A.X.: Modeling normal brain asymmetry in MR images applied to anomaly detection without segmentation and data annotation. In: Medical Imaging 2019: Computer-Aided Diagnosis, vol. 10950, p. 109500C. SPIE, San Diego, California (2019)
31. Ortiz, A., Munilla, J., Martínez, M., Gorriz, J.M., Ramírez, J., Salas-Gonzalez, D.: Parkinson's disease detection using Isosurfaces-based features and Convolutional Neural Networks. Front. Neuroinform. **13**(48), 48 (2019)
32. Ourselin, S., Stefanescu, R., Pennec, X.: Robust registration of multi-modal images: towards real-time clinical applications. In: Dohi, T., Kikinis, R. (eds.) MICCAI 2002. LNCS, vol. 2489, pp. 140–147. Springer, Heidelberg (2002). https://doi.org/10.1007/3-540-45787-9_18
33. Ribolsi, M., Daskalakis, Z.J., Siracusano, A., Koch, G.: Abnormal asymmetry of brain connectivity in schizophrenia. Front. Hum. Neurosci. **8**, 1010 (2014)
34. Ruppert, G.C.S., Teverovskiy, L., Yu, C.P., Falcao, A.X., Liu, Y.: A new symmetry-based method for mid-sagittal plane extraction in neuroimages. In: International Symposium on Biomedical Imaging: From Nano to Macro, pp. 285–288. IEEE, Chicago, Illinois, USA (2011)

35. Sarica, A., et al.: MRI asymmetry index of hippocampal subfields increases through the continuum from the mild cognitive impairment to the Alzheimer's disease. Front. Neurosci. **12**(576), 1–12 (2018)
36. Schindelin, J., et al.: Fiji: an open-source platform for biological-image analysis. Nat. Methods **9**(7), 676–682 (2012)
37. Späth, H.: One Dimensional Spline Interpolation Algorithms. CRC Press, Boca Raton (1995)
38. Thompson, P.M., et al.: Dynamics of gray matter loss in Alzheimer's disease. J. Neurosci. **23**(3), 994–1005 (2003)
39. Toga, A.W., Thompson, P.M.: Mapping brain asymmetry. Nat. Rev. Neurosci. **4**(1), 37–48 (2003)
40. Tustison, N.J., et al.: N4ITK: improved N3 bias correction. IEEE Trans. Med. Imaging **29**(6), 1310–1320 (2010)
41. Wu, W.C., et al.: Hippocampal alterations in children with temporal lobe epilepsy with or without a history of febrile convulsions: evaluations with MR volumetry and proton MR spectroscopy. Am. J. Neuroradiol. **26**(5), 1270–1275 (2005)

Temporal Image Forensics: Using CNNs for a Chronological Ordering of Line-Scan Data

Matthias Paulitsch, Andreas Vorderleitner, and Andreas Uhl$^{(\boxtimes)}$ (ID)

Department of Computer Sciences, University of Salzburg, Salzburg, Austria
`uhl@cosy.sbg.ac.at`

Abstract. We propose to use CNNs for obtaining a temporal ordering of line-scanner data. Excellent age classification accuracy is achieved only in case network training and testing is done with image patches taken from consistent spatial locations, i.e. temporal features exploited are bound to specific positions in the image. With spatially consistent patches, up to 100% classification accuracy can be achieved, whereas with spatially varying patches the accuracy stagnates at around 54%. We have also noted a result dependency on image content and have found, that a consistent patch position relative to the scanning line is not sufficient for good results.

Keywords: Temporal forensics · Image age · Chronological order · CNN

1 Introduction

In temporal image forensics [3,4], the main objective is to create a chronological sequence of pieces of evidence. A chronological sequence of images can help to deduce a causal relationship between events, which can be important in a court trial, for example. In general, images can be ordered chronologically based on the acquisition time-stamp stored in the EXIF header. However, this time-stamp is very easy to manipulate and therefore not trustworthy. For this reason, methods are required that approximate the age of an image based on traces left during the image acquisition.

One source of such traces is the image sensor. More precisely, individual pixels can become defective. These defects are known as in-field sensor defects (i.e. they develop after the manufacturing process) and they accumulate over time. In-field sensor defects are due to cosmic radiation and lead to an offset in the dynamic range of the defective pixel. Since in-field sensor defects accumulate over time, the age of image under investigation can be deduced from the detected defects. Consequently, the state-of-the-art in this field currently relies on the analysis of

---

M. Paulitsch and A. Vorderleitner–Contributed equally.

© Springer Nature Switzerland AG 2021
O. Gervasi et al. (Eds.): ICCSA 2021, LNCS 12950, pp. 147–162, 2021.
https://doi.org/10.1007/978-3-030-86960-1_11

detected pixel defects [3,4], a process not necessarily very accurate, especially when image content exhibits many textured areas.

Overall, temporal forensics have to look at subtle changes in images, eventually caused by single pixels getting defective over time. From the perspective of the requirement to analyse such subtle image properties, the fields of steganalysis (i.e., broadly speaking, steganalysis aims at finding messages hidden by steganographic methods in data) and camera authentication/identification (typically done by analyzing sensor noise properties), respectively, are closely related to temporal forensics. While the technology used in temporal image forensics is currently based on model-based approaches, the field of camera/sensor recognition has already seen the successful application of convolutional neural networks (CNN) [1,2,6,7,10]. Also in the field of steganalysis, deep learning and CNNs already received much attention from academia. Multiple networks were developed with growing success [8,9,13–15]. CNNs succeeded in extracting complex statistical dependencies from images and also achieved to improve the detection accuracy as compared to traditional techniques.

Thus, for our aim in temporal image forensics, we focused on CNNs, in particular on networks trained on finding hidden messages in images. Due to the similarity of the tasks to be conducted, the architectures of state-of-the-art steganalysis networks served us as a model for our network architecture. Basically those networks consist of three main parts, beginning with a preprocessing, continuing with a convolution part and concluding with a classification module [13–15]. All those parts can also be found in our network architecture, discussed in Sect. 2.1. Alike to steganalysis, our network should find hidden messages. In our context, however, the hidden message is any information about the age of an image. Information of that kind could be pixel defects, or other alterations occuring caused by the ageing process. This also includes information, not detectable for human eyes.

Techniques in temporal forensics currently rely on the analysis of single pixel defects [3,4] and focus on classical consumer camera (area) sensors. In this paper, we introduce the usage of convolutional neural networks for this task and consider image material acquired by a line-scanner.

In Sect. 2 we describe the structure of our network and introduce a variety of different patch selection modes. We discuss accuracy results in Sect. 3 and give an intuition on the development of the loss. Section 4 presents our concluding remarks.

2   Network

2.1   Architecture

Our network architecture is illustrated in Fig. 1. It constitutes a modified version of AlexNet [5] and can be used as a single-channel or multi-channel network. Multiple steganalysis networks inspired the fundamental construction [8,9,13–15]. For the multi-channel approach, we were inspired by the implementation of [1]. Each channel has the same structure and is based on the CNN

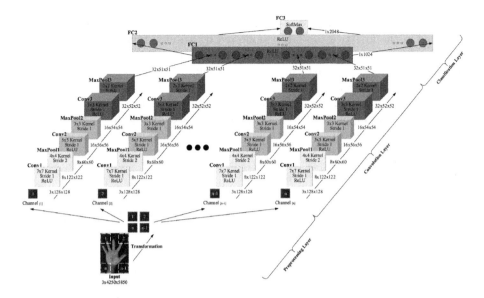

Fig. 1. Proposed CNN architecture.

network of [6]. We introduced the possibility to use one or multiple channels in the network architecture to investigate the importance of spatial position in images. In the multi-channel approach, input patches for each channel can be taken from an identical position in the image. Thus, each channel can be trained on its specific positional information by always receiving its input from the same image spatial location, while in the single-channel approach the position of the patches in the image used for training might change (depending on the patch selection strategy).

The architecture is structured in three parts, namely the preprocessing layer, the convolution layer and the classification layer. Starting with the **Preprocessing Layer (PP)**, where images are fed into the network. Depending on the chosen patch selection mode the input image is transformed into a single cropped image or into multiple cropped images of size (img-w × img-h), (e.g. 128 × 128). Further details can be found in Sect. 2.2. All resulting images are fed sequentially into the single-channel or one by one into each of the multi-channels. The **Convolution Layer (CV)** consists of a single-channel or multiple single-channels (multi-channel). Each channel starts with a 2D convolution at a kernel size of 7 × 7 and stride 1, followed by a Rectified Linear Unit (ReLU) function. Then, a 2D max pooling with a kernel size of 4 × 4 and stride 2 is applied. Every channel consists of three such blocks where the kernel size of the convolution is reduced by two and the kernel size of the max pooling is reduced by one at every block. The stride of the max pooling is then set to one for the following blocks. The image size gets reduced by each block while the feature size increases. After the last block the feature size is 32 and the image size is 51 × 51. This results in an output size of 32 × 51 × 51 for each channel. The **Classification Layer (CL)** is featured with three Fully Connected (FC) sub-layers. A linear transformation is applied in the first two sub-layers, followed by a ReLU function while the last sublayer also applies a linear transformation, but followed by a softmax function with an output size of two. This output specifies, if an image is classified as Old, or New. The input size of the first sub-layer is ch $\times$ ($32 \times 51 \times 51$) where ch denominates the number of channels. 1024 output features are produced in this process. The input size of all FC sub-layers, except the first of the three, corresponds to the output size of their preceding FC sub-layer.

2.2 Patch Selection Strategies

We followed different strategies for selecting certain image patches from scanner images, which are fed into the CNN eventually. All selection strategies described below can be applied to both versions of the network (multi-channel and single-channel), unless it is explicitly stated otherwise. Recall that running the network in multi-channel mode implies that a certain channel always receives its input from the same patch position in the image. This aims at maintaining the spatial consistency, in order to investigate the importance of spatial position coherence when investigating ageing effects. When using different patch selection strategies, the accuracy of the classification varies, as discussed in more detail in Sect. 3. All patch selection schemes use a unified patch size (i.e. 128×128).

Crop Five (CF). CF takes crops in the exact following order from the left/right top, right/left bottom and from the center position of the scanner image, marked in red in Fig. 2a. Furthermore, the single crop (SC) positions represent a selection scheme on their own, applied only in single-channel mode. For example, the center mode uses only the center patch of an image combined with the single-channel mode. The SC versions are denominated by LT (Left Top), RT (Right Top), RB (Right Bottom), LB (Left Bottom), and C (Center).

Crop Ten (CT). CT uses the same patch positions as CF and in addition also uses the horizontally flipped version of every crop, again illustrated by the red crops in Fig. 2a.

Spiral Eight (SE). SE uses the four corner crop positions, alike to CF. Additionally it uses the corners of the "inner" image of size (width $- 2 \cdot$ blocksize) $\times$ (height $- 2 \cdot$ blocksize). SE results from Fig. 2a when combining all yellow patches with the red patch positions, without the center patch.

Line Mode (LM-X). LM-X starts by cutting the scanner images into (scan-width $\times$ img-h) sized slices, resulting in several lines. These are treated as images from now on. The patches are then selected from the produced line-images. X denominates the number of patches taken from a line. Two example line-images produced by LM-X are illustrated in green in Fig. 2a.

Column Mode (CM-X). CM-X represents the opposite approach to LM-X, creating new images by cropping the scanner images into (img-w $\times$ scan-height) sized slices, resulting in columns. The patches are then again taken from the produced column-images. X denominates the number of crops, taken from a column. This behaviour is shown in blue in Fig. 2a. Note that the scan direction is from top to bottom in the images displayed. Thus, patches extracted from column-images have the identical relative position to the scan-line, and thus may be considered to have the same spatial position.

Center Block Mode (CBM-X). CBM-X is applied in combination with either CM-X, or LM-X. All patches are taken from the center area of the scanner image. Depending on the combined patch selection mode, either parts of centered columns, or centered lines are used as patches. Again X denominates the number of patches, taken from a line/column. This procedure aims at using only patches mostly consisting of scanned hand parts. The approximately covered center area is illustrated in Fig. 2b, where CBM + LM-X is colored purple and CBM + CM-X is colored cyan.

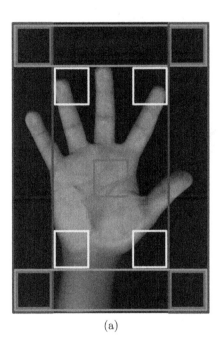 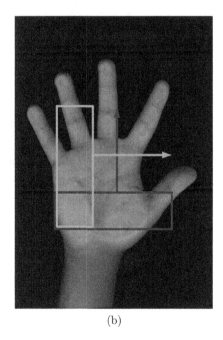

(a)   (b)

Fig. 2. (a) Different crop positions used in patch selection modes, (b) CBM crop positions. (Color figure online)

3   Experiments

3.1   Data-Set

We use a data-set that has been used before in the context of investigating ageing effects in hand-related biometrics [11,12]. We collected a database of high-resolution human palmar handprints (see Fig. 2a and Fig. 2b) from 28 members in our lab with 443 hand images in the first session (Old) captured in November 2007 and 164 hands in the second session (New) captured in October/November 2012, i.e. exhibiting a time lapse of approximately 5 years between recordings adhering to the same strict recording protocol (users were allowed to wear rings or watches and occupy an arbitrary position on the scanner as long as fingers did not touch each other). During the 5 years lapse, no images were taken with the scanner. Image acquisition was done using the same instance of a flatbed HP 3500c scanner for both sessions, recording a $216 \times 297\,mm$ area at 500 dpi resolution (resulting in 4250×5850 pixels sized images). Due to the limited illumination capabilities of traditional flatbed scanning devices, environmental surrounding light was shielded by a scanning box with a round hole for hand insertion.

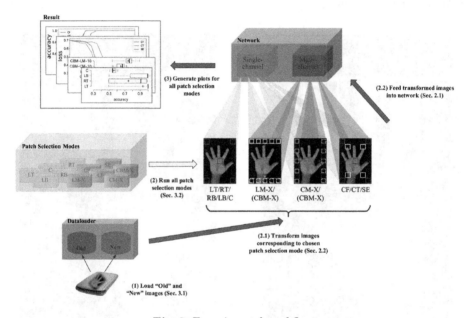

Fig. 3. Experimental workflow.

3.2   Experimental Setup

We conduct our experiments on a NVIDIA TITAN X (Pascal) 12 GB, with CUDA version 10.1. For all experiments a stochastic gradient descent optimizer with an initial learning rate of 0.01 and a momentum of 0.9 is used alike to [5]. Unlike the previous article, we set the decay to 0.005, as seen in [6]. We varied the values for learning rate, momentum, and decay, but found out that the network converged best with the selected values. The same learning rate is applied to all layers. Additionally, the bias is disabled for all convolutional layers, but remains enabled for the linear layers. No dropout is used. The weights follow a zero-mean normal distribution with a standard deviation of 0.08. We use an equal number of 164 images from the data-set for each of the two classes. The data-set is then divided into a training set, a validation set and an evaluation set with a division of 64%, 16% and 20% for a 5-fold cross-validation (5-CV). The validation set is used as feedback regarding to overfitting during the training process and the evaluation set serves for eventually testing the classification accuracy on data, which the CNN has never seen before. The batch size is set to 11 except for those experiments which are running in patch selection modes CF, CT, SE, LM-X, CM-X, CBM-X with explicitly deactivated multi-channel mode. Here the batch size is multiplied by the number of patches of the chosen selection mode. We use 19 batches for the training, four batches for the validation and five for the evaluation in each epoch. The 5-CV is repeated five times (i.e. 25 runs) with a fixed number of 15 epochs and the main results are plotted in Fig. 4. For each 5-CV repetition, the data distribution of the data-set is random. Training the network with these parameters in multi-channel mode, CT selection mode and patches of size 128×128 takes about 60 min, with convergence within 10 epochs, whereas in CF selection mode it takes about 30 min. We did not put our focus on training performance, therefore, undoubtedly this can be improved with some effort. In addition to the 5-CV with a fixed number of 15 epochs, we ran another 5-CV for some specific patch selection modes with 100 epochs (long-run) for three times to ensure the accuracy results. Figure 3 presents the workflow of our study.

3.3 Classification Accuracy

In this subsection we discuss the achieved classification accuracy with respect to the used patch selection mode, single- and multi-channel setup, respectively, and discuss the results. Furthermore, we observe the development of the loss in relation to the trend of the accuracy. In the following, it is assumed that all patch selection modes are executed in multi-channel mode, unless the operation in single-channel mode is explicitly pointed out.

The best training accuracies are achieved by the CF, CT and SE selection modes. As early as the second epoch, the loss decreases until it converges to 0.3 at epoch 8. While the loss is decreasing the accuracy increases up to 1.0. This behavior can be observed in Fig. 5a and Fig. 5b. Those three patch selection modes also achieve the best evaluation accuracies, which exactly reflect the results of the training. Figure 4 and Fig. 12 show the training and evaluation accuracies of all experiments, which are structured according to their patch selection modes. In Fig. 4a and Fig. 4b we recognize that CF, CT and SE achieve the same training and evaluation accuracy. To clarify the importance of the spatial position of the patches, we apply the same experiment in the dedicated single-channel

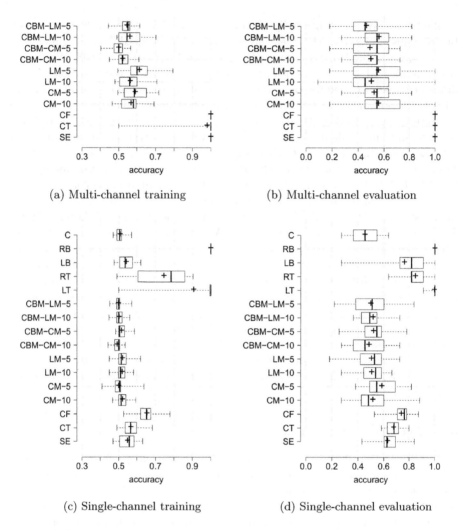

(a) Multi-channel training

(b) Multi-channel evaluation

(c) Single-channel training

(d) Single-channel evaluation

Fig. 4. Boxplot of the resulting training and evaluation accuracy for the five 5-CV runs in single-channel and multi-channel mode.

mode (where a single network is trained using patches from different spatial locations). The results are shown in Fig. 4c and Fig. 4d. The single-channel versions of CF, CT and SE perform clearly worse than in multi-channel mode. We can observe that the training and evaluation accuracy clearly do not reach the same level as the multi-channel version. Thus, we may clearly state that the position information of the patches at least contributes to high image age classification accuracy.

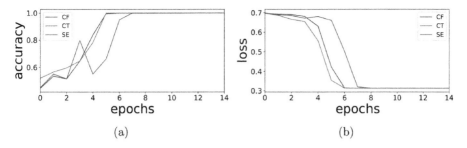

Fig. 5. (a) Comparing the training accuracies of the three best patch selection modes CT, CF and SE in multi-channel mode from an arbitrary sample run, (b) Comparing the training convergence of the three best patch selection modes CT, CF and SE in multi-channel mode from an arbitrary sample run.

Figure 6a shows the training accuracy of the CM-X and LM-X selection modes. The purpose of this experiment is to find out whether it is sufficient to use the patches from the same relative position to the scan-line, but not identical in the overall image. Patches hereby always come from the same position in the row or column, but the position of the entire row or column slides over the original scanner image. As a result, the absolute position of the patches changes continuously, but its row/column-position always stays the same. A recurring cycle is created, but of course different than in CF, CT and SE. The training and evaluation accuracies are far behind the results of CT, CF and SE. In single-channel mode, the accuracies tend to be similar, but CT, CF, and SE are still better. The loss also shows no signs of convergence over the entire epochs, as can be seen in Fig. 6b. Regarding the LM-X selection modes, it can be stated that these produce slightly better results than the CM-X modes, as can be seen in Fig. 4. The long-runs of the LM-X and CM-X in Fig. 7 show a convergence of the loss, but the results as explained before do not change. Overall, it turns out that the same relative position to the scan-line is not sufficient for decent classification (instead, a fixed position in the image is required to successfully train a network), even stronger, as we do not identify a clear difference between CM-X and LM-X, the identical position relative to the scan-line does not matter.

Furthermore, we tested a variant of LM-X, which only uses the top five lines of a scanner image to see whether the hand areas of the image are causing the poor accuracy. Those five lines mostly consist of black background. However, the results are almost exactly the same as using the entire lines of the scanner image. The same applies to a variant of CM-X, which only uses the rightmost 5 columns. For both, LM-X and CM-X, this holds for the multi-channel as well as the single-channel versions (results are not shown).

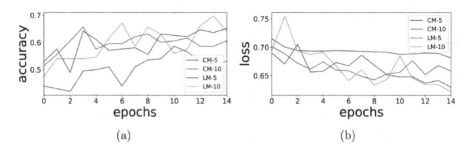

Fig. 6. (a) Comparing the training accuracies of CM-X and LM-X patch selection in multi-channel mode from an arbitrary sample run, (b) Comparing the training convergence performance of CM-X and LM-X patch selection in multi-channel mode from an arbitrary sample run.

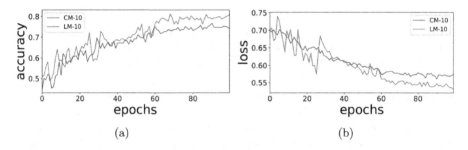

Fig. 7. (a) Comparing the training accuracies of CM-X and LM-X patch selection in multi-channel mode from an arbitrary long-run sample, (b) Comparing the training convergence performance of CM-X and LM-X patch selection in multi-channel mode from an arbitrary long-run sample.

In contrast to these variants of LM-X and CM-X, the patch selection modes CBM-CM-X and CBM-LM-X intend to use only the hand area of the scanner image. This evaluates the effects of the hand area on the accuracy of the patch selection modes LM-X and CM-X. Figure 8a and Fig. 8b present the accuracy and the loss of the training. In multi-channel mode we observe a decrease of the resulting training accuracy, while the evaluation accuracy is largely the same. In single-channel mode the training and evaluation accuracy decreases slightly, compared to the LM-X and CM-X. Thus, the line- and column oriented patch selection strategies do not turn out to be competitive to those approaches, where the spatial location of patches used in training and evaluation is fixed. Figure 9 shows the convergence of the loss up to epoch 100. Comparing the long-runs of the CBM-CM-X, CBM-LM-X to the CM-X, LM-X patch selection modes the relation does not change.

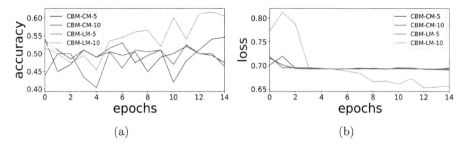

(a)   (b)

Fig. 8. (a) Comparing the training accuracies of CBM-CM-X and CBM-LM-X patch selection modes in multi-channel operation, (b) Comparing the training convergence performance of CBM-CM-X and CBM-LM-X patch selection modes in multi-channel operation, all from the same arbitrary sample run.

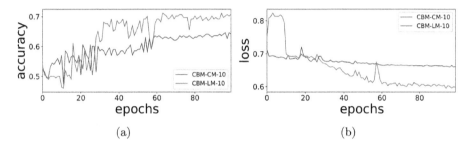

(a)   (b)

Fig. 9. (a) Comparing the long-run training accuracies of CBM-CM-X and CBM-LM-X patch selection modes in multi-channel operation, (b) Comparing the training convergence performance of CBM-CM-X and CBM-LM-X patch selection modes in multi-channel operation, all from the same arbitrary long-run sample.

Finally, we look at the single patch results - the SC selection modes show very different results for the training as well as for the evaluation accuracy. Where RB always has an accuracy of 1.0, the C selection mode performs worst in both cases (so it seems that the image content - skin texture in this case - has a clearly negative impact on age classification results). See Fig. 13 for a comparison of "Old" and "New" background patches - a stripe pattern in orthogonal direction to the scan direction is visible, with stronger and clearer patterns for the "New" class (a zoom into patches is shown). LB, RT, and LT vary widely when it comes to training accuracy. In terms of evaluation accuracy, they do not vary that much and achieve better results, but do not reach the performance of RB. The comparison between training accuracy and loss for all SC patch selection modes is shown in Fig. 10a and Fig. 10b. Due to the extreme instability of the

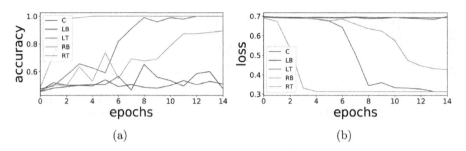

Fig. 10. (a) Comparing training accuracies of the five SC patch selection modes in single-channel operation, (b) Comparing the training convergence performance of the five SC patch selection modes in single-channel operation, all from the same arbitrary sample run.

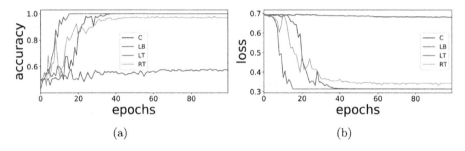

Fig. 11. (a) Comparing the long-run training accuracies of all except RB SC patch selection modes in single-channel operation, (b) Comparing the training convergence performance of all except RB SC patch selection modes in single-channel operation, all from the same arbitrary long-run sample.

resulting accuracy between training and evaluation, the SC selection modes, with the exception of RB, are not sufficient enough to make a reliable classification. The long-runs of the SC selection modes, shown in Fig. 11, reflect those results. Increasing the size of the data-set possibly has a positive impact on the reliability and the performance of the SC selection modes.

To conclude this section we want to point out that the confusion matrices of two different modes, namely RT and LM-10, show different error patterns (see Fig. 14). After looking at three arbitrary sample runs it can be observed that RT tends to classify "New" images as "Old", whereas LM-10 does not show any apparent pattern. This holds true for both, a run of 15 epochs and a long-run.

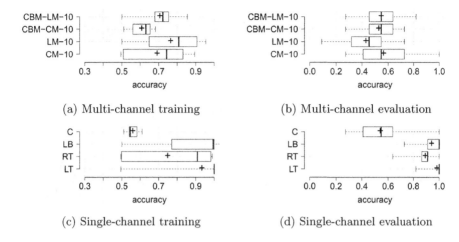

(a) Multi-channel training

(b) Multi-channel evaluation

(c) Single-channel training

(d) Single-channel evaluation

Fig. 12. Boxplot of the resulting training and evaluation accuracy for the three 5-CV long-runs in single-channel and multi-channel mode.

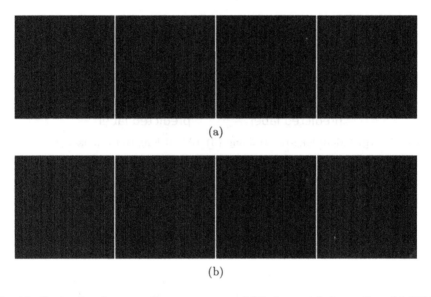

(a)

(b)

Fig. 13. Comparison between the corner crops of CF of a sample image from (a) "Old" class, (b) "New" class. (i.e. from left to right: LT, RT, RB, LB)

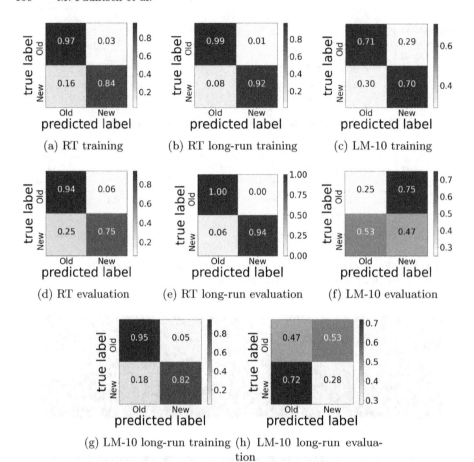

(a) RT training   (b) RT long-run training   (c) LM-10 training

(d) RT evaluation   (e) RT long-run evaluation   (f) LM-10 evaluation

(g) LM-10 long-run training (h) LM-10 long-run evaluation

Fig. 14. Confusion matrices from arbitrary sample runs out of the different patch selection modes and run lengths.

4   Conclusion

We introduced CNNs for classifying the age of scanner images. Even with a relatively small data-set, the CNN accomplished to reach high classification accuracies. We showed that maintaining the spatial consistency of patches that are fed into the network induces a clearly positive effect on the classification accuracy (i.e.: median: 100% consistent vs 54% inconsistent). This was mainly done by exploiting the benefits of multi-channel networks over single-channel networks for classifying the age of scanner images. Overall, a single network has to be trained and tested with patches from the same spatial location in the image to achieve high classification accuracy - this implies that features used for classification are local features bound to specific positions on the sensor. The single-channel versions of the network also occasionally showed decent results, but overall with a apparently lower reliability than the multi-channel versions (i.e.: highest median: 100%, lowest median: 50% of classification accuracies). Further, we were not able to exploit the line-scan nature of our data to reduce the required spatial consistency to a consistency of relative position to the scan-line. Based on our findings we see big potential for CNNs for future utilization in the field of temporal image forensics, however, a clear result dependency on image content has to be stated (i.e. uniform background leads to best results observed).

References

1. Bayar, B., Stamm, M.C.: Augmented convolutional feature maps for robust CNN-based camera model identification. In: 2017 IEEE International Conference on Image Processing (ICIP), pp. 4098–4102 (2017). https://doi.org/10.1109/ICIP. 2017.8297053
2. Bondi, L., Baroffio, L., Güera, D., Bestagini, P., Delp, E.J., Tubaro, S.: First steps toward camera model identification with convolutional neural networks. IEEE Signal Process. Lett. **24**(3), 259–263 (2017). https://doi.org/10.1109/LSP.2016. 2641006
3. Fridrich, J., Goljan, M.: Determining approximate age of digital images using sensor defects. In: Memon, N.D., Dittmann, J., Alattar, A.M., III, E.J.D. (eds.) Media Watermarking, Security, and Forensics III, vol. 7880, pp. 49–59. International Society for Optics and Photonics, SPIE (2011). https://doi.org/10.1117/12.872198
4. Jöchl, R., Uhl, A.: A machine learning approach to approximate the age of an digital image. LNCS (2020 (to appear))
5. Krizhevsky, A., Sutskever, I., Hinton, G.E.: ImageNet classification with deep convolutional neural networks. Commun. ACM **60**(6), 84–90 (2017). https://doi.org/ 10.1145/3065386. May
6. Marra, F., Poggi, G., Sansone, C., Verdoliva, L.: A deep learning approach for iris sensor model identification. Patt. Recogn. Lett. **113**, 46–53 (2018). https://doi. org/10.1016/j.patrec.2017.04.010, integrating Biometrics and Forensics
7. Mendes Júnior, P.R., Bondi, L., Bestagini, P., Tubaro, S., Rocha, A.: An in-depth study on open-set camera model identification. IEEE Access **7**, 180713–180726 (2019). https://doi.org/10.1109/ACCESS.2019.2921436
8. Reinel, T.S., Raúl, R.P., Gustavo, I.: Deep learning applied to steganalysis of digital images: a systematic review. IEEE Access **7**, 68970–68990 (2019). https://doi.org/ 10.1109/ACCESS.2019.2918086
9. Tan, S., Li, B.: Stacked convolutional auto-encoders for steganalysis of digital images. In: Signal and Information Processing Association Annual Summit and Conference (APSIPA), pp. 1–4 (2014). https://doi.org/10.1109/APSIPA.2014. 7041565
10. Tuama, A., Comby, F., Chaumont, M.: Camera model identification with the use of deep convolutional neural networks. In: Proceedings of Workshop on Information Forensics and Security (WIFS 2016). Abu Dhabi, United Arab Emirates (2016). https://doi.org/10.1109/WIFS.2016.7823908
11. Uhl, A., Wild, P.: Age factors in biometric processing. IET, pp. 153–170 (2013)
12. Uhl, A., Wild, P.: Experimental evidence of ageing in hand biometrics. In: Proceedings of the International Conference of the Biometrics Special Interest Group (BIOSIG 2013), pp. 39–50. Darmstadt, Germany (2013)

13. Xu, G., Wu, H.Z., Shi, Y.Q.: Structural design of convolutional neural networks for steganalysis. IEEE Signal Process. Lett. **23**(5), 708–712 (2016). https://doi.org/10.1109/LSP.2016.2548421

14. Yedroudj, M., Comby, F., Chaumont, M.: Yedroudj-Net: an efficient CNN for spatial steganalysis. In: 2018 IEEE International Conference on Acoustics, Speech and Signal Processing (ICASSP), pp. 2092–2096 (2018). https://doi.org/10.1109/ICASSP.2018.8461438

15. Zhang, R., Zhu, F., Liu, J., Liu, G.: Depth-wise separable convolutions and multi-level pooling for an efficient spatial CNN-based steganalysis. IEEE Trans. Inf. Forensics Secur. **15**, 1138–1150 (2020). https://doi.org/10.1109/TIFS.2019.2936913

Multichannel Color Spaces Selection for Region-Based Active Contour: Applied to Plants Extraction Under Field Conditions

Yamina Boutiche$^{(\boxtimes)}$, Nabil Chetih, Naim Ramou, Mohammed Khorchef, and Rabah Abdelkader

Research Center in Industrial Technologies - CRTI, ex CSC, P.O.Box 64, Cheraga, 16014 Algiers, Algeria
y.boutiche@crti.dz
https://www.crti.dz

Abstract. Segmentation in agriculture presents a challenging issue especially when acquisition conditions are not controlled. Segmentation algorithms should be able to deal with images that can contain shadowed regions, upper/less lighted parts and saturated pixels. Consequently, the segmentation of RGB images is more often failed. In this paper, we propose a method that can overcome some of the listed difficulties. It is based, firstly, on a transformation of RGB color space into $L^*a^*b^*$, HSV and YC_rC_b ones. Secondly, to benefit from chroma information only, a^*, H and C_b are selected and combined to construct the image to be segmented. Finally, the Piecewise Constant (PC) Chan-Vese model with fast optimization principle is adopted for the segmentation issue.

Experiments are carried out in plant images with controlled conditions and field conditions using four open datasets. In addition, an evaluation study is done using three different metrics. The obtained results show the effectiveness and the robustness of the proposed approach compared to single-channel (color index-based) approaches and the multi-channel based segmentation.

Keywords: Agricultural images · Segmentation · Plant extraction · PC active contours · Fast sweep optimization

1 Introduction

The use of image processing and analysis in agriculture is experiencing remarkable growth, where a large image processing techniques have been exploited for several issues. Automatic segmentation of plants [1–3], leaf extraction [4], leaf disease detection [5,6], irrigation control [10], fruit recognition and counting [7].

The challenging tasks in agriculture are providing algorithms and color analysis techniques that should be able to deal with the non-trivial and uncontrollable capture conditions such as shadows, noise, pixel saturation, low lighting present the real difficulties for the image processing community.

© Springer Nature Switzerland AG 2021
O. Gervasi et al. (Eds.): ICCSA 2021, LNCS 12950, pp. 163–173, 2021.
https://doi.org/10.1007/978-3-030-86960-1_12

Almost all acquisition systems provide images defined on RGB color space, where each pixel is initially represented in the three components Red, Green and Blue. The RGB color space and its derived obtained by a linear transformations are generally not suitable for color scene segmentation and analysis because of the high correlation between their three components [8]. In fact, there is non same opinion about the color space that provides the best performance in color image segmentation issue. Therefore, for each specific application, researchers try to find an optimum color space where an algorithm gives a high segmentation accuracy. Several studies have been carried out to evaluate algorithms of color images segmentation. For example, in [9] comparison studies between different color spaces in clustering-based image segmentation was carried on general nature images.

In the context of agricultural images segmentation, Hamuda et al. have done a survey and an evaluation of plant extraction based on color indices [20]. In [10] authors presented a comparisons between color models for automatic image analysis in irrigation management applications. A review on weed detection using ground-based machine vision and image processing techniques was done in [11].

In this paper, we propose a chroma combination between the three color spaces $L^*a^*b^*$, HSV and YC_rC_b. This purpose allows to benefit from color space transformation advantages, especially in term of the separation between chroma and luminance information. Then, instead of using all color space components, we select one chroma component of the above spaces which are a^*, H and C_b. Those components are combined to construct three-components image. The segmentation step is done by the Piecewise-Constant Chan-Vese model optimized by sweeping algorithm for fast convergence.

The paper is organized as follows: Sect. 2 details the stages of the proposed method. Experiments results and comparisons are given in Sect. 3. Finally, the main conclusions are done in Sect. 4.

2   Proposed Method

The present work is devoted to segment agricultural images, where they are acquisition in an uncontrolled light conditions. Figure 1 represents the flowchart of the proposed method. We develop each stage in the following subsections.

2.1   Color Space Transformations

Color spaces can be grouped into four main groups *Primary spaces, luminance-chrominance spaces, perceptual spaces* and *independent axis spaces* [12].

The great advantage of the three last groups is their ability to separate between luminance and chroma information. However, there are different opinions about the transformation that allows the best segmentation performance. Nevertheless, some comparative studies have been done in [13] and [10]. In [10] a comparative study of eleven color spaces for the problem of plant segmentation is described. The best ones were found to be HLS, HSV, YC_rC_b, YUV, $L^*a^*b^*$,

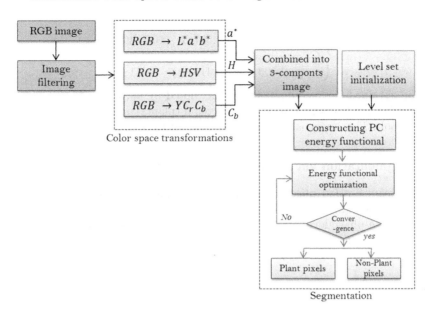

Fig. 1. Flowchart of proposed plant extraction

$L^*u^*v^*$, TSL, $I_1I_2I_3$ and XYZ. The best segmentation results are produced by the a^* channel in the $L^*a^*b^*$ color space. This later (a^*) was also used by Zhang et al., in [6], to extract cotton leaves with disease under the field environment.

In all the previous cited works, the segmentation stage is done in a single-channel image. However, to broaden the benefit from the advantages of color spaces transformation, we propose to combine three components form the three best color spaces. Consequently, the image to be segmented is a multi-channel image constructed by a^*, H and C_b channels, obtained by color space transformations $L^*a^*b^*$ in Eq. (1), HSV in Eq. (2) and YC_rC_b in Eq. (3).

$$
\begin{aligned}
L^* &= \begin{cases} 116\,Y^{\frac{1}{3}} & \text{if } Y > k; \\ 903.3\,Y & \text{if } Y \le k \end{cases} \text{ with k = 0.008856} \\
a^* &= 500(f(X) - f(Y)) \\
b^* &= 200\,(f(Y) - f(Z))
\end{aligned}
\qquad,\qquad
f(t) = \begin{cases} f(t) = t^{\frac{1}{3}}, & \text{if } t > k \\ 7.787\,t + 0.1379 & \text{if } t \le k \end{cases}
\tag{1}
$$

$$
H = \begin{cases}
\pi & \text{if } R = G = B \\[4pt]
\arccos \dfrac{\frac{1}{2}\left[(R - G) + (R - B)\right]}{\sqrt{(R - G)^2 + (R - B)(G - B)}} & \text{if } B \le G \\[10pt]
2\pi - \arccos \dfrac{\frac{1}{2}\left[(R - G) + (R - B)\right]}{\sqrt{(R - G)^2 + (R - B)(G - B)}} & \text{elsewhere}
\end{cases}
\tag{2}
$$

$$
S = \begin{cases}
0 & \text{if } R = G = B \\[4pt]
1 - \dfrac{3\min(R, G, B)}{R + G + B} & \text{elsewhere}
\end{cases}
$$

$$
V = \frac{R + G + B}{3}
$$

$$\begin{cases} Y = 0.299\,R + 0.587\,G + 0.114\,B \\ C_r = -0.169\,R - 0.331\,G + 0.500\,B \\ C_b = 0.500\,R - 0.419\,G - 0.081\,B \end{cases} \tag{3}$$

The obtained three-components image, noted by u_0, is free from luminance information and highlights well the foreground (plants) and attenuates the background (sol). This image is then used as the input to the segmentation algorithm.

2.2   Segmentation via Piecewise Constant PC Model

For the segmentation purpose, we use the famous energy-based model named Piecewise Constant Chan-Vese model in the biphase level set framework [14]. Chan-Vese bi-phase level set active contour is used for foreground/background separation. It is known by its good performance and robustness to noise. In the variational context, the PC model is defined as :

$$\begin{aligned} F(c^+, c^-, \Phi) = {} & \mu \int_\Omega \delta_\epsilon(\Phi(\mathbf{x})) |\nabla \Phi(\mathbf{x})| \, d\mathbf{x} + \int_\Omega \frac{1}{N} \sum_i^N \lambda_i^{in} \left| u_{0,i}(\mathbf{x}) - c_i^{in} \right|^2 H_\epsilon(\Phi) \, d\mathbf{x} \\ & + \int_\Omega \frac{1}{N} \sum_i^N \lambda_i^{out} \left| u_{0,i}(\mathbf{x}) - c_i^{out} \right|^2 (1 - H_\epsilon(\Phi) \, d\mathbf{x}, \quad i = 1, \dots N. \end{aligned} \tag{4}$$

where $u_{0,i}$ is the image to be segmented, $\Omega \in \mathbb{R}^2$ its domain definition and $\mathbf{x} = (x, y)$ represents the pixel coordinates. μ is scalar defining the length term weight and $\lambda_i^{in} > 0$ and $\lambda_i^{out} > 0$ weight coefficients for the error term inside and outside curve, respectively. H_ϵ is the regularized Heaviside function and δ_ϵ its derivative. $c^{in}(\Phi)$ and $c^{in}(\Phi)$ are the means inside and outside curve for each channels $(i, i = 1 \dots 3)$, they are defined as follows:

$$c_i^{in}(\Phi) = \frac{\int_\Omega u_{0,i} H_\epsilon(\Phi) d\mathbf{x}}{\int_\Omega H_\epsilon(\Phi) d\mathbf{x}}, \quad c_i^{out}(\Phi) = \frac{\int_\Omega u_{0,i} \left(1 - H_\epsilon(\Phi)\right) d\mathbf{x}}{\int_\Omega \left(1 - H_\epsilon(\Phi)\right) d\mathbf{x}} \tag{5}$$

In the present work, and to avoid the slow convergence of descent gradient optimization method, we use the sweep principal optimization algorithm. Therefor, two energies are evaluated for any movement of a pixel from inside to outside of curve as follows (see [15, 16] for more details):

$$\Delta F_{12} = \frac{1}{N} \left[\sum_i^N \lambda_i \left(u_{0,i}(\mathbf{x}) - c_i^{out} \right)^2 \frac{n}{n+1} - \sum_i^N \lambda_i \left(u_{0,i}(\mathbf{x}) - c_i^{in} \right)^2 \frac{m}{m-1} \right] \tag{6}$$

$$\Delta F_{21} = \frac{1}{N} \left[\sum_i^N \lambda_i \left(u_{0,i}(\mathbf{x}) - c_i^{in} \right)^2 \frac{m}{m+1} - \sum_i^N \lambda_i \left(u_{0,i}(\mathbf{x}) - c_i^{out} \right)^2 \frac{n}{n-1} \right], \tag{7}$$

Finally, the segmented image is deducted as follows:

$$u_{seg} = \sum_{i=1}^3 c_i^{in} \Big|_{\Phi > 0} + \sum_{i=1}^3 C_i^{out} \Big|_{\Phi < 0} \tag{8}$$

3   Experimental Results

In all experiments, the level set is initialized automatically as a rectangle that takes the image size minus 10 pixels, this is done to avoid the interaction of the user. Furthermore, as the optimization is done via sweeping algorithm, we don't need to set the constants related to variational level set models [2]. The preprocessing stage is needed where image is noised, because we omit the length term in the optimized energy (Eq. (6) and (7)) as done in [17]. This step is done by a 3×3 Gaussian filter in the R, G, B components. Otherwise, the color space transformation stage is enough to decrease the noise effect.

3.1   Qualitative Evaluations

The proposed algorithm is tested on all *black nightshade* images of the open dataset downloaded from [18]. In total, *123* images of size $4256 \times 2832 \times 3$ acquired in uncontrolled outdoor light conditions and different growth stages.

The first experiment is devoted to represent each stage results of the proposed method on an arbitrary chosen image DSC_0645 displayed in Fig. 2a. The second row Fig. 2b shows the three components a^*, H and C_b obtained by color space transformations. The combination of those components gives a three-channel image (Fig. 2c). The zero level set is initialized on the image and the three level set (Fig. 2d). After seven iterations, the algorithm convergences correctly to the plants' contours. The segmentation outcomes is represented in Fig. 2e, where the converged level set is presented, the final zero level set contour is represented, with yellow line, on the input image to be more clear. In addition, the binary segmented image (on the middle of row). Finally, the extracted object is obtained by multiplying the final binary level set by the original RGB image.

The second experiment, shown in Fig. 3, is done on three black nightshade images acquired in cloudy, overcast and very sunny days. Thereby different lighting conditions: low, medium and strong lighting. This experiment highlights the ability of the proposed method to deal with light changing, where, in the three cases, the plants are correctly extracted from the soil.

Figure 4 shows comparison results between our method and the method proposed in [19] named improved principal component analysis (PCA)-based multichannel selection C-V model (PMCV). The principle of this method is the selection of three component from RGB and HSV color spaces by using PCA, then the Chan-Vese model was applied to segmentation of the wheat leaf lesions images. When the lighting is uniform (case of DSC_{0500} image), both models converge to the plant contours correctly. However for the image with strong lighting (DSC_{0628}) where the shadow is more strongly present, the PMPC model have tendency to an over segmentation, where several part of the sol strongly lighted is classed as the plants.

(a) RGB   image
DSC_0645

(b) From left to right : a^*, H and C_b channels   (c) u_0 image to be segmented

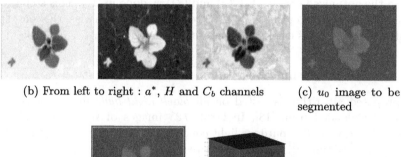

(d) Level set initialization

(e) Segmentation outcomes

Fig. 2. Outcomes of each stage of the proposed method. (a) input RGB original image, (b) The three components (a^*, H and C_b) obtained by color space transformationS. (c) The image after combination with initialized zero level set (yellow line) and corresponding level set function. (e) The algorithm outcomes: Final level set function, zero level set (yellow line) drawn on the original RGB image, segmented binary image u_{seg}, and extracted plant overlaid on the original image (Color figure online)

The last experiments are devoted to compare the proposed method performance against color index-based approaches, that are widely used in agricultural image segmentation. For that, two images from the dataset are chosen that present different light conditions and growing stages (DSC_0528 and DSC_0690). Furthermore, because there are a large color index-based formulations, we limit our comparisons to the ones classed as the best in the survey and evaluation done in [20], that are: Excess Green Index (ExG), Vegetative Index (VEG), Combined Indices 1 (COM1), Combined Indices 2 (COM2) and Excess Green minus Excess Red Index (ExGR), respectively. The segmentation

(a) Low lighting

(b) Normal lighting

(c) Strong lighting

Fig. 3. Examples of the proposed method performance on a set of images acquired in different natural light conditions. 1st row: clouded day. 2nd row: sunny day. 3th row: strongly sunny day

Fig. 4. Comparison of our method with the one based on HSV and RGB color spaces in [19].

is then applied using Otsu thresholding method. In Fig. 5, we display, for both images, the resulting image in each transformation and the segmentation results. COM1 and ExGR color index-based approaches completely fail to segment both images. However, ExG, VEG and COM2 have better performance. Compared to the proposed method, the segmentation is improved specially in terms of noise robustness and saturated pixels (due to very strong lighting).

3.2 Quantitative Evaluation of the Proposed Method Against Color Index-Based Approaches

The quantitative evaluation of the proposed method is made on three different public datasets that are:

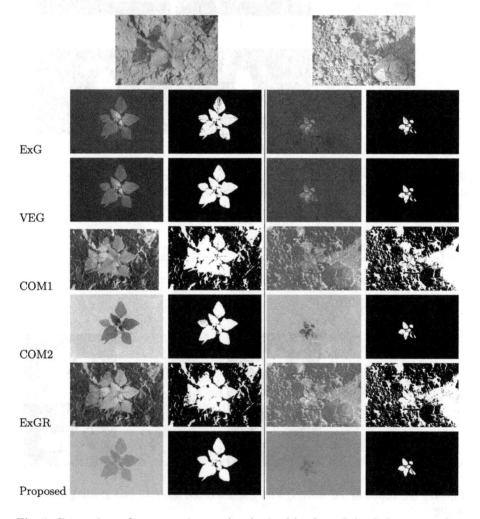

Fig. 5. Comparison of segmentation results obtained by Otsu-Color-Index approaches and the proposed one

- **CWFI (crop/weed field image) dataset** [21]: 60 RGB images (1296 × 966 pixels in resolution) acquired using ground vehicle, where the camera was shaded and artificially lit to avoid changing light conditions (controlled lighting).
- **Carrot-Weed dataset** [22]: it constitutes of 39 RGB images (1296 × 966) acquired using hand holding. They are taken under variable light conditions.
- **Wheat 2012 dataset** [23]: the acquisition system is done using camera mounted above the crop (2m). time-lapse images were taken and transmitted via 3G network. In this evaluation we use the folder named "wheat 2012" which contains 17 images (719 × 776) charactered by poor light conditions and complex background

The segmentation results are evaluated by using three different metrics Q_{seg}, *Jaccard* and *Dice*, formulated as follows:

$$Q_{seg} = \frac{TP + TN}{FN + FP + TP + TN} \tag{9}$$

$$Dice = \frac{2TP}{2 * TP + FP + FN} \tag{10}$$

$$Jaccard = \frac{Dice}{2 - Dice} \tag{11}$$

The obtained results are summarized by the mean accuracy rate of each metric and each dataset (see Fig. 6). The comparison is done against the Excess

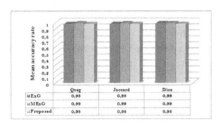

(a) CWFI dataset

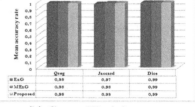

(b) Carrot-Weed dataset

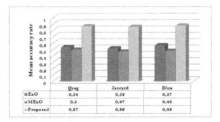

(c) Wheat 2012 dataset

Fig. 6. Quantitative evaluation of the proposed method against ExG and MExG for three different datasets using three evaluation metrics

Green Index (ExG) and the Modified Excess Green Index (MExG). This evaluation revels that the three methods give similar performance on the two first datasets (CWFI and Carrot-Weed datasets). However, the proposed method improves considerably the performance on the *wheat 2012* dataset.

4 Conclusion

This paper proposed a method to segment plants images acquired in uncontrolled outdoor lighting. Thereby those RGB images are noised, shadowed, saturated, and uneven lighting. Our method is based on a multichannel-Fast- Chan-Vese segmentation. It exploits chroma information by using three color space transformations, then a^*, H and C_b components are selected to construct the image to be segmented. Segmentation step is assured by the generalized Chan-Vese model implemented with bi-phase level set. Furthermore, the optimization of energy functional is done via sweep algorithm to ensure a fast convergence.

The obtained results allow us to see the good performance of our method in very challenging situations. Compared to color index-based approaches and PMPC model, our method improves the segmentation.

References

1. Hernández-Hernández, J.L., et al.: A new portable application for automatic segmentation of plants in agriculture. Agric. Water Manage. **183**, 146–157 (2017)
2. Boutiche, Y., Abdessalem, A., Ramou, N., Chetih, N.: Fast generalized Chan-Vese model for plant/soil segmentation to estimate percentage of ground cover in agricultural images. In: 2019 IEEE International Symposium on Signal Processing and Information Technology (ISSPIT), December 2019, pp. 1–5 (2019)
3. Guo, W., Rage, U.K., Ninomiy, S.: Illumination invariant segmentation of vegetation for time series wheat images based on decision tree model. Comput. Electron. Agric. **96**, 58–66 (2013)
4. Wang, Z., Wang, K., Yang, F., Pan, S., Han, Y.: Image segmentation of overlapping leaves based on Chan-Vese model and Sobel operator. Inf. Process. Agric. **5**, 1–10 (2018)
5. Singh, V., Misra, A.K.: Detection of plant leaf diseases using image segmentation and soft computing techniques. Inf. Process. Agric. **4**, 41–49 (2017)
6. Zhang, J., Kong, F., Wu, J., Han, S., Zhai, Z.: Automatic image segmentation method for cotton leaves with disease under natural environment. J. Integr. Agric. **17**, 180–1814 (2018)
7. Maldonado, W., Barbosa, J.C.: Automatic green fruit counting in orange trees using digital images. Comput. Electron. Agric. **127**, 572–581 (2016)
8. Busin, L., Vandenbroucke, N., Macaire, L.: Color spaces and image segmentation. Adv. Imaging Electron Phys. **151**, 65–168 (2008)
9. Jurio, A., Pagola, M., Galar, M., Lopez-Molina, C., Paternain, D.: A comparison study of different color spaces in clustering based image segmentation. In: Hüllermeier, E., Kruse, R., Hoffmann, F. (eds.) IPMU 2010. CCIS, vol. 81, pp. 532–541. Springer, Heidelberg (2010). https://doi.org/10.1007/978-3-642-14058-7_55

10. García-Mateos, G., Hernández, J.L., Escarabajal-Henarejos, D., Jaén-Terrones, S., Molina-Martínez, J.M.: Study and comparison of color models for automatic image analysis in irrigation management applications. Agric. Water Manage. **151**, 158–166 (2015)
11. Wang, A., Zhang, W., Wei, X.: A review on weed detection using ground-based machine vision and image processing techniques. Comput. Electron. Agric. **158**, 226–240 (2019)
12. Macaire, L., Postaire, J.G.: Color systems coding for color image. In: Proceedings of International Conference on Color in Graphics and Image Processing (CGIP 2000), Saint-Etienne, France, October, pp. 180–185 (2000)
13. Farid, G.-L., Jair, C., Asdrubal, L.-C., Lisbeth, R.: Segmentation of images by color features: a survey. Neurocomputing **292**, 1–27 (2018). https://doi.org/10.1016/j.neucom.2018.01.091
14. Chan, T.F., Sandberg, B.Y., Vese, L.A.: Active contours without edges for vector-valued images. J. Vis. Commun. Image Representation **11**, 130–141 (2000)
15. Boutiche, Y., Abdesselam, A.: Fast algorithm for hybrid region-based active contours optimisation. IET Image Process. **11**, 200–209 (2017). https://doi.org/10.1049/iet-ipr.2016.0648
16. Song, B., Chan, T.: A fast algorithm for level set based optimization. CAM-UCLA **2**(68), pp. 2–68 (2002)
17. He, L., Osher, S.: A fast multiphase level set algorithm for solving the Chan-Vese model. In: Proceedings in Applied Mathematics and Mechanics, vol. 7, pp. 1041911–1041912 (2007)
18. Espejo-Garcia, B., Mylonas, N., Athanasakos, L., Fountas, S., Vasilakoglou, I.: Towards weeds identification assistance through transfer learning. Comput. Electron. Agric. **171**, 105306 (2020)
19. Qiu-xia, H., Tian, J., He, D.: Wheat leaf lesion color image segmentation with improved multichannel selection based on the Chan-Vese model. Comput. Electron. Agric. **135**, 260–268 (2017)
20. Hamuda, E., Glavin, M., Jones, E.: A survey of image processing techniques for plant extraction and segmentation in the field. Comput. Electron. Agric. **125**, 184–199 (2016)
21. Haug, S., Ostermann, J.: A crop/weed field image dataset for the evaluation of computer vision based precision agriculture tasks. In: Agapito, L., Bronstein, M.M., Rother, C. (eds.) ECCV 2014. LNCS, vol. 8928, pp. 105–116. Springer, Cham (2015). https://doi.org/10.1007/978-3-319-16220-1_8
22. Lameski, P., Zdravevski, E., Trajkovik, V., Kulakov, A.: Weed detection dataset with RGB images taken under variable light conditions. In: Trajanov, D., Bakeva, V. (eds.) ICT Innovations 2017. CCIS, vol. 778, pp. 112–119. Springer, Cham (2017). https://doi.org/10.1007/978-3-319-67597-8_11
23. Guo, W., Zheng, B., Duan, T., Fukatsu, T., Chapman, S., Ninomiya, S.: EasyPCC: benchmark datasets and tools for high-throughput measurement of the plant canopy coverage ratio under field conditions. Sensors **17**, 798 (2017). https://doi.org/10.3390/s17040798

KinesiOS: A Telerehabilitation and Functional Analysis System for Post-Stroke Physical Rehabilitation Therapies

Luiz Rogério Scudeletti[1](✉) ⓘ, Alexandre Fonseca Brandão[2,3] ⓘ,
Diego Roberto Colombo Dias[3,4,5] ⓘ, and José Remo Ferreira Brega[5] ⓘ

[1] São Paulo State University, São Paulo, Brazil
rogerio.scudeletti@unesp.br
[2] Gleb Wataghin Institute of Physics, University of Campinas, São Paulo, Brazil
[3] Brazilian Institute of Neuroscience and Neurotechnology, São Paulo, Brazil
[4] Computer Department, Federal University of São João del-Rei, Minas Gerais, Brazil
[5] Interfaces and Visualization Lab., São Paulo State University, São Paulo, Brazil

Abstract. The stroke (also known as a Cerebrovascular Accident) is one of the medical conditions that most kills and incapacitates people in the world, affecting men, women and children of many different age brackets. Studies have been presented in recent years addressing the use of systems for motion capture in post stroke rehabilitation, showing that these tools could be just as efficient as the more traditional methods. In this study, we shall present KinesiOS, a system for telerehabilitation and recognition of movements for the motor and neurofunctional assessment of patients who are undergoing rehabilitation. The system tracks the joints in the human body based on their respective spatial coordinates, and then using the obtained data to construct a guide to movements in the form of a virtual skeleton, while measuring the amplitude of the movements (also known as a Range of Motion) within a certain motor action and showing the results in real time. The tracking of the joints is carried out using a Microsoft Kinect® sensor v2, while data processing, we used the C# programming language. We created the visualizations using the Windows Presentation Foundation® technology, and the data was saved in a cloud structure using the MongoDB® database. Preliminary tests performed on six healthy volunteers showed the efficiency of the system for the calculation of amplitude of movements, enabling data analysis in real time and through telemonitoring. KinesiOS is an alternative tool, portable and low-cost, compared with the traditional systems based on tracking of joints.

Keywords: Telerehabilitation · Motion capture · Kinect v2

This project has been partially supported by Huawei do Brasil Telecomunicações Ltda (Fundunesp Process # 3123/2020).

O. Gervasi et al. (Eds.): ICCSA 2021, LNCS 12950, pp. 174–185, 2021.
https://doi.org/10.1007/978-3-030-86960-1_13

1   Introduction

A stroke, or Cerebrovascular Accident (CVA), is a neurological disorder caused by the interruption of blood supply to the brain, which occurs after either a blockage or a bursting of a blood vessel. This means that oxygen and nutrients that would normally be transported to the brain are suppressed, causing damage to brain tissue, which could bring about cognitive, sensory and/or motor sequelae in the patient [1].

Data released by the World Stroke Organization [2] suggest that 13.7 million people suffer a stroke every year in the world. Out of these, some 5.5 million die as a result, while another 5 million are left with some kind of handicap, which may persist for the rest of their lives.

According to Gillen and Nilsen [3], sessions of physiotherapy and occupational therapy are essential within the process of rehabilitation of patients, as they help the strengthening of the muscles and joints that may have suffered lesions because of the diseases, and increase the chance of recovery, with the partial or total reestablishment of the individual's functional independence. Among the challenges faced by health professionals within the rehabilitation process, we have the difficulty to obtain accurate data about the progress of the patients, and also the monotony of the treatment sessions, which means that some patients undergoing rehabilitation either give up entirely or do not carry out the activities with due attention.

Due to this alarming data and the difficulties encountered, it becomes important to invest in new technologies to help the treatment of stroke. This study proposes the development of a new system for telerehabilitation, by the name of KinesiOS, in a move to help health professionals to make decisions regarding the treatment, as well as getting the patient involved in the execution of rehab activities.

2   Related Works

Brandão et al. [4] prepared a set of software packages for motion capture and virtual reality, aimed at motor and neurofunctional rehabilitation. Known as the Gesture Collection, one of the applications makes use of the Microsoft Kinect® v1 device to establish interactions between gestures of the lower limbs, meaning that the patient may make virtual navigation through cities as previously mapped by Google Street View®, using a stationary gear (simulation of a walk, but without any spatial movement). The application also allows gestural interaction based on the upper limbs, which makes it possible to put together a kind of jigsaw puzzle, with the virtual movement of the pieces with the hands, also allowing the professional to collect data regarding motor actions performed during therapy, helping the professional during diagnosis.

Virtualware's Virtual Rehab® [5] uses Kinect® v1 to track and capture patients' movements, allowing them to be immersed in a 3D environment and interact with the game. The professional defines the structure of the exercise and the game provides visual cues to the patient, eliminating the problem of forgetting the correct way to perform the exercise. In addition, the system will record the execution of movements and provide feedback to the patient and the specialist on the correct or incorrect movements made.

This paper seeks to provide a cheap and simple alternative solution, compared with the traditional systems based on monitoring of joints, using Kinect® v2, a quicker and more accurate device than the previous version, also allowing the monitoring of new body regions, such as the movements of stretching and flexion of the spine, and tweezer movements using the thumbs and the middle fingers. Moreover, the KinesiOS system was devised in a way that would allow remote service, meaning that patient and specialist do not need to be in the same place for the execution of rehabilitation exercises, which is very positive in these days of social distancing, due to the Covid-19 pandemic.

3   General View of the System

The user, positioned at a horizontal distance between 1.8 m and 2.5 m from the input device (Kinect® sensor), carries out the exercises previously entered into the system by the therapist, or exercises related to conventional therapy. The input device, which stands at an altitude of between 0.6 m and 1.2 m (above the ground), monitors the user's joints and then sends the coordinates to the system. This project uses MVC[1] architecture, so data is received first by the Controller, where the information is calculated, and the amplitude of the movements between the segments is calculated. The data is then sent on to the Model, which saves the coordinates and the respective angles in a cloud database. When necessary, the Model recalls the information from the database and then sends this information to the Controller once again, using the coordinates and generating visual feedback to the user in real time, in the application's View.

The View of the application is the reproduction of data generated for the virtual skeletons which are two: one virtual skeleton in green, which makes use of the coordinates recorded by the health worker and that is a kind of reference guide for a certain type of exercise (to be carried out by the patient), and another in red, which shows the movements of the patient, as being carried out, in real time. Figure 1 llustrates the system overview.

4   Materials and Methods

The motion capture takes place through the Kinect® v2 device [6], with an RGB camera, a depth sensor, and a multi-array microphone. To perform the monitoring of joints, the use of markers is not necessary, as the Kinect® projects a grid of infrared dots throughout the space in front, and then captures the

---

[1] MVC (or Model, View and Controller) is a software architecture standard, used for the development of user interfaces, which separates the program logic into three interconnected layers.

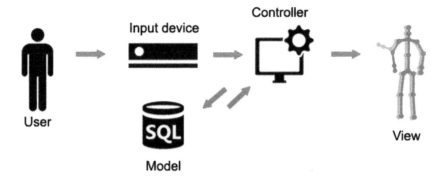

Fig. 1. KinesiOS system overview.

reflections of these points when they collide with the objects, calculating just how long each dot issued takes to be reflected back to the device, detecting joints and generating a flow of data at 30 frames per second.

Each frame has 25 lists of data used to identify the joints at a given point of reference in the 3D body. Each joint is represented by a list of coordinates, X, Y and Z, within the coordinate space of the camera, with X representing the horizontal distance, Y the vertical distance, and Z the depth of the joint in relation to the camera.

For the construction of the system, the C# programming language was used. This is the same language used for the development of the Software Development Kit (SDK) of the Kinect® v2, which ensures a greater compatibility with the device, and allows the use of Windows Presentation Foundation® (WPF) in the construction of layouts and viewings. This means that the development of a system with a modern, responsive and well usable design is now possible, thereby facilitating the use of virtual geometric forms, which shall be used for the construction of virtual skeletons [7].

The data harnessed and calculated by the system are sent to a MongoDB® database, a non-relational database aimed at documents and high performance, which is ideal to work with in this case because of the large volume of data to be generated by the system [8]. This database is available on the cloud, making it possible to access the information anywhere and at any time, requiring only that the professional is registered in the system.

4.1   System Features

The system is divided into two main modules, as shown in the flow chart in Fig. 2 the professional's module and the patient's module.

The professional's module is used by the health professional to record the exercises that shall be used by the patient during his or her rehabilitation sessions. The user must choose the capture plan, between coronal, sagittal and transversal; then select the viewing mode, which can either be 'in depth' or

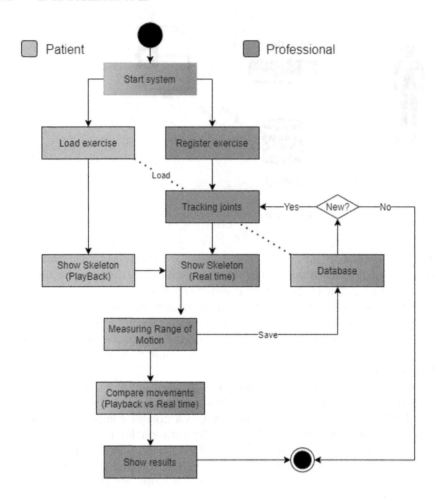

Fig. 2. KinesiOS system usage flow diagram.

'without video', the latter being to preserve the identity of the patient and also to avoid disclosure of the place where the recording is being carried out. Finally, the joints to be monitored shall then be selected. The virtual skeleton and the amplitude of the movements between joints are shown in real time, with the joints being represented by an object shaped like an Ellipse, that shall receive the coordinates X, Y and Z from the list that represents the joint being moni-tored. As this list is constantly being updated with the new position of the joint as the user moves around, the Ellipse also moves according to the new coordi-nates as received, being a kind of guide for the identification of the position of the joint. Figure 3(A) presents an example of the monitoring of the user's right shoulder, in real time of execution.

Whenever two interconnected joints, like the right shoulder and the right elbow, are selected for exhibition, there is a need to draw in the line that connects the two limbs, and for this an object of the Line type is used, having its coordinates X1-Y1 and X2-Y2 connected to the Ellipses of the monitored limbs, thereby giving rise to the virtual skeleton, as shown in Fig. 3(B).

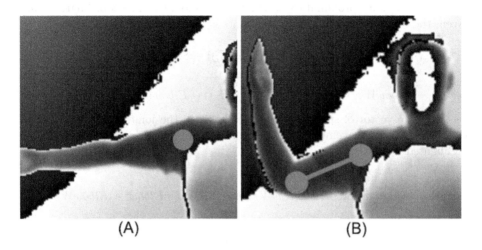

(A)   (B)

Fig. 3. Tracking spatial coordinates of the user's shoulder and right elbow.

To record the exercise, it shall suffice to select the option 'Start Recording', and then the system shall start saving the X, Y and Z coordinates and calculating the amplitude of movement of all the limbs visible on the screen, recording the information in a database in the cloud. Later on, the user may view or exclude the exercises through the 'View Recording' option. The patient's module is very similar to the professional's model, the difference being that the professional person has the option of showing the virtual skeleton as previously recorded together with the virtual skeleton of the patient, generated in real time. This shall be a guide and shall help the recovering individual with the correct execution of the movement.

4.2   Calculation of the Amplitude of Movement

Amplitude of movement (also known as a Range of Motion or ROM) is a term used to define the degree of stretching or flexion that a joint manages to reach in one or more movement plans (depending on the type of joint to be studied). The traditional method used to assess the amplitude of a movement is called goniometry, and it is normally applied by using a goniometer, a kind of ruler with two arms centrally joined by an axis, which allows the movement of the same. In the center of the device there is an inscription in degrees and as the extremities of the joint move, markings parallel to the inscriptions indicate a range of motion.

For the calculation of this amplitude of movement, the KinesiOS system there is always consideration of three joints in the body as monitored by Kinect®. For example, to calculate the amplitude of movement of the elbow, the parameter used is a vector with the position of the elbow; of the joint distal to it, which is the wrist; and the joint proximal to it, which is the shoulder. The coordinates being given, then the distance between the points is calculated, using the formula:

$$ShoulderElbow = Elbow(X_E, Y_E, Z_E) - Shoulder(X_S, Y_S, Z_S) \tag{1}$$

$$ElbowWrist = Wrist(X_W, Y_W, Z_W) - Elbow(X_E, Y_E, Z_E) \tag{2}$$

After the vector of spatial coordinates between the joints is processed, then the amplitude of movement can be found by applying the cosine law [9], as highlighted below:

$$Cos\theta = (ShoulderElbow * ElbowWrist)/(|ShoulderElbow| * |ElbowWrist|) \tag{3}$$

$$\theta = Cos^{-1}(Cos\theta) \tag{4}$$

Figure 4 illustrates the calculation of the right arm angle.

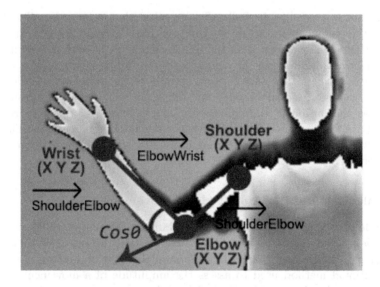

Fig. 4. Calculation of the range of motion of the right arm.

4.3   Telerehabilitation

Telerehabilitation [10] is the delivery of rehabilitation services through telecommunication networks or through the Internet, allowing the professionals to interact with the patients from a remote location, appraising and overseeing the exercises related to rehabilitation therapy. Among the advantages of this technology, we would like to highlight the fact that the patient does not have to go to the rehabilitation centre, and also the possibility that these exercises can be carried out in a residential environment, which could be a factor of motivation and is also a powerful tool at these times when social coexistence is severely restricted (social distancing due to the pandemic of Covid-19).

The system as here proposed uses an online database structure (in the cloud), on servers that can be accessed from remote locations and that receive the data as generated by the patient in real time, making it possible to transfer this data to a virtual skeleton, which reproduces the exercise that the patient is carrying out.

The professional and the patient may communicate by audio, as the system uses a special technology known as 'sockets', which sets up a link between the machines (professional and patient), thus allowing the sending and receiving of audio materials. The only need is for the computers to have microphones and speakers installed [11]. This means that the professional is able to guide and correct the movements made by the patient. In the same way, the patient can solve any doubts that may arise during the rehab sessions. Figure 5 shows the main screen of the KinesiOS system.

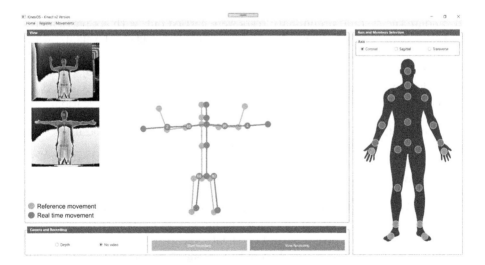

Fig. 5. Motion capture screen of the KinesiOS system. The reference movement is displayed in a green skeleton, while the skeleton in red represents the movement performed in real time by the user. (Color figure online)

4.4 Appraisal of the Exercise

To establish if an exercise has been correctly performed, two different approaches are used: identification on a frame-by-frame basis, and identification using key frames. In frame-by-frame identification, the coordinates of the user's joints are stored frame by frame, from the initial to the final position. This means that, during the execution of the movements, the angles of the joints are compared with frames from the reference exercise.

In the case of identification based on key frames, the user is advised to make the whole movement, but only the initial, middle, and end frames are considered for the appraisal. In both approaches, the professional person gets visual feedback in real time and in the form of a line graph, so that he or she may analyse the movements made, and provide guidance to the patient regarding possible corrections of posture. The execution of the exercise shall be taken as correct if the results regarding the amplitude of movements of the patient's limbs are the same as those of the reference exercise, or if they lie within the percentage of tolerable error, which is set by the professional, according to the phase of the therapy or the extent of motor impairment shown by the patient.

Apart from visual feedback, if necessary, the professional may also export the coordinates and the calculations of amplitude for all limbs monitored, to a file in XLS format, thereby permitting the use of the data in external tools, for the construction of graphs and statistical analyses.

5 Preliminary Results and Discussion

To validate the system, preliminary tests were carried out, with six participants aged between 25 and 60, men and women of different heights and weights, without any limitations on performing the chosen activities. The test sessions were carried out in a controlled environment, well lit, and without any objects blocking the field of vision of the input device.

The flow of the test comprised demonstration, training, and execution.

- Demonstration: The operation of the system is presented to the users, who are instructed about how to interact with the system;
- Training: The execution of the exercises is shown to the users; and
- Execution: The users carry out the exercises again, without any kind of help.

Figure 6 shows the results obtained on carrying out exercises for abduction and adduction of the left shoulder, where the movement starts with the arm close to the body (anatomic position) and ends with the arm at shoulder height and parallel to the ground. The initial amplitude of all users was close to 20°, while the final amplitude was around 100°, which is within the parameters considered as correct, when compared to the traditional goniometer calculation method [12].

For a second test, the data regarding amplitude of movement of a user during interaction with an application was collected, to allow the pieces of a virtual jigsaw to be put together, based on interaction with the upper limbs.

LEFT SHOULDER FLEXION RANGE OF MOTION

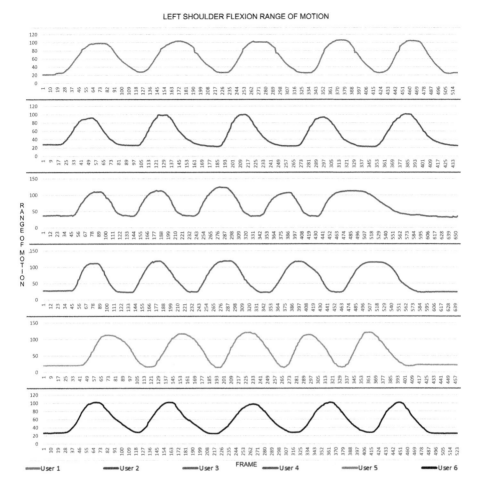

Fig. 6. Results of the amplitude of movements the left shoulder when performing the flexion exercise.

The exercise carried out when assembling this jigsaw needs movements of horizontal abduction and adduction of the right arm, and data of amplitude of movement were recorded, in relation to the right shoulder joint. In the graph highlighted in Fig. 7 one can identify the very moment when each piece of the jigsaw is placed. The pieces were separated in twelve parts, initially positioned on the left side of the application's interface, so they were controlled based on a virtual right hand (that, after monitoring of gestures, responded to the movements of the user's real hand).

This means that when the user made a move of horizontal abduction with the right arm (to pick up one of the pieces), the amplitude of movement was larger; when the user made a movement of horizontal adduction (to fit the piece in the grid) then the amplitude of movement was much less, as the reference

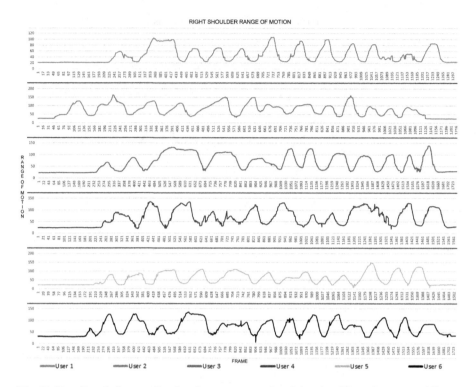

Fig. 7. Results of the amplitude of movements the right shoulder when assembling a virtual puzzle.

point for the start of calculation of the amplitude of movement was positioned on the midline of the user's body.

6   Conclusions

Applications for motion capture based on devices with infrared sensors, without markers, have been very widely used in supporting rehabilitation of patients who have had a stroke. However, the process of acknowledgement is still highly complex, especially when all the frames of a given exercise need analysis, as there is a need for a high processing cost, so that results may be generated in real time.

In this work, we seek to help the patient correct his or her own posture (bodily awareness) with the visual feedback given in real time, and supply data that could help the professional in application of the appropriate treatment, either in local or remote environment. Initial tests with healthy users were successfully conducted, showing that the application is effective, simple to use, and flexible, as it adjusts itself automatically to anatomic structures of various sizes and low costs, which makes the project feasible.

The following stages of the project shall include tests with more complex movements, also seeking to assess whether the results presented shall continue to be consistent, and if they shall allow the measurement of the progress of patients. There shall also be an assessment of how the system behaves in different environments, with different lighting, objects on the scene, and movements with occlusion of limbs, in order to gauge the limitations of the application. In addition, other types of feedback in real time, such as the speed of movement and the detection of gestures, can be implemented to give the professional more information, supporting the professional in making decisions.

References

1. Torbey, M.T., Selim, M.H.: The Stroke Book, 2nd edn, pp. 47–73. University Press, Cambridge (2013)
2. World Stroke Organization. Annual Report 2019. https://www.world-stroke.org/about-wso/annual-reports. Accessed 21 Apr 2021
3. Gillen, G., Nilsen, D.M.: Stroke Rehabilitation - A Function-Based Approach, 5th edn. Mosby Press, New York (2020)
4. Brandão, A.F., Dias, D.R.C., Castellano, G., Parizotto, N.A., Trevelin, L.C.: RehabGesture: an alternative tool for measuring human movement. Telemedicine And e-health - Mary Ann Liebert Inc. **22**(7), 584–589 (2016)
5. Virtualware, VirtualRehab System. https://evolvrehab.com/virtualrehab/. Accessed 21 Apr 2021
6. Rahman, M.: Beginning Microsoft Kinect for Windows SDK 2.0: Motion and Depth Sensing for Natural User Interfaces. Apress Press, New York (2017)
7. Yuen, S.: Mastering Windows Presentation Foundation: Build responsive UIs for desktop applications with WPF, 2nd edn. Packt Publishing, London (2020)
8. Bradshaw, S., Brazil, E., Chodorow, K.: MongoDB: The Definitive Guide: Powerful and Scalable Data Storage, 3rd edn, pp. 33–34. O'Reilly Media, California (2019)
9. Neill, H.: Trigonometry: A Complete Introduction: The Easy Way to Learn Trig, pp. 39–44. Teach Yourself, London (2018)
10. Kumar, S., Cohn, E.R.: Telerehabilitation (Health Informatics), pp. 1–11. Springer, New York (2013). https://doi.org/10.1007/978-1-4471-4198-3
11. Makofske, D., Donahoo, M.J., Calvert, K.L.: TCP/IP Sockets in C# - Practical Guide for Programmers, pp. 15–37. Morgan Kaufmann Publishers, California (2005)
12. Norkin, C.C., White, D.J., Calvert, K.L.: Measurement of Joint Motion - A Guide to Goniometry, 5th edn, pp. 66–91. F. A. Davis Company, Pennsylvania (2016)

Breast Fine Needle Cytological Classification Using Deep Hybrid Architectures

Hasnae Zerouaoui[2], Ali Idri[1,2(✉)], Fatima Zahrae Nakach[1], and Ranya El Hadri[1]

[1] Software Project Management Research Team, ENSIAS, Mohammed V University, Rabat,
Morocco
ali.idri@um5.ac.ma
[2] Modeling, Simulation and Data Analysis, Mohammed VI Polytechnic University, Benguerir,
Morocco

Abstract. Diagnosis of breast cancer in the early stages allows to significantly decrease the mortality rate by allowing to choose the adequate treatment. This paper develops and evaluates twenty-eight hybrid architectures combining seven recent deep learning techniques for feature extraction (DenseNet 201, Inception V3, Inception ReseNet V2, MobileNet V2, ResNet 50, VGG16 and VGG19), and four classifiers (MLP, SVM, DT and KNN) for binary classification of breast cytological images over the FNAC dataset. To evaluate the designed architectures, we used: (1) four classification performance criterias (accuracy, precision, recall and F1-score), (1) Scott Knott (SK) statistical test to cluster the developed architectures and identify the best cluster of the outperforming architectures, and (2) Borda Count voting method to rank the best performing architectures. Results showed the potential of combining deep learning techniques for feature extraction and classical classifiers to classify breast cancer in malignant and benign tumors. The hybrid architectures using MLP classifier and DenseNet 201 for feature extraction were the top performing architectures with a higher accuracy value reaching 99% over the FNAC dataset. As results, the findings of this study recommend the use of the hybrid architectures using DenseNet 201 for the feature extraction of the breast cancer cytological images since it gave the best results for the FNAC data images, especially if combined with the MLP classifier.

Keywords: Computer-aided diagnosis · Breast cancer · Classification · Deep convolutional neural networks · Image processing · Histological images · Hybrid architecture

1 Introduction

Breast cancer (BC) is the most commonly diagnosed cancer for women worldwide with an estimate of 2.3 million new cases in 2020 [1, 2]. It remains a global challenge, since it is the fifth leading cause of mortality worldwide, and the second cause of death for women globally [2, 3]. As the number of patients affected by the BC disease increases, it turns out to be hard for radiologists, pathologists and oncologists to accurately deal with the diagnosis process in the constrained accessible time [4]. Medical images analysis

© Springer Nature Switzerland AG 2021
O. Gervasi et al. (Eds.): ICCSA 2021, LNCS 12950, pp. 186–202, 2021.
https://doi.org/10.1007/978-3-030-86960-1_14

is one of the most promising research areas, and it provides facilities for diagnosis and decision making of several diseases such as breast cancer. Lately, more attention are paid to imaging modalities and deep learning (DL) in BC [5–7]. Therefore, interpretation of these images requires expertise and consequently several techniques have been developed and evaluated to improve and help oncologist's diagnosis. In general, deep learning techniques showed better performance in breast cancer detection by extracting the most important features of medical images, and provided high accurate features compared with classical features extraction techniques of the image processing process [6–9]. In the other hand, classical machine learning techniques gives accurate results for BC classification, are less time consuming and need fewer parameter tunning compared to the ones of the DL techniques [6, 10]. For instance, many researchers designed hybrid architectures where they combined the strengths of DL techniques for feature extraction and classical machine learning for classification [11–13]. For reference, the study [10] showed that the use of hybrid architectures combining deep CNN architectures and recurrent neural networks (RNNs) gave better results compared to CNN and deep CNN; indeed, the proposed solution combined GoogleNet and AlexNet for features extraction and LSTM (long short-term memory) for classification over the BreakHis dataset, and it achieved an accuracy of 90.5%. In [14], the authors designed a hybrid architecture using AlexNet and VGG16 DL techniques for feature extraction over the BreakHis dataset; once the features extracted, they concatenated feature vectors and apply the SVM classifier for a binary classification, the best accuracy results achieved using AlexNet and VGG16 for feature extraction and SVM for classification are: 84.87%, 89.21%, 8.65% and 86.75% for the MF 40X, 100X, 200X and 400X respectively.

This paper develops and evaluates twenty-eight hybrid architectures using four classifiers (MLP, SVM, DT and KNN) and seven of the most recent DL techniques as feature extractor (DenseNet 201, Inception V3, Inception ReseNet V2, MobileNet V2, ResNet 50, VGG16 and VGG19) for a binary BC classification since [7]: (1) they are frequently used for diagnosing breast cancer patients since 2016, and (2) they proved their strength for FE compared to the classical ones; the designed architectures are tested the FNAC dataset. We evaluate the performance of the designed hybrid architectures in terms of accuracy, precision, recall and F1-score. To the best of our knowledge, this study is the first to evaluate and compare twenty-eight hybrid architectures using seven DL techniques for feature extraction and four classifiers and uses the Scott Knott (SK) statistical test and Borda count voting method in BC image processing and classification. Note that the SK test has been widely used to comparing, clustering and ranking multiple machine learning models for parameters tunning [18, 19] in different fields such as software engineering [7, 8] and breast cancer [22]. Hence, we use the SK test since: (1) its high performance compared to other statistical tests such as Jollife [21], Calinski and Corsten [24], and Cox and Spjotvoll [25], and (2) its ability to cluster the best non-overlapping groups of machine learning techniques. Furthermore, we use the Borda Count voting method [14, 15] to rank the best SK hybrid architectures. The present study discusses three research questions (RQs):

- (RQ1): What is the overall performance of the hybrid architectures in BC binary classification?

- (RQ2): Is there any deep learning technique for feature extraction which distinctly outperformed the others when used in the hybrid architectures constructed for each classifier?
- (RQ3): Is there any hybrid architectures which distinctly outperformed the others regardless the feature extractor technique and the classifier used?

The main contributions of this empirical study are the following:

1. Designing twenty-eight hybrid architectures using four classifiers: MLP, SVM, DT and KNN, and seven DL architectures for feature extraction (FE): VGG16, VGG19, DenseNet201, Inception ResNet V2, Inception V3, ResNet 50 and MobileNet V2 in BC binary classification.
2. Evaluating the twenty-eight hybrid architectures over the FNAC dataset.
3. Comparing the performances of the twenty-eight hybrid architectures using SK clustering test and Borda Count voting method.

The rest of this paper is organized as follow. Section 2 presents the data preparation which includes data acquisition, image pre-processing and data augmentation. Section 3 presents the empirical methodology followed in this research. Section 4 reports and discusses the empirical results. Section 5 presents the threats of validity of the study. Section 6 outlines conclusions and future works.

2 Data Preparation

This section presents the data preparation process we followed for the FNAC dataset, which consists of data acquisition, data pre-processing and data augmentation.

2.1 Data Acquisition

In this subsection, we present the publicly available Fine Needle Aspiration Cytology (FNAC) dataset used in all the empirical evaluations. The images of the FNAC dataset were captured by using Leica ICC50 HD microscope using 400 resolution and 24 bits color depth and with 5 megapixels camera associated with the microscope [33]. Digitized images captured were then reviewed by experienced certified cyto-pathologists and selected a total of 212 images (113 Malignant and 99 Benign). The database can be downloaded from the link in [33].

2.2 Data Processing

In the subsection, we present the image processing we used in this study for the FNAC dataset. The first and most important step of the data preparation is to preprocess input images using different pre-processing techniques [34]. In this study, we used intensity normalization and Contrast Limited Adaptive Histogram Equalization (CLAHE). Intensity normalization is an interesting pre-processing step in image processing applications

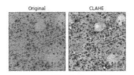

Fig. 1. (a). Original image, (c). CLAHE

and is highly used to preprocess medical images [6, 13, 35]. As shown in Fig. 1, we normalized input images to the standard normal distribution using min-max normalization of Eq. 1, and we improved the contrast using the CLAHE technique [32, 36].

$$X_{norm} = \frac{x - x_{\min}}{x_{\max} - x_{\min}} \tag{1}$$

2.3 Data Augmentation

Data augmentation was used for the training process after dataset pre-processing and has the goal to avoid the risk of overfitting [37]. Moreover, the strategies we used include geometric transforms such as rescaling, rotations, shifts, shears, zooms and flips as shown in Table 1.

Table 1. Data augmentation used

Argument	Parameter value	Description
Rescale	1/255.0	Scale images from integers 0–255 to floats 0–1
Rotation range	90	Degree range of the random rotations
Vertical shift range	0.2	The parameter value of horizontal and vertical shifts
Horizontal shift range	0.2	(20%) is a fraction of the given dimension
Shear range	0.2	Controls the angle in counter clockwise direction as radians in which our image will allowed to be sheared
Zoom range	0.2	Allows the image to be "zoomed out" or "zoomed in"
Horizontal flip	True	Controls when a given input is allowed to be flipped horizontally during the training process
Fill mode	Nearest	This is the default option where the closest pixel value is chosen and repeated for all the empty values

3   Empirical Design

This section presents the empirical design we followed in this study including: (1) performance criteria we used to evaluate the models and 10-fold cross validation to evaluate the hybrid architectures, (3) Skott Knott statistical test we used to cluster the hybrid architectures according to their accuracy values, (4) Borda Count method we used to rank the hybrid architectures of the best SK cluster according to accuracy, precision, recall, and F1-core, (5) experimental process we followed to carry out all the empirical evaluations, (6) the abbreviations we used to shorten the names of the hybrid architectures developed, and (7) the experiment configuration of the classifiers and DL feature extractors.

3.1 Performance Measures

To evaluate the performance of the Hybrid techniques, we used the followed performance measures [6]: accuracy, precision, recall and F1-score, defined in the Eq. (2), (3), (4) and (5).

$$Accuracy = \frac{TP + TN}{TN + TP + FP + FN} \tag{2}$$

$$Precision = \frac{TP}{TP + FP} \tag{3}$$

$$Recall = \frac{TP}{TP + FN} \tag{4}$$

$$F1 = 2 \times \frac{Recall \times precision}{Recall + precision} \tag{5}$$

As for the evaluation, we used 10-fold cross validation since it is easy to understand and gives better performance results compared to other evaluation techniques [71–73]. We reported the average of the performance of each of the 10 folds.

3.2 Statistical Tests

Scott Knott: The Scott-Knott (SK) algorithm is an exploratory clustering algorithm usually used in the analysis of variance (ANOVA) context. It was proposed by Scott and Knott in 1974 [38] to find distinction overlapping groups based on the multiple comparisons of treatment means. The SK algorithm is the most frequently used hierarchical clustering algorithm due to its simplicity and robustness [27–29]. In this study the SK test was performed based on the accuracy performance measure.

Borda Count: Borda count is a voting method for single winner election methods. In this technique, points are given to candidates based on their ranking; 1 point for last choice, 2 points for second-to-last choice, and so on until you are at the top. The point values for all ranks are totaled, and the candidate with the largest point total is the winner [26]. In the present study, we used Borda count technique to find the best performing hybrid architectures from the four performance measures with equal weights. This strategy was adopted to make sure that we do not favor a measure. This voting technique is widely used in SDEE, indeed [20, 30, 31], deduced that the conclusion on which a technique is the most performing depends on the chosen performance measures that may lead to contradictory results: for instance each technique may have different voting according to different performance measures. The borda count was used in this study based on four performance metrics: accuracy, precision, recall and F1-score.

3.3 Experimental Process:

Figure 2 shows the methodology we followed to carry out the empirical evaluations of this experiment. It consists of five steps described hereafters. Note that similar methodologies were used in [8, 9, 35, 36, 37]. The evaluation process involves:

1. Assess the accuracy of each variant of the four classifiers (MLP, SVM, DT, KNN) using the seven deep learning architectures as feature extractors (VGG16, VGG19, DensNet201, MobileNet_V2, ResNet50, Inception_V3, Inception_ResNet_V2) for FNAC dataset.
2. Clustering the four classifiers (MLP, SVM, DT, KNN) using Scott Knott test based on accuracy for both datasets and select the best hybrid architecture of the best SK cluster (have the best accuracy and statistically indifferent).
3. Rank the classifiers of the best SK cluster using Borda count voting system based on the four performance measures (accuracy, precision, recall and F1-score) and select the top hybrid architecture for each dataset.
4. Select the best ranked hybrid architectures for each classifier from step three and apply SK test based on accuracy.
5. Rank the best cluster of step four using Borda count based on the four performance metrics: accuracy, precision, recall and f1-score.

Fig. 2. Experimental process

3.4 Abbreviation

In order to shorten the names of the hybrid architectures, we use the following naming rules in the rest of this paper: *Classifier Deep Learning Architecture*

We shorten the name of the classifiers as follow: M for MLP, S for SVM, D for DT and K for KNN. And for the deep learning architectures we choose: V16 for VGG16, V19 for VGG19, IN for Inception V3, INR for InceptionResNet V2, Res for ResNet 50, MOB for MobilNet V2 and DEN for DensNet 201.

3.5 Experiment Configuration

To design the hybrid architectures using four classifiers (MLP, SVM, DT and KNN) and seven DL techniques for FE (VGG16, VGG19, Inception V3, DenseNet 201, ResNet 50, Inception ResNet V2 and MobileNet V2) for both datasets we used the following experiment configurations:

- All the images of the FNAC dataset were resized to 224 × 224 pixels except those of Inception_V3 and Inception_ResNet_V2 models that were resized to 299 × 299 since it is the default input size in their architectures.
- To extract the features of the input images we used the transfer learning [38] technique for FE where we downloaded the seven DL techniques pre-trained in the ImageNet dataset from the Keras library, we froze the top layers and generated a new dataset containing the features of FNAC dataset that will be the input of the four classifiers (MLP, SVM, DT and KNN).
- For binary classification we used the default configuration of the four classifiers (MLP, SVM, DT and KNN) given by the scikit-learn library of python programming language.

4 Results and Discussion

This section presents and discusses the results of the empirical evaluations of the hybrid architectures over FNAC dataset. We firstly evaluate the performances in terms of accuracy of the hybrid architectures of each classier (RQ1). Thereafter, we evaluate the impacts of the seven DL feature extraction techniques on the performances of the four classifiers in order to identify which one of them are positively influencing the classification performance (RQ2). Lastly, we compare the best hybrid architectures of the four classifiers in order to identify which hybrid architecture performed better over the FNAC dataset.

4.1 (RQ1): What is the Overall Accuracy Performance of the Hybrid Architectures in a BC Binary Classification?

To evaluate and compare hybrid architectures, we assess the accuracy values for each classifier. Note that the empirical evaluations of the hybrid architectures were performed using a computer with Processor: Intel (R) core (TM) i7-7500 CPU @ 2.90 GHZ and 12 Go in RAM running on a Microsoft Windows 10 Professional (64-bit). Python 3.7.4 was used with the two DL frameworks Keras and Tensorflow as deep learning backend, and the scikit-learn framework for machine learning. R 3.4.4 was used for statistical evaluation.

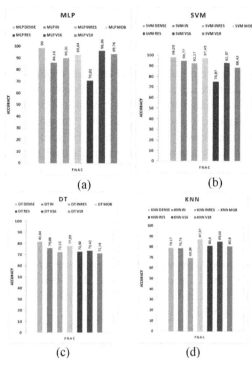

Fig. 3. Accuracy values for the twenty-eight hybrid architectures of the FNAC dataset

This subsection presents and discusses the performances of the hybrid architectures of each classifier over the FNAC dataset. Table 2 summarizes the testing accuracy values. Figures 3 shows the mean accuracy values of the 4 classifiers depending on the DL feature extractor used.

We observe that:

- For the hybrid architectures using MLP classifier the best accuracy value was obtained when using the DenseNet 201 for FE and it reached 99% and the worst accuracy value when using ResNet 50 for FE was 70.92%.
- For the hybrid architectures using the SVM classifier, the best accuracy value was obtained when using DenseNet 201 for FE and reached 98.05% and the worst accuracy value when using ResNet 50 for FE was 74.97%.
- For the hybrid architectures using the DT classifier, the best accuracy value was obtained when using DenseNet 201 for FE and reached 81.64% and the worst accuracy value when using VGG 19 for FE was 71.18%.
- For the hybrid architectures using the KNN classifier the best accuracy value was obtained when using MobileNet V2 for FE and reached 87.37% and the worst accuracy value when using Inception Inception ResNet V2 for FE was 69.54%.

4.2 (RQ2): Is There Any Deep Learning Technique for Feature Extraction Which Distinctly Outperformed the Others When Used in a Hybrid Architecture?

This section aims at evaluating the impacts of the seven DL feature extraction techniques on the performances of the four classifiers in order to identify (if exist) the feature extraction techniques that are positively impacting the classification performance. To this aim, (1) we used the SK test based on the accuracy values of the hybrid architectures of each classifier in order to cluster the techniques that have the same predictive capabilities regardless the feature extractor used, and (2) we used the Borda count method based on the four-performance metrics: accuracy, precision, recall and F1-score to rank the hybrid architectures of each classifier belonging to the best SK clusters over both the FNAC dataset. Tables 2 shows the values of the four performance measures of all the hybrid architectures. Figure 4 shows the results of SK test over the FNAC dataset. We observe that:

- For MLP, we obtained 2 clusters which implies that regardless the DL feature extractor technique used, we obtained statistically similar results in terms of accuracy. The best SK cluster contains 6 hybrid architectures using MLP with DenseNet 201, VGG16, VGG19, MobileNet V2, Inception Resnet V2, or Inception V3 for feature extraction. The last cluster contains MLP with ResNet 50.
- For SVM, we obtained 5 clusters which implies that the SVM classifier is impacted by the DL feature extractor used. The best SK cluster contains 2 hybrid architectures: SVM with DenseNet 201 or MobileNet V2. The second SK cluster contains SVM with Inception V3 as feature extractor. The third cluster, we find the architectures constructed by SVM with the DL techniques VGG16 and Inception Resnet V2. The fourth cluster includes the hybrid architecture SVM with VGG19. The last cluster contains SVM with ResNet 50 as feature extractor.
- For DT, the best cluster contains one architecture: DT with DenseNet 201. The second cluster contains the architecture combining DT and MobileNet V2. In the third cluster we have the architecture combining DT and Inception V3. For the last cluster we have the hybrid architectures constructed by the DT classifier and VGG16, ResNet 50, Inception Resnet V2 and VGG19 for feature extraction.

For KNN, 3 clusters where identified: the best SK cluster containing 2 hybrid architectures: KNN with MobileNet V2 or VGG16. The second SK cluster contains the architectures KNN with ResNet 50, VGG 19, DenseNet 201 or Inception V3. The last cluster contains KNN with Inception Resnet V2 as feature extractor.

Table 3 shows the Borda count ranks of the best SK clusters using the four classifiers over the FNAC dataset. As can be seen, the hybrid architectures designed with DensNet 201 were ranked first regardless the classifier, except for the KNN classifier. The second feature extractor is MobileNet V2 since it was ranked first with KNN, second using SVM and forth with MLP. As for the feature extractor VGG16, it was ranked second for both MLP and KNN. For the feature extractors VGG19, Inception V3 and InceptionResNet V2 they only belong to the best cluster with the MLP classifier and were ranked third, fifth and sixth respectively. Finally ReseNet 50 feature extractor underperformed compared to the others. In order to rank the impacts of each feature extractor on the classification

Table 2. Accuracy, precision, recall and F1 score values of the hybrid architectures over the FNAC dataset

Classifier	Hybrid architectures	Accuracy	Precision	Recall	F1-score
MLP	MDEN	0.99	0.9956	0.9849	0.9902
	MIN	0.8616	0.8166	0.8528	0.8259
	MINR	0.9032	0.8675	0.8355	0.8503
	MMOB	0.9305	0.8803	0.8714	0.8756
	MRES	0.7092	0.6276	0.644	0.6039
	MV16	0.9636	0.9666	0.9585	0.9623
	MV19	0.9377	0.953	0.9172	0.9343
SVM	SDEN	0.9805	0.9886	0.9729	0.9806
	SIN	0.9478	0.9563	0.936	0.9458
	SINR	0.9217	0.9489	0.8877	0.9168
	SMOB	0.9746	0.9829	0.965	0.9738
	SRES	0.7498	0.8494	0.5898	0.6938
	SV16	0.9257	0.9481	0.8982	0.9221
	SV19	0.8843	0.9304	0.8263	0.8739
DT	DDEN	0.8164	0.8379	0.8139	0.8247
	DIN	0.7589	0.7517	0.7606	0.7558
	DINR	0.7213	0.7235	0.6979	0.7086
	DMOB	0.7793	0.7692	0.7857	0.7761
	DRES	0.7258	0.7186	0.7225	0.7186
	DV16	0.7343	0.7376	0.7051	0.7201
	DV19	0.7118	0.7076	0.696	0.7011
KNN	KDEN	0.7918	0.9148	0.6684	0.7713
	KIN	0.7875	0.9791	0.58	0.7273
	KINR	0.6954	0.8941	0.4254	0.5746
	KMOB	0.8737	0.9925	0.7475	0.852
	KRES	0.8091	0.8756	0.7108	0.7826
	KV16	0.8504	0.9115	0.769	0.8332
	KV19	0.8081	0.88	0.7017	0.78

performance regardless the classifier, we count the number of occurrences of each feature extractor on the best SK clusters. In case of a tie, we refer to the borda count ranking. As can be seen in Table 4, the best performing feature extractor is DenseNet 201, since it appears 3 times in the best SK cluster and was ranked 3 times first. The following feature extractor is MobileNet V2, it appears 3 times in the best SK cluster and was ranked first, second and forth. As for the remaining DL technique VGG19, Inception V3 and Inception ReseNet V2 they appeared 1 time in the best SK cluster. Finally, Resnet 50 feature extractor underperform compared the others, and never appear in the best SK cluster. As for the sensitivity of the classifiers to the feature extraction technique used, we notice that MLP is less sensitive compared to the others. In fact, the numbers of SK clusters of MLP are 2, while those of KNN are 3, for DT 4 and finally 5 for SVM.

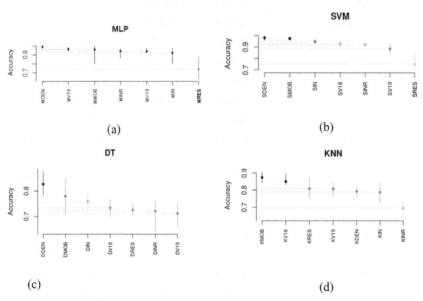

Fig. 4. Results of SK test for the hybrid architecture using the four classifiers over the FNAC dataset

Table 3. Borda count ranking fort the hybrid architectures of the best SK cluster

Classifier ranking	MLP	SVM	DT	KNN
1	MDEN	SDEN	DDEN	KMOB
2	MV16	SMOB		KV16
3	MV19			
4	MMOB			
5	MINR			
6	MIN			

Table 4. Appearance in the best SK cluster and borda count ranking

DL for feature extraction	Appearance in the best SK cluster	Borda count ranking
DenseNet 201	3	3 times first rank
MobileNet V2	3	1 time first rank 1 time second rank 1 time forth rank
VGG16	2	2 times second rank
VGG19	1	Third rank
ReseNet 50	–	–
Inception V3	1	Sixth rank
Inception ReseNet V2	1	Fifth rank

4.3 (RQ3): Is There Any Hybrid Architectures Which Distinctly Outperformed the Others Regardless the Feature Extractor and the Classifier Used?

This Section uses the SK test based on accuracy to evaluate the predictive capabilities of the best hybrid architectures of the four classifiers over each dataset (Step 4 of the empirical design). Thereafter, we discuss the ranking results of the borda count voting method based on accuracy, precision, recall and F1-score on the best SK clusters to identify the best hybrid architecture regardless the classifier over the FNAC dataset. Figure 5 shows the results of SK test over the best ranked hybrid architecture of the FNAC dataset regardless the classifier used. We obtained three clusters. The best SK cluster contains the two hybrid architectures MLP and SVM with DenseNet 201. The second SK cluster contains the KNN classifier with MobileNet V2. And the last cluster contains DT with DenseNet.

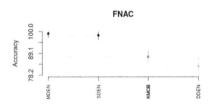

Fig. 5. Results of SK test for the best ranked hybrid architectures over the FNAC dataset

Using the Borda count voting methods on the best SK cluster, we observe from Table 5 that the hybrid architectures MLP with DenseNet 201 was ranked first, and the hybrid architecture SVM with DenseNet 201 was ranked second. As regard we conclude that MDEN hybrid architecture outperforms the others.

Table 5. Borda count ranking for the best SK cluster oft he best ranked hybrid architectures constructed over the FNAC dataset

Rank	FNAC dataset hybrid architectures
1	MDEN
2	SDEN

5 Threats of Validity

This section presents the threats to this paper's validity with respect to external and internal validity.

Internal Validity: This paper used the 10-fold cross validation method for evaluation. The main reason to use the 10-fold cross validation is that cross validation gives more stable estimations [39]. Another internal threat for this experiment is the use of transfer learning for feature extraction while using the DL techniques to extract the features of the images of the FNAC dataset.

External Validity: This study used one dataset which contain cytological images; so, we cannot generalize the obtained results for all the datasets with the same type of images and the same features. However, it will be a good benefit to redo this study using different DL techniques with another publicly or private datasets in order to confirm or refute the findings of this study.

Construct Validity: For the reliability of the classifier performances obtained, this study focused on the accuracy and other four performance criteria (accuracy, precision, recall and F1-score). The main reasons behind the choice of these performance criteria are: (1) most of the studies used them to measure the classification performance [7], and (2) the type of the data is balanced. Moreover, the conclusion was drawn by using the SK test and Borda count voting system with equal weights using these four performance criteria (accuracy, precision, recall and F1-Score). This strategy was adopted to make sure that we do not favor a particular performance criterion than another.

6 Conclusion and Future Work

The present paper presented and discussed the results of an empirical comparative study of 28 hybrid architectures using four classifiers (MLP, SVM, DT, and KNN) and seven DL techniques for feature extraction (DensNet 201, MobileNet V2, ReseNet 50, Inception V3, Inception ReseNet V2, VGG16 and VGG19), for BC Imaging classification. All the empirical evaluations used four performance criteria (accuracy, precision, recall and F1-score), SK statistical test, and Borda Count to assess and rank these 28 hybrid architecturs over the FNAC database. The main findings of this study are:

(RQ1): What is the Overall Performance of the Hybrid Architectures in BC Binary Classification?

The accuracy results of the constructed hybrid architectures were highly influenced by the DL techniques used for feature extraction. As regard, we observed that the use of DenseNet 201 for FE in the design of the hybrid methods almost gave the best results and outperformed the others regardless the classifier used in this experiment. However, ResNet 50 for FE underperformed the others regardless the classifiers used in the hybrid architectures. In respect we recommend the use of DenseNet DL technique for feature extraction for the design of hybrid architectures regardless the classifier used.

(RQ2): Is There Any Deep Learning Technique for Feature Extraction Which Distinctly Outperformed the Others When Used in the Hybrid Architectures Constructed for Each Classifier?
The hybrid architectures with DenseNet 201 for feature extraction regardless the classifier used gave the best results since they always belong to the best SK cluster for the FNAC dataset and was ranked first. Furthermore, the architectures using MobileNet V2 belong three times to the best SK followed by the architectures using VGG16 that belongs two times to the best SK cluster. Regardless the classifier, the architectures using Resnet 50 for FE never belongs to the best SK cluster and is underperforming compared to the others. As result, we highly recommend the use of DenseNet 201 to design the hybrid architectures since it gives the best results.

(RQ3): Is There Any Hybrid Architectures Which Distinctly Outperforms the Others Regardless the Feature Extractor Technique and the Classifier Used?
Step four and five of the empirical design have allowed us to distinguish the most outperforming constructed hybrid architecture regardless the classifier and for FNAC datasets. In this sense the MDEN hybrid architecture constructed using the MLP classifier and DenseNet 201 DL technique for feature extraction gave the best results since it belongs to the best cluster and was ranked first using the borda count method. The accuracy values achieved by the MDEN hybrid architecture are: 99% for the FNAC dataset.

In the future work, we are planning to investigate other public dataset to validate or refute the finding of this study. Another perspective for our research is to build homogenous and heterogeneous ensembles using different DL techniques for feature extraction for breast cancer imaging classification.

Acknowledgement. This work was conducted under the research project "Machine Learning based Breast Cancer Diagnosis and Treatment", 2020–2023. The authors would like to thank the Moroccan Ministry of Higher Education and Scientific Research, Digital Development Agency (ADD), CNRST, and UM6P for their support.

This study was funded by Mohammed VI polytechnic university at Ben Guerir Morocco.

Abbreviation

In order to shorten the names of the hybrid architectures We use the following naming rules in the rest of this paper: Classifier DeepLearningArchitecture

References

1. Metelko, Z., et al.: Pergamon the world health organization quality of life (WHOQOL): position paper from WHO. Soc. Sci. Med. **41**(10), 1403–1409 (1995)
2. Sung, H., et al.: Global cancer statistics 2020: GLOBOCAN estimates of incidence and mortality worldwide for 36 cancers in 185 countries. CA Cancer J. Clin. **71**(3), 1–41 (2021)
3. Bish, A., Ramirez, A., Burgess, C., Hunter, M.: Understanding why women delay in seeking help for breast cancer symptoms. J. Psychosom. Res. **58**(4), 321–326 (2005)
4. Zhang, G., Wang, W., Moon, J., Pack, J.K., Jeon, S.I.: A review of breast tissue classification in mammograms. In: Proceedings of the 2011 ACM Symposium on Research in Applied Computation RACS 2011, pp. 232–237 (2011)
5. Mendelson, E.B.: Imaging: potentials and limitations. Am. J. Roentgenol. **212**(2), 293–299 (2019). https://doi.org/10.2214/AJR.18.20532
6. Zerouaoui, H., Idri, A.: Reviewing machine learning and image processing based decision-making systems for breast cancer imaging. J. Med. Sys. **45**(1), 1–20 (2021)
7. Zerouaoui H., Idri A., El Asnaoui K.: Machine Learning and Image Processing for Breast Cancer: A Systematic Map. In: Rocha Á., Adeli H., Reis L., Costanzo S., Orovic I., Moreira F. (eds.) Trends and Innovations in Information Systems and Technologies. WorldCIST 2020. Advances in Intelligent Systems and Computing, vol. 1161. Springer, Cham (2020). https://doi.org/10.1007/978-3-030-45697-9_5
8. Idri, A., Chlioui, I., El Ouassif, B.: A systematic map of data analytics in breast cancer. In: International Conference Proceedings Series (2018)
9. Ouassif, E., Idri, A., Hosni, M., Abran, A.: Classification techniques in breast cancer diagnosis: a systematic literature review. Comput. Methods Biomech. Biomed. Eng. Imag. Vis. **9**(1), 50–77 (2000)
10. Yan, R. et al.: A hybrid convolutional and recurrent deep neural network for breast cancer pathological image classification. In: International Conference on Bioinformatics and Biomedicine (BIBM 2018), pp. 957–962 (2019)
11. Mendel, K., Li, H., Sheth, D., Giger, M.: Transfer learning from convolutional neural networks for computer-aided diagnosis: a comparison of digital breast tomosynthesis and full-field digital mammography. Acad. Radiol. **26**(6), 735–743 (2019)
12. Valueva, M.V., Nagornov, N.N., Lyakhov, P.A., Valuev, G.V., Chervyakov, N.I.: ScienceDirect application of the residue number system to reduce hardware costs of the convolutional neural network implementation. Math. Comput. Simul. **177**, 232–243 (2020)
13. Cordeiro, F.R., Santos, W.P., Silva-Filho, A.G.: A semi-supervised fuzzy GrowCut algorithm to segment and classify regions of interest of mammographic images. Expert Syst. Appl. **65**, 116–126 (2016)
14. Deniz, E., Şengür, A., Kadiroğlu, Z., Guo, Y., Bajaj, V., Budak, Ü.: Transfer learning based histopathologic image classification for breast cancer detection. Heal. Inf. Sci. Syst. **6**(1), 18 (2018)
15. Abdar, M., Makarenkov, V.: CWV-BANN-SVM ensemble learning classifier for an accurate diagnosis of breast cancer. Measurement **146**, 557–570 (2019)
16. Ottoni, A.L.C., Nepomuceno, E.G., de Oliveira, M.S., de Oliveira, D.C.R.: Tuning of reinforcement learning parameters applied to SOP using the Scott-Knott method. Soft Comput. **24**(6), 4441–4453 (2020)
17. Idri, A., Hosni, M., Abran, A.: Improved estimation of software development effort using Classical and fuzzy analogy ensembles. Appl. Soft Comput. J. **49**, 990–1019 (2016)
18. Mittas, N., Angelis, L.: Ranking and clustering software cost estimation models through a multiple comparisons algorithm. IEEE Trans. Softw. Eng. **39**(4), 537–551 (2013)

19. Mittas, N., Mamalikidis, I., Angelis, L.: A framework for comparing multiple cost estimation methods using an automated visualization toolkit. Inf. Softw. Technol. **57**(1), 310–328 (2015)
20. Idri, A., Bouchra, E., Hosni, M., Abnane, I.: Assessing the impact of parameters tuning in ensemble based breast cancer classification. Health Technol. (Berl) **10**(5), 1239–1255 (2020)
21. Jolliffe, I.T., Allen, O.B., Christie, B.R.: Comparison of variety means using cluster analysis and dendrograms. Exp. Agric. **25**(2), 259–269 (1989). https://doi.org/10.1017/S00144797000 16768
22. Calinski, T., Corsten, L.C.A.: Clustering means in ANOVA by simultaneous testing. Biometrics **41**(1), 39 (1985)
23. Worsley, K.J.: Confidence regions and tests for a change-point in a sequence of exponential family random variables. Biometrika **73**(1), 91–104 (1986)
24. Emerson, P.: The original Borda count and partial voting. Soc. Choice Welfare **40**(2), 353–358 (2013)
25. García-Lapresta, J.L., Martínez-Panero, M.: Borda count versus approval voting: a fuzzy approach. Public Choice **112**(1), 167–184 (2002)
26. Black, D.: Partial justification of the Borda count. Public Choice **28**(1), 1–15 (1976)
27. Simonyan, K., Zisserman, A.: Very deep convolutional networks for large-scale image recognition. In: 3rd International Conference on Learning Representations (ICLR 2015) - Conference Track Proceedings, pp. 1–14 (2015)
28. Szegedy, C., Vanhoucke, V., Shlens, J., Wojna, Z.: Rethinking the inception architecture for computer vision (2014)
29. Szegedy, C., Ioffe, S., Vanhoucke, V., Alemi, A.A.: Inception-v4, inception-ResNet and the impact of residual connections on learning. In *Proceedings of the Thirty-First AAAI Conference on Artificial Intelligence (AAAI'17)*. AAAI Press, pp. 4278–4284 (2017)
30. Saikia, A.R., Bora, K., Mahanta, L.B., Das, A.K.: Comparative assessment of CNN architectures for classification of breast FNAC images. Tissue Cell **57**, 8–14 (2019)
31. Razzak, M.I., Naz, S., Zaib, A.: Deep learning for medical image processing: overview, challenges and the future. In: Dey, N., Ashour, A., Borra, S. (eds.) Classification in BioApps. Springer, Cham (2017). https://doi.org/10.1007/978-3-319-65981-7_12
32. Kharel, N., Alsadoon, A., Prasad, P.W.C., Elchouemi, A.: Early diagnosis of breast cancer using contrast limited adaptive histogram equalization (CLAHE) and morphology methods. In: 2017 8th International Conference on Information and Communication Systems (ICICS 2017), pp. 120–124 (2017)
33. Makandar, A., Halalli, B.: Breast cancer image enhancement using median filter and CLAHE. Int. J. Sci. Eng. Res. **6**(4), 462–465 (2015)
34. Perez, L., Wang, J.: The effectiveness of data augmentation in image classification using deep learning (2017)
35. Azzeh, M., Nassif, A.B., Minku, L.L.: An empirical evaluation of ensemble adjustment methods for analogy-based effort estimation. J. Syst. Softw. **103**, 36–52 (2015)
36. Idri, A., Abnane, I., Abran, A.: Evaluating Pred(p) and standardized accuracy criteria in software development effort estimation. J. Softw. Evol. Process **30**(4), 1–15 (2018)
37. Idri, A., Abnane, I.: Fuzzy analogy based effort estimation: an empirical comparative study. In: IEEE CIT 2017-17th IEEE International Conference on Computer and Information Technology, no. Ml, pp. 114–121 (2017)

38. Chougrad, H., Zouaki, H., Alheyane, O.: Deep convolutional neural networks for breast cancer screening. Comput. Methods Programs Biomed. **157**, 19–30 (2018)
39. Xu, Y., Goodacre, R.: On splitting training and validation set: a comparative study of cross-validation, bootstrap and systematic sampling for estimating the generalization performance of supervised learning. J. Anal. Test. **2**(3), 249–262 (2018)

KLM-GOMS Detection of Interaction Patterns Through the Execution of Unplanned Tasks

Daniel Cunha[1] , Rui P. Duarte[1,2(✉)] , and Carlos A. Cunha[1,2]

[1] Polytechnic of Viseu, Viseu, Portugal
{estgv15244,pduarte,cacunha}@estgv.ipv.pt
[2] CISeD – Research Centre in Digital Services, Viseu, Portugal

Abstract. The availability of software applications has contributed to the increase in user demand, which has increased the complexity of these applications. This contributed to the adoption of automation mechanisms for the software testing process, in order to reduce coding errors and shorten the time needed to deploy a new version of the application to the user. Currently, automating the application testing process is a well-established reality and supported by many tools. However, the usability evaluation of an application requires solutions that allow to determine, in advance, the type of improvements that may be necessary in the application without the need for intensive user testing. This work deals with the automatic analysis of the impact on the user of changes in the design of an application, through the implementation of the Keystroke Level Model (KLM). Based on the execution of unplanned user interactions in a web interface, a KLM string is obtained and evaluated, providing a model that converts KLM operators and the execution time of each operator into information for designers. Moreover, performance indicators are obtained and interaction patterns are automatically defined.

Keywords: Keystroke level model · Human-centered computing · Software testing · Interaction design · User interface design

1 Introduction

Graphical User Interfaces (GUI) are available in most modern applications and represent the primary connection between the software and its users. GUI display relevant information and possible actions to users through virtual objects, or widgets, which are graphic elements such as buttons, text or edit boxes that make it intuitive to use software [1,2]. Moreover, an average of 48% of the application source code is used to implement a GUI in order to release a user-friendly software [3], and the user interaction with the software is established by performing sequences of events.

At the user level, the demand for applications is increasing and the interface between the user and the system plays an important role in the acceptance of

© Springer Nature Switzerland AG 2021
O. Gervasi et al. (Eds.): ICCSA 2021, LNCS 12950, pp. 203–219, 2021.
https://doi.org/10.1007/978-3-030-86960-1_15

the systems, as well as user's trust and satisfaction [4]. This demand focuses on usability, performance, quality of service and interface design [5]. When the interface is well understood by the users, a win-win relation is established between the user and the system; otherwise, users tend to become frustrated and more intolerant to errors. According to Hogan, L. [6], most users leave web applications when the loading wait time exceeds 2 s. In turn, when the page load time exceeds 3 s, about 40% of users tend to abandon the application.

Interface design is one of the most critical tasks in User Experience (UX) and User Interface (UI) analysis. It improves UX making the interface more intuitive and simple. That said, it is important to perform UI/UX analysis to predict the user behaviour, and, thus reduce the time spent on each interaction. To achieve this, two approaches can be considered: empirical and analytic. While the first involve users, the second uses techniques to predict user behavior. Predictive models use normalized historical data from several sources, which are processed to obtain an analysis. Example of this are the family of predictive models to compare and evaluate Goals, Operators, Methods and Selection (GOMS) of skilled, error-free performances [7]. These models are focused on the hierarchy of tasks composed of actions and cognitive operators. The simplest model of the GOMS family is the Keystroke Level Model (KLM), which predicts the execution time of a task using mouse and keyboard as its main inputs [8].

One of the current challenges in interface design is to help UI designers to improve their designs without the need to set-up a number of user testing sessions. Nevertheless this is an important part of the design process, it is a time consuming process to produce results. Moreover, UI designers make many small decisions along the way that can impact on the user interaction, and need tools that allow them to measure the quality of their design. According to Al-Megren et al. [9], it is of most importance to create methods that use the KLM model to provide guidance to introduce changes in their design and reduce the time required to complete a task. In this paper we present a KLM approach that uses the interface created by the UI designer to provide feedback about the quality of the design and proposes performance indicators on how to improve the quality of the interaction. To achieve this, we propose an algorithm based on the KLM that analyses a KLM string to detect types of interaction and, from there, identify patterns of behavior that are increasing the user task execution time, resulting in a poor interaction. The major contributions of this paper are a model of interpretation of unplanned actions by users, and performance indicators for the automatic detection of interaction patterns.

The remainder of this paper is organized in the following manner. Section 2 presents the background on KLM by identifying its applications and current trends. Section 3 presents a general view of the approach and the two main modules developed. Section 4 refers to the major conclusions that can be obtained, and provides insights on future work.

2   Background

The KLM model has been extensively used to evaluate time performance of user-computer interaction across different devices. In this section we review the KLM and its applications to related contexts. For further reading, Al-Megren et al. [9] present a survey of the KLM and recommendations are provided.

2.1   Keystroke Level Model

The GOMS model can be defined as an high-level description of the knowledge that a user has to have in order to execute tasks. It relates to the mental processing and hierarchical decomposition of predefined goals. It is a strong reference in the context of behavior analysis in what concerns the prediction of the effective usability of UIs [7].

The KLM models the decomposition of a general task obtained from the GOMS model into unit tasks, at the device level (like mouse clicks or key press). It allows the prediction of user performance, by measuring the time required for a user to complete a task. Proposed by Card et al. [8], a unit task has two parts: *acquisition* of the task and the *execution* of the task. The total time to complete a unit task is given by:

$$T_{task} = T_{acquire} + T_{execute} \tag{1}$$

At the execution level, KLM provides physical, mental and response operators with predefined time values. These operators are defined by a letter and include: K (keystroke), P (point), H (home), D (draw), B (button press), M (mental preparation for action), and R (system response). According to Card et al. [8], K is the mostly frequently used operator and represents the act of keystroke or button press (when considering $K = 0.2\,s$). P represents pointing to a specific target with a mouse. It's time varies as a function of the distance to the target, and the size of the target, according to Fitts's Law [10]. H is the action of switching the hand between keyboard and mouse. D is associated to the mouse and refers to the action of drawing with it.

To execute the physical operators, the user has to mentally prepare himself for this execution. This is represented by the mental operator, M. Since this operator is not observable from user behaviour, it is based on specific knowledge of user skills and the placement of M's is governed by a set of heuristics [8].

The final operator, R refers to the time required for the system to respond to user's actions and is typically neglected since it is assumed that the system responds in real-time. The KLM process begins with an encoding (named KLM string) containing of the physical operators (K, P, H, B, and D). After this, M's are inserted into the string according to Rule 0. Rules 1 to 4 are then applied to each M to determine if it should be deleted.

The execution time is the sum of the time for each of the operators from the final KLM string:

$$T_{execute} = T_K + T_P + T_H + T_D + T_B + T_M + T_R \tag{2}$$

Despite the advantages of using KLM, it presents several limitations. First, only expert users are considered and does not account for user errors. Second, it considers linear routine tasks; however, users can multi-task or reach a goal by different means. Finally, individual differences are not considered, once the M only assumes a single value; mental workload and fatigue are aspects of M that are not accounted by this model. KLM was validated against observed models to determine prediction accuracy. According to Card [7], the root-mean-square percentage error (RMSPE) of the model is 21%.

2.2 Related Work

Traditional methods used in the usability evaluation of an interface fall into two categories: the subjective opinion of users and experts, (mainly applying questionnaires [11,12] and inspection methods [13,14]), or using objective techniques like rules [15], and analytic modeling [16]. Nevertheless these approaches provide important tools to determine the usability of an UI, there is both cost and time needed to implement users' interaction evaluation with and acceptable coverage, increased by the need to use experts to cover for the user's faults.

Automated usability testing of web applications is a trending topic (web approaches cover 32% of the reviewed literature [17]) which presents several advantages when compared to traditional techniques [17,18]. During the last decades, predictive cognitive models, like KLM, have been applied and extended to several application domains, with focus on mobile applications and IVIS systems. Some of the most representative applications of KLM are the Cog-Tool project [19] (which was extended to web forms by the KLM Form Analyser (KLM-FA) [20]) and the SANLab-CM [21]. CogTool [19] is a tool that can produce quantitative, model-based predictions of skilled performance time from tasks demonstrated on storyboard mockups of a user interface (inspired by the DENIM project [22]). Each state if the interface is represented by a frame in the storyboard. With the storyboard created, the modeler defines the steps required to complete a task, and KLM is used to determine the time to complete a task. CogTool Explorer [23] builds upon CogTool to predict a user's goal-directed exploratory interaction with a website. Current available model-based tools require non-trivial manual work to examine forms. To tackle this problem, KLM-FA [20] extends the capabilities of existing tools for practitioners by focusing specifically on automating the analysis of web forms. On one hand they increased the automation and efficiency for evaluation tasks and minimized the required effort from CogTool. On the other hand they do not account with design support and evaluation. They also accounted for the calculation of Fitts' law, instead of using the fixed value. SANLab-CM [21] differs from these approaches since it incorporates variability as a part of its core paradigm and offers a set of tools designed to visualize and compare critical paths and how they differ within and between models. This variability is related to the execution times of an operator, which can cause the path for tasks completion to fluctuate widely.

Models focused on empirical human interaction were also developed. Instead of focusing on predictive behavior, they focus on the cognitive perception of users employing, to some extent, Artificial Intelligent (AI) techniques. These AI based tools are surveyed by Katsanos et al. [24], and cover information structure elaboration and link's appropriateness evaluation of web applications. Some of the most representative are Bloodhound [25], Method for Evaluating Site Architectures (MESA) [26], Automated Cognitive Walkthrough for the Web (ACWW) [27], InfoScent Evaluator Tool (ISEtool) [28], Combining Link Scent with Navigation Path Relevancy (CoLiDeS+) [29], and at the level of conformance to standards, the Usability Evaluation Framework (USEFul) [30].

The use of agents to extract usability metrics has later been criticized [25] since the agents do not consider the content analysis and other human-related aspects. The InfoScent Bloodhound Simulator is presented, which performs automated usability evaluation of websites through the use of information scent, and produces a usability report for the designer to identify navigation problems. This method requires three inputs: the user tasks identified by keywords, the URL address of each task target page, and the URL of the page in which exploration starts. ISEtool [28] presents an approach similar to Bloodhound and is a semi-automated tool to support the evaluation of information scent of hyperlinks' descriptions in webpages. They assumed that users have a goal and their surfing patterns are guided by information scent. Inputs provided by the designer are a description of a user goal, the URL of the webpage and select the LSA semantic similarity algorithm [31]. MESA [26] simulates navigation in a website by modeling the interoperability between the site's structure, link quality, and cognitive limitations. ACWW [27] provides an analysis of webpages for problems in navigation. It identifies four types of usability problems and produces a prediction of the mean total number of clicks for a user to select the correct link. The designer provides three types of inputs: a description of the user goal, labels of the regions of the webpage and text labels of each region. CoLiDeS+ [29] is a cognitive model of web navigation which describes a real-time step-by-step process based on positive correlations between spatial ability, path adequacy, and task performance to allow users to make navigational decisions. As inputs it requires a detailed description of the user's goal, the heading labels of subregions on each webpage, and text labels of the links in each subregion. Results provide a simulation of link selection and backtracking behavior. USEFul [30] is a web-based, nonpublically available prototype that evaluates the usability of a website by employing the Heuristic Evaluation technique which refers to the set of 242 research-based usability guidelines compiled by the author. Nevertheless, it only considers those that are straightforward to implement. Results were compared to the work of Nielsen and Tahir [32], and their framework detected on average 28.5% more violations than the referred work.

The aforementioned tools present several disadvantages. Both Bloodhound and ISEtool do not model visual search and assume a global evaluation of all hyperlinks in a webpage, whereas both ACWW and CoLiDeS+ ask from the designer to manually parse the webpage into sub regions and do not specify complex web navigation strategies. On the other hand, predictive tools simulate visual perception and provide greater insight into expected user behavior and accurate predictions of user's performance. According to Chi et al. [25] there are two types of tools that perform automated usability evaluation: those that make use of conformance to standards and those that try to predict the usage of an application. The framework proposed in this paper falls in the second category, since it generates performance indicators from a user interaction, and automatically extracts interaction patterns from a KLM string.

3   Unplanned User Interaction Framework

This section presents the process of parsing an unplanned action into user performance indicators, to assist the designer of interfaces. It presents an high level description of the architecture, and a detailed view of its components.

3.1   High-Level Architecture

At this level, a web application is designed and deployed for tests. Next, a user executes unplanned interactions (associated to a goal) in the application, which allow the extraction of performance indicators used to generate interaction patterns. To achieve this, the model is composed of two core components: the parsing module and the analysis module, as illustrated in Fig. 1.

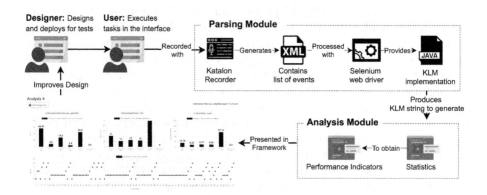

Fig. 1. General view of the framework

The parsing module captures events from the user interaction when executing a task in a web application using Katalon Automation Recorder [33]. This application stores all the interactions in a XML file which is parsed in JAVA using the Selenium WebDriver API [34], to acquire the data from the XML file. Next, the KLM [8] is applied and the corresponding KLM string is generated, resulting in the pre-processing stage of our framework.

The analysis module acts on the KLM string to extract performance indicators related to operators used in the interaction. Since user interaction is unplanned, users can reach a goal of interaction following random paths, and performance indicators are used to automatically determine interaction patterns (mouse based or keyboard based).

3.2   KLM String Extraction

The extraction of the KLM string corresponds to the output of the parsing module. To reach this goal, several steps are considered: record user interaction, process XML file, and generate the KLM string. In the first step, the user interaction is recorded with Katalon Recorder [33]. It is a browser add-on that records actions in a XML file related to elements of web applications. The XML file is composed of a set of tags associated to an event (identified as *<selenese>*) that is composed of the purpose of the action, *<command>*, containing a key or target element in the UI, *<target>*, which is given by a *<value>*. An example is presented in Fig. 2, which refers to the interaction with an email textbox.

```
<selenese>
    <command>click</command>
    <target><![CDATA[name=registrationEmail]]></target>
    <value><![CDATA[]]></value>
</selenese>
<selenese>
    <command>type</command>
    <target><![CDATA[name=registrationEmail]]></target>
    <value><![CDATA[dfghjklsdf@gmail.com]]></value>
</selenese>
```

Fig. 2. Email textbox XML tags: the user clicks the textbox and enters the email

The first *<selenese>* event is associated to the click in the textbox to activate the element, and the second relates to the writing with the keyboard. In the first event, the *<command>* returns the value *click* which corresponds to the user click in the component before writing the email. The *<value>* of the *<target>* is null. In the second event, the value of *<command>* is *type* which refers to the use of keyboard in the web element in the *<target>*. The *<value>* is given by the keys pressed. Once the XML file is generated, the second step initiates. The XML is parsed into the java implementation to extract the tags for KLM processing. This process is carried out with the Selenium WebDriver API [34]. This API provides functions suitable to extract information from the XML file, associate them to KLM operators (see Sect. 2.1), and builds the KLM string. This process is depicted in Fig. 3, and an example of application is described.

Every hierarchical *<selenese>* event is processed to obtain the *<value>* of a specific action. When a user executes a *type* *<command>* in a *<target>* of type *Id*, it is inferred that the user is interacting with a textbox. Next, the

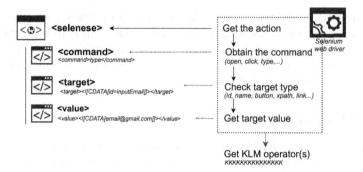

Fig. 3. XML parser: from the XML file to a KLM string

$<value>$ email@gmail.com is obtained and stored. This $<value>$ is of type *text* and is parsed in the KLM as a sequence of 15 operators K (the KLM value of keystroke). This sequence is added to the KLM string and the next $<selenese>$ event in the XML file is parsed into the KLM string. When all the events are processed, the pre-final KLM string is created. The final KLM string is generated by applying the rules to insert the mental operator, M, presented in Sect. 2.1.

To validate the accuracy of the KLM implementation, the predictions obtained with the implemented framework were compared with user testing data for the sign-up form of Cable News Network (CNN). Two interaction strategies were considered: (a) mouse based in which a user interacts with the form with a mouse, except for text entries, and (b) keyboard based, in which form fields are reached using the tab key. Ten users were asked to carry out the steps required to sign-up in the aforementioned application. No sequence of steps was defined, except for the goal of the interaction. In the framework, the following assumptions were used: (a) system response time, W, was negligible, (b) the user's hand began on the keyboard before the execution of the task, (c) Fitts' law calculations were enabled. Table 1 presents users' task execution times and the framework predicted times for CNN register form, identified as [S01] in Appendix A, along with the error rate and coefficient of covariation of the RMSE (CV-RMSE).

Table 1. Framework execution times showing means and, in parenthesis, standard deviations

Interaction	Participant's task time (ms)	Framework predicted time (ms)	Error rates of predictions (%)	CV-RMSE of predictions (%)
Keyboard based	24743 (1902.48)	25020	1.12%	7.68%
Mouse based	25679 (3522.02)	29720	15.74%	18.04%

Results show that the mean error of predictions was 8.43%, which is within the 20% margin of error reported by [35]. In general, this framework is an accurate implementation of KLM, which overestimates task time in the mouse based interaction, and underestimates in the keyboard based interaction.

3.3   Performance Indicators and Interaction Patterns

To identify potential problems in a user interaction, the KLM string is parsed into it's three visual representations: the number of operators used in the interaction, the time per operator and the timeline of events. Thus, the KLM string is interpreted as an array, and information is extracted to provide the designer visual indicators of the interaction. Figure 4 depicts an example of transforming a KLM string (*MHMPBMKKKKKKKKKKKKKKKKKKKKKK KHMPBMKKKKKKKK KKKKHMPBMPB*) into readable indicators of user performance.

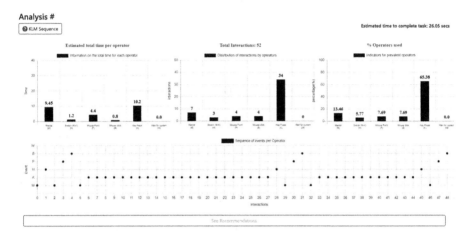

Fig. 4. Conversion of a KLM string into performance indicators for mouse and keyboard interaction

This interaction scenario was based on the task of registering a new user in [S02] using a mixed strategy of interaction (mouse and keyboard). The most prevalent operator is *K*. However, the mental operator, *M*, also consumes 9.45 *secs* of interaction time, which is high, relative to the total time of interaction. This analysis makes possible for the designer to understand that during the interaction the user had to move from the keyboard to the mouse, and again to the keyboard, which lead to an increase in the interaction time. Thus, a message was placed in the interface to the user: "did you know that you can press TAB key to change between fields?". Next, another user was asked to execute the same interaction, with the redesigned interface. As presented in Fig. 5, the user followed a keyboard-based strategy. The KLM string generated by the interaction is *MHMPBMKKKKKKKKKKKKKKKKKKKKKKKKKKKKKKKKKKKKKHMPBM PB*, and the performance indicators show that the user reduced the amount on mental and pointing interactions, which lead to a decrease in the total time required to complete a task.

Nevertheless performance indicators provide information of the quality of the interaction, they only provide a view of the task executed by the user. To

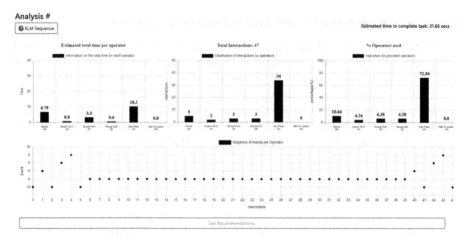

Fig. 5. Conversion of a KLM string into performance indicators for keyboard interaction resulting of changes in the web application

improve on the extraction of features, it is important to automatically detect the type of interaction that the user executed on an web interface. To implement a rule-based method, it is first required to determine patterns from the user interaction strategy. Table 2 and Fig. 6 present the results obtained by applying the interaction strategies to ten web applications (identified in Appendix A).

Results show that the percentage of time used by each operator is related to the interaction strategy. In a mouse-based strategy, the prevalent operators are M and P with 57.44% and 35.81%, in average, of the total time spent in the interaction, respectively. In a keyboard-based interaction, the operators are M (29.01%) and K (50.72%), and in a mixed combination, the prevalent operators are M (37.05%) and K (38.36%). Nevertheless similarities can be found in these two strategies, the increase in M is compensated to a decrease in K, complemented with an increase in H and P. For comparison purposes, the number of operators used (materialized in the percentage of operators) was also considered. Results show that they do not provide a precise indicator to measure the consumption of time in the execution of a task, since many operators do not imply on increase in time (cf. first scenario, where K are used in 64.38% of operators and, in time spent, they nearly correspond to the same time spent as M operator, with only 13.92% of operators used). Considering the patterns, Fig. 6 plots the percentage of time used by each operator in each interaction strategy, and its standard deviation, σ. In order to predict the interval of admissible values of operators for each type of interaction, let $O \in \{M, H, P, B, K\}$ be a general operator resulting from a user interaction, and $O_i, i = 1...n$, the value of the operator presented in Table 2. The rule is to determine $min\{O_i\}$ and $max\{O_i\}$ and verify if $O \in [min_{O_i} - \sigma_O, max_{O_i} + \sigma_O]$. If that occurs, a pattern is established, and the type of interaction is predicted, as presented in Table 3.

Table 2. Interaction strategies followed in ten sources, and corresponding percentages of time spent and, in brackets, percentage of operators used to complete a task.

	Sources	% of time per operator (% of operators used)				
		M	H	P	B	K
Sign-up (mouse & keyboard)	[S01]	35.06 (12.9)	3.89 (4.84)	17.85 (8.06)	3.24 (8.06)	39.93 (66.14)
	[S02]	36.27 (13.46)	4.60 (5.77)	16.89 (7.69)	3.07 (7.69)	39.15 (65.39)
	[S03]	37.08 (13.70)	5.49 (6.85)	15.10 (6.85)	2.74 (6.85)	39.56 (65.75)
	[S04]	37.85 (14.84)	5.31 (7.03)	19.48 (9.38)	3.24 (8.59)	34.09 (60.16)
	[S05]	37.58 (13.77)	5.27 (6.52)	16.11 (7.24)	1.46 (7.25)	39.56 (65.11)
	[S06]	42.09 (17.50)	5.34 (7.50)	19.59 (10.00)	3.56 (10.00)	29.35 (55.00)
	[S07]	36.00 (13.24)	4.74 (5.88)	16.29 (7.35)	2.96 (7.35)	40.00 (66.18)
	[S08]	38.02 (14.47)	5.12 (6.58)	16.90 (7.89)	3.07 (7.89)	36.87 (63.17)
	[S09]	36.48 (13.32)	5.40 (6.67)	14.86 (6.67)	2.70 (6.67)	40.54 (66.67)
	[S10]	34.06 (11.94)	5.04 (5.97)	13.88 (5.97)	2.52 (5.97)	44.47 (70.15)
	Average	**37.05 (13.92)**	**5.02 (6.36)**	**16.70 (7.71)**	**2.86 (7.63)**	**38.36 (64.38)**
Sign-up (keyboard - tab)	[S01]	29.67 (10.00)	2.93 (3.33)	16.11 (6.67)	2.93 (6.67)	48.35 (73.33)
	[S02]	31.17 (10.64)	3.69 (4.26)	15.24 (6.38)	2.77 (6.38)	47.11 (72.34)
	[S03]	22.13 (6.45)	3.27 (3.22)	9.01 (3.23)	1.63 (3.23)	63.93 (83.87)
	[S04]	29.78 (10.28)	4.01 (4.67)	17.65 (7.48)	2.80 (6.54)	45.73 (71.03)
	[S05]	23.19 (6.93)	2.94 (2.97)	10.79 (3.96)	1.96 (3.96)	61.10 (82.18)
	[S06]	34.06 (12.31)	3.78 (4.62)	17.35 (7.69)	3.15 (7.69)	41.64 (67.69)
	[S07]	31.87 (10.94)	4.04 (4.68)	14.83 (6.25)	2.69 (6.25)	46.54 (71.88)
	[S08]	32.19 (11.11)	4.08 (4.76)	14.99 (6.35)	2.72 (6.35)	45.99 (71.43)
	[S09]	29.34 (9.52)	4.34 (4.76)	11.95 (4.76)	2.17 (4.76)	52.17 (76.2)
	[S10]	26.73 (8.47)	3.16 (3.39)	13.06 (5.08)	2.37 (5.08)	54.65 (77.98)
	Average	**29.01 (9.67)**	**3.62 (4.01)**	**14.10 (5.79)**	**2.52 (5.70)**	**50.72 (74.80)**
Navigation (mouse)	[S11]	53.54 (35.00)	2.26 (5.00)	37.39 (30.00)	6.79 (30.00)	0.0 (0.0)
	[S06]	54.65 (35.71)	3.23 (7.14)	35.62 (28.57)	6.47 (28.57)	0.0 (0.0)
	[S12]	53.54 (35.00)	2.26 (5.00)	37.39 (30.00)	6.79 (30.00)	0.0 (0.0)
	[S13]	54.65 (35.71)	3.23 (7.14)	35.62 (28.57)	6.47 (28.57)	0.0 (0.0)
	[S03]	54.65 (35.71)	3.23 (7.14)	35.62 (28.57)	6.47 (28.57)	0.0 (0.0)
	[S07]	54.00 (35.29)	2.66 (5.88)	36.66 (29.41)	6.66 (29.41)	0.0 (0.0)
	[S14]	54.65 (35.71)	3.23 (7.14)	35.62 (28.57)	6.47 (28.57)	0.0 (0.0)
	[S01]	55.67 (36.36)	4.12 (9.09)	34.02 (27.27)	6.18 (27.27)	0.0 (0.0)
	[S02]	52.56 (34.38)	1.41 (3.13)	38.93 (31.25)	7.07 (31.25)	0.0 (0.0)
	[S10]	57.44 (37.5)	5.67 (12.5)	31.20 (25.00)	5.67 (25.00)	0.0 (0.0)
	Average	**57.44 (35.64)**	**3.13 (6.92)**	**35.81 (28.72)**	**6.51 (28.72)**	**0.0 (0.0)**

Table 3. Patterns obtained for each type of interaction

Type of interaction	Operators $\in [min\{O_i\} - \sigma_O, max\{O_i\} + \sigma_O]$				
	M	H	P	B	K
Keyboard	[18.44, 37.76]	[2.44, 4.83]	[6.33, 20.34]	[1.19, 3.60]	[34.82, 70.76]
Mouse	[51.31, 58.71]	[0.31, 6.78]	[29.20, 40.94]	[5.31, 7.44]	[0.0, 0.0]
Mouse and keyboard	[32.02, 44.15]	[3.4, 5.96]	[12.09, 21.39]	[0.92, 4.11]	[25.51, 48.37]

In order to validate these patterns, the three types of interaction were tested in ten different websites. Users were asked to follow one of three interaction strategies: keyboard, mouse, or mouse-keyboard, and the interaction is evaluated to determine if it fits the expected result. Figure 7, depicts the values for the operators within the intervals defined for each pattern of interaction, and the expected results were validated using the patterns defined in Table 3. With this approach, an unplanned interaction from a user can be automatically classified, by measuring the prevalence of operators used within the execution of a task.

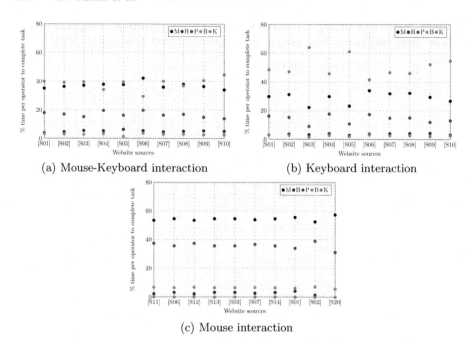

(a) Mouse-Keyboard interaction   (b) Keyboard interaction

(c) Mouse interaction

Fig. 6. Sign-up forms used for tests in different scenarios of interaction: (a) $\sigma_M = 2.05$, $\sigma_H = 0.46$, $\sigma_P = 1.79$, $\sigma_B = 0.55$, $\sigma_K = 3.89$; (b) $\sigma_M = 3.69$, $\sigma_H = 0.49$, $\sigma_P = 2.69$, $\sigma_B = 0.45$, $\sigma_K = 6.82$; (c)$\sigma_M = 1.26$, $\sigma_H = 1.1$, $\sigma_P = 2.0$, $\sigma_B = 0.36$, $\sigma_K = 0.0$

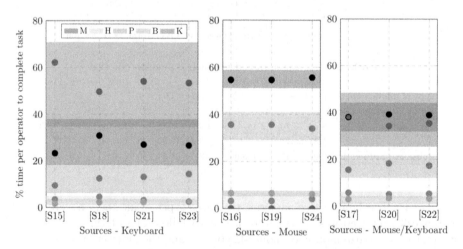

Fig. 7. Representation of scenarios for pattern fitting

4   Conclusions and Future Work

This paper presents a novel approach to assist a designer of interfaces in the quality evaluation of a designed application. It is based on unplanned user interactions that allow the user a greater degree of freedom to complete a task. Most tools available in the literature require pre-processing, for the user to follow a specific path to complete a task. The proposed framework uses KLM model in the calculation of user interaction times with web applications, and the extraction of performance indicators to allow the redesign of interfaces. By simplifying the entry model alluding to the user-oriented task (only the goal of the task to be executed is indicated), the user can execute unplanned actions in the web application in order to achieve the proposed goal. In addition, the type of interaction that the user executes on the web application is automatically detected.

Future work will follow three directions: develop a recommendation system based on the performance indicators extracted from the KLM; capture the timeline of the user facial emotions when executing a task to overlap it with the timeline of the KLM; analyse the impact of the W operator in the overall time required to complete a task. A general purpose recommendation system for designers is a key tool to obtain feedback on interface errors. At this point, this work is able to automatically detect user interactions and provide performance indicators based on patterns of interaction. From these detection, different design guidelines can be provided to designers to improve their UI, based on hybrid recommendation systems [36] that are acquired from performance indicators. Another feature that is planned is the extraction of user emotions in order to be analyze reactions when interacting with the web application to verify if the user's emotions are related to the operators obtained by the KLM. Finally, it is intended to analyze the impact of the W operator of the KLM model on the time of execution of tasks by the user. In the literature, it is suggested that W should be measured. This will allow to identify design patterns that reduce the sensitivity of UX to variations with performance.

Acknowledgements. This work is funded by National Funds through the FCT - Foundation for Science and Technology, I.P., within the scope of the project Ref. UIDB/05583/2020. Furthermore, we would like to thank the Research Centre in Digital Services (CISeD) and the Polytechnic of Viseu for their support.

A   List of Sources Used for Tests

Fourteen web applications were used to carry out the tests specified in Sect. 3.3, and are identified in Table 4: (a) mouse interactions only based on navigation; (b) keyboard interactions between fields using the Tab key; or (c) combination of mouse and keyboard interactions. Column (d) identifies web applications used for pattern fitting tests, as presented in Fig. 7.

Table 4. Websites used in tests and corresponding contexts of interaction

Label	Website	(a)	(b)	(c)	(d)
[S01]	CNN (https://edition.cnn.com/)	×	×	×	
[S02]	New York Times (https://www.nytimes.com/)	×	×	×	
[S03]	Walmart (https://www.walmart.com/)	×	×	×	
[S04]	Spotify (https://www.spotify.com/)		×	×	
[S05]	Twitter (https://twitter.com/)		×	×	
[S06]	BBC (https://www.bbc.com/)	×	×	×	
[S07]	Reddit (https://www.reddit.com/)	×	×	×	
[S08]	The Guardian (https://theguardian.com/)		×	×	
[S09]	LinkedIN (https://www.linkedin.com/)		×	×	
[S10]	GitHub (https://github.com/)	×	×	×	
[S11]	Ebay (https://www.ebay.co.uk/)	×			
[S12]	Stack Overflow (https://stackoverflow.com/)	×			
[S13]	Forbes (https://www.forbes.com/)	×			
[S14]	Reuters (https://www.reuters.com/)	×			
[S15]	Farfetch (https://www.farfetch.com/)				×
[S16]	Fox News (https://www.foxnews.com/)				×
[S17]	The Wall Street Journal (https://www.wsj.com/)				×
[S18]	The Telegraph (https://www.telegraph.co.uk/)				×
[S19]	Puma (https://eu.puma.com/)				×
[S20]	Logitech (https://www.logitech.com/)				×
[S21]	Levis (https://www.levi.com/)				×
[S22]	Gant (https://gant.com/)				×
[S23]	El Corte Ingles (https://www.elcorteingles.com/)				×
[S24]	NFL (https://www.nfl.com/)				×

References

1. Ganov, S.R., Killmar, C., Khurshid, S., Perry, D.E.: Test generation for graphical user interfaces based on symbolic execution. In: Proceedings of the 3rd International Workshop on Automation of Software Test, AST 2008, pp. 33–40. Association for Computing Machinery, New York (2008)
2. Memon, A.M.: A comprehensive framework for testing graphical user interfaces. Ph.D., University of Pittsburgh (2001). Advisors: Mary Lou Soffa and Martha Pollack; Committee members: Prof. Rajiv Gupta (University of Arizona), Prof. Adele E. Howe (Colorado State University), Prof. Lori Pollock (University of Delaware)
3. Myers, B.A., Rosson, M.B.: Survey on user interface programming. In: Proceedings of the SIGCHI Conference on Human Factors in Computing Systems, CHI 1992, pp. 195–202. Association for Computing Machinery, New York (1992)
4. Pommeranz, A., Broekens, J., Wiggers, P., Brinkman, W.P., Jonker, C.M.: Designing interfaces for explicit preference elicitation: a user-centered investigation of preference representation and elicitation process. User Model. User Adap. Inter. **22**(4–5), 357–397 (2012)
5. Delisle, S., Moulin, B.: User interfaces and help systems: from helplessness to intelligent assistance. Artif. Intell. Rev. **18**(2), 117–157 (2002)
6. Hogan, L.C.: Designing for Performance: Weighing Aesthetics and Speed. O'Reilly Media, Inc., Sebastopol (2014)
7. Card, S.K.: The Psychology of Human-Computer Interaction, pp. 49–51. Lawrence Erlbaum Associates (1983)
8. Card, S.K., Moran, T.P., Newell, A.: The keystroke-level model for user performance time with interactive systems. Commun. ACM **23**(7), 396–410 (1980)
9. Al-Megren, S., Khabti, J., Al-Khalifa, H.S.: A systematic review of modifications and validation methods for the extension of the keystroke-level model. In: Advances in Human-Computer Interaction 2018 (2018)
10. Fitts, P.M.: The information capacity of the human motor system in controlling the amplitude of movement. J. Exp. Psychol. **47**(6), 381 (1954)
11. Callahan, E., Koenemann, J.: A comparative usability evaluation of user interfaces for online product catalog. In: Proceedings of the 2nd ACM Conference on Electronic Commerce, EC 2000, pp. 197–206. Association for Computing Machinery, New York (2000)
12. Paz, F., Paz, F.A., Pow-Sang, J.A.: Evaluation of usability heuristics for transactional web sites: a comparative study. In: Information Technology: New Generations. AISC, vol. 448, pp. 1063–1073. Springer, Cham (2016). https://doi.org/10.1007/978-3-319-32467-8_92
13. Paz, F., Paz, F.A., Villanueva, D., Pow-Sang, J.A.: Heuristic evaluation as a complement to usability testing: a case study in web domain. In: ITNG 2015, pp. 546–551. IEEE Computer Society, USA (2015)
14. Yushiana, M., Rani, W.A.: Heuristic evaluation of interface usability for a web-based OPAC. Library Hi Tech **25**, 538–549 (2007). https://doi.org/10.1108/07378830710840491
15. Lim, C., Song, H.D., Lee, Y.: Improving the usability of the user interface for a digital textbook platform for elementary-school students. Educ. Tech. Res. Dev. **60**(1), 159–173 (2012)
16. Tonn-Eichstädt, H.: Measuring website usability for visually impaired people-a modified GOMS analysis. In: Proceedings of the 8th International ACM SIGACCESS Conference on Computers and Accessibility, ASSETS 2006, pp. 55–62. Association for Computing Machinery, New York (2006)

17. Fernandez, A., Insfran, E., Abrahão, S.: Usability evaluation methods for the web: a systematic mapping study. Inf. Softw. Technol. **53**(8), 789–817 (2011). Aug
18. Bakaev, M., Mamysheva, T., Gaedke, M.: Current trends in automating usability evaluation of websites: can you manage what you can't measure? In: 2016 11th International Forum on Strategic Technology (IFOST), pp. 510–514 (2016)
19. John, B.E., Prevas, K., Salvucci, D.D., Koedinger, K.: Predictive human performance modeling made easy. In: Proceedings of the SIGCHI Conference on Human Factors in Computing Systems, CHI 2004, pp. 455–462. Association for Computing Machinery, New York (2004)
20. Katsanos, C., Karousos, N., Tselios, N., Xenos, M., Avouris, N.: KLM form analyzer: automated evaluation of web form filling tasks using human performance models. In: Kotzé, P., Marsden, G., Lindgaard, G., Wesson, J., Winckler, M. (eds.) INTERACT 2013. LNCS, vol. 8118, pp. 530–537. Springer, Heidelberg (2013). https://doi.org/10.1007/978-3-642-40480-1_36
21. Patton, E.W., Gray, W.D.: SANLab-CM: a tool for incorporating stochastic operations into activity network modeling. Behav. Res. Meth. **42**(3), 877–883 (2010)
22. Lin, J., Newman, M.W., Hong, J.I., Landay, J.A.: DENIM: an informal tool for early stage web site design. In: CHI '01 Extended Abstracts on Human Factors in Computing Systems, CHI EA 2001, pp. 205–206. Association for Computing Machinery, New York (2001)
23. Teo, L., John, B.E.: CogTool-Explorer: towards a tool for predicting user interaction. In: CHI '08 Extended Abstracts on Human Factors in Computing Systems, CHI EA 2008, pp. 2793–2798. Association for Computing Machinery, New York (2008)
24. Katsanos, C., Tselios, N., Avouris, N.: A survey of tools supporting design and evaluation of websites based on models of human information interaction. Int. J. Artif. Intell. Tools **19**(06), 755–781 (2010)
25. Chi, E.H., et al.: The bloodhound project: automating discovery of web usability issues using the infoscent π simulator. In: Proceedings of the SIGCHI Conference on Human Factors in Computing Systems, CHI 2003, pp. 505–512. Association for Computing Machinery, New York (2003)
26. Miller, C.S., Remington, R.W.: Modeling information navigation: implications for information architecture. Hum. Comput. Interact. **19**(3), 225–271 (2004)
27. Blackmon, M.H., Kitajima, M., Polson, P.G.: Tool for accurately predicting website navigation problems, non-problems, problem severity, and effectiveness of repairs. In: Proceedings of the SIGCHI Conference on Human Factors in Computing Systems, CHI 2005, pp. 31–40. Association for Computing Machinery, New York (2005)
28. Katsanos, C., Tselios, N., Avouris, N.: Infoscent evaluator: a semi-automated tool to evaluate semantic appropriateness of hyperlinks in a web site. In: Proceedings of the 18th Australia Conference on Computer-Human Interaction: Design: Activities, Artefacts and Environments, OZCHI 2006, pp. 373–376. Association for Computing Machinery, New York (2006)
29. Van Oostendorp, H., Juvina, I.: Using a cognitive model to generate web navigation support. Int. J. Hum. Comput. Stud. **65**(10), 887–897 (2007)
30. Dingli, A., Mifsud, J.: Useful: a framework to mainstream web site usability through automated evaluation. Int. J. Hum. Comput. Interact. (IJHCI) **2**(1), 10 (2011)
31. Landauer, T.K., Dumais, S.T.: A solution to Plato's problem: the latent semantic analysis theory of acquisition, induction, and representation of knowledge. Psychol. Rev. **104**(2), 211 (1997)

32. Nielsen, J., Tahir, M., Tahir, M.: Homepage Usability: 50 Websites Deconstructed, vol. 50. New Riders, Indianapolis (2002)
33. Katalon: Katalon recorder v3.4: A powerful selenium ide alternative (2020). https://www.katalon.com/resources-center/blog/katalon-recorder-v3-4/
34. Selenium: Selenium webdriver (2020). https://www.selenium.dev/projects/
35. Ivory, M.Y., Hearst, M.A.: The state of the art in automating usability evaluation of user interfaces. ACM Comput. Surv. **33**(4), 470–516 (2001).
36. Burke, R.: Hybrid recommender systems: Survey and experiments. User Model. User Adap. Inter. **12**(4), 331–370 (2002)

Modelling of Moisture Effect in Safety Evaluation of Soil-Interacting Masonry Wall Structures

Emma Vagaggini, Martina Ferrini, Mauro Sassu, and Mario Lucio Puppio(⊠)

University of Cagliari, Cagliari, CA, Italy
mariol.puppio@unica.it

Abstract. This paper deals with analytical models of retaining walls interacting with soil, whose mechanical properties are influenced by penetrating moisture. Specifically, urban masonry walls are here considered. The proposed simplified procedure takes rainfalls as the intensity measure dependent on rain duration and return period. Growing levels of imbibition of the soil behind the wall determine a de-crease of the strength parameters of the soil. A geometry of the section at each step of imbibition is analyzed considering a set of soil layers with modified properties over time. The method is applied to a section of the Volterra historical urban walls named "Sperone" which was affected by a relevant collapse occurred in March 2014: a back analysis of the soil-structure section is also proposed to re-place the collapse.

Keywords: Historical urban walls · Moisture effects · Landslide risk · Volterra urban walls

1 Introduction

The effects of moisture in soils involve widespread infrastructures such as slopes, embankments, retaining walls and urban masonry walls. These last are irregular constructions situated along the external perimeter of historical city centers and built with retaining and defensive functions. In the last ten years there have been several collapses which were triggered by rainfall induced landslides as reported in [1–3]. In the presented case study, prolonged rainfalls of medium to high intensity caused long lasting wet periods in the days before the collapse: this suggests that soil moisture played a key role.

An increase of the action as hydraulic thrust and a contemporary decrease in shear strength of the soil leading to instability are well-known phenomena. The evaluation of the effects of moisture as an external action and its non-deterministic characterization remains an open issue for both traditional and bio-aggregate based masonry [4]. While other natural phenomena such as wind or snow are widely regulated by international codes, together with seismic effects on cultural heritage buildings [5, 6], no in depth details are provided for this variable [7, 8].

© Springer Nature Switzerland AG 2021
O. Gervasi et al. (Eds.): ICCSA 2021, LNCS 12950, pp. 220–234, 2021.
https://doi.org/10.1007/978-3-030-86960-1_16

Non-linear analyses taking into account an increasing level of hydraulic thrust on a section of Volterra urban walls were performed in [9]. Moisture depends not only on rainfalls but also on permanent conditions such as soil properties, drainage systems and human interferences. This has been investigated in [10] where the role of antecedent soil moisture on the critical rainfall intensity-duration that caused the landslide is considered.

The procedure here explained considers the imbibition only due to rainfalls. Starting with a short discussion on survey strategy [11–13] a section of Volterra urban walls is set up for the evaluation of the safety of the infrastructure with the increase of moisture. The SLIP model proposed by Montrasio and Valentino [14] and the results of Yoshida [15] are implemented and adapted to predict the risk of landslide.

2 State of the Art

Many studies have been conducted on the complex relationship between slopes stability and rainfalls, especially where there has been a recent increase in rainfall amounts and intensity [16–19].

Rainfalls induce changes of soil behavior: an increase of degree of saturation S_r, density γ and hydraulic thrust and a decrease of cohesion c' and angle of friction ϕ.

The effects of saturation ratio on the strength parameters of partly saturated soils were investigated in [15]. A major role is played by effective accumulative infiltration and the capacity of permeability of the soil [20]. Numerical methods can represent the problem considering many variables, however the computational burden is high and time-consuming [14].

A simplified method of analysis (SLIP model) aimed at highlighting the main aspects of the triggering event with a small amount of parameters is proposed by Montrasio and Valentino [14]. This physical phenomenon is reproduced into a mathematical model which can be described in Fig. 1.

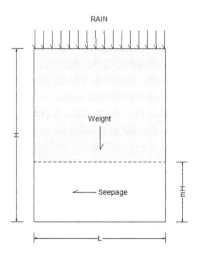

Fig. 1. Scheme of the infinite partially saturated slope of height H, as proposed in [14]

The SLIP model is applied to the case of an indefinite slope of reduced thickness and evaluates the safety factor taking into account the increase of the degree of saturation (S_R) with varying rainfall amounts. Other input parameters are the slope geometry, soil state, mechanical and hydraulic characteristics.

The water flowing vertically causes the progressive saturation of the soil until stability cannot be maintained.

The safety factor is defined applying the limit equilibrium method to an equivalent infinite slope. It is composed by two soil parts: a partially saturated one and another which is fully saturated.

Through a process of homogenization reduced shear strength parameters are obtained [6, 9]. The parameter m represents the thickness of the saturated layer and is a function of the height of rain. The model allows a quick evaluation of the decrease of the apparent cohesion and if applied to a chosen time frame provides an estimation of the safety factor.

Through the studies of Yoshida on silty sands and clays, it is possible to estimate the variations of the angle of friction. In [22] the angle is shown to decrease nearly at a constant rate with increasing saturation ratio depending on soil type and density.

The above studies are suitable to predict landslide events in a simplified and more rapid way. While the risk of these phenomena is related to their speed, to the difficulty of foreseeing their location and to the large amount of interferences also with other infrastructure that may be involved [23, 24], a swift and prompt method can help communities and authorities in the activity of monitoring and prevention.

3   Methods of Analysis

With the presence of infrastructures, the SLIP Model has been implemented to definite slopes (which contain retaining walls) in order to evaluate the reduction in shear strength due to the increase of the degree of saturation.

Moreover, to consider increasing rainfall durations, the soil layer involved in the weakening process has been discretized into horizontal strata. These are progressively affected by a process of imbibition which corresponds to a reduction of shear strength. The original application of the SLIP model discusses only superficial indefinite slopes. This hypothesis is here removed because of the "retaining" effect of the wall.

The experience of recent collapses highlights that the moisture of the soil is a recurrent condition in urban walls' failures. The degradation of the masonry due to the presence of water is not considered in this work.

The profundity and the number of chosen strata depend on the height of the wall and on the eventual presence of possible impermeable layers, as well as, in this case, on the considered rain duration. In the case of absence of imbibition, it is proposed to relate the overall height of the soil to the height of drainage. With this assumption two different areas of soil are highlighted, with (1) and without imbibition (2).

This approach is substantially different from the traditional drained and undrained conditions usually analyzed in the case of cohesive soil. The different strata are characterized by a different saturation level related to rainfall heights.

This approach can also be adapted to the case of imbibition derived from other sources as well as filtrating water or leakage of water pipes, taking into account the effective geometry of the filtrating system.

3.1 Evaluation of the Safety Factors

Considering a portion of soil of height H, which is chosen equal to 1 m, the portion mH ($m < 1$) is completely saturated. A representation of the model was already given on Fig. 1. The parameter m is directly linked to the cumulative height of rainfall in a certain duration of time and represents the percentage of saturated soil. It is given by the following [14, 21]:

$$m = \frac{\beta^* h}{n(1 - S_r)H} \tag{1}$$

where h is the amount of rainfall registered for a certain time duration; β^* represents the capacity of imbibition; H is the height of the layer altered by the rainfall; n is the soil porosity; S_r is the saturation grade.

The saturation grade is function of the initial saturation of the soil, S_{r0} which depends on the time of the year we are analyzing (higher values correspond to wet seasons such as autumn, winter or early spring [25]), and can be calculated from the following equation [14]:

$$S_r(h) = S_{r0} + \frac{\beta^* h}{nH} \tag{2}$$

The degree of saturation varies according to the amount of rainfall happening on a given time frame. This means that previous rainfalls may play an important role on the stability of the slope.

The m parameter is linked to h, rainfalls registered and to their durations meaning that m is time dependent.

However, if long periods of time (such as entire seasons) are considered, each daily increment of h would correspond to an increase of m without considering natural phenomena such as seepage or evapotranspiration: $m(t)$ is a negative exponential law as in Eqs. (3)-(7).

In this way it is possible to realistically evaluate the fluctuation of the safety factor over a long-time span [14].

$$m(t) = \sum e^{-K_T \frac{\sin \beta}{n(1-S_r)}(t-t_{0i})} \frac{h(t_{0i})}{nH(1 - S_r)} \tag{3}$$

$$F_s = \frac{\cot \beta \tan \phi'[\Gamma + m(n_w - 1)] + C'\Omega}{\Gamma + mn_w} \tag{4}$$

$$\Gamma = G_s(1 - n) + nS_r \tag{5}$$

$$n_w = n(1 - S_r) \tag{6}$$

$$\Omega = \frac{2}{\sin 2\beta \cdot H \cdot \gamma_w} \tag{7}$$

In the Eqs. (3)–(7) t_0 represents the starting instant (day in the case of this work); ß is the angle of inclination of the slope; H is the thickness of the soil layer; G_S is the soil gravity; n is the soil porosity; γ_M is the unit weight of water.

The proposed procedure takes advantage of this application to analyze backwards events of collapse which have triggered the city of Volterra. The analysis of rainfall characteristics in the area of study is the starting point. A record of the rainfalls with durations can be found on regional pluviometry charts [26].

3.2 Shear Strength Reduction

The aim of this work is to evaluate the shear strength reduction of soil layers under progressive imbibition corresponding to increasing rainfalls within 24 h. With this purpose, five layers one meter thick are considered and represent 1 h, 3 h, 6 h, 12 h, 24 h rain durations.

The coefficient m will be evaluated using Eq. (1) without considering the time dependent variable which is assumed constant throughout the application (rainfall of 24h). Moreover, the height of rain is considered dependent on the return period as well thanks to the well-known hydrology theories. The height of rain, h, is function of the rain duration d and return period t_r through:

$$h = a \cdot d^\kappa \cdot t_r^\eta \tag{8}$$

The rain duration is expressed in hours and the return period of the event in years; a (26,543), κ (0,264) and η (0,219) are dimensionless parameters given by multiple linear regression of regional rainfall records.

For our case studies they were deduced from. Equation (8) is a recurrent function of interpolation of pluviometry data. Similar equations are provided for specific zones depending on determined models of interpolation. The function is plotted on Fig. 2 for four different return periods and five rain durations in the range 1–24 h. Each increment of rainfall determines a variation of S_r which allows the calculation of the apparent cohesion expressed by the following:

$$c = c'(1 - m)^\alpha \text{ where } c' = A \cdot S_r \cdot (1 - S_r)^\lambda \tag{9}$$

where A and λ are two dimensionless coefficients depending on soil type [14, 27].

A base section and five progressively imbibed models with simplified geometries are created. These can be analyzed both with limit analyses and numerical methods with FEM geometries.

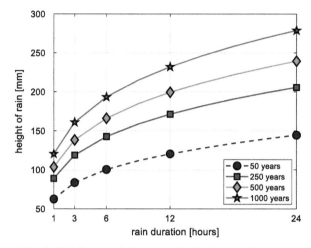

Fig. 2. Heights of rain from Eq. (8) for the site of Volterra.

The upstream soil surrounding the infrastructure is discretized into horizontal strata each representing the level of imbibition caused by increasing rain durations in terms of decrease of mechanical parameters. Figure 3 represents the most serious condition with 24 h rainfall considered.

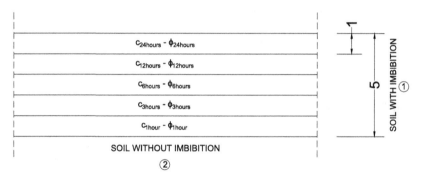

Fig. 3. Vertical section of the soil with discretization in horizontal layers. The condition shown is the worst corresponding to a rain duration of 24 h.

In the analytical models masonry is nowadays modelled also considering a parametric approach [28, 29]. This work is focused on the probabilistic effects of moisture that affects a decrease of mechanical properties of the soil only.

Both masonry and soil are modelled as continuous isotropic and homogeneous materials within the elastoplastic range and with the Mohr-Coulomb resistance criteria to only define the materials through the angle of friction and cohesion.

The simplified analysis are carried out using the SSAP2010 code [30, 31] which provides quick results through an easy interface. To evaluate the slope stability, the

Morgestern and Price method is used. The results are provided in terms of evaluation of the safety factor, FS, critic return period and rain duration.

A flow chart of the procedure is reproduced in Fig. 4.

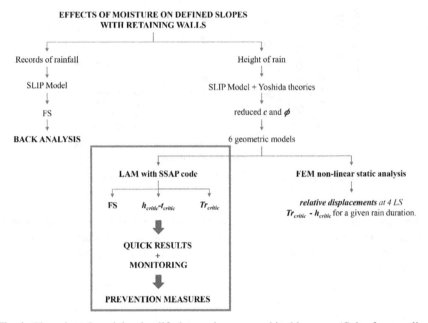

Fig. 4. Flow chart. In red the simplified procedure exposed in this paper. (Color figure online)

4 Application of the Method

The model has been applied to a section of the Volterra masonry walls, named "Sperone". The results of the simplified procedure described in 3 are here applied to a real collapse.

4.1 Case Study

On March 3rd, 2014 the Etruscan Spur close to Place "Martiri della Libertà" collapsed after some days of heavy rainfalls and an overall wet season. Indeed, in January of the same year another relevant event of collapse happened and was investigated in [7].

A picture of before and after the collapse is provided on Fig. 5 where it can also be seen how the collapse affected the entry of the parking facilities in the immediate surroundings.

Earlier than March, the spur already showed sign of incipient failure and unsuccessful interventions, such as the installation of tie-rods, took place to safeguard its stability. To prevent water from infiltrating and weaken the upstream soil, protective covers were placed. All these safeguarding measures can be seen on Fig. 6. A continuous monitoring

(a)   (b)

Fig. 5. The spur before *(a)* and after the collapse *(b)*.

Fig. 6. Evidence of the tie-rods and other protections applied on the slope.

remote system was also installed on the infrastructure providing real time information about the displacements [32].

The mechanical features of the soil are provided by on site surveys performed after both the January and March collapses (Table 1).

The main lithotype is a medium to thick sand. The masonry is made of layers of irregular porous sandstone blocks typical of the area and its characteristics are confirmed by literature reports of masonry with the same characteristics (D.M. 17/01/2018) [9]. Reports show that the "Sperone" underwent many interventions with partial reconstruction and elevation leading to a very irregular texture.

4.2 Evaluation of the Safety Factors (FS)

Through the procedure discussed in Sect. 3.1, it was possible to perform a back analysis and calculate the variation of the safety factor over 3 months including the two events of collapse. The pluviometry records are found on the Tuscany Region archive. Through Eqs. (3)–(7), it's possible to evaluate the fluctuation of FS (Fig. 7a) and of the degree of

Table 1. Mechanical parameters of the masonry and soil.

Materials	φ (°)	c' (kPa)	C_u (kPa)	γ (kN/m³)	$γ_{sat}$ (kN/m³)
Masonry	44,00	410,00	0,00	21,00	21,00
Sand 30%	37,00	0,00	0,00	20,00	22,00

saturation S_r [%] (Fig. 7b). The initial saturation grade is fixed as 75% as suggested in [25] for wet seasons.

As Fig. 7 shows, the safety factor reaches its minimum mid-February, with a sudden decrease between the 30th and 31st January where the first collapse happened. Values < 1 are recorded until mid-March. This means that on the 3rd March low values were recorded.

The graph on Fig. 7b shows an increase of saturation in correspondence of the events of collapse, with values reaching the 90% of saturation at the end of January and values slightly over than 80% at the beginning of March.

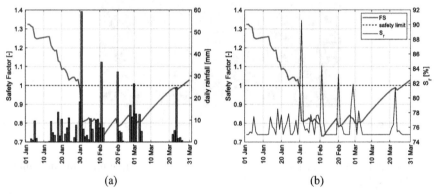

(a)   (b)

Fig. 7. Variation of FS in the period January – March 2014 and of daily rainfall amounts *(a)* or degree of saturation *(b)*.

4.3   Shear Strength Reduction

The first step consisted of the definition of the slope geometry as shown in Fig. 8.

The collapsed wall section height varies from 8.50 m to 14.00 m. A medium value of 12.10 m was considered. The top thickness of the wall was deduced from previous surveys in possession of the Municipality of Volterra. The upstream filling is almost full and the slope quite relevant.

While implementing the geometry into the SSAP2010 code it was necessary to consider the parking facilities at the base of the wall. They consist of a reinforced concrete construction approximately 4 m high. In the view of keeping the model as simple as possible it was considered as an "obstacle" not interacting with the stability analysis. The application of the method exposed in 3.2 allows the calculation of the reduced strength parameters at each increasing level of imbibition.

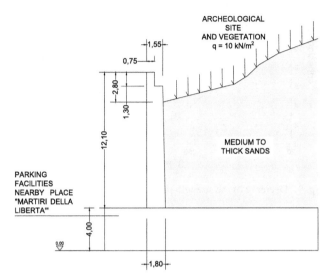

Fig. 8. Schematic geometric of the standard section.

The m parameter is calculated with Eq. (1) and depends on h, height of rain which is evaluated with Eq. (8) for four return periods and 5 rain durations. Globally, 20 values of m are calculated. The imbibition grade m also depends on the capacity of imbibition of the soil, β^*. However, assessing the percentage of water seeping into the soil and the percentage of runoff water is an open issue. In this circumstance, it has been estimated that the 30% of the total rainfall height doesn't affect the soil; thus, the value of β^* is equal to 0,70 [33, 34].

The results are shown on Fig. 9 for a return period of 250 years. The exponential interpolation of the values proves a good trend with R^2 close to 1. The cohesion reaches almost 0 with values of S_r close to 80% for rain durations of 24 h.

Once the mechanical characteristics are determined, LAM analyses are performed. First step is the analysis of the standard section. Then, 5 modified geometries corresponding to different rain durations are modelled. The process is repeated for each return period. The results are first expressed in terms of slope safety factor, FS.

The results for $t_r = 500$ years are plotted on Fig. 10.

The linear interpolation shows a good trend. The value FS = 1 is reached for rain duration close to 6 h and rainfall heights of 175 mm.

Lower return periods (50 years) show FS > 1. A decrease is expected for rain durations greater than 24 h.

The results are also presented in terms of critic return periods for each rain duration, Fig. 11, (Table 2). These correspond to values of FS < 1 which means that the stability of the slope is no longer guaranteed.

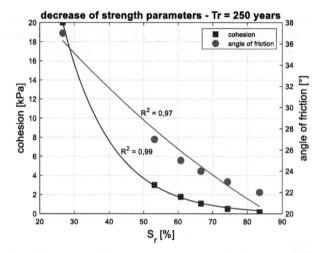

Fig. 9. Decrease of the strength parameters with the increase of Sr.

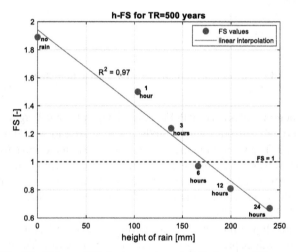

Fig. 10. Values of FS for increasing rainfalls for $t_r = 500$ years.

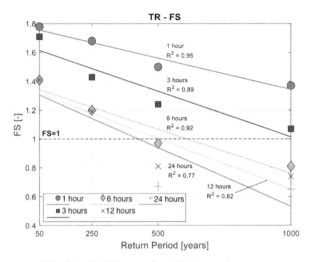

Fig. 11. TR-FS graph for 1–24 h rain durations.

Table 2. Values of critic return periods and corresponding values of h.

d [hours]	1	3	6	12	24
Tr [years]	1945	1079	623	505	432
IM - h [mm]	173	168	166	186	213

5 Conclusions

The aim of this work is to show a quick procedure to investigate the effect of moisture on soil interacting masonry wall structures. The theories for partially saturated soils and the model of Montrasio and Valentino [14] are adapted to the case of a finite slope with a retaining infrastructure. This allows the evaluation of the global safety factor for growing imbibition scenarios up to a maximum rain duration of 24 h.

The height of rain, which is the intensity measure, represents a probabilistic characterization of the meteorological event. The simplified methodology shows that moisture can significantly affect the global stability of the slope. This procedure combined with continuous monitoring techniques could help understand which are the most vulnerable sections of these extended infrastructures. The back analysis suggests a good predictability of the model, with values of FS < 1 when the relevant collapses occurred.

The results of this work highlight the role played by superficial imbibition in the stability analysis of urban walls. Indeed, the worst scenario (24-h rainfalls) considers a superficial 5 m high portion of the upstream section of the slope affected by a reduction of the strength parameters.

Some indications are provided to mitigate the vulnerability of this kind of infrastructure characterized by a strict relationship with other structures [35, 36] and with the earth [37]. A possible design strategy to mitigate the phenomena could be that of arranging drainage systems at different heights, also superficially. As well as this, superficial water management systems to avoid excessive imbibition are to be planned.

References

1. Puppio, M.L., Vagaggini, E., Giresini, L., Sassu, M.: Large-scale survey method for the integrity of historical urban walls: application to the case of Volterra (Italy). Procedia Struct. Integrity **28**, 330–343 (2020). https://doi.org/10.1016/j.prostr.2020.10.039
2. Andreini, M., De Falco, A., Giresini, L., Sassu, M.: Recenti eventi di crollo in mura storiche urbane (May), pp. 14–16, (2015).
3. Sassu, M., Andreini, M., Casapulla, C., De Falco, A.: Archaeological consolidation of UNESCO masonry structures in Oman: the sumhuram citadel of khor rori and the al balid fortress. Int. J. Archit. Herit. **7**(4), 339–374 (2013). https://doi.org/10.1080/15583058.2012. 665146
4. Sassu, M., Giresini, L., Bonannini, E., Puppio, M.: On the use of vibro-compressed units with bio-natural aggregate. Buildings **6**(3), 40 (2016). https://doi.org/10.3390/buildings6030040
5. Casapulla, C., Argiento, L.U., Maione, A.: Seismic safety assessment of a masonry building according to Italian Guidelines on cultural heritage: simplified mechanical-based approach and pushover analysis. Bull. Earthq. Eng. **16**(7), 2809–2837 (2017). https://doi.org/10.1007/s10518-017-0281-9
6. Casapulla, C., Maione, A., Argiento, L.U.: Seismic analysis of an existing masonry building according to the multi-level approach of the Italian guidelines on cultural heritage. Ing. Sismica **34**(1), 40–59 (2017)
7. Puppio, M.L., Vagaggini, E., Giresini, L., Sassu, M.: Landslide analysis of historical urban walls : the case study of Volterra, pp. 1–29 (January 2014)
8. Alvioli, M., et al.: Implications of climate change on landslide hazard in central Italy. Sci. Total Environ. **630**, 1528–1543 (2018). https://doi.org/10.1016/j.scitotenv.2018.02.315
9. Puppio, M.L., Giresini, L.: Estimation of tensile mechanical parameters of existing masonry through the analysis of the collapse of Volterra's urban walls. Frat. ed Integrita Strutt. **13**(49), 725–738 (2019). https://doi.org/10.3221/IGF-ESIS.49.65
10. Lazzari, M., Piccarreta, M., Manfreda, S.: The role of antecedent soil moisture conditions on rainfall-triggered shallow landslides. Nat. Hazards Earth Syst. Sci. (December), pp. 1–11 (2018). https://doi.org/10.5194/nhess-2018-371
11. Puppio, M. L., Vagaggini, E., Giresini, L., Sassu, M.: Large-scale survey method for the integrity of historical urban walls : application to the case of Volterra (Italy). In: VECF1 - 1st Virtual European Conference on Fracture, pp. 1–14 (2019)
12. Mistretta, F., Sanna, G., Stochino, F., Vacca, G.: Structure from motion point clouds for structural monitoring. Remote Sens. **11**(16), 1940 (2019). https://doi.org/10.3390/rs11161940
13. Pucci, A., Sousa, H.S., Puppio, M.L., Giresini, L., Matos, J.C., Sassu, M.: Method for sustainable large-scale bridge survey.pdf. In: Towards a Resilient Built Environment Risk and Asset Management, pp. 1034–1041 (2019)
14. Montrasio, L., Valentino, R.: A model for triggering mechanisms of shallow landslides. Nat. Hazards Earth Syst. Sci. **8**(5), 1149–1159 (2008). https://doi.org/10.5194/nhess-8-1149-2008
15. Yoshida, Y., Kuwano, J., Kuwano, R.: Effects of saturation on shear strenght of soils. Soils Found. **31**(1), 181–186 (1991)
16. Kristo, C., Rahardjo, H., Satyanaga, A.: Effects of variations in rainfall intensity on slope stability in Singapore. Int. Soil Water Conserv. Res. **5**, 258–264 (2017). https://doi.org/10.1016/j.swcr.2017.07.001

17. Croce, P., Formichi, P., Landi, F., Marsili, F.: Influence of climate change on extreme values of rainfall, In: IMSCI 2018 - 12th International Multi-conference on Society, Cybernetics and Informatics, Proceedings, vol. 1, pp. 132–137 (2018). https://www.scopus.com/inward/record.uri?eid=2-s2.0-85056534760&partnerID=40&md5=92e34ccc29e7fa72a35eb98256ec59ee

18. Pucci, A., Sousa, H.M., Matos, J.C.: Predicting the change of hydraulic loads on bridges: a case study in Italy with a 100-year database. In: 20th Congress of IABSE, New York City 2019: The Evolving Metropolis - Report, pp. 443–448 (2019). https://www.scopus.com/inward/record.uri?eid=2-s2.0-85074455349&partnerID=40&md5=78f7f686fc06a68bcd8820fd9431adcc

19. Puppio, M.L., Vagaggini, E., Giresini, L., Sassu, M.: Landslide analysis of historical urban walls due to rainfalls: overview of recent collapses in Italy and the case of Volterra. J. Perform. Constr. Facil. (2021). https://doi.org/10.1061/(ASCE)CF.1943-5509.0001647

20. Chen, X., Guo, H., Song, E.: Analysis method for slope stability under rainfall action, Landslides Engineered Slopes (2008)

21. Montrasio, L., Valentino, R.: Modelling rainfall-induced shallow landslides at different scales using SLIP - part II. Procedia Eng. **158**, 482–486 (2016). https://doi.org/10.1016/j.proeng.2016.08.476

22. Yoshida, Y., Kuwano, J., Kuwano, J.: Effects of saturation on shear strength of soils. Soils Found. **31**, 181–186 (1991)

23. Montrasio, L., Valentino, R., Losi, G.L.: Rainfall-induced shallow landslides: a model for the triggering mechanism of some case studies in northern Italy. Landslides **6**(3), 241–251 (2009). https://doi.org/10.1007/s10346-009-0154-7

24. Pucci, A., Puppio, M.L., Sousa, H.S., Giresini, L., Matos, J.C., Sassu, M.: Detour-impact index method and traffic gathering algorithm for assessing alternative paths of disrupted roads. Transp. Res. Rec. (2021). https://doi.org/10.1177/03611981211031237

25. Valentino, R., Meisina, C., Montrasio, L., Losi, G.L., Zizioli, D.: Predictive power evaluation of a physically based model for shallow landslides in the area of Oltrepò Pavese, northern Italy. Geotech. Geol. Eng. **32**(4), 783–805 (2014). https://doi.org/10.1007/s10706-014-9758-3

26. Settore Idrologico e Geologico Regione Toscana, Linee segnalatrici di probabilità pluviometrica, Pisa (2006)

27. Montrasio, L., Valentino, R., Terrone, A.: Application of the SLIP model. Procedia Earth Planet. Sci. **9**, 206–213 (2014). https://doi.org/10.1016/j.proeps.2014.06.023

28. Croce, P., Landi, F., Formichi, P.: Probabilistic seismic assessment of existing masonry buildings. Buildings **9**(12), 237 (2019). https://doi.org/10.3390/buildings9120237

29. Croce, P., et al.: Influence of mechanical parameters on non-linear static analysis of masonry buildings: a relevant case-study. Procedia Struct. Integrity **11**, 331–338 (2018). https://doi.org/10.1016/j.prostr.2018.11.043

30. Borselli, L.: SSAP2010 Slope Stability Analysis Program, Manuale di Riferimento (2018)

31. Borselli, L.: Muri a secco: verifiche di stabilità con software SSAP 5.0 e criterio GHB (GSI) per le strutture in roccia. (2020). https://doi.org/10.13140/RG.2.2.21048.90886/1

32. Tuscany, M.V., et al.: Early Warning GBInSAR-Based Method for Early Warning GBInSAR-Based Method for Monitoring Volterra (Tuscany , Italy) City Walls, (April 2015). https://doi.org/10.1109/JSTARS.2015.2402290

33. Franceschini, S.: Analisi critica di modelli previsionali per le frane in Emilia Romagna, University of Bologna (2012). https://doi.org/10.6092/unibo/amsdottorato/4731

34. Losi, G.L.: Modellazione spazio-temporale dei fenomeni di soil slip : dalla scala di pendio alla scala territoriale, Università degli Studi di Parma (2012)

35. Sassu, M., Stochino, F., Mistretta, F.: Assessment method for combined structural and energy retrofitting in masonry buildings. Buildings **7**(3), 71 (2017). https://doi.org/10.3390/buildings7030071

36. Stochino, F., Sassu, M., Mistretta, F.: Structural and thermal retrofitting of masonry walls: the case of a school in Vittoria (RG). In: Gervasi, O., et al. (eds.) ICCSA 2020. LNCS, vol. 12255, pp. 309–320. Springer, Cham (2020). https://doi.org/10.1007/978-3-030-58820-5_24
37. Borselli, L.: Reti in aderenza : progettazione alternativa in SSAP 5.0 per verifiche stabilità globali (LEM) (2020). https://doi.org/10.13140/RG.2.2.22444.82569

General Track 4: Advanced and Emerging Applications

Identifying the Origin of Finger Vein Samples Using Texture Descriptors

Babak Maser$^{(\boxtimes)}$ and Andreas Uhl

Multimedia Signal Processing and Security Lab, University of Salzburg, Salzburg, Austria
babak.maser@stud.sbg.ac.at, Uhl@cosy.sbg.ac.at

Abstract. Identifying the origin of a sample image in biometric systems can be beneficial for data authentication in case of attacks against the system and initiating sensor-specific processing pipelines in sensor-heterogeneous environments. Motivated by shortcomings of the photo response non-uniformity (PRNU) based method in the biometric context, we employ eight texture classification approaches, including frequency-, spatial-, and wavelet-based methods to detect finger vein samples images' origin. Besides, We use eight publicly available finger vein datasets and applying all eight novel classical texture descriptors and SVM classification in the suggested pipeline. A novel wavelet-based approach termed WMV demonstrated an excellent result for raw finger vein samples and the more challenging region of interest data among mentioned employed methods to identify sensor model. The observed results establish texture descriptors as effective competitors to PRNU in finger vein sensor model identification.

Keywords: Texture classification · Sensor identification · Image origin authentication · Finger vein recognition · PRNU · Texture descriptors

1 Introduction

Nowadays, we encounter a significant surge in the use of unattended applications of biometric systems. In many biometric modalities, a digital biometric image sensor is the core component for data acquisition, operating in the near-infrared (NIR) domain in the case of finger vein recognition. Identifying the origin of the query image is a vital task in digital image forensics. To establish the connection between a query image and the image's origin, we need to extract sensor information. Extraction of unique information of a biometric sensor can be done using various techniques and also at different levels. The aim is to extract the sensor information is to identify a sensor type and the technology used in the sensor or detect the sensor model and even the brand of the manufactured sensor. In biometrics, we also utilize the deduced information from the sensor to protect a recognition system against insertion attacks and enable the system to verify the ingenuity of image data samples. Independent biometric systems, the

© Springer Nature Switzerland AG 2021
O. Gervasi et al. (Eds.): ICCSA 2021, LNCS 12950, pp. 237–250, 2021.
https://doi.org/10.1007/978-3-030-86960-1_17

authenticity, and integrity of collected biometric data are critical to a biometric system's general stability and security issue. Attackers and Intruders always try to challenge the stability of the system. Considering an end-to-end biometric system, the Presentation Attack (PA) presents the spoofed artifacts to the sensor in contrast to the PA, Insertion Attack (IA) access directly to the feature extraction module. The feature extraction process can be done easily using another sensor, and to organize an insertion attack, the mentioned extracted information can be presented to a different biometric sensor or other modules of the recognition system. Obviously, the entire process of deducing information is done off-site. In the case of PA, the image can be manipulated to spoof the biometric system. In some complicated and advanced recognition systems, different sensors are presented. Each one may have different manufacturers or technology and models. In this complicated scenario, the interoperability of sensors is an important matter that can be impacted by how acquisition and image processing occurs on an in-sensor. Therefore, deducing sensors information leads us to utilize sensor specs to help and improve interoperability among the elements of recognition systems. And ultimately apply a sensor-tailored toolchain in a biometric system. Our work and research emphasize the feasibility of deducing sensor information in selective processing of the images and identifying the sensor model in a biometric recognition system, specifically for a finger vein (FV) sensor. We investigate the quality of captured finger vein data sample images using texture-based methods. There is a wide range of approaches to link the image's origin to the query image. One of the main approaches is exploiting the photo response non-uniformity (PRNU) [1]. However, the application of PRNU-based techniques in biometric sample data authentication has exhibited some difficulties: First, by analogy to the classical case, the PRNU fingerprint can be extracted from images of a biometric sensor and injected into forged sample images [2]. Only under certain restrictive conditions on available data can such attacks be detected or avoided (e.g., by applying the triangle test [3]). Second, authentication results with respect to biometric sensors have been reported to be widely varying at best (see e.g. [2,4,5] for iris sensor identification) and have been shown to be dependent and influenced by sensor components depicted in the images (see e.g. [6,7] for finger vein sensor identification). One reason for the difficulties is the requirement to compute the PRNU fingerprint from uncorrelated data - which is, of course, hard to satisfy given the high similarities among sample images present in biometric datasets [5,7]. Attempts to clarify this issue have not been convincing so far in the context of iris biometrics [8]. In finger vein biometrics, it has been shown that image content with a specifically structured characteristic and patterns (e.g., edge-like structures, finger positioning apparatus, illumination patterns) biases the estimated PRNU [7] as the extracted noise residual has to be independent of the image content. As a consequence, texture classification techniques have been proposed earlier to identify iris sensors at model level earlier [9,10]. However, contrasting to this present work, the authors of the former work propose rather costly techniques, i.e., improved Fisher vector encoding of denseSIFT and dense Micro-block difference features.

Therefore, due to the reasons explained, we launch a new investigation to examine the feasibility of identifying finger vein biometric sensors based on simple texture descriptors. In this work, we utilize classical descriptors, which are well known in computer vision, such as various wavelet-, frequency-, and spatial-based descriptors, respectively. Besides, we consider uncropped data samples (Original Samples), which exhibit highly correlated content, and we also study the effect in case sensor classification is done on regions of interest (ROIs) data, which do not depict special constant structures or patterns.

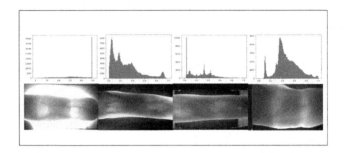

Fig. 1. Image and corresponding histogram samples of original sample images from left (a) HKPU_FV dataset, (b) UTFVP dataset, (c) SDUMLA dataset, and, (d) IDIAP dataset.

This work is structured as follows: In Sect. 2 we discuss the properties of the finger vein sample datasets as considered. Section 3 explains the conducted experiments in depth, where Subsect. 3.1 describes the used texture description methodology. Next, We discuss and analyze the experimental results in Sect. 4, and finally, we end this manuscript with a conclusion in Sect. 5.

2   Fingervein Sample Data

We considered eight different public finger vein datasets (acquired with distinct prototype near-infrared sensing devices), taking 120 samples from each dataset. As in finger vein recognition features are typically not extracted from a raw sample, but from a region-of-interest (ROI) image containing only finger vein texture, an insertion attack can also be mounted using such ROI data (in case the sensor does not deliver a raw sample to the recognition module but ROI data instead). Thus, we produced cropped ROI datasets out of the original ones (the description of the methodology is given afterward) to be able to test these data for their distinctiveness as well. Subsequently, we briefly detail the specifications of each dataset:

– **SDUMLA-HMT** [11]: Original resolution is 240×320, ROI data is 85×320 pixel. 120 images are selected from the first 30 subjects.

- **HKPU-FV** [12]: Original resolution is 256×513, ROI data is 60×390 pixel. 120 images are selected from the first 60 subjects.
- **IDIAP** [13]: Original resolution is 250×665, ROI data is 125×610 pixel. 120 images are selected from the first 60 subjects.
- **MMCBNU_6000 (MMCBNU)** [14]: Original resolution is 640×480, ROI data is 155×620 pixel. 120 images are selected from the first 20 subjects.
- **PLUS-FV3-Laser-Palmar (Palmar)** [15]: Original resolution is 600×1024, ROI data is 110×500 pixel. 120 images are selected from the first 20 subjects.
- **FV-USM** [16]: Original resolution is 480×640 pixels, ROI data is 110×280 pixel. 120 images are selected from the first 30 subjects.
- **THU-FVFDT** [15]: Original resolution is 600×1024, ROI data is 120×390 pixel. 120 images are selected from the first 120 subjects.
- **UTFVP** [17]: Original resolution is 380×672, ROI data is 140×490 pixel. 120 images are selected from the first 60 subjects.

Original sample images, as shown in Fig. 1, can be discriminated easily: Besides the differences in size (which can be adjusted by an attacker, of course), the sample images can probably be distinguished by the extent and luminance of the background. To illustrate this, we display the images' histograms above each example in Fig. 1, and those histograms clearly exhibit a very different structure. Thus, we expect texture descriptors to have an easy job of identifying the origin of the respective original sample images. The reason to use 120 images per dataset is to create the same environment and condition to have a comparable result to the PRNU-based investigations on these data (see [18]).

2.1 Generating Region of Interest Datasets

In finger vein recognition, contrasting to, e.g., fingerprint recognition, feature extraction is not applied to the entire raw sample data but instead to an ROI only [12,15]. In this ROI, only actual finger texture is contained. Depending on the setup of the system, the sensor might already extract the ROI from the raw sample. In this setting, identification of the finger vein data's origin has to be based on the ROI. Thus, it will be required under these circumstances that only finger vein texture (the ROI) is used to discriminate sensors. This is not unrealistic, as in the iris recognition case, normalized iris texture has been considered by analogy to be used for sensor identification [9,10] instead of raw iris sample data.

Different techniques have been applied to each dataset's properties to detect and segregate the finger vein region and extract a patch consisting of biometric data. For the datasets exhibiting a higher intra-variance of finger positions, we applied the following algorithm based on morphological snakes (morphological active contour without edges [19]) to extract the ROI (FV_USM, THU_FVDT, UTFVP, and MMCBNC datasets), also illustrated in Fig. 2:

1. Apply morphological snakes to the finger vein image to produce a segmented image.

2. Apply Canny edge detection and contour closing to detect the finger vein region.
3. Fill the contour.
4. Find the mass center of the filled contour and fit a line to the contour; estimate the angle (θ) of the line to the x-axis.
5. Rotate the texture area by θ degree.
6. Find the new mass center and crop the aligned original sample image.

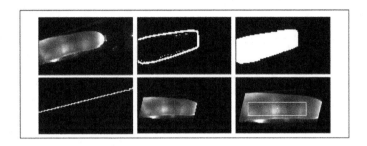

Fig. 2. ROI generation for datasets with finger position variability.

We also used this method for the Palmar dataset but replaced the morphological snakes' technique with the Chan-Vese segmentation algorithm [20].

To create the ROI for the (more straightforward) datasets HK_FV, IDIAP, and SDUMLA, we applied the following steps:

1. Apply Canny edge detection.
2. Apply a dilation operator on the detected edges.
3. Stack all images of a dataset on top of each other.
4. Fitting and cropping a rectangular (ROI) manually inside the finger vein area.

Figure 5 illustrates the results of ROI creation for a sample of each dataset (sample width has been normalized for better clarity). It gets immediately clear that discrimination is obviously more difficult based on the ROI data only. To investigate the differences between raw sample data and ROI data in more detail, we have investigated the range of luminance values and their variance across all datasets. Figures 3 and 4 display the results in the form of box-plots, where the left box-plot corresponds to the original raw sample data, and the right one to the ROI data, respectively. We can clearly see that the luminance distribution properties have been changed dramatically once we change our focus from original datasets to ROI datasets. For example, original HKPU_FV samples can be discriminated from FV_USM, MMCBNUm, PALMAR, UTFVP, and THU_FVFDT ones by just considering luminance value distribution. For the ROI data, the differences are not very pronounced anymore. When looking at the variance value distributions, we observe no such substantial discrepancy between the original

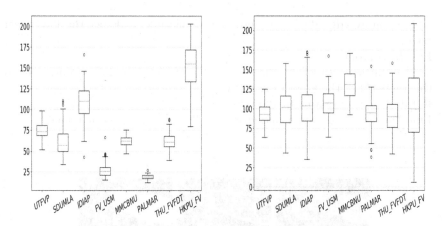

Fig. 3. Luminance distribution of original and ROI images across all datasets, respectively.

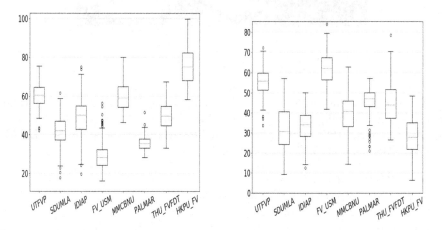

Fig. 4. Variance distribution of original and ROI images across all datasets, respectively.

sample and ROI data, still for some datasets, variance can be used as a discrimination criterion (e.g., Palmar vs. HKPU_FV in original data, FV_USM vs. HKPU_FV in ROI data). Consequently, we expect the discrimination of the considered datasets based on texture descriptors to be much more challenging when focusing on the ROI data only.

The steps to produce the cropping images is shown in Fig. 2. The size of cropped images for each dataset has been given in the Subsect. 2.

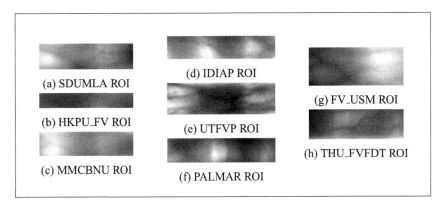

Fig. 5. ROI Samples from different datasets

3   Experimental Design

3.1   Texture Description Methodology

To discriminate sensors, we applied a number of classical yet simple approaches to produce a texture descriptor of a finger vein image. In the following subsections, we briefly describe the chosen techniques and explain how to cope with differently sized images.

Fourier Ring Filter (FRF). We generate features in the frequency domain using 2-D FFT. Independent of image size, fifteen bandpass filters split the frequency domain into equally sized bands, which are used to compute the mean and standard deviation of each ring [21] (which are used as statistical texture descriptors).

Local Binary Patterns (LBP). We use a variant of the original **Local Binary Pattern (LBP)** introduced by Ojala *et al.* [22]. This approach is called Histogram-LBP (HLBP [23]), we set the radius to 3, and the number of curricular neighborhood pixels is set to 15. The HLBP is invariant to image size if the output of the histogram for each image is normalized. Further, the number of histogram bins is fixed.

Additionally, we apply uniform LBP (ULBP) - an LBP is called uniform if the binary pattern contains at most two 0-1 or 1-0 transitions, and it has been shown that these patterns occur more frequently in natural texture (and significantly reduces feature-length vectors as the LBP histogram bins are reduced).

Image Histogram (IMHIST). We compute the image histogram and take the output as a feature vector. The IMHIST is invariant to image size by bin entry normalization and fixing the number of histogram bins.

Wavelet-Based Features. We apply 2-D wavelet decomposition using Daubechies 8-tap orthogonal filters to generate the coefficients in horizontal h, vertical v, and diagonal d directions. On every decomposition level, we compute the mean (μ) and standard deviation (std) for each of the sub-bands v, h, and d and concatenate those to get the mean and variance feature (WMV). We achieved invariance to image size by fixing the number of wavelet decomposition levels to 3.

Like WMV, we define wavelet variance (WV) by computing the variance per subband, and wavelet entropy (WE) by computing entropy per subband, respectively.

Local Entropy (LE). We slice a given image into 16 blocks (tiles) and compute the entropy from each tile. By taking a histogram of all produced entropies, a feature vector is generated. LE also is invariant to image size by fixing the number of image blocks (tiles) and bin number of the histogram.

3.2   SVM Finger Vein Texture Classification (FVTC)

The SVM classifier is trained by feeding 67% of all images then the remaining images 33% are used for the testing purpose. Images in all finger vein datasets are randomly shuffled beforehand to avoid subject-related bias. To optimize the SVM classifier and obtain the most promising hyperparameters such as C, γ, *kernel*, and *degree*, we employed a Grid Search technique with 4-fold cross-validation [24]. Also, we set the decision function to *"one vs rest"* strategy.

3.3   Evaluation Metrics

We use classical measures to rate our sensor identification task, which is a multi-class classification problem. The multi-class problem is an extension of binary classification. We use two evaluation approaches: First, receiver operating characteristic (ROC) which relates the false positive rate to the false negative rate, and second, the relation of precision and recall. For both relations, the Area Under The Curve (AUC) can be computed as a single measure.

Once we have multi-class problem, the challenging point is how to get an overall score. Often, one simply takes the average of the AUC-ROC metrics. For illustration, to calculate Recall for three-class problems, we sum up three Recalls and divide them by number of contributing classes, that is classical average. In contrast, another approach is based on summing up individual terms during the computation. In this work we obtained average of measurement metrics globally as shown in Eq. (1) and (2) (see [25, 26]).

$$Avg\ Precision = \frac{\sum_c TP_c}{\sum_c TP_c + \sum_c FP_c} \tag{1}$$

$$Avg\ Recall = \frac{\sum_c TP_c}{\sum_c TP_c + \sum_c FN_c} \tag{2}$$

Where c is the class label, TP is True Positive, FN is False Negative and FP is False Positive.

3.4   Enhancement Techniques

To investigate the impact of enhancement on sensor identification we have decided to apply an enhancement method on uncropped datasets as well as ROI datasets. we applied the following methods.

1. **Wiener Filter and CLAHE (Enh.):** To enhance the quality of the images and remove undesired noise-related artifacts, we apply a Wiener Filter [27] and also to improve the contrast of images, and we use CLAHE (Contrast Limited Adaptive Histogram Equalization). Applying these two filters is done sequentially on all images of the mentioned datasets.
2. **No Enhancement (NoEnh.):** Sample images are used as present in the datasets or as obtained after ROI computation.

4   Result Analysis

Table 1 displays the experimental results. As expected, sensor identification is easily achieved based on original samples. Image enhancement improves results (mostly slightly) in many cases. We get values >0.99 in terms of AUC-ROC and AUC Precision-Recall for ULBP and WMV for both enhancement settings, which is a perfect result.

For ROI data, results deteriorate slightly. In particular, AUC Precision-Recall for spatial domain techniques is no longer acceptable. DFT and wavelet-based descriptors, however, still result in values well above 0.9, in most cases above 0.95, which is a very good result that could not be expected given the high similarity of textures and the simplicity of our descriptors. For ROI data, there is no clear trend if enhancement as being applied is beneficial or not. In any case, similar to the original sample case, for well-performing techniques, the difference is negligible.

Tables 2, 3, 4, 5, 6, 7, and 9 show the detailed outcomes of the proposed methodology for each sensor. In the mentioned tables, all of the Avg AUC-ROC scores for uncropped images are excellent, i.e., many results give an excellent Avg AUC-ROC score of 1.0. This implies that our approach may be considered as the best choice of image' origin identification (and is a hard competitor for the PRNU-based method when it comes to sensor model identification). As is expected from the overall results, also the estimated scores for uncropped images (ROI) are excellent, at least for the best methods considered. For those, the difference in result accuracy to original samples usually is seen only at the second precision digit after the decimal point.

When explicitly comparing our best result (i.e., AUC-ROC of the WMV descriptor of 0.999 and 0.994 for original and ROI samples, respectively) to the best PRNU results shown in [18] (Table 10), We notice the original Sample's superior value and compatible value on the ROI samples for the approach suggested in this work.

Table 1. Sensor identification results.

	Descriptor	Original sample		ROI	
		No.Enh.	Enh.	No.Enh.	Enh.
Avg AUC-ROC	FRF	0.997	0.999	0.989	0.986
Avg AUC Precision-Recall		0.987	0.999	0.952	0.932
Avg AUC-ROC	HLBP	0.992	0.994	0.941	0.955
Avg AUC Precision-Recall		0.960	0.968	0.762	0.783
Avg AUC-ROC	ULBP	0.995	0.999	0.931	0.932
Avg AUC Precision-Recall		0.994	0.997	0.763	0.728
Avg AUC-ROC	LE	0.919	0.952	0.834	0.858
Avg AUC Precision-Recall		0.685	0.772	0.538	0.650
Avg AUC-ROC	ImHist	0.989	0.961	0.906	0.966
Avg AUC Precision-Recall		0.883	0.789	0.481	0.851
Avg AUC-ROC	WV	0.998	0.998	0.984	0.983
Avg AUC Precision-Recall		0.991	0.994	0.930	0.909
Avg AUC-ROC	WE	0.993	0.985	0.977	0.959
Avg AUC Precision-Recall		0.979	0.943	0.902	0.854
Avg AUC-ROC	WMV	0.999	0.999	0.982	0.994
Avg AUC Precision-Recall		0.999	0.996	0.917	0.971

Table 2. FRF exhibits how well detects and identifies different sensors, the mentioned values are Avg AUC-ROC scores

Sensor	Uncropped image		ROI	
	No.Enh	Enh.	No.Enh	Enh.
UTFVP	1.000	1.000	0.971	0.951
FV_USM	1.000	1.000	0.983	0.955
PALMAR	0.999	1.000	0.988	0.975
SDUMLA	0.988	1.000	1.000	1.000
THU_FVFDT	1.000	1.000	1.000	1.000
IDIAP	1.000	1.000	1.000	1.000
MMCBNU	0.992	0.999	0.961	0.968
HKPU-FV	1.000	1.000	1.000	1.000

Table 3. Wavelet Variance exhibits how well detects a identifies different sensors, the mentioned values are Avg AUC-ROC scores

Sensor	Uncropped image		ROI	
	No.Enh	Enh.	No.Enh	Enh.
UTFVP	0.996	1.000	0.938	0.954
FV_USM	1.000	0.997	1.000	0.992
PALMAR	1.000	0.999	0.995	0.999
SDUMLA	1.000	1.000	1.000	0.995
THU_FVFDT	0.998	0.999	0.951	0.980
IDIAP	1.000	1.000	0.996	0.990
MMCBNU	1.000	1.000	0.955	0.944
HKPU-FV	1.000	1.000	0.996	0.975

Table 4. Wavelet mean Variance exhibits how well detects and identifies different sensors, the mentioned values are Avg AUC-ROC scores

Sensor	Uncropped image		ROI	
	No.Enh	Enh.	No.Enh	Enh.
UTFVP	1.000	1.000	0.930	0.993
FV_USM	1.000	1.000	0.998	0.997
PALMAR	1.000	0.999	0.995	0.997
SDUMLA	1.000	1.000	1.000	1.000
THU_FVFDT	0.999	0.999	0.983	0.994
IDIAP	1.000	1.000	0.990	0.992
MMCBNU	1.000	1.000	0.988	0.994
HKPU-FV	1.000	1.000	0.991	0.988

Table 5. Wavelet Entropy exhibits how well detects and identifies different sensors, the mentioned values are Avg AUC-ROC scores

Sensor	Uncropped image		ROI	
	No.Enh	Enh.	No.Enh	Enh.
UTFVP	0.996	0.9960	0.972	0.969
FV_USM	0.992	0.9460	0.992	0.967
PALMAR	0.970	0.9903	0.999	0.985
SDUMLA	0.997	0.9844	1.000	0.962
THU_FVFDT	1.000	1.0000	0.948	0.889
IDIAP	1.000	1.0000	0.977	0.951
MMCBNU	0.992	0.9493	0.922	0.935
HKPU-FV	1.000	0.9997	0.965	0.992

Table 6. Histogram LBP exhibits how well detects and identifies different sensors, the mentioned values are Avg AUC-ROC scores

Sensor	Uncropped image		ROI	
	No.Enh	Enh.	No.Enh	Enh.
UTFVP	0.998	0.994	0.937	0.922
FV_USM	0.954	0.954	0.910	0.917
PALMAR	0.997	0.999	0.870	0.925
SDUMLA	0.994	0.990	0.948	0.969
THU_FVFDT	0.985	0.996	0.977	0.969
IDIAP	0.999	1.000	0.988	0.996
MMCBNU	0.997	1.000	0.901	0.929
HKPU-FV	1.000	1.000	0.978	0.977

Table 7. Uniform LBP exhibits how well detects and identifies different sensors, the mentioned values are Avg AUC-ROC scores

Sensor	Uncropped image		ROI	
	No.Enh	Enh.	No.Enh	Enh.
UTFVP	0.988	0.994	0.918	0.952
FV_USM	0.999	0.999	0.978	0.950
PALMAR	1.000	1.000	0.896	0.827
SDUMLA	1.000	1.000	1.000	0.999
THU_FVFDT	1.000	1.000	0.839	0.866
IDIAP	1.000	1.000	0.955	0.935
MMCBNU	1.000	1.000	0.821	0.880
HKPU-FV	1.000	1.000	0.990	0.986

Table 8. Image Histogram (Im Hist) exhibits how well detects and identifies different sensors, the mentioned values are Avg AUC-ROC scores

Sensor	Uncropped image		ROI	
	No.Enh	Enh.	No.Enh	Enh.
UTFVP	1.000	0.999	0.934	0.932
FV_USM	0.977	0.991	0.916	0.935
PALMAR	0.999	0.933	0.654	0.987
SDUMLA	0.999	0.976	0.950	0.960
THU_FVFDT	1.000	1.000	0.989	1.000
IDIAP	1.000	1.000	0.975	0.999
MMCBNU	0.958	0.874	0.911	0.877
HKPU-FV	1.000	1.000	0.920	0.996

Table 9. Local Entropy (LE) exhibits how well detects and identifies different sensors, the mentioned values are Avg AUC-ROC scores

Sensor	Uncropped image		ROI	
	No.Enh	Enh.	No.Enh	Enh.
UTFVP	0.939	0.917	0.651	0.825
FV_USM	0.920	0.973	0.865	0.932
PALMAR	0.928	0.951	0.783	0.901
SDUMLA	0.867	0.890	0.990	0.999
THU_FVFDT	0.993	0.998	0.985	1.000
IDIAP	0.756	0.872	0.707	0.693
MMCBNU	0.973	0.973	0.737	0.261
HKPU-FV	0.950	0.963	0.611	0.806

Table 10. Comparing results of the PRNU and the Texture Descriptors approaches on the original and the ROI sample data.

	PRNU		Texture descriptors
	NCC	PCE	WMV
Original sample	0.992	0.991	0.999
ROI	0.998	0.997	0.994

5   Conclusion

We have identified texture descriptors as being well suited for finger vein sensor *model* identification, being applied to raw sample images as well as to the more challenging finger vein ROI data. Enhancement techniques turn out to be non-decisive for classification accuracy, at least when considering top-performing techniques. Overall, but especially when considering results for ROI data, frequency-based and wavelet-based descriptors are found to perform superior to spatial domain techniques. Specifically, We introduced a novel wavelet-based texture descriptor method termed WMV that demonstrated excel on various datasets. The excellent results suggest the proposed techniques be better suited as compared to PRNU-based methods for the task investigated. Also, a fusion of both approaches seems promising, which will be subject to further investigations.

References

1. Chen, M., Fridrich, J., Goljan, M., Lukás, J.: Determining image origin and integrity using sensor noise. IEEE Trans. Inf. Forensics Secur. **3**(1), 74–90 (2008)
2. Uhl, A., Höller, Y.: Iris-sensor authentication using camera PRNU fingerprints. In: Proceedings of the 5th IAPR/IEEE International Conference on Biometrics, ICB 2012, New Delhi, March 2012, India, pp. 1–8 (2012)
3. Goljan, M., Fridrich, J., Chen, M.: Sensor noise camera identification: countering counter-forensics. In: Media Forensics and Security II, vol. 7541, p. 75410S. International Society for Optics and Photonics (2010)
4. Kalka, N., Bartlow, N., Cukic, B., Ross, A.: A preliminary study on identifying sensors from iris images. In: Proceedings of the IEEE Conference on Computer Vision and Pattern Recognition Workshops, pp. 50–56 (2015)
5. Debiasi, L., Sun, Z., Uhl, A.: Generation of iris sensor PRNU fingerprints from uncorrelated data. In: Proceedings of the 2nd International Workshop on Biometrics and Forensics, IWBF 2014, Valletta, Malta, pp. 1–6 (2014)
6. Söllinger, D., Maser, B., Uhl, A.: PRNU-based finger vein sensor identification: on the effect of different sensor croppings. In: 2019 International Conference on Biometrics (ICB), pp. 1–8 (2019)

7. Söllinger, D., Debiasi, L., Uhl, A.: Can you really trust the sensor's PRNU? How image content might impact the finger vein sensor identification performance. In: Proceedings of the 25th International Conference on Pattern Recognition (ICPR), pp. 7782–7789 (2020)
8. Debiasi, L., Uhl, A.: Techniques for a forensic analysis of the CASIA-Iris V4 database. In: Proceedings of the 3rd International Workshop on Biometrics and Forensics, IWBF 2015, Gjovik, Norway, pp. 1–8 (2015)
9. Debiasi, L., Kauba, C., Uhl, A.: Identifying iris sensors from iris images. In: Iris and Periocular Biometric Recognition, vol. 5, p. 359 (2017)
10. El-Naggar, S., Ross, A.: Which dataset is this iris image from? In: 2015 IEEE International Workshop on Information Forensics and Security (WIFS), pp. 1–6. IEEE (2015)
11. Yin, Y., Liu, L., Sun, X.: SDUMLA-HMT: a multimodal biometric database. In: Sun, Z., Lai, J., Chen, X., Tan, T. (eds.) CCBR 2011. LNCS, vol. 7098, pp. 260–268. Springer, Heidelberg (2011). https://doi.org/10.1007/978-3-642-25449-9_33
12. Kumar, A., Zhou, Y.: Human identification using finger images. IEEE Trans. Image Process. 21(4), 2228–2244 (2012)
13. Tome, P., Vanoni, M., Marcel, S.: On the vulnerability of finger vein recognition to spoofing. In: IEEE International Conference of the Biometrics Special Interest Group (BIOSIG) (September 2014)
14. Lu, Y., Xie, S.J., Yoon, S., Wang, Z., Park, D.S.: An available database for the research of finger vein recognition. In: 2013 6th International Congress on Image and Signal Processing (CISP), CISP 2013, vol. 1, pp. 410–415. IEEE (2013)
15. Kauba, C., Prommegger, B., Uhl, A.: Focussing the beam - a new laser illumination based data set providing insights to finger-vein recognition. In: Proceedings of the IEEE 9th International Conference on Biometrics: Theory, Applications, and Systems, BTAS 2018, Los Angeles, California, USA, pp. 1–9 (2018)
16. Asaari, M.S.M., Rosdi, B.A., Suandi, S.A.: Fusion of band limited phase only correlation and width centroid contour distance for finger based biometrics. Exp. Syst. Appl. 41(7), 3367–3382 (2014)
17. Ton, B.T., Veldhuis, R.N.J.: A high quality finger vascular pattern dataset collected using acustom designed capturing device. In: International Conference on Biometrics, ICB 2013. IEEE (2013)
18. Söllinger, D., Maser, B., Uhl, A.: PRNU-based finger vein sensor identification: on the effect of different sensor croppings. In: 2019 International Conference on Biometrics (ICB), pp. 1–8. IEEE (2019)
19. Chan, T.F., Vese, L.A.: Active contours without edges. IEEE Trans. Image Process. 10(2), 266–277 (2001)
20. Getreuer, P.: Chan-Vese segmentation. IPOL J. 2, 214–224 (2012)
21. Vécsei, A., Fuhrmann, T., Liedlgruber, M., Brunauer, L., Payer, H., Uhl, A.: Automated classification of duodenal imagery in celiac disease using evolved Fourier feature vectors. Comput. Meth. Program. Biomed. 95, S68–S78 (2009)
22. Ojala, T., Pietikäinen, M., Harwood, D.: A comparative study of texture measures with classification based on featured distributions. Pattern Recogn. 29(1), 51–59 (1996)
23. Boulogne, F., Warner, J.D., Yager, E.N.: scikit-image: Image processing in Python. PerrJ 2, e453 (2014)
24. Kohavi, R., et al.: A study of cross-validation and bootstrap for accuracy estimation and model selection. In: IJCAI, vol. 14, Montreal, Canada, pp. 1137–1145 (1995)

25. Ferri, C., Hernández-Orallo, J., Salido, M.A.: Volume under the ROC surface for multi-class problems. In: Lavrač, N., Gamberger, D., Blockeel, H., Todorovski, L. (eds.) ECML 2003. LNCS (LNAI), vol. 2837, pp. 108–120. Springer, Heidelberg (2003). https://doi.org/10.1007/978-3-540-39857-8_12
26. Van Asch, V.: Macro-and micro-averaged evaluation measures. In: CLiPS, Belgium, vol. 49 (2013)
27. Benesty, J., Chen, J., Huang, Y., Doclo, S.: Study of the wiener filter for noise reduction. In: Speech Enhancement. Signals and Communication Technology. Springer, Heidelberg (2005). https://doi.org/10.1007/3-540-27489-8_2

A Complex Adaptive System Approach for Anticipating Technology Diffusion, Income Inequality and Economic Recovery

Mark Abdollahian[(✉)], Yi Ling Chang, and Yuan-Yuan Lee

School of Social Science, Policy and Evaluation, Claremont Graduate University, Claremont, CA, USA
{mark.abdollahian,yi-ling.chang,yuan-yuan.lee}@cgu.edu

Abstract. Technology progress challenges equilibrium structures of human, social, cultural and behavioral systems. These are seen in 'new normals' of discontinuous market changes and nonlinear trend dynamics underpinning new individual behavior and human connectivity. Here we instantiate an agent based model integrated in a CAS framework of micro individual socio-economic transactions games empowered by technology and played on different economic network structures and market sophistication for anticipating macroeconomic and social equality outcomes. We explore scenario comparisons across GDP shocks and their impact on inequality, market behavior and recovery patterns. Several different V, U, L and W-shaped macro recovery patterns are categorized specifically dependent on such CAS structures. More sophisticated markets with preferential connectivity tend to offer lower levels of inequality and are more economically robust than less developed, randomly connected, transactional markets. High resolution computational approaches such as this can offer policymakers deeper insights and more actionable policy levers to achieve desired socio-economic outcomes in times of crisis.

Keywords: Economic growth · Technology adoption · Economic recovery · Complex adaptive system · Co-evolution · Agent-based Modeling (ABM)

1 Introduction

Technology accessibility and adoption are a critical factor for economic growth and societal knowledge transfer [1] in a fourth industrial revolution. Technological progress both destroys and creates jobs. Job destruction reduces labor required to perform tasks, automation and altering of required employment tasks that eliminate some occupations. Job creation occurs where automation complements specific job tasks, creates new industries and products, increases productivity, lowers costs and ultimately results in higher growth and income, thus boosting aggregate demand for future growth [2].

While technology if often seen as a lifeline to survival, especially in a post pandemic world, the pace of adoption and development with its accompanying life-changing possibilities is also disruptive [3]. As innovation continue to enable economic growth, it also

© Springer Nature Switzerland AG 2021
O. Gervasi et al. (Eds.): ICCSA 2021, LNCS 12950, pp. 251–262, 2021.
https://doi.org/10.1007/978-3-030-86960-1_18

intensifies market competition and changes individuals' economic employment opportunities [4]. Wealth accumulation for individuals who are capable of accessing technology easier than others fuels societal inequalities often exacerbated across different societal levels [5]. These implications drive conversations on how we can embrace technology advantages and redirect it as part of the solution for inequality challenges coupled with evolving economic complexity and increased social connectivity.

Obviously technological progress challenges equilibrium structures of human, social, cultural and behavioral systems [4] to 'new normals' often through nonlinear, discontinuous change and emergent behavior. This forces us to scientifically embrace not just transdisciplinary approaches but also multiple scale frameworks for understanding emergent human dynamics. Here we provide an integrative micro-meso-macro perspective on technology, growth and inequality that leverages computational, agent based modelling in a complex adaptive systems (CAS) simulation framework. We begin with a simplified agent based, macro economy [5] to frame individuals' economic activities and their summed GDP growth trajectories over time. We then decompose these macroeconomic activities into the requisite various meso economic activity networks and individual level decision behaviors. Integrating an agent-based approach with the added details on individual behavior and information flows can shed light on many important questions on economic equality, stability, fragility and recovery from economic shocks [6]. Leveraging such, we uncover some future research avenues at the intersection of technology, growth and income inequality that can be further calibrated to implement effective policy responses.

2 Related Work

2.1 Micro and Macroeconomics

Beginning with traditional economic models through an agent-based approach, we decompose macroeconomic technology and growth into two respective subcomponents of cross-scale human behavior: individual's micro motivations behind the adoption of technology for competitive economic gains which coevolve into networked meso societal behaviors within a CAS framework. Individual's actions are incentivized by preferences and constrained by beliefs while path dependence and information shape their strategic decisions [4] in various spaces of possibilities. Previous macroeconomic models based on equilibrium theory formally show how macro deficits can lead to economic crashes, but result often lack cross-scale explanations of micro individual behavior and meso networked relationships. Integrating an agent-based approach with the added details on individual behavior and information flows can shed light on many important questions on stability, fragility and recovery from economic shocks [6] and its impact on income inequality.

In previous work [7], we allow heterogeneous agents to play socio-economic transaction games to capture micro individual behaviors. Agents then interact under four different market typologies to simulate behavioral pattern changes driven by different societal networks of technology adoption and diffusion. For each typology, as technology proliferation increases economic connectivity and complexity, it also creates various structural changes across multiple scales of human behavior. Here we explore income inequality and economic resiliency to exogenous shocks across different technologically enabled societies.

2.2 Economic Shocks and Recovery

Endogenous technological change warrants the potential to revolutionize industries creating highly competitive markets [8] often through continuous innovations which raise living standards. However, rapid technological change can also disrupt labor markets, with heterogenous sectoral impact, which subsequently can cause income gaps to widen both across and within groups. Several societies experience such increasing income disparity and technological unemployment despite surging macroeconomic growth [2]. For example, both labor income shares and labor productivity have significantly decreased in OECD countries over the last 20 years [9, 10]. Many economists discuss technology gap approaches [11–13] which emphasizes the crucial role of technology in economic growth processes. Given the prevalence of technology across multiple economic sectors, both economists and governments actively seek to understand the consequences of economic shock and recovery process. Many look to diagnose and anticipate associated recovery shapes and duration with empirical evidence into several types: whether exogenous economic shocks result in a quick V- shaped shock and recovery pattern, or U-shaped longer term recovery, W-shaped 'double dip' relapses, or a L-shaped prolonged crisis due to a stagnation trap [14].

3 Model Design

Enabled by connectivity, technology sources of innovation and disruption proliferate faster than we realize. As humans are strategic, dynamic and adaptive beings, behaviors by design often change and evolve in emergent, nonlinear and discontinuous fashions [15]. Thus, today we operate in many environments where novel individual micro actions can have a direct impact to other agents' decision-making process and consequently drive second and higher order changes and feedbacks at other scales of human behavior [16].

We explicitly model these cross-scale, integrated behaviors: micro evolutionary, heterogenous socio-economic transaction games, meso social and market networks where such micro games are conducted with the diffusion of technology in such contexts and finally the aggregate macro sum of all activities to understand the emergence of growth, technology and inequality. In our work, agents' behavior rules are designed bottom up to increase modeling fidelity, combining network analysis and information diffusion theory for socio-economic interactions. The resulting simulation provides insights on emergent properties that arise from agents' interaction, allowing us to potentially anticipate far from equilibrium macroeconomic behavior or new emergent system orders. Agent interactions are conducted on four unique socio-economic network typologies, i.e. random, preferential, dynamic-random, and dynamic-preferential, to capture aggregate economic outcomes as seen in Fig. 1. Furthermore, it also can offer insights into more actionable policy formulation and effectiveness.

We begin with Romer and others endogenous growth Y functions [8] where firms are motivated by L labor and A technological change for the incentive of continuous K capital accumulation:

$$Y_t = K_t^{\alpha} (A_t L_{Yt})^{1-\alpha}$$

Using these macroeconomic foundations as our departure point, our simulation space is designed to capture agents' socio-economic production functions, interactions and dynamic behavioral changes on different networks within this macroeconomic context. Starting at the micro level, each agent has its own unique cognitive process of choosing strategies, with or without technology, that they believe will provide maximum individual economic returns. Of course, this also depends on whether they or counterparties possess technological advantages in any transactional game as competitive or comparative advantages. Four different possible meso level interaction games result depending on player types possessing technology or not. The intuition here is that technology can enable synergistic gains and effects in cooperative interactions while conversely maximizing individual gains in non-cooperative, asymmetric technology interactions. Thus agent decision spaces of possibilities are explicitly shaped by technology. Such symmetric technology, cooperative games often produce higher relative economic gains than asymmetric, non-technology enabled games [7].

At each iteration, agents randomly select a socio-economic transaction partner by four designed groupings: both players have technology, A_i has technology and A_j does not, A_i does not have technology and A_j has, and finally both players do not have technology. A_i strategies are adaptive over time, which affect A_{ij} pairs locally.

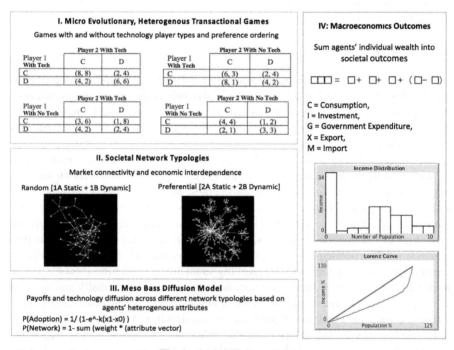

Fig. 1. CAS ABM processes

3.1 Network Typologies

In an adaptive evolving system, agents are capable of self-organizing to create or deactivate social ties which often can exhibit dynamic emergent behaviors [17]. Network formation is a complex process in which many individuals simultaneously attempt to maximize their own utility, under multiple, possibly conflicting, constraints with a tendency of homophily to avoid conflicting relationships with others [18, 19]. Our simulated economy is designed to increase interaction fidelity to capture agents' behavioral changes on different networks; creating local ties when they may be able to benefit from a cooperative relationship or dissolve social connections when they face increasing conflictual behaviors.

Agents are created with heterogeneous attributes including income, influence and perception towards technology; each agent has unique cognitive processes of choosing the strategy that best allow them to receive maximum return through socio-economic exchanges depending on whether counterparties possess technology or informational advantages. Individual's distinct cognitive characteristics and motivations contribute to creating social ties – different linkage and attachment rules, forming a collaborative social network. In this study, we include four different network typologies: static random (Erdős–Rényi) or static preferential (Barabási–Albert) network structures, as well as dynamic random and dynamic preferential networks that evolve over time. Learning and feedback loops occur as a result of personal interactions between individual and other agents and thus network formation can also dynamically evolve overtime.

Under random networks, the probability of node i has exactly k links following a binomial distribution:

$$p^k = (\frac{N-1}{k})p^k(1-p)^{N-1-k}$$

Such random networks can serve as proxies for undeveloped societies, low input factor production, opportunistic transactional markets and simpler economic structures.

Under preferential attachments, new nodes come in with m edges that prefer to attach to well-connected nodes creating various centralized hubs following a power law distribution:

$$p(i) = m\frac{k_i}{\sum_j k_j}$$

Such preferential networks help instantiate more sophisticated markets, economic production functions and value chains based on requisite factors and inputs of production or services in knowledge economies.

These various societal network typologies determine how technology diffusion and adaptation occur through the well-known Bass diffusion [20] model: with $F(t)$ as the installed base fraction of technology, $f(t)$ the change in the installed base fraction with p innovation and q imitation. The $1-F(t)$ denominator constraint creates decreasing technology diffusion at higher technology levels and over time across different network typologies.

$$\frac{f(t)}{1-F(t)} = p + qF(t)$$

3.2   Nonlinear Micro Behaviors to Simulated Macroeconomics Outcomes

At each iteration t, agents interact and perform socio-economic transaction games given differences in technology advantages which later form further connections based on various network typologies. Knowledge spreads through the network work and the rate of information diffusion varies by network type leading to the adoption of technology. As a result, agents update their income which allows us to observe change in society outcomes including inequality measured by the Gini index and Gross Domestic Product (GDP) below, where $p(c_i)$ is the probability of class c in node i.

$$Gini = 1 - \sum_{i=1}^{n} p^2(c_i)$$

Agents are adaptive, capable of learning to make strategic decisions to improve socio-economic transaction results. First, information flows and technology accessibility are designed at rates correspond to agents' income level to accurately capture inequality challenges. Based on socio-economic transactional game results, individuals can randomly choose whether to adopt technology or not depending on their ability and preference. Obviously, this has direct, second order effects across coupled meso and macro scales of behavior, including income growth, inequality, decision sets and strategy.

4   Results

Given the CAS framework described above, we instantiate several scenarios calibrated with real data for total population, export, import, investment and expenditure to our designed experiment space to capture aggregate macro changes in GDP. Simulated results allow us to explore changes in individual level games, technology diffusion and the resulting macroeconomic inequality to reveal possible recovery patterns following economic shocks. Below we explore scenario comparisons from baseline, mild and severe recessions to investigate the amplitude of economic shocks to GDP and identify underlying recovery patterns. To better understand the influence of these shocks, we leverage key control parameters of percentage of shock loss at t = 30 and 60 to systemically investigate strength of shocks and resulting system recovery patterns. For mild shock, percentage of income loss is set at 30% and for recession it is set at 80%. These values are meant to instantiate income losses, such as the 35% GDP decline in Q2 2020 due to COVID-19 pandemic, but also any other mild and severe employment shocks across heterogenous societal income groups (Table 1).

4.1   Income and Recessions

First we setup our model to instantiate various shock scenarios under heterogenous network typologies. By controlling for shock depth at $t = 30, 60$ with different socio-economic networks, we hope to offer new insights on the effects of individual ties, economic networks, market structures and social interactions. In Fig. 2, the income y-axis in each shock scenario follows different scales to allow visible variations for

Table 1. Baseline parameters

Parameters	Description	Base value
Population	Total number of agents	100
Initial percentage of using technology	Percentage of population already using technology	10%
Probability of using technology	Probability of individual have chance using technology	10%
Probability of informed	Probability of individual could receive using technology information	20%
Probability of T	Probability of individual could use technology	10%
Max Tolerance	Tolerance would impact individual preference to switch use technology	5
Nodes	Number of nodes	30
Degree	Average of nodes degree	15
Probability of rich using technology	Probability of high-income class people using technology	40%
Probability of middle using technology	Probability of middle-income class people using technology	20%
Probability of poor using technology	Probability of low-income class people using technology	10%
Probability of shock	Probability of economic shock	0%
Shock loss	Percentage of income loss from economic shock	0%

average income under different network typologies and recovery scenarios. All scenarios correctly show growth recovery, however comparisons across recovery patterns yield interesting insights. In the static random shock scenarios, recovery happens quickly regardless of shock depth due to the resiliency of randomized transactions. Conversely, in the preferential static shock scenarios, recovery to previous levels is much slower and lower due to the destruction of sophisticated preferential market linkages. Both dynamic random and dynamic preferential network shock scenarios exhibit better income recovery response levels and rates as efficient interaction patterns can reform under more agile economic structures.

4.2 Gini Index

Figure 3 displays a panel of Gini index trajectories across sampled shock scenario space considering technology adoption under various network typologies. The Gini index measures the inequality among values of a frequency distribution with a value of zero expressing perfect equality while a value of one indicates maximal inequality. In general,

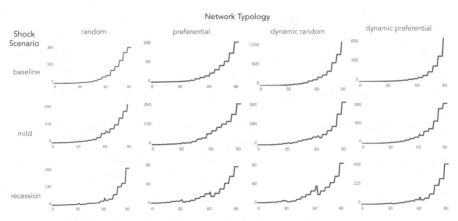

Fig. 2. Average income response across shocks and network typology

we observe an immediate inequality decrease follow by a sharp increase in income inequality, with more notable changes under preferential networks but at lower levels of inequality. This result suggests that while shocks and recession can reduce agents' income, temporarily minimizing income gaps among groups, these gaps can widen again during economic recovery. Overall, sophisticated economic structures and markets with preferential connectivity tend to offer lower mean levels of inequality than random, transactional markets and economic structures.

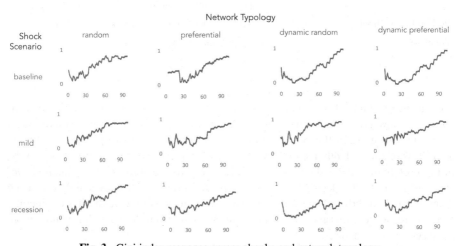

Fig. 3. Gini index response across shocks and network typology

4.3 Socio-economic Transaction Games

Figure 4 shows how agents' strategies simultaneously co-evolve over time across shock and economic network typology scenarios, driving changes with feedback processes through $t + k$ iterations. The vertical axis depicts the percentage of agent population adopting the corresponding strategy type. As t progresses and the depth of shock increases, defect strategies become dominant in all typologies, suggesting that agents will adopt different strategies that benefits themselves depending on external macro environmental influences which further impact localized path dependency of micro decision-making processes over time. However, strategy covariance is more tightly coupled in randomized network typologies given our symmetric and asymmetric technology transactional game setup. This is consistent with many other evolutionary game theory insights, experiments and findings [21–23].

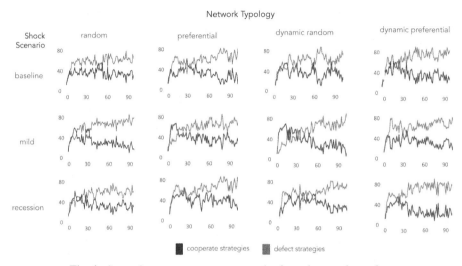

Fig. 4. Strategic game responses across shocks and network typology

4.4 Recovery

A large amount of both theoretical and empirical research discusses the relevancy between recovery path and GDP over time to identify possible recession duration and the nature of recovery to help provide evidentiary support in determining effective policy measurements under extreme uncertainties [24]. Given a particular exogenous shock, an anticipated V-shape recovery will suggest the economy may experience a sharp decline but an almost-immediate return to original growth dynamics. A U-shaped pattern indicates a relatively slower rebound compare to a V-shape paths. Next a W-shape recovery or "double-dip recession" often occurs where there may be an imbalance between fiscal and monetary policies after the first chock which further pushes the economy into a second wave of recession. Lastly, we have L-shaped recoveries which often result in structural breaks between supply and demand mechanism and are the lengthiest recession period.

In Fig. 5, we observe distinct shock and recovery geometries under different scenarios by comparing anticipated GDP trajectories from previous years mean performance. In the static random and dynamic preferential scenarios, our simulated results show a W-shaped pattern, indicating oscillatory recoveries as macroeconomic environments coupled with micro decision making and market networks are recalibrated. In the preferential network scenario, GDP recovery trends exemplifies U-shaped patterns consistent with our earlier network findings where economies with destroyed value chains take longer to recover. Finally, we detect L-shaped patterns in dynamic random markets, where sophisticated, economic value creation activities do not occur and structural breaks often persist well after an exogenous shock.

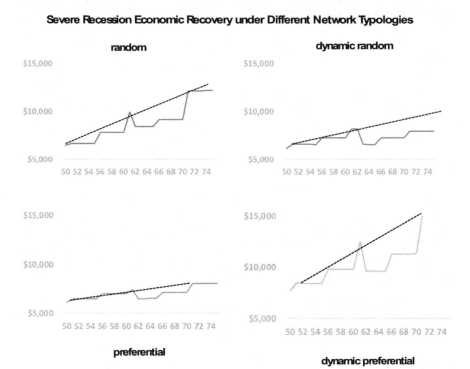

Fig. 5. GDP recovery patterns at second shock ($t = 60$).

5   Conclusions

Our initial implications include how individual strategic choices, based on localized limited resources, access to technology and incentive structures, can drive very different macro social and economic behaviors. Those macro structural outcomes feedback and shape individual agent choices and spaces of decision possibilities. This begs for us to dive deeper into understanding the nuances of technology as an enabler across multiple scales of human activity. Especially in light of shock and recovery scenarios, our approach helps illuminate the potential robustness, resiliency and adaptiveness of different economic structures, markets and societies. These have very different consequences for recovery patterns and income inequality effects enabled by technology. While not detailed here, technology asymmetries in transactional games can also drive unique emergent behaviors, new market orders with significant social and economic equality consequences.

Decomposing macroeconomic growth, technology and income equality models into their behavioral subcomponents and then re-integrating in a CAS cross-scale framework offers several scientific and policy advantages. First, modeling complex, scale variant human behavior increases not only explanatory fidelity but offers insights for bridging seemingly disparate theory from different fields via computational methods. Second, nonlinear macroeconomic outcomes and discontinuities in far from equilibrium conditions can hopefully be better anticipated and accounted for as we still continue to understand the intended and unintended second and higher order effects of quantitative easing policies for example. Third, while our approach here is just an initial foray and not fully empirically calibrated yet, it does provide policy making promise through increased descriptive, predictive and prescriptive insights in complex, dynamic human systems.

A key advantage of agent based, CAS and other cross scale modeling approaches is creating more actionable policy levers. For example, a macroeconomic model might suggest a policy option of increasing labor productivity by a certain amount to achieve desired growth targets, but does offer any suggestions on how exactly to increase such labor productivity levels. With an agent or CAS approach, one can explicitly test a priori policy 'how, who and where' answers to potential micro incentive changes as well as meso transactional network effectiveness. One general but weak critique of such approaches is modeling complexity for the sake of complexity, not specific for explanation, prediction or control. However, we know from the scientific law of requisite variety that the number of controllers or levers in a system has to be equal or greater than the number of sources of variance in the system [25] as practiced in many fields demanding systems engineering and dynamic optimization.

While simple, elegant explanatory models might help build cognitively digestible theory, policy actionability requires more detailed modeling, simulation and engineering prowess. Cross-scale modeling efforts such as the one presented here decompose complex human systems into their various subcomponents of which there is often increasing fidelity and policy lever actionability to help achieve desired socio-economic outcomes, especially in times of crisis. This seems particularly necessary in light of increasing complexity and connectivity of human activities intertwined by technology, growth and income inequality. Thus expanding scientific, empirical and policy insights into comprehensive, cross-scale human behavioral approaches is key to understanding, anticipating and hopefully shaping better societal outcomes equitably.

References

1. Schwab, K.: The fourth industrial revolution. Currency (2017)
2. Bruckner, M., LaFleur, M., Pitterle, I.: The impact of the technological revolution on labour markets and income distribution. Front. Issues 1–49 (2017)
3. Schelling, T.C.: Micromotives and Macrobehavior. WW Norton & Company, New York (2006)

4. Arthur, W.B.: Is the information revolution dead. Business **2**, 65–72 (2002)
5. Wilensky, U., Rand, W.: NetLogo Random Network model (2008). http://ccl.northwestern.edu/netlogo/models/RandomNetwork. Center for Connected Learning and Computer-Based Modeling, Northwestern Institute on Complex Systems, Northwestern University, Evanston, IL
6. Graham, C.: America's crisis of despair: A federal task force for economic recovery and societal well-being. Brookings, 23 March 2021. https://www.brookings.edu/research/americas-crisis-of-despair-a-federal-task-force-for-economic-recovery-and-societal-well-being/
7. Abdollahian, M., Chang, Y., Lee, Y.-Y.: Technology, growth and inequality: an agent-based model of micro transactional behaviors and meso technology networks for macroeconomic growth. In: Nunes, I.L. (ed.) AHFE 2021. LNNS, vol. 265, pp. 39–48. Springer, Cham (2021). https://doi.org/10.1007/978-3-030-79816-1_5
8. Romer, P.M.: Endogenous technological change. J. Polit. Econ. **98**(5, Part 2), S71–S102 (1990)
9. The Sustainable Development Goals Report 2019. United Nations, New York (2019)
10. "Decoupling of Wages From Productivity: What Implications For Public Policies?" OECD Economic Outlook, Volume 2018, Issue 2
11. Posner, M.V.: International trade and technical change. Oxford Econ. Pap. **13**(3), 323–341 (1961)
12. Gomulka, S.: Inventive activity, diffusion, and the stages of economic growth, vol. 24. Aarhus University, Institute of Economics, Aarhus (1971)
13. Singer, H.W., Reynolds, L.: Technological backwardness and productivity growth. Econ. J. **85**(340), 873–876 (1975)
14. Carlsson-Szlezak, P., Reeves, M., Swartz, P.: Understanding the economic shock of coronavirus. Harv. Bus. Rev. (2021). https://hbr.org/2020/03/understanding-the-economic-shock-of-coronavirus
15. Abdollahian, M.: Surfing the conflux: technology, information environments and great power competition convergence. In: Farhadi, A., Masys, A.J. (eds.) The Great Power Competition Volume 1, pp. 1–29. Springer, Cham (2021). https://doi.org/10.1007/978-3-030-64473-4_1
16. Camagni, R.: Technological change, uncertainty and innovation networks: towards a dynamic theory of economic space. In: Capello, R. (ed.) Seminal Studies in Regional and Urban Economics, pp. 65–92. Springer, Cham (2017). https://doi.org/10.1007/978-3-319-57807-1_4
17. Holland, J.H.: Complex adaptive systems. Daedalus **121**(1), 17–30 (1992)
18. Knoke, D., Yang, S.: Social Network Analysis. Sage Publications, Thousand Oaks (2019)
19. Voinea, C.F.: Political Attitudes: Computational and Simulation Modelling. Wiley, Hoboken (2016)
20. Bass, F.M.: The bass model. Manag. Sci. **50**(12 Supplement), 1833–1840 (2004)
21. Axelrod, R.: An evolutionary approach to norms. Am. Polit. Sci. Rev. **80**, 1095–1111 (1986)
22. Colander, D., Holt, R.P., Barkley, J.: The Changing Face of Economics: Conversations with Cutting Edge Economists. University of Michigan Press, Ann Arbor (2004)
23. Nowak, M.A.: Five rules for the evolution of cooperation. Science **314**(5805), 1560–1563 (2006)
24. Ghosh, I.: Shapes of Recovery: When Will the Global Economy Bounce Back? Visual Capitalist, 16 September 2020. https://www.visualcapitalist.com/shapes-of-recovery-when-will-the-global-economy-bounce-back/
25. Ashby, W.R.: An Introduction to Cybernetics. Chapman & Hall Ltd., London (1961)

Investigating Accuracy and Diversity in Heterogeneous Ensembles for Breast Cancer Classification

Bouchra El Ouassif[1], Ali Idri[1,2(✉)], and Mohamed Hosni[1,3]

[1] Software Project Management Research Team, ENSIAS, Mohammed V University, Rabat, Morocco
{bouchra.elouassif,ali.idri}@um5.ac.ma
[2] MSDA, Mohammed VI Polytechnic University, Ben Gueriir, Morocco
[3] MOSI, L2M3S, ENSAM-Meknes, Moulay Ismail University, Meknes, Morocco

Abstract. Breast Cancer (BC) is one of the most common forms of cancer among women. Detecting and accurately diagnosing breast cancer at an early phase increase the chances of women's survival. For this purpose, various single classification techniques have been investigated to diagnosis BC. Nevertheless, none of them proved to be accurate in all circumstances. Recently, a promising approach called ensemble classifiers have been widely used to assist physicians accurately diagnose BC. Ensemble classifiers consist on combining a set of single classifiers by means of an aggregation layer. The literature in general shows that ensemble techniques outperformed single ones when ensemble members are accurate (i.e. have the lowest percentage error) and diverse (i.e. the single classifiers make uncorrelated errors on new instances). Hence, selecting ensemble members is often a crucial task since it can lead to the opposite: single techniques outperformed their ensemble. This paper evaluates and compares ensemble members' selection based on accuracy and diversity with ensemble members' selection based on accuracy only. A comparison with ensembles without member selection was also performed. Ensemble performance was assessed in terms of accuracy, F1-score. Q statistics diversity measure was used to calculate the classifiers diversity. The experiments were carried out on three well-known BC datasets available from online repositories. Seven single classifiers were used in our experiments. Skott Knott test and Borda Count voting system were used to assess the significance of the performance differences and rank ensembles according to theirs performances. The findings of this study suggest that: (1) Investigating both accuracy and diversity to select ensemble members often led to better performance, and (2) In general, selecting ensemble members using accuracy and/or diversity led to better ensemble performance than constructing ensembles without members' selection.

Keywords: Breast cancer · Classification · Combining classifiers · Heterogeneous ensemble · Diversity measures · Voting

© Springer Nature Switzerland AG 2021
O. Gervasi et al. (Eds.): ICCSA 2021, LNCS 12950, pp. 263–281, 2021.
https://doi.org/10.1007/978-3-030-86960-1_19

1 Introduction

Breast Cancer is one of the most prominent diseases prevalent in females. It is the most common invasive cancer. In 2020, 2.3 million cases of BC were reported, which makes this cancer the most common cancer [1]. BC tumors occur when certain breast cells grow out of control, leading to the lining of the breast ducts. In addition, tumors can be classified into benign and malignant [2]. When cells are not cancerous, the tumor is benign. It would not invade nearby tissues or spread to other areas of the body (metastasize). When removed, benign tumors usually do not grow back. Malignant tumors, however, can grow and spread to other parts of the body.

Early detection of BC is the key to increase the survival rate, and the cancer is more likely to respond to effective treatment and can result in a greater probability of surviving, less morbidity, and less expensive treatment Different techniques were investigated to diagnosis BC such as ultrasound, thermography, mammography and biopsy [3]. Mammography is possibly the most intensely used medical procedure by the physicians. However, if a mammogram looks suspicious, then a biopsy is required to decide whether an abnormality is in fact a breast cancer [4]. Furthermore, if the detection accuracy of the tumor is at a low level (lower than 70%), this prevents the doctor from reporting the final result of the diagnosis. Thus, this results in a waste of time and can cause mental discomfort for the patient [5]. Accordingly, the patient faces additional tests which can be costly and demanding.

In order to provide a quick and accurate diagnosis for BC [6], various classification techniques have been investigated in the literature such as Neural Networks (ANNs), Support Vector Machines (SVMs) and K Nearest Neighbor (KNN) [7, 8]. Nevertheless, single techniques are not always the most appropriate techniques to use, since they does not achieve better performance under all situations. Actually, the performance of single techniques relies on the characteristics of the dataset [5, 8, 9]. Furthermore, each single classification technique has advantages and limitations regarding the classification tasks. In order to address this challenge, a powerful approach called ensemble classifiers has been widely investigated. They consist of combining a set of individual classifiers by means of an aggregation layer [8, 10].

One of the most important task in optimizing an ensemble learning system is to select a subset of the "best" classifiers (ensemble members) from the whole pool of classifiers, which can drive an ensemble to outperforming its members [8, 11, 12]. Otherwise, the performance of an ensemble can be worse than all or most of its members [13]. In the literature, several previous studies were carried out to identifying optimal ways to combine classifiers [14–17]. However, the selection of the ensemble members is also a crucial challenge to deal with in order to improve the ensemble performance [18]. For instance, in [15] Aytu et al., proposed a hybrid ensemble approach that employs randomized search and clustering scheme to produce an ensemble. They trained a multitude of single classifiers with different parameters, then a group of diverse classifiers is created. Classifier clusters are then created using the classification performance of single classifiers. Thereafter, two single classifiers from each cluster are selected as candidate members based on their pairwise diversity to generate the ensemble. Caruana et al. [16] proposed an ensemble selection scheme from a library of thousands of classification algorithms. In this scheme, many machine learning algorithms and parameter settings are used to build a model library. Then, a selection strategy, such as the forward step-wise selection, was used to select members that maximize the ensemble performance. In [17], Aksela proposed a method to select ensemble members based on several selection criteria such as correlation between errors, Q statistics and weighted count of errors. A measure focused on penalizing classifiers making the same error, the exponential error count approach, was identified to generate the best selections.

Performance of an ensemble learning can be influenced by many criteria including accuracy of single classifiers, number of base classifiers, combination rule, data sampling technique, and diversity of members [19]. While diversity was in general considered as the relevant criterion impacting the performance of an ensemble [20–23], other studies confirmed the opposite [24–27]. Diversity alone is a poor predictor of the ensemble accuracy" [28]. In [29], Krogh and Vedelsby underlined that members could improve the performance when they are accurate and diverse.

In the literature, several existing selection methods of ensemble members are essentially investigating one criterion: accuracy or diversity [18, 30–33]. This paper proposes a method for selecting heterogenous ensemble members for breast cancer classification, which uses both accuracy and diversity as selection criteria. Accuracy is measured in terms of recall precision, and accuracy metrics, while diversity is evaluated by means of the Q statistic diversity measure, which is one of the most popular due to its simplicity and understandability compared to other diversity measures [34].

Furthermore, we compare our proposed selection method with two existing strategies: (1) the selection of ensemble members based only on the criterion accuracy (i.e. we investigated the effect of the selection of the most accurate models from a group of seven classifiers); and (2) the selection of all the single classifiers without using any criterion. The empirical evaluations were carried out using: (1) seven single classifiers: K nearest neighbor (KNN), Multilayer Perceptron (MLP), Decision trees (DTs) and four variants of Support vector machines (SVMs) with four different kernels: Linear Kernel (LK), Normalized Polynomial Kernel (NP), Radial Basis Function Kernel (RBF), and Pearson VII function based Universal Kernel (PUK); (2) a majority voting combination rule to combine the outputs of the ensemble members. (3) three well-known available BC datasets from online repositories; (4) three performance metrics, namely accuracy, recall and precision to evaluate the constructed ensembles; and (5) the statistical test Scott-Knott and the Borda Count voting system to perform the significance tests and rank the best classifiers respectively.

The contributions of this paper are: (1) analyzing the impact of investigating both accuracy and diversity for ensemble members' selection in breast cancer classification; and (2) comparing the proposed members' selection method with two existing selection strategies.

The rest of this paper is structured as follows: Sect. 2 briefly presents the single techniques used, the ensemble concept and the existing measures of diversity. Section 3 presents an overview of related work investigating diversity in members' selection. Section 4 describes the ex.perimental design pursued in this study. The empirical findings are presented and discussed in Sect. 5. The threats to validity are given in Sect. 6. Conclusions and future works are summarized in Sect. 7.

2 Background

This section gives a summary of the single classification techniques used, the concepts of ensemble, and the measures of diversity to select ensemble members in classification.

2.1 Single Techniques

KNN: is a popular machine learning algorithm known for its simple implementation and robustness [35]. It is a non-parametric method first created in 1951 by Evelyn Fix and Joseph Hodges, and later updated by Thomas Cover. It used to solve both classification and regression problems. KNN stores all available instances and classifies new instances based on a similarity measure. To measure the similarity between its nearest neighbors, KNN uses in general the Euclidian distance.

SVMs: are powerful classification algorithms, used to solve problems of classification as well as regression. SVM was developed in the 1990s by Vladimir Vapnik [36]. It is used to classify a new unknown instance into one of the predefined classes. SVM has the ability to model complex nonlinear relationships by choosing an appropriate kernel function [37]. In fact, the Kernel function transforms the training samples so that a non-linear decision boundary is transformed to a linear equation in a higher number of dimensions [12, 38]. In this study, four variants of the SVM classifier were used. The four SVMs variants used four different kernels: Linear Kernel (LK), Radial Basis Function Kernel (RBF, Pearson VII function based Universal Kernel (PUK) and Normalized Polynomial Kernel (NP).

MLP Neural Networks: are the most frequently used feedforward neural networks due to their fast operation, ease of implementation, and smaller training set requirements, [39, 40]. They are used for both classification and regression problems [41, 42]. Their architecture consists of three types of layers: the input layer, output layer and hidden layer. The nodes present in each layer are connected to the next layer. That is the principle of feed-forward neural network; the movement information is allowed only in a forward direction. The neurons of each layer are connected to the neurons of the subsequent layer by means of weights and output signals which are a function of the sum of the inputs to neurons modified by an activation function. Generally, the neurons of the hidden layer use a nonlinear activation function, while a linear activation function is usually used for the output neurons.

DTs: are the most frequently used classification techniques, easy to use and to interpret. They can be used for both classification and regression problems [43]. DT is a tree-structured model in which internal nodes represent dataset attributes, branches represent decision rules, and each leaf node represents a class label. Depending on the task addressed, the class label could be categorical or continuous. The classification rules are described by the paths from root to leaf. In this study, the C4.5 algorithm was investigated [44].

2.2 Ensemble Classifiers

An ensemble classifiers are a powerful machine learning technique that create multiple models and then combine them by means of an aggregation rule in order to produce one optimal predictive model [8, 12]. They can be grouped into two types: Homogeneous or Heterogeneous [8, 10, 12, 45]. The Homogeneous method refers to an ensemble that combines one based learning algorithm with at least two different variants, or an ensemble that combines one base learning algorithm with one meta ensemble such as Boosting [46]. While the Heterogeneous method refers to an ensemble that combines members having different base learning algorithms. The current research is based on heterogeneous ensembles, and it adopts the majority voting combination rule to combine the decision of the individual classifiers that comprise the ensemble. Note that, the majority vote rule is the most popular and frequently used method in the literature of ensembles [47].

2.3 Measures of Diversity in Ensemble Based Classification

It is well known that the performance of an ensemble learning is impacted by diversity of its members, i.e., the degree of disagreement within the members of an ensemble [18, 25]. Diversity is loosely described as "making errors on different examples" [48, 49]. Thus, diversity has been acknowledged as a very relevant characteristic in classifiers combination. Kuncheva [28, 50] provided an analysis of ten diversity measures and classified them into two groups: Pairwise and Non-pairwise measures. Pairwise measures calculate diversity values between two base classifiers of an ensemble. The overall diversity of an ensemble can be estimated by averaging the pairwise diversity values of pairs using Q-statistic [51], double-default measure [52], and disagreement measure [50]. Non-pairwise measures, on the other hand, are used to estimate diversity among all base classifiers by accounting for all potential disagreements between them using entropy [53], generalized diversity [54], and measure of difficulty [50]. In this study, the Q statistic diversity measure was used, it is preferred over other diversity measures because of its simplicity and understandability [34].

Q statistic measure is based on Yule's Q statistic used to assess the similarity of two classifiers' outputs [50]. For two classifiers L_i and L_k, Q-statistic value is defined by Eq. 1.

$$Q_{i,k} = \frac{N^{11}N^{00} - N^{01}N^{10}}{N^{11}N^{00} + N^{01}N^{10}} \tag{1}$$

where Nab is the number of training instances for which Li gives result 'a' and Lk gives result 'b' (It is supposed that the result here is equal to 1 if an instance is classified correctly and 0 if it is misclassified). The expected value of Q for statistically independent classifiers is 0. The value of Q ranges between −1 and 1. Classifiers that appear to correctly classify the same instances will have positive Q values, while those that make errors on different instances will have negative Q values [21, 50].

3 Related Work

This section presents an overview of some related work investigating diversity in ensemble techniques.

Banfield et al. [55] proposed an algorithm termed the percentage correct diversity measure (PCDM) to construct decision trees ensemble. The proposed algorithm seeks to find the test samples for which the percentage between 10 and 90 of the single classifiers are correct. These test samples are removed from determining the ensemble diversity. The proposed technique was evaluated in twelve datasets from UCI repository using teen fold cross validation method. The empirical results suggest the effectiveness of the proposed technique.

Kadkhodaei et al. [56] proposed an entropy based approach to determine the best combination of classifiers from a pool of ten different single techniques. The evaluation of the proposed heterogeneous ensemble was evaluated on three datasets from the UCI repository. The empirical results stated that the proposed technique generates an accurate ensemble and that the time required to build it is less than the one required bagging and boosting ensemble techniques.

Nascimento et al. [57] presented a new approach for automatic selection of both base classifiers and features. The proposed approach was based on evolutionary approach composed of two genetic algorithm instances. Two proposed diversity measures were investigated in order to analyze the performance of the proposed framework. The empirical evaluations were performed using ten different classification algorithms using the bagging architecture. Five datasets from UCI repository were selected for the evaluations. The results suggested that the proposed technique was effective to generate accurate ensemble. The authors recommended to take into account other factors than diversity such as accuracy and complexity when constructing ensembles.

Lysiak et al. [58] proposed a novel approach for dynamic ensemble selection (DES) based on probabilistic measures of competence and diversity between member classifiers. The two types of ensembles were constructed: the homogeneous ensemble consisted of 20 pruned decision tree classifiers and the heterogeneous ensemble consisted of nine different classifiers. Seven public datasets were used to assess the effectiveness of the proposed approach. The Results indicated that the proposed method can eliminate weak classifiers and keep the ensemble maximally diverse. Further, the proposed DES led to better classification accuracy of the constructed ensembles compared to those generated by the DES system using only the competence measure.

4 Experimental Design

This section explains the experimental design investigated to conduct all of the empirical evaluations, including the performance metrics used, Scott–Knott (SK) test, Borda Count voting system, datasets descriptions, ensemble selection process and the abbreviations used.

4.1 Performance Metrics

The following performance metrics are used to assess the performance of single and ensemble techniques:

Accuracy, Recall and Precision defined by Eqs. 2, 3 and 4 respectively [10].

$$Accuracy = \frac{TN + TP}{TP + FP + TN + FN} \tag{2}$$

$$Recall = \frac{TP}{TP + FN} \tag{3}$$

$$Precision(Prec) = \frac{TP}{TP + FP} \tag{4}$$

where FP stands to False Positive, FN stands to False Negative TP to True Positive and TN to True Negative.

4.2 Scott-Knott Test

The Scott-Knott (SK) test is a hierarchical clustering algorithm developed by Scott and Knott (1974), is an efficient method to conduct procedures of multiple comparisons without ambiguity [59]. Compared to other statistical tests such as the Tukey test, Student– Newman–Keuls (SNK) test and t-test, the SK test is a commonly used method [60–63], it has the ability to group techniques into non-ambiguous groups [64, 65]. In this study, the SK test was used to cluster the single and ensemble techniques based on their error rates (Error rate = 1-Accuracy) and to check the significant difference between them. The ten folds cross-validation approach was used in all the experiments presented in this study.

4.3 Borda Count Voting System

The Borda count [66] is a form of single-winner election in which voters rate candidates in order of choice. The Borda count determines the winner of an election by allocating points to each candidate based on the voter's rating. After that, the system aggregates the score of each candidate based on the received points. The candidate who receives the highest score is the winner. To illustrate this process, the example in Table 1 shows the steps of Borda count to choose the winner among four candidates (Cd_1, Cd_2, Cd_3 and Cd_4) who were voted on by four voters (Vot_1, Vot_2, Vot_3 and Vot_4). Each voter Vot_i assigns a candidate to one of the positions i (1, 2, 3 or 4). As a result, we compute the

Table 1. Borda count voting system for four voters which rank four candidates according to their preferences.

Voters	VOT_1	VOT_2	VOT_3	VOT_4	POS I	Score
Cd_1	1	1	2	4	2, 1,0,1	$4 \times 2 + 3 \times 1 + 2 \times 0 + 1 \times 1 = 12$
Cd_2	3	2	3	2	0,2,2,0	$4 \times 0 + 3 \times 1 + 2 \times 2 + 1 \times 0 = 7$
Cd_3	1	2	1	4	2,1,0,1	$4 \times 2 + 3 \times 1 + 1 \times 2 + 1 \times 1 = 14$
Cd_4	4	2	3	1	1,1,1,1	$4 \times 1 + 3 \times 1 + 1 \times 2 + 1 \times 1 = 10$

vector position PI (n_1, n_2, n_3, n_4) of each candidate, where n_i is the number of times the candidate has been ranked in position i. M_i points are assigned to each position i (M_i = # ofcandidates $- i + 1$). Finally, the score of each candidate is equal to $n_i * M_i$.

4.4 Datasets Fescription

In order to evaluate the performance of the proposed techniques, three datasets obtained from the online UCI repository were investigated in this study. These datasets were the most widely used by researchers in the literature [7]. A short description of each of these datasets is reported in Table 2. Note that two of the datasets contain missing values. We simply removed them since their number was very small. Furthermore, the WPBC and Wisconsin datasets are unbalanced. The Synthetic Minority Over-sampling Technique (SMOTE) [67] was used to address this problem.

Table 2. Datasets description

Dataset	#.Attributes	Missing values?	Examples
WDBC	32	NO	569
Wisconsin	11	Yes(16)	699
WPBC	34	Yes (4)	198

4.5 Ensemble Selection Process

The first concern of constructing an effective ensemble classifier is to ensure that all individual classifiers are accurate [68]. Then we can improve the ensemble performance by rejecting weak classifiers and combining accurate members only. Toward this aim, we select ensemble members based on accuracy, referred to us Selection by Accuracy (SbA). The process of SbA is as follow:

1. Assess the performances of the N (7 in this study) single classifiers based on the three metrics: Accuracy, Precision and Recall.
2. Performing the statistical test SK based on the accuracy in order to cluster the classifiers evaluated in Step 1 into non-overlapping clusters. Each cluster includes one or more classifiers with comparable predictive abilities. The best cluster is the one with the lowest error rate value. (Error rate = 1-Accuracy); therefore, the classifiers belonging to this cluster are chosen for the next step.
3. Building an heterogeneous ensemble by combing the base classifiers belonging to the best cluster (In case of the best cluster only contains one classifier, we combine the classifiers of the two first best clusters)
4 Evaluate the performance of the heterogeneous ensemble constructed in Step 3 according to the three criteria: accuracy, recall and precision.

Since diversity of the ensemble members is relevant to improve the accuracy of an ensemble, we construct ensembles based on both accuracy and diversity, referred to us Selection by Accuracy and Diversity (SbAD). The process of SbAD is as follows.

1. Carry out the Step 1, 2 and 3 of the SbA selection process.
2. Calculate the diversity between the heterogeneous ensemble classifier constructed in Step 1 (i.e. based on accuracy criterion) and each of the remaining classifiers (i.e. classifiers that were out of the best cluster)
3. Select the single classifier with the highest diversity and include it into the SbA heterogeneous ensemble.
4. Assess the performance of the heterogeneous ensemble constructed in Step 3.
5. Repeat Steps 3 and 4 until the heterogeneous ensemble size reaches the number N-1 members.

We also combine all the seven single classifiers (DT, MLP, KNN, S-PUK, SVM-NP, SVM-RBF and SVM-LK) in one ensemble in order to compare its performance with the other constructed ensembles using SbA and SbAD. We referred to this ensemble: No Selection classifier (NSc). Figure 1 presents the experimental process we followed.

4.6 Abbreviations Used

The following abbreviation rules were used to simplify the names of ensembles

E-SingleTechnique1 SingleTechnique2

E- SingleTechnique1 SingleTechnique2SingleTechnique3.

.

.

E- SingleTechnique1 SingleTechnique2... SingleTechniqueN

It is worth noting that for ensemble techniques, we shorten the names of single classifiers as well:

KNN for K, D for DTs, M for MLP, S for SVM, SVM-PUK for P, SVM-RBF for R, S-LK for L and SVM-NP for NP.

For example, EDKLM refers to the ensemble constructed by the fusion of the four single techniques, DT, KNN, SVM-LK and MLP.

5 Empirical Results

This section discusses the empirical evaluation results of the 7 individual classifiers, the SbA and SbAD ensembles as well as the Nsc ensemble. The R software was used for statistical tests and the Waikato Environment for Knowledge Analysis (WEKA 3.9) was investigated to conduct the empirical evaluations [69].

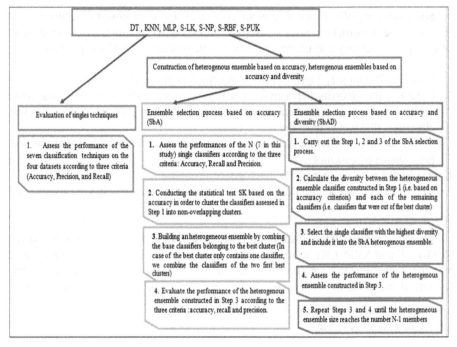

Fig. 1. Experimental process.

5.1 Individual Classifiers Evaluation

This section presents the evaluation results of the 7 classification techniques over the three datasets based on three criteria (Accuracy, Recall, and Precision). Table 3 depicts the performance values of the 7 single techniques. As shown in Table 3, we can note that in the WDBC dataset, S-LK displayed the best performance. S-PUK exhibited the best performance in the WPBC dataset. As for Wisconsin, DT was the best.

Based on the results of Table 3, we notice that there is no individual best classifier in overall datasets, and that the rankings of the same individual classifier vary depending on the dataset. This is due to the fact that the performance of single techniques depends on the characteristics of datasets (number of instances, dataset dimensionality, number of classes, etc.) [8].

Table 3. Performance results.

Classifier	WDBC			Wisconsin			WPBC		
	Acc	Prec	Recall	Acc	Prec	Recall	Acc	Prec	Recall
S-LK	**97.89**	**97.9**	**97.9**	74.66	74.8	74.7	74.66	74.8	74.7
S-NP	93.4	93.5	93.5	77.03	77	77.2	77.03	77	77.2

(*continued*)

Table 3. (*continued*)

Classifier	WDBC			Wisconsin			WPBC		
	Acc	Prec	Recall	Acc	Prec	Recall	Acc	Prec	Recall
S-RBF	92.09	92.1	92.8	65.88	65.9	66.5	65.88	65.9	66.5
S-PUK	97.54	97.5	97.5	90.88	90.9	90.9	**90.88**	**90.9**	**90.9**
DT	93.15	93.2	93.1	**96.05**	**96.1**	**96**	84.12	84.1	84.1
MLP	95.78	95.8	95.8	95.75	95.8	95.8	86.82	88.1	86.8
KNN	95.78	96.1	96.1	95.75	95.8	95.8	82.43	85	82.4

5.2 Ensembles Evaluation

Figure 2 depicts the results of the SK test carried out based on error rate overall datasets. We observed that the SK test identified 2, 4, and 2 clusters in the WDBC, WPBC and Wisconsin datasets respectively. Therefore, we constructed: (1) one SbA heterogeneous ensemble in WDBC dataset whose members were S-PUK, S-LK, KNN and DT; (2) one SbA heterogeneous ensemble in Wisconsin dataset whose members were DT, MLP and KNN; and (3) one SbA heterogeneous ensemble in WPBC dataset whose members were S-PUK, MLP, DT and KNN (since the best cluster of WPBC dataset contains only one technique we include also the techniques of the second best cluster).

For the SbAD ensembles, for each dataset we constructed ensembles based on the SbAD selection process described above (see Sect. 4.5):

- For Wisconson, we constructed 3 SbAD ensembles: EDKNM, EDKLNM and EDKLNMP

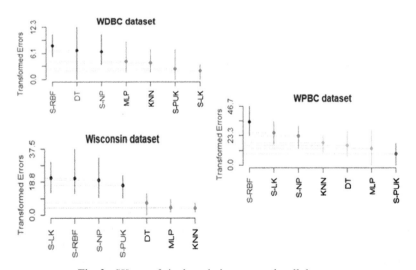

Fig. 2. SK test of single techniques over the all datasets

- For WDBC, we constructed 2 SbAD ensembles: EDLKMP and EDLKMNP
- For WPBC, we constructed 2 SbAD ensembles: EDKMPR and EDKMPRN

WDBC Dataset: Table 4 shows the performance metric values of the SbA and SbAD ensembles as well as the NSc ensemble for the WDBC dataset. We observe that the SbAD ensemble EDLKMP and the SbA ensemble come first with **97.72%**, **97%** and **97%** for accuracy, recall and precision respectively; and the SbAD ensemble EDLKMNP comes second with 97.54%, 97.5% and 97.6% for accuracy, recall and precision respectively. Note that the the SbA and SbAD ensembles outperformed the NSc ensemble.

Table 4. Performance results: WDBC dataset.

Method	Ensemble	Acc	Prec	Recall
SbDA	EDLKMP	**97.72**	**97**	97
	EDLKMNP	97.54	97.5	97.6
SbA	ELKMP	**97.72**	**97**	97
NSc		97.12	97.2	97.2

Wisconsin Dataset: Table 5 reports the performance metric values of the SbA and SbAD ensembles as well as the NSc ensemble for the Wisconsin dataset. In terms of the three performance metrics, accuracy, precision, and recall, we notice that the SbAD ensemble EDKLNMP marginally outperformed the others. It provides an accuracy, precision and recall values of 97.07%, 97.1% and 97.1% respectively.

Table 5. Performance results: Wisconsin dataset.

Method	Ensemble	Acc	Prec	Recall
SBAD	EDKNM	96.49	96.5	96.5
	EDKLNM	96.92	97	96.9
	EDKLNMP	**97.07**	**97.1**	**97.1**
SBA	EDKM	96.63	96.7	96.6
NSc		96.78	96.8	96.8

WPBC dataset: Table 6 depicts the performance metrics values of the SbA and SbAD ensembles as well as the NSc ensemble for the WPBC dataset. We observe that the ensemble SbA ensemble EDKMP outperformed all the other ensembles; it achieved an accuracy, precision and recall of 91.22%, 91.5% and 91.2% respectively. As it can be seen from Table 6, the SbA and SbAD ensembles in general outperformed the NSc ensemble.

Table 6. Performance results: WPBC dataset.

Method	Ensemble	Acc	Prec	Recall
SBAD	EDKMPR	90.54	90.8	90.5
	EDKMPRN	87.5	87.6	87.5
SBA	EDKMP	**91.22**	**91.5**	**91.2**
NSc		86.82	87.1	86.8

5.3 Comparing SbA, SbAD and Nsc Ensembles

To check the significant difference between the performances of SbA, SbAD and Nsc ensembles, the SK test was carried out based on error rate values to check whether there was a notable difference between the ensemble performances. Figure 3 displays the results of the SK test on the built ensembles for each dataset. As it can be observed in Fig. 3, in all datasets, only one cluster was identified by the SK test. This means that SbA, SbAD and Nsc ensembles show the same predictive capabilities in terms of accuracy in all datasets.

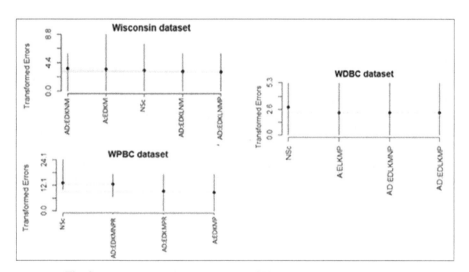

Fig. 3. Accuracy based SK test results of SbA, SBA and Nsc ensembles.

To deeply compare the predictive capabilities of SbA, SbAD and Nsc ensembles, we used the Borda Count voting system to rank them based on the three performance metrics: accuracy, precision, and recall. Table 7 displays the ranking supplied by the Borda count voting system for each dataset. We note the following:

- Except in the WPBC dataset, where the SbA ensemble EDKMP was ranked first, SbAD ensemble techniques were in general ranked at the first position in two datasets (EDLKMP in WDBC and EDKLNMP in Wisconsin).

Table 7. Borda Count ranks of the best SK cluster techniques

Rank	WDBC dataset	Wisconsin dataset	WPBC dataset
1	SbAD:EDLKMP[a]	SbAD:EDKLNMP	SbA:EDKMP
2	SbA:ELKMP[a]	SbAD:EDKLNM	SbAD:EDKMPR
3	SbAD:EDLKMNP	NSc	SbAD:EDKMPRN
4	NSc	SbA:EDKM	NSc
5		SbAD:EDKNM	

[a]The letter denotes the same rank

- In general, the SbAD and SbA ensembles outperformed the NSc ensemble overall datasets.
- The best SbAD ensemble in each dataset includes 5 to 6 single techniques (EDLKMP in WDBC, EDKLNMP in Wisconsin and EDKMPR in WPBC).
- The members DT, KNN, MLP, and S-PUK were present in all the best SbAD/SbA ensembles (EDLKMP in WDBC, EDKLNMP in Wisconsin and EDKMP in WPBC).

6 Threats to Validity

This section discusses threats to this study's validity, with regard to internal, external and construct validity.

Internal Validity: This study used a 10-fold cross validation evaluation method, which is commonly used in machine learning [10]. Another internal threat is the presence of missing values in two datasets used, which may affect the performance of a model constructed using these datasets and increase the likelihood of drawing incorrect conclusions. In this research, instead of using imputation techniques like mean imputation or expectation-maximization [70] to impute missing values, we deleted the existing missing values because their number was small, but if there are a lot of missing values, we recommend the use of imputation methods [71].

Furthermore, the majority voting was used in this study to produce the final output of the proposed ensembles. Note that, the majority voting rule is the most popular and frequently used method in the literature of ensembles [72, 73]. However, we are aware that other combination schemes such as weighted majority voting, probabilistic and weighted sum can yield different outcomes. Thus, future experiments might look into the effect of other combiners on the predictive capability of the ensembles.

External Validity: The aim of the external threat is to know whether the results of this research work can be generalized to other contexts [74]. Thus, it is crucial to figure out how broadly the findings of this study can be applied. To counteract this threat, we selected three datasets that differ in size and number of features. Moreover, this study investigated only one diversity measure, Q statistic, to calculate the classifiers' diversity. Investigating other diversity measures such as double-default measure [52], disagreement measure [50] and measure of difficulty [50] is recommended so as to generalize the findings of this study.

Construct Validity: The construct validity aims at answering the question about the measurement validity [11], or more specifically, the reliability/credibility of the measures selected to assess the performance of the techniques. We overcome this constraint by using three performance metrics (Accuracy, Precision and Recall) in order to obtain results from various perspectives. As a result, all three criteria yielded the same results. To affirm or refute the reported results, it is required to invesigate other performance metrics.

7   Conclusion and Future Work

This study assessed and compared the impacts of using accuracy and diversity instead of accuracy alone for selecting ensemble members on the performance of ensembles in BC classification. Moreover, it evaluated and compared the performances of ensembles with/out selection. We used the Q statistic measure to evaluate the diversity of ensemble members. The majority voting combination rule was used to combine the members of an ensemble. The experiments were carried out on three well-known available BC datasets from online repositories. The SK test and Borda Count were investigated to assess the significance of performance differences and to rank the ensembles respectively.

The findings were:

(1)   Investigating both accuracy and diversity to select ensemble members can improve the performance of an ensemble. This confirms the findings of [75].

(2)   Selecting ensemble members using accuracy and/or diversity in general led to better ensemble performance than constructing ensembles without members' selection. This confirms the results of Zhou et al. [76].

(3)   Individual classifiers' performance is influenced by the characteristics of the dataset as the ranks of the same technique depended on the dataset.

Ongoing work focuses on investigating other diversity measures and other combination rules to construct better heterogeneous ensembles in BC diagnosis.

References

1. breastCancer. https://www.who.int/cancer/prevention/diagnosis-screening/breast-cancer/en/. Accessed 16 Jan 2019
2. Breast Cancer Facts - National Breast Cancer Foundation. https://www.nationalbreastcancer.org/breast-cancer-facts. Accessed 11 Dec 2020
3. Luo, S.T., Cheng, B.W.: Diagnosing breast masses in digital mammography using feature selection and ensemble methods. J. Med. Syst. **36**, 569–577 (2012). https://doi.org/10.1007/s10916-010-9518-8
4. Chhatwal, J., Alagoz, O., Burnside, E.S., Burnside, E.S.: Optimal Breast biopsy decision-making based on mammographic features and demographic factors. Oper. Res. **58**(6), 1577–1591 (2010). https://doi.org/10.1287/opre.1100.0877

5. Kaushik, D., Kaur, K.: Application of data mining for high accuracy prediction of breast tissue biopsy results. In: 2016 3rd Third International Conference on Digital Information Processing, Data Mining, and Wireless Communications. DIPDMWC 2016, pp. 40–45 (2016). https://doi.org/10.1109/DIPDMWC.2016.7529361

6. Topol, E.J.: High-performance medicine: the convergence of human and artificial intelligence (2019)

7. Idri, A., Chlioui, I., El Ouassif, B.: A systematic map of data analytics in breast cancer. In: Proceedings of the Australasian Computer Science Week Multiconference (2018). https://doi.org/10.1145/3167918.3167930

8. Idri, A., Bouchra, E.O., Hosni, M., Abnane, I.: Assessing the impact of parameters tuning in ensemble based breast Cancer classification. Heal. Technol. 10(5), 1239–1255 (2020). https://doi.org/10.1007/s12553-020-00453-2

9. El Ouassif, B., Idri, A., Hosni, M.: Homogeneous ensemble based support vector machine in breast cancer diagnosis (2021).https://doi.org/10.5220/0010230403520360

10. Hosni, M., Abnane, I., Idri, A., Carrillo de Gea, J.M., Fernández-Alemán, J.L.: Reviewing ensemble classification methods in breast cancer. Comput. Methods Programs Biomed. 177, 89–112 (2019)

11. Hosni, M., Idri, A., Abran, A., Nassif, A.B.: On the value of parameter tuning in heterogeneous ensembles effort estimation. Soft. Comput. 22(18), 5977–6010 (2017). https://doi.org/10.1007/s00500-017-2945-4

12. El Ouassif, B., Idri, A., Hosni, M.: Homogeneous ensemble based support vector machine in breast cancer diagnosis - BIOSTEC 2021. In: HEALTHINF 2021 - 14th International Conference on Health Informatics, Proceedings; Part of 13th International Joint Conference on Biomedical Engineering Systems and Technologies, BIOSTEC 2021. SciTePress (2021)

13. Yang, L.: Classifiers selection for ensemble learning based on accuracy and diversity. Procedia Eng. 15, 4266–4270 (2011). https://doi.org/10.1016/j.proeng.2011.08.800

14. Alexandropoulos, S.-A., Aridas, C.K., Kotsiantis, S.B., Vrahatis, M.N.: Stacking strong ensembles of classifiers. In: MacIntyre, J., Maglogiannis, I., Iliadis, L., Pimenidis, E. (eds.) AIAI 2019. IAICT, vol. 559, pp. 545–556. Springer, Cham (2019). https://doi.org/10.1007/978-3-030-19823-7_46

15. Onan, A., Korukoğlu, S., Bulut, H.: A hybrid ensemble pruning approach based on consensus clustering and multi-objective evolutionary algorithm for sentiment classification. Inf. Process. Manag. 53, 814–833 (2017). https://doi.org/10.1016/j.ipm.2017.02.008

16. Caruana, R., Crew, G., Ksikes, A.: Ensemble selection from libraries of models. In: The Proceedings of ICML 2004 (2004)

17. Aksela, M.: Comparison of classifier selection methods for improving committee performance. In: Windeatt, T., Roli, F. (eds.) MCS 2003. LNCS, vol. 2709, pp. 84–93. Springer, Heidelberg (2003). https://doi.org/10.1007/3-540-44938-8_9

18. Butler, H.K., Friend, M.A., Bauer, K.W., Bihl, T.J.: The effectiveness of using diversity to select multiple classifier systems with varying classification thresholds. J. Algorithms Comput. Technol. 12, 187–199 (2018). https://doi.org/10.1177/1748301818761132

19. Bian, S., Wang, W.: Investigation on Diversity in Homogeneous and Heterogeneous Ensembles (2006)

20. Kuncheva, L.I., Whitaker, C.J.: Measures of diversity in classifier ensembles and their relationship with the ensemble accuracy. Mach. Learn. 51, 181–207 (2003)

21. Wang, S., Yao, X.: Relationships between diversity of classification ensembles and single-class performance measures. IEEE Trans. Knowl. Data Eng. 25, 206–219 (2013). https://doi.org/10.1109/TKDE.2011.207

22. Tsymbal, A., Pechenizkiy, M., Cunningham, P.: Diversity in search strategies for ensemble feature selection. Inf. Fusion. 6, 83–98 (2005). https://doi.org/10.1016/j.inffus.2004.04.003

23. Windeatt, T.: Diversity measures for multiple classifier system analysis and design. Inf. Fusion. **6**, 21–36 (2005). https://doi.org/10.1016/j.inffus.2004.04.002
24. Schapire, R.E.: Measures of diversity in classifier ensembles and their relationship with the ensemble accuracy. Mach. Learn. **51**, 181–207 (2003)
25. Duin, R.P.W., Tax, D.M.J.: Experiments with classifier combining rules. In: Kittler, J., Roli, F. (eds.) MCS 2000. LNCS, vol. 1857, pp. 16–29. Springer, Heidelberg (2000). https://doi.org/10.1007/3-540-45014-9_2
26. Skurichina, M., Kuncheva, L.I., Duin, R.P.W.: Bagging and boosting for the nearest mean classifier: effects of sample size on diversity and accuracy. In: Roli, F., Kittler, J. (eds.) MCS 2002. LNCS, vol. 2364, pp. 62–71. Springer, Heidelberg (2002). https://doi.org/10.1007/3-540-45428-4_6
27. Webb, G.I., Zheng, Z.: Multistrategy ensemble learning: reducing error by combining ensemble learning techniques. IEEE Trans. Knowl. Data Eng. **16**, 980–991 (2004). https://doi.org/10.1109/TKDE.2004.29
28. Kuncheva, L.I.: That elusive diversity in classifier ensembles. In: Perales, F.J., Campilho, A.J.C., de la Blanca, N.P., Sanfeliu, A. (eds.) IbPRIA 2003. LNCS, vol. 2652, pp. 1126–1138. Springer, Heidelberg (2003). https://doi.org/10.1007/978-3-540-44871-6_130
29. Krogh, A., Vedelsby, J.: Neural network ensembles, cross validation, and active learning. In: Proceedings of the 7th International Conference on Neural Information Processing Systems (1994)
30. Kuncheva, L.I., Skurichina, M., Duin, R.P.W.: An experimental study on diversity for bagging and boosting with linear classifiers. Inf. Fusion. **3**, 245–258 (2002). https://doi.org/10.1016/S1566-2535(02)00093-3
31. Narasimhamurthy, A.: Evaluation of diversity measures for binary classifier ensembles. In: Oza, N.C., Polikar, R., Kittler, J., Roli, F. (eds.) MCS 2005. LNCS, vol. 3541, pp. 267–277. Springer, Heidelberg (2005). https://doi.org/10.1007/11494683_27
32. Azizi, N., Farah, N., Sellami, M., Ennaji, A.: Using diversity in classifier set selection for Arabic handwritten recognition. Lect. Notes Comput. Sci. (including Subser. Lect. Notes Artif. Intell. Lect. Notes Bioinformatics). LNCS, **5997,** 235–244 (2010). https://doi.org/10.1007/978-3-642-12127-2-24
33. Naldi, M.C., Carvalho, A.C.P.L.F., Campello, R.J.G.B.: Cluster ensemble selection based on relative validity indexes (2013). https://doi.org/10.1007/s10618-012-0290-x
34. Kotsiantis, S., Kanellopoulos, D., Pintelas, P.: Handling imbalanced datasets: a review. GESTS Int'l Trans. Comput. Sci. Eng. **30**, 25–36 (2012)
35. Zhang, S., Li, X., Zong, M., Zhu, X., Wang, R.: Efficient kNN classification with different numbers of nearest neighbors. IEEE Trans. Neural Networks Learn. Syst. **29**, 1774–1785 (2018). https://doi.org/10.1109/TNNLS.2017.2673241
36. Vapnik, V.: Principles of risk minimization for learning theory. In: Advances in Neural Information Processing Systems (1992)
37. Schölkopf, B., Alexander, J.S.: Support Vector Machines, Regularization, Optimization, and Beyond. In: Learning with Kernels, pp. 1–27 (2001)
38. Bhavsar, H., Ganatra, A.: Radial basis polynomial kernel (RBPK): a generalized kernel for support vector machine. Int. J. Comput. Sci. Inf. Secur. **14**, 1–20 (2016)
39. Kocyigit, Y., Alkan, A., Erol, H.: Classification of EEG recordings by using fast independent component analysis and artificial neural network. J. Med. Syst. **32**, 17–20 (2008). https://doi.org/10.1007/s10916-007-9102-z
40. Übeyli, E.D.: Combined neural network model employing wavelet coefficients for EEG signals classification. Digit. Signal Process. A Rev. J. **19**, 297–308 (2009). https://doi.org/10.1016/j.dsp.2008.07.004

41. Idri, A., Khoshgoftaar, T., Abran, A.: Can neural networks be easily interpreted in soft-ware cost estimation? In: 2002 IEEE World Congress on Computational Intelligence. 2002 IEEE International Conference on Fuzzy Systems. FUZZ-IEEE 2002. Proceedings (Cat. No. 02CH37291), vol. 2, pp. 1162–1167 (2003). https://doi.org/10.1109/fuzz.2002.1006668

42. Haykin, S.: Neural networks: a comprehensive foundation (1999)

43. Wang, Y., Wang, Y., Witten, I.: Inducing model tree for continuous classes. In Proceedings of Poster Papers, 9th European Conference on Machine Learning, pp. 128–137 (1997)

44. Salzberg, S.L.: C4.5: Programs for machine learning by J. Ross Quinlan. Mach. Learn. **16**, 235–240. Morgan Kaufmann Publishers, Inc., 1993 (1994). https://doi.org/10.1007/BF0099 3309

45. Idri, A., El Ouassif, B., Hosnia, M., Abran, A.: Classification techniques in breast cancer diagnosis: a systematic literature review. Comput. Methods Biomech. Biomed. Eng. Imaging Vis. (2020)

46. Schapire, E., R.: A brief introduction to boosting (1999)

47. Sergios, T., Konstantinos, K.: Pattern Recognition, Third Edition.

48. Polikar, R.: Ensemble based systems in decision making (2006). https://doi.org/10.1109/MCAS.2006.1688199

49. Ali, K., Michael J.P.: On the Link between Error Correlation and Error Reduction in Decision Tree Ensembles (1995)

50. Kuncheva, L.I., Whitaker, C.J.: Ten measures of diversity in classifier ensembles: limits for two classifiers. IEE Colloq. 73–82 (2001). https://doi.org/10.1049/ic:20010105

51. Udny Yule, G.: On the association of attributes in statistics: with illustrations from the material of the childhood society, & c on JSTOR. Philos. Trans. R. Soc. London. A **194**, 257–319 (63 pages) (1900)

52. Giacinto, G., Roli, F.: Design of effective neural network ensembles for image classifica-tion purposes. Image Vis. Comput. **19**, 699–707 (2001). https://doi.org/10.1016/S0262-885 6(01)00045-2

53. Cunningham, P., Carney, J.: Diversity versus quality in classification ensembles based on feature selection. In: López de Mántaras, R., Plaza, E. (eds.) ECML 2000. LNCS (LNAI), vol. 1810, pp. 109–116. Springer, Heidelberg (2000). https://doi.org/10.1007/3-540-45164-1_12

54. Partridge, D., Krzanowski, W.: Software diversity: Practical statistics for its measurement and exploitation. Inf. Softw. Technol. **39**, 707–717 (1997). https://doi.org/10.1016/s0950-584 9(97)00023-2

55. Banfield, R.E., Hall, L.O., Bowyer, K.W., Kegelmeyer, W.P.: A new ensemble diversity mea-sure applied to thinning ensembles. In: Windeatt, T., Roli, F. (eds.) MCS 2003. LNCS, vol. 2709, pp. 306–316. Springer, Heidelberg (2003). https://doi.org/10.1007/3-540-44938-8_31

56. Kadkhodaei, H., Moghadam, A.M.E.: An entropy based approach to find the best combination of the base classifiers in ensemble classifiers based on stack generalization. In: 2016 4th International Conference on Control, Instrumentation, and Automation, ICCIA 2016, pp. 425–429. Institute of Electrical and Electronics Engineers Inc. (2016). https://doi.org/10.1109/ICC IAutom.2016.7483200

57. Nascimento, D.S.C., Canuto, A.M.P., Silva, L.M.M., Coelho, A.L.V.: Combining different ways to generate diversity in bagging models: an evolutionary approach. In: Proceedings of the International Joint Conference on Neural Networks, pp. 2235–2242. IEEE (2011). https://doi.org/10.1109/IJCNN.2011.6033507

58. Lysiak, R., Kurzynski, M., Woloszynski, T.: Optimal selection of ensemble classifiers using measures of competence and diversity of base classifiers. Neurocomputing **126**, 29–35 (2014). https://doi.org/10.1016/j.neucom.2013.01.052

59. Lopes Bhering, L., Cruz, D., Soares De Vasconcelos, E., Ferreira, A., Fernando, M., De Resende, R.: Alternative methodology for Scott-Knott test. Crop Breed. Appl. Biotechnol. **8**, 9–16 (2008)
60. Cox, D.R., Spjøtvoll, E.: On partitioning means into groups source. Wiley behalf Board Found. Scand. J. St. **9**, 147–152 (1982)
61. Calinski, T., Corsten, L.C.A.: Clustering means in ANOVA by simultaneous testing. Biometrics **41**, 39 (1985). https://doi.org/10.2307/2530641
62. Sharma, A., Kulshrestha, S., Daniel, S.: Machine learning approaches for breast cancer diagnosis and prognosis. In: 2017 International Conference on Soft Computing and its Engineering Applications: Harnessing Soft Computing Techniques for Smart and Better World, icSoftComp 2017, pp. 1–5. Changa, India (2018). https://doi.org/10.1109/ICSOFTCOMP.2017.8280082
63. Bony, S., Pichon, N., Ravel, C., Durixl, A., Balfourier, F.: The relationship between mycotoxin synthesis and isolatemorphology in fungal endophytes of Lolium perenne. New Phytol. **152**, 125–137 (2001)
64. Tsoumakas, G., Angelis, L., Vlahavas, I.: Selective Fusion of Heterogeneous Classiers. Intell. Data Anal. **9**, 511–525 (2005). https://doi.org/10.3233/ida-2005-9602
65. Borges, L., Ferreira, D.: Power and type I errors rate of Scott-Knott, Tukey and Newman-Keuls tests under normal and no-normal distributions of the residues. Rev. Matemática e Estatística. **21**, 67–83 (2003)
66. Rowley, C.K.: Borda, Jean-Charles de (1733–1799). In: Durlauf, S.N., Blume, L.E. (eds.) The New Palgrave: Dictionary of Economics, pp. 527–529. Palgrave Macmillan UK, London (2008). https://doi.org/10.1007/978-1-349-58802-2_148
67. Chawla, N.V, Bowyer, K.W., Hall, L.O., Kegelmeyer, W.P.: SMOTE: synthetic minority oversampling technique (2002)
68. Gu, S.: Generating diverse and accurate classifier ensembles using multi-objective optimization (2014)
69. WEKA-University of Waikato: WEKA. https://ai.waikato.ac.nz/weka/
70. Smith, B.L., Scherer, W.T., Conklin, J.H.: Exploring imputation techniques for missing data in transportation management systems. Transp. Res. Rec. J. Transp. Res. Board. **1836**, 132–142 (2003). https://doi.org/10.3141/1836-17
71. Idri, A., Abnane, I., Abran, A.: Missing data techniques in analogy-based software development effort estimation. J. Syst. Softw. **117**, 595–611 (2016). https://doi.org/10.1016/J.JSS.2016.04.058
72. Oh, S.B.: On the relationship between majority vote accuracy and dependency in multiple classifier systems. Pattern Recogn. Lett. **24**, 359–363 (2003). https://doi.org/10.1016/S0167-8655(02)00260-X
73. Kuncheva, I.L.: Combining Pattern Classifiers: Methods and Algorithms (2014). https://doi.org/10.1002/9781118914564
74. Idri, A., Hosni, M., Abran, A.: Improved estimation of software development effort using classical and fuzzy analogy ensembles. Appl. Soft Comput. J. **49**, 990–1019 (2016). https://doi.org/10.1016/j.asoc.2016.08.012
75. Hansen, L.K., Salamon, P.: Neural network ensembles. IEEE Trans. Pattern Anal. Mach. Intell. **12**, 993–1001 (1990). https://doi.org/10.1109/34.58871
76. Zhou, Z.H., Wu, J., Tang, W.: Ensembling neural networks: many could be better than all. Artif. Intell. **137**, 239–263 (2002). https://doi.org/10.1016/S0004-3702(02)00190-X

Fast Three-Dimensional Depth-Velocity Model Building Based on Reflection Traveltime Tomography and Pre-stack Time Migrated Images

D. Neklyudov[1]([⊠]), K. Gadylshin[1], M. Protasov[1], and L. Klimes[2]

[1] Institute of Petroleum Geology and Geophysics, Novosibirsk 630090, Russia
{neklyudovda,gadylshynkg,protasovmi}@ipgg.sbras.ru
[2] Charles University, 121 16, Prague, Czech Republic
klimes@seis.karlov.mff.cuni.cz

Abstract. We present a robust nonlinear traveltime inversion procedure that enables fast iterative 3D depth velocity model reconstruction. It is based on automatic grid reflection traveltime tomography. To obtain input data necessary for inversion (traveltimes of reflected waves), we utilize conventional seismic time processing products such as normal-moveout velocities and a set of time-migrated reflection horizons picked in the final prestack time migration image. Ray-based tomographic inversion fits prestack traveltimes. They are approximated by hyperbolae using the developed effective engine of grid reflection tomography. We prove the effectiveness of the proposed approach numerically using a real-life example.

Keywords: Traveltime tomography · Inverse problem · Time migration · Depth migration · Seismic exploration

1 Introduction

Nowadays, Pre-Stack Depth Migration (PSDM) has become a necessary tool in the practice of seismic data processing. Traditionally, depth migration results are called "depth seismic images" of an area under investigation. PSDM is especially important in areas with complex geological conditions in the presence of strong lateral variations of geological properties. The reliability of depth images depends entirely on the adequacy of the depth-velocity model of the medium in which PSDM is performed. Methods oriented for depth-velocity model building are based on the use of traveltimes of recorded seismic waves. Detailed reviews of existing methods that are widely used in seismic exploration can be found in the books [1–3]. Here we outline one specific feature of reflection traveltime tomography. Obtaining traveltime of reflected waves in practice is difficult (compared with traveltimes of transmitted or refracted waves observed as first arrivals within seismic records). There are two main reasons: 1) amount of seismic data acquired during modern seismic surveys oriented for reflections is enormous. It is about tens (and hundreds) of terabytes; 2) Poor quality of recorded data. Seismic records of reflected

© Springer Nature Switzerland AG 2021
O. Gervasi et al. (Eds.): ICCSA 2021, LNCS 12950, pp. 282–295, 2021.
https://doi.org/10.1007/978-3-030-86960-1_20

waves (seismograms) represent a complex pattern of interfering arrivals of various types of elastic waves. Identification of reflected arrivals, even using preliminarily processed data, is a challenging task. Reflections are difficult to distinguish because they are weak in comparison with residual coherent and random noise. These reasons exclude "manual" picking of reflection traveltimes in the practice of industrial seismic exploration. The challenge is to robustly extract traveltimes of the reflected waves from complex, noisy wavefields in an automated way.

An effective solution is provided by migration velocity analysis (MVA). MVA has now become the standard practical method for depth-velocity model building [4]. The basic idea behind MVA may be described in a simplified form as follows. Suppose some approximate initial depth-velocity model is provided. In that case, PSDM is performed for several common-offset gathers (i.e., specifically chosen subvolumes of the whole seismic dataset) when the offsets range is fixed. The set of calculated depth images corresponding to different offset are ordered into so-called Common Image Gathers (CIG). Each trace of a CIG corresponds to a depth image at a fixed lateral location in the area but obtained using a specific offset (and azimuth). If one considers a certain CIG, it consists of all migrated traces obtained using different offsets but depicting the medium's same area. If PSDM has been done using the "correct" depth-velocity model, then would align images of reflections in the CIG. The natural assumption used here is that the images of the same reflection surface should be at the same depth, regardless of the data parameters (such as offset or azimuth) from which built these images. If the depth-velocity model is incorrect, images of reflective surfaces in the CIG will have a shape different from the straight line. It is called residual moveout (RMO). The RMO analysis performed within the entire volume of the depth seismic image allows for correcting the current depth-velocity model. Each coherent event's curvature in each CIG, which is associated with a reflective surface in-depth, is recalculated into the traveltime residual of the corresponding reflected wave. (Here, the term "traveltime residuals" means the difference of traveltime of the reflected wave in the current and "correct" depth-velocity model).

Estimated reflection traveltime residuals are the input for linear traveltime tomography. As a result of traveltime inversion, a refinement of the depth velocity model is calculated. The corrected velocity model minimizes the traveltime residuals and accordingly reduces the curvature of RMO in the CIGs. The main practical advantages of MVA are: 1) estimation of reflection traveltimes performs automatically after PSDM. PSDM significantly reduces amount of random and coherent noise in the data is because of "optimal" stacking of seismic data; 2) there is no need for extended seismic horizons in the model since images on CIG are presented as a set of independent locally coherent events. Thus, traveltime tomography refines only the velocity structure of the medium. The geometry of reflectors (their shape and positions), as a rule, is not included in the inversion procedure at the early stages of velocity model building. The disadvantage of MVA is the need to repeat computationally expensive and lengthy cycles of 3D PSDM and perform re-picking at every nonlinear iteration of the velocity model update. To converge effectively, MVA requires starting from an adequate and good enough initial velocity model that provides coherent and focused CIGs for residual moveouts' picking. Robust depth-velocity model estimation approaches which may provide adequate initial guess for standard MVA have been a field of active research in recent years. The main idea behind these methods is to use products of seismic time processing such as time migration velocity [5–14] or wavefront attributes [15–17].

In this work, we propose a simple and relatively cheap algorithm to recover a reliable but simplified velocity model based on hyperbolic approximation of reflection traveltimes. We presume that such a model may be used as an initial model for standard MVA. A proper initial model allows reducing the number of computationally expensive MVA iterations at the early stages of velocity model building. Our approach is based on processing products, which are readily available at the early stages of seismic processing, namely stacking velocities and final processed "time" image provided by prestack-time migration (PSTM). The essence of the proposed approach is a robust scheme of reflection traveltimes approximation having a set of picked time migrated horizons. The advantages of the proposed algorithm are 1) it is fully automatic and based on well-known grid reflection tomography, 2) there is no need for computationally expensive two-point or dynamic raytracing, 3) it may be easily adopted for higher order traveltime approximations, 4) reflection travetimes approximation may be done azimuthally dependent.

The content of the paper is as follows. First, we present the proposed tomographic workflow and describe in more detail each step of the algorithm. We also cover some implementation details of the developed software code. Next, we present a real-life 3D example.

2 Reflection Traveltime Approximation

PSDM is always performed at the latter stages of seismic data processing workflow. In the steps preceding the PSDM, "time-images" of a target area are constructed using pre-stack time migration (PSTM). These images are used for preliminary interpretation. Main reflection horizons are usually picked there. Normal-moveout velocities V_{NMO} and velocities of time migration V_{MIG} are also always available at this stage. It is usually supposed that V_{MIG} approximate the root mean square velocity, V_{RMS}. Having these products of time processing an efficient algorithm for reflection traveltime tomography may be suggested. Below we describe the proposed algorithm in more detail. First, we specify the "ingredients" needed for workflow:

1. Initial depth velocity model. Usually, it is constructed from the available time migration velocities V_{MIG} using the Dix formula [18]. In most cases, even in areas with moderate lateral velocity variations, such a model is barely acceptable and unable to provide a reliable depth image.
2. Main reflection horizons picked in the final PSTM image (in a 3D case, it is a cube), $T_{IM}^j(X_{IM}, Y_{IM}), j = 1,..N.$ (N is several horizons.) Each horizon is defined in the form of a table, (X_{IM}, Y_{IM}, T_{IM}), where X_{IM}, Y_{IM} are the coordinates at the acquisition surface, T_{IM} is the corresponding image time. Further, we consider each surface as a set of independent locally coherent events. Note that each independent local coherent event is strictly associated with a specific reflection surface.

3. A cube of normal-moveout velocity, $V_{NMO}(X_{CMP}, Y_{CMP}, T_0)$. (Or, optionally, a set of azimuth- dependent cubes, $V_{NMO}(X_{CMP}, Y_{CMP}, T_0; \alpha)$). Here, X_{CMP}, Y_{CMP} are midpoint coordinates, T_0 is two-way traveltime along a normal ray (zero-offset traveltime), α means azimuth on the acquisition surface.

At the first stages of the proposed approach, a "numerical post-stack demigration through depth" of the picked time-migrated reflection surfaces is performed using the concept of "image ray" [19]. This term denotes an auxiliary ray that approaches the acquisition surface along the normal (see Fig. 1). Stages 1–5 described below are performed for each picked time-migrated reflection surface independently.

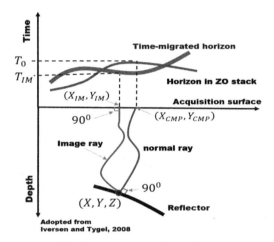

Fig. 1. Image ray and normal ray

1. For each point on the given "time-migrated" surface $T_{IM}^j (X_{IM}, Y_{IM})$, an "image" ray is traced downward within the given initial depth velocity model. It is released from the acquisition surface from the point (X_{IM}, Y_{IM}) along the normal and continues downward until the half image time $T_{IM}/2$ has expired. As a result, the position of the corresponding point (X, Y, Z) in depth is determined. It is repeated for each point of the surfaces T_{IM}^j. As a result, on performs the depth migration (of "time-to-depth conversion") of the "time" reflection horizon using the "image" rays [19]. Each point of the surface T_{IM}^j now has the corresponding position in depth in the current velocity model. The distribution of the points in depth is irregular. Note that the "image" ray provides no information about the local reflection surfaces' normal vector (contrary to normal ray tracing). All that can be said is that the "image" ray touches the reflection surface when the time $T_{IM}/2$ has passed (Fig. 1).

2. At the previous stage, we obtain the distribution of reflection horizon points in depth. Consider a depth point coordinates as a certain function $Z_j(X, Y)$ which is defined as a set of discreet values on a two-dimensional irregular grid (X, Y). It is necessary to determine numerically the surface $Z_j(X, Y)$ normal (Fig. 2). For each

fixed point in depth, its neighbors are collected within some predetermined aperture. A local plane is constructed which is closest to all selected points in the least-squares sense. Thus, the classical three-dimensional linear regression is considered locally. According to the irregular "cloud" of points (X, Y, Z) a set of local reflection surfaces is constructed. It is assumed that each local reflection surface is flat and characterized by its own normal. After time-to-depth conversion by the "image" rays using an incorrect depth-velocity model, initially smooth and regular "time" horizons might become quite irregular in depth. We assume that their local smoothness is preserved. We apply the following procedures: filtration of the values $Z_j(X, Y)$ within the regression, apertures remove outliers; local smoothing of the neighbor normal vectors within a predefined sliding window.

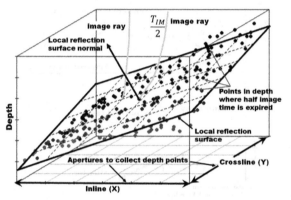

Fig. 2. Local reflection surface normal determination in depth via three-dimensional linear regressions.

3. From each local reflection surface in depth, a normal ray according to surface's normal vector is traced upward toward the acquisition surface. Thus a relation between the time-migrated horizon $T_{IM}^j(X_{IM}, Y_{IM})$ with corresponding zero-offset travel-times $T_0^j(X_{CMP}, Y_{CMP})$ is numerically constructed. Stages 1–3 are actually "numerical demigration through depth" of "time" migrated horizons picked in the PSTM cube.

4. From each reflection surface with a known normal, a fan of reflected rays is emitted. They are traced upward with certain increments of the opening angles (reflection angles) and azimuths (in depth) (Fig. 3). Each reflected ray consists of two segments satisfying Snell's law at the reflection surface. Let each of the two segments of the reflected ray arrive at the acquisition surface at certain points. These points can be thought of as "virtual" source, and receiver which have their own coordinates (X_S, Y_S) and (X_R, Y_R). Using these coordinates, reflected ray might be uniquely defined by the midpoint coordinates (X_{CMP}, Y_{CMP}), absolute offset h (distance between the source and receiver), and azimuth α. Traveltime $T_{CALC} = T_{CALC}(X_{CMP}, Y_{CMP}, h, \alpha)$ along

each reflected ray is computed in the current depth-velocity model and corresponding elements of tomography matrix, which will be used in traveltime inversion.

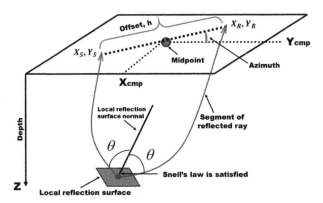

Fig. 3. Calculation of up-going reflected ray emitted from a local reflection surface in depth.

5. "Observed" reflection traveltimes should be determined in each point on the acquisition surface. At this point a reflected ray calculated at stage 4 exists. We suggest approximating actual reflection traveltime using well known 2nd order hyperbolic approximation [20, 21]:

$$T_{OBS}(X_{CMP}, Y_{CMP}, h, \alpha) = \sqrt{T_0^2(X_{CMP}, Y_{CMP}) + \frac{h^2}{V_{NMO}^2(X_{CMP}, Y_{CMP}, T_0, \alpha)}}$$

$$(1)$$

Each reflected ray is associated with its midpoint coordinates X_{CMP}, Y_{CMP} So it is easy to find the corresponding normal (zero offset) two-way traveltime $T_0 = T_0(X_{CMP}, Y_{CMP})$ (thanks to stage 3) and, therefore, to find the required value $V_{NMO}(X_{CMP}, Y_{CMP}, T_0; \alpha)$. So observed traveltime at a given point may be calculated using expression (1) since all "ingredients" are found. Such approximation is quite robust since V_{NMO} is a very stable parameter. Moreover, NMO velocities are carefully controlled. In principle, it is a purely kinematic parameter, which generally may have nothing in common with the medium's real distribution of interval velocities. Still, it always characterizes the behavior of the reflection traveltimes in the sense of regression. By definition, V_{NMO} characterizes best fits hyperbola that approximate real traveltimes of the reflected waves. More sophisticated approximations may be used here (for example, 4th order approximation described by [22, 23]). Note that NMO velocity may be additionally azimuthally dependent [20]. In that case, "observed" traveltime (1) is calculated, taking into account the reflected ray's azimuth α, i.e., it is necessary to select the NMO velocity corresponding to the given azimuth.

Finally, the input data for standard linear reflection traveltime tomography may be computed. As an input, traveltimes residuals of reflected waves are used:

$$dT(X_{CMP}, Y_{CMP}, h, \alpha) = T_{OBS}(X_{CMP}, Y_{CMP}, h, \alpha) - T_{CALC}(X_{CMP}, Y_{CMP}, h, \alpha)$$
(2)

Where T_{CALC} are calculated in stage 4 while T_{OBS} are estimated by expression (1).
6. Linear reflection traveltime tomography using reflection traveltime residuals (2) is performed.

As a result of stage 6, we determine correction to the initial velocity model. After updating of depth velocity model, steps 1–6 of the proposed scheme are repeated in the corrected model. Thus, we perform iterative refinement of the 3D depth-velocity model without involving computationally expensive PSDM procedure. After several global iterations, the velocity model's refinement must be stopped because hyperbolic approximation of reflection traveltimes may be insufficient for reliable reconstruction of the velocity model, especially in its deeper part. This result can be used as an initial model for the standard MVA, based on a much weaker assumption about the behavior of the residual moveouts picked in CIG. Note that the traveltime inversion engine (i.e., a tool responsible for the linear inverse problem solution) may be borrowed from already existing seismic data processing packages.

3   Robust Linear Traveltime Inversion

For the solution of a linear inverse problem, a tomographic system of linear algebraic equations (SLAE) is constructed during the ray tracing stage **4** and traveltime approximation stage **5**. SLAE forms a linear relationship between traveltime residuals dT and desired model update vector Δv:

$$dT = M \, \Delta v$$
(3)

Here M is a tomographic matrix with the elements representing derivatives of reflection traveltimes with respect to model parameters. Usually, the velocity model is parameterized by a 3D regular grid with constant velocity inside each voxel. Element M_{ij} of the matrix, M denotes a length of i-th reflected ray within j-th voxel. The voxels are numbered in continuous order; their number corresponds to the tomographic matrix columns. Each reflected ray corresponds to a row in the matrix M. The size of M in the isotropic case is $N_{Rays} \times (N_X \cdot N_Y \cdot N_Z)$, where N_{Rays} is a number of successfully constructed reflected rays, whereas $N_X \cdot N_Y \cdot N_Z$ is the total count of points in the subsurface velocity grid. In a typical case, the number of rows significantly exceeds the number of columns. So we arrive at a classical "grid" tomography problem [3]. Below we follow an approach proposed by [4]. Instead of the original linear system (3), a pre-conditioned system is solved:

$$LMSR\Delta v' = LdT,$$
(4)

where L is a diagonal weighting matrix of rows, R is a diagonal weighting matrix of columns. S is a spatial smoother (which acts as a 3D convolution with a triangle function). Pre-conditioned system (4) is solved using the Iterative Reweighted Least-Squares (IRLS) method [24] with the norm $l_{1.5}$. The cost function which is minimized during the solution of the inverse problem may be expressed as

$$F = \|LMSR\Delta v' - LdT\|_{1.5}^{1.5} + \lambda^2\|\Delta v'\|_2^2 = \|A\overrightarrow{x} - \overrightarrow{b}\|_{1.5}^{1.5} + \lambda^2\|\overrightarrow{x}\|_2^2 \qquad (5)$$

Where notations $A = LMSR$, $\overrightarrow{x} = \Delta v'$, $\overrightarrow{b} = LdT$ are used.

The inverse problem comes down to solving the set of typical least-squares problems with a recursively updated matrix of weights

$$A^T W_{k-1} A \overrightarrow{x_k} = A^T W_{k-1} \overrightarrow{b} \qquad (6)$$

Here diagonal weighting matrix W_k at k-th IRLS iteration is determined as $W_k = diag|r_i|^{-0.5}$, where $r_i = \sum A_{ij} - b_i$ is a residual vector after $(k-1)$-th iteration. SLAE's (6) is solved using the LSQR method [25]. The solution of the system (4) with the help of IRLS allows us to obtain solutions that are much more stable to abrupt "jumps" of input data (outliers). Outliers' impact on the final solution is a significant challenge in tomographic problems using the standard least-squares method.

It is important to highlight key features of SLAE's arising in seismic tomography that is important for the chosen realization of the code: 1) Huge dimensions of the system of equations ($> 10^6$ equations with more than 10^8 unknowns); 2) Matrix is very sparse with a relatively small number of nonzero elements (~1–3%); 3) Problem is ill-posed demanding usage of pre-conditioning and regularization procedures. For real-world applications, gigabytes of memory are needed to operate with such matrices (even taken into account their sparsity). Software implementation of 3D tomography should be initially focused on using high-performance computing systems with distributed memory (MPI - implementation). Our tomography implementation is based on using the functionality contained in the freeware library PETSc (http://www.mcs.anl.gov/petsc). The PETSc library is oriented explicitly for coding MPI-oriented programs that extensively use elements of linear algebra. PETSc has the following advantages: 1) It contains a set of MPI-oriented iterative algorithms for solving SLAE such as LSQR; 2) It is suitable for working with data arrays distributed over computing nodes (matrices, vectors) and performing any operations with such objects; 3) It provides a convenient way to deal with sparse matrices and vectors distributed over cluster nodes.

Following [4], the model updates pre-conditioning based on triangular smoothing has been implemented. Smoothing is implemented as a 3D spatial convolution with triangular functions. There are two possible strategies for smoothing.

The first strategy is "multiscale" smoothing. For each linear inverse problem, we attempt to resolve many scales progressing from large to small using a fixed framework of a single linear inverse problem. Then we apply similar "multiscale" smoothing inside all other linear iterations. Let's describe the main concept of the multiscale approach briefly. Initial problem (4) is represented as a series of the problems with "k" smoother S scales,

$$\sum_{k=1}^{N_S} [LMS_k R]\Delta v_{k-1} = d \qquad (7)$$

At the first "multiscale" iteration, $k = 1$, one fixes the biggest smoother apertures and solve the system using the IRLS method:

$$[LMS_1R]\Delta v_1 = d \tag{8}$$

When the "long-wavelength" solution $\hat{v}_1$ is obtained, the smoother aperture is decreased, and the linear system at the next scale is solved:

$$[LMS_2R]\Delta v_2 = d - [LMS_1R]\Delta\hat{v}_1 = d_2. \tag{9}$$

This stage is repeated for each given smoother scale. Note that event positions remain fixed within the current velocity model, i.e., tomographic matrix M is the same during all iterations with different smoother apertures.

This classical "multiscale" approach is proven efficient in MVA applications [4]. The remigration stage is very computationally expensive. 3D PSDM and RMO picking have to be redone for each updated velocity model.

The second strategy is "individual" smoothing for each linear inverse problem. We resolve only a specific scale of velocity variation within one solution of a single linear inverse problem. At the same time, smoothing aperture can vary for other linear iterations. The second approach consists of solving the system (8) with a fixed "individual" smoothing aperture at each linear iteration. In this case, matrix M is recalculated once again after each iteration, and we apply a new smoother aperture. In this case number of smothers, apertures equal the number of nonlinear iterations.

For our approach, where the remigration stage is relatively cheap, the second approach is the most efficient way to solve the problem. Of course, the multiscale approach may also be used. We implemented the multiscale approach and used it in all examples in this paper.

4 Real Data Numerical Example

The approach described above was applied for depth-velocity model building for a certain area located in the Kara Sea. A narrow azimuth marine survey was performed for this area. The maximum offset in the data is 3800 m. Sea depth doesn't exceed 250 m in the acquisition area. A fragment of the investigated area 40×10 km (inline/crossline) has been selected. Thus, the size of the area selected for the numerical test was equal to 400 km^2. The eight most significant reflection horizons were picked in the final PSTM cube. Tables specifying the dependence $T_{IM}^j(X_{IM}, Y_{IM})$ for each horizon were provided with a spatial step of 25×12.5 m, which corresponds to a 3D seismic survey's binning size. An example of 4 surfaces (of a total of 8) is presented in Fig. 4. The initial velocity model (shown in Fig. 5) was obtained by the standard method by converting time migration velocities to interval velocities according to the Dix formula. Discretization of the model is $25 \times 12.5 \times 10$ m (inline, crossline, depth). The total depth of the model is 7 km. Traveltime inversion was performed using a coarser grid $100 \times 100 \times 50$ m. The. The smoothers width in operator S decreased using four consecutive steps during global iterations of model refinement: initial widest smoothers are $3000 \times 3000 \times 1000$ m, the final widths are ten times smaller, i.e., $300 \times 300 \times 100$ m. It allows moving gradually from long-wavelength velocity variations to more detailed ones avoiding manifestation of instability. The total number of global nonlinear iteration was 12. While solving SLAE, the number of LSQR iteration was 20, whereas the number of IRLS (outer) iterations was equal to 10. Apertures for local reflection surface approximation (stage 2 of the proposed workflow) were chosen as 100×100 m (inline/crossline).

The recovered depth velocity model is presented in Fig. 6. The PSDM results using the initial velocity model are presented in Figs. 7a and 8a. As one can see from the CIG gathers (see Fig. 7a, where a fragment taken along the inline direction from the center of the 3D model is presented), the initial depth velocity model is far from satisfactory. At the bottom part of the images coherent events are not visible at all. It would make the use of standard MVA very difficult. In the stacked PSDM image (Fig. 8a), one can observe defocusing and poor coherency along the main reflection horizons. PSDM performed using a recovered depth velocity model (Fig. 6) allows obtaining very satisfactory results (Fig. 7b and 8b).

The results presented in this section prove the efficiency of the proposed approach. In the cases where the hyperbolic reflection traveltime approximation describes the real reflection arrivals in a sufficiently far range of offsets, the proposed algorithm could replace the standard MVA. Recall, though, that its main purpose is to be used before the standard MVA is applied to provide a reliable initial depth velocity model, which may be refined with minimum MVA iterations.

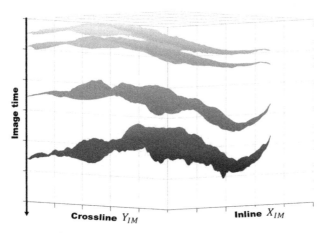

Fig. 4. Time migrated surfaces picked in PSTM cube.

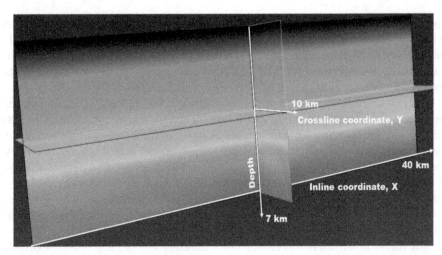

Fig. 5. Initial depth velocity model; model size is 40 × 10 × 7 km (X,Y,Z); velocity variations are within the range [1500–4400] m/s.

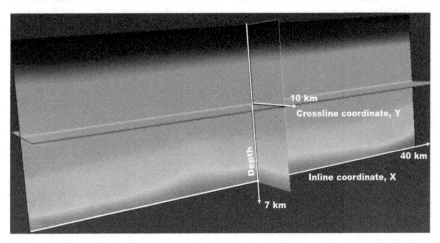

Fig. 6. Recovered depth velocity model using the proposed approach; velocity variations are within the range [1500–4400] m/s.

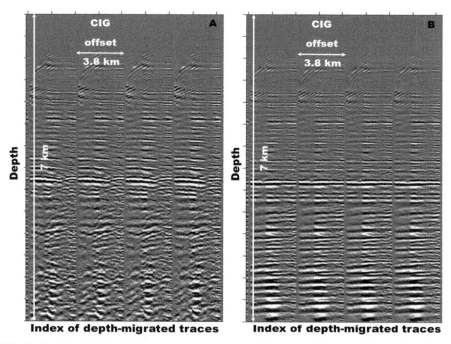

Fig. 7. Common-image gathers obtained by PSDM using (a) initial depth-velocity model; (b) recovered depth-velocity model.

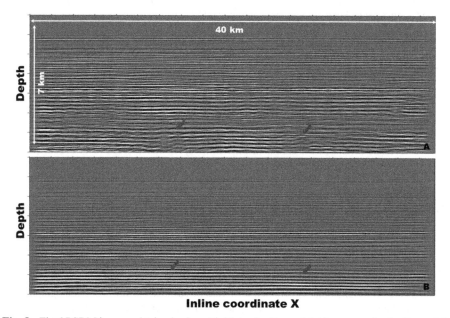

Fig. 8. Final PSDM images obtained using a) initial velocity model, b) recovered velocity model. The inline section from the middle of the models is presented.

5 Conclusions

We presented a robust nonlinear 3D reflection traveltime tomography workflow based on the use of time processing products such as NMO velocity and picked time-migrated horizons. Picking is performed within the resulting PSTM image. After numerical demigration of time-migrated surfaces, we utilized hyperbolic reflection traveltime approximation directly in the data domain. We did it instead of picking reflection traveltimes in the image domain like in standard MVA. The proposed approach enables fast iterative depth velocity model reconstruction. The reliability of the recovered model is limited due to the approximation of real reflection traveltimes, but such a model could serve as an initial velocity model for MVA. The use of an advanced initial model allows reducing the number of computationally expensive MVA iterations.

In some cases, where the hyperbolic reflection traveltime approximation can describe the real reflection arrivals in a sufficiently far range of offsets, the proposed algorithm could replace the standard MVA. The advantages of the proposed algorithm are the following: 1) there is no need for preliminary manual interpretation of seismic volumes since the algorithm is fully automatic and based on grid reflection tomography, 2) there is no need for computationally expensive two-point or dynamic raytracing, and 3) it is much cheaper computationally than standard common-image point tomography in the image domain. Real data examples confirm that depth images using the velocity model recovered by the proposed approach are more focused, and the flattening of the corresponding common-image gathers was improved.

Acknowledgments. The reported study was funded by RFBR, project number 20–55-26003, and by the Czech Science Foundation under contract 21-15272J.

References

1. Fagin, S.: Model Based Depth Imaging. SEG (1999)
2. Robein, E.: Seismic Imaging: A Review of the Techniques, their Principles, Merits and Limitations. EAGE (2010)
3. Jones I.F.: An Introduction to: Velocity model Building. EAGE (2010)
4. Woodward, V., Nichols, D., Zdraveva, O., Whitfield, P., Johns, T.: A decade of tomography. Geophysics **73**, VE5-VE11 (2008)
5. Cameron, M., Fomel, S., Sethian, J.: Time-to-depth conversion and seismic velocity estimation using time-migration velocity. Geophysics **73**, VE205-VE210 (2008)
6. Iversen, E., Tygel, M.: Image-ray tracing for joint 3D seismic velocity estimation and time-to-depth conversion. Geophysics **73**, S99–S114 (2008)
7. Santos H.B., Schleicher J., Novais A.: Initial-model construction for MVA techniques. In: Proceedings of the 2013 75th EAGE Conference and Exhibition, EAGE (2013)
8. Dell, S., Gajewski, D., Tygel, M.: Image-ray tomography. Geophys. Prospect. **62**, 413–426 (2014)
9. Lambare, G., Guillaume, P., Montel, J.P.: Recent advances in ray-based tomography. In: Proceedings of 2014 76th EAGE Conference and Exhibition, EAGE (2014)
10. Li, S., Fomel, S.: A robust approach to time-to-depth conversion and interval velocity estimation from time migration in the presence of lateral velocity variations. Geophys. Prospect. **63**, 315–337 (2015)

11. Santos, H.-B., Schleicher, J., Novais, A., Kurzmann, A., Bohlen, T.: Robust time-domain migration velocity analysis for initial-model building in a full-waveform tomography workflow. In: Proceedings of 2016 87th SEG Annual Meeting, pp. 5307–5312. SEG (2016)
12. Sadala Valente, L.S., Santos, H., Costa, J., Schleicher, J.: Time-to-depth conversion and velocity estimation by image-wavefront propagation. Geophysics **82**, U75–U85 (2017)
13. Sripanich, Y., Fomel, S.: Fast time-to-depth conversion and interval velocity estimation in the case of weak lateral variations. Geophysics **83**, S227–S235 (2018)
14. Zhao, H., Ueland Waldeland, A., Rueda Serrano, D., Tygel, M., Iversen, E.: Time-migration tomography based on reflection slopes in pre-stack time-migrated seismic data. In: Proceedings of 2018 80th EAGE EAGE Conference and Exhibition. EAGE (2018)
15. Duveneck, E.: Velocity model estimation with data-derived wavefront attributes. Geophysics **69**, 265–274 (2004)
16. Lambaré G.: Stereotomography. Geophysics, 73, VE25-VE34 (2008)
17. Gelius, L.-J., Tygel, M.: Migration-velocity building in time and depth from 3D (2D) Common-Reflection-Surface (CRS) stacking - theoretical framework. Stud. Geophys. Geod. **59**(2), 253–282 (2015). https://doi.org/10.1007/s11200-014-1036-6
18. Mesquita, L., Jorge, M., Cruz, R., Callapino, J.C., Garabito, G.: Velocity inversion by global optimization using finite-offset common-reflection-surface stacking applied to synthetic and Tacutu Basin seismic data. Geophysics **84**, R165–R174 (2019)
19. Dix, C.H.: Seismic velocities from surface measurements. Geophysics **20**, 68–86 (1955)
20. Hubral, P., Krey, T.: Interval velocities from seismic reflection time measurements. SEG (1980)
21. Yilmaz, O.: Seismic Data Analysis: Processing, Inversion, and Interpretation of Seismic Data. SEG (2001)
22. Gjoystdal, H., Ursin, B.: Inversion of reflection times in three dimensions. Geophysics **46**, 972–983 (1981)
23. Al-Chalabi, M.: Series approximation in velocity and traveltime computations. Geophys. Prosp. **21**, 783–795 (1973)
24. Alkhalifah, T.: Velocity analysis using nonhyperbolic moveout in transversely isotropic media. Geophysics **62**, 1839–1854 (1997)
25. Scales, J., Gersztenkorn, A., Treitel, S.: Fast Lp solution of large, sparse linear systems, application to seismic traveltime tomography. J. Compu. Phys. **75**, 313–333 (1988)
26. Paige, C.C., Saunders, M.A.: LSQR: an algorithm for sparse linear equations and sparse least squares. ACM Trans. Math. Softw. **8**(1), 43–71 (1982)

Synchronization Overlap Trade-Off for a Model of Spatial Distribution of Species

João Bioco[1,2,4(✉)] , Paula Prata[1,2,4] , Fernando Cánovas[3] ,
and Paulo Fazendeiro[1,2,4]

[1] Universidade da Beira Interior, UBI, Covilhã, Portugal
[2] C4 - Centro de Competências em Cloud Computing (C4-UBI),
Universidade da Beira Interior, Covilhã, Portugal
[3] Universidad Católica San Antonio de Murcia, Facultad de Ciencias de la Salud,
Murcia, Spain
[4] Instituto de Telecomunicações (IT) Covilhã, Covilhã, Portugal

Abstract. Despite of the widespread implementation of agent-based
models in ecological modeling and another several areas, modelers have
been concerned by the time consuming of these type of models.

This paper presents a strategy to parallelize an agent-based model of
spatial distribution of biological species, operating in a multi-stage synchronous distributed memory mode, as a way to obtain gains in the performance while reducing the need for synchronization. A multiprocessing
implementation divides the environment (a rectangular grid corresponding to the study area) into stage-subsets, according to the number of
defined or available processes. In order to ensure that there is no information loss, each stage-subset is extended with an overlapping section
from each one of its neighbouring stage-subsets. The effect of the size
of this overlapping on the quality of the simulations is studied. These
results seem to indicate that it is possible to establish an optimal trade-
off between the level of redundancy and the synchronization frequency.

The reported paralellization method was tested in a standalone mul-
ticore machine but may be seamlessly scalable to a computation cluster.

Keywords: Parallel programming · Multiprocessing · Agent-based
modelling and simulation · Synchronization-reducing algorithms

1 Introduction

Agent-based modelling (ABM) is an approach centered in agents (unique individuals) that interact with each other and their environment in order to accomplish a certain goal. In this modelling approach (called bottom-up modelling)
the behaviour of the system emerges by the local interactions between agents in
the environment where they exist [1]. This fact makes this modelling approach
widely applied in several areas, such as in biology, engineering, ecology, epidemiology, etc. [2,20,23,26]. One of the areas that has largely adopted agent-based

© Springer Nature Switzerland AG 2021
O. Gervasi et al. (Eds.): ICCSA 2021, LNCS 12950, pp. 296–310, 2021.
https://doi.org/10.1007/978-3-030-86960-1_21

modeling is ecology. Generally in this field, agents are described as individual organisms with behaviours and proprieties that usually change during their life cycle [18].

In the last years due to the climate changes observed in the nature, the interest in the study of species distribution has increased, in the sense of creating mechanisms for species preservation. There are some aspects that have to be taken into account when using ABM to model and simulate the distribution of species, such as: the environmental conditions that influence the life cycle of the species, the available resources (i.e., food, water), the dimension of the environment where species exist (the study area), etc. [4]. The area under study can be small or large, and the coarseness of the simulation stage can be quite heterogeneous from simulation to simulation. In the cases where the dimension of the environment is very large and the required level of detail for the simulation is high, the simulation can be quite time consuming and requires very large computation power. In this case it is necessary to implement strategies to narrow down the time requirements.

This paper presents a strategy to parallelize an agent-based model of spatial distribution of biological species, running in a multi-stage synchronous mode enabling the reduction of the processing time while assuring that there is no significant information loss. This study parameterizes the size of the overlapping section sent with each stage-subset and the number of iterations done between each global synchronization. The findings show that it is possible to attain a good trade-off between the size of the internal computation and the time of data transferring while maintaining the correctness of the algorithm.

The remaining of this paper is organized as follows: the Sect. 2 describes some studies that implemented parallelization in ABM; the Sect. 3 presents our model proposal; the experimental results are presented in the Sect. 4; in the Sect. 5 the discussion of the study is presented, followed by the concluding section.

2 Related Work

ABM in ecology was firstly applied in forest modelling, and the application to other areas of ecology began to increase in the 1990 s [12]. A particular early use of ABM was to model the recruitment of fish populations, in order to understand and assess human impacts in the mortality of fish recruitment [11]. However, two main categories of models have motivated the development of ABM in ecology based on purpose [14]. One category is mainly used to model a specific type of species, population or ecosystem with management purposes; another category aims to help ecologists to have a better understanding of the factors behind of several ecological phenomena [19]. A significant advantage of ABM in ecology concerns the study of populations by considering the individuals' behaviours and relationships. This is a more natural way of looking at population, making simulation results closer to reality [5].

The two main approaches to parallelize spatial ABMs are [22]: 1) to divide the computations at agent level where each processing element (PE) is responsible for a set of agents; 2) to divide the computations at the spatial environment level where each PE is responsible for a set of grid cells. In both cases, to handle the interaction between agents and its movement requires communication and synchronization, which constitutes the limiting factor for obtaining a scalable parallel model.

Existent proposals for parallel implementations of spatial ABMs range from multithreaded implementations in shared memory architectures [13,24] to implementations in Graphic Processing Units (GPU)s [6,7,17]. However, most of the proposals are based on distributed memory programming models which can be scaled to thousands of cores. These works include frameworks as FLAME [8,25] and Repast HPC [9,10] that use MPI for inter process communications. The work presented in [27] implements an ABM in the Apache Spark framework trying to take advantage of its in-memory computation model. In all these works the main performance bottleneck remains the communication cost.

3   Methodology

The model description follows the ODD (Overview, Design concepts, Detail) protocol [15,16]. Its main components are described in [4]. The purpose of the model is to explore different hypothesis on the space-time distribution of species. The model comprises an environment disposed in a regular rectangular grid. Each grid cell stores its suitability value (a value between 0 and 1 that sums up the environmental conditions of the cell), and the number of specimens that are present in the cell. This amount increases or decreases during the life cycle according to the suitability of the cell and the give-and-take of specimens with the neighbouring cells. Figure 1 shows in more detail the characterization of the environment.

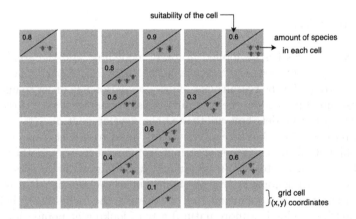

Fig. 1. Characterization of the environment.

3.1   Process Overview and Schedule

Figure 2 presents the main steps of the simulation process. It starts by initializing the environment stage (hereafter referred as patches), after setting the parameters of the model. The environment is initialized by instantiating each cell (hereafter also referred as patch) with its suitability value obtained by the combination of a set of environmental variables, and an initial quantity of species. In each iteration, to each cell is applied a birth and death rate affecting the quantity of species. These rates are conditioned by the suitability value of the cell, i.e., species are most likely to survive and reproduce in more suitable locations, on the other hand, species are more likely to disappear in less suitable locations. The spread of the species occurs through the neighbourhood of the cells. The reported model implements the Moore neighbourhood [21]. Therefore, at each iteration, each cell transfers an amount of material (quantity of species) to its neighbours, according to a spread rate. Algorithm 1 describes the updating mechanism that simulates the spatial dynamics of the spreading of a natural organism guided towards spatial self-organization (species life cycle).

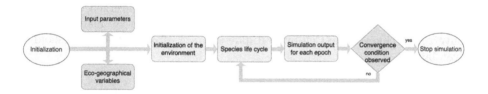

Fig. 2. General steps of the simulation process

Algorithm 1. Life cycle algorithm. The reproduce method contains the exchange policy of the cell with its neighbours depending on its birth, death and spread rates.

---

procedure $Distribution\_update(patches, previous\_patches, steps)$
    for k in $steps$ **do**
        for i, row in $patches$ **do**
            for $j, patch$ in row **do**
                $neighbours = patch.find\_neighbours(previous\_patches, i, j)$
                $patch.reproduce(previous\_patches[i][j], birth\_rate,$
                                    $death\_rate, spread\_rate, neighbours)$
            end for
        end for
    end for
    return $patches$
end procedure

The sequence of patches (different states) obtained during the iterative operation of the Algorithm 1 is called the evolution. In the sense defined in [3] evolution is the result of the simulation task, representing the process under simulation. If the process converges to a stable global state, then the algorithm evolution has a termination. If that is not the case, then the evolution is infinite, exhibiting oscillatory or chaotic behaviour. Algorithm 2 represents the evolution snapshot for a fixed amount, t, of epochs.

Algorithm 2. Evolution of the simulation task.

Initialize $patches, previous\_patches$, set $epochs, output\_interval$ and let $steps \leftarrow 1$
for t in $epochs$ **do**
    $patches = Distribution\_update(patches, previous\_patches, steps)$
    if $mod(t, output\_interval) == 0$ **then**
        $\Delta = sum(abs(previous\_patches - patches))$
        Copy $patches$ into $previous\_patches$
    end if
end for

Alternatively the evolution can be terminated once a convergence condition is observed. In that case the evolution stops, i.e. for all that matters it has converged, whenever the Cell-by-Cell difference, Δ, drops below a threshold τ given by

$$\tau = \epsilon \times N_c, \tag{1}$$

where N_c is the number of cells in the environment and ϵ stands for the amount of admissible error between any given pair of corresponding cells (or minimum distinguishability level). This threshold value can also be used as a reference value in order to analyse if two different evolutions exhibit, or not, a similar convergent behaviour.

3.2   Parallelization Strategy

The adopted parallelization strategy consists of assigning to each PE a set of grid cells. The species distribution map was divided into stage-subsets (hereafter referred as strips, since in our implementation any subset encompasses all the columns of the main stage), each one with a dimension given by the number of rows in the map divided by the number of available processes.

While pursuing high efficiency it is possible to break the equality between the initial sequential evolution and that of a decomposed parallel algorithm. Therefore in order to guarantee the equality of the sequential evolution to its parallelized version a set of correctness conditions, see [3], must be assured during the interaction between processes. Simply stated, "the problem is to organize the parallel operation in such a way that each domain interacts with the adjacent ones by exchanging data that are needed to be used in one of them for computing the next-states in the other". Moreover any given cell must be updated only once per iteration. Algorithm 3 presents the synchronous parallelization used in the reported experiments.

Algorithm 3. Synchronous parallel algorithm.

Initialize *patches, previous_patches*
Set *number_processes, overlap, steps, epochs, output_interval*
for t in *epochs* **do**
   strips $=$ *Build_strips(patches, number_processes, overlap)*
   previous_strips $=$ *Build_strips(previous_patches, number_processes, overlap)*
   for each *strip$_s$* **do**
      Run process P_s(*Distribution_update(strip$_s$, previous_strip$_s$, steps*))
   end for
   for each process P_s **do**
      Receive and fuse the P_s results into *patches*
   end for
   if $mod(t, output\_interval) == 0$ **then**
      $\Delta = sum(abs(previous\_patches - patches))$
   end if
   Copy *patches* into *previous_patches*
end for

It goes without saying that the equality between sequential and parallel versions is only assured if all the cells whose values are necessary for updating any given cell of the strips are available. In this case, using a Moore neighbourhood, this means that in order to perform a parallel iteration each strip must be extended by adding a border region (overlap) of at least one row. When dealing with parallel processes one must also be aware of the interprocess communication costs, which may be reduced by preventing frequent synchronizations between processes. This can be achieved if, instead of joining the strips after each iteration, the different stages are allowed to evolve independently in their own parallel processes for a given number of evolutive steps. Once again, the equality between versions is only achieved if the overlap between strips (or partial stages) is composed of a number of rows equal or bigger than the number of steps.

To deal with the border region, the strips are extended with a set of adjacent rows coming from each neighbouring strip. In what follows those common regions are referred as overlap, as shown in Fig. 3 where it is emphasized the domain decomposition.

In the body of the simulation cycle, the data is divided into stage-subsets. Each chunk of data (including a strip of the current distribution map, *patches$_k$*, and its overlapping sections) is sent to a process running in parallel with the ones responsible for the other data chunks. Each process will perform over its data the number of iterations given by a step parameter. At the end of the parallel phase, just the arising strips are sent back to the main process, producing the next map instance (*patches$_{k+step}$*). This means that, at this phase, the overlapping sections are discarded since their main purpose is to act as a buffer, updating the frontier cells in the starting iterations of each process. The core strips of the distribution map are fused together and the main process continues as described.

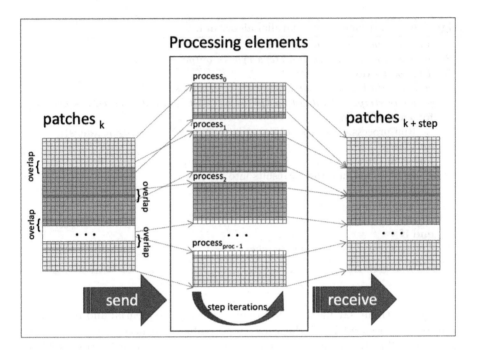

Fig. 3. Parallelization of the spatial environment. Decomposition of the study area into a set of overlapping strips for parallel processing.

4   Experimental Results

In the reported experiments were performed simulations using both sequential and parallel implementations of the model. Initially, the simulation with the sequential implementation was performed, followed by the simulation with the parallel implementation. Execution times were computed in a machine with the following hardware and software configurations: a) Operating System: Linux Ubuntu Desktop version: 18.04.5 LTS 64 bits, b) RAM: 64 GB, c) Processor: Intel® Core™ i9-9900X CPU @ 3.50 GHz × 20, d) Python version: 3.7.2.

For these experiments the spatial distribution of the african honeybee Apis mellifera in the Iberian Peninsula was simulated. The set of variables of interest for the environment was composed by four eco-geographical variables (maximum temperature of the warmest month, rainfall seasonality, average annual temperature and minimum temperature of the coldest month) which together are the input values determining the suitability of each cell.

In order to facilitate the comparison between the two implementations (sequential and parallel) the quantity of species in each grid cell was initialized by using the same seed for each simulation.

4.1 Parameterization

Table 1 shows all the parameters of the model and the values used for both sequential and parallel implementations. For the parallel implementation, 12 parameter combinations were chosen: step of 10 varying the frontier with the values (4, 6, 8, 10); step of 50 varying the frontier with the values (20, 30, 40, 50); and step of 100 varying the frontier with values (40, 60, 80, 100). Therefore, for the parallel implementation 12 different simulations were performed. For the sequential implementation the results at each timestamp were saved; and for the parallel implementation the results were saved according to the chosen step. The number of processes was fixed to 12 in order to analyse the behaviour of the algorithms when the processes were subjected to a varying workload directly related with the dimension of the data used in the experiment.

Table 1. Models' parameters.

Parameters	Value
Initial population	200 000
Number of epochs	200
Cells capacity	1000
Output generation interval	According to the step
Birth rate	0.9
Death rate	0.2
Spread rate	0.6
Type of neighbourhood	Moore
Environment dimension	1210×1940

4.2 Quasi-equality Behaviour

Aiming at the determination of the rate of degradation, resulting from the reduction of the border region, the differences between the initial sequential evolution and the parallel evolution with different process steps (period where each strip evolves independently of the remaining ones, see Fig. 3) were analysed. For each fixed number of *steps* a set of experiments were conducted for different overlap levels: 100%, 80%, 60% and 40% of the performed steps.

For a fixed number of sequential epochs (200 in the reported experiments) the number of steps has a direct influence on the number of synchronizations (with its inherent communication costs) whereas the level of overlap (determining the overall number of cells treated by each process) has a major influence on the processes' workload.

As previously noted the equality between versions is only achieved if the overlap between strips (or partial stages) is composed of a number of rows equal or bigger than the number of steps.

Table 2 shows the sum of the differences (Cell-by-Cell comparison) between each parallel combination (different steps and overlaps) and the sequential implementation in the same set of iterations.

Table 2. Cell-by-Cell differences, Δ, between the results of the sequential implementation and the results obtained from the parallel implementation in the iterations: 50, 100, 150 and 200.

Steps	Overlap	50	100	150	200
10	10	0.00067	0.00067	0.00067	0.00065
10	8	0.14313	0.11767	0.10852	0.10361
10	6	5.20148	3.82719	3.41619	3.24228
10	4	58.60432	42.44981	38.00387	36.54078
50	50	0.0	≈ 0	≈ 0	≈ 0
50	40	≈ 0	≈ 0	≈ 0	≈ 0
50	30	≈ 0	≈ 0	≈ 0	≈ 0
50	20	0.00372	0.00170	0.00099	0.000706
100	100	–	0.0	–	0.0
100	80	–	≈ 0	–	≈ 0
100	60	–	≈ 0	–	≈ 0
100	40	–	≈ 0	–	≈ 0

For $\tau < 1E^{-5}$, see (1), the error values were denoted as "≈ 0".

As expected, Table 2 shows that the differences between the sequential and parallel evolutions increase with the reduction of the level of overlap between stage-subsets. Those differences are more apparent at the earliest stages of the simulation due to the convergence of the simulation to a stable state. Interestingly enough, it should be noted a somewhat counter-intuitive observation: in this model a large number of parallel steps tends to reduce the overall inequality between the parallel and the sequential evolutions. This means that, even when there is a significant relative gap on the number of rows necessary to guarantee an equal evolution, the local convergence of the algorithm is able to circumvent this lack of data as long as enough processing steps are allowed. In the following section some performance indicators for these and other selected experiments are reported.

4.3   Performance Comparison

As a way to analyse the performance of the proposed parallel strategy the improvements on the speed of execution for several parallel implementations with different parameter combinations were measured. In the reported experiments the speedup of each parallel configuration P against the corresponding sequential version S is given by

$$S_P = T_S \,/\, T_P, \tag{2}$$

where T_S is the execution time of the sequential evolution and T_P is the execution time of the assessed parallel configuration.

The impact of varying the step and the overlap parameters for a fixed number of processes and several environment dimensions was studied. The speedups of a chosen configuration when the number of processes varies from 2 to 20 were also calculated. Figure 4 shows the speedups obtained when running the simulation for each parameter combination described in Sect. 4.1, considering three environment maps with dimensions: (1210 × 3880), (1210 × 1940) and (1210 × 970), that is, the initial map (1210 × 1940) was extended and shrunken along its second dimension. The simulations were ran by the 12 processes over 200 epochs. That number of iterations has shown to be enough to reach convergence.

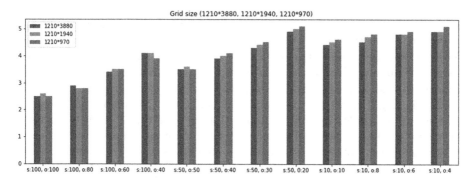

Fig. 4. Speedups obtained with maps of different number of columns, varying the number of steps s (10, 50 and 100) and for each step, the overlapping (o) size varies from 100% to 40% of the number of steps.

In Fig. 4 it is possible to observe that the speedups are approximately the same for the three maps. The speedups for the small map have values only insignificantly higher for the smallest step values. That is, increasing the number of columns has no impact in the speedup. When observing the figures with the same step value, as can be expected, the speedup increases when the overlap reduces. Less data redundancy implies less communication and less processing. Regarding the overlapping value, as seen in the previous section, it must be high enough to avoid information loss but the higher it is the worse will be the performance. In the studied cases the best speedups are obtained for the pair $steps = 50$ and $overlap = 20$ (i.e. 40% of the step value).

Then the variation of both dimensions of the map was studied, considering this pair of values. Figure 5 shows the speedups obtained when running simulations along 200 epochs in 12 processes for three maps with dimensions (605 ×

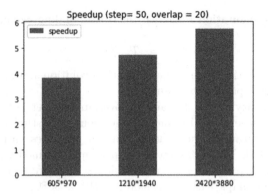

Fig. 5. Speedups obtained when increased data in both dimensions using step = 50, overlap = 20, 200 iterations and 12 processes.

970), (1210 × 1940) and (2420 × 3880). The overall parameters stay the same as before and the values of step and overlap were fixed on 50 and 20 respectively.

As can be seen the speedup increases when the dimension of the map increases. When the first dimension of the map is increased and the same number of processes is kept, the number of rows in each data strip increases. Keeping the same overlap value means less data redundancy when the size of each strip grows.

Finally, Fig. 6 shows the speedups obtained when the number of processes, p, varies from 2 to 20. The initial map with dimension (1210 × 1940) was used for a step 50 and overlap 20. The values of the remaining parameters were kept.

As can be seen from Fig. 6, the speedups increased steadily with the number of the available parallel processes to a value near of 5 times faster than the sequential implementation.

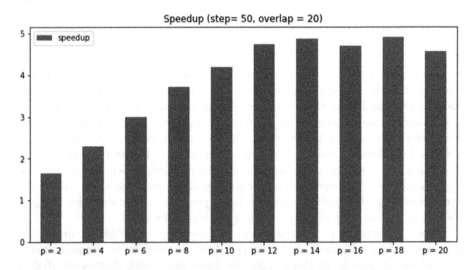

Fig. 6. Speedups obtained for the map dimension (1210 × 1940) when the number of processes, p, varies from 2 to 20, using step = 50, overlap = 20 and 200 iterations.

5   Discussion

The described parallel decomposition is able to preserve the absolute equality with
the sequential evolution, provided that a widely enough border region is shared
between contiguous stage-subsets. If the dimension of the border is less than the
number of parallel inner-process steps, potentially the equality is broken.

Figure 7 shows the evolution of the sequential implementation side-by-side
with the most different parallel evolution (step = 10; overlap = 4).

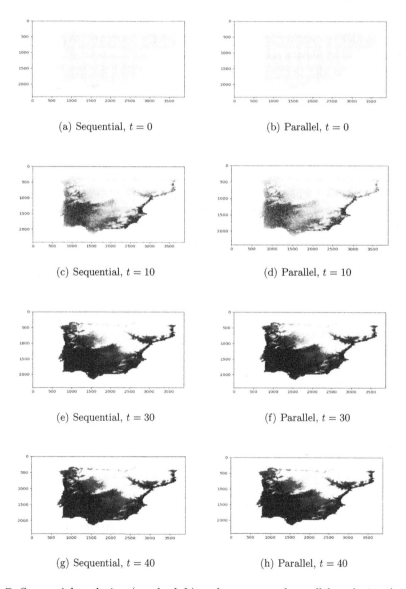

(a) Sequential, $t = 0$ (b) Parallel, $t = 0$

(c) Sequential, $t = 10$ (d) Parallel, $t = 10$

(e) Sequential, $t = 30$ (f) Parallel, $t = 30$

(g) Sequential, $t = 40$ (h) Parallel, $t = 40$

Fig. 7. Sequential evolution (on the left) and a non equal parallel evolution (on the
right) with 10 inner-process steps and only 4 overlapping rows.

It worthwhile to notice that in Fig. 7 are depicted some snapshots of the evolution only at the initial steps, period where the major differences are accounted for. As a side note, notice that at plain sight it is very difficult to spot any differences even at this earliest stages. However the simulation runs until the convergence criteria (viz. the differences between consecutive states drop below $\tau = 1E^{-5} \times N_c$, see (1), for a number of epochs) is satisfied.

On the other hand, for the same relative level of overlap, given the local convergence characteristics of this particular model, any loss of equality is less noticeable when more inner-process steps are allowed. Thus if one is ready to relax the equality constraint for the parallel version, clearly it is possible to obtain a model presenting indistinguishable results from the sequential version while achieving at least a good level of speedup.

According to the performance results, the parallel implementation had a gain, with speedups of approximately 5. Different parameters' combinations of the parallel implementation constrain both the accuracy and the performance of the model.

6   Conclusion

This paper proposes a parallelization strategy of an agent-based model of spatial distribution of species aiming at a good trade-off between the synchronization requirements and the amount of data redundancy necessary to achieve the equality between the parallel and the sequential evolution.

The adopted strategy unveils the effects that the number of parallel evolutive steps and the size of overlap between stage-subsets have in the algorithm's performance. The relation between these two parameters is explored, in order to find which are the best parameter combinations that ensure increased speedups.

According to the empirical results there is a scale opportunity to model larger problems with almost negligible errors. It was found that due to the local convergence of the model (resulting from the use of the Moore neighbourhood in the decomposed stages) it is possible to attain almost indistinguishable results from the ones of the sequential version. This is verified even when the synchronizations are scarce and the overlap is kept at a parsimonious level, hence accomplishing significant performance gains.

Thus, a clear line of future developments is related with the scale-up of the implementation to a distributed setup in order to tackle higher cardinality problems.

However, for a sounder statistical evaluation and deeper inspection of the proposal, aiming at the generalization to another kinds of models (with different, non-regular types of iterations between stage-subsets), the performance of the algorithm must be assessed over averaged metrics for multiple runs in a set of carefully chosen descriptive parameters.

Acknowledgements. This work was supported by operation Centro-01-0145-FEDER-000019 - C4 - Centro de Competências em Cloud Computing, cofinanced by the European Regional Development Fund (ERDF) through the Programa Operacional Regional do Centro (Centro 2020), in the scope of the Sistema de Apoio à Investigação Científica e Tecnológica - Programas Integrados de IC&DT. This work was also funded by FCT/MCTES through national funds and when applicable co-funded EU funds under the project UIDB/50008/2020.

References

1. Abar, S., Theodoropoulos, G.K., Lemarinier, P., O'Hare, G.M.: Agent based modelling and simulation tools: a review of the state-of-art software. Comput. Sci. Rev. **24**, 13–33 (2017)
2. An, L., Grimm, V., Turner II, B.L.: Meeting grand challenges in agent-based models. J. Artif. Soc. Soc. Simul. **23**(1), (2020)
3. Bandman, O.: Coarse-grained parallelization of cellular-automata simulation algorithms. In: Malyshkin, V. (ed.) PaCT 2007. LNCS, vol. 4671, pp. 370–384. Springer, Heidelberg (2007). https://doi.org/10.1007/978-3-540-73940-1_38
4. Bioco, J., Fazendeiro, P., Cánovas, F., Prata, P.: Parameterization of an agent-based model of spatial distribution of species. In: Miraz, M.H., Excell, P.S., Ware, A., Soomro, S., Ali, M. (eds.) iCETiC 2020. LNICST, vol. 332, pp. 251–260. Springer, Cham (2020). https://doi.org/10.1007/978-3-030-60036-5_18
5. Breckling, B., Müller, F., Reuter, H., Hölker, F., Fränzle, O.: Emergent properties in individual-based ecological models-introducing case studies in an ecosystem research context. Ecol. Model. **186**(4), 376–388 (2005)
6. Chimeh, M.K., Heywood, P., Pennisi, M., Pappalardo, F., Richmond, P.: Parallel pair-wise interaction for multi-agent immune systems modelling. In: 2018 IEEE International Conference on Bioinformatics and Biomedicine (BIBM), pp. 1367–1373. IEEE (2018)
7. Chimeh, M.K., Heywood, P., Pennisi, M., Pappalardo, F., Richmond, P.: Parallelisation strategies for agent based simulation of immune systems. BMC Bioinf. **20**(6), 1–14 (2019)
8. Coakley, S., Gheorghe, M., Holcombe, M., Chin, S., Worth, D., Greenough, C.: Exploitation of high performance computing in the flame agent-based simulation framework. In: 2012 IEEE 14th International Conference on High Performance Computing and Communication & 2012 IEEE 9th International Conference on Embedded Software and Systems, pp. 538–545. IEEE (2012)
9. Collier, N., North, M.: Parallel agent-based simulation with repast for high performance computing. Simulation **89**(10), 1215–1235 (2013)
10. Collier, N., Ozik, J., Macal, C.M.: Large-scale agent-based modeling with repast HPC: a case study in parallelizing an agent-based model. In: Euro-Par 2015. LNCS, vol. 9523, pp. 454–465. Springer, Cham (2015). https://doi.org/10.1007/978-3-319-27308-2_37
11. DeAngelis, D.L., Grimm, V.: Individual-based models in ecology after four decades. F1000Prime Rep **6**(39), 6 (2014)
12. DeAngelis, D.L., Gross, L.J., et al.: Individual-Based Models and Approaches in Ecology. Chapman & Hall, London (1992)
13. Fachada, N., Lopes, V.V., Martins, R.C., Rosa, A.C.: Parallelization strategies for spatial agent-based models. Int. J. Parallel Prog. **45**(3), 449–481 (2017)

14. Grimm, V.: Ten years of individual-based modelling in ecology: what have we learned and what could we learn in the future? Ecol. Model. **115**(2), 129–148 (1999)
15. Grimm, V., et al.: A standard protocol for describing individual-based and agent-based models. Ecol. Model. **198**(1–2), 115–126 (2006)
16. Grimm, V., et al.: The odd protocol for describing agent-based and other simulation models: a second update to improve clarity, replication, and structural realism. J. Artif. Soc. Soc. Simul. **23**(2) (2020)
17. Heywood, P., et al.: Data-parallel agent-based microscopic road network simulation using graphics processing units. Simul. Model. Pract. Theor. **83**, 188–200 (2018)
18. Huston, M., DeAngelis, D., Post, W.: New computer models unify ecological theory: computer simulations show that many ecological patterns can be explained by interactions among individual organisms. BioScience **38**(10), 682–691 (1988)
19. Lomnicki, A.: Population ecology of individuals. Monogr. Popul. Biol. **25**, 1–216 (1987)
20. Macal, C.M.: Everything you need to know about agent-based modelling and simulation. J. Simul. **10**(2), 144–156 (2016)
21. Moore, E.F.: Machine models of self-reproduction. In: Proceedings of symposia in applied mathematics. vol. 14, pp. 17–33. American Mathematical Society New York (1962)
22. Parry, H.R., Bithell, M.: Large Scale Agent-Based Modelling: A Review and Guidelines for Model Scaling. In: Heppenstall, A., Crooks, A., See, L., Batty, M. (eds.) Agent-Based Models of Geographical Systems. pp. 271-308. Springer, Dordrecht (2012) https://doi.org/10.1007/978-90-481-8927-4_14
23. Suárez-Muñoz, M., Bonet-García, F., Hódar, J.A., Herrero, J., Tanase, M., Torres-Muros, L.: Instar: an agent-based model that integrates existing knowledge to simulate the population dynamics of a forest pest. Ecol. Model. **411**, 108764 (2019)
24. Voss, A., et al.: Scalable social simulation: investigating population-scale phenomena using commodity computing. In: 2010 IEEE Sixth International Conference on e-Science, pp. 1–8. IEEE (2010)
25. Williams, R.A.: User experiences using flame: a case study modelling conflict in large enterprise system implementations. Simul. Model. Pract. Theor. **106**, 102196 (2021)
26. Wong, W.W.L., Feng, Z.Z., Thein, H.H.: A parallel sliding region algorithm to make agent-based modeling possible for a large-scale simulation: modeling hepatitis c epidemics in canada. IEEE J. Biomed. Health Inf. **20**(6), 1538–1544 (2016). https://doi.org/10.1109/JBHI.2015.2471804. Nov
27. Zhang, Q., Vatsavai, R.R., Shashidharan, A., Berkel, D.V.: Agent based urban growth modeling framework on apache spark. In: Proceedings of the 5th ACM SIGSPATIAL International Workshop on Analytics for Big Geospatial Data, pp. 50–59 (2016)

"Wave-Consistent" Ray Tracing Using Chebyshev Polynomials Representation of a Model in Three Dimensions

D. Neklyudov[(✉)] and M. Protasov

Institute of Petroleum Geology and Geophysics SB RAS, Novosibirsk 630090, Russia
{neklyudovda,protasovmi}@ipgg.sbras.ru

Abstract. We suggest solving the two-point ray tracing problem for a 3D medium using the bending method, which considers the band-limited nature of real seismic signals. It is based on a modified Fermat's principle. As a result, one may obtain more reliable ray paths and traveltimes in complex mediums. Chebyshev polynomials for model parameterization provide considerable algorithmic advantages for the two-point ray tracing based on the bending method because traveltimes and their derivatives with respect to ray parameters can be calculated analytically. Hence nonlinear conjugate gradient method can be applied efficiently for ray computation. 3D numerical experiments prove the efficiency of the proposed approach.

Keywords: Seismic exploration · Ray tracing · Ray bending · "shooting" method · Chebyshev polynomials · Fresnel volume

1 Introduction

Many technical applications require calculating traveltimes and travel paths between known locations of source and receiver of some probing signals. In seismic exploration, such a problem, known as two-point ray tracing, plays a significant role. Two-point raytracing is essential in ray-based seismic imaging techniques like Kirchhoff-type migration and seismic inverse problems such as traveltime tomography. Modern seismic exploration problems, either imaging or inversion, are often three-dimensional, where the spatial size of the computational domains may be huge. Practical experience has shown that the so-called "shooting" method for solving a two-point ray tracing problem often does not provide satisfactory results. Also, for three-dimensional cases, "shooting" requires significant computational resources. An alternative to the "shooting" method is the so-called "bending" method [1]. It is based on the modification of the initial curve that connects the source-receiver pair in such a way as to satisfy Fermat's principle, that is, to minimize the traveltime along the trajectory. In practice, it turns out that for 3D problems, the bending method is much more efficient than the shooting method.

In seismic applications, methods oriented for traveltimes and travel paths calculation usually are based on high-frequency approximations of seismic signals propagation

© Springer Nature Switzerland AG 2021
O. Gervasi et al. (Eds.): ICCSA 2021, LNCS 12950, pp. 311–321, 2021.
https://doi.org/10.1007/978-3-030-86960-1_22

(the Ray method, [2]). And they don't take into account the effects caused by the fact that (temporal) spectra of the real seismic signals are band-limited. High-frequency approximation often leads to the non-physical behavior of the constructed rays and corresponding traveltimes. After all, the ray theory has limitations on its application in complex environments [3]. Also, band-limited signals propagate not just along with rays but also within a volume surrounding the ray. The width of this zone depends on the dominant frequency of a signal. This zone is called the Fresnel volume (or zone).

Several methods are developed to go beyond the standard ray method based on the high-frequency approximation. The main idea behind these approaches is to take into account the impact of the Fresnel zone on traveltimes and ray trajectories [4–7]. In [8, 9], a promising approach called "wave-tracing" was proposed for two-dimensional situations. In this paper, we adopt "wave-tracing" for three-dimensional models using original 3D model representation via Chebyshev polynomials expansion proposed in [10, 11]. The combined approach allows, to some extent, to solve the two-point ray tracing problems taking into account the limited spectrum of the probing signals without resorting to high computational costs. It provides ray trajectories and traveltimes that give a better description of real seismic band-limited data than the standard approach.

2 Model Representation by Chebyshev Polynomials

The geological medium, a scope in seismic exploration, is usually represented as a set of sub-horizontal layers formed by sedimentary rocks. The most intensive variations of physical properties (velocity of seismic waves, density, etc.) occur in depth. Following the work [10], we assume that a 3D model of the medium (where physical parameters are dependent on three spatial coordinates x,y,z) is represented as a set of layers separated by non-planar surfaces. In each layer, a function of "slowness" $S_j(x, y)$ (the value of the reciprocal of seismic waves velocity) is specified. Slowness in each layer depends on the two lateral coordinates x, y, but does not depend on the depth, z. The function $S_j(x, y)$ is defined as two dimensional Chebyshev polynomials of 3rd order,

$$
\begin{aligned}
S_j(x, y) = C_0 + C_1 x + C_2 y + C_3 xy + C_4\left(2x^2 - 1\right) + C_5\left(2y^2 - 1\right) \\
+ C_6\left(2x^2 - 1\right)y + C_7\left(2y^2 - 1\right)x + C_8\left(4x^3 - 3x\right) + C_9\left(4y^3 - 3y\right)
\end{aligned}
\tag{1}
$$

where C_k are coefficients of decomposition defined for each layer. Each of the surfaces $Z_j(x, y)$ is described by Chebyshev polynomials by analogy with the expression (1) with the coefficients Z_k (see Fig. 1). This kind of model parametrization significantly reduces the number of required parameters compared to the standard scheme where parameter distribution is described by voxels grid. The chosen scheme provides significant computational advantages while solving raytracing problems [10, 11]. The travel times and their derivatives with respect to the rays' parameters may be expressed analytically. Hence nonlinear conjugate gradients approach or Newton-like methods of minimization may be applied very efficiently for ray computation. One more advantage of the proposed parametrization is for solving 3D traveltime tomography. It turns out that the Fréchet derivatives, i.e., a matrix of the traveltimes derivatives with respect to the model parameters (Chebyshev coefficients representing local slowness distribution within each layer and corresponding Chebyshev coefficients of the interfaces) may be expressed analytically, [10].

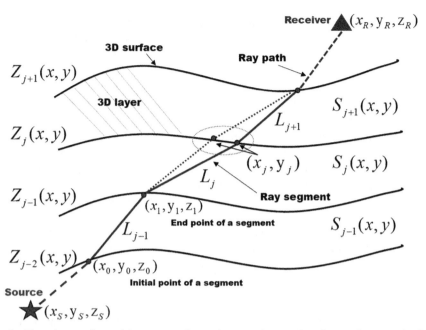

Fig. 1. The scheme of a model representation and a ray trajectory (see the text for more details).

3 Two-Points Ray Tracing Using the Bending Method

By definition, a ray is a trajectory that provides minimum traveltime of some signal between two points (Fermat's principle). Let the initial (source) and the final (receiver) points of the ray are given. Source and receiver have coordinates $(x_S, y_S, z_S), (x_R, y_R, z_R)$ respectively. Having in mind the model parametrization described above, we assume that in each layer, the ray is a straight line segment specified by the coordinates of the intersection points (x_j, y_j, z_j) and $(x_{j+1}, y_{j+1}, z_{j+1})$ with two neighboring surfaces, j and $j + 1$, $j = \overline{1, N}$. Where N is the total number of the surfaces (Fig. 1). Note that $z_j = z_j(x_j, y_j)$ because surfaces are defined analytically via Chebyshev polynomials. The ray will be fully defined if appropriate coordinates of intersection points (x_j, y_j), are known. We aim to find a proper location of $N + 1$ straight line segments to minimize traveltime calculated along a trajectory. We call a set of coordinates pairs (x_j, y_j) as ray parameters. Let T_j means traveltime within j layer. Total traveltime from source to receiver points is a sum of traveltimes along with the segments, $T = \sum_{j=0}^{N} T_j$. To construct a ray, i.e., a trajectory satisfying Fermat's principle, one must minimize the total traveltime (Fermat functional) of the form,

$$T = \sum_{j=0}^{N} T_j(x_j, y_j; x_{j+1}, y_{j+1}), \tag{2}$$

with respect to variables (x_j, y_j) while source and receiver positions are fixed. Any ray constructed in this way will satisfy Snell's law at the interfaces between the layers.

Minimization of the cost function (2) may be performed using the nonlinear conjugate gradient (CG) method. In this work, we use a conjugate gradient algorithm for solving an unconstrained minimization problem of the form "min $F(\vec{x})$" proposed in [12]. To apply the chosen approach, one needs to provide two functions. In each iteration of CG, one must be able to calculate the value of the cost function (traveltime along a current trajectory) and gradient of the cost function $\nabla_{\vec{x}} F$ depending on the parameters vector $\vec{x}$, i.e., the traveltime derivatives $\frac{dT}{dx_j}$, $\frac{dT}{dy_j}$ in our case.

Let us calculate analytical expressions for the traveltime and its derivatives along a single segment of the ray. It is necessary to calculate the curvilinear integral of the first kind of the polynomial function (1) along a straight line. To simplify notation, we assume that its first and last points (the intersection with the neighboring surfaces) have coordinates (x_0, y_0, z_0) and (x_1, y_1, z_1). Definition of a straight line in 3D space in the parametrical form is given as $x = x_0 + a \cdot l$, $y = y_0 + b \cdot l$, $z = z_0 + c \cdot l$, where a, b, c are cosines of directional angles; l is an independent parameter (distance along the line). Using the parametric representation of a line, one will obtain

$$T_j = \int_{\text{Line}} S_j(x, y) dl = \int_0^{L_j} S_j(x_0 + al, y_0 + bl) dl, \tag{3}$$

where $L_j = \sqrt{(x_1 - x_0)^2 + (y_1 - y_0)^2 + (z_1 - z_0)^2}$ is a length of a segment, $a = (x_1 - x_0)/L$, $b = (y_1 - y_0)/L$. Thus, it is necessary to integrate the polynomial expression (1) over the variable l within the interval $[0, L]$ after the appropriate substitution $x = x_0 + a \cdot l$, $y = y_0 + b \cdot l$. After some manipulations, one comes to the following expression for the traveltime along the straight ray's segment as a function of coordinates (x_0, y_0) and (x_1, y_1):

$$T_j = L_j \cdot Q, \tag{4}$$

where Q is represented as

$$\begin{aligned} Q = C_0 &+ C_1 \tfrac{1}{2}(x_1 + x_0) + C_2 \tfrac{1}{2}(y_1 + y_0) + C_3 \left(\tfrac{1}{3}(x_0 y_0 + x_1 y_1) + \tfrac{1}{6}(x_0 y_1 + x_1 y_0)\right) \\ &+ C_4 \tfrac{2}{3}(x_1^2 + x_0 x_1 + x_0^2 - 1.5) + C_5 \tfrac{2}{3}(y_1^2 + y_0 y_1 + y_0^2 - 1.5) \\ &+ C_6 \tfrac{1}{6}(x_1^2(3y_1 + y_0) + 2x_0 x_1(y_1 + y_0) + x_0^2(y_1 + 3y_0) - 3(y_1 + y_0)) \\ &+ C_7 \tfrac{1}{6}(y_1^2(3x_1 + x_0) + 2y_0 y_1(x_1 + x_0) + y_0^2(x_1 + 3x_0) - 3(x_1 + x_0)) \\ &+ C_8(x_1 + x_0)(x_1^2 + x_0^2 - 1.5) + C_9(y_1 + y_0)(y_1^2 + y_0^2 - 1.5) \end{aligned} \tag{5}$$

Note that the segment length L_j is a function of the two lateral coordinates of the intersection points $L_j = L_j(x_0, y_0, x_1, y_1)$. Chebyshev coefficients of the surfaces, Z_k, are hidden within L_j. Time derivatives (and second derivatives) are calculated analytically by differentiating the resulting expression (4). The most laborious moment is the differentiation of the factor L_j due to nonlinearity and cumbersome dependence on the parameters (Sect. 4 provides the derivation of these formulas). The total travel time along the desired ray is represented as the sum of each segment's traveltimes. The derivative of the total traveltime along the ray with respect to variable x_j (or y_j) is a sum of two traveltime derivatives in two adjacent layers, T_j and T_{j+1} (see Fig. 1):

$$\frac{dT}{dx_j} = \frac{dT_{j+1}}{dx_j} + \frac{dT_j}{dx_j} \tag{6}$$

Having analytical expressions for the derivatives of traveltimes with respect to intersection points' coordinates with the layers' surfaces, we use the nonlinear conjugate gradient method to minimize the Fermat cost functional. Nonlinear Newton method may be used as well. (It will require analytical expressions for the Hessian (matrix of the second derivatives, [10]).

The number of parameters that need to be determined for each ray is equal to the number of layers intersected by the ray multiplied by two. Ray is defined by the minimization of traveltime functional, which is carried out in two stages:

1) For a given source-receiver pair, a ray is constructed in a simplified medium model. As a "simplified" model, a model consisting of homogeneous layers separated by flat surfaces is used. It means that only the Chebyshev polynomials expansion's first coefficients differ from zero both for the slowness within the layers and for the interfaces. This ray serves as an initial guess for the next step.
2) A ray is constructed in a real model, where ten coefficients of the Chebyshev polynomials describe the slowness in the layers and the interfaces between them.

4   Traveltime Derivatives with Respect to Ray Parameters

This section shows how the first derivatives of the traveltime along a ray segment are calculated. Consider the expression (4) where Q denotes a polynomial multiplier given by the formula (5). It consists of 10 terms: the coefficients of Chebyshev polynomials of the slowness expansion within the layer. Note that in the expression L_j the depth coordinates of the intersection points $z_{0,1}$ are expressed in terms of lateral coordinates $x_{0,1}, y_{0,1}$ using the representations of surfaces by Chebyshev polynomials similar to (1). Let $Z_k^{(0)}$ and $Z_k^{(1)}$ be the coefficients of Chebyshev polynomials expansion of two adjacent surfaces (upper index correspond to surface, lower index means expansion coefficient). Using notation $p = 0, 1$ one will have,

$$
\begin{aligned}
z_p &= Z_0^{(p)} + Z_1^{(p)} x_p + Z_2^{(p)} y_p + Z_3^{(p)} x_p y_p + Z_4^{(p)}\left(2x_p^2 - 1\right) + Z_5^{(p)}\left(2y_p^2 - 1\right) \\
&+ Z_6^{(p)}\left(2x_p^2 - 1\right)y_p^2 + Z_7^{(p)}\left(2y_p^2 - 1\right)x_p^2 + Z_8^{(p)}\left(4x_p^3 - 3x_p\right) + Z_9^{(p)}\left(4y_p^3 - 3y_p\right)
\end{aligned}
\tag{7}
$$

Derivatives are expressed as.

$$
\frac{\partial T_j}{\partial \zeta_p} = \frac{\partial L_j}{\partial \zeta_p} \cdot Q + L_j \frac{\partial Q}{\partial \zeta_p},
\tag{8}
$$

where we use the notation $\zeta_p = x_p$ or $\zeta_p = y_p$.
Performing an explicit differentiation of each term, one obtains

$$
\frac{\partial L_j}{\partial x_p} = \pm \frac{1}{L_j}\left\{x_0 - x_1 + \Psi_p \cdot (z_0 - z_1)\right\},
\tag{9}
$$

Where

$$
\begin{aligned}
\Psi_p &= Z_1^{(p)} + Z_3^{(p)} y_p + 4 \cdot Z_6^{(p)} x_p y_p + Z_7^{(p)}\left(2y_p^2 - 1\right) + Z_8^{(p)}\left(12x_p^2 - 3\right) \\
\frac{\partial L_j}{\partial y_p} &= \pm \frac{1}{L_j}\left\{y_0 - y_1 + \Phi_p \cdot (z_0 - z_1)\right\},
\end{aligned}
\tag{10}
$$

Where

$$\Phi_p = Z_2^{(p)} + Z_3^{(p)} x_p + 4 \cdot Z_5^{(p)} y_p + Z_6^{(p)} \left(2x_p^2 - 1\right) + 4 \cdot Z_7^{(p)} x_p y_p + Z_9^{(p)} \left(12 y_p^2 - 3\right).$$

The sign "+" is used for $p = 0$, and the sign "−" is used for $p = 1$.

Derivatives $\frac{\partial Q}{\partial \xi_p}$ are calculated in an obvious way as derivatives of polynomials in expression (5). Thus, all the necessary components needed for the explicit calculation of the gradient are obtained. Explicit (or implicit) gradient expression is required to minimize the target functional (2) by any gradient method. Note that it is easy to obtain terms for the second derivatives, which are needed to calculate the Hessian matrix. So Newton's method of minimization may be applied straightforwardly [10].

5   Modification of Fermat Cost Function and 3D "Wave-Consistent" Ray Tracing

To consider the band-limited nature of real seismic signals, we adapt for three-dimensional cases the approach proposed in the works [8, 9] for two-dimensional two-point raytracing. The essence of the approach is that instead of the classical Fermat functional (2), a modified cost function is considered. One more term is added which characterizes the length of a ray:

$$F = \left[\sum_{j=0}^{N} \frac{T_j}{T_{SR}} + \frac{\alpha}{2} \sum_{j=1}^{N} \left(\frac{L_j}{L_{SR}} \right)^2 \right], \tag{11}$$

where T_j, L_j are traveltime and a ray segment length within the jth layer; T_{SR}, L_{SR} are auxiliary parameters used for normalization within the cost function. In our case, L_{SR} is the distance between source and receiver, T_{SR} is traveltime along this line. By minimization of (4) "wave-consistent" ray trajectory is found. The corresponding traveltime is calculated after that using explicit formulas (4).

Regularization parameter $\alpha > 0$ depends on the dominant frequency of the signal, v_0. As one can see, if $\alpha = 0$ minimization of the cost function (11) gives a "standard" ray (i.e., a ray which satisfies Fermat's principle (2)). Suppose $\alpha \to \infty$ minimization gives a straight line.

We use the following expression, which is an adaptation of formula (18) form [8] for the 3D case,

$$\alpha = \frac{1}{\sigma} \frac{Z_{SR}}{D_{SR}} \cdot \frac{1}{\sqrt{v_0 T_{SR}}}, \tag{12}$$

where $Z_{SR} = |z_S - z_R|$, $D_{SR} = \sqrt{(x_R - x_S)^2 + (y_R - y_S)^2}$, (x_S, y_S, z_S), (x_R, y_R, z_R) are the source and receiver coordinates; $\sigma = \sigma_r / \bar{r}$ is the relative standard deviation of horizontal components $r_j = \sqrt{(x_j - x_{j+1})^2 + (y_j - y_{j+1})^2}$ of the minimum traveltime ("standard") ray,

$$\bar{r} = \frac{1}{N+1} \sum_j r_j, \quad \sigma_r = \sqrt{\frac{1}{N+1} \sum_j (r_j - \bar{r})^2}, \tag{13}$$

As it was proven in [8], parameter α is defined to obtain a ray with the minimum length among all rays belonging to the Fresnel volume corresponding to the "standard" ray.

Minimization procedure used for modified functional (11) consists of three stages:

1) A straight ray is constructed for a given source-receiver pair. Auxiliary parameters T_{SR}, L_{SR} are defined.
2) A "standard" ray is constructed using the procedure described above. Regularization parameter α is defined. This ray is used as an initial guess for the last stage.
3) Minimization of the modified cost function (11) is performed. The procedure is almost the same as minimization used for a standard functional (2), with the exception that the term $\frac{\alpha}{2}\left(\frac{L_j}{L_{SR}}\right)^2$ is added for each ray segment.

6 Numerical Example

The proposed approach's advantages are illustrated here using a realistic numerical example where a situation typical for 3D crosswell traveltime tomography is modeled. The problems of crosswell traveltime tomography are essentially three-dimensional. The reason is that the boreholes in which the stimulation and registration of seismic signals occur are not strictly vertical but are significantly deviated in 3D space. Also,

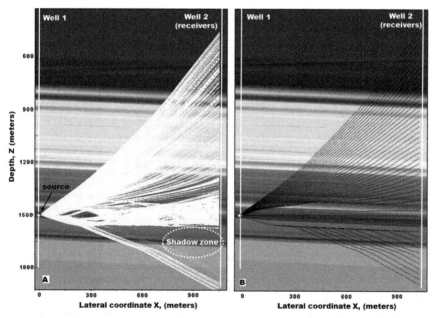

Fig. 2. Rays constructed in the realistic 3D model using two methods overlaid over a 2D section of the seismic velocities distribution: (A) "shooting" method (B) ray "bending" method. In both cases, Chebyshev polynomial representation of the model was used. (Rays projections on the plane between two wells are presented.)

when a detailed model in the vicinity of a borehole is constructed, data obtained in several wells located at a considerable distance from each other are often used. Thus, it becomes necessary to perform two-point ray tracing for large 3D models. Sources and receivers are located at various depths, so rays' trajectories become subhorizontal, i.e., their directions of propagation become close to the boundaries of geological layers. It makes considerable difficulties for two points ray tracing. The "wave-consistent" approach, which is discussed in the paper, may successfully resolve these problems. The three-dimensional model of the medium used here has $1100 \times 1000 \times 2000$ m in X, Y, and Z (depth) directions. It consists of about 1400 thin layers. Layers have a variable thickness (decreasing down to 2 m) but do not intersect with each other anywhere. All 10 Chebyshev coefficients describe each surface between the layers. Velocities in the layers vary along the lateral coordinates X, Y. Velocity variations in different layers are within the range 1200 - 5500 m/s. Two vertical boreholes are located at a distance of 1050 m from each other. It is presumed that sources of the seismic signals are located within the left well. In the right borehole, 50 receivers with a step of 30 m in depth are placed. Figure 2 shows a cross-section of the 3D velocity model in the plane of the wells. There are thin layers with rapid variations of velocity distribution. It is an unfortunate case for standard ray tracing methods, especially for the "shooting" method. Figure 2 compares the two-point ray tracing results obtained by the shooting method and the "standard" bending method. For demonstration purposes, we present the rays for one source position only. As one can see, the calculated rays' behavior differs significantly

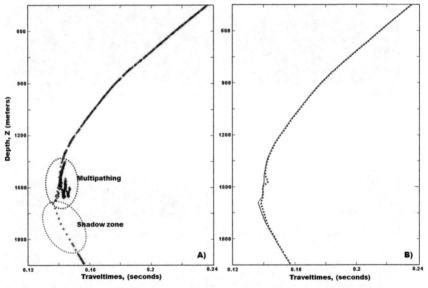

Fig. 3. A) Comparison of traveltimes calculated by the "shooting" method (shown in red) with traveltimes computed using the "standard" bending method (shown in blue); B) Comparison of traveltimes obtained by the standard bending method (red) with "wave-consistent" bending method (blue). Note that the blue line is very close to the red line. It means that the "wave-consistent" rays belong to Fresnel volumes of the "standard" rays. (Color figure online)

in the area where they are sub-horizontal. In the shooting method, many rays had to be used to illuminate the borehole where receivers are located. Rays are very irregular, which contradicts the physical nature of wave propagation in a real geological medium. There is a wide shadow zone where no rays exist. It will cause unavoidable problems during traveltime inversion.

In Fig. 3a comparison of calculated traveltimes is presented. As one can see, the rays obtained by the shooting method form "holes" while going through the high-velocity pack of layers (one of them is marked with a green oval). There are areas with multiple arrivals (marked by a blue oval), which is unfavorable since we are only interested in minimum traveltime arrivals. At the same time, the bending method gives quite adequate results if one considers traveltimes only. However, suppose one takes a closer look at the ray distribution obtained by the "standard" bending method. In that case, one will notice that the ray distribution is also quite irregular in the lower part of the figure. Rays "stick together" bend in the direction opposite the propagation direction when they go along the boundary with a sharp jump of velocities (see Fig. 4). This problem is completely resolved by "wave-consistent" ray tracing, where we presume that the dominant frequency of the signal used to define regularization parameter, is $\upsilon_0 = 100$ Hz. In Fig. 4, a comparison of "standard" rays and "wave-consistent" rays are presented. The latter demonstrates much more physical behavior from the point of view of real band limited signal propagation. It is interesting to emphasize that despite ray trajectories are pretty different, the traveltimes

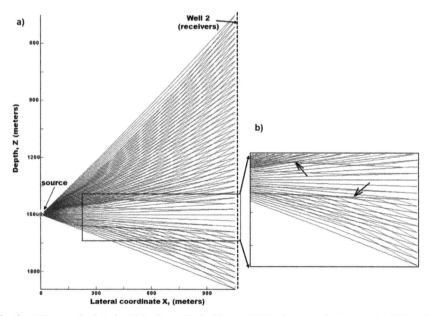

Fig. 4. a) Rays calculated within the realistic 3D model by the use of the "standard" bending method (shown in red) and "wave-consistent" bending method (shown in blue); b) Zoom of Fig. 4a. In this figure, the non-physical behavior of rays constructed by the classical bending method compared to rays obtained by "wave-consistent" bending is shown. (Projections on the plane between two wells are presented.). (Color figure online)

obtained by these two methods are very close (see Fig. 3b). The maximum difference is less than 0.002 s. This fact proves the statement we have been made above that the "wave-consistent" rays belong to Fresnel volumes of the "standard" rays. In this example, the number of parameters used in the CG –based minimization procedure is approximate ~ 1400*2. The maximum number of CG iterations does not exceed 200 and corresponds to the longest rays. CG convergence for wave-consistent ray bending is the same as for "standard" ray bending approach. Thus, the calculation for the entire set of source-receiver pairs (where the number of sources is more than 200) in a given three-dimensional model takes several minutes using a personal computer.

7   Conclusions

An algorithm for fast 3D two-point ray tracing based on the bending method is proposed, considering the effects of the band-limited spectrum of real acoustic signals. It is based on the use of a modified Fermat functional. The essence of modification is that an additional term is introduced into the standard Fermat functional that controls the ray length. As a result, the "wave-consistent" ray belongs to the Fresnel volume of the minimum traveltime ("standard") ray. A pack of layers represents the 3D model of the medium where the rays should be constructed. Physical parameters distributions within the layers and geometry of the layers' boundaries are described using laterally varying 2D Chebyshev polynomials. It gives noticeable computational advantages while solving two-point raytracing problems in three dimensions since travel times and traveltimes derivatives are calculated analytically. Modified Fermat functional responsible for "wave-consistent" ray tracing is minimized using nonlinear conjugate gradients method. "Wave-consistent" ray bending provides much more physical ray's distribution within complex 3D model than "standard" ray bending approach. Numerical examples prove the presented algorithm's high efficiency while dealing with sufficiently large and complex three-dimensional models.

Acknowledgments. The reported study was funded by RFBR and GACR, project number 20–55-26003.

References

1. Julian, B.R., Gubbins, D.: Three-dimensional seismic ray tracing. J. Geophys. Res. **43**, 95–113 (1977)
2. Babich, V.M., Buldyrev, V.S.: Asymptotical methods in problems of short-wavelength diffraction. Nauka (1972). (in Russian).
3. Ben-Menahem, A., Beydoun, W.: Range of validity of seismic ray and beam methods in general inhomogeneous media - I. General theory. Geophys. J. R. Astr. Soc. **82**, 207–234 (1985)
4. Cerveny, V., Soares, J.E.P.: Fresnel volume ray tracing. Geophysics **57**, 902–915 (1992)
5. Kravtsov, Y.A., Orlov, Y.I.: Geometrical optics of inhomogeneous media. Nauka (1980). (in Russian)

6. Lomax, A.: The wavelength-smoothing method for approximating broad-band wave propagation through complicated velocity structures. Geophys. J. Int. **117**, 313–334 (1994)
7. Vasco, D.W., Peterson, J.E., Majer, E.L.: Beyond ray tomography: wavepaths and fresnel volumes. Geophysics **60**, 1790–1804 (1995)
8. Bube, K.P., Washbourne, J.K.: Wave tracing: ray tracing for the propagation of band-limited signals: part 1 – theory. Geophysics **73**, VE377-VE384 (2008)
9. Washbourne, J.K., Bube, K.P., Carillo, P., Addington, C.: Wave tracing: ray tracing for the propagation of band-limited signals: part 2 – Applications. Geophysics **73**, VE385-VE393 (2008)
10. Washbourne, J.K., Rector, J.W., Bube, K.P.: Crosswell traveltime tomography in three dimensions. Geophysics **67**, 853–871 (2002)
11. Grechka, V.Y., McMechan, G.A.: 3D two-point ray tracing for heterogeneous, weakly transversely isotropic media. Geophysics **61**, 1883–1894 (1996)
12. Hager, W.W., Zhang, H.: A new conjugate gradient method with guaranteed descent and an efficient line search. SIAM J. Optim. **16**, 170–192 (2005)

Full Waveform Inversion in Viscoelastic Media

Vladimir Cheverda[1,2]($\boxtimes$) (iD), Ekaterina Efimova[2], and Galina Reshetova[1] (iD)

[1] Institute of Computational Mathematics and Mathematical Geophysics, Novosibirsk, Russia
kgv@nmsf.sscc.ru
[2] Institute of Petroleum Geology and Geophysics, Novosibirsk, Russia
cheverdava@ipgg.sbras.ru

Abstract. Accumulations of gas hydrates in the bottom layers pose not only a potential threat to offshore engineering structures and shipping. They can also cause the release of significant volumes of methane into the atmosphere. The most natural way to detect gas hydrates is to use seismic methods. The technology of their application on the shelf is quite well developed, and there is a wide range of instrumental and methodological solutions that ensure their practical use. Detecting gas hydrates belongs to the class of multi-parameter inverse problems. Indeed, gas hydrates in the surrounding medium lead to a change in the propagation velocity of seismic waves and an increased level of wave absorption in these areas. Thus, the correct localization of the gas hydrate accumulations requires determining those regions in space where both the propagation velocities of seismic waves and quality factor change simultaneously. This work deals with the study of the trade-off in velocities and the quality factor of seismic waves. Namely, which condition we need to separate these two parameters' perturbations by solving the dynamic inverse problem of seismic wave propagation.

Keywords: Viscoelasticity · Seismic attenuation · Generalized standard linear solid · Inverse problem · Singular value decomposition

1 Introduction

The Siberian shelf includes the Kara, Laptev, and East Siberian seas, is the most massive continental shelf on Earth. It is of strategic importance because of oil and natural gas reserves. However, it is poorly studied due to the harsh climatic conditions. State of the water and sea ice of the shelf depends on a range of climatic processes, among them are the following: the variability of atmospheric dynamics determining the processes of freezing and melting of sea ice, its drift, and water circulation in the surface layer, interaction with neighbouring regions, and the Siberian rivers. The results of observations of recent years indicate that the region has clearly expressed climatic changes that have occurred over the

Supported by Russian Science Foundation, project 20-11-20112.

O. Gervasi et al. (Eds.): ICCSA 2021, LNCS 12950, pp. 322–334, 2021.
https://doi.org/10.1007/978-3-030-86960-1_23

past decades. In particular, vivid evidence of the changes taking place due to warming is the intensification of the erosion of the Arctic shores.

The Arctic shelf is an area where permafrost sub-aqueous rocks can exist (see results in [29]). The existence of permafrost provides conditions for the formation of a zone of stability of gas hydrates in bottom sediments at shallow water depths. Hence there is a possibility of accumulation of the gas hydrates in the near bottom sediments caused by global warming. As a result, methane gas, trapping in gas hydrates, start moving to the surface and penetrate the atmosphere.

The history of the problem dates back to the early 1990 s, when at the bottom of the Barents Sea, clusters of craters were discovered, whose diameter reached one kilometre (http://earth-chronicles.ru/news/2017-06-05-105087). Subsequently, the Norwegian geologists put forward and confirmed by conducting a detailed study of the seabed the hypothesis that these craters appeared during the explosions of gas bubbles associated with the retreat of the glaciers. Subsequently, the so-called "bumps of heaving" were discovered, where, presumably, gas still accumulates as a result of the melting of gas hydrates. Analysis of the collected data showed that modern methane of the Barents Sea was formed more than thirty thousand years ago at a depth of about a kilometre below the seabed. At that time, a massive ice sheet with a thickness of more than two kilometres covered the North Seas. Its weight exerted tremendous pressure on the bottom layers, and as a result of which compressed natural gas into hydrates - crystalline compounds with water. Then, as the ice shield melted, the pressure decreased, and the gases began to accumulate in the cavities, forming massive domes in the bottom layers [31]. A further increase in water temperature will inevitably lead to the destruction of these domes and the release of methane to the surface and further into the atmosphere. A similar situation may develop on the shelf of the Kara Sea and other North Seas.

Of particular interest are the Siberian seas (Kara, Laptev, East Siberian, and Chukchi). Here the formation of gas hydrate deposits occurs in the ice ages under sub-aerial conditions as a result of lowering sea levels and the formation of permafrost in the bottom sediments. These relic hydrates can be especially sensitive to climate change. Margin relic gas hydrates Arctic shelf is estimated to 65 thousand Tg(CH4) per year [24].

An increase in the temperature of the bottom layer enhances the degradation of the permafrost. It can lead to destabilization of gas hydrates. Observation data for the years 1920–2009 in the shallow part of the shelf and the coastal zone of the Laptev Sea and the East Siberian Sea show a significant increase in the bottom temperature (up to 2.1°) starting the mid-1980s [11]. Several cases of unprecedented warming in the bottom waters were observed on the central shelf in winter [16,17].

Thus, the dynamics of bottom water temperatures near gas hydrates accumulations should be monitored especially carefully. When there is a lifting of a vast amount of gas, gas bubbles can achieve several hundred meters in diameter, posing a threat to engineering structures and shipping. Localization of such clusters and control of their state is an essential task in the light of the changing Arctic climate caused by global warming, as well as the increasing activity of geological exploration and increase in the intensity of navigation and the construction of various complex engineering structures [3]. Therefore, the study of climate change and, as a result, possible changes in temperature conditions in conjunction with the localization of clusters of gas hydrates seems to be a necessary stage for reliable prediction of risks in various regions of the shelves of the North Seas.

In the case of the development of the oil and gas industry on the shelf of the Arctic Ocean, the environmental burden on the region will inevitably increase. Exploration, production, and any type of oil transportation pose a serious threat to the natural environment. Natural purification after oil spills in the Arctic may not last for years (as in temperate regions), but for decades [27]. To provide scientific and practical support for the economic development of the Arctic shelf, it is necessary to develop criteria for the vulnerability of the region under consideration, based on the knowledge of the physical mechanisms that form the current state of its natural environment.

As follows from the above, the mathematical formulation of the problem of detecting accumulations of gas hydrates by seismic methods should base on the theory of propagation of seismic waves in viscoelastic media [28,32]. This inverse problem is multi parameters because we need to recover two families of unknowns - P- and S-wave propagation velocities and P- and S-quality factors. The theoretical proof of uniqueness of the inverse problem for viscoelastic media one can find in [30]. Here we concentrate on numerical analysis of some specific feature of multi parameter inverse problem known as coupling of parameters (see [1]). This term describes the situation when perturbations of one of the parameters, for example, absorption, appears as a perturbation of another, for example, wave propagation velocity. In more details see [20,25,26].

Below we consider the inverse problem for viscoelastic media in a linearized setting. In such media, the following inhomogeneities can occur:

- in wave propagation velocities at constant quality factor;
- in quality factor at constant wave propagation speeds;
- both in P- and S-wave propagation velocities and in P- and S- quality factors.

Let us remind, that the presence of gas hydrate inclusions increases absorption of the seismic energy, which, in its turn, leads to the increase of the wave propagation velocity [2,5,7], etc. Thus, the simultaneous increase of the phase velocity and attenuation, the latter means decrease of the quality factor, may serve as a reliable characteristic of the presence of gas hydrates.

The tool used in the paper to study the coupling of perturbations of quality factors and wave propagation velocities is the singular value decomposition of the corresponding integral operators, similar to [19]. We consider the singular value decomposition of these operators and analyze the projections of the perturbations of the required parameters onto the linear span of the elder singular vectors corresponding to the estimated condition number and a given level of noise in the data.

2 Seismic Waves Propagation in Linear Viscoelastic Media

The model of an ideal elastic medium does not consider energy absorption during the propagation of seismic waves. But the possibility of localizing areas with such absorption and its quantitative assessment provides essential additional information about the properties of the studied geological objects. The generally accepted value for describing the loss of seismic energy is the quality factor of the medium:

$$Q^{-1} = \frac{1}{2\pi} \frac{\Delta E}{E}$$

where E characterizes the elastic energy per unit volume per unit time, and δE corresponds to the energy lost due to absorption. A distinctive feature of seismic energy dissipation, which has been repeatedly confirmed as a result of field observations, is the quality factor's constancy over the entire seismic frequency range (see [22]).

In this paper, we use the τ-method to implement the Generalized Standard Linear Solid model (GSLS) to describe the propagation of seismic waves in media with absorption, see e.g. [15] and Chapter 5 of the book [14]. When using GSLS, several mechanisms are introduced that describe the relationship between deformations and stresses, which leads to the appearance of frequency dependence of the Lamé coefficients. An essential feature of GSLS is the ability to introduce the shear and longitudinal waves' quality factors. As a result, we obtain the following Lamé coefficients:

– for shear waves

$$\mu(\omega) = \mu_r \left(1 + \frac{i\omega\tau^S}{L} \sum_{l=1}^{L} \frac{\tau_l}{1 + i\omega\tau_l} \right) = \mu_r \left(1 + \tau^S S(\omega) \right) \tag{1}$$

– for longitudinal waves

$$\lambda(\omega) + 2\mu(\omega) = (\lambda_r + 2\mu_r) \left(1 + \frac{i\omega\tau^P}{L} \sum_{l=1}^{L} \frac{\tau_l}{1 + i\omega\tau_l} \right)$$
$$= (\lambda_r + 2\mu_r) \left(1 + \tau^P S(\omega) \right) \tag{2}$$

This leads to the following representation of quality factors for P- and S-waves:

$$Q_{P,S} = \frac{\sum_{l=1}^{L} \left(1 + \tau^{P,S} \frac{\omega^2 \tau_l^2}{1 + \omega^2 \tau_l^2} \right)}{\omega \tau^{P,S} \sum_{l=1}^{L} \frac{\tau_l}{1 + \omega^2 \tau_l^2}} \tag{3}$$

The τ_l determine the relaxation times for stresses in the GSLS model we are using and have the dimension of time. They should provide the desired constancy of the quality factors over a given frequency interval. At the same time, $\tau^{P,S}$ govern the value of quality factor Q for both P- and S-waves. So there are two main steps in the development of the model of a viscoelastic medium with the use of the GSLS with a given spatial distribution of the quality factors $Q_{P,S}$:

1. The computation of relaxation times providing constancy of the quality factors Q by minimization of the following functional:

$$\tau_l = argmin\|const - Q_{P,S}(\omega)\|$$

2. The calculation of the spatial distribution of τ^P and τ^S providing desired values of Q everywhere in the target domain.

Let us suppose there is the volumetric point source within a viscoelastic medium. Then in the frequency time domain we have the following system of the second order partial differential equations describing seismic waves' propagation:

$$
\begin{cases}
\omega^2 \rho u_x + \dfrac{\partial}{\partial x}\left(((\lambda + 2\mu)(1 + S^P\tau^P)divu\right. \\
\left. -2\mu(1 + S^S\tau^S)\dfrac{\partial u_z}{\partial z}\right) + \\
\dfrac{\partial}{\partial z}\left(\mu(1 + S^S\tau^S)(\dfrac{\partial u_x}{\partial z} + \dfrac{\partial u_z}{\partial x})\right) = F_1(\omega)\dfrac{\partial \delta(x - x_0)}{\partial x} \\
\dfrac{\partial}{\partial z}\left(((\lambda + 2\mu)(1 + S^P\tau^P)divu - 2\mu(1 + S^S\tau^S)\frac{\partial u_z}{\partial z})\right) + \\
\omega^2 \rho u_z + \frac{\partial}{\partial x}\left(\mu(1 + S^S\tau^S)(\frac{\partial u_x}{\partial z} + \frac{\partial u_z}{\partial x})\right) = \\
= F_2(\omega)\frac{\partial \delta(x-x_0)}{\partial z}
\end{cases}
\tag{4}
$$

To close the obtained equations, we need to add the boundary conditions:

– on the free surface $z = 0$ let us admit the vanishing of normal components of the stress tensor;
– at infinity, we suppose vanishing of the u.

3   Inverse Problem

As a result of solving the equation system (4) with boundary conditions, one has a wave field that depends on the elastic moduli λ, μ and relaxation times τ^P, τ^S. The inverse problem is to reconstruct these parameters using additional information:

$$u(x_r, z_r; \omega) = B[\lambda, \mu, \tau^P, \tau^S]
\tag{5}$$

We regard relation (5) as a nonlinear operator equation, the solution of which will be sought by the nonlinear least-squares method:

$$[\lambda, \mu, \tau^P, \tau^S] = argmin\|u(x_r, z'_r\omega) - B[\lambda, \mu, \tau^P, \tau^S]\|^2
\tag{6}$$

The first step in implementing any process of minimizing a given functional is calculating its gradient. It is necessary to select a linear increment with a small perturbation of the parameters. Here we restrict ourselves to studying the linear operator arising in this case, considering the inverse problem's linearised formulation. In this case, we will assume that we know the reference medium, determined by the parameters $\lambda_0, \mu_0, \tau_0^P, \tau_0^S$, while the actual media is slightly different from them.

Thus, the linearized formulation of the inverse problem consists of determining the perturbations of the parameters $\delta\lambda, \delta\mu, \tau^P, \tau^S$ by the corresponding perturbations of the $\delta\boldsymbol{u}$, relating by the following system of linear differential equations:

$$
\begin{cases}
\omega^2 \rho_0 \delta u_x + (\lambda_0 + 2\mu_0)\frac{\partial^2 \delta u_x}{\partial x^2} + (\lambda_0 + \mu_0)\frac{\partial^2 \delta u_z}{\partial x \partial z} + \\
+\frac{\partial^2 \delta u_x}{\partial z^2} = f_1(x, z, \omega, u^0, \delta\rho, \delta\lambda, \delta\mu, \delta\tau^P, \delta\tau^S) \\
\omega^2 \rho_0 \delta u_z + (\lambda_0 + 2\mu_0)\frac{\partial^2 \delta u_z}{\partial z^2} + (\lambda_0 + \mu_0)\frac{\partial^2 \delta u_x}{\partial x z} + \\
+\mu_0 \frac{\partial^2 \delta u_z}{\partial x^2} = f_2(x, z, \omega, u^0, \delta\rho, \delta\lambda, \delta\mu, \delta\tau^P, \delta\tau^S) \\
\omega^2 \rho_0 u_x^0 + (\lambda_0 + 2\mu_0)\frac{\partial^2 u_x^0}{\partial x^2} + (\lambda_0 + \mu_0)\frac{\partial^2 u_z^0}{\partial x \partial z} + \mu_0 \frac{\partial^2 u_x^0}{\partial z^2} = \\
= F_1(\omega)\frac{\partial \delta(\boldsymbol{x} - \boldsymbol{x}_0)}{\partial x} \\
\omega^2 \rho_0 u_z^0 + (\lambda_0 + 2\mu_0)\frac{\partial^2 u_z^0}{\partial z} + (\lambda_0 + \mu_0)\frac{\partial^2 u_z^0}{\partial x \partial z} + \mu_0 \frac{\partial^2 u_x^0}{\partial x^2} = \\
= F_2(\omega)\frac{\partial \delta(\boldsymbol{x} - \boldsymbol{x}_0)}{\partial z}.
\end{cases}
\tag{7}
$$

Right-hand sides f_i are linear functions with respect to $\delta\rho$, $\delta\lambda$, $\delta\mu$, τ^P, τ^S. Thus the formulation (5) of the inverse problem takes the following form:

$$
\boldsymbol{u}_0 + \delta\boldsymbol{u} = B[\lambda_0, \mu_0, 0, 0] + DB[\lambda_0, \mu_0, 0, 0]
\begin{pmatrix}
\delta\rho \\
\delta\lambda \\
\delta\mu \\
\tau^P \\
\tau^S
\end{pmatrix}
\tag{8}
$$

We assume that the parameters of the reference medium are known, which allows us to treat function $\boldsymbol{u}_0$ as known as well, and the linearized inverse problem looks as:

$$
\delta\boldsymbol{u} = DB[\lambda_0, \mu_0, 0, 0]
\begin{pmatrix}
\delta\rho \\
\delta\lambda \\
\delta\mu \\
\tau^P \\
\tau^S
\end{pmatrix}
\tag{9}
$$

3.1 Numerical Experiments

We will assume that the perturbations of the viscoelastic medium's parameters are within the given rectangle

$$
\Pi = \left[(x, z) \in R^2 : (-L \leq x \leq L; h \leq z \leq H) \right].
$$

To reduce system (9) to a system of linear algebraic equations, we introduce a basis of functions of two variables $\phi_{lk}(x, z) = \psi_l(x)\theta_k(z)$ in this rectangle, where

$$
\psi_l(x) = \frac{1}{\sqrt{2L}} sin\frac{\pi x}{2L}
$$

and the elementary steps $h_k(z)$. We will look for the perturbations in the form of a finite series of the functions from this basis:

$$\delta m(x, z) = \sum_{l=1}^{N_x} \phi_l(x) \sum_{k=1}^{N_z} c_{kl} \theta_k(z). \tag{10}$$

As a result, the integral Eq. (9) transforms into the system of linear algebraic equations for finding unknown coefficients. This system approximates an integral equation of the first kind with a smooth kernel; therefore, when solving it numerically, one should keep in mind that the condition number of the resulting matrix will be the higher, the more accurately the approximation is performed. Hence, we come to the situation well-known in the numerical solution of ill-posed problems: the solution becomes more unstable as approximation improves.

Therefore, a straightforward approach to this problem's numerical solution is unlikely to lead to a correct result. Consequently, we apply a regularization of the original problem by using a truncated singular value decomposition (see [19]).

3.2   Regularization by Truncated Singular Value Decomposition

Any $M \times N$ matrix possesses Singular Value Decomposition (SVD), which is a basis of right $\boldsymbol{v}_j$, left $\boldsymbol{u}_j$ singular vectors, and singular numbers σ_j, satisfying the following relations:

$$A\boldsymbol{v}_j = \sigma_j \boldsymbol{u}_j, \quad A^* \boldsymbol{u}_j = \sigma_j \boldsymbol{v}_j,$$

where A^* is the Hermitian conjugate matrix to A. Singular vectors form orthonormal bases in their spaces:

$$(\boldsymbol{u}_i, \boldsymbol{u}_k) = \delta_{ik}, \quad (\boldsymbol{v}_i, \boldsymbol{v}_k) = \delta_{ik}.$$

This opens up the possibility of looking for a solution to a system of linear algebraic equations

$$\mathcal{A}\boldsymbol{x} = \boldsymbol{f}$$

in the form of a linear combination of right singular vectors:

$$\boldsymbol{x} = \sum_{k=1}^{N} \alpha_k \boldsymbol{v}_k$$

with coefficients

$$\alpha_k = (\boldsymbol{f}, \boldsymbol{v}_k).$$

The length N of the series depends on the value of the condition number $\frac{\sigma_1}{\sigma_N}$, which in turns, depends of the noise level in the right-hand side $\boldsymbol{f}$ and the accuracy of the matrix approximation of the integral operator.

Calculation of singular value decomposition for an arbitrary medium is computationally complicated and expensive problem. At first we use matrix representation of the operator, for these purpose used 1m-mesh for target area $z \in (1000, 2000)$m, $x \in (-700, 700)$m, the characteristic functions are used as a basis (that are equal to one inside cell, and zero outside). And frequency range $(20-100$ Hz) can be divided by finite intervals with the same basis functions. Such discretisation provides with matrix approximation of the operator, by replacing the integrals by finite sums.

Then we explore coupling of parametrization of the medium. Simultaneous recovery of the parameters when the perturbation of one of the parameters falsely determines as the inhomogeneity of the others is called coupling of the parameters, and it indicates the error in the solution [1]. The ambiguity arising in the solution of the inverse problem for media with attenuation is investigated in a few papers [13, 26]. To understand whether our parametrization is coupled, heterogeneity of each parameter is located in separated subregion in target area: $\delta\lambda \neq 0$ in $z \in [1080, 1230]m$, $\delta\mu \neq 0$ in $z \in [1310, 1460]m$, $\delta\tau^P \neq 0$ in $z \in [1540, 1690]m$, $\delta\tau^S \neq 0$ in $z \in [1770, 1920]m$ (Fig. 2).

The convergence of the singular numbers σ_i to zero (Fig. 1) indicates that the approximated operator is compact. Study of advantage of truncated SVD was made for different values of the parameter N, exactly for those that correspond to the condition numbers $10^4, 10^8$ (Fig. 1) and in total the number of singular vectors does not exceed 1250.

Despite the perturbation of parameter λ was only in $z \in [1080 - 1230]m$ for the condition number 10^4, we see heterogeneities outside this interval (Fig. 3), it means that the r solutions reflect the perturbations of other parameters. A similar effect occurs when a parameter is defined. Moreover, when the problem 10^8 is conditioned (Fig. 4), the boundaries of the subregions become more clear, and the coupling of the parameters substantially decreases.

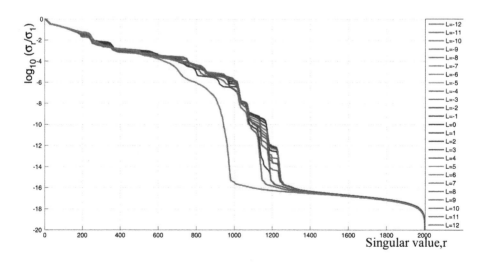

Fig. 1. Singular values in logarithmic scale.

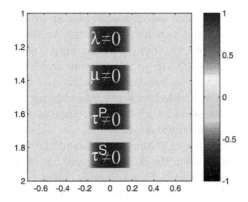

Fig. 2. Sought solution. Heterogeneities of the parameters are located in separated rectangles: $\delta\lambda$ in $z \in [1080, 1230]m$, $\delta\mu$ in $z \in [1310, 1460]m$, $\delta\tau^P$ in $z \in [1540, 1690]m$, $\delta\tau^S$ in $z \in [1770, 1920]m$.

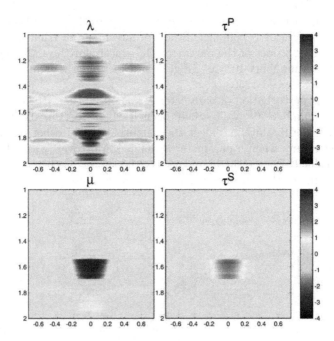

Fig. 3. Truncated SVD-solution for condition number 10^4.

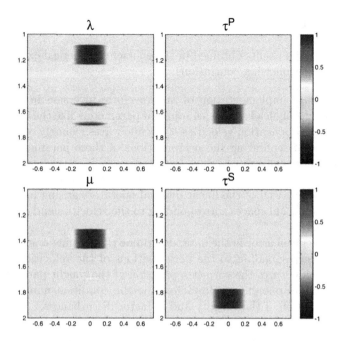

Fig. 4. Truncated SVD-solution for condition number 10^8.

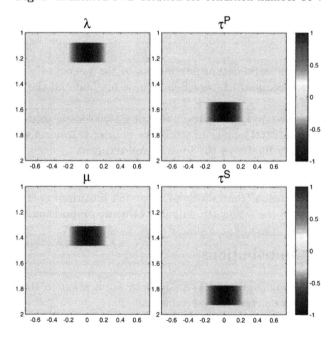

Fig. 5. Truncated SVD-solution for condition number 10^{14}.

4   Conclusions

Thus, analyzing the results obtained in the course of numerical experiments, we can formulate the following conclusions.

1. Straightforward implementation of full-waveform inversion in a viscoelastic medium inevitably leads to the coupling of parameters like the quality factors and the wave propagation velocities of seismic waves. Namely, we come to the impossibility of separating the perturbations of these parameters and their correct spatial localization.
2. To separate these perturbations, it is necessary to perform a singular value decomposition (SVD) of the linear integral operators arising in this case and to determine the subspaces corresponding to the velocities and quality factors perturbations.
3. Based on the given error in the data, determine the opening angle between the sub spaces corresponding to the perturbations of the velocities and quality factors, which ensures the correct separation of the sought parameters.
4. Using the found opening angle, determine the condition number providing correct separation of the velocity and Q-factor disturbances.
5. Based on the condition number, estimate the required accuracy of the matrix approximation of linear integral equations and the permissible error in the data.

We can describe the technique used for the solution of the problem as the follows:

- We search for the attenuation parameters of the viscoelastic medium using the Generalized Standard Linear Solid, the τ-method and the Born approximation.
- The problem of recovery the parameters of a viscoelastic medium is reduced to a system of integral equations using the Fourier transform with respect to time and the coordinates of the sources and receivers.
- We search for the characteristics of the medium as a finite Fourier series. To study the coupling between the parameters λ, τ^P, μ, τ^S we use an intermediate parametrization (that allows to construct linearized operators) and the method based on the truncated singular value decomposition.

5   Author Contributions

V.C. proposes the concept of SVD-analysis in application to the study of the coupling in multiparameters inverse problems.

E.E. derived equations connecting data variability with velocity and quality perturbations of the reference media and performed SVD computations.

G.R. has paid the main efforts to get estimations connecting data perturbations with variability of desired parameters via cond number of the matrix approximation of the operator and its apriori accuracy.

6   Funding

Galina Reshetova and Vladimir Cheverda have been supported by the Russian Science Foundation, project 20-11-20112 "Development of a modeling system for analyzing the current state and assessing the future trends in the environment of the Siberian Shelf Seas".

References

1. Assous, F., Collino, F.: A numerical method for the explanation of sensitivity: the case of the identification of the 2D stratified elastic medium. Inverse Prob. **6**(4), 487–514 (1990)
2. Asvadurov, S., Knizhnerman, L., Pabon, J.: Finite-difference modeling of viscoelastic materials with quality factors of arbitrary magnitude. Geophysics **69**(3), 817–824 (2004)
3. Bogoyavlensky V. Prospects and problems of the Arctic shelf oil and gas field development. Drilling Oil (Burenie i neft') **11**, 4–10 (2012)
4. Blanch, J.O.; Robertsson. J.O.A., Symes, W.W.: Modeling of a constant Q: Methodology and algorithm for an efficient and optimally inexpensive viscoelastic technique. Geophysics **60**(1), 176–184 (1995)
5. Carcione, J.M.: Seismic modeling in viscoelastic media. Geophysics **5**, 110–120 (1993)
6. Cheverda, V.A., Kostin, V.I.: R-pseudoinverses for compact operators in hilbert spaces:existence and stability. J. Inverse Ill-Posed Prob. **3**(2), 131–148 (1995)
7. Christensen R.M.: Theory of viscoelasticity - An introduction. 2nd edn, Academic Press, Cambridge, 364 p (1982)
8. Coleman, B.D., Noll, W.: Foundations of linear viscoelasticity. Rev. Mod. Phys. **33**(2), 239–249 (1961)
9. Curtin, M.E., Sternberg, E.: On the linear theory of visscoelasticity. Arch. Ration. Mech. Anal. **11**, 291–356 (1962)
10. Day, S.M., Minster, J.B.: Numerical simulation of attenuated wavefields using a Pade approximant method. Geophys. Res. **67**, 5279–5291 (1962)
11. Dmitrenko, I., et al.: Recent changes in shelf hydrography in the Siberian arctic: potential for subsea permafrost instability. J. Geophys. Res. **116**, C10027 (2011)
12. Dugarov G.A., Duchkov A.A., Duchkov A.D., Drobchik A.N.: Laboratory study of the acoustic properties of hydrate-bearing sediments. Sci. Rep. Facuilty Phys. Moscow Univ. (5), 1750812 (2017)
13. Efimova E.S.: Reliability of attenuation properties recovery for viscoelastic media. Open J. Appl. Sci. **3**(1B1), 84–88 (2013)
14. Fichtner, A.: Full Seismic Waveform Modeling and Inversion. Springer, Berlin (2011) https://doi.org/10.1007/978-3-642-15807-0
15. Hao, Q., Greenhalgh, S.: The generalized standard-linear-solid model and the corresponding viscoacoustic wave equations revisited. Geophys. J. Int. **219**, 1939–1947 (2019)
16. Hölemann, J.A., Kirillov, S., Klagge, T., Novikhin, A., Kassens, H., Timokhov, L.: Near-bottom water warming in the Laptev sea in response to atmospheric and sea-ice conditions in 2007. Polar Res. **30**, 6425 (2011)
17. Janout, M.A., Hölemann, J., Krumpen, T.: Cross-shelf transport of warm and saline water in response to sea ice drift on the Laptev Sea shelf. J. Geophys. Res. Oceans **118**, 563–576 (2013)

18. Janout, M., et al.: Episodic warming of near-bottom waters under the Arctic sea ice on the central Laptev Sea shelf. Geophys. Res. Lett. **43**(1), 264–272 (2016)

19. Kostov, C., Neklyudov, D., Tcheverda, V.: waveform inversion for macro velocity model recon-struction in look-ahead off-set VSP: numerical SVD-based analysis. Geophys. Prospect. **61**(6), 1099–1113 (2016)

20. Hak, B., Mulder, W.A.: Seismic attenuation imaging with causality. Geophys. J. Int. **184**, 439–451 (2011)

21. Hicks, G.J., Pratt, R.G.: Reflection waveform inversion using local descent methods: Estimating attenuation and velocity over a gas-sand deposit. Geophysics **66**(2), 598–612 (2001)

22. Liu, H.-P., Anderson, D.L., Kanamori, H.: Velocity dispersion due to anelasticity; implications for seismology and mantle composition. Geophysics **47**, 41–58 (1976)

23. McDonal, F.J., Angona, F.A., Mills, R.L., Sengbush, R.L., van Nostrand, R.G., White, J.E.: Attenuation of shear and compressional waves in Pierre shale. Geophysics **23**, 421–439 (1958)

24. McGuire, A.D., et al.: Sensitivity of the carbon cycle in the arctic to climate change. Ecol. Monogr. **79**(4), 523–555 (2009)

25. Mulder, W.A.: Velocity and attenuation perturbations can hardly be determined simultaneously in acoustic attenuation scattering. In: SEG Houston International Exposition and Annual Meeting, pp. 3078–3082 (2009)

26. Mulder, W.A., Hak, B.: An ambiguity in attenuation scattering imaging. Geophys. J. Int. **178**(3), 1614–1624 (2009)

27. Nemirovskaya, I.A.: Petroleum hydrocarbons in the ocean (in russina: Neftjanye uglevodorody v okeane). Priroda **3**, 17–27 (2008)

28. Parra, J.O., Hackert, C.: Wave attenuation attributes as flow unit indicators. Lead. Edge **21**(6), 564–572 (2002)

29. Rachold, V., et al.: Near-shore arctic subsea permafrost. Trans. EOS: Trans. Am. Geophys. Union **88**(13), 149–156 (2007)

30. Romanov, V.G.: The two-dimensional inverse problem for the equation of viscoelasticity. Siberian Math. Mag. **53**(6), 1401–1412 (2012)

31. Ruppel, C.D., Kessler, J.D.: The interaction of climate change and methane hydrates. Rev. Geophys. **55**(1), 126–168 (2017)

32. Sun, Y.F., Goldberg, D. Hydrocarbon signatures from high-resolution attenuation profiles. In: SEG Technical Program Expanded Abstracts, pp. 996–999 (1998)

33. Zener, C.: Elasticity and Anelasticity of Metals, p. 170. University of Chicago Press, Chicago (1948)

34. White, R.E.: The acuracy of estimating Q from seismic data. Geophysics **57**, 1508–1511 (1992)

35. Zhang, D., Lamoureux, M., Margrave, G., Cherkaev, E.: Rational approximation for estimation of quality q factor and phase velocity in linear, viscoelastic, isotropic media. Comput. Geosci. **15**, 117–133 (2011)

International Workshop on Advanced Transport Tools and Methods (A2TM 2021)

Issues in Modelling Traffic-Related Air Pollution: Discussion on the State-Of-The-Art

Francesco Bruzzone[1,2] and Silvio Nocera[1(✉)]

[1] Università IUAV di Venezia, Venice, Italy
nocera@iuav.it
[2] Politecnico di Torino, Turin, Italy

Abstract. Traffic-related air pollution is nowadays considered as a major externality from the transport sector. This paper conducts a literature review on three main topics: mechanisms of traffic-related air pollution formation and disposal, models in use for their simulation, and methods and models for the economic quantification of emissions. Providing an up-to-date overlook on the modelling and quantification framework of traffic-related air pollution and highlighting pros and cons of each solution with regard to the goal of the research and to the context of application, this paper aims at describing the effective knowledge made on this theme and at identifying future research directions.

Keywords: Air pollution modelling · Air pollution quantification · Traffic related pollution

1 Introduction

This paper provides a synthetic state-of-art review of methods for the analysis of the formation, disposal and quantification of traffic-related air pollution, a major negative externality from the transport sector [1]. The goal is to provide an updated overview on the modelling and quantification framework of traffic-related air pollution and to highlight strong and weak points of each solution with regard to the goal of the research and to the context of application. The brief discussion carried out within this paper should be useful for scholars and authorities to appropriately choose the most suitable methodology according to their needs, also within research efforts under the umbrella of EU-funded mobility-related projects. This paper is intended as a swift tool to support researchers in the initial stages of their work, when compiling the literature review and assessing methodological options. The paper is structured as follows: Sect. 2 looks at the formation and disposal patterns of traffic-related air pollution, while Sect. 3 discusses models to estimate those patterns. Section 4 looks at techniques and methods for the quantification of traffic-related air pollution, and Sect. 5 discusses and compares contents and draws some conclusions.

© Springer Nature Switzerland AG 2021
O. Gervasi et al. (Eds.): ICCSA 2021, LNCS 12950, pp. 337–349, 2021.
https://doi.org/10.1007/978-3-030-86960-1_24

2 Formation and Disposal of Traffic-Related Air Pollution

Road traffic, one of the main modes for both freight and passenger transport, accounts for 81% of transport-related energy consumption in Europe [2]. Large amounts of polluting substances are emitted in the atmosphere, posing severe threats to the health of human beings and of planet Earth. Primary pollutants, such as PM_{10}, $PM_{2.5}$, SO_x, NO_x and CO, constitute a direct threat to human health and to the environment, being directly responsible of premature deaths and of natural degradation [3].

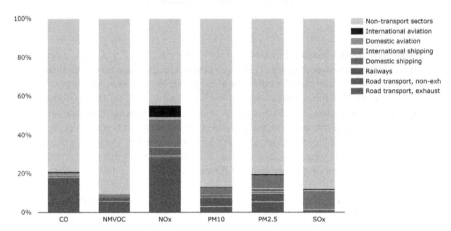

Fig. 1. Contribution of the transport sector to total emissions of the main air pollutants (Source: EEA, 2021 [5])

The contribution of the transport sector to total emissions of such elements is shrinking; however, transport still accounts for 20% of global CO, SOx and PM2.5 emissions, and for 54% of NOx emissions (Fig. 1) [1]. Emissions of GHGs (Greenhouse Gasses), with particular reference to CO_2 (carbon dioxide), CH_4 (methane) and N_2O (nitrous oxide) are growing due to human activity and in 2011 their concentrations reached 391 ppm, 1803 ppb, and 324 ppb respectively[1] [4]. Increasing GHGs emissions are directly correlated with global warming and climate change. Surface temperatures have been growing for the whole 20th century and the rate has increased in the early years of the new millennium, together with more frequent extreme weather events and other phenomena related to climate change [4, 5]. Albeit globally the transport sector accounts for approximately 14% of global yearly GHG emissions (Fig. 2), at the European level the sector is responsible for around 26% of GHG emissions, with the figure also showing an increasing trend (Fig. 3) [6, 7].

Given the differences in polluting sources and trends and the different cycles and sectors of impact of primary pollutants and GHGs, the two groups of pollutants are normally discussed, analyzed and modelled separately [3]. Some of the models adopted

---

[1] ppm stands for "parts per million"; ppb stands for "parts per billion".

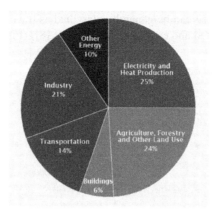

Fig. 2. Global GHG emissions by economic sector in 2014 (Source: EEA, 2021 [7])

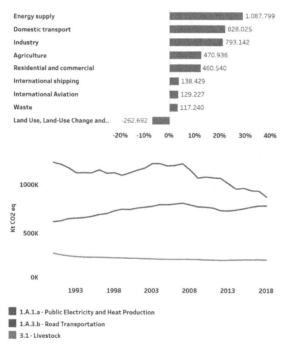

Fig. 3. GHG emissions by economic sector in the EU in 2018 and emission trend (Source: EEA, 2021 [7])

by researchers and institutions later discussed in this paper, however, are able to comprehensively evaluate emissions from the transport sectors, including both short and long-term impacts on the environment and health-related consequences.

Pollutant emissions from road traffic are generated by motorized vehicles during the various phases of the trip, and their quantity, type and emission patterns are affected by a number of factors. These can be generally classified as travel-related, driver-related, facility-related, fuel-related and environmental (Fig. 1; [8]) (Fig. 4).

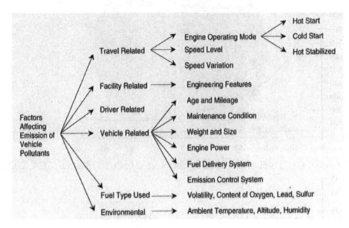

Fig. 4. Factors affecting vehicle emissions (from Sinha and Labi 2007 [8])

The sensitivity of emissions to these factors has been object of a number of report and studies. Comprehensive reviews have been conducted by [9] and by [10]. Travel-related factors include speeds and vehicle operating modes, typically cold start, hot start, and hot stabilized period [11]. Cold starts are generally acknowledged to produce higher emission rates. The consistency and smoothness of travel speed, as well as load, acceleration rates, and fuel types also impact emission rates. For most vehicle types, emissions are lower at higher speed, and rise at low speeds and when the engine is forced at high rates [12, 13]. Driver-related factors are hinged on the driver's behavior: aggressive drivers exert more severe accelerations and decelerations and require more effort to the engine, thus reducing time spent at low-emitting conditions. Facility (i.e., infrastructure) design can favor the operation of motorized vehicles at low-emitting speeds or modes, by encouraging smooth and non-aggressive driving. Infrastructural and technological interventions such as traffic flow management signal coordination, moreover, can result in emission reductions up to 25% [14, 15]. Vehicle-related factors include vehicle age, mileage, condition, weight, size, and engine power. In general terms, lighter-duty and newer vehicles emit significantly less [12]. Environmental conditions, and particularly temperatures, have an impact on emissions. Low ambient temperatures require more time for engines and emission-control systems to warm up and cause higher emissions, even in the case of hybrid-electric new generation vehicles [16].

Factors related to the environmental and meteorological conditions, as well as to the morphology of the territory in which the emitting source is located, affect the pollutants' dispersion of in the atmosphere. Atmospheric conditions, such as temperature, stability, precipitation, wind, humidity and intensity of solar radiation, impact on the temporal and spatial variation of pollutant dispersion and, therefore, on the concentration of air pollutants. Particularly relevant atmospheric phenomena ruling pollutant dispersion are the atmospheric stability (related to the change in temperature, wind speed and wind direction with height): a stable atmosphere suppresses vertical motion within its domain and thus lowers dispersion rates; thermal inversion (which entraps cold air layers and pollutants under a warmer plateau); and ceiling height: night-time and winter conditions are normally associated with lower ceilings and lower dispersion rates [8].

Topography and urban form also affect pollutant dispersion: higher morphological complexity corresponds to lower dispersion rates and thus higher pollutant concentrations, as it affects the wind speed and direction and temperatures in the area. Regardless of the urban or natural environment, steep grades and obstacles to airflow, such as building or natural elevations, contribute to the "canyon effect" and to a slower pollutant dispersion [17, 18].

The assessment of determinants, magnitude, and characteristics of pollutant dispersion can be a complex task, especially in urban areas or unregular terrains, due to the range and diversity of polluting sources, topographical and morphological features, and to the contribution of anthropic activity. The next section discusses a number of pollutant dispersion models developed by scholars to achieve the task.

3   Models for the Simulation of Traffic-Related Air Pollution Formation and Disposal

Pollutants are emitted in the atmosphere according to a number of dispersion factors including meteorological conditions (in particular wind speeds, atmospheric stability, temperature gradients), the number of emitting sources and the emission rates of such sources [8]. A number of models for assessing the emissions' impact on pollutant concentrations have been developed, among which some of the most relevant are the box model, the Gaussian plume model, and the numerical model.

The box model assumes uniform dispersion of pollutants to fill a box-shaped space, with two main factors (wind speed and mixing height) controlling dispersion patterns. Combination of multiple boxes are commonly used by scholars to improve the model's detail and solve flows as modified by the presence of buildings and other disturbing factors [19]. Box model approaches are the most basic air quality algorithm and have been often used to determine impacts from linear sources, such as highways or airport runaways; however, the analysis of morphologically complex environments through this technique is exposed to a certain degree of approximation [20].

The Gaussian Plume model uses normal distribution to describe the random wafting of plumes side to side and up and down, resulting in increased plume size with time and higher pollutant concentrations near the emitting source [8]. The Gaussian equation is used to estimate the dispersion of pollutants released from an emitting source at a steady rate, and the model is widely used as it produces similar results to those obtained with other more complex models, mathematical operations are easy to perform, it is consistent with the random nature of turbulence and dispersion, and it is often referred to in governmental guidelines [21].

Numerical models, such as for instance DANARD, RAGLAND, MROAD-2, ROADS (for further details see a review by [22]) use numerical analyses and algorithms by means of computer-based simulations to solve and analyze problems related to flows and dispersion [23, 24]. Numerical air quality models involve a three-dimensional grid of conceptual boxes that occupy the space above a transportation corridor. Emitting sources impact the overlaying "boxes" and fill them at some given rate, then including the immediately outlying areas in a continuous process. Accounting for variables such as winds, emission diffusion and emission rates, and using well-behaved functions of time, numerical models can predict the movement of polluting particles across the "boxes" with successive time increments [8]. Whereas in the past Gaussian Plume models were the most commonly used thanks to their capacity to correctly represent complex terrains and to provide good estimates of ground-level pollutant concentrations [25], numerical models have been perfectioned and have become more practical with the recent development of powerful computers. Numerical models in general have proven accurate in simulating atmospheric dispersion, recently even including meteorological conditions [26]. Silva et al. [27] found numerical models to reliably and confidently predict CO_2 emissions and fuel consumption. The EEA (European Environmental Agency) confirms that numerical models can be used to conveniently simulate transport processes [28] while the American Environmental Protection Agency (EPA) has officially adopted the AERMOD numerical model [29], which has proven reliable under a wide range of meteorological and topographical conditions [30].

4   Models and Methods for the Quantification of Traffic-Related Air Pollution

Pollutant emission models are needed to determine the impacts from the transport sector and to forecast changes due to variations to the transport system. A great number of emission models has been proposed. The most relevant ones are adopted as standard by environmental agencies (e.g. EEA, USEPA) and are available in form of online tools or software for public use, such as for instance MOBILE 6.0 and COPERT. Faris et al. [31] conducted an in-depth, extensive review of emission models, providing a number of different classifications and great-value comparisons. For deepening the topic, we suggest that the reader refers to their contribution. Basing on input variables, state-of-art models can generally be divided into microscopic or macroscopic models. At times, mesoscopic models are also indicated [31]. Conventionally, microscopic models are used in simpler simulation, with less vehicle variety and few intersections and disturbing factors, while macroscopic models are suitable for replicating the city level [32]. Table 1 synthetizes the three kinds of models that we consider in this paper, including the main source of reference for each specific model, in case one is clearly identifiable.

Single-vehicle emissions simulated through microscopic models can be aggregated to estimate network-wide measures. The most relevant microscopic models can be divided into: comprehensive modal emission model (CMEM), first debated by An et al. [33]; VT-micro, developed by Rakha and Ahn [34], which offers realistic estimation and frequent updates [35]; VETESS, adopted by the CORSIM software used by the Federal Highway Administration in the United States of America, and found to have high probability of success in real world applications [36]; VERSIT + ^{micro}, adopted by the University of Catalunya in their AIMSUN software [31]; the PHEM model, which uses an emission map, developed in terms of engine operating parameters rather than in terms of aggregate vehicle parameters, to estimate emissions microscopically, and has been

Table 1. Micro-, meso-, and macroscopic models for traffic related air pollution

Model category	Models	Source
Microscopic vehicle emission models	CNEM VT-micro VETESS VERSIT + micro PHEM EMIT MOVES INTEGRATION	An et al Rakha & Ahn Federal Highway Administration University of Catalunya Environmental Protection Agency
Mesoscopic vehicle emission models	Elemental model CONTRAM MEASURE	Herman
Macroscopic vehicle emission models	MOBILE EMFAC/MOVES CORFLO Watson COPERT CO_2MPAS	Environmental Protection Agency State of California Federal Highway Administration Watson et al European Environmental Agency European Environmental Agency

adopted in the VISSIM software [37, 38]; EMIT, that evaluates emissions depending on vehicle speed and acceleration as well as vehicle characteristics [39]; MOVES, developed by the American Environmental Protection Agency (EPA) and currently officially adopted in the United States, in its latest version MOVES3 [40]. Unlike the mentioned software, which use kinematic acceleration-speed relationship to model accelerations, INTEGRATION uses a vehicle dynamics model, computing fuel consumption for each vehicle on second-by-second basis for three operation modes (constant-speed cruise, velocity change and idling) as a function of travel speed [8].

Macroscopic models, instead, use average aggregate parameters to estimate network-wide emission rates. Some of the most relevant models using the macroscopic approach are: MOBILE (latest version 6.2 developed by EPA in 2004 [41]); EMFAC, developed in California and now replaced – together with MOBILE – by MOVES (adapted to network scale) and the TRANSIMS software [31]; CORFLO, developed by the Federal Highway Administration of the USA and characterized by fast simulation times, suitable for inexpensive computers [42]; the Watson model [43] which uses average speed to investigate fuel consumption; and COPERT (computer programme to compute emissions from road transport) developed by the EEA and widely adopted in transportation applications in Europe [44], for both emission inventories and dispersion modelling [45]. The EU has recently introduced a vehicle simulation model for CO_2 certification purposes, known as CO_2MPAS, which also contributes to the implementation of the COPERT software [46].

Finally, mesoscopic models, according to the review by Faris et al. [31], include the elemental model, CONTRAM, and MEASURE. The elemental model was proposed by Herman and adopted in the SIDRA INTERSECTION software in 1985, widely used in

Australia. CONTRAM (continuous traffic assignment model) intends to model traffic links with complex dynamics and is capable of accurate description of time-variable network conditions [47]. However, it lacks the ability to consider performance measures such as speed [48, 49]. The MEASURE (mobile emission assessment system for urban and regional evaluation) is a geographic information system GIS-based model compatible with microscopic models and tools as well [50]. It includes two major modules: start emissions and on-road emissions [51].

The monetary quantification of air pollution is a fundamental task in transport planning. According to an extensive review from Sharma and Khare [23] and to Sinha and Labi [8], three most common methods are used to determine the cost of environmental degradation due to emissions from the transport sector. A first technique consists in computing the costs for cleaning the air near the degradation source by means of physical infrastructures such as air scrubbers [52]. Air pollution costs, if quantified in this manner, can be excessive as the purchasing and operating costs of scrubbers are notable [53]. A second method describes costs as the social damage effects of air pollution, including the health care expenses due to respiratory issues, costs for repair of physical infrastructure, and compensation costs for damages to the natural environment. A review on quantification and internalization methods within this category is available in Aevermann and Schmude [54]. Finally, air pollution cost estimation can be conducted through the willingness-to-pay approach. This method has been object of a number of relevant contributions from authors exploring environmental and transportation issues, and investigates the extent to which persons and business are willing to pay to avoid a pollution-related problem. Recently, the willingness-to-pay approach has been extensively adopted to study the adoption of carbon taxes [55]. Contributions on the topic include Wang and Mullahy [56], Carlsson and Johansson-Stenman [57], Lera-López et al. [58, 59], and Istamto et al. [60].

Air pollution cost values are generally periodically defined by environmental authorities and institutions within the global effort for the fight against climate alternation. Within the European panorama, the European Commission regularly updates its Handbook on External Costs of Transport (latest version [61]) as well as a number of other documents on the internalization of transport-related externalities (like [62]). EC's documents provide air pollution cost estimates for various pollutant types, transportation modes, and operating patterns and speeds. Despite the use of internalization rates provided by public authorities has become a common practice within transport planning [63], emission estimation is not yet a well mastered task and gaps are still open for research on further approaches.

5 Discussion and Conclusions

In this paper, a short state-of-art review of methods and models for the formation, disposal and quantification of traffic-related air pollution has been presented.

In a first section, we have discussed the main factors producing traffic-related air pollution, highlighting the necessity to study different trip and engine phases (e.g. cold starts, constant-speed running, etc.) and the characteristics of trips, facilities, and the environment. It must be noted that, as discussed, the first leg of motorized journeys is that with the highest impacts, especially when conducted in the urban traffic, causing additional negative externalities to an already suffering environment [64, 65]. This inevitably calls for models and monetization criteria suitable for the analysis of urban environments and of typical urban transport operations.

In Sect. 3, we discussed models for the simulation of traffic-related air pollution formation and disposal, stressing how increasingly powerful computers have allowed for a wider use of numerical models, whereas Gaussian plume models used to be extensively used in the last decades and show good accuracy while keeping computations simple, and box models are effective in simulating simple environments and operations. Finally, we have approached the topic of the quantification and monetization of traffic-related air pollution, discussing a number of models and tools (according to the micro-, macro-, and mesoscopic classification based on input variables) and different approaches to the determination of pollution costs. Table 2 synthetizes the capacity of micro-, meso- and macroscopic emission models to effectively and precisely describe road traffic emissions.

Table 2. Accuracy and relevance to road traffic simulation in the scale of the input variable-based modelling classification (own elaboration from Faris et al. 2011)

Modelling type	Accuracy	Relevance w/r to road traffic simulation
Microscopic vehicle emission models	Relatively accurate	High
Mesoscopic vehicle emission models	More accurate than macroscopic models	Moderate
Macroscopic vehicle emission models	Less accurate than microscopic models	Low

The topic of the quantification of transport-related emissions and of proposed mitigation strategies is particularly important for transport planners, which should be able to adopt and implement the right tools to understand (first) and mitigate, then, mobility-related externalities while pursuing emission reduction strategies in harmony with global and local sustainability objectives. Sadly, governments and planning authorities are often not aware of the origin and distribution of the most impacting emitting sources [6] and are unable to estimate benefits and outcomes of potential improvements [66].

In this review, we provided an overview on different possibilities for the simulation and evaluation of pollutant emissions due to the transport sector. Future efforts within reviews on this topic should more specifically address models for each type of pollutant, on the one side; and models for the simulation of specific territorial and traffic conditions on the other side. The role of technological advancements, both from a vehicle-related perspective (e.g., alternative fuels or traction, automation), from an overall mobility-system perspective and from an ICT perspective (e.g. driving assistant, traffic management systems), should also be studied as varying conditions could alter the models' accuracy.

Despite the extensiveness of the covered topics, the aim was to provide a comprehensive yet easy to read overview of the various possible approaches, also in order to support authorities and decision makers in their approach to the topic within transport planning efforts. The understanding of the matter and the capacity to perform suitable emission evaluation and quantification in transport planning is in fact fundamental to ensure the ability of planning documents to meet global and local emission reduction and air quality targets, ensuring that policy and investments are directed towards effective measures and interventions. Whereas standard air pollution costs as published by authorities (such as the European Commission) are now widely used and have been successfully integrated in planning processes, the estimation of the emissions to be quantified is not always a straightforward, well mastered task. With this contribution, we intended to shed some light on the most common available tools for planners to adopt.

References

1. Nocera, S., Cavallaro, F., Irranca, G.O.: Options for reducing external costs from freight transport along the brenner corridor. Eur. Transp. Res. Rev. **10**(2), 53 (2018)
2. Cavallaro, F., Irranca Galati, O., Nocera, S.: Policy strategies for the mitigation of GHG emissions caused by the mass-tourism mobility in coastal areas. Transp. Res. Procedia **27**, 317–327 (2017)
3. IPCC: summary for policymakers. In: Stocker, T.F., et al. (eds.) Climate Change 2013: The Physical Science Basis. Contribution of Working Group I to the Fifth Assessment Report of the Intergovernmental Panel on Climate Change. Cambridge University Press, Cambridge (2013)
4. Nocera, S., Cavallaro, F.: The ancillary role of CO2 reduction in urban transport plans. Transp. Res. Procedia. **3**, 760–769 (2014). https://doi.org/10.1016/j.trpro.2014.10.055
5. EEA: Greenhouse gas emissions from transport in Europe — European Environment Agency (2021a). https://www.eea.europa.eu/data-and-maps/indicators/transport-emissions-of-greenhouse-gases-7/assessment
6. Nocera, S., Cavallaro, F.: A two-step method to evaluate the well-to-wheel carbon efficiency of urban consolidation centres. Res. Transp. Econ. **65**, 44–55 (2017). https://doi.org/10.1016/j.retrec.2017.04.001
7. EEA: Emissions of air pollutants from transport — European Environment Agency (2021b). https://www.eea.europa.eu/data-and-maps/indicators/transport-emissions-of-air-pollutants-8/transport-emissions-of-air-pollutants-8
8. Sinha, K.C., Labi, S.: Transportation Decision Making: Principles of Project Evaluation and Programming (2007)
9. Faiz, A., Weaver, C.S., Walsh, M.P.: Air Pollution from Motor Vehicles: Standards and Technologies for Controlling Emissions. World Bank Publications (1996)
10. Franco, V., Kousoulidou, M., Muntean, M., Ntziachristos, L., Hausberger, S., Dilara, P.: Road vehicle emission factors development: a review. Atmos. Environ. **70**, 84–97 (2013). https://doi.org/10.1016/j.atmosenv.2013.01.006
11. Reiter, M.S., Kockelman, K.M.: The problem of cold starts: a closer look at mobile source emissions levels. Transp. Res. Part D: Transp. Environ. **43**, 123–132 (2016). https://doi.org/10.1016/j.trd.2015.12.012
12. Kean, A.J., Harley, R.A., Kendall, G.R.: Effects of vehicle speed and engine load on motor vehicle emissions. Environ. Sci. Technol. **37**, 3739–3746 (2003). https://doi.org/10.1021/es0263588

13. Chong, H.S., Park, Y., Kwon, S., Hong, Y.: Analysis of real driving gaseous emissions from light-duty diesel vehicles. Transp. Res. Part D Transp. Environ. **65**, 485–499 (2018). https://doi.org/10.1016/j.trd.2018.09.015

14. Madireddy, M., et al.: Assessment of the impact of speed limit reduction and traffic signal coordination on vehicle emissions using an integrated approach. Transp. Res. Part D Transp. Environ. **16**, 504–508 (2011). https://doi.org/10.1016/j.trd.2011.06.001

15. Sharifi, F., et al.: Regional CO2 impact assessment of road infrastructure improvements. Transp. Res. Part D Transp. Environ. **90**, 1026 (2021). https://doi.org/10.1016/j.trd.2020.102638

16. Suarez-Bertoa, R., et al.: Effect of low ambient temperature on emissions and electric range of plug-in hybrid electric vehicles. ACS Omega **4**, 3159–3168 (2019). https://doi.org/10.1021/acsomega.8b02459

17. Carvalho, A.C., et al.: Influence of topography and land use on pollutants dispersion in the Atlantic coast of Iberian Peninsula. Atmos. Environ. **40**, 3969–3982 (2006). https://doi.org/10.1016/j.atmosenv.2006.02.014

18. Fang, C., Wang, S., Li, G.: Changing urban forms and carbon dioxide emissions in China: a case study of 30 provincial capital cities. Appl. Energy **158**, 519–531 (2015). https://doi.org/10.1016/j.apenergy.2015.08.095

19. Salem, N.B., Garbero, V., Salizzoni, P., Lamaison, G., Soulhac, L.: Modelling pollutant dispersion in a street network. Bound.-Layer Meteorol. **155**(1), 157–187 (2014). https://doi.org/10.1007/s10546-014-9990-7

20. Mareddy, A.R.: 5 - Impacts on air environment. In: Mareddy, A.R. (ed.) Environmental Impact Assessment, pp. 171–216. Butterworth-Heinemann (2017)

21. Miller, C.W., Hively, L.M.: A review of validation studies for the Gaussian plume atmospheric dispersion model. Nucl. Saf. 28 (1987)

22. Gokhale, S., Khare, M.: A review of deterministic, stochastic and hybrid vehicular exhaust emission models. Int. J. Transp. Manag. **2**, 59–74 (2004). https://doi.org/10.1016/j.ijtm.2004.09.001

23. Sharma, P., Khare, M.: Modelling of vehicular exhausts – a review. Transp. Res. Part D: Transp. Environ. **6**, 179–198 (2001). https://doi.org/10.1016/S1361-9209(00)00022-5

24. Sun, D., Zhang, Y.: Influence of avenue trees on traffic pollutant dispersion in asymmetric street canyons: numerical modeling with empirical analysis. Transp. Res. Part D: Transp. Environ. **65**, 784–795 (2018). https://doi.org/10.1016/j.trd.2017.10.014

25. Zannetti, P.: Gaussian models. In: Zannetti, P. (ed.) Air Pollution Modeling: Theories, Computational Methods and Available Software, pp. 141–183. Springer, US (1990)

26. Kim, G., Lee, M.-I., Lee, S., Choi, S.-D., Kim, S.-J., Song, C.-K.: Numerical modeling for the accidental dispersion of hazardous air pollutants in the urban metropolitan area. Atmosphere **11**, 477 (2020). https://doi.org/10.3390/atmos11050477

27. Silva, C.M., Farias, T.L., Frey, H.C., Rouphail, N.M.: Evaluation of numerical models for simulation of real-world hot-stabilized fuel consumption and emissions of gasoline light-duty vehicles. Transp. Res. Part D: Transp. Environ. **11**, 377–385 (2006). https://doi.org/10.1016/j.trd.2006.07.004

28. EEA: 5. Modelling — European Environment Agency (2020). https://www.eea.europa.eu/publications/TEC11a/page011.html

29. US EPA, O.: Air Quality Dispersion Modeling - Preferred and Recommended Models (2021). https://www.epa.gov/scram/air-quality-dispersion-modeling-preferred-and-recommended-models

30. Tartakovsky, D., Broday, D.M., Stern, E.: Evaluation of AERMOD and CALPUFF for predicting ambient concentrations of total suspended particulate matter (TSP) emissions from a quarry in complex terrain. Environ. Pollut. **179**, 138–145 (2013). https://doi.org/10.1016/j.envpol.2013.04.023

31. Faris, W.F., Rakha, H.A., Kafafy, R.I., Idres, M., Elmoselhy, S.: Vehicle fuel consumption and emission modelling: an in-depth literature review. IJVSMT. **6**, 318 (2011). https://doi.org/10.1504/IJVSMT.2011.044232

32. Krajzewicz, D., Behrisch, M., Wagner, P., Luz, R., Krumnow, M.: Second generation of pollutant emission models for SUMO. In: Behrisch, M., Weber, M. (eds.) Modeling Mobility with Open Data. LNM, pp. 203–221. Springer, Cham (2015). https://doi.org/10.1007/978-3-319-15024-6_12

33. An, F., Barth, M., Norbeck, J., Ross, M.: Development of comprehensive modal emissions model: operating under hot-stabilized conditions. Transp. Res. Rec. **1587**, 52–62 (1997). https://doi.org/10.3141/1587-07

34. Rakha, H., Ahn, K., Trani, A.: Development of VT-Micro model for estimating hot stabilized light duty vehicle and truck emissions. Transp. Res. Part D Transp. Environ. **9**, 49–74 (2004). https://doi.org/10.1016/S1361-9209(03)00054-3

35. Llopis-Castelló, D., Camacho-Torregrosa, F.J., García, A.: Analysis of the influence of geometric design consistency on vehicle CO2 emissions. Transp. Res. Part D Transp. Environ. **69**, 40–50 (2019). https://doi.org/10.1016/j.trd.2019.01.029

36. Zhou, M., Jin, H., Wang, W.: A review of vehicle fuel consumption models to evaluate eco-driving and eco-routing. Transp. Res. Part D Transp. Environ. **49**, 203–218 (2016). https://doi.org/10.1016/j.trd.2016.09.008

37. Fellendorf, M., Vortisch, P.: Microscopic traffic flow simulator VISSIM. In: Barceló, J. (ed.) Fundamentals of Traffic Simulation, pp. 63–93. Springer, New York (2010)

38. Tielert, T., Killat, M., Hartenstein, H., Luz, R., Hausberger, S., Benz, T.: The impact of traffic-light-to-vehicle communication on fuel consumption and emissions. In: 2010 Internet of Things (IOT), pp. 1–8 (2010)

39. Sommer, C., German, R., Dressler, F.: Bidirectionally coupled network and road traffic simulation for improved IVC analysis. IEEE Trans. Mob. Comput. **10**, 3–15 (2011). https://doi.org/10.1109/TMC.2010.133

40. US EPA, O.: Latest Version of MOtor Vehicle Emission Simulator (MOVES) (2021a). https://www.epa.gov/moves/latest-version-motor-vehicle-emission-simulator-moves

41. US EPA, O.: Description and History of the MOBILE Highway Vehicle Emission Factor Model (2021b). https://www.epa.gov/moves/description-and-history-mobile-highway-vehicle-emission-factor-model

42. Boxill, S.A., Yu, L.: An evaluation of traffic simulation models for supporting its development (2000)

43. Watson, H.C., Milkins, E.E., Marshall, G.A.: A simplified method for quantifying fuel consumption of vehicles in urban traffic. In: Presented at the Automotive Engineering Conference, 4th, 1979, Melbourne, Australia November (1979)

44. Lei, W., Chen, H., Lu, L.: microscopic emission and fuel consumption modeling for light-duty vehicles using portable emission measurement system data. Int. J. Mech. Mechatron. Eng. **4**, 495–502 (2010)

45. Boulter, P.G., McCrane, I.S., Barlow, T.J.: A review of instantaneous emission models for road vehicles (2007)

46. Mogno, C., et al.: The application of the CO2MPAS model for vehicle CO2 emissions estimation over real traffic conditions. Transp. Policy (2020). https://doi.org/10.1016/j.tranpol.2020.01.005

47. Hounsell, N., Shrestha, B., Piao, J.: Enhancing park and ride with access control: a case study of Southampton. Transp. Policy **18**, 194–203 (2011). https://doi.org/10.1016/j.tranpol.2010.08.002

48. Taylor, N.B.: The CONTRAM dynamic traffic assignment model. Netw. Spat. Econ. **3**, 297–322 (2003). https://doi.org/10.1023/A:1025394201651

49. Nguyen, J., Powers, S.T., Urquhart, N., Farrenkopf, T., Guckert, M.: An Overview of Agent-based Traffic Simulators (2021). arXiv:2102.07505 [cs]
50. Bachman, W., Sarasua, W., Hallmark, S., Guensler, R.: Modeling regional mobile source emissions in a geographic information system framework. Transp. Res. Part C Emerg. Technol. **8**, 205–229 (2000). https://doi.org/10.1016/S0968-090X(00)00005-X
51. Yue, H.: Mesoscopic Fuel Consumption and Emission Modeling (2008). Doctor of Philosophy dissertation, Virginia Polytechnic Institute and State University.
52. Vatavuk, W.M.: Estimating Costs of Air Pollution Control. CRC Press, Boca Raton (1990)
53. Gu, Y., Wallace, S.W.: Scrubber: a potentially overestimated compliance method for the emission control areas: the importance of involving a ship's sailing pattern in the evaluation. Transp. Res. Part D Transp. Environ. **55**, 51–66 (2017). https://doi.org/10.1016/j.trd.2017.06.024
54. Aevermann, T., Schmude, Jürgen.: Quantification and monetary valuation of urban ecosystem services in Munich, Germany. Zeitschrift für Wirtschaftsgeographie **59**(3), 188–200 (2015). https://doi.org/10.1515/zfw-2015-0304
55. Gupta, M.: Willingness to pay for carbon tax: a study of Indian road passenger transport. Transp. Policy **45**, 46–54 (2016). https://doi.org/10.1016/j.tranpol.2015.09.001
56. Wang, H., Mullahy, J.: Willingness to pay for reducing fatal risk by improving air quality: a contingent valuation study in Chongqing, China. Sci. Total Environ. **367**(1), 50–57 (2006). https://doi.org/10.1016/j.scitotenv.2006.02.049
57. Carlsson, F., Johansson-Stenman, O.: Willingness to pay for improved air quality in Sweden. Appl. Econ. **32**, 661–669 (2000). https://doi.org/10.1080/000368400322273
58. Lera-López, F., Faulin, J., Sánchez, M.: Determinants of the willingness-to-pay for reducing the environmental impacts of road transportation. Transp. Res. Part D Transp. Environ. **17**, 215–220 (2012). https://doi.org/10.1016/j.trd.2011.11.002
59. Lera-López, F., Sánchez, M., Faulin, J., Cacciolatti, L.: Rural environment stakeholders and policy making: willingness to pay to reduce road transportation pollution impact in the Western Pyrenees. Transp. Res. Part D Transp. Environ. **32**, 129–142 (2014). https://doi.org/10.1016/j.trd.2014.07.003
60. Istamto, T., Houthuijs, D., Lebret, E.: Willingness to pay to avoid health risks from road-traffic-related air pollution and noise across five countries. Sci. Total Environ. **497–498**, 420–429 (2014). https://doi.org/10.1016/j.scitotenv.2014.07.110
61. European Commission: Handbook on the external costs of transport (2019a). https://op.europa.eu/en/publication-detail/-/publication/9781f65f-8448-11ea-bf12-01aa75ed71a1
62. European Commission: Sustainable transport infrastructure charging and internalisation of transport externalities (2019b). https://op.europa.eu/it/publication-detail/-/publication/e0bf9e5d-a386-11e9-9d01-01aa75ed71a1
63. Bruzzone, F., Scorrano, M., Nocera, S.: The combination of e-bike-sharing and demand-responsive transport systems in rural areas: a case study of Velenje. Res. Transp. Bus. Manag. 100570 (2020). https://doi.org/10.1016/j.rtbm.2020.100570
64. Nocera, S., Pungillo, G., Bruzzone, F.: How to evaluate and plan the freight-passengers first-last mile. Transp. Policy (2020). https://doi.org/10.1016/j.tranpol.2020.01.007
65. Bruzzone, F., Cavallaro, F., Nocera, S.: The integration of passenger and freight transport for first-last mile operations. Transp. Policy **100**, 31–48 (2021). https://doi.org/10.1016/j.tranpol.2020.10.009
66. Nocera, S., Cavallaro, F., Irranca, G.O.: Options for reducing external costs from freight transport along the brenner corridor. Eur. Transp. Res. Rev. **10**(2), 5 (2018)

ITSs for Transnational Road Traffic Detection: An Opportunity for More Reliable Data

Federico Cavallaro[1]([⊠]), Ilaria De Biasi[2], and Giulia Sommacal[3]

[1] Politecnico di Torino, Interuniversity Department of Regional and Urban Studies and Planning, Viale Pier Andrea Mattioli 39, 10125 Turin, Italy
federico.cavallaro@polito.it
[2] Autostrada del Brennero SpA, Via Berlino 10, 38121 Trento, Italy
ilaria.debiasi@autobrennero.it
[3] Eurac Research, Institute for Regional Development, Viale Druso 1, 39100 Bolzano, Italy
giulia.sommacal@eurac.edu

Abstract. Reliable data collection is an unavoidable requisite for rigorous planning of transport policies that aim at achieving a more balanced modal split. This is particularly evident for long-distance road traffic, which in most cases involves two or more countries. Since each country adopts its own method for counting and classifying vehicles, the risk of using inconsistent data for common purposes is high. Intelligent Transport Systems may be a valid support for providing a standardized and effective method to solve this issue, provided that there is a preliminary agreement on the technological solutions and the location where to install them. This paper compares the methods currently adopted by different stakeholders and authorities in Italy and Austria to count transnational heavy and light vehicles along the Brenner corridor, which is one of the main European transnational axes in terms of traffic volumes. In line with this analysis, differences in the results could be noted, according to the adopted method and their position. Subsequently, a solution to achieve a more consistent dataset is proposed, based on common classification and shared technology to detect vehicles. This agreement is fundamental for policy purposes: the main aim of the Italian-Austrian Euregio Tyrol-South Tyrol-Trentino in the transport sector is to invert the current long-distance modal split by 2035, a result that can be achieved only if a correct and shared counting system is adopted.

Keywords: Traffic flow detection · ITS · Transport policies · Brenner corridor · Modal split

1 Introduction

In recent years, several transport policies have been proposed for the purpose of achieving a more balanced modal split. These policies are mainly directed at long-distance road traffic, which in most cases involves two or more countries. According to the White Paper on Transport [1], this type of journey is a typical target for a shift from road to rail, especially in delicate environmental and social contexts like the Alps [2]. Indeed,

© Springer Nature Switzerland AG 2021
O. Gervasi et al. (Eds.): ICCSA 2021, LNCS 12950, pp. 350–362, 2021.
https://doi.org/10.1007/978-3-030-86960-1_25

railway guarantees reduced externalities and less negative impacts for the communities crossed by its infrastructures. This holds particularly true for freight transport, where, on average, the economic difference in externalities between 1 tkm by road and rail is higher than 2€ [3]. By considering that in 2019 the total amount of freight that circulated by road along the Alps was 160.3 Mt and the length of transalpine corridors is more than 300 km (Brenner 415 km, Gotthard 288 km), the high external costs can be easily quantified.

Implementing modal shift policies in favour of rail in the Alps is one of the main targets of the Action Group 4 Mobility of EUSALP, the macro-regional strategy for the Alpine area [4]. However, to check progress in the modal shift, solid knowledge about road traffic volumes is a fundamental requisite. Detecting international traffic is not as trivial as it may appear, since each Country adopts its own method to classify and, consequently, to count vehicles [5]. Therefore, the risk of using inconsistent data and making analyses not based on shared values is concrete. When we refer to the transalpine transport, 5 countries (France-FR, Switzerland-CH, Austria-AT, Germany-DE and Italy-IT) and even more counting systems are considered. The latter may depend not only on the nation, but also on the data providers (e.g., regions, municipalities, companies that manage the motorways), on the scope of measurement (e.g., for fiscal or traffic monitoring reasons) and on the adopted methods (e.g., inductive loops or video detection systems).

Intelligent Transport Systems (ITS) could be a valid support for providing a standardized and effective method to solve this issue, provided that there is a preliminary transnational agreement about the technological solutions to be adopted and a shared definition of their location. The Directive 2010/40/EU [6] promotes the use of standardised information flows or traffic interfaces across borders, but in most cases alternative solutions coexist and are used alternatively.

The aim of this paper is to describe the condition of traffic detection along the Italian and Austrian sections of the Brenner corridor, the transalpine corridor with the highest traffic flows [7], and to propose a solution based on ITS to find more reliable values. The paper is structured as follows: after this introduction (Sect. 1), Sect. 2 discusses the role of ITS in traffic monitoring and management. Section 3 focuses on the Brenner corridor and identifies adopted methods for the classification of vehicles currently available. Section 4 analyses the road traffic volumes collected for the Brenner corridor, both for the Italian and Austrian motorways, showing differences in absolute values and in relation to specific subcategories of vehicles. Section 5 presents and discusses a potential solution to overcome the issue of trans-national traffic counting in this area.

2   Modern Technologies for Traffic Flow Detection

Among the wide variety of ITS and their applications both for passenger and freight transport, road traffic management systems can be helpful solutions to handle data collection [8]. Modern systems guarantee automatic classification of vehicles, which is essential for adopting adequate and immediate measures in case of unexpected impacts (e.g., road accidents, traffic congestion). According to [9], two main technologies for traffic flow detection can be identified: (i) intrusive sensors and (ii) non-intrusive sensors. The former are installed directly on or under the roadway surface (in-roadway sensors), while the latter are mounted above the roadway on sign bridges or to the side of the roadway on poles or other structures (over-roadway sensors). According to the adopted technology, sensors can detect electromagnetic energy (infrared spectrum, radio frequency spectrum, visible spectrum, microwave spectrum, or millimetre-wave spectrum) or acoustic and ultrasonic energy. In addition, the sensors can be passive (P) if they detect reflected energy from a source, or active (A), if they both emit a pulse of energy and detect reflected energy. With respect to the above specifications, inductive loops (A), magnetometers (P) and magnetic sensors (P) are intrusive sensors, while video detection systems (P), microwaves, like presence-detecting radars (A), or doppler sensors (A), acoustic sensors (P), laser radar (A), passive infrared (P), ultrasounds (A) and technology combinations (A + P) are non-intrusive sensors. The adoption of a technical solution depends on several factors, including the costs of purchase, installing, operating, and maintaining the system, the details of the traffic information to be collected and its accuracy. For instance, information about traffic flow can be gathered according to highly selective vehicle classification, to distinguish different classes based on axle counts and accurate measurement of the distances between wheels and/or their speed detection.

The inductive loop sensors are less expensive. Being installed under the road surface, they require a lane closure for their installation and maintenance, but they have a long operational life and at the same time, being invisible, may overcome issues related to the perception of drivers. Vehicle magnetic acting on loop detector parameters allows to detect and count axles correctly even if an axle is lifted. Nowadays, this solution is widely applied in traffic data collection, in part because of institutional inertia and know-how, but also for the limited costs [10]. However, the risk of collecting incomplete or inaccurate data is not negligible [11]. On the other hand, non-intrusive sensors are rapidly evolving and becoming easier to use. Compared to the previous solution, their advantage lies mainly in the minimization of interference with traffic flow and reduction of safety risk. Furthermore, they represent a valid alternative in particular sections of the line (e.g., bridges and tunnels), where pavement cutting and boring is difficult or undesirable. Current non-intrusive sensors are more accurate than inductive loop sensors for detecting volume and speed data [12]. In both cases, sensitivity to environmental conditions, traffic flow and road design can challenge accuracy, reliability, and functionality of the detectors. This preliminary defines the most suitable solutions according to the characteristics of the area to be analysed.

3   The Brenner Corridor and the Measuring Stations Located Near the Border

The Brenner corridor, which is part of the TEN-T Scandinavian-Mediterranean European Core Corridor, is a transalpine multimodal (road-rail) axis that goes from Munich (DE) to Verona (IT). This corridor has the highest freight volumes transported per year (about 53.7 Mt in 2019) [7]. Its volumes are growing consistently: an increase by 28% has been recorded since 2010. The volumes transported by road over this period are about 40 Mt (+45% compared to 2010) and are twice as high as in the second transalpine corridor in terms of volumes (Ventimiglia, 21 Mt). Thus, transalpine road freight traffic is an important component of overall volumes at Brenner.

Policies to invert the current modal share by 2035 constitute one of the pillars of the Euroregion Tirol-South Tyrol-Trentino [13], also based on the assumption that Heavy Goods Vehicles (HGVs) circulating are monitored regularly over time. In particular, for the purposes of modal shift it is fundamental to pinpoint the transalpine component. So far, reports and technical evaluations are based on Austrian data, but this simplified choice could have implications in terms of overall number of vehicles. The detection of HGVs that pass through the latest Austrian cross-border detection point but end their trip before the boundary could lead to an overestimation of the values. On the contrary, the presence of the RoLa terminal at Brennersee, which guarantees a combined transport service for the Austrian part of the infrastructure (daily connections with Wörgl) may lead to an underestimation of values, since the detection point is located before the terminal (southward direction). The ideal solution would be comparison of volumes registered at the traffic detection points that are closest to the Brenner border in AT and in IT, and to distinguish between Light Vehicles (LVs) and Heavy Vehicles (HVs). If values are similar, this would be a good indicator of the transalpine flows that circulate along the Brenner. Unfortunately, this is not the case, mainly due to the differences in counting and aggregating vehicles.

The stretches of the motorway A22 [14] that are used to compare AT and IT values are: i) the South-Tyrolean section of the A22-Autostrada del Brennero (IT) from the Brenner Toll station located near Vipiteno to the IT-AT border (16 km) and ii) the Tyrolean section of the Brennerautobahn A13 (AT) from the IT-AT border to Schönberg im Stubaital (25 km), the last toll station available in AT direction southwards. Available data are provided by A22-Autostrada del Brennero for the motorway A22 [15] and by ASFINAG for the motorway A13 [16]. State roads (i.e., SS12 in IT and B182 in AT) are not considered in our analysis. Besides the fact that they are not appropriate for long-distance travel, it would be not possible to consider them as a valid alternative: indeed, in AT there is a transit ban for vehicles above 3.5t; whereas in IT, some segments of the road are banned for vehicles above 20t (including the section between Brenner and Vipiteno, the object of our analysis).

Data are collected through different systems, which can be classified into five main categories (see Fig. 1). For each system, it is also indicated if it uses (i) intrusive and/or (ii) non-intrusive technology. In IT, A22 adopts two different methods, (ID1) toll stations and (ID2) on-road measurement systems. In AT, ASFINAG uses three different methods: (ID3) overhead sensors with triple technology, (ID4) on-road measurement systems and (ID5) toll stations. In detail:

- In IT, the solution adopted in A22 toll stations (ID1) combines intrusive (Weigh In Motion systems) and non-intrusive sensors (overhead devices that encode transit details) and classifies the vehicles according to the criterion "*assi-sagoma*" (axels-shape) [15]. It distinguishes among five classes: two for vehicles with two axles and three for vehicles or vehicle combinations with three or more axles. This system considers LVs as those belonging to ClassA–2 axles / height $\leq$ 1.3 m and HVs those belonging to Class B–2 axles / height > 1.3 m, Class3–3 axles, Class4–4 axles and

Class5–5 or more axles. In this classification, cars with a trailer and articulated trucks belong to the same Class 4.

- In the A22 on-road measurement system (ID2), the number of vehicles circulating is detected by inductive loops (intrusive sensor) located along the motorway, which collect information on instantaneous speed and vehicle type (LVs or HVs). This system classifies all vehicles ≤ 5.25 m as LVs and all vehicles > 5.25 m as HVs. This means that cars above 5.25 m are considered as HVs.

- In AT, the overhead sensors with triple technology (ID3), i.e. the TDC3–8 device (non-intrusive sensor) [17] are installed in different sites located along the A13. This system controls traffic flows thanks to three sensors installed in one device, which combines ultrasounds (to detect vehicle height profile), passive infrared (to reveal vehicle width and lane position), and doppler radar (to provide information on vehicle speed). The TDC3–8 device, mounted on overhead structures above the centre lane, classifies vehicles into 8 + 1 vehicle classes according to the standardized German TLS (Technical delivery terms for roadway stations): 1) other vehicles; 2) motorbikes; 3) cars; 4) vans; 5) cars with trailer; 6) buses; 7) HGVs; 8) HGVs with trailer; 9) HGVs articulated.

- The A13 on-road measurement system (ID4) detects the passage of vehicles through inductive loops (intrusive sensors) located along the motorway. The inductive loops classify the vehicle according to a derived shape profile, which can be compared with sample profiles. This system gathers information on instantaneous speed and vehicle type and counts vehicles on the basis of 8 + 1 vehicle classes: 1) other vehicles; 2) motorbikes 3) cars; 4) vans; 5) cars with trailer; 6) buses with more than nine seats; 7) HGVs; 8) HGVs with trailer; 9) HGVs articulated; [18].

- Overhead devices (non-intrusive sensors) are mounted in the A13 Toll station (ID5) to recognize both LVs (i.e. all vehicles with a maximum permissible weight ≤ 3.5t) and HVs (i.e. motor vehicles with a maximum permissible weight > 3.5t). HGVs and buses must be equipped with a GO-Box, an electronic device that uses microwave technology to communicate with the toll portals. The overhead device classifies the vehicles according to 4 categories: 1) Category 1: all vehicles ≤ 3.5t (i.e. cars, motorbikes and light camper vans); 2) Category 2: motor vehicles with max. 2 axles; 3) Category 3: motor vehicles with max. 3 axles; 4) Category 4: motor vehicles with 4 and more axles [16].

Along the aforesaid stretch of the Brenner motorway, five measuring stations are available (see Fig. 2). Two stations are located along the motorway A22 (Brenner Toll station and Brennero-Vipiteno) and 3 measuring stations on the motorway A13 (Brennersee/A13, Matrei am Brenner and Schönberg Toll station). The measuring stations of Brennero-Vipiteno and Matrei am Brenner use intrusive sensors based on inductive loops installed in the respective motorway sections (ID2 and ID4). The Brenner Toll station combines intrusive and non-intrusive sensors (ID1), while the Schönberg Toll station and the station of Brennersee/A13 adopt two different methods based on non-intrusive sensors: respectively, the detection toll system (ID5) and a triple technology detector system (ID3). Therefore, all technical solutions presented in Fig. 1 are visible in our study area and the information each of them provides can be compared.

Solution (ID)	Traffic flow detection	Technology*	Classification of vehicles and distinction between LVs and HVs
1	A22 Toll station (IT)	i and ii	The vehicles are classified in 5 classes: 1) ClassA-2 Axles/Height≤1.3 m (LVs); 2) Class B-2 Axles/Height>1.3 m (HVs); 3) Class3-3 Axles (HVs); 4) Class4-4 Axles (HVs); 5) Class5-5 or more Axles (HVs).
2	A22 On road measurement system (IT)	i	The vehicles are classified in 2 classes: 1) All vehicles ≤ 5.25 m (LVs); 2) All vehicles > 5.25 m (HVs).
3	A13 Triple Technology Detector System (AT)	ii	The vehicles are classified in 8+1 classes: 1) Other vehicles; 2) Motorbikes (LVs); 3) Cars (LVs); 4) Vans (LVs); 5) Cars with trailer (HVs); 6) Buses (HVs); 7) HGVs (HVs); 8) HGVs with trailer (HVs); 9) HGVs articulated (HVs).
4	A13 On road measurement system (AT)	i	The vehicles are classified in 8+1 classes: 1) Other vehicles; 2) Motorbikes (LVs) 3) Cars (LVs); 4) Vans (LVs); 5) Cars with trailer (HVs); 6) Buses with more than nine seats (HVs); 7) HGVs (HVs); 8) HGVs with trailer (HVs); 9) HGVs articulated (HVs).
5	A13 Toll station (AT)	ii	The vehicles are classified in 4 classes: 1) All vehicles ≤ 3,5t (LVs); 2) Motor vehicles with max. 2 axles (HVs); 3) Motor vehicles with max. 3 axles (HVs); 4) Motor vehicles with 4 and more axles (HVs).

* intrusive (i) and/or non-intrusive technology (ii).

Fig. 1. Main characteristics of the adopted method for classification of vehicles along the Brenner corridor.

The raw data, collected in real time by A22-Autostrada del Brennero and ASFI-NAG in each station is processed by the motorway operators and the information is then aggregated in different ways, depending on the source that publishes the data (motorway operator or Province/Land involved). The aim is to distinguish between LVs and HVs, but based on counting methods, results may vary. Figure 2 identifies for each ID the measuring station, specifying its distance from the IT-AT border and the different classification of vehicles according to the source. LVs are identified with x.1, HVs with x.2 and subclasses of HVs with x.2.y, where x stands for a letter between A and D and y is for a number, according to the number of available subclasses.

ID	Station (Nation)	Km + m	Source	LVs	HVs / HVs subclasses
1	Brenner Toll station (IT)	15+ 850	A22	**A.1**: ClassA- 2Axles/ Height≤1.3 m	**A.2**: All vehicles belonging to the classes B, 3, 4 and 5 **A.2.1**: Class B- 2 Axles/ Height >1.3 m **A.2.2**: Class 3-3 Axles **A.2.3**: Class 4 - 4 Axles **A.2.4**: Class 5 - 5 or more Axles
2	Brennero- Vipiteno (IT)	0+ 200	A22	**B.1**: All vehicles ≤ 5.25 m	**B.2**: All vehicles > 5.25 m
3	Brennersee/ A13 (AT)	5+ 800	ASFINAG Land Tyrol	**C.1**: All vehicles ≤ 3.5t **D.1** *	**C.2**: All vehicles > 3.5t **D.2**: HV-similar vehicles (such as car with trailer, van with trailer, buses with more than nine seats, HGVs with or without trailer, HVGs articulated) **D.2.1**: HVGs with or without trailer, HVGs articulated **D.2.2**: HVGs with trailer, HVGs articulated
4	Matrei am Brenner (AT)	18+ 800	ASFINAG Land Tyrol	**C.1**: All vehicles ≤ 3.5t **D.1** *	**C.2**: All vehicles > 3.5t **D.2**: HV-similar vehicles (such as car with trailer, van with trailer, buses with more than nine seats, HGVs with or without trailer, HVGs articulated) **D.2.1**: HVGs with or without trailer, HVGs articulated **D.2.2**: HVGs with trailer, HVGs articulated
5	Schönberg Toll station (AT)	23+ 400	Land Tyrol	**D.1** *	**D.2**: HV-similar vehicles (such as car with trailer, van with trailer, buses with more than nine seats, HGVs with or without trailer, HVGs articulated) **D.2.1**: HVGs with or without trailer, HVGs articulated **D.2.2**: HVGs with trailer, HVGs articulated

* LVs are not identified in the table of the traffic in Land Tyrol [17].

Fig. 2. Identification of the five measuring stations at the border section of the Brenner corridor and classification of vehicles.

4   Road Traffic Volumes Detected Along the Brenner Corridor

According to available traffic detection systems, it is now possible to quantify the number of LVs and HVs detected per day by each system and compare them. Figure 3 presents disaggregated information for the year 2019. The subdivision of vehicles in classes and subclasses is derived from the classifications identified in Fig. 2. Measuring stations in IT provide comparable data in terms of traffic for all vehicles (30,637 vehicles/day for ID1 and 30,513 for ID2), while Austrian data are generally higher: the ID3 collects 32,375 vehicles/day (+6% compared to ID1 and ID2), whereas ID4 and ID5, which are farther from the boundary and include also traffic generated or headed to the Austrian villages located until the boundary, count 37,805 and 38,276 vehicles/day (respectively, +23.5% and +25% compared to ID1 and ID2). Passing from the absolute number to the subdivision between LVs and HVs, some interesting aspects should be noted (see Fig. 4). Italian detection points registered a number of HVs higher than 11,000; in AT this value is lower than 8,000 for ID3 - the exact number depending on the source: 7,377 according to ASFINAG, 7,979 according to Land Tyrol. Even in ID4 and ID5, which presented overall higher values of vehicles (see Fig. 4), the number of HVs is lower than in IT: 7,584 according to ASFINAG and 8,343 according to Land Tyrol (ID4). As previously mentioned, this may depend on the fact that the Italian systems count all those vehicles (or vehicle combinations) that are either longer than 5.25 m, or are higher than 1.3 m, as HVs. All cars with trailers, as well as large-size cars and SUVs are therefore classified as HVs, contributing to explaining the significant differences between IT and AT values.

ID	Station (Nation)	Source	LVs per day	HVs per day HV subclasses	LVs + HVs per day
1	Brenner Toll station (IT)	A22	A.1: 19,222	A.2: 11,415 A.2.1: 3,462 A.2.2: 861 A.2.3: 366 A.2.4: 6,726	A.1+A.2: 30,637
2	Brennero-Vipiteno (IT)	A22	B.1: 19,293	B.2: 11,221	B.1+B.2: 30,513
3	Brennersee/A13 (AT)	ASFINAG Land Tyrol	C.1: 24,998 D1*: 24,396	C.2: 7,377 D.2: 7,979 D.2.1: 6,961 D.2.2: 6,548	C.1+C.2: 32,375 D1+D.2: 32,375
4	Matrei am Brenner (AT)	ASFINAG Land Tyrol	C.1: 30,225 D1*: 29,466	C.2: 7,584 D.2: 8,343 D.2.1: 7,269 D.2.2: 6,724	C.1+C.2: 37,809 D1+D.2: 37,809
5	Schönberg Toll station (AT)	Land Tyrol	D1*: n.a	D.2: n.a D.2.1: n.a D.2.2: 6,766	D1+D.2: 38,276

* LVs are calculated as the difference between Total Vehicles (D1+D2) and HV-similar vehicles (D.2).

Fig. 3. Road traffic volumes (vehicles/day) detected along the border section of the Brenner corridor – year 2019.

Repeating the same procedure for the entire period 2010–2019 produces the aggregate data presented in Fig. 5. All general comments related to the year 2019 are still valid, but some differences are also evident. In terms of total vehicles, the Italian counting systems presented higher differences in years 2010–2013, while in years 2014–2019, they registered roughly the same amount. Considering Austrian values registered at Brennersee/A13, which is the Austrian measuring station less influenced by domestic traffic, they are lower than Italian values until 2011, then they become higher. This could also

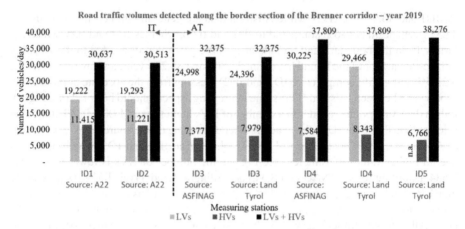

Fig. 4. LVs and HVs detected along the border section of the Brenner corridor – year 2019.

be due to the sectoral driving ban applied to the Austrian stretch of the Brenner motorway, valid only until 2011. Regarding LVs, values registered at Brennersee/A13 for year 2012–2019 are always higher than those of the two Italian measuring stations.

At a first glance, Italian data are more consistent: toll stations and road measurement systems differ by some tens or units, and this is likely due to the inaccuracy of the loops, rather than the toll system, which implies the passage from a protected toll gate. However, recalling the classification provided in Fig. 1, Italian data count all vehicles with more than 2 axles and heigh > 1.3 m (toll stations) or all vehicles longer than 5.25 m (on-road measurement system) as HVs. These categories also include passenger vehicles (with or without trailer) and buses that may not be assimilated to freight vehicles. Thus, the Italian method for classifying vehicles seems inappropriate for the purposes of this paper, and it is not possible to obtain a different classification derived from the available solutions.

On the other hand, the Austrian counting system is more rigorous in the classification of vehicles (accurate classes are used, and it is possible to consider HVs separate from passenger cars with trailers). However, some measuring stations (e.g., ID5) are quite far from the national boundary and domestic traffic or traffic with either origin or destination in the areas located along the stretch between Schönberg toll station or Matrei am Brenner and Brenner are also counted. For this reason, the counting system at Brennersee/A13, which is located after the last exit of the motorway A13 (Nößlach exit) before the Brenner Pass border, seems the best source for counting and classifying the vehicles on the Austrian side. However, there is a risk of ignoring those vehicles from the RoLa at the intermodal terminal of Brennersee (which is located close to the border, after the detection point), which provides about 20 daily connections from Brenner to Wörgl (and vice versa) during the week. From this analysis, for different reasons it is apparent that any of the systems currently available provide reliable information about the transalpine traffic flows.

ID	Station (Nation)	Km+m	Source	Classification of vehicles ***	2009 ***	2010 ***	2011 ***	2012	2013	2014	2015	2016	2017	2018	2019
1	Brenner Toll station (IT)	15+850	A22 - Brenner Motorway	LVs	17,217	17,469	17,766	17,417	17,883	17,982	18,531	19,339	19,608	19,498	19,222
				HVs	8,064	8,462	8,536	8,545	8,642	8,971	9,308	9,842	10,818	11,220	11,415
				LVs + HVs	25,281	25,931	26,302	25,961	26,525	26,953	27,839	29,181	30,426	30,718	30,637
2	Brennero Vipiteno (IT)	0+200	A22 - Brenner Motorway	LVs	17,255	18,788	18,639	17,927	18,625	18,247	18,846	19,508	19,717	19,602	19,293
				HVs	7,746	8,255	8,708	8,792	8,781	8,608	8,908	9,568	10,573	11,010	11,221
				LVs + HVs	25,001	27,043	27,347	26,719	27,406	26,855	27,754	29,076	30,290	30,613	30,513
3	Brennersee/ A13 (AT)	5+800	ASFINAG	LVs	n.a.	18,069	18,235	22,624	n.a.	21,569	22,508	23,638	25,124	25,125	24,998
				HVs	n.a.	6,162	6,279	5,624	5,589	5,813	6,026	6,374	6,786	7,202	7,377
				LVs + HVs	n.a.	24,231	24,514	28,248	n.a.	27,382	28,534	30,012	31,910	32,327	32,375
			Land Tyrol	LVs*	18,604	18,007	18,235	n.a.	n.a.	n.a.	n.a.	22,748	24,037	24,363	24,396
				HVs	6,162	6,224	6,279	n.a.	n.a.	n.a.	n.a.	7,266	7,875	7,964	7,979
				LVs + HVs	24,766	24,231	24,514	n.a.	n.a.	n.a.	n.a.	30,014	31,912	32,327	32,375
4	Matrei am Brenner (AT)	18+800	ASFINAG	LVs	24,696	n.a.	25,469	25,156	29,122	27,899	28,714	29,755	31,219	30,581	30,225
				HVs	6,063	n.a.	6,514	5,611	5,545	5,840	6,065	6,482	6,996	7,461	7,584
				LVs + HVs	30,759	n.a.	31,983	30,767	34,667	33,739	34,779	36,237	38,215	38,042	37,809
			Land Tyrol	LVs*	24,696	n.a.	25,469	24,175	n.a.	27,315	27,982	28,979	n.a.	29,787	29,466
				HVs	6,063	n.a.	6,514	6,593	n.a.	6,425	6,798	7,259	n.a.	8,257	8,343
				LVs + HVs	30,759	n.a.	31,983	30,768	n.a.	33,740	34,780	36,238	n.a.	38,044	37,809
5	Schönberg Toll station (AT)	23+400	Land Tyrol	LVs*	n.a.	n.a.	n.a.	n.a.	n.a.	n.a.	n.a.	n.a.	n.a.	n.a.	n.a.
				HVs**	4,363	4,600	4,760	4,875	4,821	5,090	5,292	5,707	6,180	6,637	6,766
				LVs + HVs	31,495	32,580	33,162	33,155	33,246	34,205	35,233	36,549	38,647	38,709	38,276

* LVs are calculated as the difference between Total Vehicles (D1+D2) and HV-similar vehicles (D.2); ** Only this data is published by Land Tyrol [17] for HVs (HV subclass D.2.2); *** Online data published by ASFINAG is only available up to 2012. To complete the dataset, iMONITRAF! data was included for years 2009-2011.

Fig. 5. Road traffic volumes (vehicles/day) detected along the border section of the Brenner corridor per year (2009–2019).

5 Conclusion: Installation of a New Traffic Measuring Station at the Italian Side of the Brenner Corridor

This paper has illustrated the methods currently adopted in IT and in AT to detect road vehicles along motorways, showing the significant differences in counting LVs and HVs. Data are managed by the respective motorway companies and road departments of the Land/Regions involved, but there is not a real sharing of such data, except for when they are published. This poses urgent issues in terms of harmonization and adoption of common values, to be used as reference to monitor the effectiveness of the policies introduced on both sides of the boundary.

Coherently with the technical analysis that we have performed in the previous sections, the TDC3–8 device (ID3) based on a classification into 8 + 1 classes (as available in the station of Brennersee/A13, see Sect. 3), provides accurate knowledge about traffic composition, allowing better differentiation between LVs and HVs. However, the lack of correspondent Italian data and its use in a point that also includes part of the domestic traffic in AT have prevented obtaining real knowledge about the transalpine vehicles circulating at Brenner.

An integration of this system, also on the Italian side, is a valid solution to guarantee more reliable counting. The idea of installing a new measuring station of this type in IT, close to the Austrian border, was proposed within the Euregio Working Group on traffic monitoring, where all the technical aspects described in this paper were discussed. It was subsequently agreed upon by the Autonomous Province of Bolzano, the Land Tyrol, the motorway A22 and ASFINAG [19]. Data gathered by the new measuring station can be easily compared with those of the station of Brennersee/A13, making accurate monitoring of the component of transit traffic crossing the Brenner corridor possible. The exact location of the new system, which classifies the vehicles in 8 + 1 classes according to TLS, is at km 0 + 130 of the A22, close to the boundary and the ID2 (km 0 + 200). Therefore, the domestic traffic is not detected, whereas the traffic from the RoLa at Brennersee is included, overcoming the main issues related to the Austrian counting stations. From a technical point of view, a new induction loop detector will be flanked by a TDC3–8 device and a system based on LiDAR sensors (2D laser scanners), which record the contour of the vehicles. All three detection systems provide traffic data according to 8 + 1 vehicle classes, allowing a comparison referred to the same point and a more coherent validation process. The installation of this system poses some management issues that are not trivial. Data collection, harmonization, and management are among them [20]. In this sense, the political agreement between local actors involved in this operation (ASFINAG, A22, Autonomous Province of Bolzano, Land Tyrol and Euregio) is an important step: it assigns the phases of data collection and harmonization to ASFINAG, with subsequent diffusion to all previously mentioned subjects. This should speed up the entire process, at the same time avoiding the risk of having different data depending on the source (a risk that cannot be ignored in this type of analysis, see Fig. 3 and Fig. 5). Regarding the economic aspects, the overall costs (labour + components) should obviously be considered in the evaluation. Costs for the system based on LiDAR sensors are almost four times higher than induction loops and TDC3–8 device, which, in turn, are of the same order of magnitude (10–15k€).

These ITS solutions could surely contribute to a more proper evaluation of the effectiveness deriving from the adoption of specific transport policies [21], in light of the ambitious objective of a modal shift mentioned in the introduction. Indeed, adequate and shared information about traffic flows can make it easier for policy makers to understand the impacts of specific transport measures and modify their mid- and long- term strategies accordingly, also considering an integration with the urban dimension [22]. Even if this is not the main purpose of our research, the proposed classification and counting of vehicles could also be helpful for daily traffic operations, especially during extraordinary weather conditions. In the geographic area in question, where extreme events are not unlikely (e.g., intense snowfalls in wintertime, severe rainstorms in summertime [23]), monitoring of hourly and daily vehicles that circulate could contribute to managing circulation and introducing appropriate temporary measures to reduce accident risks. Furthermore, this system could support traffic management in case of congestions. Again, this is a condition that is not negligible along the Brenner corridor, especially related to the transboundary section, mostly caused by the high number of HGVs. Thanks to rigorous detection of vehicles, adequate signals or even temporary closures of the toll stations could be foreseen, and information about parking areas located along the motorway could be provided.

The passage from technical to political issues is particularly relevant along the Brenner corridor, where in recent years the single provinces have often introduced specific measures valid only for their territories, but whose effects have also impacted neighbouring regions [24]. Sharing the vehicle counting system (at least at transalpine level) and using common data could be a first step towards integrated policies, which should be the real overall objective of this initiative.

References

1. EC, European Commission. White Paper. Roadmap to a Single European Transport Area – Towards a competitive and resource efficient transport system. EUR-Lex (2011). http://eur-lex.europa.eu/legal-content/EN/TXT/PDF/?uri=CELEX:52011DC0144& from=EN, Accessed 05 Mar 2021
2. Alpine Convention. Air quality in the Alps. Report on the state of the Alps (2020). https://www.alpconv.org/fileadmin/user_upload/Organization/AC/XVI/RSA8_final_draft_ACXVI.pdf, Accessed 05 Mar 2021
3. Cavallaro, F.: Policy implications from the economic valuation of freight transport externalities along the Brenner corridor. Case Stud. Transp. Policy **6**(2018), 133–146 (2018). https://doi.org/10.1016/j.cstp.2017.11.008. Accessed 05 Mar 2021
4. EUSALP, EU Strategy for the Alpine Region – Action Group 4 Mobility. Final report on external costs in mountain areas (2017). https://www.alpine-region.eu/results/study-external-costs-mountain-areas, Accessed 05 Mar 2021
5. TMS-DG07. Harmonising EU ITS Services, Deployment Guideline on Traffic Management Services - Traffic management plan for corridors and networks (2015). https://portal.its-platform.eu/index.php?q=filedepot_download/1729/5383, Accessed 05 Mar 2021
6. EU, European Union. Directive 2010/40/EU of the European Parliament and of the Council of 7 July 2010 on the framework for the deployment of Intelligent Transport Systems in the field of road transport and for interfaces with other modes of transport. EUR-Lex (2010). https://eur-lex.europa.eu/legal-content/EN/ALL/?uri=CELEX%3A32010L0040, Accessed 05 Mar 2021
7. Lückge, H., Heldstab, J., Maibach, M., Sommacal, G., Dianin, A., Skoniezki, P.: iMONITRAF! Annual Report 2020. New policy scenarios re-confirm the need for an ambitious and coordinated modal shift policy (2021). http://www.imonitraf.org/, Accessed 05 Mar 2021
8. Guerrero-Ibáñez, J., Zeadally, S., Contreras-Castillo, J.: Sensor technologies for intelligent transportation systems. Sensors (MDPI) **18**(4), 1212 (2018). https://doi.org/10.3390/s18041212, Access 03 Mar 2021
9. Klein, L.A.: ITS Sensors and Architectures for Traffic Management and Connected Vehicles. Taylor & Francis. Published by CRC Press in 2017 (2017). ISBN 9781138747371.
10. Gajda, J., Piwowar, P., Sroka, R., Stencel, M., Zeglen, T.: Application of inductive loops as wheel detectors. Transp. Res. Part C Emerg. Technol. **21**(1), 57–66 (2012). https://doi.org/10.1016/j.trc.2011.08.010, Access 05 Mar 2021
11. Nocera, S., Cavallaro, F.: Economic evaluation of future carbon dioxide impacts from Italian highways. Procedia Soc. Behav. Sci. **54**, 1360–1369 (2012). https://doi.org/10.1016/j.sbspro.2012.09.850
12. Yu, X., Prevedouros, P.D.: Performance and challenges in utilizing non-intrusive sensors for traffic data collection. Adv. Remote Sens. **2013**(2), 45–50 (2013). https://doi.org/10.4236/ars.2013.22006. 05 Mar 2021
13. EGTC, European Grouping of Territorial Cooperation. Resolution 01/2018 of the European Region Tyrol-South Tyrol-Trentino. Strategy for a modal shift of transport from road to rail on the Brenner axis (2018). http://www.europaregion.info/downloads/Beschluss_EVTZ_deliberazione_GECT_15.01.2018_DEF(1).pdf, Accessed 05 Mar 2021
14. Cavallaro, F., Maino, F., Morelli, V.: A new method for forecasting CO2 operation emissions along an infrastructure corridor. Eur. Transp. Trasporti Europei **55**(4) (2013). ISSN 1825–3997

15. A22 motorway. A22 Brenner Motorway Operator (Autostrada del Brennero S.p.A) – Homepage (2021). https://www.autobrennero.it/en/, Access 05 Mar 2021
16. ASFINAG. Austrian Motorway and Expressway Network Operator (Autobahnen- und Schnellstraßen-Finanzierungs-Aktiengesellschaft) – Homepage (2021). https://www.asfinag.at/, Accessed 05 Mar 2021
17. ADEC. ADEC Technologies AG – Homepage (2021). https://adec-technologies.ch/en/, Accessed 05 Mar 2021
18. Land Tirol. Traffic in Tyrol, Traffic reports (2021). https://www.tirol.gv.at/verkehr/publikationen-statistiken/publikationen-verkehr-mobilitaet/, Accessed 05 Mar 2021
19. EGTC, European Grouping of Territorial Cooperation. European Region Tyrol-South Tyrol-Trentino - Final report on the state of implementation of the Resolution n. 14 concerning strategic traffic measures for the reduction of air and noise pollution on the motorway section of the Brenner corridor, published in August 2019 (2019)
20. Cappelli, A., Nocera, S.: Freight modal split models: data base, calibration problem and urban application. WIT Trans. Built Environ. **89**, 369–375 (2006). https://doi.org/10.2495/UT060371
21. EC, European Council. Digitalisation of transport - Council conclusions (2018). https://www.consilium.europa.eu/en/press/press-releases/2017/12/05/digitalisation-of-transport-council-conclusions/, Accessed 05 Mar 2021
22. Nocera, S., Cavallaro, F.: A two-step method to evaluate the well-to-wheel carbon efficiency of urban consolidation centres. Res. Transp. Econ. **65**, 44–55 (2017). https://doi.org/10.1016/j.retrec.2017.04.001
23. Cavallaro, F., Ciari, F., Nocera, S., Prettenthaler, F., Scuttari, A.: The impacts of climate change on tourist mobility in mountain areas. J. Sustain. Tour. **25**(8), 1063–1083 (2017). https://doi.org/10.1080/09669582.2016.1253092
24. Nocera, S., Cavallaro, F., Irranca Galati, O.: Options for reducing external costs from freight transport along the Brenner corridor. Eur. Transp. Res. Rev. **10**(2), 1–18 (2018). https://doi.org/10.1186/s12544-018-0323-7

A Probability Estimation of Aircraft Departures and Arrivals Delays

Ivan Ostroumov[1]([✉]) [iD], Nataliia Kuzmenko[1] [iD], Olga Sushchenko[1] [iD],
Maksym Zaliskyi[1] [iD], Oleksandr Solomentsev[1] [iD], Yuliya Averyanova[1] [iD],
Simeon Zhyla[2] [iD], Vladimir Pavlikov[2] [iD], Eduard Tserne[2] [iD], Valerii Volosyuk[2] [iD],
Kostiantyn Dergachov[2] [iD], Olena Havrylenko[2] [iD], Oleksandr Shmatko[2] [iD],
Anatoliy Popov[2] [iD], Nikolay Ruzhentsev[2] [iD], Borys Kuznetsov[3] [iD],
and Tatyana Nikitina[4] [iD]

[1] National Aviation University, Kyiv, Ukraine
{ostroumovv,nataliiakuzmenko,sushoa,maximus2812,
avsolomentsev}@ukr.net, ayua@nau.edu.ua
[2] National Aerospace University H.E. Zhukovsky, Kharkiv, Ukraine
{s.zhyla,v.pavlikov,e.tserne,v.volosyuk,k.dergachov,
o.havrylenko,o.shmatko,a.v.popov}@khai.edu
[3] State Institution "Institute of Technical Problems of Magnetism",
National Academy of Sciences of Ukraine, Kharkiv, Ukraine
[4] Kharkiv National Automobile and Highway University, Kharkiv, Ukraine

Abstract. Delays in air transport systems have a significant impact on the safety of aviation. We propose to use a trajectory of airplanes to estimate time of airplane delay. Surveillance data is obtained using the Automatic Dependent Surveillance-Broadcast receiver. A network of software defined radios is used to receive position reports of particular airspace user transmitted by airplane transponder of Mode 1090ES. Linear regression with a spline function is used for airplane trajectory approximation and time synchronization. Delays are estimated based on rapid changing of airplane altitude with respect to runway elevation in comparison with a scheduled time. Statistical data of airplane take-off time, landing time, and flight duration based on numerous historical flights is analyzed. Probabilities of airplane delays during take-off and landing are estimated with a help of the Kernel density function. Proposed approach for probability estimation of aircraft departures and arrivals delays can be useful in air traffic management and airline planning for efficient usage of aviation transport system.

Keywords: Transport · Aviation · Air traffic delays · Big data · Automatic dependent surveillance-broadcast · Software defined radio · Airplane trajectory · Linear regression · Spline function · Kernel density estimator · Probability density function

1 First Section

Aviation is referred to one of the most developed transport domains. Numerous advantages of air transportation lead to a continuously increasing amount of carried cargo

© Springer Nature Switzerland AG 2021
O. Gervasi et al. (Eds.): ICCSA 2021, LNCS 12950, pp. 363–377, 2021.
https://doi.org/10.1007/978-3-030-86960-1_26

and passengers. It increases the number of flights and amount of aircraft which use airspace simultaneously [1, 2]. Current air traffic system operates in highly congested conditions, due to limited capacity of runways, airports, and flight routes net-work [3]. Operation of air traffic system in highly congested conditions may be easily affected by numerous factors which can lead to delays in transport services [4]. Airplane departure and arrival delays play an important role in the safety of transport system.

Airplane delays have huge economical and safety effects. An air traffic has al-ways a highly congested schedule. Many low-cost airlines would like to minimize time spent in airport facilities as less as possible. Thus, an unexpected delay causes delays for all further scheduled flights. On the other side, unexpected delays of one connecting flight cause serious problems to continue the journey and significant losses in airline profit. Also, delays have dramatical influence on aviation safety. Each flight is preplanned by the flight plan. Operation of air traffic system within scheduled time is connected with the low-risk environment due to planned traffic capacity at different airspace volumes. For example, delayed departure of an airplane may be a reason to put an airplane at holding pattern during arrival at a busy airport due to losing a scheduled landing window. All of that significantly reduce the safety of air transportation. Thus, operation of all airport services has to be provided based on delays minimization criteria. In common case, the safety of air transportation is based on accurate following of predefined trajectory [5–7] withing the defined time.

Statistical analysis of delays helps to identify the main factors which affect air trans-portation service to minimize its influence into the future air traffic system [8]. Numerous studies consider data processing of delays based on departure and arrival times avail-able in the air traffic control system [9–11]. In this case, the data processing includes histogram analysis and data fitting by Gaussian probability density function. But, numer-ous factors affecting departure and arrival times with different probability of occurrence taking the assumption about Gaussian noise may degrade performance of estimation.

In our study, we would like to propose an approach for airplane departure and arrival delay estimation based on trajectory data processing. Surveillance data can be eas-ily obtained under the Automatic Dependent Surveillance-Broadcast concept. Statisti-cal analysis of flight based on numerous trajectories helps to identify delays exactly. The probabilities of airplane delays can be estimated based on the probability density function.

2   The Main Causes of Delays in Air Transportation

A statistic of delays is dramatically increased over the European airspace [12]. A con-tinuously increasing tendency is the same for both departures and arrivals. Statistics of average delay per departure and arrival in more than 5 min of a total amount of flights is represented in Fig. 1 [12, 13].

Based on data in Fig. 1 the amount of delayed aircraft is constantly increasing until 2018, were reached its maximal value for departure in 48.4% and 42.8% for the arrival of total flights over Europe. It means that about a half of total flights was delayed in more than 5 min. Also, based on this statistics, the amount of delayed airplanes during departures is always more than for arrivals.

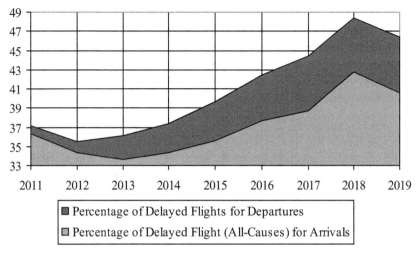

Fig. 1. Statistics of delayed flights for departures and arrivals.

According to the International Air Transport Association (IATA) regulation, it is possible to classify the causes of delays into the main seven groups [12]:

– Airline. Consists of delays caused by passenger and baggage carrying side; cargo loading; airplane and ramp handling; technical problems and aircraft equipment failure; damage of airplane part; flight operations and human factor.
– Airport. Includes delays initiated by airport facilities; restrictions at departure and arrival airports.

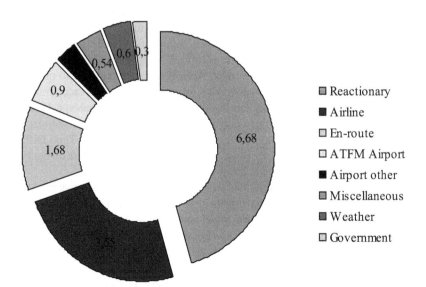

Fig. 2. An average minutes of delay per flight per categories for 2018.

– En-route. Air traffic flow management (ATFM) due to Air Traffic Control (ATC) en-route demand; ATC staff and ATFM equipment.
– Governmental. Includes delays from security and immigration reasons.
– Weather.
– Miscellaneous.
– Reactionary. Is a result of the late arrival of aircraft, passengers, cargo, or crew.

Due to airplane operation in quite a busy environment, any significant primary delay causes a reactionary one. For example, an airplane arrival delay automatically cases delays for all of the further flights of the same airplane during the scheduled time. Average time of delay per flight caused by different factors is represented in Fig. 2 [12].

3 Trajectory Data Processing

Nowadays, the Automatic Dependent Surveillance-Broadcast (ADS-B) concept is considered as the most prospective surveillance technology in the future. According to ADS-B, each airspace user has to be equipped with equipment for transmitting a data message continuously. ADS-B message includes user position report which includes actual coordinates in Lattitude, Longitude, and barometric Altitude (LLA) measured by on-board sensor [14]. The most frequently used on-board equipment of ADS-B is the Air Traffic Control Radar Beacon System (ATCRBS) or airplane transponder of Mode 1090ES [15]. An ADS-B message can be easily received by multiple Software Defined Radios (SDRs) and decoded to obtain exact airplane coordinates [16–18]. There are multiple networks of ADS-B receivers over the globe including national networks of Air Navigation Service Providers (ANSP). A radio waves of Very High-Frequency (VHF) band attenuate in airspace. The service area of each ADS-B ground station depends on transmitting power, a gain function of receiving antenna system, SDR performance, attenuation in the troposphere, multipath from high-altitude construction, and relief. Thus, maximum operational range of ADS-B communication line is limited to 400 NM, approximately. A network of ground stations is used to get stable airplane tracking over the whole flight [19]. Received and decoded ADS-B messages are sent to the data collecting and processing unit automatically. Also, some ground receivers are accompanied by a receiver of Global Navigation Satellite System (GNSS) for time synchronization (see Fig. 3). Received messages from one airplane by different ground stations, which are synchronized by time, allow measuring of airplane coordinates by Time Difference of Arrival (TDOA) method [20, 21]. Thus, some networks of ADS-B data do not provide receiving data from on-board sensor, but provide airplane localization.

According to airplane operation rules, an airplane transponder is active immediately after undocking the airplane. According to that, the ADS-B data sequence includes airplane path during taxing, take-off, en-route, landing, and taxing in destination airport up to airplane docking to particular gate. Usage of ADS-B transponder for ground operations helps the airport system to provide an exact tracking and traffic control within ground facilities [22].

The whole sequence of traffic data (ground and air) is useful to detect the exact departure, take-off, landing, and arrival phases of an airplane [19]. Departure time can

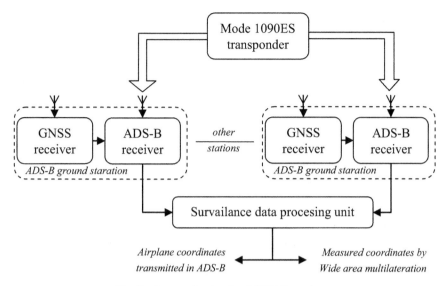

Fig. 3. A ground network of ADS-B receivers.

be obtained from the moment of initial airplane coordinates changing, based on starting moving from the gate. Time of take-off and landing can be estimated based on criteria of equal altitudes of airplane and runway. Arrival time is a time of stopping movement after landing.

In order to estimate the probability of airplane delays, we have to use various realizations of a particular flight. Regular airline connections are scheduled and usually are performed by the same flight-plan and a particular time of a day. Thus, statistical data processing of airplane trajectory from different realizations of particular flight connection creates statistics of airplane departures and arrivals. Obtained statistics can be used to estimate delays based on scheduled time. Also, the results of the statistical analysis give a possibility for efficient planning to reduce delays in future flights.

Transmitted message from on-board equipment is not synchronized in time and each user can use different settings for a time interval between transmissions. The majority of ATCRBS transponders uses reporting frequency in 1 Hz, but some of them can vary the time between transmission in the range from 2 s to 17 s. Also, some messages may be corrupted and lost during transmission. Thus, finally obtained sequence of coordinates is not synchronized. Interpolation of data can be used at the stage of the post-flight data processing to obtain a synchronized sequence of trajectory data [23].

For trajectory data processing we propose to use linear regression with a spline function. Spline interpolation is currently one of the most accurate methods of approximation. The widespread use of splines is due to their high approximation accuracy, which is very important factor in solving practical problems [24, 25].

The spline function is a piece-wise smooth cubic polynomial function, the first and second derivatives of which are continuous in different parts of the curve. Splines are functions that are "glued" from "pieces" of polynomials. The spline curve is given by a set of coordinates of points called control points and indicate the general shape of the curve. According to these points, a piecewise continuous parametric polynomial function is selected. An interpolation function is a curve created by joining all control points with polynomial parts.

B-spline is the most commonly used type of spline functions. They are formed by approximating a set of control points. The degree of the B-spline polynomial can be set regardless of the number of control points, and they also allow local control over the shape of the curve. In addition to local control, splines allow you to change the number of control points used to construct the curve without changing the degree of the polynomial.

Cubic B-splines, i.e. splines of the third order, were selected as the most effective taking into account the following factors:

1. Polynomials of lower degrees give very low flexibility in controlling the shape of the curve. First-order B-splines do not give a satisfactory smoothness of the approximating curve. Second-order B-splines give a smooth curve, but there is a problem at the points of connection of the curve segments. Second-order B-splines do not guarantee satisfactory continuity at the points of union of the curve segments. The use of cubic B-splines, which are C^0, C^1, and C^2 continuous can solve this problem.
2. Higher polynomials require more time in the computational process and can lead to unwanted approximation jumps.
3. Cubic B-splines allow obtaining sufficient continuity required for processing experimental data.

Therefore, despite of some difficulties in the calculations, in comparison with other methods for application in approximation problems, the choice of cubic B splines is explained by the higher accuracy of its results.

Spline function for interpolation can be represented in the next form:

$$S(t) = \sum_{j=1}^{N+3} B_{j,m}(t)P_j, 0 \leq t \leq T, \tag{1}$$

where $B_{j,m}(t)$ is a basis function of B-spline; P_j matrix of control points.

A Cox-Debour relation of different m order is used as a basis function of spline [23, 26]:

In case $m = 1$, then

$$B_{j,1}(t) = \begin{cases} 1, \tau_j \leq t \leq \tau_{j+1} \\ 0, \tau_j > t > \tau_{j+1} \end{cases}$$

In case $m \geq 2$, then

$$B_{j,1}(t) = \frac{t - \tau_j}{\tau_{j+m-1} - \tau_j} B_{j,m-1}(t) + \frac{\tau_{j+m} - t}{\tau_{j+m} - \tau_{j+1}} B_{j+1,m-1}(t), \tag{2}$$

where τ is a node matrix.

Control points are calculated based on input statistics, which includes results of not-synchronized measurements of airplane coordinates. The trajectory of an airplane is transformed from LLA to a local Cartesian reference frame. We use the Noth-East-Down (NED) local frame with a reference point in departure airport coordinates. The total matrix of airplane trajectory includes a correspondent time of measurements and can be represented in the following form:

$$D = \left[X_{NED}^T, Y_{NED}^T, Z_{NED}^T, T^T \right], \tag{3}$$

where X_{NED}, Y_{NED}, Z_{NED} are matrices of airplane coordinates, T is a matrix of time of receiving ADS-B message.

Then, Eq. (1) can be represented in the following matrix form:

$$D = BP, \tag{4}$$

$$B = \begin{bmatrix} B_{1,\mathrm{m}}(t_1) & B_{2,\mathrm{m}}(t_1) & \cdots & B_{n,\mathrm{m}}(t_1) \\ B_{1,\mathrm{m}}(t_2) & B_{2,\mathrm{m}}(t_2) & \cdots & B_{n,\mathrm{m}}(t_2) \\ \vdots & \vdots & & \vdots \\ B_{1,\mathrm{m}}(t_n) & B_{2,\mathrm{m}}(t_n) & \cdots & B_{n,\mathrm{m}}(t_n) \end{bmatrix}$$

where P is a matrix of control points, B is a matrix of basis functions.

The elements of the regression matrix $B_{n,m}$ depend on the time variable, called the regressor or predictor variable. In the general case, the elements of $B_{n,m}$ form linearly independent columns. Then, for known basic functions (2), control points can be obtained from the solution of this equation by least squares method in matrix form:

$$P = \left(B^T B \right)^{-1} B^T D. \tag{5}$$

The Least-Squares Method minimizes the sum of standard deviations relative to the values of the matrix of control points P. After obtaining the control points (5) and the basic functions of Eq. (2), to extrapolate the navigation parameter for the required time, we use Eq. (1) in matrix form:

$$S = BP \tag{6}$$

Interpolated airplane trajectory is used to detect the time of airplane departure and arrival.

4   Probability Density Function Fitting

Numerous flight realizations of some particular connection, based on the scheduled time of gate departure and arrival makes possible to estimate statistics of delays. Factors that take action into delay within the same connection and scheduled time have the same nature in the most cases. Thus, probabilities of delays are estimated based on statistical data of departure and arrival times for particular flight connection separately. Also, a few factors action depends on airport capacity which varies in time of the day, day of the week or holidays.

Each of the factors has the probabilistic nature of the action. In common case, it is possible to assume the Normal (Gaussian) distribution of summarized action of all of these factors. Normal Probability Density Function (PDF) can be represented in the following form:

$$\rho(y) = \frac{1}{\sigma\sqrt{2\pi}}exp\left(-\frac{(y-\mu)^2}{2\sigma^2}\right), \tag{7}$$

where σ is a mean squared deviation, μ is a mean value.

Normal PDF (NPDF) gives a rough estimation due to the high rate of rare factors occurrence which shifts scheduled time much far than mean value. In some cases, a Double or Triple Univariate Generalized Error Distribution (DUGED, TUGED) can be used to fit a PDF [27, 28]. TUGED takes into account the appearance of rare delays for a long time. In general, TUGED can be represented by the following form:

$$\rho(y) = \frac{\alpha}{2a_1 b_1 \Gamma(b_1)}exp\left(-\left|\frac{y-b_1}{a_1}\right|^{b_1^{-1}}\right) + \frac{\beta}{2a_2 b_2 \Gamma(b_2)}exp\left(-\left|\frac{y-b_2}{a_2}\right|^{b_2^{-1}}\right)$$
$$+ \frac{1-\alpha-\beta}{2a_3 b_3 \Gamma(b_3)}exp\left(-\left|\frac{y-b_3}{a_3}\right|^{b_3^{-1}}\right), \tag{8}$$

where a_1, a_2, a_3 are scale factors; b_1, b_2, b_3 are shape coefficients; μ_1, μ_2, μ_3 are mean values; α and β are weight coefficients.

However, the best fitting property of statistics dataset with unknown factors action can be obtained by Kernel Density Estimation (KDE) [11, 29, 30]. KDE is a PDF estimator that does not require an assumption about a family of the underlying function. In common case, KDE of an identically distributed random sample of X data can be represented in the following form [31]:

$$\rho_{KDE}(x) = \frac{1}{nh^d}\sum_{i=1}^{n} K\left(\frac{x-X_i}{h}\right) \tag{9}$$

where $K(y)$ is a kernel function; h is a smoothing bandwidth ($h > 0$); X is a statistical data set.

Based on an assumption of the normal action of each factor, we use the Gaussian kernel function:

$$K(y) = \frac{exp\left(-0.5|y|^2\right)}{\int exp\left(-0.5|y|^2\right)dy}, \tag{10}$$

A special attention should be taken to choose a smoothing bandwidth. Incorrect values of h can lead to under or over smoothing. Both of them lead to significant errors in PDF. Thus, h should be chosen based on some criteria of optimality. Basically, h can be chosen based on minimization of mean square error of KDE [31]:

$$h_{AMISE} = \left(\frac{4\mu_K}{n\sigma_K^4 \int |\nabla^2 p(x)|^2 dx}\right)^{\frac{1}{d+4}}, \tag{11}$$

where μ_K and σ_K are mean and mean squared deviation of kernel function; $\nabla^2 p(x)$ is a Laplacian of the function $p(x)$.

Also, h can be chosen to get a balance between bias and variance in the under smoothing approach [31]:

$$h \approx n^{-\frac{1}{d+4}}. \tag{12}$$

Different confidence bands can be easily estimated based on the obtained PDF. A probability of delay at some particular time (t_x) is an important value in air traffic scheduling:

$$P[x \geq t_x] = \int_{t_x}^{+\infty} \rho(x)dx. \tag{13}$$

5 Numerical Demonstration

Numerical demonstration of statistical analysis of delays is based on ADS-B data. We use ADS-B trajectory data obtained from network of SDRs. The data includes only ADS-B transmitted airplane position reports. We use a dataset of connection between Kyiv Borispol International (UKBB) departure airport and Frankfurt International (EDDF) arrival airport. The dataset includes 107 unique flights between Sep 12, 2020 and Jan 23, 2021. Take-off is scheduled at 4:30 UTC and landing at 7:20 UTC. Flights are mostly performed by A319 and A321 airplanes. Results of statistical analysis of input statistics are represented in Fig. 4 and Fig. 5 for a frequency of take-off and landing, correspondently. We use a 5-min interval for histogram bins construction.

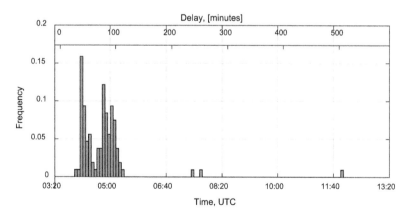

Fig. 4. Frequency of take-off in Borispol departure airport.

Based on the scheduled time and available statistics, we estimate delays for aircraft departure and arrivals. The histogram of flight duration, represented in Fig. 6, includes the time between aircraft take-off and landing.

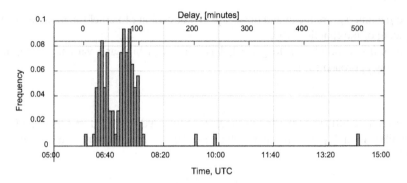

Fig. 5. Frequency of landing in Frankfurt arrival airport.

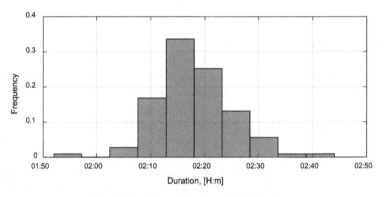

Fig. 6. Statistics of airplane flight duration.

Results of PDF estimation for flight duration, take-off, and landing time are presented in Fig. 7, Fig. 8, Fig. 9. Obtained results for KDE fit input statistics more accurately than NPDF.

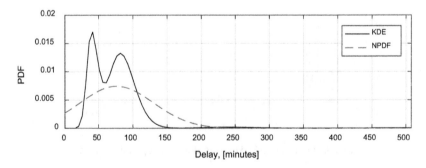

Fig. 7. PDF of take-off in Borispol airport.

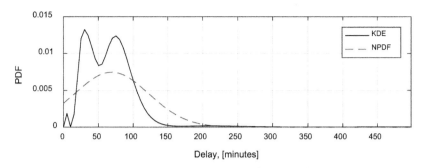

Fig. 8. PDF of landing in Frankfurt airport.

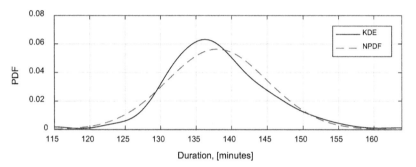

Fig. 9. PDF of flight duration.

Also, a Cumulative Distribution Function (CDF) has been estimated. Results of probability to be late at defined time t_x are estimated by (13) and represented in Fig. 10, Fig. 11, Fig. 12.

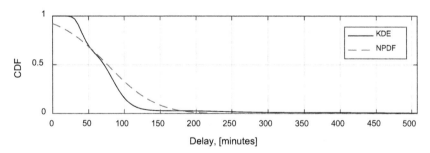

Fig. 10. CDF of take-off in Borispol airport.

Correct choosing of a bandwidth is an important parameter for KDE value. KDE estimated based on different bandwidth values including optimal one, based on a minimal error of approximation (11) is represented in Fig. 13.

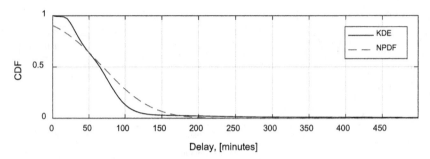

Fig. 11. CDF of landing in Frankfurt airport.

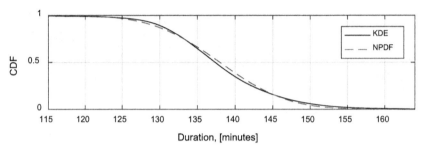

Fig. 12. CDF of flight duration.

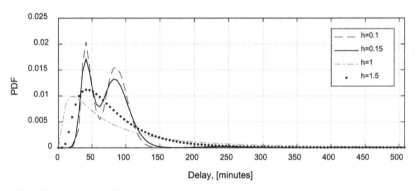

Fig. 13. KDE with different bandwidth values for take-off data in Borispol airport.

Results of mean and mean standard deviation values for each data set, together with optimal bandwidth are presented in Table 1.

Represented values of CDFs have a significant practical application. It shows probability/risk to be delayed for various time intervals. Obtained probabilities are useful in air traffic planning for ATC and airline representatives. Moreover, only one flight connection should be used for statistics collection in order to minimize ATC error influence.

Table 1. Mean and mean standard deviation of delay and flight duration

Time	Mean value	Mean Standard Deviation	Bandwidth
Delay of take-off	76	53	0.15
Delay of landing	69	53	0.19
Duration of flight	137	7	0.02

6 Conclusion

Airplane delay is an important problem of current and future air traffic system. Delays are caused by multiple factors, some of which are quite difficult for simulation due to the low probability of their occurrence. The obtained results show a practical side of delays estimation based on statistical data processing of airplane trajectory obtained by collecting position reports from on-board equipment by the ADS-B concept. Also, linear regression with a B-spline function is used for trajectory fitting to provide data time synchronization. Accumulated statistics of numerous trajectories for unique flight connection have been used in statistical analysis.

Usage of some assumptions about PDF of delays does not fit statistical data very well, due to influence of numerous factors and different forms of PDF. Normalization in this case leads to significant mistakes. Estimation of PFD with a kernel density function gives more accurate fitting in comparison with NPDF and TUGED. Obtained PDFs for flight duration, take-off, and landing delays are used for estimation of airplane risk to be delayed. Probability estimation of an airplane delay is an important step for efficient traffic configuration and safety of aviation.

References

1. Aviation Benefits report 2017, IHLG (2017)
2. Accident Statistics, ICAO. https://www.icao.int/safety/iStars/Pages/Accident-Statistics.aspx. Accessed 15 Jan 2021
3. Safety Report 2018, 55th edn. International Air Transport Association, Geneva (2019)
4. Eurocontrol. https://www.eurocontrol.int/our-data. Accessed 18 Jan 2021
5. Ostroumov, I.V., Kuzmenko, N.S.: Risk assessment of mid-air collision based on positioning performance by navigational aids. In: 6th International Conference on Methods and Systems of Navigation and Motion Control, pp. 34–37. IEEE, Kyiv Ukraine (2020)
6. Ostroumov, I.V., Kuzmenko, N.S.: Risk Analysis of Positioning by Navigational Aids. In: International Conference Signal Processing Symposium, pp. 92–95. IEEE, Krakow Poland (2019).
7. Ostroumov, I.V., Kuzmenko, N.S.: An area navigation RNAV system performance monitoring and alerting. In: 1st International Conference System Analysis & Intelligent Computing, pp. 211–214. IEEE, Kyiv Ukraine (2018)
8. Airport Handling Manual (AHM), IATA (2020)
9. Mueller, E., Chatterji, G.: Analysis of aircraft arrival and departure delay characteristics. In: Aircraft Technology, Integration, and Operations (ATIO) Technical Forum, p. 5866, AIAA (2002)

10. Cao, Y., Zhang, L., Sun, D.: An air traffic prediction model based on kernel density estimation. In: American Control Conference, pp. 6333–6338, IEEE, Washington DC USA (2013)
11. Yufeng, T., Ball, M.O., Jank, W.S.: Estimating flight departure delay distributions – a statistical approach with long-term trend and short-term pattern. J. Am. Stat. Assoc. **103**(481), 112–125 (2008)
12. CODA digest. All-Causes Delay and Cancellations to Air Transport in Europe, Eurocontrol (2019)
13. CODA digest. All-Causes Delay and Cancellations to Air Transport in Europe, Eurocontrol (2015)
14. Certification Specifications and Acceptable Means of Compliance for Airborne Communications, Navigation and Surveillance. CS-ACNS. European Union Aviation Safety Agency. Annex I to ED Decision 2019/011/R, EASA (2019)
15. Technical Provisions for Mode S Services and Extended Squitter, Doc 9871, First Edition ICAO (2008)
16. Piracci, E.G., Galati, G., Pagnini, M.: ADS-B signals reception: A Software Defined Radio approach. In: Metrology for Aerospace (MetroAeroSpace), pp. 543–548. IEEE (2014)
17. Calvo-Palomino, R., Ricciato, F., Repas, B., Giustiniano, D., Lenders, V.: Nanosecond-precision time-of-arrival estimation for aircraft signals with low-cost SDR receivers. In: 17th ACM/IEEE International Conference on Information Processing in Sensor Networks (IPSN), pp. 272–277. IEEE (2018)
18. Ostroumov, I.V., Kuzmenko, N.S.: Interrogation rate measurements of distance measuring equipment in air navigation system. In: 2nd International Conference on System Analysis & Intelligent Computing (SAIC), pp. 1–5, Kyiv Ukraine, IEEE (2020)
19. Sun, J., Ellerbroek, J., Hoekstra, J.: Large-scale flight phase identification from ads-b data using machine learning methods. In: 7th International Conference on Research in Air Transportation, pp. 1–8, Philadelphia, USA, TUDelft (2016)
20. Nijsure, Y.A., Kaddoum, G., Gagnon, G., Gagnon, F., Yuen, C., Mahapatra, R.: Adaptive air-to-ground secure communication system based on ADS-B and wide-area multilateration. IEEE Trans. Veh. Technol. **65**(5), 3150–3165 (2015)
21. Jan, S.S., Jheng, S.L., Tao, A.L.: Wide area multilateration evaluation test bed using USRP based ADS-B receiver. In: 26th International Technical Meeting of the Satellite Division of the Institute of Navigation, pp. 274–281, ION (2013)
22. Tran, T.N., Pham, D.T., Alam, S.: A map-matching algorithm for ground movement trajectory representation using A-SMGCS data. In: International Conference on Artificial Intelligence and Data Analytics for Air Transportation (AIDA-AT), pp. 1–8. IEEE (2020)
23. Tarasevich, S., Ostroumov, I.V.: A light statistical method of air traffic delays prediction. In: 2nd International Conference on System Analysis & Intelligent Computing (SAIC), pp. 1–5, IEEE (2020)
24. Ude, A., Atkeson, C.G., Riley, M.: Planning of joint trajectories for humanoid robots using B-spline wavelets. In: International Conference on Robotics and Automation 3, pp. 2223–2228. IEEE (2000)
25. Delahaye, D., Puechmorel, S., Tsiotras, P., Féron, E.: Mathematical models for aircraft trajectory design: a survey. Air Traffic Management and Systems, pp. 205–247. Springer, Tokyo (2014). https://doi.org/10.1007/978-4-431-54475-3_12
26. Biagiotti, L., Melchiorri, C.: B-spline based filters for multi-point trajectories planning. In: International Conference on Robotics and Automation, pp. 3065–3070, IEEE (2010).
27. Ostroumov, I.V., Marais, K., Kuzmenko, N.S., Fala, N.: Triple probability density distribution model in the task of aviation risk assessment. Aviation **24**(2), 57–65 (2020)

28. Tsymbaliuk, I., Ivashchuk, O., Ostroumov, I.: Estimation the risk of airplane separation lost by statistical data processing of lateral deviations. In: 10th International Conference on Advanced Computer Information Technologies (ACIT), pp. 269–272, IEEE, Deggendof Germany (2020)
29. JooSeuk, K., Scott, C.: Robust kernel density estimation. Mach. Learn. Res. **13**(1), 2529–2565 (2012)
30. Zhixiao, X., Yan, J.: Kernel density estimation of traffic accidents in a network space. Comput. Environ. Urban Syst. **32**(5), 396–406 (2008)
31. Yen-Chi. C: A tutorial on kernel density estimation and recent advances. Biostatistics Epidemiol. **1**(1), 161–187 (2017)

Changing Mobility Behaviors: A State of Art Analysis

Federico Moretti, Massimiliano Petri[✉] [iD], and Antonio Pratelli

Department of Industrial and Civil Engineering, University of Pisa, Pisa, Italy
f.moretti7@studenti.unipi.it, {m.petri,a.pratelli}@ing.unipi.it

Abstract. Sustainable Urban Mobility Plans are today implemented in all most important European Urban Areas. The change in mobility habits towards sustainable transport modes is one their main objective. Methodology to achieve this goal are differents and go from restriction and payments for access to urban area (the stick) to rewarding systems (the carrot). The analysis carried out on the use cases shows how in recent years the 'carrot action type' has seen a great development even if, to date, there are few truly applicative and correctly applied solutions. The article highlights different use cases describing their pros and cons with a special attention to rewarding framework and technologies.

Keywords: Sustainable mobility · Transport access restriction · Mobility reward · Smartphone APP

1 Introduction

In this first part we want to introduce those elements of the mobility state of the art at European level useful in the subsequent considerations relating to the modifying mobility behavior systems. The data that are proposed, to clarify the situation of the transport sector in Europe with a greater focus on road transport, were collected from the publications that Eurostat, the statistical office of the European Union, made available in 2020 [1, 2].

In 2017, passenger cars accounted for 82.9% of passenger land transport in the EU, with coaches, buses and trolley buses and trains both accounting for less than a tenth of all traffic measured in passengers/sqkm and 9.4% and 7.8% respectively (see Fig. 1).

As for passenger cars, 11 of the 23 EU Member States for which 2018 data are available, more than 50% of the cars were petrol (see Fig. 2). Cyprus reported the highest percentage (81.7%), followed by Finland (73.7%) and Denmark (68.3%). Diesel cars exceeded the 50% mark in Lithuania (67.7%), France (65.8%), Latvia (59.1%), Luxembourg (58.9%), Spain and Portugal (both 56, 1%), Austria (55.8%), Ireland (55.5%), Belgium (54.7%) and Croatia (50.7%). Alternative fuels made a significant contribution in Poland (15.9%), Italy (8.6%), Lithuania (8.2%) and Latvia (7.3%).

Moreover, in 2018, it was estimated [2] that more than half (51.6%) of all passenger cars in the EU-27 were at least 10 years old, compared to a scant 11.8% of those less than two years old.

© Springer Nature Switzerland AG 2021
O. Gervasi et al. (Eds.): ICCSA 2021, LNCS 12950, pp. 378–393, 2021.
https://doi.org/10.1007/978-3-030-86960-1_27

Given also the high motorization rate of the European population, rather than acting on incentives to change its type (electric, hybrid, others), there is the need to decrease the vehicle density and encourage the use of alternative transport modes to the car.

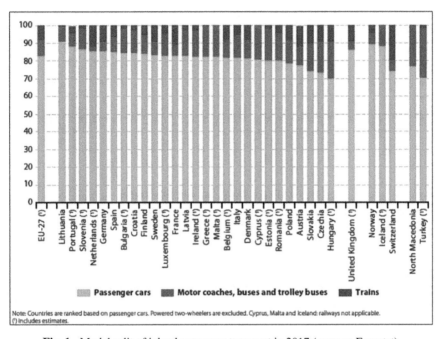

Fig. 1. Modal split of inland passenger transport in 2017 (source: Eurostat).

Going into the Italian situation, the annual study carried out by the ISFORT institute [4] provides a more detailed picture and indicates that occasional journeys are almost 40% of the total and the urban modal split sees private mobility (by car or motorcycles) down from 59 (in 2008) to 55%s (in 2019) while extra-urban mobility has an unsustainable share of travel equal to over 88% and constant over the last 10 years. It is, therefore, clear that it is particularly necessary to intervene in the traffic flows entering in the cities from external areas. Given the almost impossibility of acting on existing infrastructures (for limits in road network or in disposable resources), the rewarding policies in this context acquire a significant weight and are opposed to, or better integrate with, the tolling and charging policies related to urban accessibility.

The next paragraph presents the methodologies used to manage urban accessibility and mobility behaviors, with particular attention to the interventions following the important features now underlined.

2   Mobility Management Policies

The European Commission, in the Communication no. 433 "Greening Transports" at the European Parliament in July 2008, underlines the importance to identify methodologies for calculating user mobility costs. Commission highlighted the necessity to

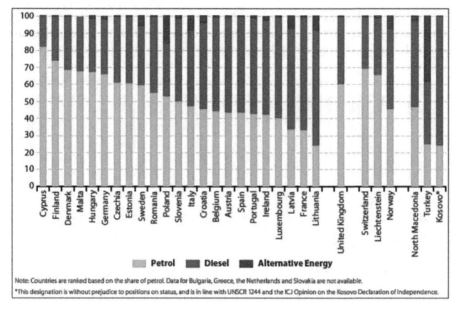

Fig. 2. Passenger cars by fuel type (%) in 2018 (source: Eurostat).

evaluate indirect prices that society must support and the need for incentives to engage behaviors that can generate lower costs for society. This is a first input to introduce mobility management measures such as charging company employees for parking vehicular access, rewarding commuters using park & ride or other. Moreover, in many European cities there is practically the impossibility of acting on existing infrastructures, so soft mobility management policies in this context acquire a significant weight and rewarding ones are opposed, or rather (and better) integrated, with tolling and charging policies. It is important to calibrate mobility management policies because we need to guarantee accessibility to residents even in restricted traffic areas and avoid, with excessively restrictive policies, the citizens urban centers abandonment. Unfortunately, these are dynamic already existing in some inhabited centers such as Cagliari, for example, where the excessive restriction on residents accessibility (only in some daily time periods) led to the transfer of larger family groups, where the presence of children required greater flexibility in daily travel.

Moreover, recent statistics [4] shows relevant suggestions to design and implement correctly mobility management policies. For example, from 2008 to 2019 in Italy trips within urban areas saw a decrease in the share in private cars from 58.8% to 54.9% with an increase in sharing mobility of almost 2% as well as for public transport. In the same period, extra-urban journeys ending in urban areas remained practically unchanged, with a share of private cars near 85%. It is therefore important to manage more efficiently the trips accessing urban areas, often commuting ones.

Policies to manage mobility go from **accessibility restriction** to **reward systems.** In the next two paragraphs we describe both.

2.1   Accessibility Restriction Mobility Management Policies

These former policies have already seen many application areas and can be applied on a point-base framework, such as for the Bridge Tolls in Motala and Sundsvall in Sweden [5] where a tax was introduced in February 2015, and it is still in operation, which all vehicles have to pay for every time they use the two bridges. The toll is active 24 h a day and a fixed fee is charged for crossing through the automatic checkpoints.

Access restrictions can be cordons-based, with the restriction applied for crossing a dividing line between two city areas. Examples of representative applications are the Stockholm and Gothenburg congestion charge.

The Stockholm congestion charge is a charging system (still working) applied to vehicles passing through checkpoints, both inbound and outbound, placed on a cordon. The congestion charge was applied on a permanent basis on 1 August 2007, after a seven-month testing period between 3 January 2006 and 31 July of the same year.

The main purpose was to reduce traffic congestion and improve the environmental condition in central Stockholm, as well as to allocate the collected revenues to new infrastructure supply. The charge zone initially included the whole Stockholm city center, but from 1 January 2016 it was also extended to the access and exit ramps of the two Essingeleden motorway junctions to reduce traffic jams during peak periods and, consequently, in a reduced way, even on the surrounding roads. Operation is simple: there are unattended electronic control points at all bead inputs. Vehicles passing through them are identified by automatic license plate recognition thanks to cameras, laser detectors, antennas and information signals mounted on a series of portals at each control point. There are no payment booths and this is done through a monthly report relating to the identifications. The idea is that with a traditional toll booth, a substantial percentage of the toll, instead of being used for infrastructure investments, would be destined for staff, which is avoided here. The system is not exclusively aimed at traveling by car to the city center; the congestion charge is applied both at the entrance and at the exit of the interested area. The tax amount depends on the control point (if it is on the motorway or if it concerns the cordon of the center of Stockholm) and on the time of day when a citizen enters or leaves the congestion zone. There is no charge on Saturdays, public holidays and days before holidays or between 18:30 and 06:29. Some vehicle classes are exempt from congestion charge such as: emergency service vehicles, certain types of buses, motorcycles, military vehicles, etc.

Stockholm has seen a drop in traffic volumes across the cordon of around 25%. Of these, about 10 percentage points represented business trips and a further 6% were due to changes in personal trip (change of destination, reduction of frequencies and planning of multiple activities in a single trip). Commercial vehicle drivers have also changed their routes so to decrease the number of cordon crossings. It should be noted that a good part of the traffic throw the checkpoints was occasional, therefore, not all transits took place for work reasons and then it was possible to adapt mobility behaviors to these limitations.

Gothenburg's congestion charge, which is still working, was introduced on 1 January 2013, covering the entire center of Gothenburg and the main E6 road through the city. The operation is very similar to the congestion charge applied in Stockholm. Here, too, different rates are applied depending on the time slot, it is charged in both directions (both inbound and outbound) when the cordon is crossed and it is applied to all vehicles, even the least polluting ones. In addition, a new rule has been introduced that attributes the maximum amount to those who pass through two checkpoints in one hour. There is no charge on Saturdays, days before holidays and holidays or between 18:30 and 05:59. About 8 months after its introduction, a reduction in traffic volumes in transit at the cordon was observed, stabilizing at 12%.

The last type of restriction is area-based where the restriction is applied for entry by car within an area during a certain period of time and with a specific vehicle. These restrictions are simple to understand and easy to implement. The London Congestion Charge is the best known example of this vehicle access restriction regime, together with the Milan Area C and the Singapore ERP.

The London congestion charge was introduced in 2003, within the London Inner Ring Road, an area enclosed by a series of main roads surrounding central London and which forms the boundary of the restricted traffic zone. The goal was to make vehicular traffic more fluid, obtain funds to invest for public transport network greater efficiency, improve the quality of life and above all reduce congestion and pollution in the delimited area or in the city center. The London Congestion Charge is a daily rate of £ 11.50, between 7 am and 6 pm, Monday to Friday which drivers of certain motor vehicles must pay in order to access the parts of London designated as congestion charge zones and located in the city center. Some vehicles, such as taxis, buses, commercial vehicles do not have to pay the toll, while residents receive a 90% discount.

The tax is still present and has led to a reduction of about 20% of vehicles subject to full toll even if it has been accompanied by an increase in entrances not subject to any tolls (taxis and private rental vehicles). A significant modal change was also obtained also because the revenues were used to support sustainable mobility. In particular, vehicle entrances for transit only increased from 29% of the total in 2002 to 37% in 2015 (mostly buses) while car transits decreased, from 46% to 36% in the same period. The share of bicycle admissions doubled from 1% to 2%.

London has shown that it is possible to achieve significant changes in travel patterns in a relatively short time frame.

The Milan Area C is a large area (the ZTL includes about 8.2 square kilometers and 77,000 residents) of the Milan historic center with access restrictions for some types of vehicles. It coincides with the limited traffic zone (ZTL) "Cerchia dei Bastioni" and is bordered by 43 gates with cameras, of which 7 are for the exclusive use of public transport.

Introduced in 2012 and active, the rate is applied to all vehicles entering the city center on weekdays (excluding Saturdays) from 7:30 to 19:30. Each vehicle that enters the paid area must pay 5 € to be able to enter and exit for the whole day. Electric vehicles, mopeds and motor vehicles (motorcycles, tricycles and quadricycles), hybrids with certain specifications, public utility vehicles, police and emergency vehicles, buses and taxis are exempt. Residents inside the restricted area also have to pay to reach their home but have 40 free admissions per year and a discounted rate of € 2 from the 41st access onwards. As a result, there was a decrease in traffic of 35.9% and it was observed that occasional entrances are absolutely prevalent, with 45.65% of vehicles entering only one day and 86.99% of vehicles entering for less than 10% (<13 days) of the 125 days considered. Systematic movements are extremely limited, i.e. 2.58% of vehicles entered the ZTL. It is interesting to note that, for the category of residents, 85.8% did not exceed the 40 free admissions provided for by the provision, while 7.5% made systematic admissions ($\geq$63 days of admissions).

Electronic Road Pricing (ERP) was implemented by the Singapore Land Transportation Authority in September 1998 to replace the previous "Area Licensing Scheme" and was the world's first electronic toll collection system.

The charge takes place thanks to an IU (In-vehicle Unit - On-board unit) device, connected to a CashCard, which is placed in the lower right part of the windshield. When a vehicle crosses the portal, the installed sensors communicate with the UI via a short-range communication system and the driver receives the message of the deducted amount on the LCD screen of his UI.

The results were a reduction in road traffic of nearly 25,000 vehicles (13% of total traffic) during rush hour, with an increase in average road speed of around 20%.

An interesting example of restriction policies is the Greater Manchester congestion charge. The Greater Manchester Congestion Charge is reported because the congestion charge was not introduced due to strong resistance from a large part of the population, who saw the intervention as one of the many impositions to make citizens pay.

In 2008, two curbs were proposed: one outside that surrounded the main urban core and the other that covered the center of Manchester (for a total of 210 kmq). The Greater Manchester Transport Innovation Fund was rejected in a referendum at the end of the same year (December 2008). It was proposed that vehicles entering the area bounded by the M60 motorway would be charged £ 2.00 during the morning rush hour, plus £ 1.00 also when crossing the internal cordon. If the exit from each cordon had taken place in the evening, an additional £ 1 was charged (these rates were those proposed in 2007).

Inbound rates would be applied between 7:00 and 9:30, and outbound rates between 16:00 and 18:30. There would be no charge during the central hours of the day, after 6:30 pm, on weekends and for travel during non-congestion hours: leaving the city in the morning or entering it in the evening.

The charge payment had to be carried out through a prepaid "tag and beacon" system (tag that sends signals within its range and communicates with the receiver made up of antennas, devices, etc.), connected directly to the driver's account, to the overcoming of the seams. Casual visitors to Manchester without a tag could have paid via the call center or the internet for a surcharge. Exempt from the fee were motorcycles, black taxis and private rental cars. Presumably in the spring of 2022, a Clean Air Zone will be introduced by order of the government. This will be different from a Traffic Restriction Zone where all or most of the vehicles will be subject to payment for transit, but will hit with a daily rate, only vans, buses, coaches, taxis, private hire vehicles, minibuses and heavy commercial vehicles that do not they will meet emission standards. Private cars, motorcycles and mopeds will not be included.

So, this last example show the needs to integrate restriction policies also with other types like rewarding ones, described in the following paragraph.

2.2 Rewarding Mobility Management Policies

Reality shows that people are more motivated when they are rewarded rather than when they are punished, as in this way positive thinking is associated with the behavior itself [6]. As reported in their article Dogterom, Ettema and Dijst [7], road pricing methods and other toll measures are attractive from a theoretical point of view but are often perceived as controversial due to their lack of social acceptability. Therefore, the introduction of tools able of allowing the management of transport demand in a politically and socially acceptable way is a factor of primary importance.

One of the few comparisons between pricing and reward systems at a scientific level [8] shows how the theoretical pricing solution must make each user pay what it costs the company. To pursue this hypothesis, taxes of different levels should be applied for each user and for each road line, an element that is difficult to apply and has led pricing policies to simplified solutions that are less socially acceptable.

Another work that compares reward systems (in particular the Tradable Mobility Permits-TMP) and pricing systems is that of De Palma, Proost, Seshadri and Ben-Akiva (9). In it, the search for the social optimum was simulated, that is, the condition for which total welfare is maximized. The results show that, in real conditions, where the uncertainty, represented in the model by an increase in the random term of the logit behavioral formulation (e.g. variations in queues, travel times and more), is not zero, TMP systems maximize more global welfare and consumer surplus (positive difference between the price that an individual is willing to pay to receive a specific good or service and the market price of the same good). It is interesting to note that an equity measure was carried out (represented by the Gini coefficient) with more equitable values closer to zero and where equity is assessed with respect to the different income classes. In this case, the Gini index is always higher for payment systems, indicating how they have a lower level of equity.

The preference of road users towards reward systems was also found by Parag, Capstick and Poortinga [10] who investigated preferences on the carbon emissions reduction in a population sample, verifying how the rewarding systems have an higher conviction capacity than payment systems [11].

In fact, the reward, together with the other toll/fare interventions, is an excellent tool able of changing the behavior of citizens towards sustainable mobility. The reward, to be effective and achieve the goal, cannot ignore human behavior and must take into account the relating research conducted in the field of sociology, psychology and marketing.

As for the intervention strategies aimed at modifying the behavior itself, "The Fogg Behavior Model (FBM)" is very interesting, a behavior modification model proposed by B.J. Fogg [12] who states that behavior is composed of three different factors: **motivation, ability** and **triggers**. According to this model, to be successful in changing behavior, each person must be motivated, have the ability to execute the behavior, and have a trigger to execute it (for example, in the GOODGO reward system [13] there are prizes for three age groups, in order to motivate also the elderly who, if placed in competition with the highest level of mobility of young people, could lose motivation to participate). Fogg defines several motivators: such as pleasure or pain, hope or fear, social acceptance or rejection. Skill, on the other hand, refers to the perception of self-efficacy in performing a target behavior. The elements that characterize an high capacity or on the contrary simplicity to perform a behavior could be time, money, physical effort, brain cycles, social deviance, non-routine, etc.

Fogg states that an activity that can be done easily requires little motivation and a trigger element is enough while, when an activity is tiring, such as changing one's behavior, the motivating element must grow (see Fig. 3). In the field of mobility, this model consolidates the position of rewarding, in fact, when the goal is to change the behavior of citizens towards sustainable solutions, a long and tiring process, the motivating element must grow and this consists precisely in introducing prizes.

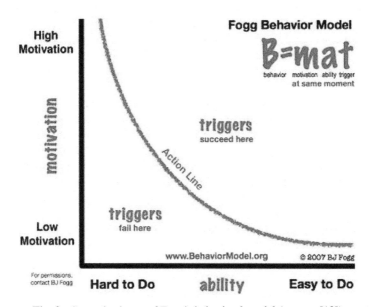

Fig. 3. General scheme of Fogg's behavioral model (source: [12])

Motivation to change

A first intervention can concern the communicative level. Already the knowledge of the change possibility and the presence of an alternative mobility network are elements of information which, in themselves, constitute a stimulus to change, even more so if connected to reward elements [14].

The identification of achievable individual goals can also be part of incentive mechanisms together with feedback on the goal achieved and their sharing; knowing that one's virtuous behavior will be made public makes it more pleasant to continue with such attitudes.

Barsky et al. [15] report that it is much more impactful on the user to receive information in a negative format than a positive one; for example, receiving the quantification of the CO_2 avoided Kg leads to more incisive results than showing the emitted Kg. Tools to activate the stimulus can be: posters, radio/TV advertising, newsletters, magazines, guides, websites or informative and interactive material, videos, social networks such as Facebook groups, Twitter, LinkedIn, events, meetings, workshops, smartphone apps etc.

Trigger to change

This point concerns the need to give people access to the resources necessary to change behavior. This can be done by providing economic sums or by providing alternative systems (for example innovative accessibility systems) to existing ones that do not increase individual spending levels. The physical or social context of people must be restructured in order to activate or facilitate the desired behavior.

Ability to change

They must be provided goals quickly achievable. Furthermore, people are likely to adopt a new behavior if they see that other individuals have managed to achieve the same goal. Along the same lines is the theory belonging to Skinner [16] which states that complex behavior is learned gradually through the modification of simpler behaviors and that imitation and reinforcement play important roles as individuals learn by duplicating the behaviors observed in others and rewards are essential to ensure desirable behavior repetition.

The self-image communicated to others is an important incentive element, as it leads the citizen to be seen as a high social stature person. Often the use of monetary rewards goes against this principle as the citizen loses the social goal towards the more venal economic one. Furthermore, the virtuous behaviors of friends, acquaintances and colleagues can influence individuals and, so, the use of social networks, to spread sustainable behaviors, acquires a high importance. Ampt (2003) also highlighted how communicating one's own virtuous behavior to others involves a mental process of strengthening commitment to sustainability.

In conclusion, giving up daily "bad" habits can be painful and for this there is a need to be rewarded for every small step, especially at the beginning of this long process.

Another key element is to promote behavior change; personalized communications and tailored messages are effective and are a good way to incentivize behavioral change and improve its impact. This approach can be used to generate an emotional response such as fear, hope or anxiety and to promote the comparison between real and ideal behaviors.

Equally significant is to understand how to maintain a certain stable behavior over time, and therefore the results obtained, after the first active phase of intervention and change without continuous support. In fact, experience shows that in most cases, the effects on behaviors inherent to lifestyles are often more pronounced in the initial phase of the intervention. Without further support, the positive effects tend to diminish or fade over time also due to some factors such as climate or seasonal and/or personal changes. Hence, maintaining behavior change requires an understanding of the surrounding context that can facilitate or prevent this process.

In order to change a given behavior in the future, it is necessary to know the behavior of users and past alternatives, in fact, to achieve the desired change, the user must be aware of his present attitude and possible future alternatives. People are rarely aware of their routine behavior, since most of the actions that compose it are performed unconsciously and therefore the difficulty arises in considering other possibilities.

In this respect, technology offers countless advantages. One way to effectively influence user behavior is to communicate sustainable alternatives including the ability to monitor oneself, set goals, be able to compare and be rewarded for achieving target behaviors.

The stage of change reached by users is important, as it affects the success of the new behavior promoted. The stages of change, therefore, highlight how ready and willing people are to behavior change. The phase-of-change model proposed by Prochaska [17] (see Fig. 4) describes the five stages people go through on their path to change. The process starts with an individual who has no interest in changing his behavior (no = pre-contemplation), begins to contemplate change (perhaps = contemplation), prepares to change it (prepare/plan = preparation), begins a new behavior (doing = action) and maintaining it (continuing = maintaining).

The "action" and "maintenance" phases are of particular importance in changing mobility behavior, because the initiation process versus the maintenance process is very different (making the initial decision versus continuing effort). Initial attraction or a willingness to try a new behavior may be enough to encourage the transition to the "action" phase, but it is important to assume that someone will continue a behavior just because they have started doing it. In fact, creating new habits and breaking old ones is incredibly difficult and there is a high relapse chance.

For a less complex service such as a single journey on public transport or even a monthly pass (low complexity due to less time spent on commitment) the mobility buyer's decision process occurs more quickly. This makes public transport users more susceptible to new solutions for their mobility needs than car owners who have a high financial loss to amortize.

It is therefore essential to consider that the success of the behavior change results from an understanding of the target group and its objectives and needs in terms of mobility; for example, by introducing car-pooling as an element of reward, allowing the use, even if minor, of one's car and, therefore, not contrasting with previous purchasing choices.

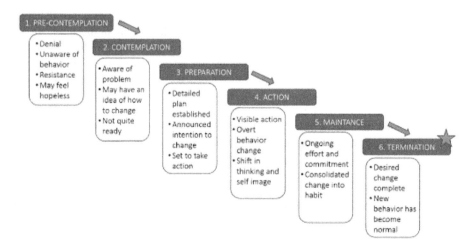

Fig. 4. Prochaska's Six Stages of Change [13]

It is very important to understand how people can be motivated to change behavior and what are the effective mechanisms that can keep users motivated once the novelty effect of such a system wears off. Planning tools, which automatically consider multiple users by offering decision support for multiple objectives, by making users aware of their past behavior and attempting to integrate personal information with spatio-temporal information, are needed to link complex objectives with simple activities. For this reason, a growing range of smart apps and price comparison tools are increasingly being made available to travelers to manage their travel in order to reduce costs and save time.

2.3 Rewarding Mobility Technologies

Many online tools, apps, ITS (Intelligent Transportation System) technologies have been created and integrated into web platforms to encourage sustainable mobility behaviors using rewarding policies. The technical term of these technologies is *Captology*, acronym for Computers As Persuasive TecnOlogy which has been becoming a new research field since the 2000s [12]. This field studies the use of different computer-based technologies as a means of persuasion to change behavior (websites, smartphones, video games, smart environments, virtual reality and more).

ICT (Information and Communications Technology) tools and especially smartphone apps are the technologies most used to monitor people's mobility behavior, collect data (for example, useful for the Administration to plan mobility), increase the motivation of users through competition and reward those who move in a sustainable way. Many examples show that smartphone is one of the most important platforms for modifying human change [17].

Analyzing all the characteristics of the individual reward projects developed would be excessive and, therefore, for brevity reasons, the analysis made (see Tab. 1) on the apps used in the most important reward projects is reported, in order to provide a general picture of the current implementation state.

The high number of reward platforms developed highlights the great interest that this system is enjoying, while the detailed analysis highlights how many of the applications have been developed internally at research projects and have been applied only at a prototype level with the decommissioning of the system. at the end of the project.

Often the applications are not designed with the target to provide a real decision support and incentive to modal change but only exercises and technological developments not adequately designed to be used in a massive and effective way.

The detailed analysis of the few platforms applied to date highlights a general superficiality in the management of the reward modules, with the following elements:

- problems from the point of view of privacy (in some Apps you can see the names and surnames of all the members in the ranking and not the nicknames);
- bugs in the registration system or in the tracking system;
- absence of verification of the declared transport modes, with consequent invalidation of the rewarding objectives
  Elements of interest are the connection with the APP personalization systems (Avatar, possibility to invite friends, to compete with friends, to publish their own photos or

Table 1. Main features of the main rewarding Apps (PT = Public Transport)

Name	Country	N° downl	Monitored modes	Note	Year	Current State	Start Stop
Muv	many	>100	multimodal	avatar, weekly competition, multipliers	2020	freezes on registrat	manual
Biklio	many	>1.000	bike	Trace proj. shops checkpoints	2018	impossible to enter	manual
Positive Drive	Holland	>1.000	multimodal	Trace Project	2018	absent in APP store	manual
BetterPoint	Italy, UK	>50.000	multimodal, car-pooling	medal, prize draw, user union	2016	no initiatives currently active	manual
GoodGo	Italy	>1.000	multimodal car-pooling	Integrated with bike antitheft system	2017	working	manual
WeCity	Italy	>10.000	multimodal unique PT	modes certification, survey on path	2015	Active with bugs in registration	mixed
Kultur-Token	Austria	>100	multimodal	owned by the Austrian government	2019	suspended for covid	auto
Pin-Bike	Italy	>1.000	bike, car-pooling	certified paths	2018	working	auto
IMove	many	unk	unk	second level maas	2019	absent in APP store	unk
Smart	many	>10.000	multimodal	personal and group competitions	2015	working	auto
Enschede Fietsstad	Holland	>10.000	bike	complete dashboard	2015	working	auto
Sharing Cities	Italy,UK Portugal	unk	bike, car-pool., PT, car-sharing	also waste recycling and energy cons.	2015	not working	unk

(continued)

Table 1. (*continued*)

Name	Country	N° downl	Monitored modes	Note	Year	Current State	Start Stop
Travel Smart Rewards	Singapore	unk	rail	prize draws or credits	2010	absent in APP store	unk
Alternativ	Israel	>50	multimodal	complex rules for identifying modes and stops	2016	absent in APP store	auto
Nuride	USA	> 10.000	car-pooling	money to the passage of users in front of activities	2003	absent in APP store	unk
TripZoom	Hooland UK, Sweden	>100	multimodal	social networks, friends rankings, O / D travel info	2014	impossible to enter	auto
Commuter Greener	Sweden	unk	multimodal	unk	2015	absent in APP store	unk
Love To Ride	Italy, UK	>10.000	bike	google fit, bike shops, add photos, stories	2007	working	manual
Travel Watcher	Holland	unk	multimodal	automatic travel classification system	2010	absent in APP store	auto
CO2 Fit Chargers	Germany	>10.000	multimodal	google fit, vehicle setting, invite friends, healthy nutrition	2018	working	auto
Vivibici	Italy	>50.000	bike	rewards with mobile minutes	2015	working	manual
Play&Go	Italy	>1.000	multimodal	planned trip entry	2016	working	manual

(*continued*)

Table 1. (*continued*)

Name	Country	N° downl	Monitored modes	Note	Year	Current State	Start Stop
La'Zooz	Israel	unk	multimodal	collaborative platform, zooz bitcoin coin	2017	absent in APP store	unk

travel stories or to connect to social networks) or the connection with external apps (e.g. Google Fit or Google Health).

Among the results of the various experiments it should be emphasized that, despite many apps do automatic route tracking (Table 1 last column), it emerges that users do not like continuous monitoring but rather prefer to decide for themselves when to be tracked. This approach is also beneficial in terms of sustainability, as it avoids excessive battery consumption.

Another element of interest is the use of multipliers, a typical gamification feature. They provide extra points in case of special conditions that occur. MUV, one of the platform using multipliers, has three different types of multipliers:

- Weather conditions: rewards users who move when it rains or snows;
- Rush hours: rewards users who move sustainably during peak traffic hours;
- Frequency of use: rewards users who play for multiple consecutive days.

In other ways, credits/points can also be collected by inviting friends, answering a survey, in order to encourage the dissemination of tools or the collection of important data from the point of view of mobility behavioral analyzes.

Moreover, one of the problems of rewarding systems is the involvement of occasional users, users who do not have the possibility to collect points but, in some projects such as Betterpoints, players participate in real prize draws or virtual raffles to win physical prizes.

3 Conclusions and Perspectives

In general, the lesson learned from the analysis of the rewarding best practices state of the art allows to extract some basic elements to be provided to each application:

- provide awareness of their mobility choices thanks to self-monitoring;
- provide positive and instructive feedback;
- reward users both extrinsically and intrinsically;
- create rankings to fuel healthy competition between participants and push them to do better;
- create the emulation effect and disseminate good mobility practices by publishing the achievements on social networks;
- the importance of the role of gamification;

- the importance of the prizes choice;
- the importance of tying the prizes to the territory in which the reward system is implemented.

In all the experiences viewed, good results were obtained (often in qualitative terms) but when the rewarding ceased there was a slow return to previous mobility habits. This must push to build reward systems that are not one-off applications in order to raise sustainable mobility awareness (as often happens with tenders linked to state funding or fixed-term projects) but must enter into the daily planning of mobility, for example entering between the Sustainable Urban Mobility Plan measures.

Finally we need to underline that data relating to users who move on foot, by bicycle or using other sustainable means are generally extremely difficult to collect both for public administrations and for most private data providers because these mainly depend on the willingness of individual citizens to monitor their own mobility behaviors and cannot be collected through technological devices (urban sensors, recording systems, etc.). Since decision makers, but also local communities of interest, do not have the information of those who move sustainably within the urban space, the effectiveness of the policies that should aim to motivate those same actors is reduced. Furthermore, the unavailability of these data leads to political and infrastructural planning focused on the means of transport on which the most information is available, i.e. the most polluting means.

So rewarding systems and the related data collecting are, in this sense, a double way to incentive sustainable mobility.

References

1. Eurostat - "Energy, transport and environment statistics, 2020 Edition". https://ec.europa.eu/eurostat/documents/3217494/11478276/KS-DK-20-001-EN-N.pdf/06ddaf8d-1745-76b5-838e-013524781340?t=1605526083000
2. Eurostat - "Key figures on Europe, STATISTICS ILLUSTRATED, 2020 Edition". https://ec.europa.eu/eurostat/documents/3217494/11432756/KS-EI20-001-EN-N.pdf/6b9097d9-ea05-a973-d931-c08334db979b?t=1602776043000
3. European Environment Agency - "Air quality in Europe - 2020 report". https://www.eea.europa.eu/publications/air-quality-in-europe-2020-report
4. ISFORT, 2020. 17° Rapporto Audimob sulla mobilità degli italiani. https://www.isfort.it/wp-content/uploads/2020/12/RapportoMobilita2020.pdf
5. https://www.transportstyrelsen.se/en/road/road-tolls/Infrastructure-charges-in-Motala-and-Sundsvall/
6. EPOMM-European Platform on Mobility Management, 2017. Rewarding behaviour change. http://www.epomm.eu/newsletter/v2/eupdate.php?nl=0317_2&lan=enDogterom et al. 2017
7. Dogterom, N., Ettema, D., Dijst, M.: Tradable credits for managing car travel: a review of empirical research and relevant behavioural approaches. Transp. Rev. 37(3), 322–343 (2017). https://doi.org/10.1080/01441647.2016.1245219
8. Tillema, T., Ben-Elia, E., Ettema, D.: Road pricing vs. peak-avoidance rewards: a comparison of two Dutch studies, paper presented to the 12°. In: World Conference on Transport Research, Lisbon (2010)

9. De Palma, A., Proost, S., Seshadri, S., Ben-Akiva, M.: Tools versus Mobility Permits: a comparative analysis, report prepared for the French Ministry of Ecology, Sustainable Development and Energy (2016)

10. Parag, Y., Capstick, S., Poortinga, W.: Policy attribute framing: a comparison between three policy instruments for personal emissions reduction. J. Policy Anal. Manage. **30**, 889–905 (2011)

11. Pratelli, A., Petri, M.: SaveMyBike – a complete platform to promote sustainable mobility. In: Lecture Notes in Computer Science (including subseries Lecture Notes in Artificial Intelligence and Lecture Notes in Bioinformatics) (2019)

12. Fogg, B.J.: Persuasive Technology: Using Computers to Change What We Think and Do, 1st edn. Morgan Kaufmann, San Francisco (2002)

13. Nicotra, I., Petri, M., Pratelli, A., Souleyrette, R.R., Wang, T.(: Mobility impacts of the second phase of covid-19: general considerations and regulation from Tuscany (Italy) and Kentucky (USA). In: Gervasi, O., et al. (eds.) ICCSA 2020. LNCS, vol. 12250, pp. 255–268. Springer, Cham (2020). https://doi.org/10.1007/978-3-030-58802-1_19

14. Ampt, E.: Understanding voluntary travel behaviour change, articolo presentato alla 26°. In: Australian Transport Research Forum, Wellington (2003)

15. Barsky, Y., Galtzur, A.: Integration of Social Inventives Aimed to Promote Behavioural Change, in Civitas Tel Aviv-Yafo report, WP 8 Skinner B.F., 1953. Science and human behaviour, The Free Press (2016)

16. Prochaska, J.O., Norcross, J.C., DiClemente, C.C.: Changing for good. New York: Morrow. Released in paperback by Avon, 1995. (Hungarian 2009, Polish 2008, Hebrew 2006, Japanese 2005) (1994). ISBN: 0-380-72572-X

17. Wouter, B., Koolwaaij, J., Peddemors, A.: User behaviour captured by mobile phones, Novay, Brouwerijstraat 1, 7523 XC Enschede, The Netherlands (2014)

Economic Risk Evaluation in Road Pavement Management

Vittorio Nicolosi[1]([✉]) [iD], Maria Augeri[1] [iD], Mauro D'Apuzzo[2] [iD], Luis Picado Santos[3],
Azzurra Evangelisti[2] [iD], and Daniela Santilli[2]

[1] University of Rome "Tor Vergata", via del Politecnico 1, 00133 Rome, Italy
augeri@ing.uniroma2.it, dapuzzo@unicas.it
[2] University of Cassino and Southern Lazio, Via G. Di Biasio 43, 03043 Cassino, Italy
nicolosi@uniroma2.it, daniela.santilli@unicas.it
[3] CERIS, Instituto Superior Técnico, Universidade de Lisboa, Av. Rovisco Pais,
1049-001 Lisbon, Portugal
luispicadosantos@tecnico.ulisboa.pt

Abstract. Pavement maintenance is essential to prevent the deterioration of asset value and satisfy all stakeholders' expectations. However, the budgets are often insufficient to keep the road pavement at optimum levels. Therefore, a decision-making process ought to be used to prioritize different maintenance activities to optimize the fulfillment of the pre-defined goals. At the same time, there is a growing need to integrate risk management into asset management, and therefore into the Pavement Management System. It is the best way to understand risk in decision-making at the program and organizational levels. This paper examines how risk-based pavement management practices can be implemented. The idea is to identify the best combination of maintenance actions given budget constraints, also considering budget risk reduction within a multiobjective optimization process. As far as economic risk assessment is concerned, probabilistic LCCA with Monte Carlo Simulation was used to investigate the risk of budget exceeding in Pavement Management Systems as a secondary criterion for choosing the optimal maintenance strategy on a road network. The method allows assessing epistemic uncertainties regarding discount rate and materials, man-power, transportation, and equipment rental costs. Outputs were able to show the possible variability of maintenance strategies costs. Moreover, probability density functions provide for establishing the most economically advantageous solutions (lower mean value) and for the riskiest ones (greater standard deviation). The optimal strategy might be selected by minimizing the probability of budget exceeding. The innovation of this research is in the introduction of the quantitative economic risk analyses into pavement management, with the aim to integrate epistemic and aleatory uncertainties in the process.

Keywords: Pavement management · Budget risk management · Multi-criteria decision-making

O. Gervasi et al. (Eds.): ICCSA 2021, LNCS 12950, pp. 394–410, 2021.
https://doi.org/10.1007/978-3-030-86960-1_28

1 Introduction

Roads are the core of an integrated transport system and their performance is essential for all citizens in terms of mobility, quality of life, economic competitiveness and sustainable development. Furthermore, road infrastructures are a huge economic public asset in many countries, for example the European road network consists of 5.5 million Km and, according to European Road Federation "ERF" it represents an estimated value of over € 8,000 billion.

Since road networks have an important instrumental function and a significant asset value, it is necessary to maintain them in order to ensure adequate performance and their durability.

Asset management has been widely accepted by road organizations as a means to deliver a more efficient and effective approach to management of highway infrastructure assets through longer term planning, ensuring that standards are defined and achievable for available budgets.

The transportation infrastructure system includes many assets: pavements, signals, lighting, bridges, right-of-way, and all other roadway related structures. Thus, an asset management system can include various individual component management systems such as a pavement management system (PMS), bridge management system (BMS), etc., all coordinated together. Asset management can only be efficient if component management systems are efficient, properly integrated into the process and coordinated. Particularly, road pavements represent one of the most important infrastructure elements, thus needs to be managed in such a way that supports broad asset management approach.

Analyses for prioritization of interventions usually consider average values for the variables. However, the need to evaluate the risk associated with the planning decisions undertaken is becoming increasingly evident, as this can produce variability of used and/or available economic resources and variability of expected performance.

For this reason, systems that support resource allocation decisions should be implemented with a Risk Management (RM) process, in order to ensure the most effective use of limited funds to maintain assets. Today, the leading international transportation organizations, manuals and standards consider risk assessment and management an important components of the asset management process (ISO 2015). Although they pay adequate attention to the definition of risk in asset management, little guidance is given on how risk management can be applied to road asset management.

All activities from management, identification and prioritization of works to the establishment of budgets have risks and opportunities associated with them. The most common use of the term "risk", when applied to transportation infrastructure, refers to the risk of failure of a transportation asset, but there are several different types of risks that can impact the management of pavements in a highway network, such as: funding uncertainty, variability of costs, regulatory changes, improper maintenance treatment selection, inaccurate pavement condition forecasts, severe weather events and evolving technology.

Risks affecting transportation asset management programs can be generally grouped in 4 categories:

• Economic Risks - increasing material prices, rising interest rates, budget uncertainty or shortfall;

• Hazard and external Risks - natural events, such as earthquakes, floods, climate changing, hurricanes;

• Strategic Risks - environmental standards, stakeholder demands, public opinion;

• Operational Risks: weak program management, lack of management support, poor asset inventories, maintenance failures.

The analytical definition of risk, or "Risk Rating", usually adopted is:

$$R = L * C$$

where "L" is the likelihood for an event to occur, and "C" is its consequence.

Approaches managing risk range from qualitative assessments of likelihood and consequence at the enterprise level to quantitative, probabilistic approaches such as scenario analysis, simulation, and other approaches to predictive modelling. Current practices provide for qualitative risk analyses, and most road organizations found in "Risk Register" the perfect tool to integrate the Risk Management in Management Systems at Network Level.

Qualitative risk analysis cannot be integrated in the multi-criteria optimization approaches required in the modern asset management system.

The objectives of this research project is to:

• Introduce a quantitative enhanced technique to evaluate economic risk related to road pavement;

• Develop a procedures for incorporating risk in the multi-objective optimization tools of the PMS;

• Demonstrate the effectiveness of the proposed methodology by applying it to a case study.

The article is structured on 5 sections: after this introduction, Sect. 2 describes the analytical methods to deal with economic risk and uncertainties. Subsequently, Sect. 3 analyses the stochastic modeling to determine life-cycle cost of maintenance strategies and Sect. 4 presents a case study. Finally, Sect. 5 summarizes the features of the proposed methodology and the results obtained.

2   Analytical Methods to Deal with Economic Risk and Uncertainties

Aging infrastructure, increasingly sophisticated stakeholder demands and low funding availability have created a need for increasingly efficient tools to manage road networks and to take wise decisions of investing founds. Net present value (NPV), pay back period (PBP), internal rate of return (IRR) are some of widely used tools to evaluate alternative infrastructure investment options. In this paper, life-cycle cost analysis (LCCA) has been used to evaluate alternative maintenance options, as it considers every cost category throughout the life of the maintenance strategies, and better represents the effective use of funds in its totality.

LCC can be defined as "the sum of the total direct, indirect, recurring, nonrecurring, and other costs estimated to be incurred in the design, development, production, operation, maintenance, support and final disposition of a system over its anticipated useful life span" (U.S. Department of Energy 2011). There are numerous costs associated to its calculation, such as construction costs, user's costs, maintenance and repair costs, residual value, and replacement costs.

Many methods can be used to analyze life-cycle costs of infrastructure rehabilitation alternatives (Hudson et al. 1997).The AASHTO Guide (AASHTO 1993) recommends the use of the Present Worth Method to evaluate construction/rehabilitation alternatives for infrastructure projects, and particularly pavement and highway projects. It involves discounting all future costs to the present using a specific discount rate. The main advantage of using this method is that it is straightforward and simple.

Using the Present Worth Method, life cycle cost model can be applied through the following equation, that represents the whole life cycle cost for a maintenance strategy:

$$LCC_S = \sum_{x=1}^{N} LCC_{x,s} = \sum_{x=1}^{N} \left[\sum_{t=0}^{T} PWF_t \cdot \rho_{s,i,x,t} \cdot CM_{i,x,s,t} + \sum_{t=0}^{T} PWF_t \cdot CU_{x,s,t} - PWF_T \cdot SV_{x,s} \right] \quad (1)$$

Where:

LCC_s is the total present worth of life-cycle cost for the maintenance strategy s in the analysis period (from year 1 to T),

$LCC_{x,s}$ is the present worth of life-cycle cost for the maintenance strategy s in the road section x,

PWF_t is the present worth factor of future amount at time t at the discount rate r $PWF_t = 1/(1+r)^t$,

$CM_{i,x,s,t} = UCM_i \cdot Q_{x,i,s,t}$ is the cost of maintenance intervent i on the road section x for the maintenance strategy s, in the year t of the analysis period,

UCM_i is the unit cost of maintenance intervent i.

$Q_{x,i,s,t}$ is the estension of intervent i on the road section x at time t for the strategy s.

$\rho_{s,x,i,t}$ is a bolean variabile $= 1$ if in the maintenance strategy s the intervent "i" is applied on the road section "x" at time t, $= 0$ otherwise.

$CU_{x,s,t}$ is the user cost on the road section x for the maintenance strategy s in the year t of the analysis period (random variable).

$SV_{x,s} = min \left[1; \frac{TF_{x,s}}{T} \right] \cdot \sum_{t=0}^{T} \rho_{s,i,x,t} \cdot CM_{i,x,s,t}$ is the remaining service life value for maintenance strategy s on the road section x at the end of analysis period (year T).

TF is the time to failure (i.e. the remaing life at the end of period of analysis "T" for the maintenance strategy s on the road section x).

Some of the quantities in the formulation are random variables (i.e. UCM_i, r) so the result obtained is also a random variable. In the evaluation of the life cycle costs of the maintenance strategies, the variations of the costs sustained by the users are often difficult to calculate and they are of less interest for the managers therefore, in order to make simpler the problem, at least in a first approach, they have been neglected.

Therefore the Eq. (1) is simplified as follows:

$$LCC_{x,s} = \sum_{t=0}^{T} PWF_t \cdot \rho_{s,i,x,t} \cdot CM_{i,x,s,t} - PWF_T \cdot SV_{x,s} \quad (2)$$

To estimate the stochastic variability of the LCC parameter on the road section for the maintenance strategy s, the variabilities of the quantities that appear in the Eq. (2) have been analyzed.

The Life Cycle Cost (LCC) of the construction and maintenance actions cannot be univocally defined, as it is affected by risk and uncertainties. Therefore, it needs to be treated by a Risk Management process, in order to get useful outputs to the decision-making process.

Usually, the best guessed values for each input variable are selected and then are computed in a single deterministic result. On the other hand, this approach often excludes information that could improve the decision, even if a sensitivity analysis is conducted (FHWA 1998). Monte Carlo Simulation (MCS), as a probabilistic method, can take the haziness of the variables into account and produces a random variable as a finale Net Present Value for LCC, with a probability distribution function (PDF).

Interpretation of risk analysis goes beyond a simple comparison of which alternative on average costs less. In the example of Fig. 1, two NPVs of maintenance strategies, considered as random variables, are represented by their cumulative distribution function (CDF).

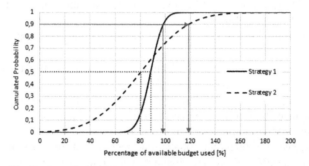

Fig. 1. Cumulative risk profiles of two maintenance strategies

Considering the budget that not be exceed with a probability of 90%, from the cumulative distribution curves clearly emerges that the best solution for a risk-based approach is strategy 1 although strategy 2 has an average budget used lower than strategy 1. This because the chances of funding overrun for strategy 2 are lower than strategy 1.

It have to be highlight that LCCAs have been rarely implemented in practice (Wu, et al, 2015) because they are usually a large-scale problem involving many input parameters with a high-level uncertainty. Besides, the time-horizon of the analysis plays a fundamental role and its increasing could leads to unreliable output. While it is possible to represent through the described process the variability of the parameters, it is difficult to fully represent the vagueness of the long-term forecasts (e.g. investigate the whole life-cycle cost of a pavement can still lead to unreliable results). Therefore, it is advisable to restrict the use of probabilistic LCCA to time-horizon of 4/5 years. However, this does not represent a limit for the application in PMS, as the time analysis for maintenance program is usually 3 to 5 years. In the next section a methodology to determine LCC of the maintenance strategies by a stochastic approach is presented.

3   Stochastic Modeling to Determine LCC of Maintenance Strategies

The LCC accuracy of strategies is strongly conditioned by the unit cost of the interventions. A deterministic approach obscures risk associated with road pavement treatment selection and inhibits the capability of the agency or administration to mitigate the budget risk. Preservation costs are often unpredictable, as they are affected by uncertainties and should be treated probabilistically. An attempt to assess risk on unit costs have to be made for each of the considered interventions. To perform a probabilistic analysis, it is necessary to apply some variations to the factor affecting LCCs, attributing to them a theoretical probability distribution. It is possible to do this by two ways:

1. Defining probabilistic variations on the basis of a large data sample gathered e.g. historical data, data from bid tabulations or management systems (Pittenger et al. 2012);

2. Choosing a priori the most suitable probability distribution and calibrating it based on a small amount of data or qualitative information.

The framework of the proposed methodology is reported in Fig. 2 and it is explained in detail in the following subsections.

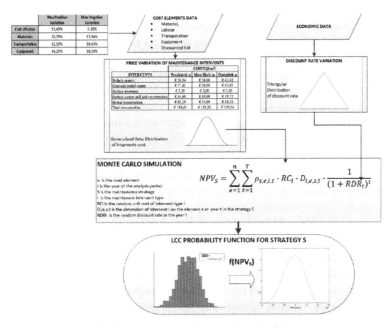

Fig. 2. Framework of proposed methodology

3.1   Probabilistic representations of the Factors Influencing the Cost of Interventions

The normal distribution is commonly assumed to be the appropriate distribution for data and is further justified by the central limit theorem. However, selection of normal distributions can introduce, in some cases, bias into the LCCA (Tighe 2009) for various reasons. Firstly, normal distribution has not the capability to model distribution with a finite range, as it is defined from $(-\text{``}\infty\text{''})$ to $(+\text{``}\infty\text{''})$; secondly, most of the costs would tend to yield an asymmetric shape with a positive skew, so normal distribution is not suitable. For the second reason, the lognormal distribution should be more appropriate. The compromise between the two distributions, which also makes it possible to define the probability distribution in a finite interval, is the Generalized Beta Distribution (GBD).

Therefore, in this study, a generalized beta random variable for the stochastic representation of factors that influence the cost of interventions (i.e., materials, equipment, labor, and contractor profit) was used. The actual costs of the administration are also affected by the discount applied by construction companies for maintenance activities (tender discount). This discount can obviously be at most equal to the profit of the construction company, which is normally 10%, so it is plausible to imagine that the discount can fluctuate between 10% and 0%. The unit cost is:

$$UCM_i = (LC_i + MC_i + TC_i + EC_i) \cdot d \tag{3}$$

Where:

LC_i is the labor cost for the maintenance intervent i,
MC_i is the material cost for the maintenance intervent i,
TC_i is the transport cost for the maintenance intervent i,
EC_i is the equipment cost for the maintenance intervent i,
d is the tender discount.

It can be assumed that unit cost can be represented through a Generalized Beta Distribution (GBD) with the following probability density function (PDF):

$$f(z) = \frac{1}{B(a, b)} \cdot \frac{(z - p)^{a-1} \cdot (q - z)^{b-1}}{(q - p)^{a+b-2}} \tag{4}$$

$$B(a, b) = \int_p^q \left(\frac{z - p}{q - p}\right)^{a-1} \cdot \left(\frac{q - z}{q - p}\right)^{b-1} dz \tag{5}$$

where "a" and "b" are shape parameters and "p" and "q" are respectively the minimum and maximum value, as $f(p) = 0$ and $f(q) = 0$.

The availability of data allows to obtain the probability density functions in rigorous way, by applying the method of moments for estimating all the parameters of GBD (Zaven et al. 2000).

However if only limited data are available but, the parameters of the distribution can be equally estimated (Johnson and Kotz 1970). As matter of fact, starting from the estimate of "optimistic cost" p, "most likely" m, "pessimistic" q values, the mean and variance of the distribution can be calculated as follow (Hendirckson 2008):

$$a = \frac{(\mu - p)^2 \cdot (q - \mu) - \sigma^2 \cdot (\mu - p)}{\sigma^2 \cdot (q - p)} \tag{6}$$

$$\sigma^2 = \frac{1}{36} \cdot (q - p)^2 \tag{7}$$

If mean, variance and extreme values "p" and "q" are known, the shape parameters of GBD can be obtained as follows (D'Apuzzo and Nicolosi 2010):

$$a = \frac{(\mu - p)^2 \cdot \cdot (q - \mu) - \sigma^2 \cdot (\mu - p)}{\sigma^2 \cdot (q - p)} \tag{8}$$

$$b = \frac{(\mu - p)^2 \cdot (q - \mu) - \sigma^2 \cdot (q - \mu)}{\sigma^2 \cdot (q - p)} \tag{9}$$

So, considering the unit cost of each type of maintenance work as a generalized beta random variable and estimating the most probable value, the optimistic one and the pessimistic one, the probability function of unit costs can be obtained, applying the relations represented above. Then, multiplying the random variable "discounted unit cost" of the intervention by the extent of the intervention on each section "x" (i.e. the surface of the pavement on which to carry out the intervention) the random variable cost of the intervention i on section x can be achieved as:

$$f\left(CM_{i,x,s,t}\right) = f(UCM_i) \cdot Q_{x,i,s,t} \tag{10}$$

3.2 Probabilistic Representations of the Discount Rate

The discount rate in analyses is normally assumed to be constant throughout the analysis period, although the historical data indicate that it is not particularly stable. Since it is difficult to predict the average value of the discount rate during the analysis period, it is appropriate to consider this parameter as a random variable in the risk analysis. In this study, a symmetric triangular probability distribution was used to describe the discount rate random variable, as suggested by FHWA (FHWA 1998). Thus, the probability distribution can be uniquely determined once the minimum and maximum values are estimated during the analysis period.

3.3 Probabilistic representations of the Life Cycle Cost

Life-cycle cost assessment contains complex functions of random variables so it is not possible to estimate the final probability distribution a priori. Therefore, a Monte Carlo simulation (MCS) has been employed to obtain the final distribution of the cost on the life cycle.

After the probability distribution of the unit costs of the single interventions has been determined a vector of m = 1000 random numbers that follow the wanted distribution is generated through MCS for each maintenance intervention. For the generic intervention "I", the cost matrix [UCM$_I$] can be, for example:

$$[UCM_I] = \begin{pmatrix} UCM_{I1,1} & \cdots & UCM_{I1,m} \\ \vdots & \ddots & \vdots \\ UCM_{IN,1} & \cdots & UCM_{IN,m} \end{pmatrix}_{Nxm} \tag{11}$$

Similarly, from the known probability distribution of the discount rate, it is possible to generate, through a random sampling, 1000 values of r (discount rate) and then a matrix of values of the Present Worth Factor (PWF) for each year of the analysis period:

$$[PWF] = \begin{pmatrix} PWF_{m=1,t=1} & \cdots & PWF_{m=1,t=5} \\ \vdots & \ddots & \vdots \\ PWF_{m=1000,t=1} & \cdots & PWF_{1000,t=5} \end{pmatrix}_{mxt} \tag{12}$$

Replacing these data in the Eq. (2) it is possible to obtain, for every strategy of maintenance "s", 1000 values of the LCC of which it is possible to calculate the average and the statistical variance. Therefore applying the theorem of the central limit it is reasonable to assume that the distribution of the total LCC is a normal random variable whose parameters are known using the statistical estimates of the average and the variance carried out on the base of the data obtained from the Monte Carlo simulation.

Thereafter, they are combined with working areas for each section, obtaining in the end the Net Present Value (NPV) of the strategy.

The risk analysis lets to different outcomes such as total NPV of the solutions and Yearly NPV for each year of analysis. As the probabilistic LCCA does not provide a single result, every parameter can be useful to the decision-making process, in order to select the best strategy selection. Consequently, a method based on the probability of exceeding the budget limits is proposed, with the aim to consider whether the probabilistic LCCA can be regarded as the most suitable tool to investigate the financial risk.

In the following section, a case study will be analyzed with the aim to show how the budget risk minimization can be introduced in the probabilistic calculation of LCC at Network Level.

4 Case Study: Choice of M&R Strategy with Budget Risk Reduction

The methodology was adopted to test the effectiveness and the incidence of probabilistic Life Cycle Cost Calculation, overall decision-making process; risk is assessed through the evaluation of Epistemic Uncertainties regarding M&R unit costs interventions for the Italian context.

The study consisted in the identification of the optimal maintenance strategies for the main road network of IX Municipality in Rome that involved 46 road sections with a total length of 23 km.

The time period of the analysis is 5 years and three strategies are considered for the sections maintenance:

- Strategy 1: Do Nothing;
- Strategy 2: Functional and Preservation Maintenance;
- Strategy 3: Structural Maintenance.

Decision trees express the trigger values of their realization. The evolution of pavement performance and the calculation of work quantities, are obtained by Highway Development and Management Software "HDM-4".

A multiobjective optimization method has been used for prioritizing different maintenance activities in order to achieve pre-defined goals, optimizing the use of the available budget. One of the biggest difficulties in multiobjective optimization method is the large number of the feasible solutions (Pareto optimal set or its approximation), which makes it hard for the Decision Maker to select the best solution. To support interaction with the decision maker for identifying the best combination of maintenance actions, this paper refers to a new methodology named "Interactive Multiobjective Optimization-Dominance Rough Set Approach" (IMO-DRSA), already used in a previous research. For more details about the method see (Augeri et al. 2019).

In order to select the most suitable solution starting from the feasible solutions of the multi-objective optimization, in this paper a new element is introduced, in fact additional information comes from minimizing the risk of exceeding the budget was added.

For the analytical formulation of the problem, the following variables was considered:

- $x_{i,u}$ is decision variable, and it is equal to 1 if the strategy u is applied to the section, 0 otherwise;
- S is the set of all the possible strategies to be applied on road section i;
- $C_{i,u}$ is the cost of maintenance strategy u on section i;
- C_{TOT} is the total budget constraint in the planning period;
- C_{YEAR} is the yearly budget constraint in the planning period;
- $c_{i,u,y}$ is the cost of maintenance strategy u, applied on section i, in the year y, as a discount rate will be considered;
- l_i is the length of section i;
- $L = \sum_i^n l_i$ is the total length of the network;
- $t = 5 years$ is the planning period.

The constraints of the optimization problem are:

$$\sum_{i=1}^{n} \sum_{u=1}^{s} \sum_{y=1}^{t} c_{i,u,y} \cdot x_{i,u} \leq C_{TOT}, \; C_{TOT} = 4.034.300 \; € \qquad (13)$$

$$\sum_{i=1}^{n} \sum_{u=1}^{s} c_{i,u,y} \cdot x_{i,u} \leq C_{YEAR}, \qquad C_{YEAR} = 1.049.000 \; € \qquad (14)$$

The yearly budget is obtained from the total one, divided by the years of planning period and increased of its 30%.

The objectives are mathematically formulated by the objective functions described in Table 1. These have to be maximized.

Where:

li is the length of the road section i and is the total length of the road network;

$\overline{PCI}_{i,u,y}$ is the annual average value of pavement condition index (ASTM D5340 standard) on the road section i, in the year y of the analysis period obtained applying treatment provided by strategy u;

$\overline{PSI}_{i,u,y}$ is the annual average value of the Present serviceability Index on section i, in the year y of the analysis period obtained applying treatment provided by strategy u;

Table 1. Objective functions

Pavement structural condition during the planning time-span	User comfort during the planning time-span:
$$F_1(x_{i,u}) = E = \dfrac{\sum\limits_{i=1}^{n}\sum\limits_{u=1}^{s}\sum\limits_{y=1}^{t}\left(\overline{PCI}_{i,u,y}\cdot l_i \cdot x_{i,u}\right)}{L}$$	$$F_2(x_{i,u}) = C = \dfrac{\sum\limits_{i=1}^{n}\sum\limits_{u=1}^{s}\sum\limits_{y=1}^{t}\left(\overline{PSI}_{i,u,y}\cdot TI_i \cdot l_i \cdot x_{i,u}\right)}{L}$$
Average safety improvement during the planning time-span:	percentage of road network in sufficient serviceability condition during the planning time-span:
$$F_3(x_{i,u}) = \Delta AC = \dfrac{\sum\limits_{i=1}^{n}\sum\limits_{u=1}^{s}\left[\left(\sum\limits_{y=1}^{t}\overline{AC}_i\cdot CMF_{i,u,y}\right)\cdot x_{i,u}\right]}{L\cdot t}$$	$$F_4(x_{i,u}) = PSIS = \dfrac{\sum\limits_{i=1}^{n}\sum\limits_{u=1}^{s}\left(l_i\cdot x_{i,u}\cdot\sum\limits_{y=1}^{t} PSIS_{i,u,y}\right)}{L\cdot t}$$
percentage of road network in good serviceability condition during the planning time-span	percentage of road network in sufficient structural condition during the planning time-span:
$$F_5(x_{i,u}) = SIG = \dfrac{\sum\limits_{i=1}^{n}\sum\limits_{u=1}^{s}\left(l_i\cdot x_{i,u}\cdot\sum\limits_{y=1}^{t} PSIG_{i,u,y}\right)}{L\cdot t}$$	$$F_6(x_{i,u}) = PCIA = \dfrac{\sum\limits_{i=1}^{n}\sum\limits_{u=1}^{s}\left(l_i\cdot x_{i,u}\cdot\sum\limits_{y=1}^{t} PCIS_{i,u,y}\right)}{L\cdot t}$$
percentage of road network in good structural condition during the planning time-span:	
$$F_7(x_{i,u}) = PCIG = \dfrac{\sum\limits_{i=1}^{n}\sum\limits_{u=1}^{s}\left(l_i\cdot x_{i,u}\cdot\sum\limits_{y=1}^{t} PCIG_{i,u,y}\right)}{L\cdot t}$$	

TI is the traffic impact factor (TI = 1 if AADT $\leq$ 10000 veic/day TI = 2 if 10000 < AADT $\leq$ 20000 a nd TI = 3 if AADT > 20000).

ΔACi is the expected number of accidents on site i without maintenance activity implementation.

As mentioned earlier, three strategies were considered for the maintenance of the road sections. Each of them includes different interventions that can be applied, with different costs associated (Table 2).

The fluctuation of the unit costs was assessed and for each of the considered interventions the impact of the four factors that make up the final cost was evaluated. These factors are: cost of labor, materials, transportation, equipment rental.

Starting from ISTAT (Italian Institute of Statistic) data, positive and negative maximum variation of costs for a period of 5 years were evaluated obtaining percentages in Table 3.

Table 2. Unit cost of maintenance intervention

Interventions	Unit Cost €/m^2	Strategies
Pothole repair	50	Do Nothing
Surface patching	50	Do Nothing
Crack and pothole repair	50	Preservation/Structural Maintenance
Chip seal	2,97	Preservation/Structural Maintenance
Surface course reconstruction	29,57	Preservation/Structural Maintenance
Surface course reconstr. and reprofi	38,08	Preservation/Structural Maintenance
Partial reconstruction	53,61	Structural Maintenance
Total reconstruction	160,5	Structural Maintenance

Table 3. Maximum variations of unit cost factors for a 5-year period of analysis

	Max positive variation	Max negative variation
Cost of labor	13,43%	5,38%
Materials	22,59%	13,94%
Transportation	12,12%	10,43%
Equipment rental	14,25%	18,50%

Subsequently, the variations were applied to the cost factors that determine the interventions final price, in order to obtain the extreme values. Results are reported in Table 4 for only one intervention to show how the final price is obtained. Instead, in Table 5 there is an overview of the final prices for all interventions.

Table 4. Variation applied to cost factors for "partial reconstruction"

Partial reconstruction	Price (€/m^2)	Percentage	Maximum value (€/m2)	Minimum value (€/m2)
Cost of Labor	12	23%	14,03	11,72
Materials	27	50%	32,63	22,91
Transportation	4	7%	4,06	3,24
Equipment Rental	11	21%	12,57	8,97
Final price	**54**		**63,29**	**36,10**

To perform the probabilistic analysis, it is necessary to attribute to these costs a theoretical probability distribution. So, using a distribution function that is better suited to the variability of the cost, the variation of price index of road construction in Italy was analyzed.

Table 5. Price Variation of the interventions

Interventions	Price ($€/m^2$)	Maximum value ($€/m2$)	Minimum value ($€/m2$)
Pothole repair	50	58,94	42,42
Surface patching	50	58,85	42,50
Crack and pothole repair	50	57,61	43,05
Chip seal	2,97	3,50	2
Surface course reconstruction	29,57	34,66	19,75
Surface course reconstr.and reprofil	38,08	44,61	25,42
Partial reconstruction	53,61	63,29	36,10
Total reconstruction	160,5	189,41	109,14

Following the framework illustrated in Sect. 3.1, the probability density functions of unit costs for all the considered interventions were calculated. Results are shown in Table 6.

Table 6. Parameters of GBD for unit costs of maintenance work.

Intervention	Mean [$€/m^2$]	Std Dev	a	b
Pothole repair	50,23	2,75	3,77	4,20
Surface patching	50,23	2,72	3,77	4,21
Crack and pothole repair	50,11	2,43	3,88	4,12
Chip seal	2,92	0,25	4,62	2,93
Surface course reconstruction	28,79	2,49	4,60	3,00
Surface course reconstruction and reprofiling	37,06	3,20	4,60	2,99
Partial reconstruction	52,31	4,53	4,57	3,10
Total reconstruction	156,76	13,38	4,56	3,13

To carry out the risk analysis, for each solution (strategy) obtained by multi-objective optimization problem, 1000 scenarios have been generated by Monte Carlo Simulation (MCS). To this purpose, a Matlab code was developed and fitted to the case study.

For explanatory purpose, the obtained outputs are showed for one of the strategies obtained by optimization process, named S1. The first output is the histogram of the scenarios, from which a normal probability distribution has been obtained by means of a statistical inference, and then the cumulative distribution function has been determined (Fig. 3).

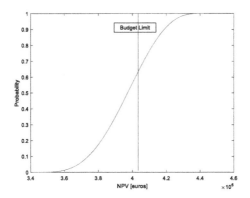

Fig. 3. NPV Cumulative Distribution Function of Strategy S1

The result achievable from these representations is the probability that, for the strategy considered, the budget limit is not exceeded. For the case in analysis, this turns out to be 63%.

The second constraint considered in the multi-objective optimization problem takes the limitation on the Annual Budget into account. Thus, the annual budget risk was investigated for each year of analysis: the functions related to the Yearly NPV were derived in the same way as done for the Total NPV. So, for each strategy, the probability of not exceeding the annual budget was evaluated as shown in Fig. 4 for strategy S1.

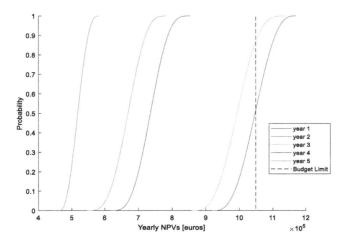

Fig. 4. Yearly NPVs Cumulative Distribution Functions

The parameter Annual Risk Rating (ARR) can be introduced, in order to establish an annual risk factor. It can be evaluated by the following expression:

$$ARR = \frac{1}{t} \cdot \sum_{y=1}^{t} \left(P\left(Yearly_{NPV_y} \geq Annual_{Budget} \right) \right) \tag{15}$$

All the information and parameters obtained for each strategy have to be compared to identify a risk-based solution.

Once the variable "cost of intervention I" on section 1 at time t is known, it is possible to associate to each maintenance strategy, besides the performance indicators, also a random total cost (sum of the variable cost of the various interventions chosen). For the theorem of the central limit, the sum of the variable costs of the interventions of the program p will tend to a normal aleatory variable, with average equal to the sum of the average costs and variance equal to the sum of the variances, having hypothesized that the costs are between them independent. This hypothesis is not entirely true but can be considered acceptable.

All the information and parameters obtained for each strategy can be compared to identify a risk-based solution.

The criterion for the optimal strategy selection is:

$$\min(P_i, ARR_i) \tag{16}$$

The choice can be supported by the comparison of cumulative distribution functions of the strategies, as reported in Fig. 5 for some of the obtained strategies, as example.

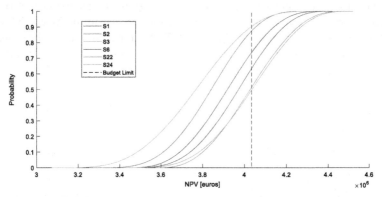

Fig. 5. NPV cumulative distribution functions of some analyzed strategies

The analysis of these curves can be useful to the decision.

The widest curve, which corresponds to Strategy 3, is the one with the greatest amplitude (standard deviation) and therefore representative of the riskiest solution. The worst-case scenario of strategy 3 is not remarkably different from the worst-case scenario of strategy 2; on the other hand, strategy 3 best-case scenario is considerably lower than the strategy 2 one.

Consequently, the riskiest solution appears to be the most economically advantageous, as the most optimistic scenarios belong to it. Finally, the choice of strategy 3 offers the possibility to "take advantage of the risk": despite the solution presents the greatest uncertainty and variability on the final cost, it is the one that minimizes the budget risk.

5 Conclusion

An efficient Asset Management System has to include risk analyses to ensure more informed and transparent choices in resource allocation. Qualitative analyses can surely improve the knowledge of the risk levels involved in maintenance planning; however, they are not sufficiently detailed to improve the decision-making process. Instead, quantitative risk analyses provide more comprehensive results, as the probability theory can assimilate epistemic and aleatory uncertainties.

This paper aims to examine how budget risk reduction can be considered to identify the best combination of maintenance actions given budget constraints within a multiobjective optimization process.

Probabilistic LCCA with Monte Carlo Simulation was used to investigate the risk of budget exceeding in Pavement Management Systems as a secondary criterion for choosing the optimal maintenance strategy on a road network. The method allowed to assess epistemic uncertainties regarding costs and discount rate. Outputs were able to show the possible variability of maintenance strategies costs. Moreover, probability density functions provided for establishing the most economically advantageous solutions (lower mean value) and the riskiest ones (greater standard deviation). Thus, the optimal strategy might be selected by minimizing the probability of budget exceeding. The proposed framework can be used within any multi-objective optimization method.

References

1. ISO 55001 "Asset management – Management systems – Requirements", 2015
2. U.S. Department of Energy. (2011). Cost Estimating Guide. Washington D.C.
3. Hudson, W., Haas, R., Uddin, W.: Infrastructure Management. McGraw-Hill, New York. (1997)
4. AASHTO. Guide For Design Of Pavement Structures, pp I49-I51. (1993)
5. FHWA. Life-Cycle Cost Analysis in Pavement Design. Washington, D.C.: U.S. Department of Transportation (1998)
6. Wu, D., Yuan, C., Liu, H.: A risk-based optimization for pavement preventive maintenance with probabilistic LCCA: a Chinese case. Internation Journal of Pavement Engineering (2015)
7. Pittenger, D., Gransberg, D., Zaman, M., Riemer, C.: Stochastic LCCA for Pavement Preservation Treatment; pp 45–51.: Transportation Research Board No. 2292. Washington D.C. (2012)
8. Tighe, S.: Guidelines for Probabilistc Pavement Life Cycle Cost Analysis. Washington D.C.: Transportation Board Research Record: Journal of the Transportation Research Board No. 2093, pp 84–92 (2009)
9. Zaven, A., Dudewiez, E.: Fitting Statistical Distribution: The Generalized Lambda Distribution and Generalized Bootstrap Methods. Chapman and Hall/CRC, New York (2000)
10. Johnson, N., Kotz, S.: Continuous Univariate Distributions-2; Chapter 24. John Wiley & Sons. (1970)

11. Hendirckson, C.: Project Management for Construction: Fundamental Concepts for Owners, Engineers, Architects and Builders; Chapter 11. Departmen of Civil Engeneering, Carnagie Mellon University, Pittsburgh (2008)
12. D'Apuzzo, M., Nicolosi, V.: A new methodology for stochastic modelling of pay factors in hot-mix asphalt pavements. Road materials and pavement design **11**(sup1), 559–585 (2010)
13. Augeri, M.G., Greco, S., Nicolosi, V.: Planning urban pavement maintenance by a new interactive multiobjective optimization approach. Eur. Transp. Res. Rev. **11**(1), 1–14 (2019). https://doi.org/10.1186/s12544-019-0353-9

Urban Air Mobility: A State of Art Analysis

Dominique Gillis[2(✉)], Massimiliano Petri[1] 🄳, Antonio Pratelli[1], Ivana Semanjski[2,3], and Silvio Semanjski[4]

[1] Department of Industrial and Civil Engineering, University of Pisa, Pisa, Italy
{m.petri,a.pratelli}@ing.unipi.it
[2] Department of Industrial Systems Engineering and Product Design, Ghent University, Ghent, Belgium
{Dominique.Gillis,Ivana.Semanjski}@UGent.be
[3] Industrial Systems Engineering (ISyE)—Flanders Make@Ghent University, Kortrijk, Belgium
[4] Seal Aeronautica S.L., Barcelona, Spain
silvio.semanjski@sealaero.com

Abstract. Urban Air Mobility (UAM) is an aerial component of urban mobility system which integrates an emerging transport mode, Unmanned Aerial Vehicles (UAV), also known as drones, into multimodal urban mobility context. UAM has potential to bring new services related to both passengers and logistic/freight mobility (like passenger carrying air taxis or small package delivery drones) as well as enable better resilience in emergency situations resulting due to various causes (for example, traffic accidents, traffic congestion, catastrophic events and others). Such new services, accelerated thanks to the recent introduction of Vertical Take-Off Landing (VTOL) capable vehicles, able to need less air space to take-off or landing, have the possibility to transform the way people move within, around and between urban areas by shortening commute times, bypassing ground mobility congestion and enabling specific and oriented point-to-point flight across the cities. As currently the UAM integrations are still hindered by a number of constraints, in this paper we provide an overview of UAM legislative frameworks, with particular focus on European Union (EU), together with the overview of the most relevant UAM case studies and the potential new UAM services.

Keywords: Urban Air Mobility · Sustainable mobility · Passenger mobility · Logistic and freight delivery

1 Introduction

In recent years, Unmanned Aerial Vehicles (UAV), also known as drones, have gained much relevance. The initial applications of UAV technologies included a somewhat limited number of tasks, such as aerial photography or mapping, that were either costly or inconvenient to be performed using traditional aerial vehicles. However, with the advances concerning the payload of UAVs and relevant technologies, the number of possible applications have rapidly expanded paving the way for future UAV based services. This is even more highlighted in the context of a number of catastrophic and

© Springer Nature Switzerland AG 2021
O. Gervasi et al. (Eds.): ICCSA 2021, LNCS 12950, pp. 411–425, 2021.
https://doi.org/10.1007/978-3-030-86960-1_29

emergency events (due to climate change, to the recent pandemic and other causes). Hence, it has become evident that UAVs will make up an increasingly relevant component of the mobility system, when it comes to transportation of both fright and humans, in the years to come.

Nonetheless, the initial pilot studies and demonstrator use cases indicate that there are still a number of challenges associated with the wider adoption of UAV based services. For one, UAVs will need to fit into existing, both ground and aviation, systems to ensure ongoing safety and public acceptance. These systems have been established for a while and are underpinned by a number of regulations and standards, which are reflected in international conventions, national legislations, regulations, and practices. Integrating UAVs services and operations into these systems without compromising the safety or security will present a significant challenge for many years to come. This is especially pronounced in urban areas where UAVs represent Urban Air Mobility (UAM) component of urban mobility system, characterised by densely populated areas and highly interlinked mobility system networks and human and freight mobility patterns. Associated challenges as legal and regulatory, environmental, certification, public perception and infrastructure constraints become even more evident in such context, hindering the potential use and future development of UAM.

In order to support ongoing UAM development efforts, in this paper, we provide an overview of UAM legislative frameworks, with particular focus on European Union (EU). This section is followed by the overview of the most relevant case studies, starting from inter-governmental urban air transport collaborative programmes and continuing with the analysis of the individual national, regional or local applications, within the EU and wider. This is followed with the brief discussion on UAM and potential new UAM services.

2 Legal and Regulatory Assessment

As the EU agency responsible for regulating air operations and aircraft airworthiness certification, the European Union Aviation Safety Agency (EASA) has initiated a rule-making process which as a goal has regulated VTOL aircraft use in both production and operations, through development of regulatory building blocks to enable safe VTOL operation and new air mobility in Europe. The first building block is a complete set of dedicated technical specifications in the form of a special condition for VTOL aircraft. *This special condition addresses the unique characteristics of these products and prescribes airworthiness standards for the issuance of the type certificate, and changes to this type certificate, for a passenger-carrying VTOL aircraft in the small category. Certification with this small category Special Condition applies to an aircraft with a passenger seating configuration of 9 or less and a maximum certified take-off mass of 3 175 kg (7 000 lbs) or less.*

As a part of the work on the second building block, in 2020 the EASA has published the Proposed Means of Compliance with the Special Condition VTOL *to address the applicant's requests for clarification of EASA's interpretation of these objectives and of possibilities how to demonstrate compliance with them.*

As a part of work on the third building block, also in 2020 the EASA has published proposed methods on how to certify hybrid or electric air taxis (VTOL) [Ref.]. *This*

Special Condition has been developed to support Applications received by the Agency for the certification of Electric and / or Hybrid Propulsion Systems. This Special Condition is articulated so as to provide objective based certification requirements which are independent of the propulsion system design or architecture. The type of technology used in the propulsion system will be addressed in the Acceptable Means of Compliance.

3 Actual UAM Application State of Art

In this section, we describe the application of UAM regarding both passenger mobility than logistic/freight mobility cases and relative to daily mobility or emergency services. The description is divided in two part: a first one describes the inter-governmental urban air transport collaborative programmes while the second part describe briefly some case studies dividing them for geographical area, starting from European Case Studies [3] and 'navigating' the rest of the world through Far East and Australasia, America and Middle East.

3.1 Inter-governmental Urban Air Collaborative Programmes

At european level, starting in May 2018, the European Union's Urban Air Mobility (UAM) Initiative, supported by the European Commission[1], brings together cities and regions, citizens, industries, SMEs, investors, researchers and other smart city actors to join a unique partnership. This initiative is part of the European Innovation Partnership in Smart Cities and Communities (EIP-SCC) and it is aligned with ongoing and future SESAR Joint Undertaking (SESAR JU) funded studies, including demonstrations, on drone traffic management in Europe moving one step closer towards the European Commission's U-space vision for ensuring safe and secure access to airspace for drones. With demonstrable benefits to citizens and their approval, developing a market for drones and drone services will create jobs and growth in Europe. Particularly in urban areas, civil drones could be a way to address mobility needs such as emergency needs and traffic congestion; the latter currently costs more than €100 billion a year in the EU alone.

In June 2018, the EU gave the European Aviation Safety Agency (EASA) new powers to set EU-wide rules for urban air mobility. Following EASA: "The so-called new Basic Regulation formalises EASA's role in the domain of drones and urban air mobility, enabling the Agency to prepare rules for all sizes of civil drones and harmonize standards for the commercial market across Europe", covering the full spectrum of aviation landscape with the possibility for EASA and European Member States to work closer together in a flexible way.

---

[1] Commissioner for Transport Violeta Bulc said: "Drones offer exciting opportunities for new services and business models, particularly in our cities. At the same time we need to ensure that drones operations taking place above our heads are safe, secure, quiet and clean. In that regard, the Urban Air mobility initiative is an important demonstration project involving several European cities to address these challenges and plan for the future. It will also contribute to the EU's U-Space, which is a flagship project of the European Union to manage air traffic at low level.".

In February 2018, the UK NESTA Challenge Prize Centre foundation, that designs and runs challenge prizes that help solve pressing problems that lack solutions, organizes a challenge involving five cities (Bradford, London, Preston, Southampton and the West Midlands) to design how drone technology can be used to support their local needs. In September 2018 NESTA published a research on the potential market of drones in the UK elaborated inside the Flying High project underlining the enormous prospects for future commercial development.

In October 2019, the European Commission and the European Investment Bank (EIB) announced the launch of a "European Drone Investment - Advisory Platform" to support innovation and investment in drones. The initiative aims to improve access to EU support in this field, and to develop a better understanding of the market to improve investment in this emerging field.

Going in the world far-east side, the Japan Prime Minister, in 2017, designs the "Road for the Aerial Industrial Revolution" with a japanese UTM consortium to create a new industrial drone innovation space with communication and collision avoidance technologies with the possibility to incorporate japanese systems into 'international standards" like the NASA's UTM.

In 2018 the Civil Aviation Authority of Singapore - CAAS, the EASA and Airbus have agreed to develop together safety standards and regulatory requirements for unmanned aircraft systems (UAS) in urban environments. The MoU established a framework for the expert knowledge exchange as well as operational and technological assessments for the deployment of UAS in urban environments, such as last-mile deliveries or safety information and learning outcomes from the applied trials.

Moreover, in February 2018, Singapore Government designs the One-North R&D hub to research about the application of drone in urban areas and for test-bedding innovative UAS technologies. Airbus, ST Aerospace and NTU's Air Traffic Management Research Institute have soon come on-board of the One-North drone research.

Finally, going in the American continent, in 2017 the U.S. Department of Transportation (USDOT) and the Federal Aviation Administration (FAA) have launched the Unmanned Aircraft System (UAS) Integration Pilot Program (IPP) to test and evaluate the integration of civil and public drone operations into national airspace system. The IPP Lead Participants evaluated operational concepts and possibility, including night operations, flights over people and beyond the pilot's line of sight, package delivery, detect-and-avoid technologies and the reliability and security of data links between pilot and aircraft. Fields that could see immediate opportunities from the program include commerce, photography, emergency management, agricultural support and infrastructure inspections.

The IPP program ended in October 2020 and, to continue work on the remaining challenges of UAS integration, the FAA launched a new program called BEYOND in the same month. The challenge areas to study in the program regard the Beyond Visual Line of Sight (BVLOS) operations that are repeatable, scalable and economically viable with specific emphasis on infrastructure inspection, public operations and small package delivery, the leveraging industry operations to better analyze and quantify the societal and economic benefits of UAS operations and Focusing on community engagement efforts to collect, analyze and address community concerns.

The DriveOhio Research Program is a partnership started in 2018 between Driveohio's UAS Center and the Ohio State University College of Engineering. The Program foresees the experimentation on 33 road corridors of traffic monitoring with drones, with an experimentation of communication between air and ground vehicles/ITS sensors. Unmanned aircraft will monitor traffic and accident response along the corridor in conjunction with the state's current fixed-location traffic camera system, so interacting with sensors and communication equipment along the corridor to feed data into the Traffic Management Center.

Nasa's Aeronautical Research Mission Directorate (ARMD) in 2018 has updated its strategic plan to also include research regarding Urban Air Mobility. It started with a consultant's research on Urban Air Mobility Market Study [1] describing legal and regulatory assessment, societal barriers, weather barriers, air-shuttle and air-taxi market analysis and air ambulance. In 2019 NASA has signed Space Act Agreements with 17 companies in the aviation industry to advance plans for the first in a series of technology demonstrations known as the Urban Air Mobility (UAM) Grand Challenge (see Challenge framework in Fig. 1). The final results are expected in 2022.

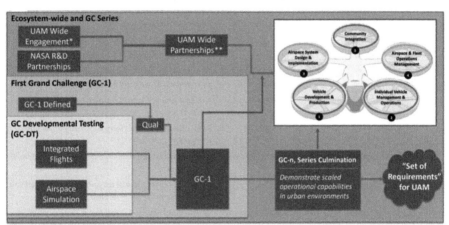

Fig. 1. Partnership and grand challenge series overview ([1]) source

3.2 Worldwide UAM Case Studies

Some recent UAM case studies are summarized in Tables 1 and 2 dividing european from extra-european applications (all cases relate to the period 2017–2021). In these tables, the general safety and flight infrastructure testing cases have not been included as well as cities that joined the UAM, the NESTA or other initiative but only cases regarding application of future UAM services [2, 4, 5, 8–11].

Moreover there are three cross-border area partnerships regarding respectively:

- France-Belgium cross-border region between Kortrijk (Belgium) and Tournai and Lille (France);
- Euregio project between Enschede-Netherlands and Munster-Germany, experiment the use of UAM in the field of emergency services for public safety;
- MAHHL (Maastricht and Heerlen-Netherlands, Aachen-Germany, Hasselt and Liege-Belgium).

Table 1. Some European UAM case studies (O = operative applications)

Country	Area	Field	Description
Austria	Linz	Vehicle technology	Trail flights to evaluate the implementation in urban areas
Belgium	Antwerp	Security	Drones to monitor by the local police
	Antwerp	Health	Medical supplies transport by drone
		Port	Drone for surveillance flight (terminal container and oil spill inspection)
		Civil	Drone for high voltage line mapping and pylon inspection
	Bruxelles	Emergency	Drones for emergency services
	Ghent	Health	Ambulance drone application
	St Truiden	Many	The area is a 'drone port' to conduct tests in dedicated area
Estonia	Tallinn	Logistic	International drone flight between Tallinn and Helsinki
Finland	Helsinki	Emergency	Create a temporary flight restriction area around an accident scene
		Logistic	Wing Company start delivery goods with drone based on an App-O
	Gulf of Finland	Passenger/Security	Maritime traffic surveillance and drone taxi flight from airport to city
France	Nouvelle- Aquitaine	Many	Air goods delivery, tourist transport (based in Bordeaux city) or to cover needs of isolated people
	Paris	Vehicle technology	Test area at Pontoise-Cormeilles-en-Vexin airfield, starting in summer 2021. Focussing on 5 areas: vehicle development, ground infrastructure, airspace integration, operations and public acceptance
	Toulouse	Many	Drone and remote sensing for environm. Monitoring appl., urban services (traffic manag., air quality or infrastr. Monit.), emergency and logistic services
Bulgaria	Plovdiv	Many	Public transport, ambulance services, public safety and small rapid goods del
Germany	Bad Neustadt	Logistic	Drone delivery of pizza-O
	Northern Hesse	Logistic	Logistic across the airports chain of Kassel Calden and Frankfurt am Main
	Frankfurt	Passenger/Logistic	Airport Fraport AG start developing ground infrastructure and operations for air taxi service and drone for operational airport purpose
	Hamburg	Many	Emergency medical delivery, large infrastructure inspection (for ex. Bridges, turbines)

(*continued*)

Table 1. (*continued*)

Country	Area	Field	Description
	Ingolstadt	Health/ Passenger	Pilot programme for air taxi and organ transport
		Logistic	Drones on the national railway network to record storm damage or vegetation on the tracks-O
	Munich	Passenger	Air taxi from airport's labcampus to the city center (Lilium, 2020)
Iceland	Reykjavik	Logistic	Aha and Flytrex food delivery-O
Italy	Rieti	Many	Drone to parcel delivery, road traffic patrol, professional photography, railway and power lines monitoring, search and rescue, airport operations and firefighting
Switzerland	All Country	Passenger	Air taxi to develop last mile trips from railway stations to destination
	Berne, Zurich, Lugano	Health	Drone to bring goods and bloods from laboratory to hospital
	Zurich	Logistic	Drone delivery of e-commerce goods
The Netherl	Amsterdam	Logistic	Domino uses drone for pizza delivery
	Leiden	Passenger	Airbus forms Airbus Urban Mobility division with the ambition to offer passenger drone flights at a cost that is competitive with traditional ground taxi service (Downing, 2019; Airbus, 2020)
United Kingdom	London	Logistics	London residential complex implements rooftop "vertiports", encouraging drone delivery services (Stouhi, 2019)
		Civic/ Safety	Drone for safer infrastr. Inspections and for the capital's emergency services
	Cambridge	Logistics	Amazon carried out its first successful drone delivery operation (2016), unveiled the latest version of its Prime Air delivery drone and announced the intention to launch a delivery service using drones (Vincent and Gartenberg, 2019)
	Bradford	Safety/ Health	Drones To Help Disaster Response, Digital Health, Surveying And Community Safety
	Oxford	Many	Drone-based services to include roof and building surveying, land mapping, aerial photography and filming
	Preston	Civil	Drone for utilities and building inspection (O) and exploring its use for flood management and road networks upgrading

Referring to the MAHHL cities (see Fig. 2), they consider better transport and logistical connections within the region as well as better connections to cross regional transport hubs with the use of drones and Air Taxis. A roadmap considering the inclusion of the mobility plans within the cross border region of the different aircrafts has been created at the official kick-off-event. The map focuses on the inclusion of UAS in rescue missions, transport of medical goods as well as transporting medical specialists. The Institute of Flight System Dynamics currently works on many projects according to the roadmap. One already finished project is GrenzFlug, where in cooperation with the city of Aachen, the first fully automated cross border rescue mission has already been performed recently. Currently further projects for supporting rescue workers within the region are VISION and FALKE. The project SafirMed which started in December 2020 focusses on transporting medical goods.

Several European projects specifically focus on application of UAM for emergency applications. AiRMOUR project focuses on the use of UAM for doctors and medical supplies, in order to develop a UAM toolbox for aviation and urban authorities. The project includes the cities of Luxemburg (simulations), Stavanger, Helsinki and Nord-Hessen (demonstrations). The main objective of MOBNET project is to locate isolated victims during natural disasters and situations of emergency such as earthquakes, hurricanes or large snowstorms. MOBNET will also help first responder services to find lost people in general. AMBULAR has a different focus, with the goal of developing an eVTOL air ambulance. SAFIR-Med (Safe and Flexible Integration of Advanced U-Space Services for Medical Air Mobility) aims at real demonstrations in collaboration with hospitals, focusing on demonstrating the operational safety level.

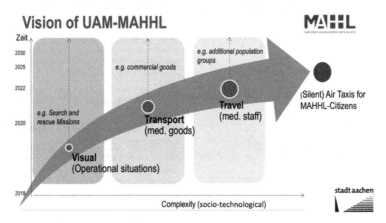

Fig. 2. Partnership and grand challenge series overview ([1]) source

Moreover, in the CASCADE project, started in 2018 and always ongoing, a partnership of five english universities (Bristol, Cranfield, London, Manchester, Southampton), are implementing some case studies, described in Fig. 3 (to which we need to add two recent ones related to the deliver NHS supplies to the Isle of Wight and to the ojin of machine-learning and drone technology to monitor populations of the Critically Endangered Kordofan giraffe at Bénoué National Park):

All the previous case studies show that there is a growing demand for urban services based on air mobility worldwide. In particular, the use of drones sees most of the current applications, with a more mature market, like for operative services regarding private distribution logistics. Applications relating to passenger transport are almost exclusively in the field of experimentation due to the greater complexity of both application and legislation (laws and regulations).

Significant legal and regulatory, weather, certification, public perception and infrastructure constraints exist to date. High service cost is a severe limitation, especially for

Table 2. Some extra-European UAM case studies (O = operative applications)

Country	Area	Field	Description
Australia	*Camberra Logan*	*Logistic*	*Wing Company start delivery goods with drone based on an App - O*
Brazil	*Sao Paulo*	*Passenger*	*Helicopter to bring people out from urban congestion*
Canada	*Calgary*	*Civic*	*Drone to collect mapping data for urban plan*
	Québec	*Vehicle technology*	*Trail flights*
China	*Muyang*	*Health*	*Drones equipped with speakers are being used to criticize people in public without a Covid19 mask*
	Hezhou	*Security /logistics*	*Transport of medical supplies and personnel*
	Shanghai	*Logistic*	*Drone to delivery food for 100 local restaurant on 17 different routes*
	Dongguan, Guandong	*Logistic*	*DHL uses a drone landing on an intelligent cabinet for delivery service*
Japan	*Changsha and Hohhot*	*Logistic*	*JD.com with Rakuten to use drones and UGVs in unmanned goods delivery service*
	Fukushima	*Security*	*After fukushima earthquake, unmanned flying vehicles were used to explore the disaster site, and specifically the nuclear plant*
Jordan	*Dead Sea*	*Security*	*Drone for the arab league summit to ensuring security of world leaders*
Mexico	*Ensenada*	*Security*	*Drone for daily police surveill. With a 10% decrease in crime and 30% reduction in home burglary after six-months pilot application-o*

(*continued*)

Table 2. (*continued*)

Country	Area	Field	Description
New Zealand	*Canterbury, Queenstown*	*Passenger*	*Air-taxi into the air-navigation service*
	Tararua	*Logistic*	*Drones to capture images of slips in the district's network of rural roads and where streams have scoured their banks to threatening roads*
Pakistan	*Karakoram Mountains (Himalayas)*	*Security*	*Application of a drone to locate and rescue a lost mountainier*
Singapore	*All Country*	*Many*	*UAS for solutionsas surveillance, inspection, package and marittime deliv*
South Korea	*Seoul*	*Logistic*	*Drone for disaster monitoring and logistic transport*
Turkey	*Zigana mountain region*	*Security*	*Experiment for use of drones in searching and location victims in a mountain area*
United Arab Emir	*Dubai*	*Many*	*Air-taxi service, drone delivery and traffic surveillance to incident detection*
USA	*Virginia*	*Logistic*	*Wing Company start delivery goods with drone based on an App - O*
	North Carolina	*Health*	*Drones to monitor residents' adherence to imposed protection measures (Covid-19)*
	Nevada	*Health*	*UAS to time-sensitive delivery of life-saving medical equipment (defibrillators for example)*

(*continued*)

Table 2. (*continued*)

Country	Area	Field	Description
	Florida	*Agricult*	*Low-altitude aerial applications to control/survey the mosquito population*
	Calif./NY	*Security*	*UAV for police search and rescue operations or access hard-to-reach crime scene*
	North Dakota	*Many*	*UAS to linear infrastr inspections, crop health monitoring and emergency response*
	Ohio	*Logistic*	*Drone commercial delivery services - O*
	California	*Passenger*	*Uber is trying to develop air-taxi with four-seats*

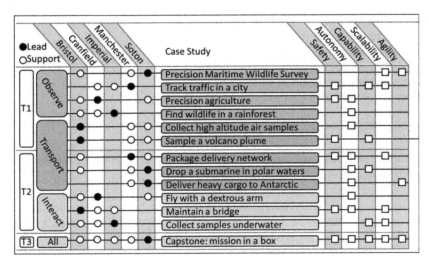

Fig. 3. Cascade project case studies ([6, 7]) source:

passenger transport where also problems due to weather events are bigger (both from damage than comfort side). Current battery technology creates a barrier like also battery weight and recharging times. Moreover noise introduces an environmental impact that could affect community acceptance. Other problems are lack of existing infrastructure, the habit of using traditional transport modes, the competitions of other innovative mobility (electric and autonomous cars for ex.).

4   Impacts of UAM on Sustainable Mobility

In general UAM case studies include application at different scales, depending on the support and services involved going from local areas, like port application, city center areas, to suburban/edge city areas and to inter-urban areas.

The following table presents **all potential UAM applications** coming from the case studies analysis (Table 3).

The case studies analyzed underline different UAM scopes:

- Decongestion of road traffic;
- Improve mobility level of services;
- Reduce transport time;
- Decrease pollution;
- Reduce strain on existing public transport networks;
- Reduce traffic accidents;
- Improve capacity of smart city management;
- Improve city security (decrease in crime and home burglary).

The use of drone has generally shown, in respect to classic mobility means, the following advantages:

- Cost reduction;
- Increase flexibility, e.g. to reach less accessible areas, more robust in case of infrastructure deficiency;
- Decrease of atmospheric mission;
- Highest planning reliability;
- More suitable for transport of small parts;
- About 5% of the maintenance costs of a car;
- 80–85% of energy-saving compared to electric cars;
- Decrease in transportation time (about 75%);

The exact evaluation and measurement of these impacts is however complicated by various considerations:

- The reduction of emissions is valid in the local context, but should also be considered on a more global level (e.g. include impact of energy production) and on a lifecycle basis (e.g. include battery production, replacement and recycling);
- Also the infrastructure required for a UAM system needs to be considered: hub stations, charging locations, maintenance, …
- Many aspects are largely depending on the local context and the exact design of the system (number and density of hubs), the vehicle properties (battery capacity, charging time, weight capacity) and regulations (e.g. which trajectories are available for flights, are trajectories combined with flows of motorized traffic, …)
- Large fleet sizes may be needed to obtain a system, which is economically viable. Similarly, large numbers of UAM flights (and thus fleet size) are needed to realize significant impact on e.g. urban congestion level. This also raises questions on the potential capacity of an urban UAM system.

Table 3. Potential UAM markets

Category	Potential UAM case study	Type
Air commute	*Airport Shuttle*	
	Air Taxi	*Passenger*
	Bus last-miles connection	
	Train last-miles connection	
First response	*Air ambulance*	
	Police/security event service	*Logistic +*
	Firefighter quick response	*Passenger*
	Humanitarian aid intervention	
Company level UAM	*Company shuttle*	
	Office-to-office travel	*Passenger*
	Inter-office/client delivery	
Events	*Pick-up and drop-off people*	*Passenger*
Entertainment/Media	*Extreme sport application* *Film/TV/Reporting* *Tourism Sightseeing*	*Logistic*
Logistic/Good del	*Air freight delivery service* *Air freight warehouses service*	*Logistic*
Construction/Real estate	*Aerial inspection and survey* *Replace external building operations (painting/window washing) in extreme environm* *Access to electrical wires and poles*	*Logistic*
Healthcare services	*Pick-up and drop-off patient for remote areas* *Delivery of urgent needed medical items*	*Passenger + Logistic*
Daily public services	*Replace winter snow plow and salt trucks* *Replace trash collection (hazardous waste disposal)* *School Busses service* *Smart City Monitoring (congestion, safety, etc..)*	*Passenger + Logistic*
Agriculture	*Reach remote flocks* *Reach remote farmland for harvesting goods* *Distribute seeds / other in the fields*	*Logistic*

- The effects may be compensated by rebound effects: the freed capacity on the road network may cause induced traffic; UAM as a last-mile solution may create a higher demand for deliveries; the resulting stimulation of e-commerce will increase the pressure on urban logistics.

Public acceptance is another major issue, because of negative or uncertain perception of safety, visual impact, noise hindrance, privacy issues, … This puts even more Importance in test projects to demonstrate and evaluate the effects of UAM.

For this reason, applications in emergency service (urgent interventions, deliveries of medical equipment and/or staff) seem most promising. On one hand, UAM offers clear benefits compared to classic motorized transport (speed, reliability), while the impact is limited because of the lower frequence of flights. The use for transport of passengers or goods suffers higher resistance, because of the large fleet size and number of flights needed to become profitable and to realize the expected impact on sustainability.

With the opening of EU drones services market in 2019, it is important to remember the "Drones Helsinki Declaration" that stressed the need to work on:

- Legal requirements for operations and airspace and U-Space services;
- An effective standard setting process;
- Further investment in demonstrators and the establishment of a European U-Space Demonstrator Network.

The Declaration also underlined a commitment to safe, secure, green drone operations that also respect privacy.

The future and the actual development of demonstration projects in close collaboration with local and/or regional authorities need to follow the aforementioned needs. Infact, one of the major needs to reach the objectives of UAM real application consists in the preparation of a permanent, definitive and clear regulation framework, overcoming the current condition of continuous updating and modification of UAM rules and legislation.

Acknowledgement. This paper is supported by European Union's Horizon 2020 research and innovation programme under grant agreement No. 101007134, project AURORA (Safe urban air mobility for European citizens).

References

1. Reiche, C., Goyal, R., Cohen, A., et al.: Urban Air Mobility Market Study. National Aeronautics and Space Administration – NASA. Permalink, 2018 November. https://escholarship.org/uc/item/0fz0x1s2
2. Urbanmobilitynew.com Document: Global Urban Air Mobility Report. Permalink (2019). https://www.urbanairmobilitynews.com/wp-content/uploads/2019/03/Global-Urban-Air-Mobility-Report-7-March-2019.pdf
3. Gianmarco, S., Walid, N., et al.: En-Route to Urban Air Mobility. On the fast track to viable and safe on demand air-services, Altran Company. Permalink (2020). https://www.altran.com/as-content/uploads/sites/27/2020/03/en-route-to-urban-air-mobility.pdf

4. European Commission: Mobility and Transport, Permalink (2020). |https://ec.europa.eu/tra nsport/home_en
5. Canergie Endowment for International Peace: 2021 Mapped: Drone Privacy Law around the World, Electronic Frontier Foundation, AlgorithmWatch. For the full research behind the map, visit: bit.ly/DroneProvacyLaws (2021)
6. David, T., Hine, D., Schellenberg, B., et al.: S. CASCADE Open Aircraft Project: University of Bristol VTOL Drone Development AIAA Scitech 2021 Forum. AIAA 2021–1930, January 2021
7. Ferraro, M., et al.: CASCADE project, presentation at the Connected Places Catapult Drone Workshop (2019)
8. Lilium (2020): Website. https://lilium.com/about-us. Accessed 15 Apr 2021
9. Vincent, J., Gartenberg, C.: Here's Amazon's new transforming Prime Air delivery drone. The Verge (2019). https://www.theverge.com/2019/6/5/18654044/amazon-prime-air-delivery-drone-new-designsafety-transforming-flight-video. Accessed 15 Apr 2021
10. Stouhi, D.: London Development to Offer First Dedicated Drone Port in the UK, Arch Daily (2019). https://www.archdaily.com/911950/london-development-to-offer-first-dedica ted-drone-port-in-the-uk/. Accessed 15 Apr 2021
11. Airbus (2020): Website. https://www.airbus.com/innovation/zero-emission/urban-air-mob ility.html. Accessed 15 Apr 2021

Path Planning Approach for a Quadrotor Unmanned Aerial Vehicle

César A. Cárdenas R.[1]($\boxtimes$), V. Landero[2], Ramón E. R. González[3],
Paola Ariza-Colpas[4], Emiro De-la-Hoz-Franco[4],
and Carlos Andrés Collazos-Morales[1]($\boxtimes$)

[1] Vicerrectoria de Investigaciones, Universidad Manuela Beltrán, Bogotá, Colombia
{cesar.cardenas,carlos.collazos}@docentes.umb.edu.co
[2] Universidad Politécnica de Apodoca, Apodaca, Nuevo León, Mexico
[3] Departamento de Física, Universidade Federal Rural de Pernambuco, Recife,
Pernambuco, Brazil
[4] Departamento de Ciencias de la Computación y Electrónica,
Universidad de la Costa, Barranquilla, Colombia

Abstract. A path planning method for an unmanned aerial system type quadrotor is proposed in this work. It is based on Dubins curves. Therefore, different points (initial and ending) are set for generation of several paths. Additionally, to validate the proposed model a computational resource is applied. Also, some flight dynamics limits and orientation angles computations are considered to be able to determine a simplified Dubins model. Dubins paths are commonly divided into low, medium and high altitude gains. It will depend on the altitude established for the start and end points and other configurations.

Keywords: Dubins · VTOL · Flight dynamics · Path planning

1 Introduction

Motion planning involves path as well as trajectory planning. Path planning is a very important matter regarding unmanned autonomous vehicles. It can be defined as a search of a feasible path starting at any initial point to a final one in which a collision-free environment can be obtained. Hence, kinematic limitations of the system can be met. It is significant to point that there is a difference between path and trajectory planning. The first one considers the path geometry. The other one is more related to how a path evolves over time. Thus, in this research a path planning with 2D Dubins curves is developed and proposed. Moreover, other 3D paths are also generated similar to the ones studied in [15,16]. Dubins paths designed for aerial vehicles are indeed more difficult due to the altitude changes. It looks like short paths will be able reduce consumption of fuel, travel time, energy and even life cycle of the platform. An unmanned aircraft flying at a steady altitude might save fuel. To determine an appropriate path planning approach, a vehicle and an aeronautics or aerospace

© Springer Nature Switzerland AG 2021
O. Gervasi et al. (Eds.): ICCSA 2021, LNCS 12950, pp. 426–439, 2021.
https://doi.org/10.1007/978-3-030-86960-1_30

reference frame have to be defined. Also, a 2D path planning option could help to a path with an unchanging altitude. So, path planning is a very remarkable issue to ensure that an unmanned aircraft can travel considerable distance missions completely. [13,17–19].

2 Dubins Curves

As it is presented in [8], to be able to find a short path between two points, a mixture of lines and circular segments is needed. As stated earlier, Dubins paths can be classified as low, medium, or high altitude gains. Therefore, both altitude changes and steady altitude are looked into. There are three possibilities for a body to travel. These are straight lines, left and right curves. The resulting paths can be either CCC or CSC. C describes a circular arc in which the radius is ρ or R_{min}. S shows a straight line. This is method known as the Dubins path. Hence, some of path configurations that can be got from a Dubins are: $D = LSL, LSR, RSR, RSL, LRL, RLR$. The arcs should be a left (L) and right (R) turn illustration and (S) is a straight line. The first point is for Dubins path generation is to establish what type of path will be employed. [9,11]. Similary, other path planning options can be found in [1,4,5].

3 Dubins Aircraft Model

An airplane should be considered to move with a continuous velocity v and a steady altitude h. Thus, g is neglected since an aerial vehicle will be taken. So, the 2D system is as follows:

Dubins' car model: Dubins' aerial vehicle model:

$$
\begin{bmatrix} \dot{x} \\ \dot{y} \\ \dot{\psi} \end{bmatrix} = \begin{bmatrix} cos\ \psi \\ sin\ \psi \\ u \end{bmatrix} \qquad \begin{bmatrix} \dot{x} \\ \dot{y} \\ \dot{\psi} \end{bmatrix} = \begin{bmatrix} v\ cos\ \psi \\ v\ sin\ \psi \\ \omega \end{bmatrix} \tag{1}
$$

where ω is the turning rate of the aircraft, v is the velocity, x and y are the inertial position and ψ is the heading angle.

3.1 Dubins Paths

For Dubins paths generation that can be feasible, it is necessary to consider vehicle constraints. The minimum turning radius is one of the most important ones. Besides the starting and ending points, the orientation of the angle is also crucial. Hence, if the curvature constraint is denoted as κ, the path planning equation may be written as:

$$
P_i(x_i,\ y_i,\ \psi_i) \xrightarrow{\kappa} P_f(x_f,\ y_f,\ \psi_f) \tag{2}
$$

If an unmanned vehicle flyes around a circle at constante velocity, the circle radius and turning angular velocity can be described as follows:

$$R_{min} = \frac{v^2}{g\sqrt{n^2 - 1}} \tag{3}$$

To be able to obtain the smallest turning radius, the highest load factor and a low velocity are required. Likewise, to get an important turning rate, the largest possible load factor is also needed. The minimum velocity is selected. At any velocity the maximum load factor for a continuous turning flight is limited by the available thrust. [2,7,12].

$$\omega = \frac{g\sqrt{n^2 - 1}}{v} \tag{4}$$

In which n is the load factor and g is the gravity acceleration. The equations described previously are based on the aircraft dynamic performance. There is another possible way to calculate the minimum turning radius. It is shown in [11] as:

$$\rho_{min} = \frac{v^2}{g \tan \phi_{max}} \tag{5}$$

The following expression is a manner to show the equations applied to compute start and finish circles based on the vehicle performance:

for $C_i(x_{ci}, y_{ci})$:

$$C_{x_{ci}} = x_i + R_{min} \cos \left(\psi_i + \frac{\pi}{2}\right) \tag{6}$$

$$C_{y_{ci}} = y_i + R_{min} \sin \left(\psi_i + \frac{\pi}{2}\right) \tag{7}$$

and for $C_f(x_{cf}, y_{cf})$:

$$C_{x_{cf}} = x_f + R_{min} \cos \left(\psi_f + \frac{\pi}{2}\right) \tag{8}$$

$$C_{y_{cf}} = y_f + R_{min} \sin \left(\psi_f + \frac{\pi}{2}\right) \tag{9}$$

Similarly, the angle ψ_L is the angle measured from the straight line segment $\chi_{(ci-cf)}$ to the y axis is given as follows:

$$\psi_L = \psi_{ci-cf} = \tan^{-1} \left(\frac{y_{cf} - y_{ci}}{x_{cf} - x_{ci}}\right) \tag{10}$$

The shortest path is estimated by matching the distance between the center of the circles. The distance between the center of the circles CR_i and CR_f is calculated as:

$$\chi_{(ci-cf)} = \sqrt{(x_{cf} - x_{ci})^2 + (y_{cf} - y_{ci})^2} \tag{11}$$

The total length of the Dubins path is given by:

$$\chi_{Dubins} = \chi_{arci} + \chi_{tangent} + \chi_{arcf} \tag{12}$$

3.2   Dubins Airplane Paths

There is an option to get a 3D model. A significant analysis associated to disturbances is that wind is not considered in the equations of motion when the Dubins aircraft model is utilized. Thus, the kinematics model can be described as:

$$\begin{bmatrix} \dot{x} \\ \dot{y} \\ \dot{z} \\ \dot{\psi} \end{bmatrix} = \begin{bmatrix} v \cos \gamma \cos \psi \\ -v \cos \gamma \sin \psi \\ v \sin \gamma \\ \omega \end{bmatrix} \tag{13}$$

where v is the airspeed, ψ is the heading angle and γ is the angle of flight. The relationship between the yaw angle ψ and the roll or bank angle ϕ is as follows:

$$\dot{\psi} = \frac{g}{v} \tan \phi \tag{14}$$

where g is the gravity acceleration. Another hypothesis taken into consideration is that the well-tuned autopilot. So, the airframe model can be given as follows: [3,10,11].

$$\left(\begin{matrix} \dot{x} = v \cos \psi \cos \gamma \\ \dot{y} = -v \sin \psi \cos \gamma \\ \dot{z} = v \sin \gamma \\ \dot{\psi} = \frac{g}{v} \tan \phi \end{matrix} \right) \tag{15}$$

The height of the vehicle could be modified if the flight path angle γ is changed. Moreover, the physical aircraft restrictions set some limitations to both roll and flight path angles. Such restrictions are shown as: [14].

$$|\phi^c| = \leqslant \bar{\phi} \tag{16}$$
$$|\gamma^c| = \leqslant \bar{\theta} \tag{17}$$

Table 1. Dubins parameters for quadrotor paths

Dubins Path	R_{min}	γ_{max} (rad)	ψ_i	ψ_f	H_i	H_f	S_{Dubins}
RSR Low altitude	5 m	1.22	0.34	4.01	100	170	210.5 m
RSR High altitude	5 m	1.22	0.34	4.712	100	300	218.62 m
RSL Low altitude	5 m	1.22	5.75	−6.8	150	200	145.07* m
RSL High altitude	5 m	1.22	5.2	−6.8	350	450	147.85* m
LSR Low altitude	5 m	1.22	1.22	1.22	100	350	153.2 m
LSR High altitude	5 m	1.22	1.22	1.22	200	350	230.82 m
LSL Low altitude	5 m	1.22	1.22	−2.35	350	90	149.8 m
LSL High altitude	5 m	1.22	−1.22	−2.35	350	100	233.18 m

3.3 The Shortest Dubins Paths

In the RSR case, the total length is as follows:

$$
\begin{aligned}
\chi_{Dubins1} = {} & \varphi_i R_{min} \left[\left(\psi_L - \frac{\pi}{2} \right) - \left(\psi_i - \frac{\pi}{2} \right) \right] \\
& + \chi_{tangent(ci-cf)} + \varphi_f R_{min} \left[\left(\psi_f - \frac{\pi}{2} \right) - \left(\psi_L - \frac{\pi}{2} \right) \right]
\end{aligned}
\tag{18}
$$

where φ points how the path is moving along the arcs of the circles. The directions can be clockwise or counter-clockwise as shown in Fig. 1.

In the RSL case (Fig. 4), the total length is:

$$
\begin{aligned}
\chi_{Dubins2} = {} & \sqrt{\chi^2_{(c_i - c_f)} - 4R^2_{min}} + R_{min} \left[(\psi_L + \varphi_i + \psi_l) - \psi_L \left(\psi_i - \frac{\pi}{2} \right) \right] \\
& + R_{min} \left[(\psi_L + \pi) - \left(\psi_f + \frac{\pi}{2} \right) \right]
\end{aligned}
\tag{19}
$$

With respect to LSR path (Fig. 3), the total length is given by:

$$
\begin{aligned}
\chi_{Dubins3} = {} & \sqrt{\chi^2_{(c_i - c_f)} - 4R^2_{min}} + R_{min} \left[\left(\psi_i + \frac{\pi}{2} \right) - (\psi_L + \varphi_i \psi_l) \right] \\
& + R_{min} \left[\left(\psi_f - \frac{\pi}{2} \right) - (\psi_l + \psi_L + \varphi_i \psi_l - \pi) \right]
\end{aligned}
\tag{20}
$$

The LSL case looks similar to the RSR path (Fig. 2). Thus, the total length can be given as follows:

$$
\begin{aligned}
\chi_{Dubins4} = {} & \varphi_i R_{min} \left[\left(\psi_f + \frac{\pi}{2} \right) - \left(\psi_L + \frac{\pi}{2} \right) \right] \\
& + \chi_{(c_i - c_f)} + \varphi_f R_{min} \left[\left(\psi_L + \frac{\pi}{2} \right) - \left(\psi_f + \frac{\pi}{2} \right) \right]
\end{aligned}
\tag{21}
$$

Table 2. Flight angle computation

Flight Path Angle γ
RSR Low altitude = 0.75 rad
RSR High altitude = 0.73 rad
RSL Low altitude = 0.33 rad
RSL High altitude = 0.6 rad
LSR Low altitude = 0.9 rad
LSR High altitude = 0.57 rad
LSL Low altitude = 0.78 rad
LSL High altitude = 0.77 rad

Table 3. Roll angle computation

Roll angle ϕ
$\phi = 1.53$ rad

Table 4. Pitch angle computation

Pitch Angle θ
RSR Low altitude = 1.75 rad
RSR High altitude = 1.73 rad
RSL Low altitude = 1.33 rad
RSL High altitude = 1.6 rad
LSR Low altitude = 1.9 rad
LSR High altitude = 1.57 rad
LSL Low altitude = 1.78 rad
LSL High altitude = 1.77 rad

Table 5. Yaw angle computation

Yaw Angle ψ
RSR Low altitude = 1.52 rad
RSR High altitude = 1.54 rad
RSL Low altitude = 0.85 rad
RSL High altitude = 0.85 rad
LSR Low altitude = 0.84 rad
LSR High altitude = 0.84 rad
LSL Low altitude = 0.82 rad
LSL High altitude = 1.13 rad

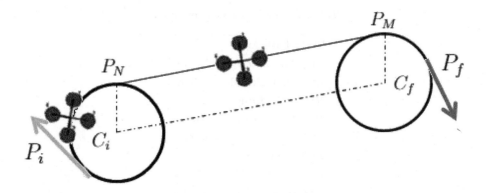

Fig. 1. RSR dubins path

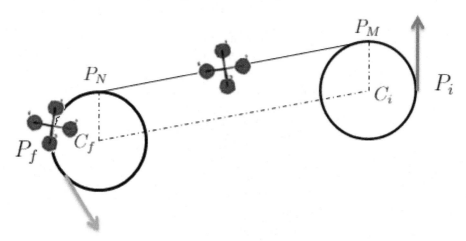

Fig. 2. LSL dubins path

3.4 Helical Paths

A helical path can be described as follows:

$$p(t) = C_h + \begin{pmatrix} R_h \; cos \; (\Pi_h \; t + \psi_h) \\ R_h \; sin \; (\Pi_h \; t + \psi_h) \\ -t \; R_h \; tan \; \theta_h \end{pmatrix} \tag{22}$$

Where $p(t) = [p_x \; p_y \; p_z]^T$ is the path position, and $C_h = [C_x \; C_y \; C_z]$ is the helix center. Hence, The initial helix position is given by:

$$p(0) = C_h + \begin{pmatrix} R_h \; cos \; \psi_h \\ R_h \; sin \; \psi_h \\ 0 \end{pmatrix} \tag{23}$$

Where R_h is the radius, $\Pi = +1$ denotes that the helix rotates in a clockwise direction and $\Pi = -1$ is the opposite direction. θ_h is the helix path angle.

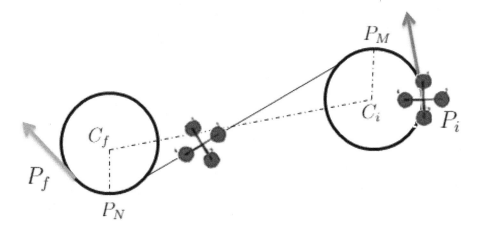

Fig. 3. LSR dubins path

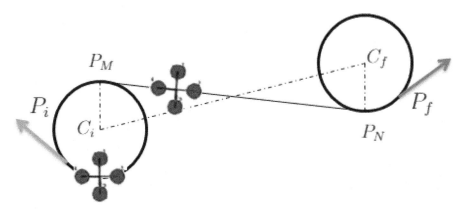

Fig. 4. RSL dubins path

Therefore, a helical path can be expressed as:

$$\mathcal{P}_h = (C_h \ \psi_h \ \Pi_h \ R_h \ \theta_h) \tag{24}$$

Several values for θ_{max}, ψ_{max}, ψ_{min}, H_i, H_f and R_{min} are given (Table 1). Possible Dubins paths for a quadrotor are generated and presented in the following figures. These types of paths applicable when a a quadrotor is flying at a steady speed and when roll and pitch angles have certain constraints. The velocity considered is about 15 m/s. The minimum radius is decreased to avoid large arcs and keep VTOL features. This particular detail show the way to compute a maximum flight path angle γ of 1.22 radians. It is a close constraint value as in [14]. In addition, Eq. 10 is used to find the flight path angle values (Tables 2, 3, 4). In the same way, pitch and roll angles (ϕ θ, ψ) are computed by applying the equations that follow:

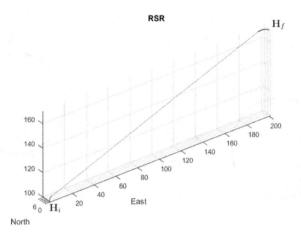

Fig. 5. RSR dubins path low altitude

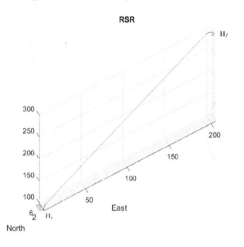

Fig. 6. RSR dubins path high altitude

$$\theta = \alpha_a + \gamma$$

$$\phi = tan^{-1}\frac{v^2}{R_{min}} \qquad (25)$$

and the heading angle is found by using Eq. 30. These calculations are done just for the Dubins paths that are a suitable choice for a quadrotor with a + configuration. Dubins airplane paths are more complicated because of the altitude variable included. The path cases are determined by the difference of the altitude between the start and final positions. These are described as low, medium and high altitude [6,14]. H is be taken as z for altitude regarding to start and final configurations or altitude difference. Low and high alternatives are only taken in this research work. The path climbs and descends from H_i to

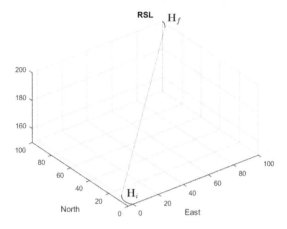

Fig. 7. RSL dubins path for low altitude

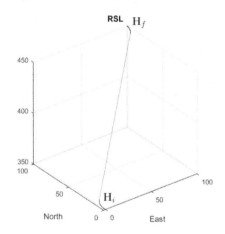

Fig. 8. RSL dubins for high altitude

H_f when γ is varied. So, it can be as follows (Table 5):

$$\gamma = tan^{-1} \left(\frac{H_f - H_i}{\chi_{tangent}} \right) \tag{26}$$

4   Results

As it is given in Table 1, different values for θ_{max}, ψ_{max}, ψ_{min}, H_i, H_f and R_{min} are determined. (R_{min}, γ_{max} and ϕ are constant). A low level autopilot is included into the computational resource with common aeronautics symbols. It changes within commands of straight lines and helical paths. Also orientation regulation is also adjusted. All Dubins paths are performed for both altitude cases (low and high). A helix parameter is considered as well. However, it is

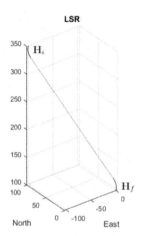

Fig. 9. LSR dubins path low altitude

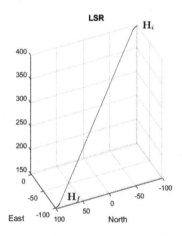

Fig. 10. LSR dubins path high altitude

not really accomplished because of the set values and path conditions. It can suggested that the RSL is the shortest path. Quadrotor attitude parameters are established. Thus, after calculating flight angles, all of seem very small without significant variations. Also, the pitch and yaw angles are also very similar specially, the RSL ones at low altitudes. The generated paths are shown in the following graphs (Figs. 5, 6, 7, 8, 9, 10, 11 and 12).

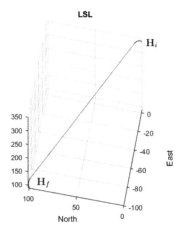

Fig. 11. LSL dubins path low altitude

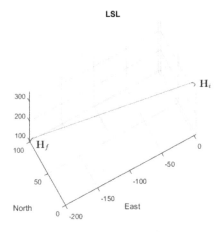

Fig. 12. LSL dubins path high altitude

5 Conclusions

It can be inferred that a quadrotor can truly fly in RSR, RSL, LSL and LSR path types keeping a constant minimum radius and velocity. This method has been implemented as an useful way to measure the shortest path with the maximum curvature for an unmanned aerial airplane based on different estimated points. Additionally, a modification of the Dubins car model is established. Thus, another way for turning flight might be achieved just by taking a constant velocity. Therefore, very similar paths with no Dubins curves can be performed. It is observed that helical paths are not easily got or are not really necessary to get the initial and final points. This factor is because of the VTOL characteristics of this type of airplane. Likewise, it appears to be that the integrated autopilot needs proper adjustments. Attitude angles can be computed from basics flight dynamics equations.

References

1. Alexis, K Dr.: Dubins Airplane. http://www.kostasalexis.com/dubins-airplane. html. (2017)
2. Anderson, J.D.: Aircraft performance and design. McGraw-Hill international editions: Aerospace science/technology series. WCB/McGraw-Hill, New York (1999). ISBN: 9780070019713. https://books.google.com.co/books?id=PwtO7aiwbBwC
3. Cárdenas, C.A., Grisales, V.H., Collazos Morales, C.A., Cerón-Muñoz, H.D., Ariza-Colpas, P., Caputo-Llanos, R.: Quadrotor Modeling and a PID Control Approach. In: Tiwary, U.S., Chaudhury, S. (eds.) IHCI 2019. LNCS, vol. 11886, pp. 281–291. Springer, Cham (2020). https://doi.org/10.1007/978-3-030-44689-5_25
4. Beard, R.W., McLain, T.W.: Small Unmanned Aircraft: Theory and Practice. Princeton University Press, New Jersey (2012)
5. Cárdenas Ruiz, C.A.: Performance study of the flight control and path planning for a UAV type Quadrotor. In: Ingeniería Mecatrónica (2018)
6. Chitsaz, H., LaValle, S.M.: Time-optimal paths for a Dubins airplane. In: Decision and Control, 2007 46th IEEE Conference on, pp. 2379–2384. IEEE (2007)
7. Collazos, C., et al.: State estimation of a dehydration process by interval analysis. In: Figueroa-García, J.C., López-Santana, E.R., Rodriguez-Molano, J.I. (eds.) WEA 2018. CCIS, vol. 915, pp. 66–77. Springer, Cham (2018). https://doi.org/10.1007/978-3-030-00350-0_6
8. Dubins, L.E.: On curves of minimal length with a constraint on average curvature, and with prescribed initial and terminal positions and tangents. Am. J. Math. **79**(3), 497–516 (1957)
9. Grymin, D.: Development of a novel method for autonomous navigation and landing of unmanned aerial vehicles. In: (2009)
10. Li, W., et al.: A 3D path planning approach for quadrotor UAV navigation. In: Information and Automation, 2015 IEEE International Conference on, pp. 2481–2486. IEEE (2015)
11. Lugo-Cárdenas, I., et al.: Dubins path generation for a fixed wing UAV. In: Unmanned Aircraft Systems (ICUAS), 2014 International Conference on, pp. 339–346. IEEE (2014)
12. Cárdenas, R.C.A., et al.: Mathematical Modelling and Identification of a Quadrotor. In: Gervasi, O., et al. (eds.) ICCSA 2020. LNCS, vol. 12249, pp. 261–275. Springer, Cham (2020). https://doi.org/10.1007/978-3-030-58799-4_19
13. Omar, R.B.: Path planning for unmanned aerial vehicles using visibility line-based methods. PhD thesis. University of Leicester (2012)
14. Owen, M., Beard, R.W., McLain, T.W.: implementing Dubins airplane paths on fixed-wing UAVs*. In: Valavanis, K.P., Vachtsevanos, G.J. (eds.) Handbook of Unmanned Aerial Vehicles, pp. 1677–1701. Springer, Dordrecht (2015). https://doi.org/10.1007/978-90-481-9707-1_120
15. Pharpatara, P.: Trajectory planning for aerial vehicles with constraints. Theses. Université Paris-Saclay; Université d'Evry-Val-d'Essonne, September (2015). https://tel.archives-ouvertes.fr/tel-01206423
16. Poyi, G.T.: A novel approach to the control of quad-rotor helicopters using fuzzy-neural networks. In: (2014)

17. Shanmugavel, M., et al.: Path planning of multiple autonomous vehicles. In: (2007)
18. Antonios Tsourdos, Brian White, and Madhavan Shanmugavel. Cooperative Path Planning of Unmanned Aerial Vehicles. vol. 32. Wiley, Chichester (2010)
19. Valavanis, K.P., Vachtsevanos, G.J. (eds.): Handbook of Unmanned Aerial Vehicles. Springer, Dordrecht (2015). https://doi.org/10.1007/978-90-481-9707-1

International Workshop on Advances in Artificial Intelligence Learning Technologies: Blended Learning, STEM, Computational Thinking and Coding (AAILT 2021)

Understanding Factors Influencing Infusion and Use of an Online Collaboration Tool: A Case of a Higher Education Institution in Lesotho

Pakiso J. Khomokhoana⬤ and Okuthe P. Kogeda(✉) ⬤

Department of Computer Science and Informatics, University of the Free State,
Bloemfontein, South Africa
{khomokhoanap,kogedapo}@ufs.ac.za

Abstract. It is imperative for organisations adopting new technologies to investigate factors that may encourage or discourage users from using such technologies in the initial stages of adoption. This may help in identifying, among others, training needs to focus on in further encouraging users to continue to use the technology. This study investigated the above phenomenon by specifically focusing on the Online Collaboration Tool (OCT) at a higher education institution in Lesotho. The study followed a quantitative approach where data was collected from a sample of 216 respondents through a questionnaire. The data was analysed using the Statistical Package for the Social Sciences (SPSS). Theoretically grounded on the Unified Theory of Acceptance and Use of Technology model, the study found that performance expectancy and effort expectancy positively influence the behavioural intention of students to use the online collaborative tool (OCT); while social influence, facilitating conditions and behavioural intention do not have a significant influence on the students' behaviour to use OCT. The results suggest that students actually use technology irrespective of whether they think they will learn or not learn from using it.

Keywords: Online Collaboration Tool · Infusion of technology · Unified Theory of Acceptance and Use of Technology · Technology acceptance models · Higher education institutions · Technology adoption · Learning management systems

1 Introduction

Information and Communications Technology (ICT) has become an integral and accepted part of people's daily lives [36, 40]). As such it may not be surprising that some members of the millennial generation and most members of generation Z are almost always busy with their technological gadgets - at work, home, school, play, on the go and when they go to sleep [26]. Seamlessly, the use of technology helps both profit-making and non-profit making organisations in running their operations [2]. The profit making organisations benefit in that through using these innovations, they are able

© Springer Nature Switzerland AG 2021
O. Gervasi et al. (Eds.): ICCSA 2021, LNCS 12950, pp. 443–458, 2021.
https://doi.org/10.1007/978-3-030-86960-1_31

to conduct business transactions irrespective of time and location (*e.g., e-commerce, m-commerce and e-procurement*) [34], hence, maximising profits. The latter organisations benefit in such that they are able to expedite service delivery, hence, satisfying customers [23].

Higher educational institutions are not an exception to using technology. As an example, South African institutions of higher learning such as University of Cape Town (UCT), University of Kwazulu Natal (UKZN) and University of the Free State (UFS) have the online application facility for new and returning students. Furthermore, UCT uses an online collaboration and learning environment named Vula [7]. This environment is used to support UCT courses as well as other UCT-related groups and communities. The UFS uses a learning management system called Blackboard [8]. However, one higher education institution in Lesotho had been lagging behind its counterparts in either using the collaborative environment or learning management systems. It was only at the beginning of the academic year 2014/5 that this institution, for the first time in its lifetime, acquired an Online Collaboration Tool (OCT).

According to the developers of OCT [31], it offers a rich set of collaboration features to enable secure communication and sharing of resources within the community of the institution. It also helps students and educators to form an efficient online teaching and learning ecosystem where course resources can be shared and accessed conveniently. OCT extends beyond course management by providing a comprehensive collaborative environment that can facilitate collaboration between researchers sharing similar research interests and staff members within non-academic units working toward common goals. Finally, OCT purportedly enhances productivity by enabling personnel to locate and view information related to their roles and responsibilities faster and easily. According to Franzoni et al. [17], systems similar to OCT should also allow educators to track how frequent students utilise them, as well as, to check the activities that students perform on those systems. In this way, educators are able to continuously monitor the learners.

What obtains in practice is sometimes not always what was planned. Several benefits, though limited, have been reported since the inception of the tool usage. These benefits can be classified into fast and easy communication, sharing of resources and cost cutting measures. In relation to communication, students are able to communicate (*email and/or chat*) with their educators and vice versa; employees are able to communicate among themselves; and administrative offices are able to make announcements (*e.g., invitations to seminars, meetings or conferences, notifications of new appointments, etc.*) through OCT. In relation to sharing resources, educators are able to upload course materials (*e.g., course outlines, spreadsheets for student marks, lecture notes, etc.*) on OCT and students access them from there; and administrative offices are able to share necessary documents with the community of the institution through uploading them on OCT (*e.g., strategic plan, almanac of events, lecture time table, notices, etc.*). According to Moteete (personal communication, 16 September 2015), massive printing of documents had reduced because prior to using OCT, every message that was meant for consumption of the community of the institution, had to be printed and posted on the notice boards. She further indicated that it was even worse that such wasted paper is not even recycled as there is not such a service at the institution.

As a way to integrate OCT with Integrated Tertiary Software (ITS) – a corporate Computer Information System used for administrative functions at the selected institution of higher education in Lesotho, [31] had proposed loose coupling. In defining and creating this integration, they had proposed to use the Simple Object Access Protocol (SOAP) (see Fig. 1). However, during the first year of adoption of OCT, the intended integration had not happened (see a dashed line labelled "not implemented?" in Fig. 1). The implication was that, for OCT to be updated with data *(for example, details of student registration)*, the OCT team had to be given data retrieved from ITS on an external storage *(physically)*. As a result, that caused a lot of inconvenience to users.

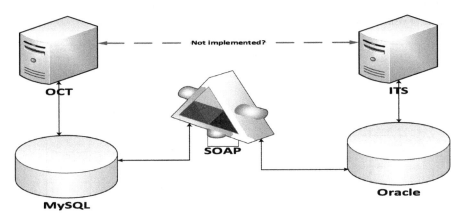

Fig. 1. Proposed integration model between OCT and ITS [Source: Adapted from [31] (p. 3)].

Notwithstanding the benefits that resulted from the use of OCT, statistics indicated that not a lot of intended audience was actually using the tool. Table 1 shows usage data for the majority of courses in Moshoeshoe[1] Department for the academic year 2014/5 *(August 2014 to May 2015)*. Note that a few courses have been excluded from the list in Table 1 because they were not linked to any educator(s) in the period in question due to either one or more of the following reasons: Course(s) being taught by part timers; educator(s) who resigned; or not being offered that academic year. Moshoeshoe (See footnote 1) Department had 10 full-time educators *(since individuals teach more than one course)* in the selected time period. This implies that the average number of visits to OCT by Moshoeshoe (See footnote 1) Department educators from August 2014 to May 2015 was approximately 29 times. The average number of students enrolled into the Moshoeshoe (See footnote 1) Department courses visited OCT 0.03 times in the selected period. This number of visits was far below the number of visits by the educator of, and students enrolled into the BC [3] course *(199 and 1949 visits respectively)* (see Table 1). This study, therefore, is aimed at investigating the infusion *(the degree to which technology is deeply integrated into an area or department)* of the tool and its use by both students and educators at the selected higher education institution. Based on the Unified

---

[1] A specific name of the department has been withheld for anonymity purposes, and throughout the paper, reference to the department will be used as such.

Theory of Acceptance and Use of Technology (UTAUT), the paper specifically examines the factors that determine the use and/or non-use of OCT at the chosen institution.

In the remainder of this paper, a theoretical framework guiding this study together with a review of relevant background literature are presented in Sect. 2. This is followed by a discussion of the research design and methods in Sect. 3, and a presentation of the results and interpretation in Sect. 4. The significance of the study and limitations in using OCT are provided in Sect. 5, while conclusions and recommendations for future research are presented in Sect. 6.

Table 1. Usage data on Moshoeshoe[1] department courses for the academic year 2014/15 on the number of visits to OCT.

Course Code	Total # of students enrolled	# of Educators Linked to course	Total # of visits	# of visits by Educator(s)	# of visits by students	Average visits per Educator	Average visits per student
A1[0]	307	1	664	1	663	1	2
B1[0]	234	1	399	0	399	0	2
A2[0]	120	1	115	0	115	0	1
B2[0]	116	1	40	5	35	5	0
C2[0]	89	1	17	5	12	5	0
A3[0]	95	1	45	1	44	1	0
A3[1]	92	1	98	0	98	0	1
B3[0]	44	1	551	51	500	51	11
B3[1]	87	1	50	0	50	0	1
B3[2]	106	2	25	7	18	4	0
B3[3]	85	1	2148	199	1949	199	23
C3[0]	46	1	40	1	39	1	1
A4[0]	66	1	16	0	16	0	0
B4[0]	98	1	8	2	6	2	0
A4[0]	70	1	2	0	2	0	0
A4[1]	70	1	8	0	8	0	0
B4[0]	95	1	35	0	35	0	0
C4[0]	24	1	2	0	2	0	0
C4[1]	24	1	4	0	4	0	0
B4[1]	24	1	63	3	60	3	3
B4[2]	12	2	247	40	207	20	17
Total averages	1904	23	4577	315	4262	292	63

Legend: For the sake of anonymity, the module codes have been coded - The first alphabet represents a module; the second number represents level of study; and a number in square brackets represents the number of modules for the level of study.

2 Theoretical Framework

Associated with infusion and use of technological innovations are several models. These include, but are not limited to Motivational Model (MM) [11], Technology Acceptance

Model (TAM) [12], Diffusion of Innovations (DOI) Theory [32] and Theory of Planned Behaviour (TPB) [1]. It was not until 2003 that the eight models, including the just stated, were evaluated by [42]. The evaluation gave birth to the Unified Theory of Acceptance and Use of Technology (UTAUT) model. As its underlying theoretical framework, this study adopted this model. The model was chosen for this study because it explains at least 70% of user's behavioural intention to use a technological innovation, while the previous models only explain 30 to 60% of the user's intention [42].

2.1   UTAUT Model

In the UTAUT model, the key constructs that determine the behavioural intention to use and the actual use behaviour to use some technological innovation are defined. As the moderating and mediating factors may be inherent in one's actual use of some named innovation, the model also considers such factors [42]. This study, however, focuses on only the original four constructs to determine their impact on the behavioural intention and actual behavioural use of OCT. Similar to other studies [2, 24], the moderating factors (gender, age, experience and voluntariness of use) have been excluded in this study. Figure 2 depicts the theoretical framework of this study as adapted from the UTAUT model.

Performance Expectancy and Behavioural Intention
Performance expectancy (PE) is a conviction that users will gain benefits from the use of a system [42]. The benefits that may accrue from using a new system include increased productivity, increased performance, efficiency and time saving [35]. As per researchers in [42], performance embraces the concepts of perceived usefulness, extrinsic motivation, relative advantage, outcome expectations and job-fit. According to [41], perceived usefulness has been identified as the principal predictor in adopting a technology. Davis [12] (p. 320) defines perceived usefulness as "the degree to which a person believes that using a particular system would enhance his/her job performance". Behavioural intention (BI) is a measure of the strength of one's intention to perform a specified behaviour [42].

Several studies [14, 20, 24] have found that there is a significant positive relationship between performance expectancy and behavioural intentions to use named technological systems. These systems include 3G Mobile technology, E-government Services, Wireless Technology, and Digital Library.

H1: Performance expectancy has a positive effect on the behavioral intention to use OCT.

Effort Expectancy and Behavioural Intention
Venkatesh et al. [42] define effort expectancy (EE) as the degree of ease of use that is associated with the system. EE is composed of such constructs as perceived ease of use (PEOU), ease of use (EOU) and complexity. PEOU is a degree to which an individual believes that using a system would be free of effort. Complexity is the degree to which an individual perceives a system as relatively difficult to understand and use. EOU is the degree to which an individual perceives using an innovation as being difficult to use

[12]. A myriad of studies have found that effort expectancy positively influences user's behavioural intention to use technological systems [9, 15].

H2: Effort expectancy has a positive effect on the behavioral intention to use OCT.

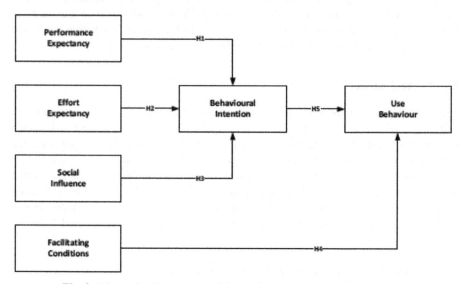

Fig. 2. Theoretical Framework of the study [Adapted from [42] (p. 447)]

Social Influence and Behavioural Intention

Social influence (SI) plays a key role in better understanding adoption logistics of any technology enabled or computerised system [42]. It is defined as the extent to which an individual perceives that people important to him/her believe that s/he should use the new system [42]. Social factors are defined as "the individual's internalisation of the reference group's subjective culture and specific interpersonal agreements that the individual has made with others in specific situations" [37] (p. 126). The influence could be from either friends, supervisors, colleagues or children. According to [22], social influence is a complex and delicate concept that can have an impact on people's decisions to use technology related systems. Several studies have found that social influence positively affects user's intention to use technological systems [3, 16, 18].

H3: Social influence has a positive effect on the behavioral intention to use OCT.

Facilitating Conditions and Use Behaviour

Facilitating conditions (FCs) are the objective factors in the environment that observers agree make an act easy to accomplish [42]. In the context of information systems, an example of a facilitating condition that can influence the usage of a system is provision of support to computer users [37]. In using OCT, it may be necessary to investigate whether the institution provides all the necessary resources to allow users to effectively use it. The resources may include expertise, infrastructure, policies, funding, support services, and culture. Many studies have found that facilitating conditions positively affect user's behavioural intention to use technological systems [27, 43, 44].

H4: Facilitating conditions have a positive effect on the actual behavioural to use OCT.

Behavioural Intention and Use Behaviour

People have a tendency to form attitudes and intentions toward a new technology before they actually use it [1]. The attitudes and intentions come as a result of fearful expectations formed by perceptions, social influences and traditional or habitual behaviour [4]. Several studies have found that behavioural intention positively affects user's actual intention to use technological systems [13, 19, 30]. In using OCT, it may be necessary to examine how attitudes and intentions that users develop impact their actual use of the system.

H5: Behavioural intention has a positive effect on the actual behavioral to use OCT.

3   Research Methods

3.1   Design

This study was grounded on the UTAUT model [42] as the research design. Within this design, a quantitative approach [33] was followed. Data was collected by means of using a questionnaire. The study population consisted of senior students from a selected institution of higher education in Lesotho. The study was based on a convenient sample [33] of 216 respondents. The sample consisted of 84 first year, 116 third year and 16 fourth year students enrolled in the courses offered by Moshoeshoe[1] Department at the selected institution. The respondents were purposefully [5] drawn from five programs of study offered in this department. The choice of these groups was driven by the fact that they were familiar and had, at least once, used OCT (*actual use of OCT is one construct tested in the study*).

3.2   Data Collection

Two hundred and sixteen structured questionnaires were distributed to our respondents. Of these, 155 were filled-in and returned. Hence, the return rate was 72%, which according to [25] (p. 126) exceeds the minimum acceptability level of 70% for questionnaires. Of the 155 respondents, five did not indicate their gender, hence, making a valid percentage of 61% (*females*) and males accounting for the rest. Age-wise, 61% of the respondents were between 18 and 23 years inclusively; thirty-percent were between 24 and 29 years and those 30 years old and above accounted for less than 10%.

Prior to respondents filling-in the questionnaire, they were given a consent letter, which indicated the purpose of the study and the approximate time (*10 to 15 min*) to fill-in the questionnaire. The respondents were also assured that by filling-in the questionnaire, they were giving the researchers consent to use their information for research purposes only and that their responses would be treated in a manner confidential and their privacy would be protected to the maximum extent allowable by law. They were also notified that their response or non-response would not, in any way, affect their academic performance.

3.3 The Instrument

The structured questionnaire developed had two sections – Section A was capturing the demographic details of respondents, while Section B focused on the experiences of respondents with the use of OCT. Section B had 23 statements where respondents had to indicate the extent to which they agreed or disagreed with such statements about the use of OCT. The statements were testing mainly the five constructs namely: performance expectancy, effort expectancy, social influence, facilitating conditions and behavioural intention in relation to the use of OCT, on a four-point Likert scale (1–2–3–4). This scale was used in order to maximise reliability by not including a neutral choice for respondents [10].

3.4 Measurement and Reliability

Upon capturing data into the Statistical Package for Social Sciences (SPSS) and in order to test whether the scale used in this research was consistently reflecting the construct(s) being measured, reliability analysis was conducted. This led to deleting some items forming a construct. All the items were adapted from [42] (see Fig. 2). Several items were combined to form appropriate constructs.

The PE construct was measured using three items [Cronbach's alpha (α) = 0.75]. These items were adapted from [42]. The items were measured on a four-point Likert scale ranging from (1) strongly agree to (4) strongly disagree. Sample items were: "I find OCT useful in my learning"; "Using OCT enables me to accomplish tasks more quickly"; and "Using OCT increases my chances of performing well academically".

The EE construct was measured using five items ($\alpha = 0.86$). These items were adapted from [42]. The items were measured on a four-point Likert scale ranging from (1) strongly agree to (4) strongly disagree. The five used sample items were: "My interaction with OCT is clear and understandable"; "I find OCT easy to use"; "OCT is user friendly"; "Learning to use OCT is easy"; and "OCT is easy to interact with".

The SI construct was measured using five items, of which two had to be deleted (*I would use OCT if I needed to; and I use OCT because many people use it*) to improve on the scale (*α value moving from 0.42 to 0.64*). All the items were adapted from [42]. The items were measured on a four-point Likert scale ranging from (1) strongly agree to (4) strongly disagree. The three used sample items were: "People who influence my behaviour think that I should use OCT"; "Other people expect me to use OCT"; and "People important to me expect me to use OCT".

The FCs construct was originally measured using three items, of which one had to be deleted (*Technicians are available to assist me with the difficulties I encounter with OCT*), hence, leading to better internal reliability (*α value moving from 0.47 to 0.62*). The BI construct was measured using six items ($\alpha = 0.88$). All the items were adapted from [42]. The items were measured on a four-point Likert scale ranging from (1) strongly agree to (4) strongly disagree. The two used sample items were: "I have the resources necessary to use OCT" and "I have the knowledge necessary to use OCT".

Use behaviour (UB) was measured using two items as follows: (*1) How often do you log onto OCT from home? and (2) How many times do you log onto OCT per day?* The first item was adapted from [6]. Both of these items used four categories (*Never, Once,*

Twice and > 2 times), which were respectively given codes 0, 1, 2, and 3. Gender was coded "0" for males and "1" for females. Age (*in years*) was categorised into four ranges [(18–23), (24–29), (30–35), and above 35], and these ranges were respectively coded 1, 2, 3, and 4. Year of study was ranging from "Year 1" through "Year 4" and were coded 1, 2, 3 and 4 respectively. Experience with computers (*years*) was categorised into four ranges [(0–5), (6–10), (11–15), and above 15], and these ranges were respectively coded 1, 2, 3, and 4. The first time use of OCT was given four labels [*Never Used, Semester 1 (2014), Semester 2 (2014), Semester 1 (2015)*] and their respective codes were 0, 1, 2, and 3.

4 Results and Interpretation

In this study, data was analysed through using SPSS. The tests employed are zero order correlations and regression analysis. The latter was used due to limitations associated with zero order correlations. One main limitation is that simple correlation does not regulate the counterfeit relationships that may come as a result of such other variables as demographic.

Table 2 shows a correlation matrix that indicates means, standard deviations and bivariate correlations of variables. All the means of the main variables center around 2 (*agree*), which implies that on average most of the respondents agreed with the statements that were tested in this study. Table 3 indicates the results for the linear regression analysis. The presentation below is based on information from Tables 2 and 3.

4.1 Hypothesis 1

Hypothesis 1 had predicted that performance expectancy has a positive effect on the behavioural intention to use OCT. Correlation analysis initially found that PE really had a positive and significant effect on BI ($r = 0.56$, $p \leq 0.01$). When all other variables were controlled, the correlational finding was confirmed ($\beta = 0.26$, $p \leq 0.01$). This significant relationship suggests that students were eager to use OCT because they believed they could benefit a lot from its use. This finding resonates with findings of previous studies [14, 20, 24]. Findings in regard to Hypothesis 1 suggest that the digital native students are definitely willing to embrace the new technologies. However, there may be various reasons why this is the case. Most of the students in this generation are interested in just playing around with the technology and related gadgets. This is to such an extent that some of them even spend more time with their technological gadgets than they have to, at times playing games and visiting social networking sites. In the same spirit, some can be attracted by the content that is related to their school work, hence, take advantage of that to perform well in their studies. Another reason that attracts students into being glued up with the technological gadgets is that they are a generation that is lazy to read – they would rather listen to an audio text than to read the actual text. Hence, it is not surprising that they hold a belief that they can always benefit from the use of the technology related systems.

4.2 Hypothesis 2

A similar pattern of results (to H1) was obtained in relation to hypothesis 2 since it was correlationally found that effort expectancy had a positive and significant effect on the behavioural intention to use OCT ($r = 0.68$, $p \leq 0.01$). After controlling for all the other variables, the correlational finding was confirmed ($\beta = 0.34$, $p \leq 0.01$). This relationship proposes that, of the factors that attracted students to use OCT and kept them using it was the ease of use [12, 29, 39] that was associated with it.

4.3 Hypothesis 3

A different pattern of results (from H1 and H2) was obtained in relation to hypothesis 3 because it was correlationally found that social influence had a positive and significant effect on the behavioural intention to use OCT ($r = 0.30$, $p \leq 0.01$). Upon controlling for all the other variables, the correlational finding was not confirmed as student intentions to use OCT under social influence were advancing away from zero ($\beta = 0.08$, $p \geq 0.05$). This relationship suggests that students did not use OCT because they had been influenced by people around them. Even though students may be eager to experiment with the technological gadgets and related application(s), they may as well be discouraged from using the technology/system because they have a perception that it is not easy to use. Immediately after forming that perception, they can completely stop using the system/technology or they may only concentrate on the aspects of the technology/system that excite them.

As an example, if the OCT has the chat component, students might just use OCT for chatting purposes only and forget about the principal aim for which OCT was designed and implemented. Therefore, this implies that educators as well must be brought on board to ensure that they give assignments or activities that may require students to use OCT in dealing with such. This might go a long way in helping users of a named system to benefit academically or otherwise from using such a system. This is interesting because the study finds that, though students are attracted to using OCT and keep using it, they have not been influenced by anyone to do so. This implies that a lot of benefits can be reaped from using the system if all the relevant parties that must take part are effectively doing so.

4.4 Hypothesis 4

In relation to hypothesis 4, the same pattern of results to hypothesis 3 was found. This is because it was initially found that facilitating conditions had a positive and significant influence on the actual behaviour to use OCT ($r = 0.28$, $p \leq 0.01$). However, when all the other variables were controlled, the facilitating conditions had no influence on the actual behaviour to use OCT ($\beta = -0.03$, $p \geq 0.05$). This relationship suggests that students felt that the environment was conducive for them to really use OCT to achieve their objectives. Prior training (as part of facilitating conditions) is equally important because for one to use the system/technology effectively s/he has to be equipped with the right skills including knowing what the system or technology is aimed at. This might help individuals to prioritise the use of the system by focusing on the secondary aspects

Table 2. Correlation Matrix (*means, standard deviations and bivariate correlations of variables*).

Variable	Mean	δ	1	2	3	4	5	6	7	8	9	10	11
1. PE	2.26	0.68	1.00										
2. EE	2.09	0.72	0.59**	1.00									
3. SI	2.25	0.66	0.25**	0.25**	1.00								
4. FCs	1.99	0.74	0.41**	0.65**	0.23**	1.00							
5. BI	2.01	0.70	0.56**	0.68**	0.30**	0.57**	1.00						
6. Gender	0.61	0.49	0.07	0.04	0.07	-0.05	0.78	1.00					
7. Age	1.51	0.75	-0.20*	-0.22**	-0.15	-0.13	-0.27**	0.14	1.00				
8. YOS	2.59	0.60	-0.30**	-0.27**	-0.06	-0.13	-0.39**	0.08	-0.34**	1.00			
9. EWCs	1.82	0.89	0.08	0.01	-0.03	0.17*	-0.03	0.14	-0.05	-0.00	1.00		
10. FTUoT	1.82	0.92	-0.14	-0.25**	-0.00	-0.13	-0.18*	0.11	-0.14	-0.25**	-0.06	1.00	
11. UB	1.28	-0.08	0.39**	0.37**	0.24**	0.28**	0.42**	0.08	0.07	0.27**	-0.10	0.11	1.00

Note: **= correlation significant at 0.01 level (2-tailed); *= correlation significant at 0.05 level (2-tailed); δ=Standard Deviation; PE=Performance Expectancy; EE=Effort Expectancy; SI=Social Influence; FCs=Facilitating Conditions; BI=Behavioural Intention; YOS=Year of Study; EWCs=Experience With Computers; FTUoT=First Time Use of OCT; LFfH=Login Frequency from Home; LF/D=Login Frequency per Day.

Table 3. Linear regression analysis results.

	Model 1	Model 2	Model 3
Variable	Behavioural Intention	Use Behaviour	Use Behaviour
Performance expectancy	0.26**	-0.21*	-0.18
Effort expectancy	0.34**	-0.16	-0.12
Social Influence	0.08*	-0.18*	-0.15
Facilitating Conditions	0.17*	-0.03	0.01
Gender	-0.03	0.09	0.07
Age	-0.08	-0.06	-0.05
Year of study	-0.21**	0.16	0.12
Experience with computers	-0.09	-0.09	-0.09
First time use of OCT	0.04	-0.03	-0.03
Behavioural intention	-	-	-0.15
R	0.79	0.50	0.49
R^2	0.62	0.25	0.24

Note: **= correlation significant at 0.01 level (2-tailed); *= correlation significant at 0.05 level (2-tailed). Values of Regression Coefficient (β) represented here are the standardised ones.

at a later stage when the imperative business is done. It is also important to note that the digital native students are not concerned about whether the institution where they are studying has computer resources or not. This is because almost all students have computer gadgets that are Wireless Access Point (WAP)-enabled (iPad, laptop, iPhone, etc.) at their disposal. Therefore, whether the institution has enough computers, acceptable and easily available internet connectivity, sufficient Wi-Fi hot-spots, acceptable number of technicians or not, students still continue to use the related technology/systems. They would rather purchase data bundles from a service provider or a configured modem so that they can connect to the Internet and use a related system or technology (*i.e., to download learning materials, assignments, test or assignment feedback*) irrespective of time and location (*any time anywhere*).

4.5 Hypothesis 5

Hypothesis 5 had predicted that behavioural intention has a positive effect on the actual use of OCT. Correlation analysis initially found that BI really had a positive and significant effect on UB (r = 0.42, p ≤ 0.01). This initial finding was not confirmed when all the other variables were controlled (β = −0.15, p ≥ 0.05). This relationship suggests that the behavioural intention of students to use OCT did not have any influence on their actual behaviour to use it. In testing Hypothesis 5, three procedures were followed. The mediator variable (behavioural intention) was first regressed on the independent variables PE, EE and SC. The dependent variable (use behaviour) was then regressed on the same three independent variables (PE, EE, and SC). Then the same dependent variable was regressed on both the three independent variables and a mediator (BI). Students of this age are independent, techno-savvy, entrepreneurial, hardworking and thrive on flexibility [38]. They also have easy access to cell phones, personal pagers, and computers [21]. These factors make them to, whether they have intentions or not, use the technology/systems at their disposal. This is attributed to the fact that the students who participated in this study indicated no intentions to use OCT, but at the end of the day, they actually used it.

5 Significance of the Study

The study contributes to literature and in practice. In literature, it contributes to the body of knowledge in the field of the use of technology for teaching and learning at higher education institutions. It also helps in understanding the perceptions held by students at higher education level with regard to the use of technology in teaching and learning. Furthermore, the UTAUT model seems not to have been employed in the setting where the study was carried out, but has been applied many times in the Western setting. Therefore, considering the culture of people (e.g., Basotho) where this study was conducted such as their thinking abilities, social interactions, resources at their disposal and several other aspects, this study has produced interesting results.

One of the top-executives of the selected institution asserts "the good news is that staff and students seem to have embraced OCT as evidenced by the use of the system following the training programme" [28] (p. 1). However, as evidenced above, this did not seem to be what is on the ground at least in the first academic year upon acquiring and using OCT. Hence, in practice, top management of the selected institution is highly likely to benefit from the findings of this paper. Moreover, the OCT designers and developers may also benefit from the findings of this paper. OCT provides both the mobile view and full view. However, students were struggling to complete assignments using their cellphones because of some challenges encountered with the mobile view. Students normally encountered problems in that they were not even sure when their submissions had gone through (i.e., no notification of submission). As a result they (students) had to physically consult with their educators to confirm their submissions. This causes unnecessary workload on educators, and in responding to the continuous monitoring call [17], they would not even have to wait for students to ask, but proactively make the designers aware of the need for such functionalities to be incorporated in OCT.

6 Conclusions and Future Work

The aim of this study was to examine the factors that determine the use and/or non-use of OCT at the selected institution. Having followed the UTAUT model, the study concludes that, of the investigated hypotheses, performance expectancy (PE) and effort expectancy (EE) positively influence the behavioural intention to use the online collaborative tool (OCT); while social influence (SI), facilitating conditions (FCs) and behavioural intention (BI) do not have a significant influence on the students to use OCT. The findings indicate that students actually use technology irrespective of whether they think they will learn or not learn from using it. However, they are interested only in some components that each finds interesting to use. The students completely ignore the other system/technology components that they are not interested in. Hence, future studies may investigate ways to make students use most of the components of the system/technology.

Furthermore, in as far as using technology in the learning process is concerned, students do not have a problem because they already have technological gadgets. Perhaps the story would be different if they had to go an extra mile to secure such gadgets. This is why the findings of this study are in agreement with [42] that the behavioural intention does not mediate the direct relationship between facilitating conditions and the actual behavior to use the system. However, the findings implicate that behavioural intention does not have to mediate the relationship between social influence and the actual behavior to use the system/technology.

The findings of this study imply that technology plays a vital role in the learning of students if used effectively. It also suggests that all parties that must partake in using the system/technology must do so in order that its benefits are realised. This is irrespective of whether cultures, thinking abilities, social interactions and resources are concerned. A follow-up study to investigate the status-quo at the institution can be conducted as far as the use of OCT is concerned.

References

1. Ajzen, I.: The theory of planned behavior. Org. Behav. Hum. Decis. Proceses **50**(2), 179–211 (1991). https://doi.org/10.1016/0749-5978(91)90020-T
2. Attuquayefio, S.N., Addo, H.: Using the UTAUT model to analyze students' ICT adoption. Int. J. Edu. Dev. Infor. Comm. Technol. **10**(3), 75–86 (2014)
3. Ayaz, A., Yanartaş, M.: An analysis on the unified theory of acceptance and use of technology theory (UTAUT): acceptance of electronic document management system (EDMS). Comput. Hum. Behav. Reports **2**, 1–7 (2020). https://doi.org/10.1016/j.chbr.2020.100032
4. Bagozzi, R.P., Davis, F.D.D., Warshaw, P.R.: Development and test of a theory of technological learning and usage. Hum. Relat. **45**(7), 659–686 (1992). https://doi.org/10.1177/001872679 204500702

5. Bernard, H.R.: Research methods in Anthropology. Rowman and Littlefield Publishers, New York (2017)
6. Brown, R., Ogden, J.: Children's eating attitudes and behaviour: a study of the modelling and control theories of parental influence. Health Educ. Res. **19**(3), 261–271 (2004). https://doi.org/10.1093/her/cyg040
7. Centre for Innovation in Learning and Teaching: Vula - What is Vula? Vula Features. Training & Resources (2021). http://www.cilt.uct.ac.za/cilt/vula Accessed 11 Feb 2021
8. Centre for Teaching and Learning: Blackboard (2021). https://www.ufs.ac.za/ctl/home-page/ctl-registration-support-2021/blackboard. Accessed 14 Apr 2021
9. Chen, P.Y., Hwang, G.J.: An empirical examination of the effect of self-regulation and the unified theory of acceptance and use of technology (UTAUT) factors on the online learning behavioural intention of college students. Asia Pacif. J. Edu. **39**(1), 79–95 (2019). https://doi.org/10.1080/02188791.2019.1575184
10. Chyung, S.Y.Y., Roberts, K., Swanson, I., Hankinson, A.: Evidence-based survey design: the use of a midpoint on the likert scale. Perfor. Improv. **56**(10), 15–23 (2017). https://doi.org/10.1002/pfi.21727
11. Davis, F.D., Bagozzi, R.P., Warshaw, P.R.: Extrinsic and intrinsic motivation to use computers in the workplace. J. Appl. Soc. Psychol. **22**(14), 1111–1132 (1992)
12. Davis, F.D.: Perceived usefulness, perceived ease of use, and user acceptance of information technology. MIS Q. Manag. Infor. Syst. **13**(3), 319–339 (1989). https://doi.org/10.2307/249008
13. Deng, S., Liu, Y., Qi, Y.: An empirical study on determinants of web based question-answer services adoption. Online Infor. Rev. **35**(5), 789–798 (2011). https://doi.org/10.1108/14684521111176507
14. Dhaha, I.S.Y.A., Ali, A.Y.S.: Mediating effects of behavioral intention between 3G predictors and service satisfaction. Malays. J. Comm. **30**(Special Issue), 107–128 (2014). https://doi.org/10.17576/jkmjc-2014-30si-07
15. Donmez-Turan, A.: Does unified theory of acceptance and use of technology (UTAUT) reduce resistance and anxiety of individuals towards a new system? Kybernetes **49**(5), 1381–1405 (2019)
16. Durak, H.Y.: Examining the acceptance and use of online social networks by preservice teachers within the context of unified theory of acceptance and use of technology model. J. Comput. High. Educ. **31**(1), 173-209 (2019). https://doi.org/10.1007/s12528-018-9200-6
17. Franzoni, V.: Artificial intelligence visual metaphors in E-Learning interfaces for learning analytics. Appl. Sci. **10**(20), 2–25 (2020). https://doi.org/10.3390/app10207195
18. Hoque, R., Sorwar, G.: Understanding factors influencing the adoption of mHealth by the elderly: an extension of the UTAUT model. Inter. J. Med. Infor. **101**, 75–84 (2017). https://doi.org/10.1016/j.ijmedinf.2017.02.002
19. Im, I., Hong, S., Kang, M.S.: An international comparison of technology adoption: testing the UTAUT model. Infor. Manag. **48**(1), 1–8 (2011). https://doi.org/10.1016/j.im.2010.09.001
20. Kim, S., Lee, K.H., Hwang, H., Yoo, S.: Analysis of the factors influencing healthcare professionals' adoption of mobile electronic medical record (EMR) using the unified theory of acceptance and use of technology (UTAUT) in a tertiary hospital. BMC Med. Infor. Decis. Mak. **16**(1), 1–13 (2016). https://doi.org/10.1186/s12911-016-0249-8
21. Lancaster, L.C., Stillman, D.: When generations collide: who they are. Why they clash. How to solve the generational puzzle at work. Harper Collins Publishers, Inc., New York (2003)
22. Lee, Y., Lee, J., Lee, Z.: Social influence on technology acceptance behavior: self-identity theory perspective. ACM SIGMIS Database: DATABASE Adv. Infor. Sys. **37**(2–3), 60–75 (2006). https://doi.org/10.1145/1161345.1161355

23. Ma, S.: Fast or free shipping options in online and omni-channel retail? The mediating role of uncertainty on satisfaction & purchase intentions. Inter. J. Logist. Manag. **28**(4), 1099–1122 (2017). https://doi.org/10.1108/IJLM-05-2016-0130

24. Maqableh, W., Mhamdi, C., Al Kurdi, B., Salloum, S.A., Shaalan, K.: Studying the social media adoption by university students in the United Arab Emirates. Inter. J. Infor. Technol. Lang. Stud. **2**(3), 83–95 (2018)

25. McClelland, S.B.: Organizational Needs Assessments: Design, Facilitation and Analysis. Quorum Books, Westport, London (1995)

26. McCrindle, M., Fell, A.: Understanding Generation Z: Recruiting, Training and Leading the Next generation. McCrindle Research Pty Ltd., Norwest, AU (2019)

27. Mosweu, O., Bwalya, K.J., Mutshewa, A.: A probe into the factors for adoption and usage of electronic document and records management systems in the Botswana context. Infor. Dev. **33**(1), 97–110 (2017). https://doi.org/10.1177/0266666916640593

28. Moteete, M.: A quarterly newsletter of the National University of Lesotho: Maiden issue. Quarterly newsletter (2014). https://thuto.nul.ls/access/content/group/communications and marketing/NUL News- Maiden issue June to September 2014.pdf. Accessed 12 Nov 2019

29. Preece, J., Rogers, Y., Sharp, H.: Interaction Design: Beyond Human-Computer Interaction, 4th edn. John Wiley & Sons Inc, New Delhi (2015)

30. Pynoo, B., Devolder, P., Tondeur, J., Van Braak, J., Duyck, W., Duyck, P.: Predicting secondary school teachers' acceptance and use of a digital learning environment: a cross-sectional study. Comput. Hum. Behav. **27**(1), 568–575 (2011). https://doi.org/10.1016/j.chb.2010.10.005

31. Ramaboli, A.L., Mphatsi, L., Keta, K.: Thuto: Educate, learn and collaborate (Unpublished Project proposal) (2014)

32. Roger, E.M.: Diffusion of Innovations, 4th edn. Free Press, New York (1995)

33. Saunders, M.N.K., Lewis, P., Thorhill, A.: Research Methods for Business Students, 8th edn. Pearson Education Inc, Harlow (2019)

34. Selyer, C., Mugova, C.: An Investigation into the impact of E-commerce, M- commerce and modern technology on the translation industry in South Africa. J. Manag. Admin. **1**, 1–23 (2017)

35. Srinivasan, S.S., Anderson, R., Ponnavolu, K.: Customer loyalty in e-commerce: an exploration of its antecedents and consequences. J. Retail. **78**(1), 41–50 (2002). https://doi.org/10.1016/S0022-4359(01)00065-3

36. Sullins, J.: Information technology and moral values. In: Zalta, E.N. (ed.) The stanford encyclopedia of philosophy (2021). https://plato.stanford.edu/archives/spr2021/entries/it-moral-values/. Accessed 25 Mar 2021

37. Thompson, R.L., Higgins, C.A., Howell, J.M.: Personal computing: toward a conceptual model of utilization. MIS Q. Manag. Info. Sys. **15**(1), 125–142 (1991). https://doi.org/10.2307/249443

38. Tulgan, B., Martin, C.A.: Managing Generation Y: Global Citizens Born in the Late Seventies and Early Eighties. HRD Press, Amherst, MA (2001)

39. Tullis, T., Albert, B.: Measuring the User Experience: Collecting, Analyzing, and Presenting Usability Metrics, 2nd edn. Morgan Kaufmann, Waltham, MA (2013)

40. Uwameiye, B.E.: Application of information and communication technology (Icts) in the effective teaching and learning of Home Economics in secondary schools in Nigeria. Europ. Sci. J. **11**(7), 1857–7881 (2015)

41. Venkatesh, V., Davis, F.D.: A theoretical extension of the technology acceptance model: Four longitudinal field studies. Manag. Sci. **46**(2), 186–204 (2000). https://doi.org/10.1287/mnsc.46.2.186.11926

42. Venkatesh, V., Morris, M.G., Davis, G.B., Davis, F.D.: User acceptance of information technology: toward a unified view. MIS Q. **27**(3), 425–478 (2003)

43. Wentzel, J.P., Diatha, K.S., Yadavalli, V.S.S.: An application of the extended technology acceptance model in understanding technology-enabled financial service adoption in South Africa. Dev. Southern Africa **30**(4–5), 659–673 (2013). https://doi.org/10.1080/0376835X. 2013.830963

44. Wu, Y.L., Tao, Y.H., Yang, P.C.: The use of unified theory of acceptance and use of technology to confer the behavioral model of 3G mobile telecommunication users. J. Stat. Manag. Syst. **11**(5), 919–949 (2008). https://doi.org/10.1080/09720510.2008.10701351

45. Zhou, T., Lu, Y., Wang, B.: Integrating TTF and UTAUT to explain mobile banking user adoption. Comput. Hum. Behav. **26**(4), 760–767 (2010). https://doi.org/10.1016/j.chb.2010. 01.013

The Model of Curriculum Constructor

Gulmira Bekmanova[1,2]([⊠]), Aizhan Nazyrova[1], Assel Omarbekova[1], and Altynbek Sharipbay[1,2]

[1] L.N.Gumilyov Eurasian National University, Nur-Sultan, Kazakhstan
[2] Nuclear University MEPhI, Moscow, Russia

Abstract. The curriculum constructor model based on ontology that allows establishing connection between curriculum objectives, learning outcomes, expertise and disciplines is introduced in this paper. Thus, it makes possible to consider the compliance of the Education Programme with the learning outcomes and expertise, as well as, it helps check the curriculum goal achievements, the learning outcome development and mastering the competences. The constructor lets create different subsets with different properties and set various individual trajectories.

Keywords: e-learning · Ontological model · Artificial intelligence · Curriculum development · Learning outcomes · Competences · Curriculum goal

1 Introduction

The main issue of this research is to create an estimation model of achievement of curriculum objectives. Curriculum design consists of several stages. The Education Programme goal achievement is carried out through the Education Progamme learning outcomes. Each learning outcome is associated with one or more expertise, while the expertise are associated with a set of disciplines. To achieve the curriculum goal in the context of programme disciplines is ensured by the student's competency mastery. However, each stage requires the verification of the goal achievement, learning outcome, and expertise. Frequently, it happens that not all learning outcomes or achievements are achieved resulting the lack of achievement formal verification at each stage that leads to the difficulties of assessing the Education Programme goal achievement. Besides, in case of the errors in the Education Programme design it is impossible to check them at the design stage. A different approach to design the Education Programme that allows all verifications and checks incorporate into the Education Programme model is required in that regard. That is why the development of the curriculum goal achievement model estimation with the verification is the foundation for the Intellectual Learning Management System.

2 Related Works

There are various work dedicated to the development and assessment of the Academic programme. Thus, in the paper [1] researchers have introduced a new way of executing intellectual data analysis for the development instructions and the academic programme

© Springer Nature Switzerland AG 2021
O. Gervasi et al. (Eds.): ICCSA 2021, LNCS 12950, pp. 459–470, 2021.
https://doi.org/10.1007/978-3-030-86960-1_32

assessment. Authors propose the association rules of the analysis technique in order to identify a set of rules that manage the relationship between two main elements: an academic education programme and student outcomes.

The result-based training is considered as a learning model in the paper [2] that creates an emphasis in assessing student performance according to the outcomes that lie in knowledge, skills and behavior. The study revealed that this teaching form is widely used in technical higher institutions around the world. At the time of the study the authors have emphasizes that the outcome lies in what learners can do rather than what they know.

It is shown that the MOOC provides an opportunity to assess student learning in a technically mediated online environment in the papers [3, 4]. An analysis has been carried out from MOOC or from several MOOCs with different numbers of students to carry out the research in the paper [3]. The study has found that MOOC course design can have a significant impact on course student engagement rates. The model of intelligent MOOC presented in the paper [4] blended and any e-learning at its designing, using the knowledge base, ontological model of discipline, and their relevant question-answer system and intelligent search. The separate important part of each MOOC is the intelligent assessment of knowledge and achievement of announced training results. The suggested MOOC model is more effective for distance, blended and any e-learning.

Perspectives of artificial intelligence influence on teaching technologies have been considered in the paper [5]. Four different categories of education landscape have been considered in the study course. They are: individual learning content, modern teaching methods, score multiplying technology, and relationship among students, as well as teachers. The researchers have presented the research outcome stating that Artificial Intelligence (AI) is able to change the learning trajectory. The knowledge assessment is an important artificial intelligence task that is often based on natural language processing. In the paper [6] empirical research is studied with the aim of assessing knowledge that can be applied without human participation.

The authors of the paper [7] have created an intelligent assessment system that is considered a collective and joint project designed for the construction intensive, heuristic and adaptive part that is considered a subsystem in the concept of teaching students. This study assumes a three-part structure based on standard training flows, and explains the function and construction of each part. The researchers have analyzed the use of the electronic assessment in the field of computer teaching, and identified factors that contribute to effective electronic knowledge assessment in the paper [8]. The authors of the paper [9] have provided the research overview through the systematic analysis that considers the use of AI in the higher education.

They have reviewed over 2,656 publications from 2007 to 2018. 146 papers have been used for final generalization according to the exact aspects of introduction and exclusion. The research results introduced in the AIEd publications show that they belong to informatics and STEM, as well as point out the fact that quantitative methods are more often used in empirical research. In the paper [10] the research authors studied and discovered the methods and ways of dynamic generating the test tasks that are based on the student knowledge examination using methods grounded in neural networks. These methods can help ensure flexibility in the controlling alternative adaptation that adjusts to personal knowledge level. As a result, this procedure will enable the teacher to acquire the most representative assessment of the student knowledge. In the paper [11] authors have presented a tool for Assessing Infrastructure Service Level Environment (AISLE).

The purpose of this tool is to improve the application of artificial intelligence methods in assessing student's understanding of a specific teaching topic with the concept map support. They have used the XML parsing method to perform the required evaluation. A two-layer feedback neural network has been used in order to construct the analyzer.

The authors of the paper [12] have studied the intelligent software system introduction for e-learning that is focused on increasing the student motivation and academic achievements in computer programming courses. The created intellectual approach seeks to provide students with appropriate materials based on teaching degree.

The authors have provided the efficient scheme of fuzzy automatons for the intelligent recognition of students' textual responses in the paper [13]. The recommended system adapts to unintentional mistakes made by learners, as well as simulates a benevolent human evaluator. The authors of the paper [14] have analyzed the latest neural network scheme in order to intelligently analyze the student interaction in the field of study. The model is used for a typed one-word text solution at the end of the teaching. The authors have considered an approach in the paper [15] that is based on k-variable fuzzy finite state automation for the implementation of a scoring system for short answers. The presented method attempts to imitate student behavior in the context of made mistakes that may be knowledge-based or unintentional in nature. The developed methodology has been described using sample assessments from a test conducted on a group of students.

The authors on the paper [16] have developed a technology that allows integrating existing partially effective tools with general learning management systems (LMS) in accordance with the standard. The designed and developed system can be used as modules for existing systems or LMS. Universities interested in the system can use this tool as a module.

The scholars of the paper [17] have also presented an improved assessment system technology that can be used both for teaching skills and knowledge. A general technologically advanced grading system for online logic courses has been developed taking into account the e-learning standards and specifications.

The scholars in the paper [18] have focused on the triangular membership function, where the two inputs have been two different system results, namely the finance and business statistics principles, while the output has been the performance value. The authors have applied the method to a sample of twenty-two first-year students at the University of Malaysia.

The authors [19] have reviewed the student performance on two separate tests on the same subject "Electrical Control" to measure student performance. Then they have used the classical methods to compare the assessment results with the obtained results. Their sample consists of twenty students of laboratory courses in control methodology in Turkey.

The authors of the paper [20] have sought to assess collectively the student group knowledge and skills. They have created a fuzzy model in order to assess the student group knowledge and abilities, in which the evaluated student characteristics have presented in fuzzy subsets variant of a set of linguistic meanings determining their work. Student profiles have been for quantitative/qualitative study results of student performance in groups.

The authors of the paper [21] have applied the fuzzy logic to determine the student knowledge level in a certain knowledge area using the response time to test questions and the student's score as metrics.

The authors have presented a fuzzy neutral inference system for modeling a student in the context of a web-based intelligent learning system in the paper [22]. The proposed model has been implemented and tested on simulated student data.

The authors of the paper [23] have considered an approach to student knowledge assessment that is based on fuzzy modeling neural network. A fuzzy neural diagnostic process model has been proposed in order to establish the student data. The model can be simply adapted to the individual teacher judgment. This approach can be applied in order to implement the disclosed student model that will be interactively corrected by the teacher. Scientists [24] suggest an integrated approach to fuzzy sets in order to assess the learning outcomes aimed at learners. They apply the fuzzy set foundations in order to find out the fuzzy concept for an individual proposal, and also use the fuzzy set method to establish the evaluation criteria for their corresponding values.

The authors [25] have created the Cogito software package, in which they have used the neural network to search for patterns in "noisy" data from paper and pencil, and also connected them with Perry models and Reflexive intellectual development judgments.

The researchers [26] have given the latest strategy for assessing education achievements during the teaching process of students, in which each question is assigned by the linguistic variables of a fuzzy set depending on its importance, difficulties and problems through the membership functions.

The scientists have presented a fuzzy logical diagnostic model that is called the profile of teaching students in the paper [27]. In the process of research, the authors of the current paper [28] considered and researched different ways and methods, which where, as a result, utilized for the evaluation of capstone projects.

The authors [29] gave an example of an educational domain model. During the development of the current paper, the authors [30] suggested a variety of methods which would analyze and calculate the students' activity with the intent of improving the students' performance. As a result of the research it was determined that the most effective and at the same time practical method happens to be the analysis of study, meaning the constant analysis and comparison of the results using algorithms.

During the research of the educational methods, the author of the current paper, [31] opened a course for future engineers which had a participation of 27 students. By analyzing the completed course in question, the students were satisfied with the high quality educational material, and in addition, were able to give feedback for improving the course. The author of this article [32], as a result of his research, came to the conclusion that the theoretical educational part of the programs not only develops the development of educational programs, but also simultaneously allows expanding the possibilities of the learning process. In the course of the study, the authors of this scientific work [33] consider the concepts of educational programs that are formed on the basis of the results. The authors [34] of the article gave an example of curriculum development using the artificial intelligence (AI) backpropagation method. In the work [35] for the design of curricula, the design of the PLM course was considered. The article [36] considered the use of aggregated profile clustering to assess the curriculum of a higher educational institution. The authors [37] conducted a study in the field of the educational process. And also the structure of the analysis of the curriculum is proposed for the study to test the proposed approach. In the article [38], the authors gave an example for developing a computer science curriculum at a secondary liberal arts university in the context of cybersecurity. This article [39] is dedicated to modulating a new curriculum for integration into the existing curriculum. This article [40] aims to redesign the curriculum for Industry 4.0.

3 Curriculum Constructor

The Curriculum is being developed by the Academic Committee members. Furthermore the curriculum is being approved by the administration undergoing the examination and the review and approval procedure at the University collegiate bodies.

Currently, the constructor functions are as follows (Fig. 1).

Users of the system:

1) the Academic Committee members.
2) Administration staff.
3) External expert.
4) Internal expert.
1. A member of the academic committee develops CURRICULUM, enters general data: task, type, degree (bachelor, master, doctor), time period of study, distinctive characteristics, language of study, date of approval by the academic council, availability of accreditation, formed results of research.
2. The composition of the index of disciplines by a member of the academic committee.
3. The creation of the CURRICULUM template by a member of the academic committee: enters mandatory subjects to the educational program.
4. After the creation of the CURRICULUM template by am employee, a member of the academic council chooses the subjects and fills out CURRICULUM.
5. A employee reviews the created CURRICULUM. In case everything is correct, it is sent for examination. In case of the opposite, the CURRICULUM is sent for revision.
6. A roster of experts is available, an employee assigns these experts to particular CURRICULUM. The experts review the CURRICULUM. Examination is done according to a particular template. If the CURRICULUM is correct, it is sent for confirmation, if not, it is sent for revision.
7. An employee downloads the CURRICULUM, prints it, and announces at the academic council.

An employee creates necessary reports for CURRICULUM. The staff generates the necessary reports in the CURRICULUM.

The disadvantage of this constructor is that it is impossible to consider the compliance of the Education Programme with the learning outcomes and expertise, as well as to check the curriculum objective achievements, the learning outcome development and mastering the expertise.

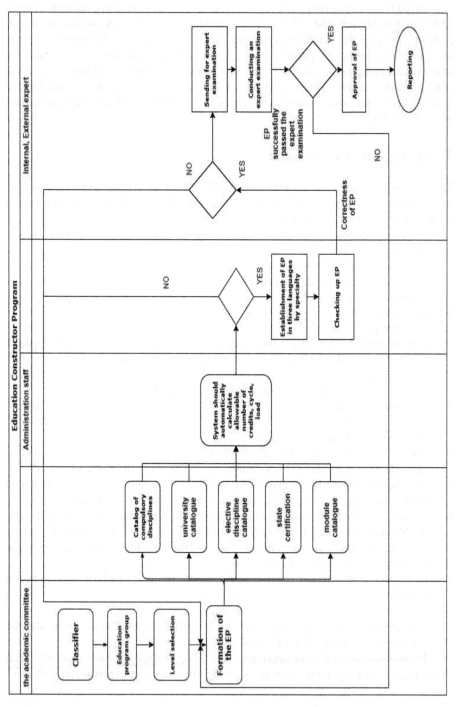

Fig. 1. Current curriculum constructor's model

4   Development of Curriculum with Using Ontological Model

Development of curriculum consists of several stages that are described in Fig. 2.

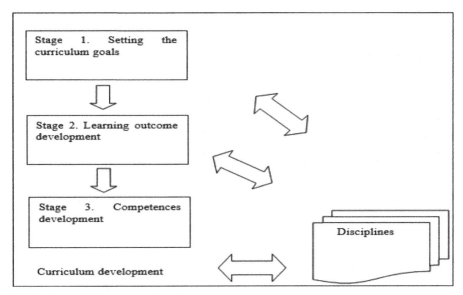

Fig. 2. Curriculum development stages

It is crucial to correlate the competences with the learning outcomes, and the learning outcomes with the programme objectives.

Stage 1. Setting the curriculum objectives:

Curriculum objectives are aimed at training specialists of high qualification level without a category, the second category and the first category. It is necessary to complete a number of tasks including the targeted creation of students' contingent, the specialized theoretical and practical training of the students during the learning process focused on the modern employer requirements. Hence, the Programme goal is designated as GDP, the learning outcomes as R, and the expertise as C.

The requirements for the goal are as follows: brevity, comprehensibility, connection with the learning outcomes.

Stage 2. Curriculum learning outcomes:

The learning outcomes are created both at the Curriculum level of higher education and at the individual module or academic discipline level.

The learning outcome requirements:

1. Demonstrating the attained knowledge in the field of research;
2. Implementing the attained knowledge, being able to form theses and arguments, being able to solve problems in the field of research on a professional level;
3. Perform data collection on the bases of social, ethical, and scientific reasonings;

4. Being able to implement theoretical and practical skills into solving problems in the field of research;
5. Independence for further training in the field of research;
6. Using the methods of scientific researches and academic writings in the field of research;
7. Utilizing the attained knowledge and understanding of facts, theories, and phenomena in the field of research;
8. Following the principles and customs of academic integrity.

The results should be achievable, measurable and related to the competences.

Stage 3. Curriculum competences:

Competence is the ability to put into practice the knowledge, skills and abilities acquired in the learning process in professional activities.

Competences must be short, achievable, measurable and related to the disciplines.

The learning achievements can only be verified if the learning process, learning validation and student's knowledge evaluation are fully formalized.

The curriculum goal can be understood in general terms as the combination of all learning outcomes of the programme.

$$G_{DP} = R_1 \& R_2 \& R_3 \& \ldots \& R_n \tag{1}$$

where $R = \{R_1, R_2, R_3, \ldots, R_n\}$ a set of learning outcomes.

However, in a particular case, these learning outcomes can be divided into mandatory learning outcomes of the Education Programme, and they can be variable.

Mandatory core learning outcomes, as the name suggests, are mandatory and included in full. Thus, the subset of the mandatory learning outcomes can be represented as $R = \{R_{c1}, R_{c2}, R_{c3}, \ldots R_{cn}\}$.

Varieties of the learning outcomes can be represented as $Rv = \{R_{v1}, R_{v2}, R_{v3}, \ldots R_{vn}\}$. However, there is a formal need to impose restrictions on minimum necessary and sufficient variable learning outcomes for obtaining a degree. But, it is impossible at this stage, the restrictions are introduced at the stage of formulating competences while verifying the compliance with the disciplines. Then the formula (1) can be rewritten as (2):

$$GDP = R \& R_v \tag{2}$$

Many of the competences can be described as (3):

$$C = \{C_1, C_2, \ldots, C_n\} \tag{3}$$

Many of the competences can be represented as (4):

$$D = \{D_1, D_2, \ldots, D_n\} \tag{4}$$

The example of ontology illustrating the relationship between objectives, learning outcomes, competences and disciplines is represented in Fig. 3.

The term "ontology" (the word "ontology" comes from the Greek "ontos" - existing and "logos" - concept, teaching, reason), proposed by R. Goklenius, appeared in 1613.

Fig. 3. Example of curriculum ontological model

At present, the methods of knowledge engineering, the young science of the extraction, structuring, presentation and processing of knowledge are becoming more and more popular [41]. The ontological model is built in Protégé tool [42].

This use of ontology to create the Curriculum model makes it possible to establish a connection between the curriculum goal, learning outcomes, competences, and disciplines. It allows creating different subsets with different properties and defining various individual trajectories.

There are 15 options for developing curriculum using the ontological modeling through various combinations of possible options.

1. Formulating the Curriculum objectives;
2. Formulation/choice (among existing numbers) of the relevant objectives of the learning outcomes and linking them to the goal;
3. Formulation/ choice (among existing numbers) of the competences and linking them to the learning outcomes;
4. Creation/ selection of existing disciplines with descriptions and keywords, and linking disciplines to the competences.

5   Conclusion

In this paper, an attempt is made to create the Curriculum constructor model based on the ontological modeling. The links between the curriculum goal, learning outcomes, competences and disciplines are considered while developing the Curriculum Model. The existing curriculum constructor does not provide an opportunity to check the compliance of the Education Programme with the goal, learning outcomes and competences, as well as to verify the goal achievement, the learning outcome development and mastering the competences. The lack of linkages subsequently precludes analysis and assessment of the learning outcomes and results of competence mastering.

6 Further Work

Further work prospects, in our opinion, lie in creating a kernel of a new knowledge-based learning management system. The Curriculum as the basic document of the educational process and its correct design plays a major role in providing high-quality education. The next modules of the system are the intelligent knowledge assessment module and the module of MOOC. Connection of these modules will significantly expand the traditional LMS services for students and professors.

References

1. Yahya, A., Osman, A.: Using data mining techniques to guide academic programs design and assessment (2019)
2. Rathy, G.A., Sivasankar, P., Gnanasambandhan, T.G.: Developing a knowledge structure using outcome based education in Power Electronics Engineering. Procedia Comput. Sci. **172**, 1026–1032 (2020). https://doi.org/10.1016/j.procs.2020.05.150
3. Redecker, C.: European framework for the digital competence of educators: DigCompEdu. – Joint Research Centre (Seville site) (2017), JRC107466
4. Bekmanova, G., Omarbekova, A., Kaderkeyeva, Z., Sharipbay, A.: Model of intelligent massive open online course development. In: Gervasi, O., et al. (eds.) ICCSA 2020. LNCS, vol. 12250, pp. 271–281. Springer, Cham (2020). https://doi.org/10.1007/978-3-030-58802-1_20
5. Chassignol, M., Khoroshavin, A., Klimova, A., Bilyatdinova, A.: Artificial intelligence trends in education: a narrative overview. Procedia Comput. Sci. **136**, 16–24 (2018). https://doi.org/10.1016/j.procs.2018.08.233
6. Matveeva, T., Galiullina, N.: An empirical investigation of language model based reverse turing test as a tool for knowledge and skills assessment. https://doi.org/10.28995/2075-7182-2020-19-696-707
7. Shen, R.M., Tang, Y.Y., Zhang, T.Z.: The intelligent assessment system in Web-based distance learning education 31st annual frontiers in education conference. In: Impact on Engineering and Science Education. Conference Proceedings (Cat. No. 01CH37193). IEEE (2001). T. 1. C. TIF-7
8. Sitthiworachart, J., Joy, M., Sutinen, E.: Success factors for e-assessment in computer science education E-Learn: world conference on E-Learning in corporate, government, healthcare, and higher education. In: Association for the Advancement of Computing in Education (AACE), pp. 2287–2293 (2008)
9. Zawacki-Richter, O., et al.: Systematic review of research on artificial intelligence applications in higher education–where are the educators?. Int. J. Educ. Technol. High. Educ. **16**(1), 1–27 (2019)
10. Petrovskaya, A. et al.: Computerization of learning management process as a means of improving the quality of the educational process and student motivation. Procedia Comput. Sci. **169**, 656–661 (2020)

11. Jain, G.P., et al.: Artificial intelligence-based student learning evaluation: a concept map-based approach for analyzing a student's understanding of a topic. IEEE Trans. Learn. Technol. **7**(3), 267–279 (2014)
12. Kose, U., Arslan, A.: Intelligent e-learning system for improving students' academic achievements in computer programming courses. Int. J. Eng. Educ. **32**(1), 185–198 (2016)
13. Chakraborty, U.K., Roy, S.: Fuzzy automata inspired intelligent assesment of learning achievement IICAI, 1505–1518 (2011)
14. Chakraborty, U.K., Roy, S.: Neural network based intelligent analysis of learners' response for an e-Learning environment. In: 2010 2nd International Conference on Education Technology and Computer, vol. 2, pp. V2–333-V2–337. IEEE (2010)
15. Chakraborty, U., Konar, D., Roy, S., Choudhury, S.: Intelligent evaluation of short responses for e-learning systems. In: Satapathy, S.C., Prasad, V.K., Rani, B.P., Udgata, S.K., Raju, K.S. (eds.) Proceedings of the First International Conference on Computational Intelligence and Informatics. AISC, vol. 507, pp. 365–372. Springer, Singapore (2017). https://doi.org/10. 1007/978-981-10-2471-9_35
16. Hettiarachchi, E., et al.: A standard and interoperable technology-enhanced assessment system for skill and knowledge acquirement CSEDU, vol. 2, pp. 157–160 (2012)
17. Hettiarachchi, E., et al.: A technology enhanced assessment system for skill and knowledge learning CSEDU, vol. 2, 184–191 (2014)
18. Ishak, I.: Application of fuzzy logic to student performance in calculation subjects. In: Proceedings of the 4th National Symposium & Exhibition on Business & Accounting (2015)
19. Gokmen, G., et al.: Evaluation of student performance in laboratory applications using fuzzy logic. Procedia-Soc. Behav. Sci. **2**(2), 902–909 (2010)
20. Voskoglou, M.G.: Fuzzy logic as a tool for assessing students' knowledge and skills. Educ. Sci. **3**(2), 208–221 (2013)
21. Iskander, M. (ed.): Innovations in E-Learning, Instruction Technology, Assessment and Engineering Education. Springer Science & Business Media (2007)
22. Ali, M., Ghatol, A.: A neuro-fuzzy inference system for student modeling in web-based intelligent tutoring systems. In: Proceedings of International Conference on Cognitive Systems, pp. 14–19 (2004)
23. Stathacopoulou, R., et al.: Neuro-fuzzy knowledge processing in intelligent learning environments for improved student diagnosis. Inf. Sci. **170**(2–4), 273–307 (2005)
24. Zadeh, L.A.: Fuzzy logic. Computer **21**(4), 83–93 (1988)
25. Weon, S., Kim, J.: Learning achievement evaluation strategy using fuzzy membership function 31st Annual Frontiers in Education Conference. In: Impact on Engineering and Science Education Conference Proceedings (Cat. No. 01CH37193), vol. 1, pp. T3A-19. IEEE (2001)
26. Samarakou, M., et al.: Application of fuzzy logic for the assessment of engineering students. In: 2017 IEEE Global Engineering Education Conference (EDUCON), pp. 646–650. IEEE (2017)
27. Karthika, R., Deborah, L.J., Vijayakumar, P.: Intelligent e-learning system based on fuzzy logic. Neural Comput. Appl., 1–10 (2019)
28. Milani, A., Suriani, S., Poggioni, V.: Modeling educational domains in a planning framework. In: ACM International Conference Proceeding Series, vol. 113, pp. 748-753 (2005). https://doi.org/10.1145/1089551.1089687
29. Sasipraba, T., et al.: Assessment tools and rubrics for evaluating the capstone projects in outcome based education. Procedia Comput. Sci. **172**, 296–301, ISSN 1877–0509 (2020). https://doi.org/10.1016/j.procs.2020.05.047
30. Srimadhaven, T., Chris Junni, A.V., Naga, H., Jessenth Ebenezer, S., Shabari Girish, S., Priyaadharshini, M.: Learning analytics: virtual reality for programming course in higher education. Procedia Comput. Sci. **172**, 433–437, ISSN 1877–0509 (2020). https://doi.org/10. 1016/j.procs.2020.05.095

31. Lueny, M.: An undergraduate engineering education leadership program. is it working? outcomes of the second phase. Procedia Comput. Sci. **172**, 337–343, ISSN 1877–0509 (2020). https://doi.org/10.1016/j.procs.2020.05.169

32. Taylor, P.H.: Introduction: curriculum studies in retrospect and prospect. New Directions in Curriculum Studies **33**, 9–12 (2018). https://doi.org/10.4324/9780429453953-1

33. Young, M.: Curriculum theory: what it is and why it is important. [Teoria do currículo: O que é e por que é importante] Cadernos De Pesquisa. **44**(151), 191–201 (2014). https://doi.org/10.1590/198053142851

34. Jadhav, M.R., Kakade, A.B., Jagtap, S.R., Patil, M.S.: Impact assessment of outcome based approach in engineering education in India. Procedia Comput. Sci. **172**, 791–796, ISSN 1877–0509 (2020). https://doi.org/10.1016/j.procs.2020.05.113

35. Somasundaram, M., Latha, P., Saravana Pandian, S.A.: Curriculum design using artificial intelligence (AI) back propagation method. Procedia Comput. Sci. **172**, 134–138, ISSN 1877–0509 (2020). https://doi.org/10.1016/j.procs.2020.05.020

36. Kulkarni, V.N., Gaitonde, V.N., Kotturshettar, B.B., Satish, J.G.: Adapting industry based curriculum design for strengthening post graduate programs in Indian scenario, Procedia Comput. Sci. **172**, 253–258 (2020), ISSN 1877–0509. https://doi.org/10.1016/j.procs.2020.05.040

37. Priyambada, S.A., Mahendrawathi, E.R., Yahya, B.N.: Curriculum assessment of higher educational institution using aggregate profile clustering. Procedia Comput. Sci. **124**, 264–273, ISSN 1877–0509 (2017). https://doi.org/10.1016/j.procs.2017.12.155

38. Bendatu, Y., Yahya, B.N.: Sequence matching analysis for curriculum development. Jurnal Teknik Industri, 17 (2015). https://doi.org/10.9744/jti.17.1.47-52

39. Cao, P.Y., Ajwa, I.A.: Enhancing computational science curriculum at liberal arts institutions: a case study in the context of cybersecurity. Procedia Comput. Sci. **80**, 1940–1946, ISSN 1877–0509 (2016). https://doi.org/10.1016/j.procs.2016.05.510

40. Rodriguez, J.: Modularization of new course for integration in existing curriculum. Procedia Comput. Sci. **172**, 817–822, ISSN 1877–0509 (2020). https://doi.org/10.1016/j.procs.2020.05.117

41. Ellahi, R.M., Khan, M.U.A., Shah, A.: redesigning curriculum in line with industry 4.0. Procedia Comput. Sci. **151**, 699–708, ISSN 1877–0509 (2019). https://doi.org/10.1016/j.procs.2019.04.093

42. Bekmanova, G., Ongarbayev, Y.: Flexible model for organizing blended and distance learning. In: Gervasi, O., et al. (eds.) ICCSA 2020. LNCS, vol. 12250, pp. 282–292. Springer, Cham (2020). https://doi.org/10.1007/978-3-030-58802-1_21

43. https://protege.stanford.edu/

Inference Engines Performance in Reasoning Tasks for Intelligent Tutoring Systems

Oleg A. Sychev⬤, Anton Anikin$^{(\boxtimes)}$⬤, and Mikhail Denisov⬤

Volgograd State Technical University, Lenin Ave, 28, Volgograd 400005, Russia
anton.anikin@vstu.ru

Abstract. The use of formal knowledge representation models in intelligent tutoring systems often requires logical reasoning on these models by predefined rules. This process can be time and memory consuming, so finding effective software reasoners for different applications is an important research field. This problem is relevant for cognitive and constraint-based intelligent tutoring systems. We performed a comparative study of various software reasoners (Pellet, Apache Jena inference subsystem, Apache Jena SPARQL query processor, SWI-Prolog with semweb package, Closed World Machine, and Answer Set Programming solvers Clingo and DLV) for solving tasks specific to intelligent tutoring systems using three formal models with different properties and corresponding rule sets created for intelligent tutoring systems in introductory programming education domain. We compared features of rule-definition formalisms for different approaches and measured run and wall time, average CPU load, and peak RAM usage based on the count of inferred RDF triples. The experiments show that Apache Jena infers the solution quicker than other reasoners on the majority of tasks but consumes a significant amount of memory, while Clingo performs significantly better for combinatorial problems.

Keywords: Intelligent tutoring systems · Constraint-based ITS · Inference engines · Performance · Resource description framework · Answer set programming

1 Introduction and Related Work

E-learning is an application of computational science where technologies improve the learning process. It concerns developing software tools for students' automated assessment, students' self-assessment, and teacher's self-evaluation [18]. Using artificial intelligence is a promising approach in e-learning systems. Intelligent models and algorithms allow providing immediate feedback, based on the learning analytics, during online learning [13,16,17], checking the correctness of the learner answers [30], generating the learning tasks, and managing the learning process in an adaptive manner [11,14,22].

© Springer Nature Switzerland AG 2021
O. Gervasi et al. (Eds.): ICCSA 2021, LNCS 12950, pp. 471–482, 2021.
https://doi.org/10.1007/978-3-030-86960-1_33

The intelligent tutoring systems (ITS) development implies the use of knowledge modeling for these systems. The use of formal knowledge-representation models in modern cognitive, rule-based, and constraint-based intelligent tutoring systems requires logical inference. The rules and facts for it can be implemented as OWL/RDF ontologies using OWL/SWRL rules, allowing using reasoners like Pellet and Jena [25]. So, this combination of models and reasoning mechanisms can be applied to different tasks in the ITS implementation. E.g., in [29] authors use a set of SWRL rules to find an order of evaluation of C-language expressions and capture fault reason for incorrect students' answers. A similar approach was applied to verify algorithm execution tracings [28] in ITS for programming learning. Other approaches include using Prolog language [31] and Answer Set Programming [9] (ASP) [10, 21]. For example, authors of AFFLOG [15] use ASP for learning course planning according to task difficulty and student affection.

There are ontology benchmarks, such as the recent OWL2Bench [26]. These benchmarks, however, focus on reasoning within the semantics of the various OWL/OWL2 profiles without considering arbitrary, user-defined rules. Also, they usually do not investigate problems specific to intelligent tutoring systems.

Research investigating production rule engines seldom deal with RDF or OWL. The OpenRuleBench study [24] examines eight rule engines, only one of which (Jena) handles RDF/OWL. Again, this test examines the performance of three features specific to this type of reasoners, each involving several rules (using both real-world and synthetic datasets). The results of this study are outdated; since then, the approaches have evolved, and new tools such as Clingo have emerged.

In [25], the authors investigate three reasoners (including Jena) on six rulesets for forward-chaining reasoning tasks with data represented as RDF. The scope of the rules is real-world (products and recommendations for customers), and rulesets range in size from 1 to 24 rules. The subject of the study is close to our own and includes, in addition to performance, reasoning strategies, supported rule languages and their expressiveness, and the use of built-in functions.

This study aims to cover more reasoners dealing with RDF and also emerging problem solvers like Clingo, and test them on problems related to intelligent tutoring systems that are often more complex than regular semantic-web problems, having up to a hundred rules within a problem domain. So, the reasoning engine used in ITS should perform complex reasoning while grading students' answers in real time which can be a time- and memory-consuming task.

We considered seven inferences engines.

1. `Pellet` 2.3 [27] is an OWL reasoner supporting SWRL [6] built-ins.
2. Apache `Jena` [3] - RDF/OWL framework for Java including rule reasoner.
3. Apache Jena ARQ [1] - a `SPARQL` query processor capable running batches of SPARQL Update [5].
4. SWI-`Prolog` [31] with the semweb (rdf11) package [4] for handling RDF data alongside prolog native predicates representing the domain rules.
5. `CWM` (Closed World Machine) [8] - ancient reasoner for RDF written in Python 2 with its own rule syntax supporting some useful built-ins.

6. `Clingo` [20] - Answer Set Programming (ASP) tool that combines `gringo` grounder and `clasp` solver [19] (handles data as ASP facts, no native support for RDF).
7. `DLV` [2,7] - another ASP solver using a dialect of ASP slightly different from Clingo's one.

`CWM` (Closed World Machine) and `DLV` (an ASP solver) performed poorly during the first evaluation and were excluded from the following experiments. The extensive study included five software reasoners.

We aimed to compare reasoners in terms of their suitability for constraint-based intelligent tutoring systems. To do it, we used three sets of rules scaling the size of 5 datasets representing inference tasks common for intelligent tutoring systems:

- **Program Traces** domain [28] with low inter-connectivity (about 6 triples per subject), originally written in SWRL: builds the correct trace according to the given algorithm and determines errors in the given trace (3 datasets of traces of 5–20 lines).
- **Expressions** [29] with moderate inter-connectivity (about 40 triples per subject), originally in SWRL: finds errors in the order of evaluation of a C programming language expression and constructs a correct partial order DAG (the dataset includes a concatenation of expressions with 2–13 operands).
- **DEC** (Discrete Event Calculus) [15] with high inter-connectivity (about 75 triples per subject), originally in ASP: calculates the state of each fluent (changeable entity) at each point in time provided that events occur (in the generated dataset, each fluent maps to two events).

2   Method

For this study, we converted three selected rule sets, representing real-word ITS problems, to the rule syntax of all studied reasoners. The input facts were represented as RDF triples. For ASP reasoner, these were emulated by a ternary predicate `t`, while RDF terms - as a string or term literals. Inferred facts were converted back to RDF triples to extend the initial ontology. For each of the selected reasoning methods, it was necessary to convert the original SWRL rules to the native format since no reasoner other than Pellet supports SWRL with built-ins. The conversion of rules between different formalisms mainly required writing rules in different syntaxes. Some built-ins of the original rule language like regexp matching may require polyfill as syntax extension. Because of the disjunctive nature of a rule head in ASP, each rule with multiple fact assertions in its head breaks down into multiple rules asserting one fact per rule.

While the SWRL semantics assume basic RDFS inference rules, the rdf11 SWI-Prolog library supports this, while the simple Jena rules, SPARQL, and ASP solvers do not. Two of the were rules added to prepared rulesets, namely the `subClass` and `subProperty` propagation.

The solution method for each domain implemented in SWRL, is effective, but not optimal due to the limitations of SWRL. SWRL follows Open World Assumption, which is why it does not support default negation and does not provide the means to modify (update/delete) an ontology (incremental approach). Most of the compared reasoners have more rule-language capabilities than SWRL. However, for a more fair comparison of reasoners, the original solution was not changed, i.e. the semantics of the rules have not changed. This also made it possible to convert the rules programmatically, taking into account only the differences in the syntax of the equivalent constructs but not the semantics. When making comparisons, we nevertheless must keep in mind that the specific capabilities of different resonators may give them an edge on some tasks.

The semantics of the original rulesets are consistent with the output, i.e., using the deduced facts in the reasoning process. Reasoning ends when there is nothing more to display (forward chaining mechanism).

Two reasoners - Prolog (native backward chaining) and SPARQL (no immediate access to just inserted triples) do not support forward chaining. For these, we iteratively ran a batch of all the available rules (without regard to actual applicability to the data) until the size of triple store stopped increasing (thus, the order of the rules may affect the number of complete iterations).

In Prolog, a rule batch is a predicate with N definitions for N rules, ending with the `fail` command each. In SPARQL, a rule batch for N rules is a big query of N semicolon-concatenated INSERT..SELECT queries. This implementation is principally close to breadth-first search and shows results comparable to the hybrid RETE algorithm (Jena) and ASP (Clingo).

2.1   Measurement

The following metrics were recorded with each reasoner invoked as a standalone process:

1. Input problem size.
2. The numbers of input, inferred, and output triples.
3. `Wall time` – the running time from the start of the reasoner's application process to its completion, including i/o, rules and data parsing, and results serialization (in the case of Clingo called using the `clyngor` Python library, the time of grammar-based parsing of the printout is also included).
4. `Exclusive time` – the pure reasoning time excluding any preparation and input/output (this metric is relevant when the calling code is in the same environment – e.g., calling Jena from java code).
5. `CPU load` – average CPU usage in percent per core (similarly to Linux's `top` utility).
6. `Memory` – a peak RAM consumption defined as the average between RSS (working set) and VMS (pagefile, private bytes). The interval for collecting the indicators from the running process was 0.1 s.

The Python code organising the evaluation (using Owlready2 library [23] to handle RDF data) was not included in time measurements. All measurements were performed on a computer with 4-cores CPU AMD A10-5745M, 8 Gb RAM, no CUDA, Windows 8.1; power profile - high applications performance.

3 Results and Discussion

3.1 Rule Definition Formalisms and Reasoners Capabilities

The inference engines (reasoners) below were studied: Pellet 2.3 (supports SWRL formalism), Apache Jena inference engine (custom rule syntax), SWI-Prolog with the semweb (rdf11) package (Prolog), Apache Jena ARQ (SPARQL Update), Clingo (Answer Set Programming).

All considered rule formalisms except SWRL follow Closed World Assumption that is sensible for local reasoning [12]. All the reasoners allow triple matching and inferring new relations and types of individuals, although SWRL and ASP do not allow retracting and modifying existing ones. Also, SWRL does not allow creating new classes and individuals.

In RDF, a data unit is represented as a triple, while Prolog and ASP support native nesting of data structures.

Rule composition is not implied by the rule formalisms, but SPARQL allows SELECT queries to be nested, and Prolog predicates are reusable.

Besides, Jena supports a kind of rules that create other rules intended to speed up the inference.

Count, sum, and other aggregates are available in SPARQL and ASP; Prolog allows implementing them. A some sort of optimized transitivity checking is presented in SPARQL as "property paths", and Prolog can utilize recursion. The concept of integrity constraints (applied to any answer produced) exists only in ASP.

It is not possible to deactivate or prioritize rules, although it is possible to order rules in synthetic iterative approaches for Prolog and SPARQL.

Finally, all the rule languages except SPARQL provide a way to introduce custom built-ins.

Table 1 gives a summary of comparison of rule languages (formalisms) used by the studied inference engines.

3.2 Reasoners Performance Tests

For CWM, we ran two tests with the minimal problems, which took 10–20 times longer than using Pellet, one of the slowest reasoners.

For DLV, five tests were conducted, showing very low RAM consumption (about 4 MB) and the sharp increase in the time to solve a problem (6–10 times with a 2-fold increase in the problem size). It was clear that DLV is not scalable to big problems.

So, these two reasoners were discarded from further research because of too poor performance.

Table 1. Features of rule definition formalisms

Feature	SWRL	JENA rules	Prolog	SPARQL Update	ASP (Clingo)
Open/Closed World Assumption	OWA	CWA	CWA	CWA	CWA
Arithmetic built-ins	Yes	Yes	Yes	Yes	Yes
Assert relations	Yes	Yes	Yes	Yes	Yes
Retract relations	No	Yes	Yes	Yes	No
Creating individuals	No	Yes	Yes	Yes	Yes
Creating classes	No	Yes	Yes	Yes	Yes
Data structures	Plain	Plain	Nested	Plain	Nested
Rule composition	No	No	Yes	Nested SELECTs	No
Negation support	No	Yes	Yes	Yes	Yes
Aggregates	No	No	User-defined	Built-in	Built-in
Integrity constraints	No	No	No	No	Yes
Transitivity check	Assert all	Assert all	Recursion	Property path	Assert all
Custom builtins	Yes (Java)	Yes (Java)	Yes (Prolog)	No	Yes (Lua)
Prioritize rules	No	No	Ordering only	Ordering only	No

CPU load shows that Clingo and Prolog engines used only one core, while Java-based engines (Pellet, Jena, and SPARQL) used multithreading, but their memory consumption was significantly higher.

The two groups of reasoners showed significantly different results: the engines based on logic programming (Clingo, Prolog) versus Java engines with Rete algorithm or ontology querying (Pellet, Jena, SPARQL). The reasoners from the first group run on a single CPU core and spend less RAM (Fig. 4, 5). The second group of reasoners uses multithreading (2..3 cores on 4-cored CPU) and consumes a lot of memory even for tiny problems.

Prolog (running iteratively) shows low memory consumption on all tasks (Fig. 4). It has the fastest dry start and best time on light load tasks (up to 3k..5k inferred triples on Traces domain) (Fig. 3). Then its running time increases faster than that of Jena and Clingo, however, in the Expressions domain it outperformed SPARQL on large tasks as well.

Pellet reasoner consistently performed worse than all the others in terms of time and memory on all rule sets. Prolog demonstrates low memory usage regardless of the task size (in triples). It is the best reasoner for small reasoning tasks (up to about 4000 of inferred triples in the Traces domain, see Fig. 1), but it scales poorly with the task size, though it still outperforms SPARQL in the Expressions domain.

Jena turned out to be the quickest reasoner for low inter-connectivity domains, but its memory consumption rises quickly with the task size (see Fig. 2a); the reasoning can fail because of insufficient memory. For the models with a high number of links per entity (like DEC domain and, to some extent, Expressions domain), Clingo outperforms other reasoners while maintaining relatively low memory consumption.

To better assess run time and memory consumption, we plotted them on the same graph (see Fig. 6; the lower value, the better). Colored lines connect the measurements of different reasoners on the same input. The vertex in the lower-left corner of any figure points to the best reasoner for this situation. It shows that in any domain Prolog steeply increases execution time, while Jena steeply increases memory consumption when the task size rise while Clingo becomes the best reasoner for large tasks.

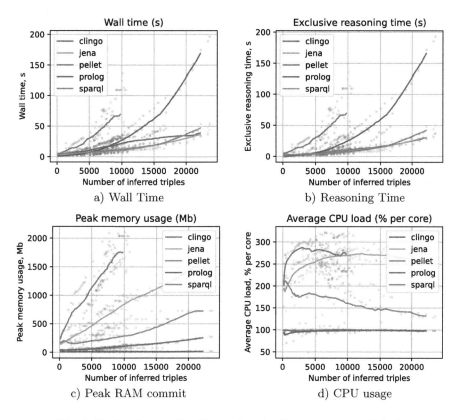

Fig. 1. Evaluation results: Traces domain (low inter-connectivity).

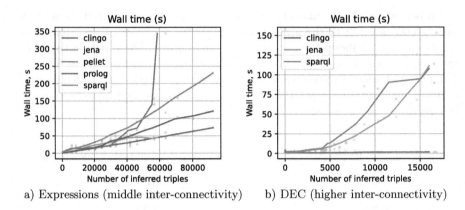

Fig. 2. Evaluation results: Wall Time in two domains.

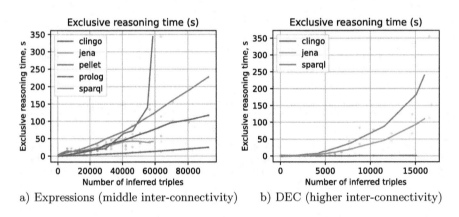

Fig. 3. Evaluation results: Exclusive Reasoning Time in two domains.

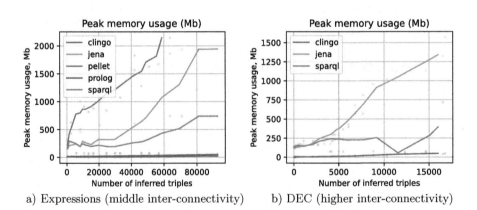

Fig. 4. Evaluation results: Peak RAM commit in two domains.

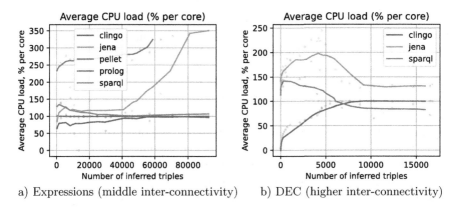

a) Expressions (middle inter-connectivity)   b) DEC (higher inter-connectivity)

Fig. 5. Evaluation results: CPU usage in two domains.

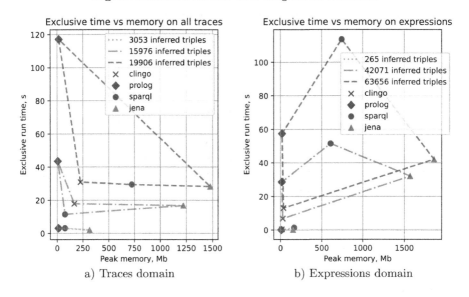

a) Traces domain   b) Expressions domain

Fig. 6. Time and memory consumption in two domains.

4   Conclusion

We compared several semantic rule engines – RDF reasoners (Pellet, Jena inference engine, SPARQL, and SWI-Prolog with rdf11) and ASP solver Clingo – as candidates for ITS backend using three different rule sets used in modern Intelligent Tutoring Systems.

CWM, released in 2000 and last updated in 2008, is no more relevant. The latest DLV (released in 2012) showed low suitability for something bigger than trivial problems.

We evaluated five rule engines on three real-world formal models related to Intelligent Tutoring Systems: Program Traces, Programming Language Expressions, and Discrete Event Calculus.

Prolog provides the fastest solution for small problems, but its efficiency drops quickly with the problem size.

Jena reasoner performs better for rule sets with low inter-connectivity but consumes much memory.

Clingo outperforms other reasoners for high inter-connectivity models, especially those implying combinatorial search while keeping memory consumption low. Its disadvantages are the lack of ability to modify the existing facts, poor RDF support, and using only one CPU core per task. Developing a multithreaded version of Clingo can make it the best reasoner for most tasks.

The options shown most stable behaviour concerning time/memory ratio are Clingo and iterative SPARQL approach.

For well-parallelable problems (especially implying combinatorial search) like DEC, Clingo shows promising speed keeping low memory consumption, although it does not provide RDF interface and disallow facts modification.

In the case of limited hardware resources, Prolog and ASP (Clingo) can be considered, because they require less space both in installation (single executable vs. java virtual machine with JARs) and in operation.

Our results can be used for selecting an inference engine for ITS according to its formal model properties. For example, they were used to choose the inference engine for the intelligent tutor "How it Works: Algorithms" [30]: Jena reasoner was chosen as the best-performing reasoner for the programming traces domain. Complex systems may use different reasoners for different models. Generally, Clingo and Jena reasoner are the best choices among modern inference engines.

Acknowledgment. The reported study was funded by RFBR, project number 20-07-00764 "Conceptual modeling of the knowledge domain on the comprehension level for intelligent decision-making systems in the learning".

References

1. ARQ - A SPARQL Processor for Jena. https://jena.apache.org/documentation//query/. Accessed 30 Apr 2021
2. DLV System. http://www.dlvsystem.com/dlv/. Accessed 30 Apr 2021
3. Jena - a free and open source Java framework for building Semantic Web and Linked Data applications. https://jena.apache.org. Accessed 30 Apr 2021
4. library(semweb/rdf11): The RDF database. https://www.swi-prolog.org/pldoc/man?section=semweb-rdf11. Accessed 30 Apr 2021
5. SPARQL Update. A language for updating RDF graphs. W3C Member Submission 15 July 2008. https://www.w3.org/Submission/SPARQL-Update/. Accessed 30 Apr 2021
6. SWRL: A Semantic Web Rule Language Combining OWL and RuleML. W3C Member Submission 21 May 2004. https://www.w3.org/Submission/SWRL/. Accessed 30 Apr 2021

7. Adrian, W.T., et al.: The ASP System DLV: Advancements and Applications. KI - Künstliche Intelligenz, pp. 177–179 (2018). https://doi.org/10.1007/s13218-018-0533-0

8. Berners-Lee, T.: Cwm: General-purpose data processor for the semantic web. http://www.w3.org/2000/10/swap/doc/cwm (2000). Accessed 30 Apr 2021

9. Brewka, G., Eiter, T., Truszczyński, M.: Answer set programming at a glance. Commun. ACM **54**(12), 93–103 (2011). https://doi.org/10.1145/2043174.2043195

10. Calegari, R., Ciatto, G., Mascardi, V., Omicini, A.: Logic-based technologies for multi-agent systems: a systematic literature review. Autonomous Agents Multi-Agent Syst. **35**(1), 1–67 (2020). https://doi.org/10.1007/s10458-020-09478-3

11. Chang, M., D'Aniello, G., Gaeta, M., Orciuoli, F., Sampson, D., Simonelli, C.: Building ontology-driven tutoring models for intelligent tutoring systems using data mining. IEEE Access **8**, 48151–48162 (2020). https://doi.org/10.1109/access.2020.2979281

12. Cortés-Calabuig, A., Denecker, M., Arieli, O., Van Nuffelen, B., Bruynooghe, M.: On the local closed-world assumption of data-sources. In: Baral, C., Greco, G., Leone, N., Terracina, G. (eds.) LPNMR 2005. LNCS (LNAI), vol. 3662, pp. 145–157. Springer, Heidelberg (2005). https://doi.org/10.1007/11546207_12

13. Demaidi, M.N., Gaber, M.M., Filer, N.: OntoPeFeGe: ontology-based personalized feedback generator. IEEE Access **6**, 31644–31664 (2018)

14. Dermeval, D., Albuquerque, J., Bittencourt, I.I., Isotani, S., Silva, A.P., Vassileva, J.: GaTO: An ontological model to apply gamification in intelligent tutoring systems. Frontiers Artif. Intell. **2**, July 2019. https://doi.org/10.3389/frai.2019.00013. https://doi.org/10.3389/frai.2019.00013

15. Dougalis, A., Plexousakis, D.: AFFLOG: A Logic Based Affective Tutoring System. In: Kumar, V., Troussas, C. (eds.) ITS 2020. LNCS, vol. 12149, pp. 270–274. Springer, Cham (2020). https://doi.org/10.1007/978-3-030-49663-0_31

16. Franzoni, V., Biondi, G., Milani, A.: Emotional sounds of crowds: spectrogram-based analysis using deep learning. Multimed. Tools Appl. **79**(47–48), 36063–36075 (2020)

17. Franzoni, V., Milani, A., Mengoni, P., Piccinato, F.: Artificial intelligence visual metaphors in e-learning interfaces for learning analytics. Appl. Sci. **10**(20), 7195 (2020)

18. Franzoni, V., Pallottelli, S., Milani, A.: Reshaping higher education with e-studium, a 10-years capstone in academic computing. In: Gervasi, O., et al. (eds.) ICCSA 2020. LNCS, vol. 12250, pp. 293–303. Springer, Cham (2020). https://doi.org/10.1007/978-3-030-58802-1_22

19. Gebser, M., Kaufmann, B., Neumann, A., Schaub, T.: *clasp*: a conflict-driven answer set solver. In: Baral, C., Brewka, G., Schlipf, J. (eds.) LPNMR 2007. LNCS (LNAI), vol. 4483, pp. 260–265. Springer, Heidelberg (2007). https://doi.org/10.1007/978-3-540-72200-7_23

20. Gebser, M., Kaminski, R., Kaufmann, B., Schaub, T.: Multi-shot ASP solving with clingo. CoRR abs/1705.09811 (2017)

21. Janhunen, T.: Cross-Translating Answer Set Programs Using the ASPTOOLS Collection. KI - Künstliche Intelligenz **32**(2-3), 183–184 (2018). https://doi.org/10.1007/s13218-018-0529-9

22. Kultsova, M., Anikin, A., Zhukova, I., Dvoryankin, A.: Ontology-based learning content management system in programming languages domain. Commun. Comput. Inf. Sci. **535**, 767–777 (2015). https://doi.org/10.1007/978-3-319-23766-4_61

23. Lamy, J.B.: Owlready: ontology-oriented programming in Python with automatic classification and high level constructs for biomedical ontologies. Artif. Intell. Med. **80** (2017). https://doi.org/10.1016/j.artmed.2017.07.002
24. Liang, S., Fodor, P., Wan, H., Kifer, M.: OpenRuleBench. In: Proceedings of the 18th International Conference on World Wide Web - WWW 2009. ACM Press (2009). https://doi.org/10.1145/1526709.1526790
25. Rattanasawad, T., Buranarach, M., Saikaew, K.R., Supnithi, T.: A comparative study of rule-based inference engines for the semantic web. IEICE Trans. Inf. Syst. E101.D(1), 82–89 (2018). https://doi.org/10.1587/transinf.2017swp0004. https://doi.org/10.1587/transinf.2017swp0004
26. Singh, G., Bhatia, S., Mutharaju, R.: OWL2Bench: a benchmark for OWL 2 reasoners. In: Pan, J.Z., Tamma, V., d'Amato, C., Janowicz, K., Fu, B., Polleres, A., Seneviratne, O., Kagal, L. (eds.) ISWC 2020. LNCS, vol. 12507, pp. 81–96. Springer, Cham (2020). https://doi.org/10.1007/978-3-030-62466-8_6
27. Sirin, E., Parsia, B., Grau, B.C., Kalyanpur, A., Katz, Y.: Pellet: a practical OWL-DL reasoner. J. Web Semantics 5(2), 51–53 (2007). https://doi.org/10.1016/j.websem.2007.03.004. https://www.sciencedirect.com/science/article/pii/S1570826807000169, software Engineering and the Semantic Web
28. Sychev, O., Denisov, M., Anikin, A.: Verifying algorithm traces and fault reason determining using ontology reasoning. In: 19th International Semantic Web Conference on Demos and Industry Tracks: From Novel Ideas to Industrial Practice, ISWC-Posters 2020, vol. 2721, pp. 49–53 (2020). http://ceur-ws.org/Vol-2721/paper495.pdf
29. Sychev, O., Penskoy, N.: Ontology-based determining of evaluation order of c expressions and the fault reason for incorrect answers. In: 19th International Semantic Web Conference on Demos and Industry Tracks: From Novel Ideas to Industrial Practice, ISWC-Posters 2020, vol. 2721, pp. 44–48 (2020). http://ceur-ws.org/Vol-2721/paper494.pdf
30. Sychev, O., Denisov, M., Terekhov, G.: How it works: Algorithms - a tool for developing an understanding of control structures. In: Proceedings of the 26th ACM Conference on Innovation and Technology in Computer Science Education V. 2. ACM, June 2021. https://doi.org/10.1145/3456565.3460032
31. Wielemaker, J., Schrijvers, T., Triska, M., Lager, T.: Swi-prolog. Theory and Practice of Logic Programming **12**(1–2), 67–96 (2012). https://doi.org/10.1017/S1471068411000494

International Workshop on Advancements in Applied Machine-learning and Data Analytics (AAMDA 2021)

A Comparison of Anomaly Detection Methods for Industrial Screw Tightening

Diogo Ribeiro[1], Luís Miguel Matos[1], Paulo Cortez[1(✉)], Guilherme Moreira[2], and André Pilastri[3]

[1] ALGORITMI R&D Centre, Department of Information Systems, University of Minho, 4804-533 Guimarães, Portugal
id6336@alunos.uminho.pt, {luis.matos,pcortez}@dsi.uminho.pt
[2] Bosch Car Multimedia, Braga, Portugal
Guilherme.Moreira2@pt.bosch.com
[3] EPMQ - IT CCG ZGDV Institute, 4804-533 Guimarães, Portugal
andre.pilastri@ccg.pt

Abstract. Within the context of Industry 4.0, quality assessment procedures using data-driven techniques are becoming more critical due to the generation of massive amounts of production data. In this paper, we address the detection of abnormal screw tightening processes, which is a relevant industrial task. Since labeling is costly, requiring a manual effort, we focus on unsupervised approaches. In particular, we assume a low-dimensional input screw fastening approach that is based only on angle-torque pairs. Using such pairs, we explore three main unsupervised Machine Learning (ML) algorithms: Local Outlier Factor (LOF), Isolation Forest (iForest) and a deep learning Autoencoder (AE). For benchmarking purposes, we also explore a supervised Random Forest (RF) algorithm. Several computational experiments were held by using recent industrial data with 2.8 million angle-torque pair records and a realistic and robust rolling window evaluation. Overall, high quality anomaly discrimination results were achieved by the iForest (99%) and AE (95% and 96%) unsupervised methods, which compared well against the supervised RF (99% and 91%). When compared with iForest, the AE requires less computation effort and provides faster anomaly detection response times.

Keywords: Autoencoder · Deep learning · Industry 4.0 · Isolation Forest · One-class classification · Random Forest · Unsupervised learning

1 Introduction

The current competition market increases the pressure for industrial companies to improve their productive processes (e.g., increasing efficiency and reducing costs). Within this context, a key aspect is the reduction of assembly errors during the production of products. Following the Industry 4.0 revolution, most

© Springer Nature Switzerland AG 2021
O. Gervasi et al. (Eds.): ICCSA 2021, LNCS 12950, pp. 485–500, 2021.
https://doi.org/10.1007/978-3-030-86960-1_34

modern factories make use of automation and robots that are interconnected with data sensors. While fully autonomous robots are used in some production plants, there are still industrial tasks that require a human operator. In effect, several companies use assembly machines that combine the flexibility of robotic arms with the guidance of human operators. In particular, handheld screwdrivers are often used in assembly factories, presenting the advantage of an automatic adaption to different torque profiles. During the screw tightening process, data is collected in real-time, generating several instances with multiple features, including angle-torque pairs. Once the operation is finished, the full screw response curve is presented to the operator, who needs to accept or reject it (due to lack of quality issues) in a small amount of time. While defects are rare in this domain, they are related with a diverse range of situations, such as floating screws, stripped screws and other situations. Often, the defect is detected too late down the production chain, which results in extra production times and costs.

A common approach to detect industrial screw tightening anomalies is to use a defect catalog that contains a curated set of normal and failure examples. Each time a new screw tightening is executed, the obtained angle-torque curve is compared with a predefined set of rules. However, this expert system detection method is rather static, requiring a manual collection and labeling of data. Thus, there is a potential to automate and improve the detection task by using Machine Learning (ML) algorithms. Within our knowledge, this approach has been scarcely studied. The research work more closely related with our approach was based on a Long Short-Term Memory (LSTM) neural network that uses annotated angle-torque screw responses [5]. Yet, such LSTM assumes a supervised learning that is rather impractical in several industrial settings. In effect, labeled data is costly (involving manual effort) and assembly companies typically produce big data. Thus, using a unsupervised learning for industrial screw tightening anomaly detection would result in a more valuable and automated approach.

One of the challenges that anomaly detection has to address is that boundaries between normal data and abnormal data are often not clearly defined [4]. Moreover, the scarcity and diversity of anomalous data makes it a non trivial task that is typically addressed by using an unsupervised or one-class learning [19]. Under the one-class learning approach, the training datasets only contain "normal" data points and the anomaly (often termed outlier) is detected when new data is considered distant from the training learning space.

In this paper, we explore a low-dimensional input approach for the anomaly detection of industrial screw tightening processes that only assumes angle-torque pairs. To model the data, we adapt three main unsupervised algorithms that are only trained with normal screw tightening process instances (thus one-class): Local Outlier Factor (LOF), Isolation Forest (iForest) and a deep learning Autoencoder (AE). For comparison purposes, we also test a supervised (two-class) Random Forest (RF). To evaluate the learning methods, we use a dataset with 2.8 million angle-torque pairs that were recently collected from an automotive multimedia assembly company. Several computational experiments were conducted by using a realistic rolling window evaluation that simulates several training and testing iterations through time. For each iteration, both the anomaly detection performance and the computational effort were measured.

This paper is organized as follows. Section 2 presents the related work. Next, Sect. 3 describes the collected industrial assembly data, anomaly detection learning models and evaluation procedure. Then, Sect. 4 presents the experimental results. Finally, Sect. 5 discusses the main conclusions and future work directions.

2   Related Work

The topic of anomaly detection has been researched in several fields of study [6]. It is often regarded as a one-class task, since anomalous patterns are typically scarce in most datasets. Several learning methods have been proposed for an unsupervised anomaly detection, such as [1,3]: Local Outlier Factor (LOF), One-Class Support Vector Machine (OC-SVM) and Isolation Forest (iForest). Following the success of Deep Learning, Autoencoders (AE) have been increasingly adopted for anomaly detection tasks [21]. When compared with conventional methods (e.g., LOF, OS-SVM and IF), AEs tend to provide faster training times, thus are capable of handling a larger amount of training data. In particular, the AE model can be updated to include new instances without retraining the full model [14], which is highly relevant when high velocity data is generated.

Regarding the specific screwing tightening use case, the research literature involving ML approaches is scarce. For the automatic identification of behaviour patterns on blind fasteners installation, [7] presents a kernel density-based pattern methodology to classify good or bad examples. In [18], Support Vector Machine (SVMs) with different kernel functions (e.g., linear and polynomial) were used to monitor screw fastening processes on the cover of hard disks based on the driver motor data. The obtained data-driven models were then used to detect incomplete screwing processes. This work differs from ours as it correlates motor power on a time series instead of the actual handled screw driver output (e.g., angle and torque). In [15], another data-driven analysis was conducted on thread fastening tasks that used a robotic manipulator. The problem was addressed by a supervised SVM classifier, aiming to estimate how much of the screw has been inserted by using the thread vibration from a torque sensor. As explained in Sect. 1, the most similar related work was performed by Cao et al. [5], where a supervised learning LSTM was used to infer the quality of the assembly by analyzing the screwing tightening angle-torque curve. The main novelty of our approach is that we use a pure unsupervised one-class learning, which is more suited for high volume industrial screwing data.

3   Materials and Methods

3.1   Industrial Data

Our case study is related with a major automotive multimedia assembly company, where a new fastening process starts with the insertion of several sub-parts (the total number varies from product to product) into the assembly jig. Then a scanner reads the imprinted barcode and the machine automatically loads the screw tightening program. Next, the human operator proceeds with an alignment of the matching pair and initiates the task. Using the help of the handheld screwdriver, the operator is guided to fasten each screw in a specific order, leaving the remaining technical details (e.g., torque and angle) to be handled automatically by the machine. Screw after screw, the operator is presented with a "Good Or Fail" (GOF) result, which is calculated automatically by the assembly machine based on its internal configuration. Each time the tightening is finished, the assembly system generates a local .csv file with all process details. All files are then imported into a centralized big data server cluster. The new data is then made available via the export of a .json file. Until this point, we have no control over the generated data stream.

In this work, we have designed a Python virtual environment that is capable of retrieving the industrial data by using Apache Spark queries to the central cluster. The raw data variables of the retrieved data are summarized in Table 1. Let i denote a screw fastening procedure for some product unit, where $i \in \{1, 2, ..., N\}$ and N represents the total number of screw tightening processes. For each i process, there is a real-time generation of several $k \in \{1, 2, 3, ..., K_i\}$ values, where K_i denotes the total number of observations and each product unit can produce a different K_i value. For a particular i and k value, the machine automatically registers the angle ($\alpha_{i,k}$) and torque ($\tau_{i,k}$) measurements. Typically, each screw fastening produces hundreds of angle-torque pairs, such as exemplified in Fig. 1. It should be highlighted that while there is no direct temporal variable, the angle ($\alpha_{i,k}$) attribute can be used as sequential or temporal measure of the fastening, since in almost cases its value increases as the procedure continues.

A tightening process is composed of four main steps $s \in \{1, 2, 3, 4\}$ (Table 2). In order to succeed at combining sub-parts, every screw must sequentially meet the transition conditions of each step within the parameterized time. The steps are exemplified in Fig. 1, where the short blue curve denotes the initial rotation (step $s = 0$) used to catch the screw and the remainder curve colors ($s = 1$ – red, $s = 2$ – green, $s = 3$ –purple and $s = 4$ – orange) denote the four main fastening steps. For each step, the screw fastening tool stores the step number, total angle and torque values. At the beginning of step $s = 4$, the fastening tool computes an estimate of the *DTM Clamping Angle* (*dtmclampt*). Once the process is nearly finishing, the real *DTM Clamping Angle* is measured (*dtmclampa*). Then, the final GOF result is computed by the screw assembly machine (attribute *screwgof*).

Table 1. Description of the industrial screw tightening data variables.

Variable	Description
	Hundreds of values per screw fastening (k):
angle ($\alpha_{i,k}$)	Value for the angle (e.g., $-208.1°$)
torque ($\tau_{i,k}$)	Value for the torque (e.g., 35.6 Ncm)
gradient	Torque gradient for two consecutive angle values (e.g., 20)
	Four values per fastening (one for each step):
stepnr	Screwing step number ($s \in \{1,2,3,4\}$)
tstepangle	Total step angle (e.g., $990°$)
tsteptorque	Total step torque (e.g., 39.3)
	One value per fastening (i):
dtmclampa	Value for actual DTM Clamp (e.g., 79.1)
dtmclampt	Value for predicted DTM Clamp (e.g., 30.2)
timestamp	Timestamp (e.g., "2020-10-08 06:30:51")
product	Product family identifier (e.g., "2222111")
serial	Product serial number (e.g., "11111")
screwnr	Screw number ($\in \{1, 2, ..., 8\}$)
screwenergy	Total energy required (e.g., 19959)
totalangle	Total angle (e.g., $2993°$)
totaltorque	Total torque (e.g., 30000 Ncm)
screwgof (y_i)	Screwing process "Good" (1) or "Fail" (0)

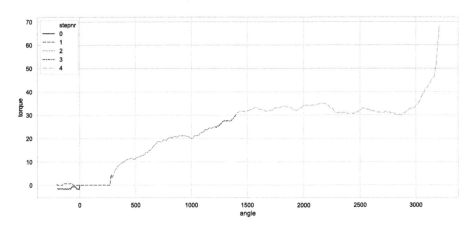

Fig. 1. Example of a normal ("Good") screw tightening.

In this work, we adopt the last variable (*screwgof* attribute) as the supervised label of the screwing process. Thus, this variable, denoted as y_i for the i-th screw tightening process, is used to select the instances for learning methods (e.g., only normal examples are fed into the one-class methods) and also to define

Table 2. Description of the screw fastening process steps.

Step	Description	Transition condition
Simple torque ($s = 1$)	Serves as a transition to the next step when the thread starts to be formed	Achieve either the transition torque or angle within the parameterized time range
Angle ($s = 2$)	Fixed number of turns conducted	Min Angle < Angle Target < Max Angle within stipulated time
Torque ($s = 3$)	Apply a fixed torque value	Min Torque < Torque Target < Max Torque within stipulated time
Seating control ($s = 4$)	Apply a torque value from the screw seating value to the part	Clamping Torque < Max Seating Torque; Min Total Angle < Total Angle < Max Total Angle; Min Total Torque < Total Torque < Max Total Torque

the normal and abnormal labels used to evaluate the models (when using test "unseen" data). The values were directly computed by the assembly machine software tool, which includes a rather rigid internal set of rules that define a normal fastening cycle (based on a catalog of normal and abnormal examples). It should be noted that in a few cases, this assembly tool provides a wrong labeling, such as classifying a fastening cycle as normal when it actually was a defect. However, since we have big data and it is costly to manually verify all fastening cycles, we assume here that the provided y_i values are a reasonable oracle for testing the proposed anomaly detection methods (which do not use any labels).

In terms of preprocessing, we grouped all records by product family (*product*), serial number (*serial*) and screw number (*screwnr*, maximum of 8 screws per individual unit). This grouping allows a fast identification of all k values related with a tightening process (i), which is useful for anomaly detection evaluation purposes. The resulting dataset corresponds to around two days of screw tightening processes, collected in November of 2020, and it includes a total of 2,853,967 entries related with $N = 6,162$ fastening cycles. On average, there is around $\overline{K_i} = 463$ records (angle-torque pairs) per cycle. Moreover, the dataset is highly unbalanced. In effect, there are 74 defects that correspond to only 2% of the screw tightening cycles.

3.2 Anomaly Detection Angle-Torque Approach

There are two screw tightening anomaly detection goals. The first and most important goal is to automatically detect if there has been a failure in the screw fastening cycle (y_i = "Fail"). Once a failure has been detected, the secondary goal is to automatically identify where (within the full cycle) the defect took place. The secondary goal is needed to fine tune the detection details, providing additional information that helps the human operator to accept the machine learning result.

In this work, we assume that each anomaly detection model outputs a decision score (d_i), such that the higher the score, the more probable is the chance that an anomaly has occurred for the i-th tightening screw process. In order to achieve both goals simultaneously, the anomaly detection methods are fed with the all individual and more instant k values (one at the time). This means that each i-th screw fastening produces a total of $k \in \{1, 2, ...K_i\}$ examples that are used for training or testing purposes, resulting in $d_{i,k}$ anomaly decision scores.

To compute the overall decision score d_i (thus perform the main goal), the individual scores are averaged, thus $d_i = \sum_k d_{i,k}/K_i$. For distinct i values, the resulting scores (d_i) can be compared with the target variable (y_i), allowing the computation of the Receiver Operating Characteristic (ROC) curve [8]. If class labels are needed, then a fixed Th threshold value needs to be set, such that the i-th output is interpreted as an anomaly if $d_i > Th$. The specific Th can be selected by using historical unlabeled data (e.g., by adopting a given percentage above the maximum AE reconstruction error) or by selecting the best specificity-sensitivity trade-off from a ROC curve that is generated when using a labeled validation set. As for the secondary goal, if fine-grained target values $(y_{i,k})$ were available, then the individual decision scores $(d_{i,k})$ could be used to generate ROC curves or other classification performance measures. Since our dataset does not contain any fine-grained target labels, we will adopt a fixed threshold to demonstrate the applicability of the secondary goal but without robustly studying it. Moreover, given that we use pure unsupervised ML models, the threshold will be set using historical unlabeled data (see Sect. 4).

Regarding the input features, the industrial screw tightening data includes several attributes with different granularities (Table 1): fine-grained – angle $(\alpha_{i,k})$, torque $(\tau_{i,k})$ and gradient; step-grained – step number, total step angle and torque; and one per each i-th tightening – such as *dtmclampa*, *screwenergy* and y_i. To select the input variables (and also to tune the AE models, see Sect. 3.3), we conducted preliminary experiments by considering only the training data of the first rolling window iteration, thus related with the oldest data records (Sect. 3.4). This oldest training data was further split into training (to fit the models, with $W - T$ tightening screw processes) and validation sets (with T tightening cycles). In particular, we initially tested anomaly detection models that were fed with all possible input variables (e.g., angle, torque, gradient, total step angle, energy). However, the obtained models provided substantially worst results when compared with the low-dimensional angle-torque input approach. Moreover, the full input anomaly detection models required much more memory and computational effort. In some cases, such as when using LOF, the computational training process even halted due to an out-of-memory issue. Based on these results, all screw anomaly methods described in this work assume just two input values: the simpler angle-torque pairs $(\alpha_{i,k}, \tau_{i,k})$ for each (i, k) example.

3.3   Learning Models

Three unsupervised learning algorithms were selected for the empirical comparison: LOF, iForest and a AE. These methods are only trained with normal cases (one-class approach). For benchmarking purposes, we selected the popular Random Forest (RF) method. However, we note that RF is a supervised learning method that, contrary to the one-class methods, requires labeled data for the training procedure. Thus, RF was trained with both normal and abnormal angle-torque instances.

To implement the methods we used the `scikit-learn` (for RF, LOF and iForest) [16] and `TensorFlow` (for AE) [9] Python modules. At an initial stage, we also explored the OC-SVM implementation of `scikit-learn`. However, our training sets are too large for the algorithm, which did not compute results in an affordable time (e.g., the first rolling window iteration execution was halted after 40 h of execution time). In order to provide a fair comparison, and also reduce the computational effort, in the experiments we assumed in general the default `scikit-learn` and `keras` hyperparameter values. Before feeding the models, all angle-torque data values were first normalized, where the numeric values were transformed to fit into a [0,1] scale by using a max-min normalization, via the `MinMaxScaler` procedure of the `sklearn` module.

Unlike the other anomaly detection methods, LOF is a density-based algorithm that heavily depends on K-nearest neighbors [3]. It is designed as local because it depends on how well isolated an object is from the surrounding neighborhood. Instead of classifying an object as being an outlier or an inlier, an outlier factor is assigned describing up to which degree the object differs from the rest of the dataset. It should be noted that LOF can be used to detect both local and global outliers. Moreover, it can be trained with multi-class instances. In this paper, in order to provide a fair comparison, we only use normal examples to fit the model, where the outlier factor score is directly used as the anomaly decision score ($d_{i,k}$).

Isolation Forests (iForest) take advantage of two major characteristics of anomalous points: they are present in fewer quantities and are also numerically different to normal instances. This means that anomalous instances are prone to be more easily isolated (separated from the rest of the instances) than normal points. Using this principle, an anomaly detection method can be based on the construction of a tree structure to isolate every single instance and then evaluate their normality. Due to their nature and tendency to be quickly isolated, abnormal instances are closer to the root of the tree. This characteristic, also known as Isolation Tree or iTree, forms the basis of anomaly detection for this model. An iForest [13] is based on an ensemble of iTrees for a given dataset. Anomalies are the instances that are closer to the root of the tree and will have shorter average path lengths. The `scikit-learn` provides an anomaly score that ranges from $\hat{y}_{i,k} = -1$ (highest abnormal score) to $\hat{y}_{i,k} = 1$ (highest normal score). Thus, we rescaled these scores into the [0, 1] anomaly probability range by computing $d_{i,k} = (1 - \hat{y}_{i,k})/2$.

Autoencoders (AE) are unsupervised learning techniques that efficiently compresses and encode data to a lower-dimensional representation [10]. The includes an encoding stage, where compression occurs by reducing the input data (the number of features describing the dataset) into a latent space (defined by a bottleneck layer), and a decoding stage, where the model tries to reconstruct the original data from the latent space. When adapted for our anomaly detection case, the training algorithm is only fed with normal instances and it attempts to generate two output values $(\hat{\alpha}_{i,k}, \hat{\tau}_{i,k})$ that are identical to its inputs $(\alpha_{i,k}, \tau_{i,k})$. When a new (i, k) instance is evaluated, we compute the Mean Absolute Error (MAE) [1]:

$$MAE_{i,k} = (|\alpha_{i,k} - \hat{\alpha}_{i,k}| + |\tau_{i,k} - \hat{\tau}_{i,k}|)/2 \qquad (1)$$

This reconstruction error is used as the decision score $d_{i,k} = MAE_{i,k}$, where higher reconstruction errors should denote a higher anomaly probability. In terms of the deep AE architecture, we assume a dense AE with two inputs $(\alpha_{i,k}, \tau_{i,k})$, two output nodes and a fully-connected structure that includes a stack of layers in both the encoder and decoder components. All hidden layers use the popular ReLU activation function, while the output nodes assume the logistic function (all inputs are normalized $\in [0,1]$). Every hidden layer is also attached with a *Batch Normalization* layer, which reduces the internal covariate shift, discards the need of dropout layers and normalizes the layer inputs for each batch of data that passes through it [12]. In order to define the number of layers and units per layer, we performed several trial-and-error preliminary experiments. These experiments assume the same oldest training data used to set the input variables (Sect. 3.2). The tested architectures use a bottleneck layer with 1 hidden node (to define the latent space) and a varying range of hidden layers. The best results, measured in terms of the lowest reconstruction error for the validation set, were achieved for an AE with 20 layers (including the input, Batch Normalization and output layers), as shown in Fig. 2. In this work, the selected AE is trained with the Adam optimization algorithm using the MAE loss function, a total of 100 epochs and early stopping (with 5% of the training data being used as the validation set).

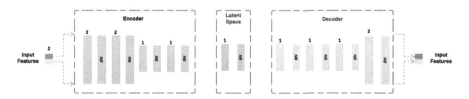

Fig. 2. The adopted dense AutoEncoder (AE) structure (the numbers denote the number of hidden units per layer and the term BN represents a batch normalization layer).

As explained in Sect. 2, the AE model can be adapted to dynamic environments when there are frequent data updates. In [14], two neural network learning modes were compared: reset and reuse. When new data arrives and the neural network is retrained, the former assumes a random weight initialization, while the latter uses the previously trained weights as the initial set of weights. In this work, we compare both learning modes when executing the rolling window evaluation.

Finally, RF is a supervised learning algorithm that works as an ensemble of decision trees that form a "forest". Each tree depends on the values of a randomly sampled input vector and a bagging selection of training samples [2]. RF can be used for both regression and classification tasks. In this work, we used RF to output anomaly class probabilities ($\in[0.0,1.0]$), which are used as the decision score ($d_{i,k}$).

3.4   Evaluation

To compare the anomaly detection ML models, we execute the realistic rolling window procedure [14,20]. This procedure is more robust than the popular random holdout train-test split, since it realistically simulates an anomaly detection through time, producing several training and test updates, as exemplified in Fig. 3. A fixed training window with W examples is first set. Then, in the first iteration ($u = 1$), the model is trained with the oldest W instances. Once the model is fit, it performs T ahead predictions. Then, it simulates the passage of time by "rolling" the training and test sets by assuming a temporal jump step (S). The oldest S instances are discarded from the training set, being replaced by S more recent ones. A new model is then fit, allowing to perform T new subsequent predictions, and so on. In total, there are $U = (D - (W + T))/S$ model updates (training and test iterations), where D is the total dataset size. To split the data, in this work we adopt the screw tightening granularity, where all k instances for a particular i tightening process are always kept together, thus $D = N = 6,162$. After consulting the industrial experts, we adopted the values $W = 5,000$, $T = 500$ and $S = 33$, which produces a total of $U = 20$ model updates.

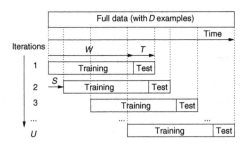

Fig. 3. Schematic of the adopted rolling window procedure.

To evaluate the anomaly detection methods, we adopt the popular ROC [8], which is computed by using the target labels (anomaly or normal) and the predicted anomaly decision scores ($d_{i,k}$) for the T tightening test examples of each u rolling window iteration. The ROC represents the full discrimination performance of a binary classifier when considering all possible Th threshold values, plotting one minus the specificity (x-axis), also known as False Positive Rate (FPR), versus the sensitivity (y-axis) or True Positive Rate (TPR). The overall discrimination quality is measured by the Area Under the Curve $AUC = \int_0^{Th_{max}} ROCdTh$, where Th_{max} denotes the maximum threshold value (in this work, $Th_{max} = 1$ for the iForest, AE and RF methods). This AUC metric has two main advantages [17]. Firstly, when the data is highly unbalanced (which is our case), the interpretation of the goodness of the metric values does not change. Secondly, the quality of the AUC values can be interpreted as: 50% - performance of a random classifier; 60% - reasonable; 70% - good; 80% - very good; 90% - excellent; and 100% - perfect. Given that the rolling window produces several ROC curves and AUC values (one for each iteration), we aggregate the results by computing the median (AUC, training and inference time) and average (AUC) values, using the Wilcoxon non parametric statistic [11] to check if paired AUC differences are statistically significant. All experiments were executed using a 2.3 GHz Intel Core i9 processor. Besides the predictive decision scores, for each anomaly detection method we recorded the computational effort, measured in terms of the average (over all u iterations) total training and anomaly screw tightening inference times (both in s).

4   Results

Table 3 summarizes the rolling window evaluation results, while Fig. 4 shows the evolution of AUC values through the rolling window iterations. The supervised method (RF) provides a high median and average AUC scores (99% and 91%), which sets a high comparison standard for the unsupervised one-class methods. Regarding LOF, it corresponds to the fastest training method, requiring an average of 18.3 s for each rolling window iteration. However, this method also provides the worst AUC values (median of 80% and average of 75%). When comparing both AE learning modes, reuse and reset, the former provides better median and average AUC results, and lower average computational training times (explained by the usage of the early stopping procedure). The best one-class median and average results are obtained by the iForest (99%). As shown in Table 3, the median AUC differences are only significant when comparing RF, iForest or the AEs with LOF. As for the average AUC values, iForest is significantly better than other methods, while AE reuse is significantly better than LOF and AE reset. Figure 4 confirms that the two best one-class performing models continuously provide an excellent (almost perfect) discrimination along the distinct evaluation iterations. Given these results, we recommend the two unsupervised methods iForest and AE resuse, since they do not require labeled data (as RF). In particular, iForest obtains the highest discrimination capability. As for AE reuse, the method produces an almost similar performance while requiring much less computational effort. In effect, when compared with iForest, the AE reuse training is around 7 times faster. Moreover, the AE reuse anomaly inference time is half of the one required by iForest and one third of the response time required by RF.

Table 3. Rolling window screw tightening anomaly detection results (best values in **boldface** font; selected models in *italic* font).

	AUC		Train	Inference
	Median	Avg.	Time (s)	Time (s)
LOF	0.80	0.75	**18.30**	0.12
RF	**0.99**⋆	0.91⋆	25.33	0.12
iForest	**0.99**⋆	**0.99**†	168.18	0.08
AE reset	0.93⋆	0.92⋆	29.64	**0.04**
AE reuse	0.95⋆	0.96◇	23.00	**0.04**

⋆ – Statistically significant under a paired comparison with LOF.
† – Statistically significant under a paired comparison with all other methods.
◇ – Statistically significant under a paired comparison with LOF and AE reset.

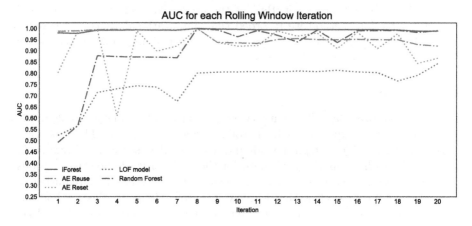

Fig. 4. AUC evolution for the screw tightening anomaly detection methods.

As an example, Fig. 5 plots the individual ROC curves for the selected models during the third rolling window iteration. The curves overlap in the sensitive (top right) region, with the AE reuse model slightly presenting a better specificity (bottom left region) in this example.

To demonstrate the secondary goal (identification of the anomaly process angle-torque regions), we selected the iForest and AE reuse models that were obtained in the same rolling window iteration and a defect example from the test set related with a torque not reached error. Then, we defined a threshold value (Th) that corresponds to the average plus one standard deviation of the training decision scores (a different Th value was used for each method). Figure 6 shows the obtained anomaly points that correspond to the highest decision scores, such that $d_{i,k} > Th$. It is interesting to note that both models signal several identical

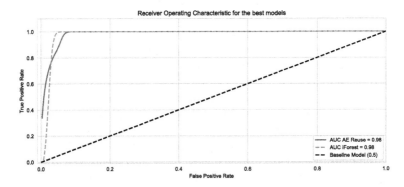

Fig. 5. ROC curves for the best two models for the rolling window iteration $u = 3$ (dashed black line denotes the performance of a random baseline classifier).

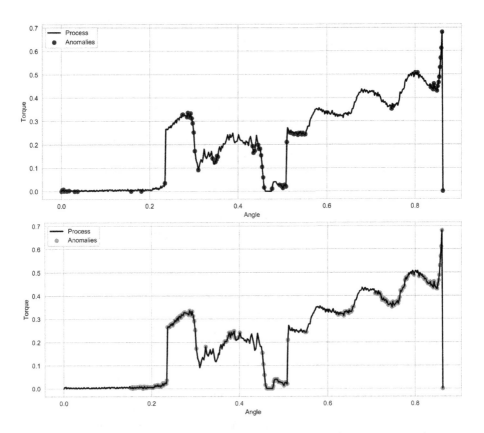

Fig. 6. Anomaly points identified by iForest (top, blue points) and AE reuse (bottom, orange points) for a screw fastening defect process. (Color figure online)

abnormal regions, such as the three abrupt torque decays (set at the normalized x-axis angle values of 0.30, 0.46 and 0.86).

The obtained results were shown to the industrial experts, who considered them very positive. Besides the high quality AUC results, the iForest and AE reuse models allowed to detect one real defect example that was labeled as "normal" by the screw assembly expert system tool. Thus, there is a potential for a better screw tightening defect detection by the proposed data-driven models. Moreover, the secondary goal demonstration example (Fig. 6) was valued as interesting and useful, particularly in the end part of the process (right $x-$axis region) when there is an abrupt torque decay. The experts highlighted that this is one of the defect behaviors. Nevertheless, they also mentioned that there are other types of screw tightening failures (at more initial stages), which will be analysed in future work.

5   Conclusions

Following the Industry 4.0 phenomenon, there is currently a vast abundance of digital data that reflect several stages of the manufacturing process. By using these data, analytic tools based on machine learning methods can potentially provide valuable production insights. In particular, within the electronic component assembly industry, the automatic detection of screw tightening processes can improve productivity and reduce costs. The assembly of screws is currently a semi-automated procedure, involving both a human operator and robotic machines. Thus, typically each tightening procedure requires a two-stage inspection, first by the machine and then by the operator. Moreover, current screw tightening failures, as identified by the machine tool, are often based on a defect catalog comparison, which uses a rigid set of rules defined by an expert system. Thus, there is room for improvement by using machine learning algorithms, creating a data-driven model that can more quickly adapt to changes in the assembly environment (e.g., assembly of new products).

In this paper, we focus on an industrial screw tightening anomaly detection task by using an unsupervised approach, which does not require any labeled data and thus it is more easy to automate. Our approach assumes that only normal (thus one-class) angle-torque input pairs are used to train the models. Using recently collected data, from an automotive electronics assembly company, we collected around 2.8 million angle-torque pairs, related with 6,162 screw fastening cycles. Four different learning algorithms were compared: unsupervised – Local Outlier Factor (LOF), Isolation Forest (iForest) and a dense Autoencoder (AE, under two learning modes: reset and reuse); and supervised - Random Forest (RF), used for benchmarking purposes. The algorithms were trained with individual angle-torque observations (two input variables), allowing to output an anomaly decision score for each angle-torque pair. The anomaly detection methods were compared by using a realistic rolling window evaluation, which simulated 20 training and test iterations through time. Overall, the best unsupervised learning results were obtained by the iForest and AE reuse. Both methods provide an excellent anomaly discrimination capability that is identical or better than the supervised RF (median of 99% and average of 91%) benchmark. In effect, iForest produces the best unsupervised anomaly detection result (median and average of 99%), slightly outperforming the AE reuse (median of 95%, average of 96%) approach. However, it should be noted that the AE reuse method requires much less computational effort, which is relevant in this industrial domain, since it generates high velocity data.

In future work, we intend to explore more deep learning architectures, such as based on an AE with Long Short Term Memory (LSTM) layers, which can capture directly sequential data [5]. Another interesting research direction is to extend the research study by comparing the proposed data-driven approach with the screw assembly expert system tool (main goal) and also evaluate the data-driven model capability to correctly identify specific defect angle-torque regions (secondary goal), which however would require the collection of a more costly fine-grained human labeled data.

Acknowledgments. This work is supported by: European Structural and Investment Funds in the FEDER component, through the Operational Competitiveness and Internationalization Programme (COMPETE 2020) [Project n 39479; Funding Reference: POCI-01-0247-FEDER-39479]. The authors also wish to thank the automotive electronics company staff involved with this project for providing the data and the valuable domain feedback.

References

1. Alla, S., Adari, S.K.: Beginning Anomaly Detection Using Python-Based Deep Learning. Apress, Berkeley (2019). https://doi.org/10.1007/978-1-4842-5177-5_8
2. Breiman, L.: Random forests. Mach. Learn. **45**(1), 5–32 (2001). https://doi.org/10.1023/A:1010933404324
3. Breunig, M.M., Kriegel, H., Ng, R.T., Sander, J.: LOF: identifying density-based local outliers. In: Chen, W., Naughton, J.F., Bernstein, P.A. (eds.) Proceedings of the 2000 ACM SIGMOD International Conference on Management of Data, May 16–18, 2000, Dallas, Texas, USA, pp. 93–104. ACM (2000). https://doi.org/10.1145/342009.335388
4. Cao, N., Lin, Y.R., Gotz, D., Du, F.: Z-glyph: visualizing outliers in multivariate data. Inf. Vis. **17**(1), 22–40 (2018). https://doi.org/10.1177/1473871616686635
5. Cao, X., Liu, J., Meng, F., Yan, B., Zheng, H., Su, H.: Anomaly detection for screw tightening timing data with LSTM recurrent neural network. In: 15th International Conference on Mobile Ad-Hoc and Sensor Networks, MSN 2019, Shenzhen, China, December 11–13, 2019, pp. 348–352. IEEE (2019). https://doi.org/10.1109/MSN48538.2019.00072
6. Chandola, V., Banerjee, A., Kumar, V.: Anomaly detection: a survey. ACM Comput. Surv. **41**(3), 15:1–15:58 (2009). https://doi.org/10.1145/1541880.1541882
7. Diez-Olivan, A., Penalva, M., Veiga, F., Deitert, L., Sanz, R., Sierra, B.: Kernel density-based pattern classification in blind fasteners installation. In: Martínez de Pisón, F.J., Urraca, R., Quintián, H., Corchado, E. (eds.) HAIS 2017. LNCS (LNAI), vol. 10334, pp. 195–206. Springer, Cham (2017). https://doi.org/10.1007/978-3-319-59650-1_17

8. Fawcett, T.: An introduction to ROC analysis. Pattern Recognit. Lett. **27**(8), 861–874 (2006). https://doi.org/10.1016/j.patrec.2005.10.010
9. Gulli, A., Pal, S.: Deep learning with Keras. Packt Publishing Ltd, Birmingham (2017)
10. Hinton, G., Salakhutdinov, R.: Reducing the dimensionality of data with neural networks. Science **313**(5786), 504–507 (2006). https://doi.org/10.1126/science.1127647
11. Hollander, M., Wolfe, D.A., Chicken, E.: Nonparametric Statistical Methods. Wiley, Hoboken (2013)
12. Ioffe, S., Szegedy, C.: Batch normalization: accelerating deep network training by reducing internal covariate shift. In: Bach, F.R., Blei, D.M. (eds.) Proceedings of the 32nd International Conference on Machine Learning, ICML 2015, Lille, France, 6–11 July 2015. JMLR Workshop and Conference Proceedings, vol. 37, pp. 448–456. JMLR.org (2015)
13. Liu, F.T., Ting, K.M., Zhou, Z.: Isolation forest. In: Proceedings of the 8th IEEE International Conference on Data Mining (ICDM 2008), December 15–19, 2008, Pisa, Italy, pp. 413–422. IEEE Computer Society (2008). https://doi.org/10.1109/ICDM.2008.17
14. Matos, L.M., Cortez, P., Mendes, R., Moreau, A.: Using deep learning for mobile marketing user conversion prediction. In: International Joint Conference on Neural Networks, IJCNN 2019 Budapest, Hungary, July 14–19, 2019. pp. 1–8. IEEE (2019). https://doi.org/10.1109/IJCNN.2019.8851888
15. Matsuno, T., Huang, J., Fukuda, T.: Fault detection algorithm for external thread fastening by robotic manipulator using linear support vector machine classifier. In: 2013 IEEE International Conference on Robotics and Automation, Karlsruhe, Germany, May 6–10, 2013, pp. 3443–3450. IEEE (2013). https://doi.org/10.1109/ICRA.2013.6631058
16. Pedregosa, F., et al.: Scikit-learn: Machine learning in python. J. Mach. Learn. Res. **12**, 2825–2830 (2011). http://dl.acm.org/citation.cfm?id=2078195
17. Pereira, P.J., Cortez, P., Mendes, R.: Multi-objective grammatical evolution of decision trees for mobile marketing user conversion prediction. Expert Syst. Appl. **168**, 114287 (2021). https://doi.org/10.1016/j.eswa.2020.114287
18. Ponpitakchai, S.: Monitoring screw fastening process: an application of SVM classification. Naresuan Univ. Eng. J. (NUEJ) **11**(2), 1–5 (2016). https://doi.org/10.14456/nuej.2016.11
19. Ruff, L., et al.: Deep one-class classification. In: Dy, J.G., Krause, A. (eds.) Proceedings of the 35th International Conference on Machine Learning, ICML 2018, Stockholmsmässan, Stockholm, Sweden, July 10–15, 2018. Proceedings of Machine Learning Research, vol. 80, pp. 4390–4399. PMLR (2018). http://proceedings.mlr.press/v80/ruff18a.html
20. Tashman, L.J.: Out-of-sample tests of forecasting accuracy: an analysis and review. Int. J. Forecast. **16**(4), 437–450 (2000). https://doi.org/10.1016/S0169-2070(00)00065-0
21. Zhou, C., Paffenroth, R.C.: Anomaly detection with robust deep autoencoders. In: Proceedings of the 23rd ACM SIGKDD International Conference on Knowledge Discovery and Data Mining, Halifax, NS, Canada, August 13–17, 2017. pp. 665–674. ACM (2017). https://doi.org/10.1145/3097983.3098052

Object Detection with RetinaNet on Aerial Imagery: The Algarve Landscape

C. Coelho[✉], M. Fernanda P. Costa[ID], L. L. Ferrás[ID], and A. J. Soares[ID]

Centre of Mathematics, University of Minho, Campus de Gualtar, 4710-057 Braga, Portugal
cmartins@cmat.uminho.pt, {mfc,llima}@math.uminho.pt
https://www.cmat.uminho.pt

Abstract. This work presents a study of the different existing object detection algorithms and the implementation of a Deep Learning model capable of detecting swimming pools from satellite images. In order to obtain the best results for this particular task, the RetinaNet algorithm was chosen. The model was trained using a customised dataset from Kaggle and tested with a newly developed dataset containing aerial images of the Algarve landscape and a random dataset of images obtained from *Google Maps*. The performance of the trained model is discussed using several metrics. The model can be used by the authorities to detect illegal swimming pools in any region, especially in the Algarve region due to the high density of pools there.

Keywords: Computer vision · Neural networks · Deep learning · Object detection · RetinaNet

1 Introduction

The main objective of this work is to detect swimming pools from satellite images, especially in the Algarve region. Due to the high density of pools that exist in this Portuguese region, some may be illegal and actively contribute to tax avoidance. Based on the work developed in [4] and [18], the Deep Learning (DL) methods are the most suitable to perform this type of detection.

In [4], both two-stage and single-stage DL algorithms were studied and compared. The two-stage group includes Convolutional Neural Networks (CNN) [12], Region-based CNN (RCNN) [13], Fast RCNN [14] and Faster RCNN [15]. These algorithms showed very high performance on the proposed task, but very high computational cost and slow prediction speed, due to a first stage that extracts the objects of an image, followed by a second stage in which these objects are classified.

Supported by FCT – Fundação para a Ciência e a Tecnologia, through projects UIDB/00013/2020 and UIDP/00013/2020 of CMAT-UM.

The single-stage group includes You Only Look Once (YOLO) [16], Single-Shot Detector (SSD) [17] and RetinaNet [1]. These algorithms classify the entire image and are faster than two-stage detectors. The remarkable speed of YOLO makes it to be preferred for real-time detection at the expense of accuracy [16]. In comparison with YOLO, SSD is slower but its approach improves the detection of objects of different sizes, in an image. The common problem of YOLO and SSD is the presence of class imbalance when the target images have high background presence, resulting in low positive classifications, as in the case of swimming pools [1]. The RetinaNet algorithm has been presented in the literature [1] to address class imbalances and improve detection of smaller objects, achieving a performance comparable with two-stage algorithms, but with the speed of single-stage algorithms.

Since satellite images are characterised by the presence of a massive amount of background, and objects can have different scales, one can conclude, based on the studied DL methods, that the RetinaNet method seems to be the most appropriate to solve the problem proposed in our study.

Having these ideas in mind, in this work we present the development of three RetinaNet models capable of detecting swimming pools in satellite images with high accuracy.

After this Introduction, the paper is organised as follows. The RetinaNet method is briefly described in Sect. 2. Then, to obtain and test the RetinaNet models, we created four datasets that are presented in Sect. 3, one for training and three for testing. The three RetinaNet models are trained and the results obtained with the testing datasets are presented and discussed in Sect. 4. The work ends with the main conclusions presented in Sect. 5.

2   RetinaNet

The RetinaNet algorithm is a single-stage detector that brought two new improvements over YOLO and SSD, namely Focal Loss [1] and Feature Pyramid Network [2], allowing it to match the performance of two-stage detectors without sacrificing speed.

To address possible class imbalances between the background class and the class under investigation, RetinaNet uses a modified version of the cross-entropy function called Focal Loss,

$$FL(p, y) = -\sum_t y_t(1 - p_t)^\gamma \log p_t \tag{1}$$

where t is the class index, y_t the class label such that $y_t = 1$ if the object belongs to class t and $y_t = 0$ otherwise,

p_t is the predicted probability of the object being of class t and $\gamma \in [0, 5]$. As suggested by the authors [1], a value of $\gamma = 2$ was used in our work.

The Focal Loss function defined in (1) gives lower importance to easily classified examples by using the modulating factor, γ, associated with a new term,

$(1 - p_t)$. In this way, easily classified examples, *i.e.* with high prediction probability (above 0.6), receive a value close to zero, resulting in very little or no learning, and the focus is on hard classified examples [1].

Multiple scale image processing can be used to detect objects with different scales in a single image, but is computationally expensive and time consuming.

As anticipated above, RetinaNet uses a backbone architecture called Feature Pyramid Network. This is a feature extractor that generates multiple multiscale feature maps, focusing on both accuracy and speed [2].

Several algorithms can be used as extractors, and the ResNet was chosen in this work. The ResNet architecture allows to add layers without compromising accuracy [8]. Therefore, different network depths were analysed to determine the one that gives the best results. Specifically, we analysed the following ResNet architectures, namely ResNet50, ResNet101 and ResNet150 with 50, 101 and 150 layers respectively [9].

In summary, the RetinaNet architecture, shown in Fig. 1, consists of: (a) a bottom-up path with a backbone network called Feature Pyramid Network, which computes multiple feature maps at different scales of an entire image; (b) a top-down path that upsamples feature maps from higher pyramid layers and associates equally sized top-down and bottom-up layers through lateral connections; (c) a classification subnetwork responsible for object classification; (d) a regression subnetwork to improve bounding box estimation [2].

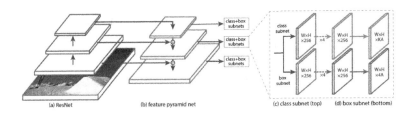

Fig. 1. RetinaNet network architecture composed of four main components: a) bottom-up pathway; b) top-down pathway; c) classification subnetwork; d) regression subnetwork; adapted from [2]. (Color figure online)

The RetinaNet also includes several algorithms to optimise the weights, *i.e.* the parameters, of the neural network, allowing for customisation. By default, RetinaNet uses Adaptive Moment Estimation (Adam) as it has been shown to be quite efficient in training neural networks [3].

3   Datasets

Four datasets were created in this work. One dataset for training, denoted by $\mathcal{D}_{tr}$, and three datasets for testing, denoted by $\mathcal{D}_{test1}$, $\mathcal{D}_{test2}$, $\mathcal{D}_{test3}$, respectively. The proposed testing datasets are composed of images with increasing difficulty for swimming pool detection. In particular, $\mathcal{D}_{test1}$ is constructed based on a dataset from *Kaggle* and consists of images similar to those used in the training dataset. Moreover, $\mathcal{D}_{test2}$ consists of *Google Maps* images from the Algarve region, in Portugal. This region was chosen because of the high density of swimming pools in this area. Finally, $\mathcal{D}_{test3}$ consists of images with swimming pools and similar looking objects taken from *Google Maps*. These testing datasets are used to illustrate the performance differences among the three RetinaNet models.

3.1   Training Dataset - $\mathcal{D}_{tr}$

A new dataset for training swimming pool detection models was created from a dataset retrieved from *Kaggle*. The original dataset contains 3748 training satellite images with bounding box annotations and labels for classifying cars (labelled 1) and swimming pools (labelled 2). The images are of the RGB type and have 224×224 pixels [11].

For each training image that contains objects of one of the classes, car or swimming pool, there is a corresponding file in XML format (EXtensible Markup Language) written in the PASCAL Visual Object Classes (PASCAL VOC) structure that contains important information, such as the file name of the corresponding image, the size of the image, the recognised object class and the coordinates of the bounding box. An example of an annotation file from the training dataset is the following:

```
<?xml version="1.0"?>
<annotation>
    <filename>000000012.jpg</filename>
    <source>
        <annotation>ArcGIS Pro 2.1</annotation>
    </source>
    <size>
        <width>224</width>
        <height>224</height>
        <depth>3</depth>
    </size>
    <object>
        <name>2</name>
        <bndbox>
            <xmin>149.53</xmin>
            <ymin>196.11</ymin>
            <xmax>193.97</xmax>
            <ymax>224.00</ymax>
        </bndbox>
    </object>
    <object>
        <name>2</name>
        <bndbox>
```

```
          <xmin>120.24</xmin>
          <ymin>212.77</ymin>
          <xmax>158.87</xmax>
          <ymax>224.00</ymax>
        </bndbox>
    </object>
</annotation>
```

From this example of the annotation file, we can see that the image has two pools represented by the two object sections with class label 2 and the coordinates of the bounding boxes.

Since in this work only the detection of swimming pools was of interest, it was necessary to discard the information related to the car class. Therefore, as a first step, a *Python* script (*Py1*) was developed to remove from the dataset the images and the annotation files that contained only cars. The script *Py1* implements the following strategy.

Open each training annotation file and do:

- Count the number of objects of each class, pool (denoted by 2) and car (denoted by 1);
- If there are only objects of class car, then delete the image and annotation file.

For samples with both classes, the car objects annotation data must be removed. Therefore, as a second step, another *Python* script (*Py2*) was developed to implement the following steps.

Open each training annotation file and for each object block do:

- If the label is 1, then exclude the object block.

After applying *Py1* and *Py2* scripts to the complete training dataset of *Kaggle*, the new training dataset D_{tr} was obtained. Dataset D_{tr} contains a total of 1993 training images with their corresponding XML annotation files.

However, the RetinaNet algorithm requires a single annotation file in Comma-Separated Values (CSV) with a specific format. Therefore, in a third step, another *Python* script (*Py3*) was implemented to perform the conversion by extracting the information from the XML files and merging all objects, one per line. The required format for each line is:

```
path/to/image.jpg,x1,y1,x2,y2,class_name
```

After the conversion, an example of some rows from the new CSV annotations file is:

```
000000012.jpg,149.53,196.11,193.97,224.00,pool
000000012.jpg,120.24,212.77,158.87,224.00,pool
000000014.jpg,211.71,156.41,224.00,196.16,pool
```

3.2 Testing Dataset from *Kaggle* - $\mathcal{D}_{test1}$

The dataset $\mathcal{D}_{test1}$ was created based on a testing dataset from *Kaggle* containing 2703 satellite images without any kind of annotation files [11]. Therefore, in order to compare and analyse the performance of the three RetinaNet models, it was mandatory to annotate all testing images. Thus, a *Python* script (*Py4*) was developed to help with this task by reading the coordinates of mouse clicks as bounding box coordinates and extracting the data inherent to the image, *i.e.* size and filename, and writing an XML annotation file for each image. After obtaining the XML files, it is known that the $\mathcal{D}_{test1}$ contains 2703 satellite images with a total of 620 swimming pools distributed over 524 images. These images are similar to those used to create the $\mathcal{D}_{tr}$ training dataset.

3.3 Testing Dataset from Algarve's Region - $\mathcal{D}_{test2}$

A new dataset was created with images from the Algarve region taken from *Google Maps*. We point out again that this region has been chosen essentially for several reasons. First, the main objective of this work is to detect illegal swimming pools, and the Algarve region has a high density of swimming pools. Second, there is a large amount of noise in this area, namely semi-hidden swimming pools and similar looking objects, which makes the problem much more interesting.

To speed up the process of sequentially capturing these sample images, a *Python* script (*Py5*) was also developed in view of automating the process by cutting a large image into smaller images, and allows full-screen printouts of the software to be obtained manually. When using this script, one can choose the number and type of divisions or images, which can be all of the same size or of a random size.

To analyse the performance of the training models in a variety of scenarios mimicking real-world use of satellite imagery, both equal and random size imagery were considered. Therefore, after applying *Py5*, the script *Py4* was applied to create the corresponding XML files for each image.

The new test dataset $\mathcal{D}_{test2}$ is a collection of 289 images of different sizes, zoom percentages and image quality. In total, 298 swimming pools are distributed among 172 examples, the others being negative examples with absence of swimming pools. Some selected examples are shown in Fig. 2. It can be seen that the images of $\mathcal{D}_{test2}$ were intentionally not carefully selected. The goal of using $\mathcal{D}_{test2}$ was to make more difficult the task of finding swimming pools by the training models in real case scenarios.

3.4 Testing Dataset from *Google Maps* - $\mathcal{D}_{test3}$

In order to examine the limitations of the trained models in depth, a few cropped images were carefully selected from *Google Maps*. These images were taken at different zoom percentages and from areas with different image quality. The eight images selected to create this test dataset are shown in Fig. 3.

Fig. 2. Algarve's landscape dataset examples in scale. The various samples have different zoom percentages, sizes and image quality. (Color figure online)

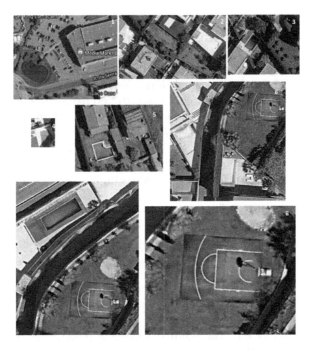

Fig. 3. Eight test images taken from *Google Maps* with different sizes and resolutions. The images are numbered in yellow on the top right corner. (Color figure online)

These images were carefully selected because they have objects that closely resemble swimming pools, and also swimming pools with an unusual appearance. This selection was made with the intention of fooling the detectors and quantifying their possible limitations. Below are the characteristics of each image:

- **image 1** - 433 × 340 pixels, presence of a lake and a blue tag;
- **image 2** - 442 × 258 pixels, encloses two swimming pools and a pool look-alike glass structure;
- **image 3** - 298 × 241 pixels, a pool without one of the most common characteristics, the blue colour;
- **image 4** - 100 × 118 pixels, very small and high zoom image;
- **image 5** - 366 × 270 pixels, unusual swimming pool body;
- **image 6** - 1351 × 1364 pixels, very high-quality 3-dimensional picture with two pools and a basketball court;
- **image 7** - 1099 × 1333 pixels, the same court as in image 6 and a green swimming pool;
- **image 8** - 732 × 613 pixels, the basketball court seen in images 6 and 7 but with higher zoom percentage.

To generate the XML annotation files for each of these images, the script *Py4* was applied. The new test dataset D_{test3} thus contains a total of 8 images with their corresponding XML annotation files.

4   Training and Testing of RetinaNet Models

In this section, we present the training parameters and metrics used to analyse the performance of the three trained RetinaNet models with different backbone architectures, namely the ResNet50, ResNet101 and ResNet150, using the testing datasets D_{test1}, D_{test2} and D_{test3}. For each trained model, in order to calculate the metrics *accuracy*, *precision*, *sensitivity* and *specificity*, confusion matrices, with values of True Negative (TN), True Positive (TP), False Negative (FN) and False Positive (FP), had to be created. In addition, Type I and Type II errors were also calculated following the ideas in [5].

- **True Positive (TP):** The expected and the predicted values are positive (meaning that, there is a swimming pool in an image and the model identified it). For each swimming pool correctly detected by the model, 1 is added to the TP value;
- **True Negative (TN):** The expected and the predicted values are negative (meaning that, there are not swimming pools in an image and the model does not detect any in the image). For each image correctly classified as not having pools, 1 is added to the TN value;
- **False Positive (FP):** The predicted value is positive but the expected is negative (meaning that, there is not a swimming pool in the image but the model identifies a part of the image as one). For each pool detected but not being one, 1 is added to the FP value;

- **False Negative (FN):** The predicted value is negative but the expected is positive (meaning that, there are swimming pools in the image but the model is not able to detect them). For each object, wrongly classified as being a swimming pool, 1 is added to the FN value.

4.1 Training

The three RetinaNet models were trained using the training dataset presented in Sect. 3. It is well known that deep neural networks have a large number of unknown parameters that are used to learn different features and map the input to the desired output. Therefore, the task of finding the optimal values for all these parameters, *i.e.* weights, requires a huge labelled training dataset. This can lead to a huge computation time and requires the use of supercomputers. To overcome this problem, one can use the weights of similar pre-trained models as a starting point for training the three RetinaNet models with ResNet50, ResNet101, and ResNet150, see [6] and [7].

Nowadays, there are several neural network models trained using the *ImageNet* dataset (a project that provides a dataset consisting of a large collection of images with human annotations created by academics, containing more than 14 million images of more than 20000 classes and bounding box annotations for 1 million images). The weights learned by some neural networks, *e.g.*, ResNet, InceptionV3, MobileNet, etc., are available to the community for download and use in *Keras, Tensorflow*, and in other deep learning libraries.

In this work, we used the weight files of pre-trained ResNets available in *Keras* using *ImageNet* as initial weights for training the three RetinaNet models with ResNet50, ResNet101, and ResNet150. We recall that the total number of weights to be initialised and optimised in these backbone models is 25636712 in ResNet50, 44707176 in ResNet101, and 60419944 in ResNet150 [10]. We used the RetinaNet algorithm options by default, with the following exceptions: i) three different depth ResNet backbone architectures, by default ResNet50; ii) initialisation of the weights using the weight files available in *Keras*; iii) a stack size of 1; iv) a learning rate of 10^{-5} for the optimisation algorithm, by default Adam; v) a maximum number of iterations of 500. We note that using a batch size of 1 means that each time an image traverses the entire network, the weights are updated by the Adam algorithm.

The RetinaNet algorithm requires two Comma-Separated Values (CSV) files with specific formats as input data. A CSV file with the annotations developed with the script *Py3*, and a CSV file in class mapping format with the labels corresponding to the classes. The mapping file must contain the target classes and a corresponding ID number (starts at 0), one per line. In our case, the CSV mapping file looks as follows:

pool , 0

Thus, after preprocessing the training dataset D_{tr}, the three RetinaNet models were trained with ResNet50, ResNet101 and ResNet150. As with the training dataset, a CSV file containing the annotations, developed using the *Py3* script, and a CSV file in class mapping format were also used for training. The training was performed in a PC Intel Core i5-8365U Processor, CPU 4.10GHz, with 8GB RAM and integrated graphics. Note that, the usage of a dedicated graphics card would drastically reduce training time, allowing to further train the networks in search of even better weights approximations.

In the next subsections, the validation of the trained models is performed using the developed test datasets $\mathcal{D}_{test1}$, $\mathcal{D}_{test2}$ and $\mathcal{D}_{test3}$.

4.2   Results Obtained with $\mathcal{D}_{test1}$

As expected, the results obtained with the dataset $\mathcal{D}_{test1}$ are excellent for the three considered architectures, see Table 1. It can be seen that the accuracy of each model is high and remarkably close, about 0.95, indicating that the classifiers are correct in most cases. All RetinaNet models have high precision, above 0.95, suggesting that an image labelled as positive is indeed positive, as confirmed by the low number of false positives in the table. Also, type I and II errors are small and very close to each other, indicating low false positive and false negative numbers, respectively. The model using a ResNet152 backbone architecture is better at detecting the class under study, showing lowest number of false negatives, but the performance differences between the three architectures are insignificant.

Table 1. Confusion matrix values and computed evaluation quantities for each of the trained models when using $\mathcal{D}_{test1}$

	TP	TN	FP	FN	Accuracy	Precision	Sensitivity	Specificity	Type I	Type II
ResNet50	471	2200	18	149	0.9412	0.9632	0.7600	0.9919	0.0081	0.2400
ResNet101	480	2285	7	140	0.9495	0.9856	0.7742	0.9969	0.0031	0.2258
ResNet152	503	2241	6	117	0.9571	0.9882	0.8113	0.9973	0.0027	0.1887

The results are good due to the large number of elements in the dataset and also to the fact that the testing images are similar to the training images.

To illustrate the difficulties of each model in detecting swimming pools, two tests are now performed, considering the more complex datasets $\mathcal{D}_{test2}$ and $\mathcal{D}_{test3}$.

4.3   Results Obtained with $\mathcal{D}_{test2}$

The Algarve landscape dataset was given to all RetinaNet models and the obtained results are shown in Table 2.

It is important to keep in mind that there are a total of 298 swimming pools in the dataset and that the images represent the worst case scenario of the models. From Table 2, we can see that none of the three models were able to detect the entirety of the positive examples, see the table column True Positive,

Table 2. Confusion matrix values and computed evaluation quantities for each of the trained models using the Algarve's landscape dataset.

	TP	TN	FP	FN	Accuracy	Precision	Sensitivity	Specificity	Type I	Type II
ResNet50	131	63	161	167	0.3716	0.4486	0.4396	0.2813	0.7187	0.5604
ResNet101	154	97	83	144	0.5251	0.6498	0.5168	0.5389	0.4611	0.9461
ResNet152	276	112	29	22	0.8838	0.9049	0.92617	0.7943	0.2057	0.07383

TP, with the ResNet152 model having by far the highest number of correct detections. Impressively, the ResNet152 model has a high sensitivity value, which means that no swimming pool passes unnoticed, which is also indicated by the low error of type II. This is a very important result, as inspecting an object incorrectly classified as a swimming pool causes less losses than overlooking an illegal construction. The models with a ResNet50 and ResNet101 have such low performance results that they should not be considered for real swimming pool detection. On the other hand, the model with a ResNet152 shows very high accuracy and precision, indicating that the classifier is correct in most cases and if an object is labelled as positive, it is indeed positive. This conclusion is also recognised by the number of false positives reported, corroborated by the lowest type I error.

From this analysis, it can be concluded that the RetinaNet model using ResNet152 backbone architecture outperforms by far the other models using ResNet50 and ResNet101 and shows high performance results. This means that the RetinaNet with a ResNet152 can be used in practise without hesitation.

4.4   Results Obtained with $\mathcal{D}_{test3}$

The results obtained with the three RetinaNet models are detailed and discussed in depth below.

Model Using ResNet50. The images in Fig. 3 were fed to the RetinaNet model using a ResNet50. Objects detected as swimming pools were enclosed by a blue bounding box, as shown in Fig. 4.

Some comments concerning the results obtained for each image are listed below:

- **image 1** - The model did not detect a swimming pool, although the image contains a lake and a blue sign;
- **image 2** - As expected, two swimming pools were detected and the glass structure did not fool the model;
- **image 3** - The model could detect the unclean swimming pool;
- **image 4** - Several swimming pools were detected instead of one. This can be explained by the very low resolution and size of this image;
- **image 5** - Again, three detections were made for a single pool. The confusion can possibly be explained by the unusual shape;

Fig. 4. Eight test images from *Google Maps* with different sizes and resolutions. The objects detected by RetinaNet+ResNet50 are surrounded by blue boxes. The images are numbered in the upper right corner. (Color figure online)

- **image 6** - The model recognised the swimming pools and did not confuse the basketball court with a swimming pool;
- **image 7** - Unlike image 3, the unclean pool was not recognised. This could explain why the lake in image 1 was not detected. Also, the previously unrecognised yard (image 6) was classified as a pool this time. The difference between the images suggests that the model is sensible to the zoom;
- **image 8** - The test performed on this image supports the statement presented in the previous point. The higher zoom helped to fool this model.

Model Using ResNet101. The images in Fig. 3 were fed to the RetinaNet model built on top of a ResNet101. The detected pools were enclosed by a blue bounding box, as shown in Fig. 5.

Some comments concerning the results obtained for each image are listed below:

- **image 1** - The model did not detect a swimming pool, although the image contains a lake and a blue sign;

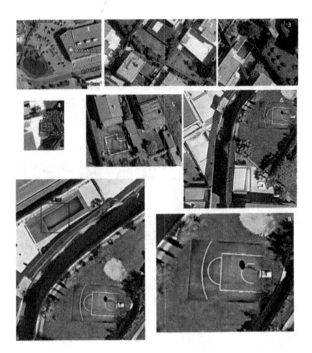

Fig. 5. Eight test images from *Google Maps* with different sizes and resolutions. The objects detected by RetinaNet+ResNet101 are surrounded by blue boxes. The images are numbered in the upper right corner. (Color figure online)

- **image 2** - The model failed to detect a swimming pool, performing worse than the ResNet50 model for this image;
- **image 3** - The model was unable to detect the unclean swimming pool;
- **image 4** - Multiple swimming pools were detected instead of one. Again, this can be explained by the very low resolution and size of this image;
- **image 5** - Similar to the previous model, three detections were made for a single swimming pool. The confusion could be explained by the unusual shape;
- **image 6** - The model recognised the swimming pools and did not mistake the basketball court by a swimming pool;
- **image 7** - All objects were correctly classified. The only difference between this image and image 3 is the quality. Image 7 has higher quality and 3-dimensional textures;
- **image 8** - The higher zoom percentage had no effect on detection.

Model Using ResNet152. Finally, the images in Fig. 3 were passed to the RetinaNet model built on top of a ResNet152. The detected swimming pools were enclosed by a blue bounding box, as shown in Fig. 6.

Some comments concerning the results obtained for each image are listed below:

Fig. 6. Eight test images from *Google Maps* with different sizes and resolutions. The objects detected by RetinaNet+ResNet152 are surrounded by blue boxes. The images are numbered in the upper right corner. (Color figure online)

- **image 1** - The model did not detect a swimming pool, although the image contains a lake and a blue sign;
- **image 2** - The ResNet152 model was able to correctly classify the objects in this image;
- **image 3** - The unclean pool was correctly detected;
- **image 4** - Unlike the previous models, the swimming pool in this image was not detected;
- **image 5** - Unlike the ResNet50 and ResNet101 models, the ResNet152 model correctly detected a single pool in this image;
- **image 6** - It correctly identified the pools and was not fooled by the blue basketball court;
- **image 7** - Both the pool and the court were correctly classified;
- **image 8** - As seen before, the basketball court did not fool the model.

The results obtained with the three models are now summarised.

- The models with ResNet50 and ResNet101 are not able to detect unusually coloured pools, green as in the tested examples;
- The results show that the model with ResNet50 is sensitive to the zoom of the image, indicating a wrong classification of objects when the zoom percentage is high;

- Classification of exceptionally small images is difficult for the model with ResNet152.

The model with the best performance was RetinaNet with a ResNet152 backbone architecture. In fact, it was able to correctly detect all swimming pools, with the exception of one in the smallest image 4. This result is not surprising, since the superiority of the model with a ResNet152 has already been shown by the results with $\mathcal{D}_{test2}$. Also, this last test data set, $\mathcal{D}_{test2}$, in addition to the results with $\mathcal{D}_{test3}$, proves that the models with a ResNet50 and ResNet101 are highly error-prone.

5 Conclusions

The aim of this work was to develop a model capable of detecting swimming pools in satellite images using a Deep Learning approach. To achieve this, several object detection algorithms were investigated and compared. The concluding remarks of this work are:

- We developed a dataset for training, $\mathcal{D}_{tr}$, and for testing, $\mathcal{D}_{test1}$, by modifying an existing one, filtering swimming pool annotations and creating label files for the testing images. To simulate real-world usage, *e.g.* by government agencies, 289 images of the Algarve region, in Portugal, were extracted from *Google Maps* resulting in a new dataset, $\mathcal{D}_{test2}$. In addition, to closely examine the limitations of the trained networks, eight cropped images from *Google Maps* were carefully selected to fool RetinaNet, $\mathcal{D}_{test3}$.
- Three RetinaNet models were trained with the training samples of the modified dataset, $\mathcal{D}_{tr}$, each one with a different depth backbone architecture: ResNet50, ResNet101, and ResNet152. To evaluate the models, the testing sets ($\mathcal{D}_{test1}$, $\mathcal{D}_{test2}$, $\mathcal{D}_{test3}$) were given to the models and a comparison between the predictions and the expected values was performed using confusion matrices that allowed to compute five performance metrics.
- The confusion matrices along with the values of the performance metrics showed that the RetinaNet model with a ResNet152 backbone architecture is the best at detecting swimming pools and achieved the highest performance in all three testing sets. Furthermore, this model was not fooled by the $\mathcal{D}_{test3}$ samples and only showed difficulty in correctly predicting very small and poor quality images.

References

1. Lin, T., Goyal, P., Girshick, R., He, K., Dollár, P.: Focal loss for dense object detection. In: Proceedings of the IEEE International Conference on Computer Vision, pp. 2980–2988 (2017)
2. Lin, T., Dollár, P., Girshick, R., He, K., Hariharan, B., Belongie, S.: Feature pyramid networks for object detection. In: Proceedings of the IEEE conference on computer vision and pattern recognition, pp. 2117–2125 (2017)

3. Kingma, D., Ba, J.: Adam: a method for stochastic optimization. In: International Conference on Learning Representations (2014)
4. Coelho, C.: Machine Learning and Image Processing. Master's thesis, to appear in Repositorium, University of Minho (2020). http://repositorium.sdum.uminho.pt
5. Powers, D.: Evaluation: from precision, recall and F-measure to ROC, informedness, markedness and correlation. J. Mach. Learn. Technol. **2**, 37–63 (2011)
6. Zabir, M., Fazira, N., Ibrahim, Z., Sabri, N.: Evaluation of pre-trained convolutional neural network models for object recognition. Int. J. Eng. Technol. **7**(3), 95–98 (2018)
7. Salahat, E., Qasaimeh, M.: Recent advances in features extraction and description algorithms: A comprehensive survey. In: Proceedings of the IEEE International Conference on Industrial Technology, pp. 1059–1063 (2017)
8. He, K., Zhang, X., Ren, S., Sun, J.: Deep residual learning for image recognition. In: Proceedings of the IEEE conference on computer vision and pattern recognition, pp. 770–778 (2016)
9. Keras RetinaNet. https://github.com/fizyr/keras-retinanet. Accessed 4 May 2021
10. Keras Applications. https://keras.io/api/applications. Accessed 4 May 2021
11. Swimming Pool and Car Detection. https://www.kaggle.com/kbhartiya83/swimming-pool-and-car-detection. Accessed 4 May 2021
12. Goodfellow, I., Bengio, Y., Courville, A.: Deep Learning. MIT Press, Cambridge (2016)
13. Girshick, R., Donahue, J., Darrell, T., Malik, J.: Rich feature hierarchies for accurate object detection and semantic segmentation. In: Proceedings of the IEEE Computer Society Conference on Computer Vision and Pattern Recognition, pp. 580–587 (2013)
14. Girshick, R.: Fast R-CNN. In: Proceedings of the IEEE International Conference on Computer Vision, pp. 1440–1448 (2015)
15. Ren, S., He, K., Girshick, R., Sun, J.: Faster R-CNN: towards real-time object detection with region proposal networks. IEEE Trans. Pattern Anal. Mach. Intell. **39**(6), 1137–1149 (2015)
16. Redmon, J., Divvala, S., Girshick, R., Farhadi, A.: You only look once: Unified, real-time object detection. In: Proceedings of the IEEE conference on computer vision and pattern recognition, pp. 779–788 (2016)
17. Liu, W., et al.: SSD: single shot multibox detector. In: Leibe, B., Matas, J., Sebe, N., Welling, M. (eds.) ECCV 2016. LNCS, vol. 9905, pp. 21–37. Springer, Cham (2016). https://doi.org/10.1007/978-3-319-46448-0_2
18. Coelho, C., Costa M., Ferrás L., Soares A.: Development of a machine learning model and a user interface to detect illegal swimming pools. In: SYMCOMP 2021 5th International Conference on Numerical and Symbolic Computation Developments and Applications, pp. 445–454. Évora (2021)

Identifying Anomaly Detection Patterns from Log Files: A Dynamic Approach

Claudia Cavallaro and Elisabetta Ronchieri$^{(\boxtimes)}$

INFN-CNAF, Bologna, Italy
{claudia.cavallaro,elisabetta.ronchieri}@cnaf.infn.it

Abstract. Context: Services that run in a data center can be configured to store information about their behaviour in specific logs. They contain huge amount of data that makes difficult to manually understand run-time properties of services. Therefore, to facilitate their analysis it is important to use dynamic solutions to quickly parse log files, detect anomaly and diagnose problems in order to react promptly.

Objectives: We want to build a model for determining anomaly detection in a certain period of time. Furthermore, we wish to identify machine learning techniques that support us determining problems patterns according to the messages available in logs.

Method: We have selected machine learning techniques, such as Invariant Mining model, natural language process and autoencoder, able to work with messages and identify anomaly patterns. According to the data available we have decided to study monthly data and detect samples with a higher frequency of problems. We have scanned the various logs to search the services with a wrong behaviour in the same period of time to recognize past anomalies in the data center and code these behaviours.

Results: The results are promising. We have obtained an average of F-measure metric over 86%.

Conclusion: Our model aims at quickly recognizing problems and solving them. It helps site administrators to better understand the run services, code anomalies and crosscheck different messages in the same time slot.

Keywords: Log analysis · Log mining · Log parsing · Anomaly detection · Machine learning

1 Introduction

A data center has to gather information about computer systems states. It is a common practice to have programs logging events that provide insights on the service activity and report their internal state, allowing to detect anomalies. Logs are semi-structured texts, usually appended to a file, with the '.log' extension, that grows in size and becomes very large, therefore the fact that system administrators analyze systems' health according to log files does not scale. This leads

© Springer Nature Switzerland AG 2021
O. Gervasi et al. (Eds.): ICCSA 2021, LNCS 12950, pp. 517–532, 2021.
https://doi.org/10.1007/978-3-030-86960-1_36

to develop solutions that automate the processing of logs, reducing human intervention. Software components and applications produce heterogeneous log files. There are services that use extremely flexible logging methods in their syntax, like syslog [22]. Log files usually do not contain the same type of information: 'syslog' event logs a system activity, while 'crond' event logs the CRON entries that show up in 'syslog'. Each file tends to describe a partial view of the whole data center. The stored information can contain the time and date of specific event to log exactly what happened and does use a simple formatting. Furthermore, the services can use different key words to express normal or erroneous behaviour. Logging analysis involves processing large amount of data that are not easy to read and understand manually. This can make difficult to perform log summarizing. Non-automatic processes for trouble shooting [40] are discouraged in large computing system. Several studies have proposed approaches to scale up log analysis [20] , moving the error problems from manual operation to automated operation. However, they still may require log processing and conversion of log files into a format that can be understood by analysis tools. Some methods are used to compact the information into more readable formats, such as '.csv' and '.json', and to also automatically extract any suspicious information, like the cause of an 'error' message. Quite often it is not feasible just selecting the file and retyping the file extension as '.csv' or '.json', because the transformed file could be wrongly reformatted. Furthermore, the file usually contains daily service information, so the number of data can be over $60 - 120\ K$ rows, penalizing the usage of some analysis tools, because they will likely become unresponsive when they perform, e.g., selecting and adding new variables in the data sets. Before starting any analysis it is essential to decide which variables have to be included in the resultant data sets. Spreading the data across multiple columns is another aspect to consider to organise your data set into a manageable format. The resultant files can be used to identify trends and unusual activity that is beneficial for both short- and long-term data center management. Machine Learning techniques are a solution to identify anomaly detection patterns automatically, predicting the failure of the machine.

In this paper, we focus on determining problems patterns by combining the results obtained by different machine learning (ML) techniques. Log files contain a considerable amount of texts, therefore we have used modern Natural Language Processing (NLP) methods [1,33] to map the log files' words to a high dimensional metric space in order to define site administrators' actions. The aim is to cluster and- if possible- classify the various system events. Furthermore, having to work with unsupervised data, we have used a machine (deep) learning technique called autoencoder [56]: the decision taken by and information available to site experts are not stored in any local framework. We have also used an invariant mining technique [37] that supports, for example, detection of failures and anomalies. During the log analysis, we have adopted the invariant mining model, because it is a general approach that does not rely on the nature of the data and it does not require any meaningful knowledge of the domain or constant rules. Furthermore, it does not require training labels and messages are not grouped with respect to distance only. The proposed approach is repeatable in other contexts and domains. With minimum setup effort and the usage of machine learning tools, it is possible to automatically extract relevant information about system state. Through experiments, we illustrate the potential benefits of our approach by answering our research questions.

The reminder of this paper is as follows. Section 2 introduces the background of log analysis and provides related works. Section 3 introduces the rational of our approach. Section 4 provides some promising experimental results. Section 5 concludes the paper with future work.

2   Log Mining and Related Works

Existing commercial tools, such as Elk [18], Splunk [45], Loggly [36] or Over Ops [41], are mainly used for Big Data visualization, but are not appropriate for anomaly detection. Moreover, utilities such as Salsa [30] or Log Enhancer [55] for diagnosability are not good on large volume data, because they are heavy. Log mining is an approach that uses statistical and ML algorithms in structured event list to extract knowledge from logs, discovering a model inference and detecting outliers in a system.

Log Compression. Preliminary phases to log mining are log compression, which uses, for example, dictionary-based and statistic-based techniques or Jaccard similarity [28], and log parsing, which collects clustering, heuristics, frequent pattern mining and evolutionary algorithms. Frequent pattern mining identifies frequencies, correlations and causalities between sets of items in transactional databases: among the most used algorithms, Apriori [46] and FP-growth [23] can be mentioned. The Apriori algorithm is based on a level structure and it is useful to search for the most frequent schemes [5] from a large dataset, to quickly exclude non-recurring patterns. FP-growth bases its operation on the construction of a tree structure that models the considered dataset. In particular, the frequent item sets are extracted by going through the tree, without therefore having to read the database several times.

Log Parsing. In the preprocessing techniques category, also called log abstraction, there is the log parsing activity used to abstract and simplify the initial data. Among the clustering methods for log parsing, we can mention Simple Logfile Clustering Tool (SLCT) [49], Hierarchical Event Log Organized (HELO) [21] and Iterative Partitioning Log Mining (IPLoM) [38]. In addition, POP [25], a parallel log analyzer, is suitable for large-scale data processing. During the log parsing process, metrics are sometimes used, for example LevenShtein distance [34] or Longest Common Subsequence [48] in Spell [14].

Data Transformation. In the preprocessing phase, there is also data transformation. Through this process the log set is transformed into an appropriate form, such as an histogram matrix of events or a binary transformation, which becomes the input in the next data mining step. El-Masri et al. [17] compare some aspects, such as scalability and efficiency, of the main log abstraction techniques.

Log Analysis. Logs can be analyzed in a static way (off-line method) or dynamically (on-line method), but the latter method is time consuming. The former is less precise but also faster. Hybrid analysis is a mixture of the two. The most popular on-line log parsing approaches are SHISO [39], LenMa (which are based on clustering), Drain [26] (based on heuristics) and Logram [9] (based on frequent pattern mining). A more detailed description of clustering and log parsing methods can be found in the survey [27]. Log analysis techniques can be divided into the following branches: failure detection techniques that include, for instance, decision tree and associative rule learning; anomaly detection techniques that comprises agglomerative or divisive hierarchical clustering, nearest neighbor approach, chi-squared test and Naive Bayes (NB) classifier [13], and failure diagnosis that includes the SherLog [54] tool and Decision Tree (DT) [42] approaches.

Machine Learning Techniques. Some traditional ML algorithms are applied in literature, including Principal Component Analysis (PCA) [53], LogCluster [50], Support Vector Machine (SVM) [51] and mixtures of Hidden Markov models. Some Deep Learning approaches for this scope are DeepLog (LSTM-RNN) [15], LogGAN (CNN) [52], Aarohi (a failure prediction method) [10], AirAlert (a framework based on Bayesian network) [7] and TCFG (Time Weighted Control Flow Graphs) [29]. Machine Learning techniques are divided into classification techniques (such as SVM and Bayesian networks), clustering techniques (including K-means [35], Self-Organizing Feature Map, SOFM [32] and DBSCAN [44]) and statistical techniques (for example PCA). A clustering algorithm divides data into groups, according to the criterion that the elements belonging to the same cluster are very similar. Outliers, which are those that do not belong to any cluster, often contain useful information on the anomalous characteristics of the system as they are generally different from other data. To estimate how much the results of the algorithms implemented in the log mining process correspond to reality, one can proceed in several ways. A first strategy is to measure true and false positives, true and false negatives and plotting the curve receiver operating characteristic (ROC). Other procedures, used to evaluate clustering methods, include internal evaluation (given by the Davies-Bouldin index [11] or Dunn index [16]), and external evaluation (with Rand measure [43], F-measure or Jaccard index).

Related Works. Breier and Branisova [3], employing Apache Hadoop framework, process log files in parallel. Their method, based on the generation on dynamic rules, spots anomalies through MapReduce. The parallel implementation, compared with A-priori and FP-Growth algorithms, has accelerated the process of detecting security breaches in log records.

Borghesi et al. [2], through a deep learning technique, perform anomaly detection in high performance computing systems. The Examon infrastructure (the code is available on Github [19]) dataset is explored through a neural network called autoencoder. It is trained with the Adam optimizer [31], by using an off-line approach. Finally, the metric used to evaluate the accuracy of the obtained results is F-measure. Layer et al. [33] use NLP to predict the operator's action for the CMS experiment [8] workflow handling. They employ the information of the computing operators for taking the decision in case of failures, stored in the experiment framework, during the ML training phase. Bertero et al. [1] also use NLP to identity anomaly detection. They emulate different types of errors that refer to CPU consumption, misuse of memory, abnormal number of disk accesses, network packet loss and network latency.

The results presented in previous papers show that the optimal procedure for analyzing logs is the combination of, first, heuristics or filtering and, next, machine learning steps. In this article, we have chosen to use invariant mining and autoencoder techniques, because we are interested in the relationships between different messages and source hostnames. We have also applied NLP to identify anomaly categories that can be used to label the log entries and contribute to defining administrators' operations. Tools, like SLCT, have the disadvantage of being based on searching only for the most frequent types of messages in the log file, neglecting the less frequent ones. For anomaly detection, however, it may be necessary to find rare types of messages, so this option has been omitted. Furthermore, the application of clustering methods in the log core analysis phase would lead to the risk of losing significant results on the anomalies to be considered individually, if they have been aggregated with others.

3   Approach

The approach proposed as the contribution of this paper is presented in Fig. 1. Once data has been collected and the log files characteristics have been identified, we have dedicated to the following phases: data preprocessing and transformation that have allowed us to build a set of datasets to train and build our anomaly detection models.

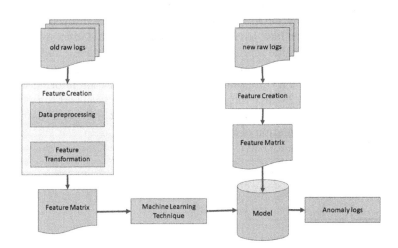

Fig. 1. General approach overview.

Source Data. The log files examined in this study are related to a set of services running at INFN Tier-1 data center [12], used by the large hadron collider experiments [4]. Low level services are shared among the highlighted groups, such as crond and sudo. The log files mainly belong to Linux system services, such as the software utility crond, the free and open-source main transfer agent postfix and the standard for message logging syslog. Table 1 summarizes the first 30 filenames according to their frequency, that is the number of times a value of an unique filename occurs. The suffix-type frequencies of the various available files are summarized as follows: .gz 10869, .log 3562759, .manifest 5, .meta 6, .pending 1, .txt 2.

Table 1. The top 30 log files per frequency.

Filename	Frequency	Filename	Frequency	Filename	Frequency
sudo.log	378781	systemd.log	107700	userhelper.log	21380
puppet-agent.log	368530	mmfs.log	72620	nslcd.log	20544
run-parts.log	365734	rsyslogd.log	70210	neutron_linuxbridge.log	8572
crontab.log	348896	kernel.log	65938	runuser.log	6859
crond.log	347708	logrotate.log	62531	cvmfs_x509_validator.log	6031
sshd.log	303919	syslog.log	47330	cvmfs_x509_helper.log	5399
anacron.log	287419	yum.log	43301	srp_daemon.log	4938
postfix.log	175558	fusinv-agent.log	42125	edg-mkgridmap.log	4083
auditd.log	120473	root.log	37345	libvirtd.log	3328
smartd.log	109441	gpfs.log	31000	dbus.log	3301

Log files. Each of these log files contains a different amount of lines. They contain numerical and textual data that describe system states and run time information. Each log entry includes a message that contains a natural-language text (i.e. a list of words) describing some events. Logs are generally generated by logging statements inserted, either by software developers in source code or by system administrators in configuration files, to record particular events and software behaviour. Each log entry in the log file represents a specific event. Figures 2 and 3 show two different log entry samples composed of a log header and a log message: the former is generally composed of a timestamp, a custom-configuration information (such as the hostname in Fig. 2 where the service runs and a log level verbosity in Fig. 3) and the name of the service the message is associated to; the latter is just the message that contains information of the logged event.

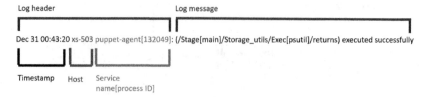

Fig. 2. A log entry sample from the puppet-agent service.

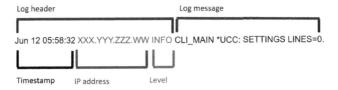

Fig. 3. A log entry sample from the puppet-agent service.

Figures 4 and 5 show two examples of the log message of a log entry, characterized by a natural-language text which interpretation is difficult because there is not an official standard defining the message format. The text is usually composed of different fields called dynamic and static: a dynamic field (DF) is a string or a set of strings that are assigned at run time; a static field (SF) does not change during events. Such fields can be delimited by different separators, such as a comma, a white space, a parenthesis. Logging practice is scarcely well documented. This activity mainly depends on human expertise [24]. They often have to analyze a large volume of information that may be unrelated to the problematic scenarios and lead to overwhelming messages [6].

Fig. 4. A log message fields with just one dynamic field and static field.

Data Preprocessing. During this phase the log files change format and turn into '.csv' files. In addition the following variables are added to each file: date, time, timestamp, hostname, internet protocol (ip) address, service name, process identifier (id), component name, message (msg). The hostname and ip couple are not always both available, especially when the service runs on a virtual machine. Each file is related to a particular service that runs on a well-known machine in the data center. Its location is got by a local database and included into the resulting file. During this phase we have tackled some site administration peculiarities: the same service called in lower or capital letters; the process identifier included in the service logging file; the service name included (or not) in the log message; the process identifier included (or not) in the log message; the logging filename included typo error. Before performing any cleaning operations, we have excluded meaningless services' log files, especially those with a small number of events. In the remaining logs, the following changes have been applied in the message field: the removal of unwanted texts, such as punctuation, non-alphabets,

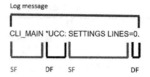

Fig. 5. A log message fields with a sequence of dynamic and static fields.

and any other kind of characters that are not part of the language by involving regular expressions; the exclusion of non-English characters; the cancellation of stopwords, that is frequent general words, like 'of, are, the, it, is', with a low meaning. According to the amount and types of services, in this phase we have started to trace the types of log events, such as assert, fail, error and debug, and to identify anomaly key terms that can be used to classify the reason of the problems in the service. This part of the study is still ongoing requiring experts' check. However, Table 2 summarizes few examples of message lines that describe a wrong service behaviour.

Table 2. Examples of message lines for the crond log file.

log event msg	log event type	anomaly key term
.. reset error counters	error	reset
.. failed create session connection time out	fail	connection

Features Transformation. Data transformation includes the creation of a new dataset that includes a matrix, whose dimension and values change according to the machine learning technique used to build the anomaly detection model. The resulting dataset can contain either binary data or numerical data. In case of binary data there is a numerical value of 1 for features that are present in the log event and a numerical value of 0 otherwise. In case of NLP usage, the tokenization phase (see Fig. 3) determines the element of the matrix: the message string is split up in single words. Once transformed the message events into a matrix of features, in case of problem with large dimension, it is possible to apply rules to reduce data, e.g., according to the uniqueness of the messages. For NLP, low frequency words or words that are not important for the meaning of the anomalies are filtered out. For a matter of time computation, the event count matrix has been created for each month, in which its elements indicate the occurrences of all messages given in input and relating to the hostnames from which they come in that time window.

Table 3. Examples of message strings split up in single words for the crond log file.

log event msg	..	connection	..	error	..	fail	..	time	..
.. reset error counters	..	0	..	1	..	0	..	0	..
.. failed create session connection time out	..	1	..	0	..	1	..	1	..

Machine Learning. Starting from the event count matrix, NLP, autoencoder and invariant mining techniques have been considered. NLP includes the word2vec [47] unsupervised algorithm, a popular embedding approach by Google to process natural language. word2vec takes a text corpus as input and produces a high-dimensional euclidean space with each unique word in the corpus. Each word of an event is mapped to nearby points in the vector space. This technique is able to produce meaningful word embeddings: similar words end up close, words that are not related to each other end up far away in the embedding space. Following the same approach to all log files, it is possible to form clusters of similar anomaly events and use traditional supervised classifiers to e.g. determine anomalous and normal behaviour of the services.

Autoencoder is a type of neural network that can be used to detect anomalies at the service level. With respect to Borghesi et al. [2], in this study we have created a set of separate autoencoder models, one for each service in the system. Each model is trained by using a loss function to ensure that the output is close to the input.

Invariant mining model is based on the fact that a number of initial operations on the files corresponds to the same number of final operations on the same or a similar type, for example the number of jobs arrived with the number of jobs started, or the number of jobs arrived and number of jobs completed. Linear program invariant is then a predicate that always contains the same values in different normal operations, such as opening and closing files. Invariants reflect the underlying correlation between variables and the properties of the execution path. Whenever an invariant is broken during a system operation, the anomaly in the log corresponding to these variables is detected because it is considered a sign of a probable malfunction. The final result is a vector of length equal to the number of unique messages given in input, and composed of values "0" and "1". Each element corresponds to a log message, in which "1" indicates that it is labeled as anomalous, "0" instead as non-anomalous.

Some messages, erroneously labeled as anomaly, could be log patterns that at the beginning of a month correctly close the operations of the end of the previous month, but are not linked to the messages that precede them if we have separated the dataset by month. These boundary cases can be assessed individually, considering a limited time window e.g. 24 h before and after, or taking into account the average closing time of execution flows. Once the logs with anomalies have been identified automatically, through this strategy we can trace the causes and files that potentially triggered the errors. This could be useful for us to know in advance what the alarm messages are in real time and to act in time to prevent problems from occurring, that is, to do predictive maintenance.

Performance Metrics. To access the performance metrics of the ML techniques we have considered the following metrics. *Precision* is the measure of the model's performance with respect to false positives (FP), which are the messages identified as anomalies, when no anomalies occurred. False negatives (FN), on the other hand, are messages mislabeled with non-anomaly, which should instead be indicated as anomalous. From the expression: $precision = \frac{TP}{TP+FP}$, where true positives (TP) are messages that correctly report an occurred anomaly, it is evident that high precision values indicate a low rate of FP. *Recall* measures model's performance with respect to false negatives, according to this formula: $recall = \frac{TP}{TP+FN}$. f-measure, expressed by: $f1 = 2 \times \frac{precision \times recall}{precision+recall}$, is the harmonic mean of precision and recall. It is less affected by extreme values; the closer is its value to 1, the better the model performs.

4   Results

For this study, we have created a set of GitLab projects to store code we have implemented so far to perform the various phases of the presented approach. The collected and preprocessed data contains reserved information, therefore at the moment of the conference we have preferred to keep private our GitLab projects. We have developed our analysis on jupyter notebooks for python3 programming language by leveraging data analysis python libraries, such as pandas, nltk, and scikit-learn.

In the following we are going to show the effectiveness of the presented approach by showing the results obtained for a subset of services (see Table 1). The logs analyzed, after their preparation with the preprocessing phase, relate to the period June–October 2020. The causes of the anomalies detected are various: we can mention, for example, bugs related to the memory or silent corruptions of the data, jobs aborted due to field validation file, software downs or errors due to some users who use the data center services. We have grouped the extracted anomalies into specific message templates, so that they can be used as warnings in future system states without waiting further data acquisition time. Our approach is off-line: however, an on-line method is simulated by comparing the anomalous results of one month logging files with both the previous and the consecutive periods, and specific ML evaluation metrics are used. Next, the success is assessed through a manual check of the most frequent messages, in order to assign labels to certain messages and compare the experimental results with the real and expected ones. The precision, recall and F-measure performance metrics of ML techniques have been calculated through the data of two consecutive months. Table 4 shows the mean values of the ML performance metrics. We can assert that these models are effectively able to describe the anomalies.

Table 4. Mean values of the ML performance metrics.

Method	Precision	Recall	F-measure
Invariant mining	0.827	0.904	0.864
NLP word2vec	0.899	0.932	0.912
Autoencoder	0.819	0.925	0.895

All the methods highlight the same period of time for the anomalies due to the presence of problem key terms in the files. Messages, quite explanatory, labeled as anomalous by the methods are for example: "failing disk", "left power supply failure", recovered: "medium error disk", "internal queue full", "Total Queue full", "process abort" and "Disk Failed: Abort".

In the following we present details of the different methods.

Invariant Mining Model. The invariant mining technique is able to detect the intrinsic linear characteristics of workflows linked to a machine, in particular used to automatically collect the frequent patterns of error messages that generate a problem in the Tier-1 data center. The message-hostname pair has been chosen, because it is the most significant in the search for anomalies, and because other variables such as the ip address could be obtained from the hostname source. The percentage of anomalous messages, calculated between the number of unique suspicious messages of each month and the total number of monthly log messages, varies from a minimum of 0.06% to a maximum of 10.3%.

Figure 6 shows some examples of anomaly messages detected from logs of Tier-1, where the histograms report the number of occurrences per day. Their frequency is high at particular dates, and therefore can be associated with some problems. These patterns come from two specific hostnames and ips, which we indicate for privacy "host_a, ip_XXX.XXX.XXX.XY" and "host_b, ip_XXX.XXX.XXX.XZ", and therefore, this information is useful for monitoring the future operations of the specific associated machines.

Natural Language Processing Method. To establish the word2vec training set, we use the concatenation of 300 log files that constitute the basis of our model training. Therefore, the message corpus is relatively small. We have use the NLP method avoiding any optimization of the computation in order to get the maximum number of information. The output of the word2vec is a file containing the coordinates of distinct words of our training corpus.

Autoencoder Method. An autoencoder is a feed-forward multi-layer neural network with the same number of input and output neurons. We have used it to learn a more efficient representation of data while minimizing the corresponding error.

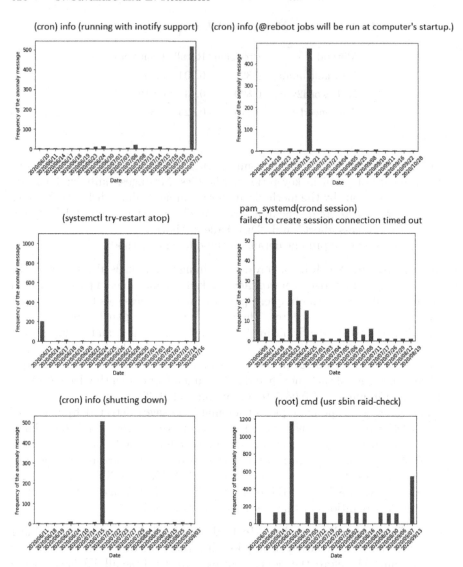

Fig. 6. Some anomaly messages detected with invariant mining.

5 Conclusions

The anomalies in logs can be related to permanent or transient errors. Performing manual diagnostics in a data center is not feasible, given the large-scale increase in unstructured information produced by machines in operation. For a human operator, even by dividing the data analysis for small time intervals, it would be very time-consuming. Most of the traditional tools on error analysis are based on verifying the standard behavior of the system by knowing a priori the behavior of specific software, but it is therefore linked to the specificity of the system and to the knowledge of the operators. If new software is used, manual analysis cannot be accurate after a short period of use, because in that case it is not possible to outline all the different normal system behaviors and tag the other unknowns as errors. If it were done for all the prototypes of unknown messages this could lead to a categorization of false positives.

In this work, the log mining is performed on monthly time slots and on the frequencies of each anomalous message, which are automatically extracted together with the hostnames registered to them. The results are promising. We have obtained an average of F-measure metric over 86%. A subset of services has been used during the study, therefore we aim at improving our study by considering all the available datasets.

Acknowledgements. This work has been supported by the IoTwins project N. 857191.

References

1. Bertero, C., Roy, M., Sauvanaud, C., Trédan, G.: Experience report: log mining using natural language processing and application to anomaly detection. In: 28th International Symposium on Software Reliability Engineering (ISSRE 2017). p. 10p. Toulouse, France (October 2017). https://hal.laas.fr/hal-01576291
2. Borghesi, A., Bartolini, A., Lombardi, M., Milano, M., Benini, L.: Anomaly detection using autoencoders in high performance computing systems. In: Proceedings of the AAAI Conference on Artificial Intelligence, vol. 33, pp. 9428–9433 (July 2019). https://doi.org/10.1609/aaai.v33i01.33019428
3. Kim, K.J. (ed.): Information Science and Applications. LNEE, vol. 339. Springer, Heidelberg (2015). https://doi.org/10.1007/978-3-662-46578-3
4. Breskin, R.V.A.: The CERN Large Hadron Collider: Accelerator And Experiments, vol. 2, CMS, LHCb, LHCf, And Totem. CERN (2009)
5. Cavallaro, C., Vitrià, J.: Corridor detection from large GPS trajectories datasets. Appl. Sci. **10**(14), 5003 (July 2020) https://doi.org/10.3390/app10145003
6. Chen, B., Jiang, Z.M.J.: Characterizing and detecting anti-patterns in the logging code. In: Proceedings of the IEEE/ACM 39th International Conference on Software Engineering (ICSE), pp. 71–81. IEEE Press (2017). https://doi.org/10.1109/ICSE.2017.15
7. Chen, Y., et al.: Outage prediction and diagnosis for cloud service systems. In: The World Wide Web Conference on - WWW 2019, ACM Press (2019). https://doi.org/10.1145/3308558.3313501
8. Collaboration, T.C., Chatrchyan, S., Hmayakyan, G., Khachatryan, V., Sirunyan, A.M., et al.: The CMS experiment at the CERN LHC. J. Instrum. **3**(08), S08004–S08004 (2008) https://doi.org/10.1088/1748-0221/3/08/s08004
9. Dai, H., Li, H., Chen, C.S., Shang, W., Chen, T.H.: Logram: efficient log parsing using n-gram dictionaries. IEEE Trans. Softw. Eng. 1 (2020). https://doi.org/10.1109/tse.2020.3007554
10. Das, A., Mueller, F., Rountree, B.: Aarohi: making real-time node failure prediction feasible. In: 2020 IEEE International Parallel and Distributed Processing Symposium (IPDPS), IEEE (May 2020) https://doi.org/10.1109/ipdps47924.2020.00115

11. Davies, D.L., Bouldin, D.W.: A cluster separation measure. IEEE Trans. Pattern Anal. Mach. Intell. PAMI-1(2), 224–227 (1979). https://doi.org/10.1109/tpami. 1979.4766909

12. dell'Agnello, L., et al.: Infn tier–1: a distributed site. EPJ Web Conf. **214**(08002), 01 (2019). https://doi.org/10.1051/epjconf/201921408002

13. Domingos, P., Pazzani, M.: On the optimality of the simple bayesian classifier under zero-one loss. Mach. Learn. **29**(2/3), 103–130 (1997). https://doi.org/10. 1023/a:1007413511361

14. Du, M., Li, F.: Spell: Streaming parsing of system event logs. In: 2016 IEEE 16th International Conference on Data Mining (ICDM), IEEE (December 2016) https:// doi.org/10.1109/icdm.2016.0103

15. Du, M., Li, F., Zheng, G., Srikumar, V.: DeepLog : anomaly detection and diagnosis from system logs through deep learning. In: Proceedings of the 2017 ACM SIGSAC Conference on Computer and Communications Security, ACM (October 2017) https://doi.org/10.1145/3133956.3134015

16. Dunn, J.C.: Well-separated clusters and optimal fuzzy partitions. J. Cybern. **4**(1), 95–104 (1974). https://doi.org/10.1080/01969727408546059

17. El-Masri, D., Petrillo, F., Guéhéneuc, Y.G., Hamou-Lhadj, A., Bouziane, A.: A systematic literature review on automated log abstraction techniques. Inf. Softw. Technol. **122**, 106276 (2020) https://doi.org/10.1016/j.infsof.2020.106276

18. ELK: Elasticsearch. https://www.elastic.co/elk-stack (2021). Accessed 11 Jun 2021

19. Examon: Examon HPC Monitoring. https://github.com/EEESlab/examon (2021). Accessed 11 Jun 2021

20. Farshchi, M., Schneider, J.G., Weber, I., Grundy, J.: Experience report: anomaly detection of cloud application operations using log and cloud metric correlation analysis. IEEE Trans. Softw. Eng. (2015). https://doi.org/10.1109/ISSRE.2015. 7381796

21. Gainaru, A., Cappello, F., Trausan-Matu, S., Kramer, B.: event log mining tool for large scale HPC systems. In: Jeannot, E., Namyst, R., Roman, J. (eds.) Euro-Par 2011. LNCS, vol. 6852, pp. 52–64. Springer, Heidelberg (2011). https://doi.org/10. 1007/978-3-642-23400-2_6

22. Gerhards, R.: The syslog protocol. In: RFC. RFC Editor (2009)

23. Han, J., Pei, J., Yin, Y.: Mining frequent patterns without candidate generation. ACM SIGMOD Record **29**(2), 1–12 (2000). https://doi.org/10.1145/335191. 335372

24. He, P., Chen, Z., He, S., Lyu, M.R.: Characterizing the natural language descriptions in software logging statements. In: Proceedings of the 33rd ACM/IEEE International Conference on Automated Software Engineering ASE, pp. 178–189 (2018). https://doi.org/10.1145/3238147.3238193

25. He, P., Zhu, J., He, S., Li, J., Lyu, M.R.: Towards automated log parsing for large-scale log data analysis. IEEE Trans. Dependable Secure Comput. **15**(6), 931–944 (2018). https://doi.org/10.1109/tdsc.2017.2762673

26. He, P., Zhu, J., Zheng, Z., Lyu, M.R.: Drain: An online log parsing approach with fixed depth tree. In: 2017 IEEE International Conference on Web Services (ICWS), IEEE (2017) https://doi.org/10.1109/icws.2017.13

27. He, S., He, P., Chen, Z., Yang, T., Su, Y., Lyu, M.R.: A survey on automated log analysis for reliability engineering. ArXiv (September 2020)

28. Jaccard, P.: The distribution of the flora in the alpine zone. New Phytol. **11**(2), 37–50 (1912). https://doi.org/10.1111/j.1469-8137.1912.tb05611.x

29. Jia, T., Yang, L., Chen, P., Li, Y., Meng, F., Xu, J.: LogSed: anomaly diagnosis through mining time-weighted control flow graph in logs. In: 2017 IEEE 10th International Conference on Cloud Computing (CLOUD), IEEE (2017). https://doi.org/10.1109/cloud.2017.64

30. Tan, J., Pan, X., Kavulya, S., Gandhi, R., Narasimhan, P.: Salsa: analyzing logs as state machines (cmu-pdl-08-111). In: First USENIX Workshop on the Analysis of System Logs, WASL 2008, San Diego, CA, USA, Proceedings. Carnegie Mellon University (2008). https://doi.org/10.1184/R1/6619766

31. Kingma, D.P., Ba, J.: Adam: a method for stochastic optimization. In: International Conference on Learning Representation (ICLR) (2015)

32. Kohonen, T.: Self-organized formation of topologically correct feature maps. Biol. Cybern. **43**(1), 59–69 (1982). https://doi.org/10.1007/bf00337288

33. Layer, L., et al.: Automatic log analysis with NLP for the CMS workflow handling. In: 24th International Conference on Computing in High Energy and Nuclear Physics (CHEP 2019), p. 7 (November 2020) https://doi.org/10.1051/epjconf/202024503006

34. Levenshtein, V.I.: Binary codes capable of correcting deletions, insertions and reversals. Doklady Akademii Nauk SSSR **163**(4), 845–848 (1965)

35. Lloyd, S.: Least squares quantization in PCM. IEEE Trans. Inf. Theor. **28**(2), 129–137 (1982). https://doi.org/10.1109/tit.1982.1056489

36. Loggly: Loggly - log management by loggly. https://www.loggly.com (2021). Accessed 11 Jun 2021

37. Lou, J.G., Fu, Q., Yang, S., Xu, Y., Li, J.: Mining invariants from console logs for system problem detection. In: Proceedings of the 2010 USENIX Conference on USENIX Annual Technical Conference, USENIXATC 2010, p. 24, USENIX Association, USA (2010)

38. Makanju, A., Zincir-Heywood, A.N., Milios, E.E.: A lightweight algorithm for message type extraction in system application logs. IEEE Trans. Knowl. Data Eng. **24**(11), 1921–1936 (2012). https://doi.org/10.1109/tkde.2011.138

39. Mizutani, M.: Incremental mining of system log format. In: 2013 IEEE International Conference on Services Computing, IEEE (June 2013) https://doi.org/10.1109/scc.2013.73

40. Oliver, R.: What supercomputers say: a study of 5 system logs. In: Proceedings of the 37th Annual IEEE/IFIP International Conference on Dependable Systems and Networks (DSN 2007), IEEE Press (2007). https://doi.org/10.1109/DSN.2007.103

41. OverOps: OverOps Continuous Reliability Solution. https://www.overops.com/ (2021). Accessed 11 Jun 2021

42. Quinlan, J.: Simplifying decision trees. Int. J. Man-Mach. Stud. **27**(3), 221–234 (1987). https://doi.org/10.1016/s0020-7373(87)80053-6

43. Rand, W.M.: Objective criteria for the evaluation of clustering methods. J. Am. Stat. Assoc. **66**(336), 846–850 (1971). https://doi.org/10.1080/01621459.1971.10482356

44. Sander, J., Ester, M., Kriegel, H.P., Xu, X.: Density-based clustering in spatial databases: The algorithm dbscan and its applications. Data Min. Knowl. Discov. **2**(2), 169–194 (1998). https://doi.org/10.1023/a:1009745219419

45. Splunk: Splunk platform. http://www.splunk.com (2005-2021). Accessed 11 Jun 2021

46. Srikant, R., Agrawal, R.: Mining generalized association rules. Future Gener. Comput. Syst. **13**(2–3), 161–180 (1997). https://doi.org/10.1016/s0167-739x(97)00019-8

47. Tomas, M., Ilya, S., Kai, C., Greg, C., Jeffrey, D.: Distributed representations of words and phrases and their compositionality. In: Proceedings of the 26th International Conference on Neural Information Processing Systems (NIPS 2013), NIPS 2013, pp. 3111–3119. Curran Associates Inc., Red Hook, NY, USA (2013)

48. Ullman, J.D., Aho, A.V., Hirschberg, D.S.: Bounds on the complexity of the longest common subsequence problem. J. ACM **23**(1), 1–12 (1976). https://doi.org/10.1145/321921.321922

49. Vaarandi, R.: Mining event logs with SLCT and LogHound. In: NOMS 2008–2008 IEEE Network Operations and Management Symposium, IEEE (2008). https://doi.org/10.1109/noms.2008.4575281

50. Vaarandi, R., Pihelgas, M.: LogCluster - a data clustering and pattern mining algorithm for event logs. In: 2015 11th International Conference on Network and Service Management (CNSM), IEEE (November 2015) https://doi.org/10.1109/cnsm.2015.7367331

51. Vapnik, V.: The Nature of Statistical Learning Theory. Springer, New York (1995). https://doi.org/10.1007/978-1-4757-3264-1

52. Xia, B., Bai, Y., Yin, J., Li, Y., Xu, J.: LogGAN: a log-level generative adversarial network for anomaly detection using permutation event modeling. Inf. Syst. Front. **23**(2), 285–298 (2020). https://doi.org/10.1007/s10796-020-10026-3

53. Xu, W., Huang, L., Fox, A., Patterson, D., Jordan, M.I.: Detecting large-scale system problems by mining console logs. In: Proceedings of the ACM SIGOPS 22nd symposium on Operating systems principles - SOSP 2009. ACM Press (2009). https://doi.org/10.1145/1629575.1629587

54. Yuan, D., Mai, H., Xiong, W., Tan, L., Zhou, Y., Pasupathy, S.: SherLog: error diagnosis by connecting clues from run-time logs. ACM SIGARCH Comput. Architect. News **38**(1), 143–154 (2010). https://doi.org/10.1145/1735970.1736038

55. Yuan, D., Zheng, J., Park, S., Zhou, Y., Savage, S.: Improving software diagnosability via log enhancement. In: Proceedings of the Sixteenth International Conference on Architectural Support for Programming Languages and Operating Systems - ASPLOS 2011. ACM Press (2011). https://doi.org/10.1145/1950365.1950369

56. Zhang, C., et al.: A deep neural network for unsupervised anomaly detection and diagnosis in multivariate time series data. ArXiv arXiv:1811.08055 (2019)

A Cloud-Edge Orchestration Platform for the Innovative Industrial Scenarios of the IoTwins Project

Alessandro Costantini[1(✉)], Doina Cristina Duma[1], Barbara Martelli[1],
Marica Antonacci[2], Matteo Galletti[1], Simone Rossi Tisbeni[1],
Paolo Bellavista[3], Giuseppe Di Modica[3], Daniel Nehls[4],
Jean-Christian Ahouangonou[5], Cedric Delamarre[5], and Daniele Cesini[1]

[1] INFN -CNAF, Bologna, Italy
`alessandro.costantini@cnaf.infn.it`
[2] INFN-Bari, Bari, Italy
[3] Università di Bologna, Bologna, Italy
[4] Fraunhofer FOKUS, TU, Berlin, Germany
[5] ESI GROUP, Rungis, France

Abstract. The concept of digital twins has growing more and more interest not only in the academic field but also among industrial environments thanks to the fact that the Internet of Things has enabled its cost-effective implementation. Digital twins (or digital models) refer to a virtual representation of a physical product or process that integrate data from various sources such as data APIs, historical data, embedded sensors and open data, giving to the manufacturers an unprecedented view into how their products are performing. The EU-funded IoTwins project plans to build testbeds for digital twins in order to run real-time computation as close to the data origin as possible (e.g., IoT Gateway or Edge nodes), and whilst batch-wise tasks such as Big Data analytics and Machine Learning model training are advised to run on the Cloud, where computing resources are abundant. In this paper, the basic concepts of the IoTwins project, its reference architecture, functionalities and components have been presented and discussed.

Keywords: Digital twins · Digital models · IoT · Edge · Cloud

1 Introduction

Many projects have been developed under the European Commission financial project Horizon 2020, to sustain and promote research in IOT, edge and cloud technologies in the scientific and industrial fields. The European Open Science Cloud Hub [1] contributes to the development of a Hub for European researchers and innovators to discover, access, use and reuse a broad spectrum of resources for advanced data-driven research. The above investments, which could be perceived as cost-prohibitive, are now becoming a business necessity for innovative

© Springer Nature Switzerland AG 2021
O. Gervasi et al. (Eds.): ICCSA 2021, LNCS 12950, pp. 533–543, 2021.
https://doi.org/10.1007/978-3-030-86960-1_37

companies, in both the manufacturing and facility management domains, with particular attention to the rich ecosystem of EU small and medium enterprises (SMEs).

This is also the case of the H2020 IoTwins project that, by designing a framework for a seamless, straightforward and loose integration of already developed and deployed industrial software components running in typical long-lived industrial test-beds, it aims to lower the barriers for building edge-enabled and cloud-assisted intelligent systems and services based on big data for the domains of manufacturing and facility management.

The IoTwins project propose a highly distributed and hybrid digital twin model, which provides for an integration of simulative and data-driven models to feed AI services. This will be tackled by harmonising standards to enable interoperability, and by developing an easy-to-use service layer that facilitates and decreases the cost of integration and deployment.

IoTwins leverage on the experience and the outcomes from past EU funded projects such as DEEP-Hybrid-DataCloud (DEEP-HDC) [2] aimed at bridging cloud and intensive computing resources to explore large datasets for artificial intelligence, deep, and machine learning, eXtreme-DataCloud (XDC) [3] aimed at developing scalable technologies for federating storage resources and managing data in highly distributed computing environments and INDIGO-DataCloud [4] aimed at developing a computing platform, deployable on multiple hardware, and provisioned over hybrid infrastructures.

In this paper the IoTwins project platform is presented and discussed in Sect. 2, where the proposed architecture and the platform capabilities derived form the use case driven analysis are described. In Sect. 3 the IoTwins software stack and the related IoT, Edge and Cloud service components are treated, including the interaction among them. Section 4 summarises the activity performed and draft the related conclusions.

2   IoTwins Project Platform

2.1   Proposed Architecture

IoTwins will provide during its activity the needed support to the users for building and running their services by deploying a platform providing a single point of access to heterogeneous computing, high capacity storage and interconnected resources.

The platform will deliver interfaces to data analytics and AI techniques, physical simulation, optimizations and virtual lab services. In addition, the IoTwins platform will apply well established and emerging standards to enable communication- and data-level interoperability to 12 different use cases coming from (but not limited to) the industrial IoT domain.

The proposed computing architecture is drafted in Fig. 1.

The figure shows the different computing layer of the platform and the different interactions among them, including the interfaces between IoT and Edge layers and between Edge and Cloud layers. In particular we have:

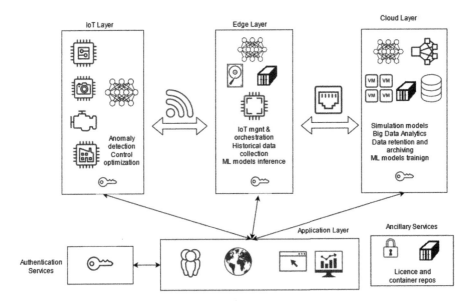

Fig. 1. IoTwins high level architecture

- Internet of Things (IoT) layer
- Interface between IoT and Edge layers
- Edge Layer
- Interface between Edge and Cloud layers
- Cloud Layer
- Application Layer

2.2 Use Case Requirements

The IoTwins User is the main stakeholder that will benefit from the services offered by the IoTwins platform. In the scope of the project, the IoTwins User role is played by Testbeds, who have contributed with their use cases to identify the requirements of the platform.

In Table 1, the different use cases are categorized according to the infrastructure where computing tasks are run. Furthermore, the platform will provide the User with services to build "data processing chains", i.e., a collection multiples services, distributed on at least two different infrastructures, chained together to serve the specific computing needs.

The "IoT-related use cases" refer to the action of configuring and running computing tasks on data sensed by sensors that the IoT device is equipped with. Depending on the need and the capabilities of the device, computing tasks may consume data on-the-fly or data at rest.

"Edge-related use cases" configure data processing allows the User to set-up an environment for data elaboration. Configuration involves actions like a) setting up data sources address, type, format, b) choosing and setting up the

Table 1. Mapping among IoTwins infrastructure, use case and their computational tasks

Reference infrastr.	Use case flow	Computational task
IoT-related	Configure data processing and run data processing	Data on-the-fly or data at rest
Edge-related	Configure data processing, Run bulk-data processing, Run data stream processing and Run ML Model	Data stream processing
Cloud-related	Configure data processing, Run bulk-data processing, Run data stream processing, Run ML training and run simulation	Data streams from either IoT or Edge or data-at-rest stored in the cloud

parameters of a specific computing task, c) selecting the destination of the data output, etc. Run bulk-data is the action of invoking tasks that elaborate data at rest, i.e., data locally stored, while Run data stream processing refers to the elaboration of live data coming from IoT devices. Finally, the Run ML model is a sample data stream processing use case.

In "Cloud-related use cases" instead, most of use cases available in the Edge are available in the Cloud too. The User can configure and run computing tasks on data streams coming from either IoT or Edge, or on data-at-rest stored in the Cloud. For instance, the User can request to run Simulations (Run Simulation use case) or to train ML models (Run ML training case) on bulk data residing in the Cloud. Trained ML models can then be moved to the Edge where, fed by data streams coming from IoT devices, will execute (see Edge's Run ML Model use case).

Out of the project scope, potential Users will be players of the manufacturing and facility/infrastructure management sectors that intend to benefit of the IoTwins platform services to implement a digital twin-based management of their business processes.

2.3 Platform Capabilities

The IoTwins platform is proposed as an open framework enabling Users to configure and run data processing tasks on IoT, Edge and Cloud respectively. It manages digital twins in the Cloud-to-Things continuum, built on top of open-source software and tools, allowing third-party software to integrate by way of open application programming interfaces (API).

The main principles that drive the design of the IoTwins platform are openness and software re-use respectively.

Starting from this principle, in fact, the platform will provide different services listed hereafter.

Data Handling. As one of the Digital Twins' cornerstones, Data Handling becomes extremely important in order to represent (or even forecast) the monitored asset. It is divided into three different consecutive steps: Data gathering, Data transportation and Data storage.

Data Gathering is performed by the sensing devices of the testbeds. The variety of sensors involved in the testbeds is quite wide. Thus, the platform has to be able to adapt itself, in order to enable the gathering of different data types and sources.

Data Transportation provides the protocols used to implement communication in the IoTwins architecture depend on the level (IoT, edge, cloud/backend). Currently, data transportation is performed by adopting standard (or de facto standard) technologies to enable all required data transportation and, at the same time, minimize the number of different technologies used by finding common particularities.

Data storage solutions (Data Models (D.M) and database) will be used either as a preliminary storage or as long terms storage for the testbeds.

Computation. Computation is requested to be done at all levels: IoT, Edge and Cloud.

At the IoT level few computation tools may be needed. The tools may be provided either by the data collection tool or as a service of the platform. Light and quick calculation may be done at the IoT level for some types of devices

At the Edge level operation like filtering, data aggregation or Big Data Analytics may be needed.

Cloud level will be mostly dedicated to off-line computations. Simulations and ML training will be performed at this level. Obviously, all the above listed computations (IoT & EDGE levels) may be performed here.

Anonimization. Anonymisation consists of replacing/encrypting/deleting elements directly or almost directly identifying individuals, generally does not protect individuals. Anonymisation of data is a compromise between data and individual protection on the one hand, and the possibility to process data in an interesting way on the other.

Due to the fact that the most of the testbed comes from industries, those data need to be kept under strict confidentiality ruled. So a process of anonymization that prevents from whatsoever association between data, their elaboration and inferenced conclusions is needed before sending the data out.

Encryption. Encryption converts a plain text into a ciphertext using a secret key. Encryption may be used as an alternative to anonymization for those testbeds that operates on sensitive data. Encrypted channels for data transfers from the IoT level to the cloud is, in general, requested by all testbeds. Obviously, further investigation is needed in order to clarify if data should be encrypted also during access and whether it should be encrypted in the Edge before being uploaded to the Cloud.

Moreover, two obstacles have to be tackled to applying encryption:

1. The computational cost of the encryption algorithms leading to a performance slow down and a higher energy consumption
2. the difficulty of processing over encrypted data.

If efficient solutions exist to deal with the first obstacle, enabling efficient processing over encrypted data is one the biggest challenge in the security field that the project will try to address.

3 IoTwins Orchestration Platform

3.1 Authentication

The authorization and authentication infrastructure service (AAI) in IoTwins is provided among the above mentioned stack layers by the INDIGO Identity and Access Management Service (INDIGO-IAM). INDIGO-IAM is a service developed by the INDIGO-DataCloud EC project and maintained for the foreseeable future by the INFN partner.

INDIGO-IAM is provided through multiple methods (SAML [5], OpenID Connect [6] and X.509 [7]) by leveraging on the credentials provided by the existing Identity Federations (i.e. IDEM [8], eduGAIN [9], etc.). The support to Distributed Authorization Policies and Token Translation Service will guarantee selected access to the resources as well as data protection and privacy.

INDIGO-IAM is based on the OpenIDConnect and OAuth2.0 protocols and supports two types of users:

- Normal users: these users can login with the IAM, register client applications that use the IAM for authentication, link external accounts to their account (when allowed by the configuration),
- Administrators: these are users that have also administration privileges for an IAM organization.

The service exposes the OpenID Connect/OAuth dynamic client registration functionality offered by the MitreID OpenID Connect server libraries. In OAuth terminology, a client is an application or service that can interact with an authorisation server for authentication/authorization purposes and a new client can be registered in the IAM in two ways: (1) using the dynamic client registration API (2) via the IAM dashboard (which simply acts as a client to the API mentioned above). After registration the client can obtain a token using APIs or already available scripts and GUIs.

The INDIGO-IAM service has been successfully integrated with many off-the-shelf components like Openstack, Kubernetes, Atlassian JIRA and Confluence, Grafana and several middleware services that can be used in the IoTwins architecture.

3.2   Cloud Service Components

The Cloud Level Software provides services aimed to orchestrate the full stacks (IoT, Edge and Cloud) and controls which software is running on which hardware.

To properly address the requests coming from the use cases and, at the same time, to be flexible enough to interact with the most popular choices of the computer centers, without interfering in the operation of their facilities, the virtualization of resources becames a key word.

Taking care of such needs, set of PaaS core components that have been deployed (within the activities of the INDIGO-DC project [4] and subsequent evolutions [2,3]) as a suite of small services using the concept of microservice. Referred to a software architecture style, this term identify complex applications composed of small independent processes communicating with each other via lightweight mechanisms like HTTP resource APIs. The modularity of microservices makes the approach highly desirable for architectural design of complex systems, where many developers are involved.

The base components and functionalities of the INDIGO PaaS are briefly described and their interrelations are shown in Fig. 2

- **PaaS Orchestrator** [10] is the core component of the PaaS layer. The requests related to application or service deployment coming from the user are expressed using TOSCA [11], the OASIS [12] standard to specify the topology of services provisioned in IT infrastructures. Such requests are grouped and organized in the TOSCA template that is processed by the INDIGO Orchestrator. The Orchestrator is able to implement a complex workflow aimed at fulfilling a user request using information about the health status and capabilities of underlying IaaS and their resource availability, QoS/SLA constraints, the status of the data files and storage resources needed by the service/application. This process allows to achieve the best allocation of the resources among multiple IaaS sites.
- **Managed Service/Application (MSA) Deployment Service** [13] is in charge of scheduling, spawning, executing and monitoring applications and services on top of one or more Mesos clusters.
- **Infrastructure Manager (IM)** [14] is in charge to deploy complex and customized virtual infrastructures on different IaaS Cloud deployment, providing an abstraction layer to define and provision resources in different clouds and virtualization platforms. IM enables computing resource orchestration using OASIS standard languages (i.e. TOSCA protocol). Moreover, it eases the access and the usability of IaaS clouds by automating the VMI (Virtual Machine Image) selection, deployment, configuration, software installation, monitoring and update of the virtual infrastructure,
- **Data Management Services** is a collection of services that provide an abstraction layer for accessing data storage in a unified and federated way. These services provide a unified view of the storage resources together with the capabilities of importing data, schedule transfers of data from different

sources. The support of high-level storage requirements, such as flexible allocation of disk or tape storage space and support for data life cycle is achieved by the adoption of the INDIGO CDMI Server [15]. The CDMI server has been extended in the INDIGO project to support Quality-of-Service (QoS) and Data Life-cycle (DLC) operations for multiple storage back-ends like dCache [16], Ceph [17], IBM Spectrum Scale [18], TSM [19], StoRM [20] and HPSS [21],

– **Monitoring Service** [22] is in charge of collecting monitoring data from the targeted clouds, analysing and transforming them into information to be consumed by the Orchestrator,

– **CloudProviderRanker** [23] is a rule-based engine that allows to manage the ranking among the resources that are available to fulfil the requested services. The Service uses the list of IaaS instances and their properties provided by the Orchestrator in order to choose the best site that could support the user requirements,

– **Cloud Information Provider** [24] generates a representation of a site's cloud resources, to be published inside INDIGO Configuration Management DataBase (CMDB),

– **Configuration Management DataBase** [25] is a REST service storing all the information about the cloud sites available and providing details such as the images and containers they support. It is used as an authoritative source of information for matchmaking and orchestration of VM and containers.

– **QoS/SLA Management Service** [26] it allows the handshake between a user and a site on a given Service Level Agreement (SLA) and it describes the Quality of Service (QoS) for a group or specific user both in the PaaS as a whole or over a given site.

3.3   Edge Service Components

The Edge Level Software provides services aimed to provide communication with both the IoT level and the Cloud level by providing methods to manage the provided functionality (apps) dynamically. In the IoTwins architecture, the applications are packed into OCI containers. The Edge Service Orchestration thus has to provide functionality to download, start, stop and configure OCI [27] containers.

In such respect, the IoTwins platform adopted the Apache Mesos [28] to host the Edge Service Orchestration. Apache Mesos is an open-source project to manage computer clusters which has two framework: Marathon[29] and Chronos[30]. Marathon is a production-grade container orchestration platform for Mesosphere's Datacenter Operating System (DC/OS) and Apache Mesos. Marathon is a framework (or meta framework) that can launch applications and other frameworks and can also serve as a container orchestration platform which can provide scaling and self-healing for containerized workloads. Chronos, insted, is a fault-tolerant scheduler that runs on top of Apache Mesos that is used for job orchestration.

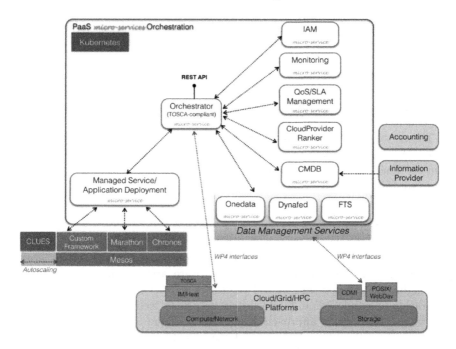

Fig. 2. Graphic schema showing the interrelations among INDIGO PaaS components.

Mesos CLI or UI can be used. Docker containers can be launched using JSON definitions that specify the repository, resources, number of instances, and command to execute. Scaling-up can be done by using the Marathon UI, and the Marathon scheduler will distribute these containers on slave nodes based on specified criteria. Autoscaling is supported. Multi-tiered applications can be deployed using application groups. Containers can be scheduled without constraints on node placement, or each container on a unique node (the number of slave nodes should be at least equal to the number of containers). High availability for Mesos and Marathon is supported using Zookeeper. Zookeeper provides election of Mesos and Marathon leaders and maintains cluster state. Host ports can be mapped to multiple container ports, serving as a front-end for other applications or end users. Marathon continuously monitors the number of instances of a Docker container that are running. If one of the containers fail, Marathon reschedules it on other slave nodes. Auto-scaling using resource metrics is available through community-supported components only. Local persistent volumes (beta) are supported for stateful applications such as MySQL. When needed, tasks can be restarted on the same node using the same volume. The use of external storage, such as Amazon EBS, is also in beta. At the present time, applications that use external volumes can only be scaled to a single instance because a volume can only attach to a single task at a time.

Starting from the available Mesos frameworks: Marathon, which allows to deploy and manage LRS, and Chronos, which allows to execute jobs, the INDIGO project has added new interesting functionalities in Mesos that make them suitable for explitation: the elasticity of a Mesos cluster so that it can automatically shrink or expand depending on the tasks queue; the automatic scaling of the user services running on top of the Mesos cluster and a strong authentication mechanism based on OpenID-Connect.

Thanks to the integration between the Cloud orchestration layer and the Edge orchestration layer provided by the INDIGO-DC components, is it possible, from the PaaS Orchestrator to deploy and made available services on the Edge layer (see Fig. 2).

4   Conclusion

This paper reports on the effort made as an outcome of the IoTwins project to define and deploy a computing and software architecture to be proposed as a highly distributed and hybrid digital twin model, which provides for an integration of simulative and data-driven models to feed AI services.

The use case driven approach permitted to taking into account the heterogeneity of the different use cases and at the same time to defines standard based interfaces for functionality blocks to allow for further development and integration of 3rd party tools.

Furthermore, following a proven software development approach, the requirement collection and analysis phase was followed by applying techniques aimed to define a general high-level platform that has been presented and described.

The present work lays the foundation to build a prototype platform which then can be deployed and tested with the project's use case testbeds. Experiences and discussions resulting from these prototyping will directly influence further development iteratively.

Acknowledgements. This work was partially supported by EU H2020 IoTwins Innovation Action project (g.a. 857191).

References

1. European Open Science Cloud Hub, web site, last view April 2021. https://www.eosc-hub.eu/
2. DEEP-Hybrid-DataCloud, web site, last view April 2021. https://deep-hybrid-datacloud.eu/
3. eXtreme-DataCloud, web site, last view April 2021. http://www.extreme-datacloud.eu/
4. INDIGO-DataCloud, web site, last view April 2021. https://repo.indigo-datacloud.eu/
5. Security Assertion Markup Language, web site, last view April 2021. https://it.wikipedia.org/wiki/Security_Assertion_Markup_Language
6. OpenID Connect, web site, last view April 2021. https://openid.net/connect/

7. X.509, web site, last view April 2021. https://en.wikipedia.org/wiki/X.509
8. IDEM Federation, web site, last view April 2021. https://idem.garr.it/en/federazione-idem-en/idem-federation
9. eduGAIN interfederation service, web site, last view April 2021. https://edugain.org/
10. INDIGO Orchestrator, web site, last view April 2021. https://github.com/indigo-dc/orchestrator
11. TOSCA, web site, last view April 2021. http://docs.oasis-open.org/tosca/TOSCA/v1.0/os/TOSCA-v1.0-os.html
12. OASIS, web site, last view April 2021. https://en.wikipedia.org/wiki/OASIS_(organization)
13. Managed Service/Application, web site, last view April 2021. https://github.com/indigo-dc/mesos-cluster
14. Infrastructure Manager, web site, last view April 2021. https://indigo-dc.gitbook.io/im/
15. INDIGO Cloud Data Management Interface, web site, last view April 2021. https://www.snia.org/cdmi
16. dCache, web site, last view April 2021. https://www.dcache.org
17. CEPH, web site, last view April 2021. https://ceph.io
18. IBM Spectrum Scale, web site, last view April 2021. https://www.ibm.com/docs/en/spectrum-scale/5.0.4?topic=overview-spectrum-scale
19. IBM Spectrum Protect, web site, last view April 2021. http://www-03.ibm.com/software/products/it/spectrum-protect
20. Storage Resource Manager, web site, last view April 2021. http://italiangrid.github.io/storm/documentation/functional-description/1.11.2/
21. High Performance Storage Systems, web site, last view April 2021. https://www.hpss-collaboration.org
22. High Performance Storage Systems, web site, last view April 2021. https://github.com/indigo-dc/Monitoring
23. CloudProviderRanker, web site, last view April 2021. https://github.com/indigo-dc/CloudProviderRanker
24. Cloud Info Provider, web site, last view April 2021. https://github.com/EGI-Federation/cloud-info-provider
25. Configuration Management DataBase, web site, last view April 2021. https://github.com/indigo-dc/cmdb
26. QoS/SLA Management Service, web site, last view April 2021. https://github.com/indigo-dc/slam
27. Open Container Initiative, web site, last view April 2021. https://opencontainers.org/
28. Mesos, web site, last view April 2021. http://mesos.apache.org/
29. Apache Marathon framework, web site, last view April 2021. https://mesosphere.github.io/marathon/
30. Apache Chronos framework, web site, last view April 2021. https://mesos.github.io/chronos/

Unsupervised Learning Model to Uncover
Hidden Knowledge from COVID-19 Vaccines Literature

Tasnim Gharaibeh[✉] and Elise de Doncker

Western Michigan University, Kalamazoo, MI 49008, USA
{tasnim.gharaibeh,elise.dedoncker}@wmich.edu

Abstract. Severe acute respiratory syndrome coronavirus 2 (or SARS-CoV-2) has spread globally, causing a pandemic with, so far, more than 152 million infections and more than three million deaths (as of May 2021). In order to address the COVID-19 pandemic by limiting transmission, an intense global effort is in the development of a safe and effective vaccine, which generally requires several years of pre-clinical and clinical stages of evaluation as well as strict regulatory approvals. However, because of the unprecedented impact of COVID-19 worldwide, the development and testing of a new vaccine are being accelerated. There are currently some authorized, not yet approved, vaccines to fight COVID-19, besides other ones in clinical evaluation or in a pre-clinical stage, and many more being researched. In this work, we used natural language processing and a machine learning model to predict good candidate vaccines. We built an unsupervised deep learning model (CVW2V) to produce word-embeddings using Word2vec from a corpus of published articles, selectively focusing on COVID-19 candidate vaccines that appeared in the literature, to identify promising target vaccines according to their similarity with approved and authorized vaccines.

Keywords: Unsupervised learning · Word embeddings · Word2vec · COVID-19 vaccines

1 Introduction

Machine learning approaches are usually divided into three categories, supervised, unsupervised, and reinforcement learning. Supervised learning deals with labeled data, data that contains both the inputs and the desired outputs, to train the model. It uses linear regression, neural networks, nearest neighbor, or naive Bayes algorithms to predict or classify data. Reinforcement learning allows machines or software agents to automatically decide the required behavior within a specific situation to maximize its performance. Using Q-Learning, the temporal difference (TD), or deep adversarial networks, reinforcement learning algorithms can be applied in robotic hands, computer played board games, and self-driving cars. Unsupervised learning allows the model to work on its own to find structure and discover patterns and information that was previously undetected; mine for rules; and summarize and group the data points, without the need to supervise or lead the model and without any previous insertion knowledge. It mainly deals with the unlabeled data as input [1]. Clustering and association are two types

© Springer Nature Switzerland AG 2021
O. Gervasi et al. (Eds.): ICCSA 2021, LNCS 12950, pp. 544–559, 2021.
https://doi.org/10.1007/978-3-030-86960-1_38

of unsupervised learning. The association rules establish relations among data objects inside large databases, whereas clustering splits the dataset into groups based on their similarities.

Word embedding is one of the most important and versatile examples of unsupervised learning models, leading to a type of language modeling or word representation that allows words with similar meanings to be understood by machine learning algorithms [2]. As a representation of words obtained from an unlabeled large corpus, by embedding both semantic and syntactic meanings, word embedding can be represented as:

$$V \rightarrow R^D : w \rightarrow \vec{w}$$

mapping a word w from a vocabulary V to a real-valued vector $\vec{w}$ in an embedding space with dimensionality D [33]. To compute the similarities between vectors in the embedding space, the cosine similarity function can be used, which is defined as

$$similarity(w1, w2) = \frac{\vec{w}1 \cdot \vec{w}2}{\|\vec{w}1\| \, \|\vec{w}2\|}$$

According to this function, the nearest neighbors of a word w are given as a list of words $v \in V \backslash \{w\}$, sorted in descending order [33].

This is an important method commonly used in tasks of modern natural language processing (NLP), such as semantic analysis [2], retrieval of information [3], dependency parsing [4–6], query answering [7, 8], and computer translation [7, 9, 10]. In these techniques, individual words are represented as real-valued vectors, often of tens or hundreds of dimensions, in a predefined vector space. Words that are closer in the vector space are expected to be similar in meaning. Each word is mapped to one vector. The vector values are learned in an approach that resembles a neural network, and hence the technique is often lumped into the field of deep learning. Word embedding methods vary and can be grouped into different categories. For example, they can be classified as either paradigmatic models or syntagmatic models, based on the word distribution information [11, 12]. The most important aspect of the syntagmatic model is that words co-occur in the text region, but for the paradigmatic model, it is the similar context that matters. Another grouping depending on how word embedding is generated divides models into two classes, matrix factorization and sliding-window sampling. The first method is based on the word co-occurrence matrix, where word embeddings are obtained from matrix decomposition. In sliding-window sampling, data sampled from sliding windows is used to predict the context word [13]. Word embedding has many important applications. It can be applied in idiomaticity analysis, sentiment analysis, syntax analysis, part of speech (POS) tagging, named entity recognition, as well as textual entailment [14]. Various word embedding models are available, such as Word2vec, Glove, and fastest.

The Word2vec method is based on a shallow neural network with two layers that takes a large corpus of text as its input and produces a vector space, typically of several hundred dimensions, with remarkable linear relationships called analogies. These analogies allow for math operations such as vec("king") - vec("man") + vec("woman") $\approx$ vec("queen") [20]. As an unsupervised learning technique, and a machine learning algorithm used to draw inferences from datasets consisting of input data without labeled responses, it comprises two techniques: CBOW (continuous bag of words) and Skip-gram. CBOW predicts the probability of a word given a context, while Skip-gram predicts the context given a word.

These models in machine learning have great benefits and a significant impact in many areas of science and technology, including medical research, because the models created by machine learning can learn and generalize the patterns and similarities within the available huge data and can draw conclusions from previously unseen data [15]. On the other hand, the increase in knowledge and understanding of diseases, vaccines, and drugs has been associated with a growth in information. Such big data is ready to deliver the benefits of the application of machine learning (ML) in diagnosis, prognosis, drug development, and vaccine discovery.

Vaccine discovery methods have been costly, and it may take many years to develop an appropriate vaccine against a specified pathogen. In the COVID-19 case, vaccines are essential in reducing morbidity and mortality especially since the virus establishes itself in the population. Thus, the development of a new vaccine should be accelerated to save lives. When most people in a community are vaccinated against a disease, the ability of the pathogen to spread is limited. This also provides indirect protection to people who cannot be vaccinated. Machine learning can accelerate the discovery of effective vaccines, or suggest components of a vaccine after understanding the viral protein structure, and assist medical researchers scouring tens of thousands of related scientific papers [16, 17].

In this research we employ word embedding, as an unsupervised machine language and NLP model, to predict potential vaccines according to their similarity with the authorized vaccines, without any prior explicit insertion knowledge about vaccines, treatment, or biomedical information about COVID-19.

2 Background and Previous Work

2.1 Vaccine Development

A pathogen is a bacterium, virus, parasite, or fungus that can cause disease within the body. Each pathogen has its unique subparts; one important factor is the antigen; which causes the formation of antibodies. The antibodies produced in response to the pathogen's antigen are an important part of the immune system. Usually, vaccines contain weakened or inactive parts of a particular antigen that train the immune system to recognize and fight pathogens, which greatly reduces the risk of infection. The immune system can safely recognize the pathogen when it is encountered naturally; by delivering an immunogen, a specific type of antigen elicits an immune response [15].

To fight virus-infected cells, the host immune system produces antibodies on the surface of B-cells or attacks the virus directly using T-cells. To assist T- and B-cells in attacking and binding with invaders, the human leukocyte antigen (HLA) gene encodes MCH-I and MCH-II proteins, which present epitopes as antigenic determinants [18].

To ensure the safety and efficiency of the new vaccine, many actions and steps are taken, such as pre-clinical studies, phase I clinical trial, phase II clinical trial, phase III clinical trial, phase IV post-marketing surveillance, and human challenge studies. In the pre-clinical studies, the vaccine is tested first in animal studies for efficacy and safety.

Then in phase I clinical trial, small groups of healthy adult volunteers receive the vaccine to test for safety. After that in phase II clinical trial, people who have characteristics for whom the new vaccine is intended are given the vaccine. The next step of testing the vaccine is by giving it to thousands of people and that is in phase III clinical trial. After the vaccine is approved and licensed, many studies are conducted to study future effects of the vaccine in the population in the long term [19].

Since the outbreak of this novel coronavirus, different machine learning approaches have been used to predict potential effective vaccines.

2.2 Machine Learning and the Development of COVID-19 Vaccines

To accelerate the vaccine design process by predicting the design of a multi-epitope vaccine for COVID-19, Zikun et al. implemented a supervised deep learning model (DeepVacPred) in silico. From the available SARS-CoV-2 spike protein sequence, their model was able to predict 26 candidate vaccine subunits, by combining the in silico immunoinformatic and deep neural network strategies. In order to construct a multi-epitope vaccine for the SARS-CoV-2 virus, they identified the top 11 from the 26. According to the best 11 candidates, they suggested a 694aa (consisting of 694 amino acid residues) multi-epitope vaccine that contains 16 B-cell epitopes, 82 CTL epitopes, and 89 HTL epitopes, as a promising vaccine to fight the SARS-CoV-2 viral infection. They also concluded that the designed vaccine can be used successfully with the recent RNA mutations of the virus by investigating the SARS-CoV-2 mutations for RNA [24].

Since the T-cell and B-cell epitopes could be used to produce good vaccines as well as recognize neutralizing antibodies [28], Fast et al. tried to identify 2019-nCoV T-cell and B-cell epitopes based on viral protein antigen presentation and antibody binding properties, using computational tools from structural biology and two supervised neural networks (MARIA, a multimodal recurrent neural network [29] and NetMHCpan4, a trained neural network in [30]). Their results show good antigen presentation scores for 405 potential viral peptides, for both human MHC-I and MHC-II alleles. Their models also identified two potential neutralizing B-cell epitopes close to the 2019-nCoV spike protein receptor-binding domain (440–460 and 494–506). They also concluded that the spike protein for the SARS-CoV-2 virus could be a potential vaccine candidate. This conclusion was the result of their analysis of mutation profiles of 68 viral genomes spread through four continents that showed no mutations are present near the spike protein receptor-binding domain and d 96 coding-change mutations occurred only in regions with good MHC-I presentation scores [28].

By using the same method, the recurrent neural networks (RNN), and a training set consisting of alpha, beta, gamma, and delta coronavirus spike sequences, Crossman simulated the spike protein sequences of coronaviruses for the previously known sequences and tested their characteristics. This simulation could help in the development of vaccine design to predict alternative possible spike sequences that could arise in the future [31].

In another study that introduces a computational framework from several computational approaches such as reverse vaccinology and immunoinformatics tools with deep learning, Abbasi et al. predicted a potentially suitable protein vaccine candidate against SARS-CoV-2. Reverse vaccinology uses the expressed genomic sequences to find new potential vaccines, whereas immunoinformatics tools such as PSORTb, FungalRV, SignalP, TargetP, IEDB, BLASTp, ProtParam, Vaxijen, etc. were modified to classify proteins into positive and negative datasets. As a result, one protein (Spike S- Surface Glycoprotein with accession no. QHD43416.1) was shortlisted as a potential vaccine candidate out of the 10 proteins of SARS-CoV-2. Out of 41 B-cell epitope sequences that were predicted using BcePred, only 15 peptides were identified to be highly antigenic as specified by the Vaxijen server, where a peptide "DLCFTNVY" is predicted as the highest-ranking peptide. For T-cell epitopes, they consider peptides with a 100% conservancy rate, so 45 epitopes were identified by the ProPred-I tool to be conserved out of the 46 predicted MHC class-I binding epitopes, where KIADYNYKL was found with the highest antigenicity score and 90 out of 94 MHC- II allele binding T-cell epitopes predicted by the ProPred tool with VKNKCVNFN as the highest antigenicity score [32].

3 Methods

3.1 Data Collection and Processing

From Science Direct at Elsevier, we obtained around 54,427 full-text published papers and articles primarily focusing on COVID-19 in the years 2019, 2020, and 2021. In order to perform the text mining process, we used Science Direct application programming interfaces (APIs) (https://dev.elsevier.com/), interactive APIs, and some Python (3.6) code [21]. We searched the Science Direct database using specific search terms as a first step in cleaning the data. These terms combined "COVID-19 OR 2019-nCoV OR COVID19 OR nCoV-19 OR Sars-CoV-2" with vaccines, biomedical, medication, antibodies, immunity, immunology, and drugs. Extensive cleaning processes were needed to reduce the noise and unwanted words in the corpus, speed up model training, and improve model efficiency. We deleted words with a length of more than 20 letters; usually, these words are URLs or descriptions related to the metadata. Then we converted all uppercase characters in the corpus to lowercase, so Vaccine and vaccine will have one vector. Finally, we deleted the stop words and punctuation from the corpus. We started with a data file of size 2.74 GB and obtained a reduced file of size 2 GB.

3.2 Word2vec Training

To train our model (CVW2V), a combination of the Word2vec implementation in genism [22] and the open-source code implemented by Vahe Tshitoyan et al. [23] were used with some modifications. Several parameters and tuners that affect both training speed and efficiency are recognized by Word2vec such as word count, vector size, workers, negative sampling, the model type (CBOW or skip-gram), learning rate, and window size. We have more than 298 million different words in the data, but we consider only 1,679,911 words. These are words that occurred more than five times because words that appear in a billion-word corpus only once or twice or less than 5 times are probably uninteresting typos and garbage. Furthermore, there is not enough information to include any practical training on those terms, so it is better to ignore them. The vector size, size of the embedding, in CVW2V is 200. Higher dimensionality increases the quality of word embedding, but after reaching some point, the benefit would decrease [20]. Larger size values need more data from training but can lead to better (more precise) models. Usually, fair vector size values are in the tens to hundreds.

For the model type, CVW2V uses a skip-gram model for the training with negative sampling loss. With negative sampling, we randomly select just a small number of "negative" words (let's say n = 15) for which to update the weights. In this context, a "negative" word is one for which the neural network should output a 0. Negative sampling speeds up the training by decreasing the computational burden and also increases the efficiency of the resulting word vectors by having each training sample only modify a small percentage of the weights, rather than all of them [29]. According to Tshitoyan's result, it was found that skip-gram with negative sampling loss (n = 15) performed best [23]. With respect to Google code [25], the author indicated that CBOW is faster, while skip-gram is slower, but it does a better job for infrequent words because Skip-Gram seeks the prediction of the context given a word instead of the prediction of a word given its context like CBOW.

The window size parameter in CVW2V is 8, which relates to the maximum distance within a sentence between the actual and the predicted word, or how many words before or after a given word would be included as context words of the given word. The initial learning rate was 0.01 and it would drop linearly to 0.0001 as training progresses in 30 epochs. The typical range for the learning rate is between 1 and 10^{-6}. The learning rate aims to decide what proportion of the observed error should be used to change weights. The model will oscillate by setting the learning rate too high and will converge too slowly to a solution by setting it too low. A learning rate of 0.01 usually works with most networks [26]. To optimize the computation, we used a hierarchical SoftMax training algorithm, which uses a binary tree to represent all words in the vocabulary [27].

4   Results and Discussion

4.1   Verification

To verify CVW2V results, some general known words and phrases related to COVID-19 vocabulary such as SARS, vaccines, wearing a mask, dry cough were chosen as seeds for the model to test the relatedness between these words and their similarities depending on human judgment as shown in Table 1. For the first word, SARS, the top similar words are sarscov, sarscov1, sars_coronavirus, middle_east_respiratory_syndrome, sarscov2, severe_acute_respiratory_syndrome, mers_sars, corona_virus, infection, and pandemic. The first eight similarities are similar names for COVID-19, whereas infection and pandemic are more general references related to COVID-19. Similar words for vaccines are vaccine_candidates, inactivated_vaccines, mrna_vaccines, immunization, live_attenuated, viral_vector, proteinbased_vaccines, subunit_vaccines, rabies_vaccine, and hpv_vaccine. Most of these are types of vaccines. Wearing a mask, as a protective means for COVID-19, has washing_hands, avoid_touching_face, wearing_gloves, frequent_handwashing, maintaining_social_distance, staying_home, face_covering, personal_hygiene, avoiding_crowds, and cough_etiquette as the top 10 similarities and it is obvious that all of these are also protective measures against COVID-19. As a symptom of COVID-19 we chose dry cough, resulting in the top 10 similarities of short-ness_breath, fever_cough, dyspnoea, chills, runny_nose, chest_tightness, sore_throat, difficulty_breathing, myalgias, and diarrhea_abdominal_pain. Thus the similarities found are also symptoms of COVID-19.

4.2 Similarities for Some Vaccines

Vaccines are being developed with different technologies, some well-known and others completely new for human vaccines, such as peptide and nucleic acid technologies.

Whole Virus Vaccine. These are live-attenuated and inactivated vaccines. Live-attenuated vaccines use a weakened (or attenuated) pathogen that causes disease and that creates a strong and long-lasting immune response. Because these vaccines are so close to the real natural infection, only one or a maximum of two doses are needed to give lifetime immunity against the pathogen. They are used to protect against rotavirus, yellow fever, and chickenpox [34]. The inactivated vaccines use the dead version of the pathogen. Therefore, these vaccines do not provide immunity protection as strong as activated vaccines, so several doses are needed over time to get lasting immunity. They are used to immunize against hepatitis a, flu, polio, and rabies [34]. Sinopharm and CoronaVac Are examples of the inactivated virus vaccines for COVID-19.

Sinopharm. Also known as BBIBP-CorV and developed by Sinopharm, the National Pharmaceutical Group Corporation (CNPGC) in China, with 86% efficacy against the COVID-19 infection [36]. As the most similar words for Sinopharm, CVW2V found sinovac, coronavac, inactivated_vaccine_candidate, ad5ncov, sputnik_v, cov-axin, codagenix, adenovirusvectored, adults_aged_18–59_years, and wibp_vaccine (see Fig. 1).

Sinovac is a biopharmaceutical company in China that focuses on the research, devel-opment, manufacturing, and commercialization of vaccines. This company developed the CoronaVac inactivated virus COVID-19 vaccine. Ad5-nCoV, or Convidicea, was the first novel coronavirus vaccine for COVID-19 in China. It is a viral vector-based vaccine. Both of these vaccines are under emergency use authorization status in China, Mexico, Pakistan, Hungary, and Chile [37]. Sputnik-V (or Gam-COVID-Vac) is a viral vector vaccine for COVID-19 developed by the Gamaleya Research Institute of Epidemiology and Microbiology in Russia. Registered on 11 August 2020 by the Russian Ministry of Health, followed by an interim analysis from the trial was published on February 2021, indicating 91.6% efficacy. Over a billion doses of that vaccine were requested by Russia, Argentina, Belarus, Hungary, Serbia and the United Arab Emirates for immediate dis-tribution globally [38]. Covaxin is another inactivated virus-based COVID-19 vaccine being developed in India, by Bharat Biotech in collaboration with the Indian Council of Medical Research. After the second dose, it showed 81% intermediate effectiveness in preventing COVID-19 in those that had not previously been infected [39]. Codagenix is a clinical-stage synthetic biology company that uses programs to recode the genomes of viruses in order to create live-attenuated vaccines or viruses that can be used to pre-vent viral infections. It generated COVI-VAC, a single-dose, intranasal, live-attenuated

Table 1. The top 10 first similarities for SARS, vaccines, wearing a mask, and dry cough.

SARS	Vaccines	Wearing mask	Dry cough
sarscov	vaccine_candidates	washing_hands	shortness_breath
sarscov1	inactivated_vaccines	avoid_touching_face	fever_cough
sars_coronavirus	mrna_vaccines	wearing_gloves	Dyspnoea
middle_east_respiratory_syndrome	immunization	frequent_handwashing	Chills
sarscov2	live_attenuated	maintaining_social_distance	runny_nose
severe_acute_respiratory_syndrome	viral_vector	staying_home	chest_tightness
mers_sars	Proteinbased_vaccines	face_covering	sore_throat
corona_virus	subunit_vaccines	personal_hygiene	difficulty_breathing
infection	rabies_vaccine	avoiding_crowds	Myalgias
pandemic	hpv_vaccine	cough_etiquette	diarrhea_abdominal_pain

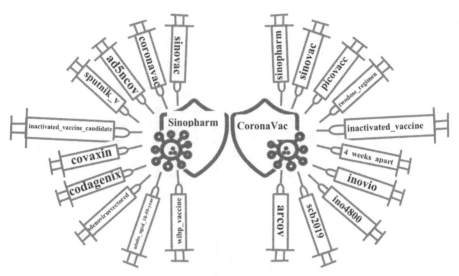

Fig. 1. The similar words for the inactivated virus vaccines for COVID-19, Sinopharm and CoronaVac

vaccine against COVID-19. In Phase I, COVI-VAC showed good results in safety and protection after just one dose in the gold standard animal model and now it is being evaluated in the clinical phase. It is designed to produce immunity against all SARS-CoV-2 proteins to protect against a range of SARS-CoV-2 strains [41]. Adenovirus vectored is a vector based vaccine, that adds a gene for the coronavirus vaccine into a modified version of a chimpanzee adenovirus that can enter human cells but not replicate inside, so the vaccine can target the spike proteins that SARS-CoV-2 uses to enter human cells [42]. The wibp_vaccine is one of two inactivated virus COVID-19 vaccines developed by Sinopharm [36].

CoronaVac. Is an inactivated virus COVID-19 vaccine developed by the Chinese company Sinovac Biotech. It has been in Phase III clinical trials in several countries. It does not need to be frozen, and can be transported and refrigerated at 2–8 °C (36–46 °F), as traditional flu vaccines are kept [37]. When it was used as a seed in our model, the top similar words Were sinopharm, inactivated_vaccine, sinovac, picovacc, twodose_regimen, 4_weeks_apart, inovio, ino4800, scb2019, and arcov (see Fig. 1).

Picovacc is the previous name for CoronaVac. twodose_regimen and 4_weeks_apart according to the treatment plan that specifies the amount and the schedule for the vaccination. Inovio, or ino4800, is a DNA vaccine INO-4700 against MERS CoV that is currently in preparation to initiate a Phase 2 vaccine trial [43]. The scb2019 subunit vaccine candidate is developed by Clover Pharmaceuticals [45]. Arcov is the first mRNA vaccine in China and may be ready in May for final stage trials overseas [44].

RNA or mRNA Vaccine. MRNA vaccine development uses a new technology, based on proteins to trigger an immune response. They have several benefits compared to other types of vaccines, including shorter manufacturing times and, because they do not contain a live virus, no risk of causing disease in the person getting vaccinated. Moderna and pfizer-BioNTech are examples of mRNA vaccines for COVID-19. Under an Emergency Use Authorization (EUA) in December 2019, the FDA has authorized their emergency use to prevent or reduce the COVID-19 infection in individuals of age 16 years and older. We used these two vaccines as seeds to our model to search for similar candidate vaccines.

Moderna. The top similarities for Moderna are pfizer_biontech, mrna1273, Mrnabased_vaccine, encapsulated_mrna, Johnson_johnson, Astrazeneca, Curevac, Arcturus, Gsk, and Gx19 (see Fig. 2).

The mrna1273 similarity is another name for Moderna. CureVac COVID-19 is an mRNA COVID-19 vaccine candidate developed by CureVac N.V. and the Coalition for Epidemic Preparedness Innovations (CEPI). It is different from the current mRNA COVID-19 vaccines, Pfizer–BioNTech and Moderna. CureVac uses unmodified RNA, while the other two use nucleoside-modified RNA. It is currently in Phase III clinical trials, as of April 2021 [46]. Arcturus, or ARCT-02,1 is an mRNA COVID-19 vaccine candidate developed by Arcturus Therapeutics and currently in phase II [47]. Gsk, or VAT00002, is a protein subunit COVID-19 vaccine candidate in phase II and developed by Sanofi Pasteur and GSK [48]. The gx19 DNA COVID-19 vaccine candidate is developed by Genexine and in phase II status [49].

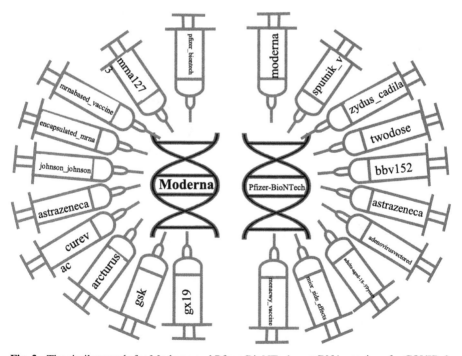

Fig. 2. The similar words for Moderna and Pfizer-BioNTech as mRNA vaccines for COVID-9.

Pfizer-BioNTech. The similar words found for pfizer-bioNTech are astrazeneca, moderna, sputnik_v, zydus_cadila, twodose, bbv152, adenovirusvectored, adults_aged_18–59_years, minor_side_effects, and menacwy_vaccine (see Fig. 2).

zydus_cadila is an Indian multinational pharmaceutical company that develops a DNA plasmid-based COVID-19 vaccine named ZyCoV-D, which is expected to get emergency use authorization in May or June [50]. BBV152, or Covaxin, is an inactivated virus-based COVID-19 vaccine. It was developed by the Indian Council of Medical Research and Bharat Biotech [51]. The MenACWY vaccine is a single injection given into the upper arm that protects against four meningococcal bacteria strains that cause meningitis and blood poisoning: A, C, W, and Y [52].

Viral Vector. Is a type of vaccine that uses a modified version of one virus as a vector. This vector is a DNA molecule used as a vehicle to artificially carry foreign genetic material into another cell. Even though it has two strange objects for the body, this type of vaccine does not cause infection with either the virus used as the vector or the source of the antigen. The genetic material it delivers does not integrate into a person's genome [40]. Johnson & Johnson's Janssen and Oxford–AstraZeneca are examples of viral vector-based COVID-19 vaccine.

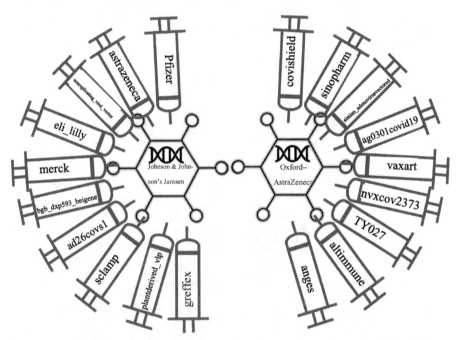

Fig. 3. The similar words for viral vector-based COVID-19 vaccine, Johnson & Johnson's Janssen and Oxford–AstraZeneca.

Johnson & Johnson's Janssen. Is a one dose viral-based COVID-19 vaccine developed by Janssen Vaccines company. We used Johnson & Johnson's Janssen as a seed to our model. pfizer, astrazeneca, nonreplicating_viral_vector, eli_lilly, merck, bgb_dxp593_beigene, ad26covs1, sclamp, plantderived_vlp, and greffex are some similar words for this vaccine (see Fig. 3).

eli_lilly is an antibody-based treatment for COVID-19 with an Emergency Use Authorization from the Food and Drug Administration [53]. Merck is a pharmaceutical company working on developing an oral medication for COVID-19. It is in phase II clinical trial with promising early findings [54]. BGB-DXP593 is a neutralizing antibody, or protein-based therapy, under investigation for SARS-CoV-2 in a clinical trial for participants with mild-to-moderate COVID-19 by BeiGene [55]; ad26covs1 is another name for Johnson & Johnson's Janssen vaccine. Sclamp, or V451, is a subunit_based COVID-19 vaccine developed by the University of Queensland and the Australian pharmaceutical company CSL Limited, and it is still in Phase I [56]. The plantderived_vlp vaccine is a different type of COVID-19 vaccine. Instead of using a viral vector, it uses living plants as bioreactors to produce non-infectious versions of viruses, called Virus-like Particles, or VLPs. This type of vaccine was developed by the Canada-based biopharma Medicago and GlaxoSmithKline. It is currently in Phase 3 clinical trials [57]. Greffex is a vaccine based on an adenovirus viral vector, currently in Pre-Clinical phase and developed by Greffex, a genetic engineering company [58].

Oxford–AstraZeneca. Is a viral vector-based COVID-19 vaccine developed by Oxford University and AstraZeneca. According to CVW2V, covishield, sinopharm, anges, ag0301covid19, vaxart, nvxcov2373, simian_adenovirusvectored', TY027, altimmune are interesting similar words for Oxford–AstraZeneca (see Fig. 3).

The word covishield is another name for Oxford–AstraZeneca. Anges, or AG0302-COVID-19, is a DNA candidate vaccine developed by AnGes Inc. in Japan and currently in Phase II/III [60]. Vaxart is a nonreplicating viral vector oral vaccine platform for COVID-19 in Phase I [59]; nvxcov2373 is a protein-based COVID-19 vaccine [35]. The simian_adenovirusvectored vaccine is used as a vector in the vector-based vaccine instead of the human adenovirus, because of its advantages over the human adenovirus. TY027 is a treatment for patients with COVID-19 to slow the progression of the disease and accelerate recovery, in addition to its potential to provide temporary protection against infection with SARS-CoV-2 [61]. Altimmune is a single dose viral vector COVID-19 vaccine that triggers a broad immune response – neutralizing IgG, mucosal IgA, and T-cells and it is still in Phase I [62].

5   Conclusions and Future Work

To stop the pandemic, wearing masks and social distancing are required but not enough; we need to use all prevention tools, especially vaccines. COVID-19 vaccines, like other vaccines, work with our immune system, to prepare our body to fight the virus if we are exposed. As an addition to our work in [63] on drug discovery using Word2vec for COVID-19 treatment, we trained an unsupervised word embedding model (CVW2V) to search for potential candidate vaccines in the COVID-19 literature according to their similarities with authorized or most promising existing ones. These studies rely on the property that words sharing similar surrounding words are semantically close. After verifying CVW2V with the terms SARS, vaccines, wearing a mask, and dry cough as general words and their similarities, CVW2V was provided with currently authorized vaccine names such as Sinopharm and CoronaVac as inactivated whole virus-based vaccines, Moderna and Pfizer-BioNTech as mRNA vaccines, and Johnson & Johnson's Janssen and Oxford–AstraZeneca as viral vector-based COVID-19 vaccine. As a result, without any explicit insertion knowledge about COVID-19 or vaccines, or any supervision on the words or objects and their meanings, the CVW2V model was able to suggest several potential target vaccines for clinical investigation and support other ones in different clinical experimental phases according to their high similarity with currently authorized and most promising vaccines.

In future work, we plan on applying this approach in different areas, such as the materials science field. Since the data types may not be the same, the word embedding parameters need to be tuned. Different combinations of parameter values for the word embedding training could be changed to improve the approach, such as negative sampling distribution, number of epochs, window size [64], minimum word count, and vector size [65].

Acknowledgments. The authors would like to thank Dr. Alvis Fong and Dr. Pnina Ari-Gur for their valuable suggestions in the development of this work. Furthermore, we thank the anonymous reviewers for their valuable feedback and suggestions.

References.

1. Usama, M., et al.: Unsupervised machine learning for networking: techniques, applications and research challenges. IEEE Access. **7**, 65579–65615 (2019)
2. Yu, L.-C., et al.: Refining word embeddings using intensity scores for sentiment analysis. IEEE/ACM Trans. Audio Speech Lang. Process. **26**(3), 671–681 (2018)
3. Manning, C.D., Raghavan, P., Schütze, H.: Introduction to Information Retrieval, Cambridge University Press, vol. 39 (2008)
4. Chen, W., et al.: Distributed feature representations for dependency parsing. IEEE/ACM Trans. Audio, Speech Lang. Process. **23**(3), 451–460 (2015)
5. Ouchi, H., et al.: Transition-based dependency parsing exploiting supertags. IEEE/ACM Trans. Audio, Speech, Lang. Process. **24**(11), 2059–2068 (2016)
6. Shen, M., et al.: Dependency parse reranking with rich subtree features. IEEE/ACM Trans. Audio Speech Lang. Process. **22**(7), 1208–1218 (2014)
7. Zhou, G., et al.: Learning the multilingual translation representations for question retrieval in community question answering via non-negative matrix factorization. IEEE/ACM Trans. Audio Speech Lang. Process. **24**(7), 1305–1314 (2016)
8. Hao, Y., et al.: An end-to-end model for question answering over knowledge base with cross-attention combining global knowledge. In: Proceedings of the 55th Annual Meeting of the Association for Computational Linguistics (Volume 1: Long Papers). (2017)
9. Zhang, B., et al.: A context-aware recurrent encoder for neural machine translation. IEEE/ACM Trans. Audio Speech Lang. Process. **25**(12), 2424–2432 (2017)
10. Chen, K., et al.: A neural approach to source dependence based context model for statistical machine translation. IEEE/ACM Trans. Audio Speech Lang. Process. **26**(2), 266–280 (2018)

11. Sun, F., et al.: Learning word representations by jointly modeling syntagmatic and paradigmatic relations. In: Proceedings of the 53rd Annual Meeting of the Association for Computational Linguistics and the 7th International Joint Conference on Natural Language Processing (Volume 1: Long Papers) (2015)
12. Lai, S., et al.: How to generate a good word embedding. IEEE Intell. Syst. **31**(6), 5–14 (2016)
13. Yin, W., Schütze, H.: Discriminative phrase embedding for paraphrase identification. In: Proceedings of the 2015 Conference of the North American Chapter of the Association for Computational Linguistics: Human Language Technologies (2015)
14. Li, Y., Yang, T.: Word Embedding for Understanding Natural Language: a Survey. Studies in Big Data, pp. 83–104 (2017)
15. Mellet, J., Pepper, M.S.: A COVID-19 vaccine: big strides come with big challenges. Vaccines. 9(1), 39 (2021)
16. Beck, B.R., et al.: Predicting commercially available antiviral drugs that may act on the novel coronavirus (SARS-CoV-2) through a drug-target interaction deep learning model. Comput. Struct. Biotechnol. J. **18**, 784–790 (2020)
17. Zhavoronkov, A., et al.: Deep learning enables rapid identification of potent DDR1 kinase inhibitors. Nat. Biotechnol. **37**(9), 1038–1040 (2019)
18. Keshavarzi Arshadi, A. et al.: Artificial Intelligence for COVID-19 Drug Discovery and Vaccine Development. Frontiers in Artificial Intelligence. 3, (2020).
19. Center for Biologics Evaluation and Research: Vaccine Development – 101. https://www.fda.gov/vaccines-blood-biologics/development-approval-process-cber/vaccine-development-101
20. Mikolov, T., Chen, K., Corrado, G., et al.: Efficient estimation of word representations in vector space. arXiv preprint arXiv:1301.3781 (2013)
21. ElsevierDev: ElsevierDev/elsapy. https://github.com/ElsevierDev/elsapy. Accessed 22 Apr 2021
22. gensim: topic modelling for humans. https://radimrehurek.com/gensim_3.8.3/index.html. Accessed 22 Apr 2021
23. Tshitoyan, V., et al.: Unsupervised word embeddings capture latent knowledge from materials science literature. Nature **571**(7763), 95–98 (2019)
24. Yang, Z., et al.: An in silico deep learning approach to multi-epitope vaccine design: a SARS-CoV-2 case study. Sci. Rep. **11**, 1 (2021)
25. "Google Code Archive - Long-term storage for Google Code Project Hosting". code.google.com. Retrieved 22 October 2020.
26. Strandqvist, W.: Neural Networks for Part-of-Speech Tagging (Dissertation) (2016). http://urn.kb.se/resolve?urn=urn:nbn:se:liu:diva-129296
27. Rong, X.: Word2vec Parameter Learning Explained. ArXiv Preprint ArXiv:1411.2738 (2014)
28. Fast, E., Altman, R.B., Chen, B.: Potential t-cell and b-cell epitopes of 2019-ncov. (2020). Goldberg, Y. and Levy, O., "word2vec Explained: deriving Mikolov et al.'s negative-sampling word-embedding method", <i>arXiv e-prints</i> (2014)
29. Chen, B., et al.: Predicting HLA class II antigen presentation through integrated deep learning. Nat. Biotechnol. **37**, 1332–1343 (2019)
30. Jurtz, V., Paul, S., Andreatta, M., Marcatili, P., Peters, B., Nielsen, M.: NetMHCpan-4.0: Improved Peptide–MHC Class I Interaction Predictions Integrating Eluted Ligand and Peptide Binding Affinity Data. J. Immunology **199**, 3360–3368 (2017)
31. Crossman, L.C.: Leveraging Deep Learning to Simulate Coronavirus Spike proteins has the potential to predict future Zoonotic sequences (2020)
32. Abbasi, B.A., Saraf, D., Sharma, T., Sinha, R., Singh, S., Gupta, P., Sood, S., Gupta, A., rawal, kamal: Identification of vaccine targets & design of vaccine against sars-cov-2 coronavirus using computational and deep learning-based approaches. (2020).

33. Schnabel, T., Labutov, I., Mimno, D., Joachims, T.: Evaluation methods for unsupervised word embeddings. In: Proceedings of the 2015 Conference on Empirical Methods in Natural Language Processing (2015)
34. The different types of COVID-19 vaccines. https://www.who.int/news-room/feature-stories/detail/the-race-for-a-covid-19-vaccine-explained
35. Pollet, J., Chen, W.-H., Strych, U.: Recombinant protein vaccines, a proven approach against coronavirus pandemics. Adv. Drug Deliv. Rev. **170**, 71–82 (2021)
36. Wee, S.-lee, Qin, A.: China Approves Covid-19 Vaccine as It Moves to Inoculate Millions, https://www.nytimes.com/2020/12/30/business/china-vaccine.html.
37. Corum, J., Zimmer, C.: How the Sinovac Vaccine Works. https://www.nytimes.com/interactive/2020/health/sinovac-covid-19-vaccine.html
38. Logunov, D.Y., et al.: Safety and efficacy of an rAd26 and rAd5 vector-based heterologous prime-boost COVID-19 vaccine: an interim analysis of a randomised controlled phase 3 trial in Russia. The Lancet. **397**, 671–681 (2021)
39. Bharat Biotech-Vaccines & Bio-Therapeutics Manufacturer in India. https://www.bharatbiotech.com/covaxin.html. Accessed 22 Apr 2021
40. Understanding Viral Vector COVID-19 Vaccines. https://www.cdc.gov/coronavirus/2019-ncov/vaccines/different-vaccines/viralvector.html?CDC_AA_refVal=https%3A%2F%2Fwww.cdc.gov%2Fvaccines%2Fcovid-19%2Fhcp%2Fviral-vector-vaccine-basics.html. Accessed 22 Apr 2021
41. Codagenix Home. https://codagenix.com/. Accessed 27 Apr 2021
42. Dutta, D.S.S.: What are Adenovirus-Based Vaccines? https://www.news-medical.net/health/What-are-Adenovirus-Based-Vaccines.aspx
43. Modjarrad, K.,et al.: Safety and immunogenicity of an anti-Middle East respiratory syndrome coronavirus DNA vaccine: a phase 1, open-label, single-arm, dose-escalation trial. Lancet Infectious Diseases **19**, 1013–1022 (2019)
44. China's mRNA COVID-19 vaccine may start late-stage trial in May - state media. https://www.reuters.com/business/healthcare-pharmaceuticals/chinas-mrna-covid-19-vaccine-may-start-late-stage-trial-may-state-media-2021-04-13/. Accessed 27 Apr 2021
45. COVID-19 S-Trimer (SCB-2019) Vaccine. https://www.precisionvaccinations.com/vaccines/covid-19-s-trimer-scb-2019-vaccine. Accessed 28 Apr 2021
46. Celonic and CureVac Announce Agreement to Manufacture over 100 Million Doses of CureVac's COVID-19 Vaccine Candidate, CVnCoV. https://www.curevac.com/en/2021/03/30/celonic-and-curevac-announce-agreement-to-manufacture-over-100-million-doses-of-curevacs-covid-19-vaccine-candidate-cvncov/. Accessed 28 Apr 2021
47. Ascending Dose Study of Investigational SARS-CoV-2 Vaccine ARCT-021 in Healthy Adult Subjects – Full Text View. https://www.clinicaltrials.gov/ct2/show/NCT04480957. Accessed 2 May 2021
48. Study of Recombinant Protein Vaccine Formulations Against COVID-19 in Healthy Adults 18 Years of Age and Older - Full Text View. https://www.clinicaltrials.gov/ct2/show/NCT04537208. Accessed 2 May 2021
49. Philippidis, A.: Genexine - GX-19. https://www.genengnews.com/covid-19-candidates/genexine-gx-19/. Accessed 2 May 2021
50. Dey, A., Chozhavel Rajanathan, T.M., Chandra, H., Pericherla, H.P.R., Kumar, S., Choonia, H.S., Bajpai, M., Singh, A.K., Sinha, A., Saini, G., Dalal, P., Vandriwala, S., Raheem, M.A., Divate, R.D., Navlani, N.L., Sharma, V., Parikh, A., Prasath, S., Rao, S., Maithal, K.: Immunogenic Potential of DNA Vaccine candidate, ZyCoV-D against SARS-CoV-2 in Animal Models. (2021).
51. Ella, R., et al.: Safety and immunogenicity of an inactivated SARS-CoV-2 vaccine, BBV152: a double-blind, randomised, phase 1 trial. Lancet. Infect. Dis **21**, 637–646 (2021)

52. Pizza, M., Bekkat-Berkani, R., Rappuoli, R.: Vaccines against Meningococcal Diseases. Microorganisms. **8**, 1521 (2020)
53. Commissioner, O.of the: Coronavirus (COVID-19) Update: FDA Authorizes Monoclonal Antibodies for Treatment of COVID-19. https://www.fda.gov/news-events/press-announcem ents/coronavirus-covid-19-update-fda-authorizes-monoclonal-antibodies-treatment-covid-19-0. Accessed 2 May 2021
54. Miller, K.: Merck Oral COVID-19 Drug Shows Promise in Early Trials. https://www.verywe llhealth.com/merck-oral-covid-19-drug-clinical-trial-5115909. Accessed 2 May 2021
55. BGB-DXP593. https://go.drugbank.com/drugs/DB16357. Accessed 2 May 2021
56. A Study on the Safety, Tolerability and Immune Response of SARS-CoV-2 Sclamp (COVID-19) Vaccine in Healthy Adults - Full Text View. https://www.clinicaltrials.gov/ct2/show/NCT 04495933
57. Dhama, K., et al.: Plant-based vaccines and antibodies to combat COVID-19: current status and prospects. Hum. Vaccin. Immunother. **16**, 2913–2920 (2020)
58. The Future of Genetic Engineering. https://www.greffex.com/. Accessed 2 May 2021
59. New Data from Vaxart Oral COVID-19 Vaccine Phase I Study Suggests Broad Cross-Reactivity against Other Coronaviruses. https://investors.vaxart.com/news-releases/news-rel ease-details/new-data-vaxart-oral-covid-19-vaccine-phase-i-study-suggests. Accessed 2 May 2021
60. Phase II / III Study of COVID-19 DNA Vaccine (AG0302-COVID19) - Full Text View. https:// www.clinicaltrials.gov/ct2/show/NCT04655625. Accessed 2 May 2021
61. Efficacy and Safety of TY027 a Treatment for COVID-19 in Humans. https://www.cen terwatch.com/clinical-trials/listings/259289/efficacy-and-safety-of-ty027-a-treatment-for-covid-19-in-humans/. Accessed 2 May 2021
62. Safety and Immunogenicity of AdCOVID in Healthy Adults (COVID-19 Vaccine Study) - Full Text View. https://clinicaltrials.gov/ct2/show/NCT04679909. Accessed 4 May 2021
63. Gharaibeh, T., de Doncker, E.: Unsupervised Learning with Word Embeddings Captures Knowledge from COVID-19 Literature. CSCI 2020 (Dec. 2020), IEEE CPS, Accepted
64. Caselles-Dupré, H., Lesaint, F., Royo-Letelier, J.: Word2vec applied to recommendation. In: Proceedings of the 12th ACM Conference on Recommender Systems. (2018)
65. Yildiz, B., Tezgider, M.: Learning quality improved word embedding with assessment of hyperparameters. In: Schwardmann, U., et al. (eds.) Euro-Par 2019. LNCS, vol. 11997, pp. 506–518. Springer, Cham (2020). https://doi.org/10.1007/978-3-030-48340-1_39

Short Papers

Short Papers

Substrate Soil Moisture Impact on Green Roof Performance for an Experimental Site in Tomblaine, France

Mirka Mobilia[1], Roberta D'Ambrosio[1(✉)], Remy Claverie[2], and Antonia Longobardi[1]

[1] Department of Civil Engineering, University of Salerno, Via Giovanni Paolo II, 132, 84084 Fisciano, Salerno, Italy
robdambrosio@unisa.it

[2] Cerema Est, TEAM Research Group, 71 rue de la grande haie, 54510 Tomblaine, France

Abstract. During the last decades, urbanization completely changed the layout of the cities, increasing the imperviousness and reducing infiltration, evapotranspiration and groundwater recharge. The direct consequence of this is the enhancement of floods in urban context. It is widely known that Nature Based Solutions such as Green Roofs may represent a successful solution in supporting traditional drainage systems in the management of an ever-increasing stormwater runoff.

To the purpose, an improved knowledge of the hydrological behavior of such infrastructures and the relation between their retention capacity and climate or design variable is essential to enhance their effectiveness.

Starting from the analysis of 56 rainfall-runoff events, this paper aims at assessing the behavior of a 98 m^2 experimental green roof, located in Tomblaine (Northern-Eastern France), with specific reference to the relation between runoff retention capacity, soil moisture content prior to the triggering rainfall event and the rainfall event properties. For the specific case study, the ANOVA analysis pointed out that soil moisture content appeared to be the best predictor of retention capacity both under moderate and severe rainfall. However, a weak correlation between the two variables was detected with a maximum 0,67 coefficient reached by a single regression of the low retention group.

Keywords: Green roofs · Soil moisture content · Retention capacity

1 Introduction

Land use changes and soil sealing gradually transformed the hydrological cycle and increased urban flooding risk [1]. In this context, Nature Based Solutions appear fundamental strategies for a sustainable management of stormwater runoff at the city scale [2–4]. Among them, green roofs (GRs) are valuable solutions that allows an optimal retrofitting of traditional roofing, reducing stormwater runoff and bringing numerous additional benefits for both nature and humans [5–7].

Specifically referring to the role of GRs in the mitigation of urban flooding, these infrastructures seem to be effective in the reduction of runoff peak flow and volumes [5].

© Springer Nature Switzerland AG 2021
O. Gervasi et al. (Eds.): ICCSA 2021, LNCS 12950, pp. 563–570, 2021.
https://doi.org/10.1007/978-3-030-86960-1_39

Nevertheless, according to literature, both climate and design variables seem to affect the GRs behaviour. For example, higher substrate depths and spread or light rains are undoubtedly the best conditions for their efficiency [8, 9], as well as the right choice of vegetation [10].

However, the retention performance of a GR can vary on a large extent and this is mainly due to the interplay of technological features and geographical contexts of each infrastructure [11, 12]. Researchers worldwide discuss about the existence of a relationship between retention capacity of GRs and some specific hydrological variables, such as the initial soil moisture content or the antecedent dry period to an event but so far, contrasting results came out [13, 14].

In the current paper, the performance of an experimental green roof of 98 m² located in Tomblaine, Northern France, was investigate through an event scale analysis that involved 56 rainfall-runoff events occurred during the period July 2014 – December 2015. The study aimed at providing additional explanation on the hydrological behaviour of the GRs in a geographical context characterized by a temperate climate and at improving the prediction of the retention capacity using design and climate-related variables.

2   Materials and Methods

2.1   Case Study

The in situ experimental GR, set up in July 2011, is located in Tomblaine (Fig. 1a), northeast of France (48° 40′ 48″ N, 6° 13′ 9″E, 202 m a.s.l.). The infrastructure is placed above a 6-m high building with a traditional flat roof [15]. Even if the overall area occupied by the GR is of about 600 m², this study analysed only a part of it, characterized by a 98 m² area (Fig. 1b) independent from the rest of the roof.

The GR only includes two layers: a vegetation and a substrate layer. The vegetation, as often occurs for these systems, is composed by different kind of sedum (Sedum album, Sedum reflexumlarix, Sedum reflexum germanium, Sedum sexangulare and Sedum floriferum). It does not exceed the 15-cm height, including 5 cm of roots. The substrate, ranging between a 7.5 to a 9.5-cm depth, is made by pozzolana (80%) and organic fraction (20%) (Fig. 1c). The pozzolana, widely used for its physical-hydraulic properties (low bulk density, mechanical resistance and high storage) is made by particles ranging from 3 to 6 mm, for 80% of the composition, and from 7 to 15 mm for the remaining 20%. The organic fraction is instead composed of peat dust and maritime pine bark in equal shares.

The runoff overflows from the top of the roof and is collected in gutters and conveyed through a pipe in a tipping bucket system, closed in a metal box and fixed to an outside wall, in order to measure it (Fig. 1d, just one of the three systems in the picture refers to the investigated roof).

2.2   Case Study Climate and Hydrological Data Characterization

The city of Tomblaine is characterized by a temperate climate. It comes under Cfb class according to the Köppen-Geiger climate classification that include temperate-climate

Fig. 1. a. Localization of Tomblaine, the city where the installation is placed; b. in situ experimental green roof; c. detail of the growing media (pozzolana and organic fraction); d. runoff measurement system (just one of the three systems refers to the roof analyzed).

areas with wet summers, temperature of the hottest month below 22 °C and at least four month with temperatures higher than 10 °C. The average temperature in Tomblaine is 10.6 °C while the annual rainfall amount is around 897 mm (Fig. 2).

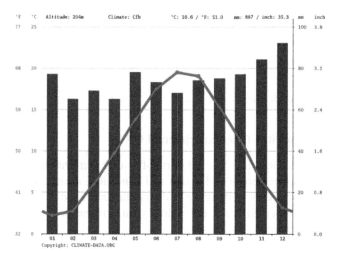

Fig. 2. Tomblaine mean monthly temperature and rainfall. Picture from https://en.climate-data.org/

From July 2014 to December 2015, rainfall, runoff and soil moisture content (SWC) values were continuously measured and collected (Fig. 3). During the monitoring period (almost one year and half) the total rainfall amount was 875.1 mm while the total runoff registered was 256.3 mm, defining a long-run retention capacity of 71%. Soil moisture content reached the maximum value, 45.93%, in January while the lowest value, 6.89%, in June; the mean was 35.87%. Provided the high frequency of rainfall occurrences and the absence of a drainage system to remove subsurface water stored into the GR substrate, the soil moisture temporal profile is characterized by long period of stationarity with no significant changes in soil moisture values. Soil water content depletion typically occurs during the summer period, but it quickly recovers after precipitation events.

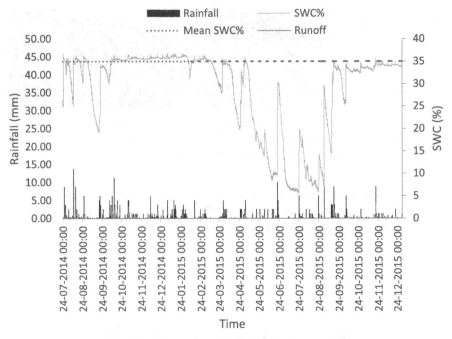

Fig. 3. Rainfall, runoff and SWC monitoring over time.

2.3   Retention Capacity of the Experimental GR

The current analysis concerned 56 rainfall-runoff events extracted from the mentioned database. Rainfall characteristics (duration and cumulate), mean substrate soil water content registered before the occurrence of the rainfall event and cumulate runoff were assessed. For each observed event, the hydrological performance in term of retention capacity (RC) was computed as follows:

$$1-C* = RC \tag{1}$$

Where $C*$ is the discharge coefficient, computed as the ratio between the cumulative observed runoff depth and the cumulative observed rainfall depth at the event scale.

3   Results and Discussions

Soil moisture content of the selected rainfall-runoff events varies from a minimum of 7.28% to a maximum of 44.59% with a mean value of 30.46%. Probably due to the building technology of the experimental roof assessed, few values of soil water content below 20% registered and only during the summer season. Much more frequent are SWC major than 35%. Rainfall also varies a lot both in terms of cumulate (ranging from 0.20 to 60.20) and duration (ranging from 1 to 77 h). Retention Capacity shows a great variability, ranging from 1 to 100%, with a mean value of 76% representative of a good overall performance of the experimental roof.

Table. 1 Minimum, maximum and average values for pre-event soil water content (SWC), cumulate rainfall (rain) and rainfall duration (duration).

	Min	Max	Average
SWC (%)	7.28	44.85	30.46
Rain (mm)	0.20	60.20	10.08
Duration (hr)	1.00	77.00	15.64

Longobardi and co-workers [14] demonstrated how SWC prior to a triggering rainfall event can be used to identify different hydrological behavior that a single GR can exhibit depending on SWC thresholds. With the objective to assess how the soil moisture content prior to the occurrence of the rain could affect the hydrological behavior of the presented case study, the relationship between the retention capacity and soil moisture content was investigated for each of the rainfall-runoff events. Looking at Fig. 4 it is possible to observe, for the specific case study, that for SWC percentage below 35% the GR is able to store and retain the entire rainfall amount. Hence, in those conditions, the runoff is zero and subsequently the RC is 100%.

When SWC approaches values larger than 35%, RC largely varies from a minimum value of about 1% to maximum of about 100% apparently with no relationship with SWC values.

Especially for what concerns events characterized by SWC > 35%, a distinction can be made with reference to the RC values. Two distinct groups can be highlighted, laying respectively above (high retention group) and below (low retention group) a RC threshold of about 50% (Fig. 4).

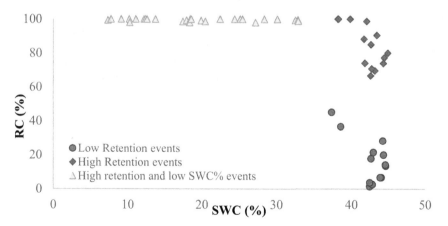

Fig. 4. Relation between retention capacity and soil moisture content: identification of groups.

For each group, events SWC, cumulate rainfall, duration and RC were represented as a box plot (Fig. 5). Differences between the two groups are mostly detected in terms of rainfall cumulate and duration. The high retention group of events (RC > 50%, SWC

> 35%), are mostly characterized by rainfall cumulate and duration below the mean value (10 mm and 16 h respectively in Table 1); the group of low retention events (RC never exceeding 50%, SWC > 35%), are characterized by rainfall cumulate and duration above the mean value. A larger variability in terms of cumulate rainfall and duration, compared to the high retention group, is a distinctive characteristic of the low retention group.

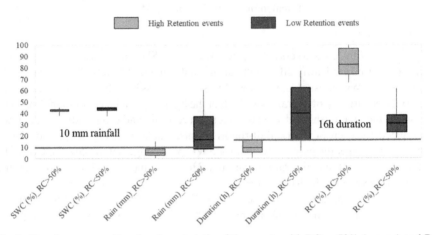

Fig. 5. Boxplots representing the characteristic of the events with RC > 50% (orange) and RC < 50% (blue) (Color figure online).

In order to identify an empirical data-based model for RC prediction, the relationship between RC and initial soil water content SWC, cumulate rainfall and rainfall duration, was investigated in a multiple regression framework. The one-way analysis of variance (ANOVA) is a common technique that can be used to compare means of two or more independent groups in order to determine whether there is statistical evidence that the associated population means are significantly different [16]. The ANOVA analysis results are illustrated in the following, for $\alpha = 5\%$. Both in the case of high retention (Table 2) and low retention (Table 3) group, SWC appears as the only significant variable (p-value $< \alpha$). The correlation coefficients equal 0,44 and 0,49 respectively for the high and low retention groups, providing an indication for a weak correlation. Also the case of a single regression of RC against SWC was considered. This led to an improvement in the correlation coefficient, especially for the low retention group for which it increased up to 0,67.

Table. 2 ANOVA analysis for the high retention events group. ANOVA refers to a multiple regression model of RC versus SWC, cumulate rain and rain duration ($\alpha = 0{,}05$).

	Coefficients	Standard error	Stat t	p-value
SWC	−4,21	1,60	−2,63	0,03
Rain	−0,06	0,22	−0,29	0,78
Duration	−0,03	0,15	−0,17	0,87

Table. 3 ANOVA analysis for the low retention events group. ANOVA refers to a multiple regression model of RC versus SWC, cumulate rain and rain duration ($\alpha = 0{,}05$).

	Coefficients	Standard error	Stat t	p-value
SWC	−4,04	1,42	−2,85	0,02
Rain	−0,10	0,72	−0,15	0,89
Duration	−0,26	0,40	−0,64	0,54

4 Conclusions

In this paper, the relationship between the retention capacity of an experimental GR located in Tomblaine (France) and the initial (to the triggering rainfall event) substrate soil moisture content SWC was investigated. The results showed that the GR hydrological response is mainly driven by SWC. An SWC threshold of about 35% was found to delimit (SWC < 35%) events for which, regardless for the rainfall properties, no runoff occurred. If SWC > 35%, it was possible to distinguish between low and high retention events. The first occurred for rainfall cumulate and duration well above the relevant average values.

In contrast to other studies, the rainfall cumulate and duration seemed not to affect the performance of this specific experimental roof in terms of retention. In an ANOVA analysis, SWC appeared the only significant variable to predict RC, but the correlation coefficients was rather weak. The results reported in this research are likely related to the specific building technology adopted for the construction which does not see a drainage layer. Further research supported by a conceptual model framework is foreseen to achive a better prediction of the retention properties.

References

1. Mobilia, M., Longobardi, A.: Using SAR satellite imagery for potential green roof retrofitting for flood mitigation in urban environment. In: GISTAM (Accepted)
2. Bell, S.: Urban water sustainability. Constructing Infrastructure for Cities and Nature. Routledge, London and New York (2018)

3. Brears, R.C.: Blue and green cities: the role of blue-green infrastructure in managing urban water resources. Macmillan Publishers Ltd, London (2018)
4. Sartor, J., Mobilia, M., Longobardi, A.: Results and findings from 15 years of sustainable urban storm water management. Int. J. Safety Security Eng. **8**(4), 505–514 (2018)
5. D'Ambrosio, R., Longobardi, A., Mobilia, M., Sassone, P.: Sustainable strategies for flood risk management in urban areas. Enhancing city resilience with Green Roofs. UPLanD – J. Urban Plann Landscape Environ. Des. **5**(2), 87–98 (2021)
6. Ercolani, G., Chiardia, E.A., Gandolfi, C., Castelli, F., Masseroni, D.: Evaluating performances of green roofs for stormwater runoff mitigation in a high flood risk urban catchment. J. Hydrol. **566**, 830–845 (2018)
7. Mobilia, M., Longobardi, A.: Smart stormwater management in urban areas by roofs greening. In: International Conference on Computational Science and Its Applications. Springer, Cham, pp. 455–463 (2017)
8. Carbone, M., Nigro, G., Garofalo, G., Piro, P.: Experimental testing for evaluating the influence of substrate thickness on the sub-surface runoff of a Green Roof. Appl. Mech. Mater. **737**, 705–709 (2015)
9. Soulis, K.X., Ntoulas, N., Nektarios, P.A., Kargas, G.: Runoff reduction from extensive green roofs having different substrate depth and plant cover. Ecol. Eng. **102**, 80–89 (2017)
10. Cristiano, E., Urru, S., Farris, S., Ruggiu, D., Deidda, R., Viola, F.: Analysis of potential benefits on flood mitigation of a CAM green roof in Mediterranean urban areas. Build. Environ. **183**, 107179 (2020)
11. Nawaz, R., McDonald, A., Postoyko, S.: Hydrological performance of a full-scale extensive green roof located in a temperate climate. Ecol. Eng. **82**, 66–80 (2015)
12. Mobilia, M., D'Ambrosio, R., Longobardi, A.: Climate, soil moisture and drainage layer properties impact on green roofs in a Mediterranean environment, frontiers in water-energy-nexus—nature-based solutions. Adv. Technol. Best Pract. Environ. Sustain. 169–171 (2019)
13. Chenot, J., Gaget, E., Moinardeau, C., Jaunatre, R., Buisson, E., Dutoit, T.: Substrate composition and depth affect soil moisture behavior and plant-soil relationship on Mediterranean extensive green roofs. Water **9**(11), 817 (2017)
14. Longobardi, A., D'Ambrosio, R., Mobilia, M.: Predicting stormwater retention capacity of green roofs: an experimental study of the roles of climate, substrate soil moisture, and drainage layer properties. Sustainability **11**(24), 6956 (2019)
15. Bouzouidja, R., Rousseau, G., Galzin, V., Claverie, R., Lacroix, D., Séré, G.: Green roof ageing or Isolatic Technosol's pedogenesis? J. Soils Sediments **18**(2), 418–425 (2016). https://doi.org/10.1007/s11368-016-1513-3
16. Lepore, E., Aguilera Benito, P., Piña Ramírez, C., Viccione, G.: Indoors ventilation in times of confinement by SARS-CoV-2 epidemic: a comparative approach between Spain and Italy. Sustain. Cities Soc. **72**, 103051 (2021)

Usable Privacy in Location-Based Services: The Need for a Regulatory Framework

Björn Beringer(✉), Per Ole Uphaus, and Harald Rau

Institute for Media Management, Ostfalia University of Applied Sciences, Salzgitter, Germany
{bjo.beringer,pe.uphaus,h.rau}@ostfalia.de

Abstract. Facing the last decades of information technology, the importance of 'usable privacy and security' (UPS) increased significantly. Previous studies have uncovered a tremendous gap in UPS-related research, especially concerning location-based services (LBS). This contribution aims to reduce this gap by identifying factors leading to a more user-centric design of LBS. It relies on guided interviews with international experts either with a scientific or an entrepreneurial background. The findings support solutions equipping the user with more control of owned data, thus ensuring more self-determination and decision-making authority affecting personal information. The study makes use of a qualitative content analysis in order to identify relevant factors that reveal which data security-related aspects are important by designing location-based services. The findings lead to two key outcomes: Firstly, the implementation of easier to use data security tools in applications based on LBS often comes with an information overload affecting the user. Secondly, a regulatory framework for third party vendors, users, and providers to address the lack of knowledge for user-friendly LBS turns out to be the key requirement for successful implementation of UPS.

Keywords: Location-based services · Usable privacy and security · Regulatory framework

1 Introduction

Dealing with aspects of data security in particular, users are increasingly faced with difficulties in understanding what is happening with their data and how to still maintain control [1, 2]. Research identified several problems regarding the use of mobile applications and the possibility of data control [3], involving the compatibility of data security, data protection and user-friendliness in applications using location-based services (LBS). In this regard, Abbas et al. favor a framework that accounts for the complexity of LBS applications in regulatory terms [1]. Given that, this contribution examines factors supporting usable data security while designing location-based services. By that, new insights into the relationship between location-based services (LBS) and Usable Privacy and Security will be generated, and factors influencing user-friendly data security in location-based technologies will be identified. Doing so, this study represents an important contribution in current research about regulatory considerations and ethical

© Springer Nature Switzerland AG 2021
O. Gervasi et al. (Eds.): ICCSA 2021, LNCS 12950, pp. 571–579, 2021.
https://doi.org/10.1007/978-3-030-86960-1_40

dilemmas [1]. For that purpose, interviews with international experts were conducted and analyzed. As an outcome, the opportunities and risks of UPS implementation in LBS will be summarized in several hypotheses for further research in the field. Above all, the results show a clear requirement for a framework covering socio-ethical, regulatory, and technological factors.

2 Theoretical Background

Mobile applications are increasingly becoming an essential part of people's everyday lives [4]. Depending on the user's context, LBS applications are able to deliver relevant, contextual information tailored to the current situation [5]. Data transmission of information tailored to the user's location often takes place via networks such as Wi-Fi as well as via mobile networks (GSM). Discussing the providers' part, servers receive the corresponding requests from the user and initiates the processing – depending on the service [6], to be exact: content and data servers respond to the user requests in a context-sensitive response way, and in some cases they also draw on the data of third-party providers [7]. During the use of LBS, data is generated, that is collected, stored, and processed on these servers of providers *and* third-party providers [8]. Data security is usually represented by means of a server with appropriate security functions and legally secured by data protection regulations [9]. Due to the access of the LBS provider or any third-party providers to the user's location and the collection of personal data, there are a number of security risks that may impact LBS usage [8]. Common security risks while using LBS that are frequently discussed in literature include the accessibility of personal information such as addresses, consumption characteristics, and places visited and corresponding times of stay for third parties without the user's permission [10, 11]. The increased risk of compromised privacy in LBS is considered to make the protection of privacy the greatest challenge within these technologies [12]. Despite their individual nature by focusing on sensitive security risks, LBS can be grouped, using the concept of 'location privacy' [13, 14].

Privacy is the *most frequently* considered issue when it comes to concerns during the use of mobile applications – especially location-based applications [15, 16]. This would include (but is not limited to) acts of extortion, tracking, surveillance, or intrusion [17, 18]. How exactly security risks are to be reduced and how new technologies can be created to be more user-friendly, and – at the same time – more secure, is summarized under the catchphrase of 'Usable Privacy and Security' (UPS). Saltzer and Schroeder were the first to mention a computer systems' easiness of use while also providing the highest possible level of security for the user [19]. Therefore, it seems to be important to integrate usability testing cycles into the design process of applications instead of testing and adapting them after development [20]. With these findings, they laid the foundation for today's 'Usable Privacy and Security' research. According to Naqvi and Porras, it is relevant to create a 'match' between system-internal development processes and the user's cognition [21]. The incorporation of the user's values into the design of safety systems enables the user's security-relevant needs and values to be met while harmonizing all security precautions [21]. Implementing this endeavor would presume a kind of education of the users and/or the incorporation of human aspects into the design process for security systems [21]. A fundamental approach in UPS research today is to analyze the development of behavioral and activity-based safety systems – and one of the most relevant tasks then will be the reduction of safety tasks or prompts on users [22]. According to Atzeni et al., users always strive to pursue a goal or satisfy a need when using applications [23]. Therefore, systems should be designed in such a way that these goals can be achieved as quickly as possible, and in a secure way.

In summary: security aspects should be designed to be as user-friendly as possible; they must not hinder the user's search for information in order to avoid potential mistakes in the human interaction with the system [23]. Consequently, the target for researching this field is to identify ways to develop secure *and* 'usable' systems – systems, that do not restrict or frame users, while at the same time taking their individual needs into account [22, 24].

3 Methodology

In order to identify factors influencing possible correlations between user-friendly data security and LBS, guided interviews with experts were conducted and subsequently analyzed, making use of a qualitative content analysis. The sample consisted of 21 international and multidisciplinary participants from academia as well as from industry. All business representatives interviewed are coming with an entrepreneurial approach. An overview of the entire sample can be found in Table 1. To generate a sound set of qualitative data, all interviews were based on a guideline comprising the following over-arching and for the coding process predefined dimensions, which provided the structure and ensured comparability of the individual interviews [25]: *meaning of location, security risks for location-based services, usability aspects, barriers of implementation*, and *future expectations* in this field. This contribution will focus explicitly on the barriers going along with security risks, as this dimension turned out to be particularly insightful. The identified barriers offer several invitations for further research and they provide deep insights how to implement UPS in location-based services. The interviews were recorded, transcribed, and coded using the QDA software MAXQDA. The interpretation made use of Mayring's qualitative content analysis – an approach considered particularly suitable for a systematic, rule- and theory-guided analysis of text-based interviews [26]. Except for the previously mentioned dimensions – adopted from literature, and implemented directly into the interview guideline – the derivation of all the further upcoming categories followed an inductive approach. The used routine determined the abstraction level as well as the process of coding, evaluation, and identification of context units. Finally, the derivation of categories from the transcript's semantic content resulted in a category system of various categories and subcategories. With the coding being handled by only one of the authors, Krippendorff's Alpha test was used to estimate the intracoder reliability with exemplary interviews at the beginning, in the middle and at the end of the coding process. The results indicate acceptable values for the interviews tested in the beginning ($\alpha = 0.7275$), in the middle ($\alpha = 0.8426$) and in the end ($\alpha = 0.6956$) of the entire encoding process, indicating reliable coding. Further while validity was ensured by revising the categories during and after the encoding process [26].

Table 1. Sample overview

Abbrevation	Position	Organization	Area of expertise	Country
P1	Ph.D	University	Privacy and Security	USA
P2	Dr.	Institute	Privacy and Security	Germany
P3	Ph.D	Institute	Movement data in GIS	Austria
P4	Prof. Dr.	University	Media law	Germany
P5	Prof. Dr.		Locative media	USA
P6	Prof. Dr.		Digital Technologies	Singapore
P7	Prof. Dr.		Geo-Information	Austria
P8	App Designer	Company	Critical application software	Germany
P9	Dr.	University	Privacy and Security	USA
P10	Prof. Dr.		Media sociology	USA
P11	Dr.		Socio-technical systems	Australia
P12	Prof. Dr.		Privacy and Security	USA
P13	Ph.D		Privacy and Data Protection	Germany
P14	Prof. Dr.		Social Implications of Technology	USA
P15	Prof. Dr.		Geo-Information Science	Belgium
P16	Prof. Dr.		Data protection	Germany
P17	CTO	Company	Data protection consulting	Germany
P18	CEO	Non-Profit-Organization	Data protection	Germany
P19	CEO	Company	Trustworthy Technologies	Australia
P20	CEO		Software development	Germany
P21	CEO		Geo Marketing and media planning	Germany

4   Results

The final analysis of the interviews resulted in a fairly complex category system. The evaluation of individual code relations subsequently provided in-depth insights into the interrelationship between several categories. Clear advantages of UPS incorporation in LBS were revealed: above all, the category *data combination* (e.g., visited locations, time spent, usage of camera) is of interest, because strong ties between UPS and LBS's usage for *marketing*, *profiling*, and *tracking* became apparent. Connecting a two-sided view to the whole development process for applications seems to result in an added value for both areas, UPS as well as LBS. Some statements of the participants could be summarized to the conclusion that three overarching categories may form a kind of 'framework' for user-friendly data security in LBS applications. To name them: *socio-ethical*, *regulatory* and *technological* factors. *Socio-ethical* factors often show the connection between particularly sensitive data – such as one's personal address, timeslots at which a user leaves the house – and data security risks. This was especially mentioned in the category of *GPS use*. Furthermore, *regulatory* factors play a significant role in the case of LBS-based profiling. The same applies to the delivery of location-based advertising depending on the user's location-based identification. When it comes to technological factors, the user identification seems to be of particular importance – more precisely: the creation of movement profiles is to be seen as a risk. Further findings clearly reveal several obstacles by combining user-friendly data security with LBS usability aspects. During the analysis, barriers from the providers' as well as the users' perspective could be identified. On the providers' side, the development effort required turns out to be a decisive barrier. As P1 puts it: "On the other hand, it's a curse because if you talk to developers, they absolutely hate it" (P1). This is accompanied by operational costs that providers have to spend, as well as accompanying business interests of third-party providers based on business relationships. According to participants P2, P3 and P4, nearly all third parties involved are more interested in collecting user data than in giving users more options for data security. On the user's side, the participants primarily observe obstacles arising from the *antipattern* phenomenon in connection with location-based advertising (P2, P3). In this regard, antipattern refers to "software design shortcomings that can negatively affect software quality, in particular maintainability" [27] instead of supporting user-friendliness. Another obstacle not to be neglected is a lack in the know-how of users – especially in the case of location-based advertising. Additionally, barriers in the form of an information overload when users themselves become responsible for data management (e.g., in order to provide more control about personal data) were frequently mentioned (P1, P6, P7). P9 summarizes the participants' basic understanding of data management quite aptly, stating "more information, more transparency and something else that goes hand in hand with that, obviously, is control" (P9). Despite the obvious possibility of an increase in control, the analysis shows that security aspects may also turn out to be conflicting or complicating factors – especially in the case of location-based profiling. Collectively, the participants endorse the call for more regulation and guidance by designing the aforementioned framework. "I do think it comes back to creating a framework, [a] code that can help guide the way [in] which [the LBS-application] is developed" (P10). The following hypotheses summarize the outcome of the content analysis:

1. The successful integration of UPS factors into LBS seems to require a regulatory framework covering the entire value chain of LBS. This includes all stakeholders, at governmental, economic, and civil society levels.

2. Embedding UPS factors into LBS applications increases the ability to control data security regulations. As a result, more responsibility and expertise on behalf of the user is mandatory.
3. By involving the user in the configuration of user-friendly systems, the danger of information overload arises. In this scenario, users may favor previously known systems that they perceive to be easier to use.
4. User-friendly data security in LBS applications is primarily prevented by antipatterns. The resulting obstacles may primarily affect the field of location-based advertising.
5. More data security for the user in LBS applications contradicts the business interests of third-party providers, who are mainly interested in collecting as much data as possible.

5 Discussion

For user-friendly data security to be part of the LBS application, the participants of the interview consider a current focus on effort and cost. However, the now obvious business interests of third parties can be outlined as a surprising result. From the participants' point of view, third parties in general are less interested in user-friendly data security than in gaining the most comprehensive collection and storage of user data possible. For these companies "the database itself is of course the treasure [...] with which they will likely want to earn money in the future" (P4) in order, for example, to create movement profiles with regard to advertising strategies or marketing activities. The collection and analysis of data could be regarded as the focus of LBS usage, this was also confirmed in a study by Fischer, who examined tracking of location data in different LBS application scenarios (mobility, tourism and event, out-of-home advertising, and consumer research) [28]. A regulatory framework could provide a possible solution for this issue, as this study's findings suggest an overarching need for some kind of regulation concerning the application of user-friendly security to location-based services. According to the participants, this framework should include socio-ethical, regulatory and technological factors reflecting the requirements of a value chain for the joint benefit of all stakeholders involved. First attempts of such an approach can be found in the research of Abbas et al. [1]. This way, the users' lack of knowledge could be compensated, and a clear guideline for service providers, third-party providers and users could be established. Another advantage of user-friendly data security, according to the participants, could be an increased use of LBS in general, going along with significantly more control over personal data and its use. Consequently, user-friendly data security is expected to increase the app retention in mobile applications that use LBS technologies [4]. In the case of commercial use, this could be directly linked to an increased customer loyalty.

6 Conclusion

The study investigated factors influencing user-friendly data security in location-based services. The findings indicate an overarching need for a framework regulating the compatibility of user-friendly security and location-based services (P10). This framework should include socio-ethical, regulatory and technological factors aligned with the interests of all stakeholders involved. With regard to possible security risks, it was revealed that both profiling and tracking [29] can involve a considerable risk. In terms of security risks, associations with criminally motivated use of LBS became apparent, as also mentioned by Michael and Michael [30]. Several barriers were identified to interfere with user-friendly data security in LBS usage: A significant additional expense (costs for personnel and development) on the part of the providers prevents LBS and UPS from being combined. As pointed out by several business representatives, user-friendly data security seems to be not compatible with a commercially oriented collection of user data. It could also be shown that this commercially oriented actors apparently expect regulatory restrictions, and, with that, governments incentivizing user-friendly data security. Strategic communication and the concept of privacy by design [31], on the other hand were said to play a subordinate role and considered to be not significant. Potential factors having a positive influence on the data management also had been identified. An increased control over one's own data and the possibility of data minimization are two of them. From the scientists' point of view, it is precisely these two aspects that have a key function in harmonizing user-friendly data security and LBS. However, above all, currently, the user side appears to be characterized by antipatterns. At the same time, an increasing differentiation of usage can overwhelm the end user. Despite clear obstacles mentioned by all participants, there are outstanding advantages that exert a clearly positive effects on user-friendly data security and LBS: On the one hand, extensive legal regulations, such as the GDPR (general data protection regulation) [32], are taking effect to minimize security risks and obstacles – as also evidenced by the anchoring of the principles of privacy by design in the GDPR [31]. On the other hand, an increased app usage in general could be highlighted by the use of user-friendly data security in LBS. This impact can be considered a possible success factor for the progression towards a framework for the compatibility of user-friendly data security in location-based services.

7 Limitations and Future Research

Although this study was conducted with great diligence, it was also subject to some limitations. With regard to the chosen method, the effect of social desirability may have had an impact on interviews [33]. The authors deliberately chose to apply the approach of Mayring [26] in order to evaluate all interviews. In this context, however, aligning with alternative, more internationally recognized methods for qualitative content analyses might be advisable in order to comply with the overarching topic's international character in future studies.

Further research in this area should deal with the design of a framework for in-depth results. In this context, it would be useful to investigate how this framework could be implemented under the GDPR and what relationships exist between the stakeholders. The results could supplement the set of rules of the GDPR with regard to location-based applications [1]. The international focus of this work is essentially to be regarded as a benefit. In order to deepen this perspective, it could prove useful to investigate and compare possible correlations between user-friendly data security and LBS in different (national) jurisdictions. For example, a study in the context of the 'California Consumer

Privacy Act' [34], a law protecting consumer rights in California, often mentioned by the participants, would appear reasonable. Based on the results of this work, the consideration of user-friendly data security in commercially used mobile applications, especially in location-based services, offers extensive opportunities for further research. Not least because of the further development of location-based services [35], this field of research can be expected to become even more relevant in the future.

References

1. Abbas, R., Michael, K., Michael, M.G.: The regulatory considerations and ethical dilemmas of location-based services (LBS). Info Technol. People **27**(1), 2–20 (2014)
2. Michael, K., Michael, M. G.: Innovative Automatic Identification and Location-Based Services. IGI Global, Hershey, PA (2009)
3. Cao, Y., Takagi, S., Xiao, Y., Xiong, L., Yoshikawa, M.: PANDA: policy-aware location privacy for epidemic surveillance. Proc. VLDB Endowment **13**(12), 3001–3004 (2020)
4. Heinemann, G.: Die Neuausrichtung des App- und Smartphone-Shopping. Springer Gabler, Wiesbaden (2018)
5. Parviainen, P., Tihinen, M., Kääriäinen, J., Teppola, S.: Tackling the digitalization challenge: how to benefit from digitalization in practice. Int. J. Inf. Syst. Proj. Manag. **5**(1), 63–77 (2017)
6. Singhal, M., Shukla, A.: Implementation of location based services in android using GPS and Web services. Int. J. Comput. Sci. Issues **9**(2), 237–242 (2012)
7. Wendt, A., Brand, M., Schüppstuhl, T.: Semantically enriched spatial modelling of industrial indoor environments enabling location-based services. In: Schüppstuhl, T., Tracht, K., Franke, J. (eds.): Tagungsband des 3. Kongresses Montage Handhabung Industrieroboter, pp. 111–121. Springer, Heidelberg (2018)
8. Zhou, T.: Understanding location-based services users' privacy concern. Internet Res. **27**(3), 506–519 (2017)
9. Dao, D., Rizos, C., Wang, J.: Location-based services: technical and business issues. GPS Solutions **6**(3), 169–178 (2002)
10. Asuquo, P., et al.: Security and privacy in location-based services for vehicular and mobile communications: an overview, challenges, and countermeasures. IEEE Internet Things J. **5**(6), 4778–4802 (2018)
11. Chen, L., et al.: Robustness, security and privacy in location-based services for future IoT: a survey. IEEE Access **5**, 8956–8977 (2017)
12. Farouk, F., Alkady, Y., Rizk, R.: Efficient privacy-preserving scheme for location based services in VANET system. IEEE Access **8**, 60101–60116 (2020)
13. Tefera, M.K., Yang, X., Sun, Q.T.: A survey of system architectures, privacy preservation, and main research challenges on location-based services. KSII Trans. Internet Inf. Syst. **13**(6), 3199–3218 (2019)
14. Aditya, P., Bhattacharjee, B., Druschel, P., Erdélyi, V., Lentz, M.: Brave new world. ACM SIGMOBILE Mob. Comput. Commun. Rev. **18**(3), 49–54 (2015)
15. He, Y., Chen, J.: User location privacy protection mechanism for location-based services. Digital Commun. Netw. Accepted Manuscript (2020)
16. Parmar, D., Rao, U.P.: Dummy generation-based privacy preservation for location-based services. In: ICDCN 2020: Proceedings of the 21st International Conference on Distributed Computing and Networking, pp. 1. ACM, New York, NY (2020)
17. Abul, O., Bayrak, C.: From location to location pattern privacy in location-based services. Knowl. Inf. Syst. **56**(3), 533–557 (2018). https://doi.org/10.1007/s10115-017-1146-x

18. Sun, G., et al.: Location privacy preservation for mobile users in location-based services. IEEE Access **7**, 87425–87438 (2019)
19. Saltzer, J., Schroeder, M.: The protection of information in computer systems. In: Proceedings of the IEEE 1975, vol. 63, pp. 1278–1308 (1975)
20. Zurko, M.E., Simon, R.T.: User-centered security. In: Haigh, T., Hosmer, H (eds.) Proceedings of the 1996 workshop on New security paradigms – NSPW '1996, the 1996 workshop, pp. 27–33. ACM Press, New York, NY (1996)
21. Naqvi, B., Porras, J.: Usable security by design: a pattern approach. In: Moallem, A. (ed.) HCII 2020. LNCS, vol. 12210, pp. 609–618. Springer, Cham (2020). https://doi.org/10.1007/978-3-030-50309-3_41
22. Cranor, L.F., Buchler, N.: Better together: usability and security go hand in hand. IEEE Secur. Priv. **12**(6), 89–93 (2014)
23. Atzeni, A., Faily, S., Galloni, R.: Usable security. In: Khosrow-Pour, M. (ed.): Advanced Methodologies and Technologies in System Security, Information Privacy, and Forensics, pp. 348–359 (2018)
24. Alt, F., von Zezschwitz, E.: Emerging trends in usable security and privacy. i-com **18**(3), 189–195 (2019)
25. Helfferich, C.: Leitfaden- und Experteninterviews. In: Baur, N., Blasius, J. (eds.) Handbuch Methoden der empirischen Sozialforschung, pp. 669–686. Springer VS, Wiesbaden (2019). https://doi.org/10.1007/978-3-658-21308-4_44
26. Mayring, P.: Qualitative content analysis. Forum Qual. Sozialforschung **1**(2), 20 (2000)
27. Fontana, F.A., Dietrich, J., Walter, B., Yamashita, A., Zanoni, M.: Antipattern and code smell false positives: preliminary conceptualization and classification. In: 2016 IEEE 23rd international conference on software analysis, evolution, and reengineering (SANER), pp. 609–613. IEEE, Piscataway, NJ (2016)
28. Fischer, B.: Mit einer Smartphone-App kontinuierlich Standortdaten erheben und Insights generieren. In: Keller, B., Klein, H.-W., Wachenfeld-Schell, A., Wirth, T. (eds.) Marktforschung für die Smart Data World, pp. 71–83. Springer, Wiesbaden (2020). https://doi.org/10.1007/978-3-658-28664-4_6
29. Bareth, U., Kupper, A.: Energy-efficient position tracking in proactive location-based services for smartphone environments. In: 2011 IEEE 35th Annual Computer Software and Applications Conference, pp. 516–521. IEEE, Piscataway, NJ (2011)
30. Michael, K., Michael, M.G.: The social and behavioural implications of location-based services. J. Location Based Serv. **5**(3–4), 121–137 (2011)
31. Cavoukian, A.: Understanding how to implement privacy by design, one step at a time. IEEE Consum. Electr. Mag. **9**(2), 78–82 (2020)
32. Schiering, I., Mester, B.A., Friedewald, M., Martin, N., Hallinan, D.: Datenschutz-Risiken partizipativ identifizieren und analysieren. Datenschutz und Datensicherheit **44**(3), 161–165 (2020)
33. Reuband, K.-H.: Soziale Erwünschtheit und unzureichende Erinnerung als Fehlerquelle im Interview: Möglichkeiten und Grenzen bei der Rekonstruktion von früherem Verhalten – das Beispiel Drogengebrauch. Zentralarchiv für Empirische Sozialforschung **23**, 63–72 (1988)
34. Aridor, G., Che, Y.-K., Nelson, W., Salz, T.: The economic consequences of data privacy regulation: empirical evidence from GDPR. In: SSRN Electronic Journal (2020)
35. Göll, N., Lassnig, M., Rehrl, K.: Location Based Services im Tourismus – Quo Vadis? In: Egger, R., Jooss, M. (eds.): mTourism – Mobile Dienste im Tourismus, pp. 27–44. Gabler, Wiesbaden (2010)

A Covid-19 Data Analysis During the First Five Months in Senegal

Hamidou Dathe[1,2] and Python Ndekou Tandong Paul[2(✉)]

[1] Sorbonne Université, UMMISCO, UMI 209, 93143 Bondy, France
hamidou.dathe@ucad.edu.sn
[2] Department of Mathematics and Computer Science,Cheikh Anta Diop University,
Dakar, Senegal
python.paul@ucad.edu.sn

Abstract. In this paper, we present an analysis of Covid-19 data during
the first 138 days in Senegal. The data come from the website of the
Ministry of Health and Social Action. We build a database with R and
Python scripts helping to carry out data analysis for answering questions
that support decision-makers. We compute the basic reproduction rate
with the collected data using the SIRD model approach (susceptible,
infected, recovered, death). Results show that there are two phases in the
evolution of infected cases, 4.75% are between the first day of the Covid-
19 outbreak and the 50th day, 95.25% are between the 51st day and
the 138th day corresponding to the period of the reopening of airports.
The average value of the basic reproduction rate during the first 138
days is 2.77. We observed a strong correlation between infected cases
and community infected cases. The obtained results could help decision-
makers to evaluate the statistical evolution of infected persons.

Keywords: Data analysis · Covid-19 disease · Statistical analysis ·
Basic reproduction rate · UML language · Mathematical modeling

1 Introduction

The Covid-19 pandemic has officially appeared in Senegal since March 2, 2020,
[2]. On April 26, 736 cases of infection and nine deaths were confirmed [2].
For fighting against the spread of covid-19, Senegal Government had canceled
international flights since Friday, March 20, 2020 [2]. On March 14, 2020, the
public authority closed schools, prohibited public events, canceled events involving more than 50 people [2]. On March 23, 2020, Senegal Government declared
a state of emergency throughout the territory and extended it for 30 days on
April 4. On April 19, the Government imposed the mandatory wearing of masks
in public and private services, shops and transports [2]. Each day, Senegal Government presents collected data as follows: the number of infected cases, the
number of tested cases, the number of the imported cases, the number of community cases, the number of infected cases under treatment, and the number of

© Springer Nature Switzerland AG 2021
O. Gervasi et al. (Eds.): ICCSA 2021, LNCS 12950, pp. 580–588, 2021.
https://doi.org/10.1007/978-3-030-86960-1_41

deaths cases. Data mining models have been used to evaluate the features of the covid-19 outbreak in several countries in the world [4]. For fighting against the transmission of infectious diseases, one can use data mining technics. Jahanbin et al. [5] have used social networks mining [10] in the management of the infectious disease systems. Huang and et al. have worked on the clinical features of patients infected with the covid-19 in China [6]. The study of the epidemiological characteristics of the novel coronavirus diseases has helped public authorities all over the world to develop new strategies for fighting the covid-19 disease [7–9]. The objectives of this work are: collect the Covid-19 disease data, develop a database containing these data, carry out statistical analysis of Covid-19 data, find correlations between statistical variables, examine statistical variations in the different critical periods during the evolution of Covid-19 infection cases.

2   Basic Mathematical Model

For computing the average basic reproduction rate with the real data, we first formulate the covid-19 mathematical model and carry out mathematical analysis that helps to compute the value of R_0. We use the compartments S,I,R,D (Susceptible, Infectious, Recovered, Death) [1] during the covid-19 infection period. We present the ordinary differential equations describing the covid-19 transmission as follows:

$$
\begin{cases}
\dfrac{dS}{dt} = \lambda - \mu S - \dfrac{\beta I S}{N} \\[2mm]
\dfrac{dI}{dt} = \dfrac{\beta I S}{N} - \gamma I - \mu I - \alpha I \\[2mm]
\dfrac{dR}{dt} = \gamma I - \mu R \\[2mm]
\dfrac{dD}{dt} = \alpha I
\end{cases}
\tag{1}
$$

where the susceptible population is characterized by the death rate μ, the susceptible recruitment rate λ, the infectious contact rate β, the death rate due to the disease α, the recoved rate γ, the total population N. The Mathematical analysis shows that the disease-free equilibrium (DFE) point is: $(S, I, R, D) = \left(\dfrac{\lambda}{\mu}, 0, 0, 0\right)$. We used the next generation matrix approach described by Diekmann et al. [3] for the computation of R_0. We have found that:

$$R_0 = \frac{\beta}{(\mu + \alpha) + \gamma}$$

when posing $R_0(t) = \dfrac{\beta \dfrac{\beta I(t) S(t)}{N}}{(\mu + \alpha) \dfrac{\beta I(t) S(t)}{N} + \gamma \dfrac{\beta I(t) S(t)}{N(t)}}$

assume that NIP(t) = number of infected persons at time t, NDP(t) = number of death persons at time t, NRP(t) = number of recovered persons at time t and let $R_1(t)$ be $R_1(t) = \dfrac{NIP(t)}{NDP(t) + NRP(t)}$.

If $R_1(t) < 1$ there are insufficient new infection, and the covid-19 cannot invade the population. When $R_1(t) > 1$, the covid-19 disease may persist in the environment. We have used $R_1(t)$ formula for computing the basic reproduction rate using collected data.

3   Data Modeling

3.1   Development of the Database

Like several African countries, Senegal had been preparing for the Covid-19 outbreak and developed a plan to fight against this disease. Hospitals and hotels had been recommended for the treatment of infected individuals. Covid-19 test centers have also been identified and placed at the disposal of the Ministry of Public Health. From the appearance of the first cases of the infection of this disease, the following data were published by the Minister of Public Health each day: the number of tested cases, the number of infected cases, the number of infectious contact cases, the number of deaths, the number of recovered cases, the number of infectious community cases. We used a UML (unified modeling language) class diagram (Fig. 1) to show the relationships between the interacting entities in Senegal. The UML model helps to build the Covid-19 database. The dataset used in this study comes from the different test centers and is updated every day. It's possible to see all the data on the Ministry of Public Health (http://www.sante.gouv.sn) website. In the first step of the data treatment, data were collected in each test center with the detection of errors. We carried out different scripts for plotting the data. The study of the correlation between two or several variables helps to know the intensity of the relationship between variables. We have computed for each couple of variables (x, y) the coefficient of linear correlation and have obtained the matrix of correlation. Mathematical models used to compute correlations between two statistical variables x and y are presented as follows:

$$
\begin{cases}
rcol = \dfrac{cov(x, y)}{\sigma_x \sigma_y} \\[2ex]
cov(x, y) = \dfrac{1}{p} \sum_{i=1}^{p} x_i y_i - (E(x)E(y)) \\[2ex]
\sigma_x^2 = \dfrac{1}{p} \sum_{i=1}^{p} n_i x_i^2 - (E(x))^2, \\[2ex]
\sigma_y^2 = \dfrac{1}{p} \sum_{i=1}^{p} n_i y_i^2 - (E(y))^2.
\end{cases}
\tag{2}
$$

p is the size of data, n_i is the size of x_i. rcol is the linear correlation between x and y, σ_x is a standard deviation of x, the standard deviation helps to measure the dispersion of the data.

3.2   Description of Algorithms

In this part, we present the following algorithms (Table 1) used to compute the statistical results: mean(X, n) computes the average of data belonging to X. var(X, n) computes the variance of data belonging to X. cov(X, Y, n) computes the covariance between X and Y.

Table 1. Algorithms used for data analysis

Algorithm 1	Algorithm 2	Algorithm 3
function mean(X,n):real	function var(X,n):real	function cov(X,Y,n):real
1. s1 = 0	1.s2 = 0	1. s3 = 0
2. for i = 1 to n do	2. for j = 1 to n do	2. for k = 1 to n do
3 s1 = s1 + X[i]	3. s2 = s2 + X[j]**2	3. s3 = s3 + X[k] * Y[k]
4. endfor	4. endfor	4 endfor
5. xb1 = s1/n	5. xb2 = s2/n	5. xb3 = s3/n
6. return(xb1)	6. V = xb2-(mean(X,n)**2)	6. CV = xb3-(mean(X,n) * mean(Y,n))
7. endfunction	7. return(V)	7. return(CV)
	8. endfunction	8. endfunction

3.3   Matrix of Linear Correlation Coefficients

We suppose that x_i represents the data value at the intersection of row i and column x, y_i represents the data value at the intersection of row i and column y. The database file of Covid-19 data is called Covid-19.csv, using the R software language, the following command has helped to record data into the dataframe. dataframeData<-read.csv(Covid-19.csv, sp = ";", decimal = "."). The database has p fields, the dataframe also has p columns, Data[, i] allows to get all data from column i. A correlation matrix is a set of values allowing to measure the intensity of the relationship between variables. The correlation matrix contains the linear correlation coefficients between each variable and others. We used the function rcorr () coming from the R language to compute the Pearson and Spearman correlations (Table 2).

3.4   Results and Discussion

We collected data during 138 days. We observed that there were 20 days without death cases and recovered cases. There were two phases in the evolution of

Table 2. correlations between variables during the first 138 days of the presence of Covid-19

	InfectedC	ImportedC	ContactC	CommunityC	RecoveredC	DeathC
InfectedC	1.00	−0.004	0.98	0.71	0.67	0.60
ImportedC	−0.004	1.00	−0.087	0.028	0.037	0.12
ContactC	0.98	−0.08	1.00	0.59	0.64	0.53
CommunityuC	0.71	0.028	0.59	1.00	0.58	0.62
RecoveredC	0.67	0.037	0.64	0.58	1.00	0.50
DeathC	0.603	0.121	0.53	0.62	0.50	1.00

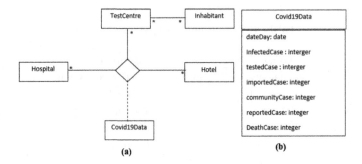

Fig. 1. UML class diagram showing different sources of data represented by a graphic formalism(a) and the Covid19 data class used for creating the database containing all data concerning the spread of the coronavirus 2019 in Senegal(b)

infected cases: 4.75 % of infected cases were between the first day and the 50th day, 95.25% of infected cases were between the 51st day and the 138th day (Fig. 2). There were three phases in the evolution of infectious imported cases: 49.72 % of infectious imported cases were between the first day and the 33rd day, 6.43 % of the infectious imported cases were between the 34th day and the 100th day, 43.85% of infectious imported cases were between the 101st day and the 138th day (Fig. 3). There were two phases in the evolution of infectious contact cases: 3.98% of infectious contact cases were between the first day and the 50th day, 96.02 % of infectious contact cases were between the 51st day and the 138th day (Fig. 4).

Concerning infectious community cases, we observed two scenarios: the first scenario in two phases showed that there were 3.71 % of infectious community cases between the first day and the 50th day, 96.29% of infectious community cases between the 51st day and the 138th day (Fig. 5). A second scenario showed that there were 3.71% of infectious community cases between the first day and the 50th day, 22.88% of infectious community cases were between the 51st day and the 90th day, 73.41% of infectious community cases were between the 91st day and the 138th day (Fig. 5). For the recovered cases, we observed two scenarios: the first scenario in two phases showed that there were 4.14% of recovered cases between the first day and the 50th day, 95.86% of recovered cases were between the 51st day and the 138th day (Fig. 5).

The second scenario in three phases showed that there were 6.08% of recovered cases between the first day and the 60th day, 43.17% of recovered cases between the 61st day and the 100th day, 50.75% of recovered cases between the 101st day and the 138th day (Fig. 6). For death cases, we observed two scenarios: a first scenario showed that there were 3.06% of death cases between the first day and the 50th day, 95.94% of death cases between the 51st day and the 138th day (Fig. 7). A second scenario showed that there were 11.66% of death cases between the first day and the 70th day, 20.24% of death cases between the 71st day and the 100th day, 68.1% of death cases between the 101st day and the 138th day (Fig. 7). The Computation of daily infection rates showed that the average infection rate (infected cases/number of tests) between the first day and the 31st day was 29.54%, the average infection rate between the 32nd day and the 138th day was 8.68% (Fig. 8). For the computation of the basic reproduction number, we observed four phases during 119 days: between the first and the 11th day, the average basic reproduction rate was 4.77, between day 12 and day 31 the average basic reproduction rate was 1.48, between the day 32 and day 46 the average basic reproduction rate was 7.87, between the day 47 and day 119 the average basic reproduction rate was 1.7 (Fig. 9).

The average basic reproduction rate between the first day and the 138th day was 2.72, explaining the persistence of Covid19 disease. The matrix of correlations between variables showed that there was a very strong correlation between infected cases and infectious contact cases, a strong correlation between infected cases and infectious community cases.

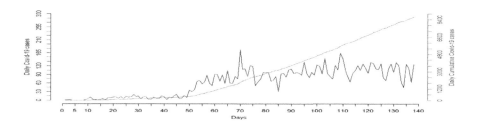

Fig. 2. Evolution of infected cases

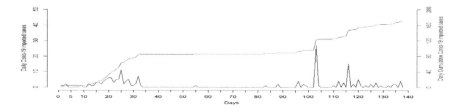

Fig. 3. Evolution of imported cases

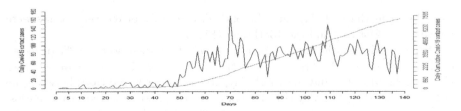

Fig. 4. Evolution of contact cases

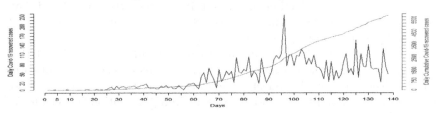

Fig. 5. Evolution of community cases

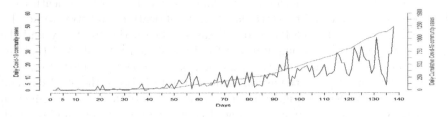

Fig. 6. Evolution of recovered cases

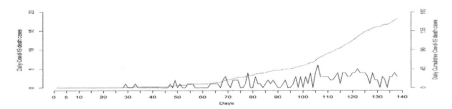

Fig. 7. Evolution of the death cases

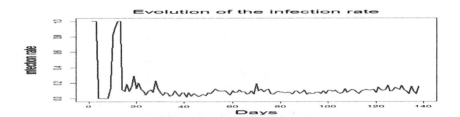

Fig. 8. Evolution of infection cases

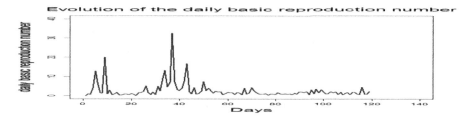

Fig. 9. Evolution of daily basic reproduction rate. The average basic reproduction rate is 2.72

We computed the rate q of the population that can be infected at the end of the covid-19 in Senegal and observed that this rate could be 90.54% of the total population.

4 Conclusion

In this study, we collected data from the different test centers of the covid-19 disease in Senegal. With those data, we have built a database that allowed us to carry out the statistical analysis. Results coming from the statistical analysis helped the decision-makers to evaluate the dynamics of covid-19 infection. During 138 days, corresponding to the 138 first days of the covid-19 presence disease in Senegal, we observed that imported cases of covid-19 had a high impact on the growth of the infected cases. We also observed that when the community infected cases and contact infected cases increase at the same time, we had exponential growth of infected cases. Due to the difficulty of the Ministry of Public Health and Government to evaluate the real impact of each barrier measure, the graphics that we have drawn could help to estimate the effects of each type of measure barrier (mask, social distancing, hand washing). With the statistical results, the Ministry of public health could evaluate the efficiency of barrier measures during the first 138 days of the presence of the covid-19 in Senegal.

References

1. Ross, R.: An application of the theory of probabilities to the study of a priori pathometry. Part I. Proc. R. Soc. London **92** (638) (1916). https://doi.org/10.1098/rspa.1916.0007
2. Covid19 Cases number. http://www.sante.gouv.sn/Actualites/
3. Diekmann, O., Heesterbeek, J.A., Metz, J.A.: On the definition and the computation of the basic reproduction ratio R0 in models for infectious diseases in heterogeneous populations. J. Math. Biol. **28**(4), 36582 (1990). ISSN 0303-6812. https://doi.org/10.1007/BF00178324
4. Li, X.Q., Cuomo, R., Purushothaman, V.: Data mining and content analysis of the Chinese social media platform weibo during the early COVID-19 outbreak: retrospective observational infoveillance study. JMIR Public Health Surveill. **6**(2), e18700 (2020). PMID: 32293582 PMCID: PMC7175787 https://doi.org/10.2196/18700

5. Jahanbin, K., Rahmanian, F., Rahmanian, V., Sotoodeh, J.A.: Application of Twitter and web news mining in infectious disease surveillance systems and prospects for public health. GMS Hyg. Infect. Control **14**, 1–12 (2019)

6. Huang, C., Wang, Y., Li, X., Ren, L., Zhao, J., Hu, Y., et al.: Clinical features of patients infected with novel coronavirus in Wuhan. China. Lancet (2019). https://doi.org/10.1016/S0140-6736(20)30183-5

7. Nishiura, H., Jung, S.M., Linton, N.M., Kinoshita, R., Yang, Y., Hayashi, K., et al.: The extent of transmission of novel coronavirus in Wuhan, China. J Clin Med. (2020). https://doi.org/10.3390/jcm9020330

8. Majumder, M., Mandl, K.D.: Early transmissibility assessment of a novel coronavirus in Wuhan, China. SSRN (2020). https://doi.org/10.2139/ssrn.3524675

9. Wang, C., Horby, P.W., Hayden, F.G., Gao, G.F.: A novel coronavirus outbreak of global health concern. Lancet. **395**(10223), 470–473 (2020). [Europe PMC free article] [Abstract] [CrossRef] [Google Scholar]. https://doi.org/10.1016/S0140-6736(20)30185-9

10. Guellil, I., Boukhalfa, K.: Social big data mining: a survey focused on opinion mining and sentiments analysis. In: Conference of ISPS 2015: 12th International Symposium on Programming and Systems, Algiers. IEEE (2015). https://doi.org/10.1109/ISPS.2015.7244976

Dataset Anonimyzation for Machine Learning: An ISP Case Study

Lelio Campanile[1], Fabio Forgione[2], Fiammetta Marulli[1],
Gianfranco Palmiero[2], and Carlo Sanghez[2(✉)]

[1] Universitá degli Studi della Campania "L. Vanvitelli", Dip. Matematica e Fisica,
Caserta, Italy
{lelio.campanile,fiammetta.marulli}@unicampania.it
[2] GAM Engineering srlu, Caserta, Italy
{fforgione,gpalmiero,csanghez}@gameng.it
https://www.gameng.it

Abstract. Internet Service Providers technical support needs personal data to predict potential anomalies. In this paper, we performed a comparative study of forecasting performance using raw data and anonymized data, in order to assess how much performance may vary, when plain personal data are replaced by anonymized personal data.

Keywords: ISP · Cryptography · Customer Premise Equipment · WISP · Logistic regression · Random forest · Attribute suppression · Character masking · Hash · Pseudo-anonymization · Data generalization

1 Introduction

Preservation of privacy in data mining and machine learning has emerged as an absolute prerequisite in several practical scenarios, especially when the processing of sensitive data is outsourced to an external third party. Currently, privacy preservation methods are mainly based on randomization and/or perturbation, secure multiparty computations and cryptographic methods.

In [7] is proposed a strategy that leverages partial homomorphic property of some crypto-systems to train n simple machine learning models on encrypted data. The basic scenario discussed involves three parties, represented by: 1)multiple Data Owners, which provide encrypted training examples; 2) the Algorithm Owner (or Application), processing them to adjust the parameters of its models; 3) a semi-trusted third party, which provides privacy and secure computation services to the Application, when operations that are not supported by the homomorphic cryptosystem occur. More precisely, their focus is on the use of multiple-key crypto-systems, and on the impact of the quantization of real-valued input data, required before encryption.

In [10], the authors design a protocol framework to perform classification tasks in a privacy-preserving manner. To demonstrate the feasibility of their proposal, two protocols, supporting Naive Bayes classification, are implemented and

© Springer Nature Switzerland AG 2021
O. Gervasi et al. (Eds.): ICCSA 2021, LNCS 12950, pp. 589–597, 2021.
https://doi.org/10.1007/978-3-030-86960-1_42

the heavy computational load of conventional fully homomorphic encryption-based privacy-preserving protocols, is overcome by adopting different optimization techniques. This method differs from other existing techniques because it requires no intermediate interactions between the server and the client while executing the protocol, except for the mandatory interaction to obtain the decryption result of the encrypted classification output. As a result of this minimal interaction, the proposed method results in a relatively stable solution. Furthermore, in [8], is presented the *CryptoDL* framework, thus offering new techniques to provide solutions for applying deep neural network-based algorithms to encrypted data.

In the work we propose, we address the problem of evaluating the performance degradation when a classification task is performed over the same data samples but several different machine learning algorithms and various anonymization methodologies are adopted.

2   Related Work

The main field of application of the studies introduced in the previous section, is focused on the personal privacy requirements of each individual, thus preserving as much information as possible and when it applied to a large quantity of data preserving also good performance.

For the purposes of the analysis, there is the need to limit the risks of disclosure to an acceptable level, while retaining the information and thus the usefulness. When using data sets containing sensitive information and data that should not be disclosed, it becomes necessary to use anonymization techniques on the raw data.

In [9], various privacy techniques are compared, evaluating the strengths and weaknesses of the different techniques. An interesting insight that focuses on the possibility of pseudo-anonymizing data and subsequently de-pseudonymizing it, is presentend in [2]; in that case, healthcare patients data are used and the proposed technique enables searches on anonymous aggregated data, then reporting the results to the individual patient.

Several national authorities in different countries have issued interesting guidelines related to privacy and anonymization and pseudo-anonymization of personal data [4], [12], [11].

A more complex approach to pseudo-anonymization of data is offered by the *K-anonimization* technique, which is investigated in [14] and [13].

However, most of the anonymization and pseudo-anonymization techniques result to be inefficient, when evaluated in terms of performance; furthermore, they introduce an overhead to the operations that need to be performed on the data, including machine learning operations. The authors of the work described in [5], face the performance problems and propose a framework that addresses the data privacy preservation in an efficient way. In [6], the authors address, in particular, the topic of the application of machine learning algorithms on anonymized data, carefully concerning with the problems related to the accuracy and performance of the resulting models.

3   Case Study

In this section we describe the case study on which is based our study on performance decay. It consists in *Flyber*[1], a Wireless Internet Service Provider (WISP) based on Teverola, Italy, that offers its customers free technical support to solve problems related to the quality of signal reception. Since the technology is wireless over open bands, the quality of the connection depends mainly on radio interference. To check the health status of the Customer Premise Equipments, Flyber adopts a monitoring system based on two servers: the first one is the Dude[2] network monitor by MikroTik and, the second one, is cnMaestro by Cambium Networks [1]. Since most clients use Cambium CPEs, we focused only on cnMaestro related data. The monitoring system produces a daily report with 25 fields. An extract of a daily report is showed in (Fig. 1).

Fig. 1. cnMaestro on premises performance 24 h system daily report.

Flyber staff uses a customer relationship management system based on a custom software solution. In Fig. 2 is shown an extract of records.

The first part of the work consists in classifying the records according to two conditions: with and without reporting by the customer. In the first step, we obtained a model able to classify a given condition on the CPE, by using personal data without anonimyzation. Since it was a classic binary classification, we considered the logistic regression and the random forest algorithms. The second step consisted in transforming the data set through the anonymization of personal data and, finally, the last step was to obtain a model able to classify a given condition on the CPE using data from anonimous data set, in order to assess performance changes. The considered time frame was between 21st December 2020 and 23rd March 2021. During this time frame, we collected 68 reports, each including daily data incoming from about 600 antennas and we received 1580 reports to Flyber's customer care. The experiment was therefore conducted on about 40,000 samples. As regards the samples of the reports to customer care, the experiment was conducted by cleaning the data set from administrative issues, or referred to non-Cambium equipment.

---

[1] www.flyber.it.

[2] https://mikrotik.com/thedude, [Online; accessed 02-May-2021].

1	Open issue - M⬛⬛⬛LA,2020-12-07 10:50:42+01:00
2	Open issue - ST⬛⬛OLO,2021-01-22 11:59:08+01:00
3	Open issue - G⬛⬛RAFFAELE,2020-09-11 14:46:47+02:00
4	Open issue - R⬛⬛2021-03-03 17:13:31+01:00
5	Open issue - N⬛⬛AKI ABBAS,2020-09-28 09:08:46+02:00
6	Open issue - ER⬛⬛,2021-02-09 13:00:54+01:00

Fig. 2. CRM records with personal data.

4   Discussion and Results

4.1   Raw Dataset

The experiment was conducted considering a data set with 43780 rows × 25 columns. In the following, a description of the columns that include personal data, is provided.

- **MAC** is the MAC address of the CPE.
- **Device Name** is the configured device name of the AP, used for identifying the device in an NMS such as the Cambium Network Services Server (CNSS). It includes personal data.
- **COD. CLIENTE** is the customer ID.
- **COMUNE INSTALLAZIONE** is the ZIP code of customer.

The initial data set was further cleaned by deleting useless, null or dirty records. In fact, sometimes the CPE communicates incorrect data, or the monitoring system adds null values when the CPE does not respond in time. By doing so, the number of samples was reduced to about 17.000.

Classification of Data Set with Personal Data Using the Logistic Regression: the first classification tests were not satisfactory. Indeed, the system was able to correctly classify less than one out of five anomalies.

Classification of Data Set with Personal Data Using the Random Forest: the first classification tests were satisfactory. Indeed, the system was able to correctly classify more than one out of three anomalies. The results are summarized in the Table 1. In the first column of the table we reported the total number of samples, in the second column we reported the percentage of samples with issues, in the third column of the table we reported the percentage of success in identifying the critical samples using logistic regression. Finally, in the last column we reported the success rate we achieved using the random forest on the non-anonymous data set. We could have improved the performance of the algorithms further or changed the algorithm, but the goal here is to measure the performance drop by anonymizing the data. So, we have selected the random forest classification algorithm.

Table 1. Data set with personal data: logistic regression (LR) VS Random forest (RF)

Samples	% Issues	% of success by LR	% of success by RF
17.000	3,2%	20%	37%

4.2   Anonymous Data Set

Our intent was to follow good practices in terms of data anonymization. We knew that we would reduce the original information present in the data set, but the question was how much we would lose in terms of performance. We have chosen to use different approaches to anonymize the data.

Details below.

– **Attribute suppression:** refers to the removal of the personal part from data. We simply deleted the columns with personal data from the data set.
– **Character Masking:** refers to the changing some characters of a personal data record by using a 'x'.
– **Pseudo-anymization by HASH:** refers to the replacement of personal data with anonymous data. We simply used a hash function on strings containing personal data. The new value was used instead of the old one.
– **Generalization:** refers to the converting personal data in an range of values. We simply grouped personal values in order to keep the rough information, but accepting the loss of detail.
– **Generalization and k-anonymity:** refers to the converting data in an range of values as seen before, but in this case we deliberately deleted those samples that had no other k-similar. We did not use K-anonymization due to the size of our data set [3].

Results from Attribute Suppression. We have simply deleted columns that include personal data from the data set. By applying the classification algorithm we obtained a success rate of 31,79%. In fact, during our tests, the system was able to correctly classify 55 out of 173 samples.

Results from Character Masking. We have replaced the last three octets of the MAC with the 'x' character. We have replaced the last four characters of the 'Device name' column with the 'x' character. We have replaced the last two characters of the 'COMMON INSTALLATION' column with the 'x' character. Finally, we have replaced the last three characters of the 'COD. CUSTOMER' with the character 'x'. In this way, we have made the data set completely anonymous. Below is the code block that performs the masking.

```
1 XY_data_ctc['MAC_MASK']=\
2 XY_data_ctc['MAC'].str[:-8]+'xx:xx:xx'
3
4 XY_data_ctc['Device Name_MASK']=\
```

```
5  XY_data_ctc['Device Name'].str[:-4]+'xxxx'
6
7  XY_data_ctc['COMUNE INSTALLAZIONE_MASK']=\
8  XY_data_ctc['COMUNE INSTALLAZIONE'].str[:-2]+'xx'
```

By applying the classification algorithm we obtained a success rate of 35,26%. In fact, during our tests, the system was able to correctly classify 61 out of 173 samples. The critical issue here is the time that is lost in transforming personal data via code. In fact, on the ARM Linux system used for the inference tests, about 0,158 s are lost to anonymize the 400 CPEs data. In our case study, samples from 2000 CPEs are collected every second, so considering the time lost to anonymize the data using masking function, we lose 0,785s for each step.

Results from Pseudo-Anonymization. We simply applied a hash function directly to the personal data. Thus, the number of columns and features did not change. We have modified the MAC, Device_Name, COD_CLIENTE, COMUNE_INSTALLAZIONE columns using Python code. The code is shown below.

```
1  import hashlib
2
3  def encrypt_string(hash_string):
4      sha_signature = \
5          hashlib.sha256(hash_string.encode()).hexdigest()
6      return sha_signature
7
8  MACRIPTO = []
9
10 for row in XY_data_ctc['MAC']:
11     temp = (str(row))
12     encMessage = encrypt_string(temp)
13     MACRIPTO.append(encMessage)
14
15 XY_data_ctc['MAC_CRIPTO_HASH'] = MACRIPTO
16
17 DeviceNameCRIPTO = []
18
19 for row in XY_data_ctc['Device Name']:
20     temp = (str(row))
21     encMessage = encrypt_string(temp)
22     DeviceNameCRIPTO.append(encMessage)
23
24 XY_data_ctc['Device Name_HASH'] = DeviceNameCRIPTO
25
26 CODCLIENTECRIPTO = []
27
28 for row in XY_data_ctc['COD.CLIENTE']:
29     temp = (str(row))
30     encMessage = encrypt_string(temp)
```

```
31      CODCLIENTECRIPTO.append(encMessage)
32
33  XY_data_ctc['COD. CLIENTE_HASH'] = CODCLIENTECRIPTO
34
35  CCRIPTO = []
36
37  for row in XY_data_ctc['COMUNE INSTALLAZIONE']:
38      temp = (str(row))
39      encMessage = encrypt_string(temp)
40      CCRIPTO.append(encMessage)
41
42  XY_data_ctc['COMUNE INSTALLAZIONE_HASH'] = CCRIPTO
```

By applying the classification algorithm we obtained a success rate of 37,57%. In fact, during our tests, the system was able to correctly classify 65 out of 173 samples. The critical issue here is the time that is lost in transforming personal data via code. In fact, on the ARM Linux system used for the inference tests, about 0,4778 s are lost to anonymize the 400 CPE data. In our case study, samples from 2000 CPEs are collected every second. Considering the time lost to anonymize the data using a function, we lose too much time at every step. The wasted time is 2,389 s at every step.

Results from Generalization. We have eliminated the last octet present in the 'MAC' column, we have grouped the values present in the 'CUSTOMER CODE' column into groups of 500, we have eliminated the last two digits from the ZIP code present in the 'COMMON INSTALLATION' column and finally we have considered only the first letter present in the values of the 'DEVICE NAME' column. In this way, the number of features has changed. Indeed, the non-anonymous data set included more than 1200 characteristics, while the grouped data set has 311 characteristics.

The code is shown below.

```
1   XY_data_ctc['MAC_GEN']=XY_data_ctc['MAC'].str[:-2]
2   XY_data_ctc['Device Name_GEN']=\
3   XY_data_ctc['Device Name'].str[:1]
4   XY_data_ctc['COD. CLIENTE_GEN'] =\
5   XY_data_ctc['COD. CLIENTE'].str[3:]
6   XY_data_ctc['COD. CLIENTE_GEN']=\
7   pd.to_numeric(XY_data_ctc['COD. CLIENTE_GEN'])
8   XY_data_ctc.loc[XY_data_ctc['COD. CLIENTE_GEN'] \
9   <= 499 ,['COD. CLIENTE_GEN']]    = 10000
10  XY_data_ctc.loc[XY_data_ctc['COD. CLIENTE_GEN'] \
11  <= 999 ,['COD. CLIENTE_GEN']]    = 20000
12  XY_data_ctc.loc[XY_data_ctc['COD. CLIENTE_GEN'] \
13  <= 1499 ,['COD. CLIENTE_GEN']]    = 30000
14  XY_data_ctc.loc[XY_data_ctc['COD. CLIENTE_GEN'] \
15  <= 1999 ,['COD. CLIENTE_GEN']]    = 40000
16  XY_data_ctc.loc[XY_data_ctc['COD. CLIENTE_GEN'] \
17  <= 2499 ,['COD. CLIENTE_GEN']]    = 50000
```

```
18  XY_data_ctc.loc[XY_data_ctc['COD. CLIENTE_GEN'] \
19  <= 2999 ,['COD. CLIENTE_GEN']]  = 60000
20  XY_data_ctc['COMUNE INSTALLAZIONE_GEN']=\
21  XY_data_ctc['COMUNE INSTALLAZIONE'].str[:-2]
```

By applying the classification algorithm we obtained a success rate of 35,26%. In fact, during our tests, the system was able to correctly classify 61 out of 173 samples. The critical issue here is the time that is lost in transforming personal data via code. In fact, on the ARM Linux system used for the inference tests, about 0,17 s are lost to anonymize the 400 CPE data. In our case study, samples from 2000 CPEs are collected every second. Considering the time lost to anonymize the data using this algorithm, we lose too much time at every step. The wasted time is 0,85 s at every step.

Table 2. Attribute suppression (AS) VS Character Masking (CM) VS Pseudo-anonymization (PS) VS Generalization (GE)

Algorithm	Attribute suppression	Character masking	Pseudo-anonymization by HASH	Generalization
Success rate	31,79%	35,26%	37,57%	35,26%
Loss of data?	NO	NO	NO	NO
Time issue?	NO	YES We lose 0,785 s for each step	YES We lose 2,389 s for each step	YES We lose 0,85 s for each step

5   Conclusion and Future Work

In this work, we explored the impact of adopting anonymization and pseudo-anonymization in a real-word application, when a Machine Learning algorithm is performed on anonymized data. Our experiments were focused on computing performance and accuracy, when data anonymization is applied or not. The preliminary results bring evidence of certain differences among the adoption of different techniques, as it is summarized in Table 2. As evidenced from the results provided, the best performance was obtained by applying hashing techniques, but the time spent removing personal data was a big issue. The future direction of this work consists in improving the effectiveness of the anonymization, evenly adopting the k-anonymization technique, and to enhance the computing performance, by distributing the computational load on multiple computational devices.

References

1. Cambium Networks. https://www.cambiumnetworks.com/products/software/cnmaestro-management/. Accessed 02 May 2021
2. Aamot, H., Kohl, C.D., Richter, D., Knaup-Gregori, P.: Pseudonymization of patient identifiers for translational research. BMC Med. Inform. Decis. Mak. **13**(1), 1–15 (2013). https://doi.org/10.1186/1472-6947-13-75
3. Aggarwal, C.C.: On k-anonymity and the curse of dimensionality. VLDB **5**, 901–909 (2005)
4. Article 29 Data Protection Working Party: Opinion 05/2014 on Anonymisation Techniques. Working Party Opinions (April), 1–37 (2014). http://ec.europa.eu/justice/data-protection/index_en.htm%0Aec.europa.eu/justice/article-29/documentation/opinion-recommendation/files/2014/wp216_en.pdf
5. Ghinita, G., Karras, P., Kalnis, P., Mamoulis, N.: Fast data anonymization with low information loss. In: 33rd International Conference on Very Large Data Bases, VLDB 2007 - Conference Proceedings, pp. 758–769 (2007)
6. Goldsteen, A., Ezov, G., Shmelkin, R., Moffie, M., Farkash, A.: Anonymizing machine learning models. arXiv (2020)
7. González-Serrano, F.-J., Amor-Martín, A., Casamayón-Antón, J.: Supervised machine learning using encrypted training data. Int. J. Inf. Secur. **17**(4), 365–377 (2017). https://doi.org/10.1007/s10207-017-0381-1
8. Hesamifard, E., Takabi, H., Ghasemi, M., Wright, R.N.: Privacy-preserving machine learning as a service. In: Proceedings on Privacy Enhancing Technologies, vol. 2018, no. 3, pp. 123–142 (2018)
9. Murthy, S., Abu Bakar, A., Abdul Rahim, F., Ramli, R.: A Comparative study of data anonymization techniques. In: Proceedings - 5th IEEE International Conference on Big Data Security on Cloud, BigDataSecurity 2019, 5th IEEE International Conference on High Performance and Smart Computing, HPSC 2019 and 4th IEEE International Conference on Intelligent Data and Security, IDS 2019, pp. 306–309 (2019). https://doi.org/10.1109/BigDataSecurity-HPSC-IDS.2019.00063
10. Park, H., Kim, P., Kim, H., Park, K.W., Lee, Y.: Efficient machine learning over encrypted data with non-interactive communication. Comput. Stan. Interf. **58**, 87–108 (2018)
11. Rifaut, A.: Office of inspector general, health care compliance association: guidance note: guidance on anonymisation and pseudonymisation. In: 2011 4th International Workshop on Requirements Engineering and Law, RELAW 2011, Proceedings - Held in Conjunction with the 19th International Requirements Engineering Conference (June), pp. 1–54 (2019). https://oig.hhs.gov/compliance/101/files/HCCA-OIG-Resource-Guide.pdf
12. Singapore, P.D.P.C.: Guide to basic data anonymisation techniques. In: Published 25 January 2018. Personal Data Protection Commission Singapore (PDPC) (January), pp. 1–39 (2018). https://www.pdpc.gov.sg/-/media/Files/PDPC/PDF-Files/Other-Guides/Guide-to-Anonymisation_v1-(250118).pdf
13. Sweeney, L.: k-anonymity: a model for protecting privacy. Int. J. Uncertain. Fuzziness Knowl. Based Syst. **10**(5), 557–570 (2002). https://doi.org/10.1142/S0218488502001648
14. Zhong, S., Yang, Z., Wright, R.N.: Anonymization of customer data, vol. 1, pp. 139–147 (2005)

Traffic Modelling Through a LSTM Variational Auto Encoder Approach: Preliminary Results

Stefano Chiesa and Sergio Taraglio[✉]

ENEA, Robotics Laboratory, ENEA, Rome, Italy
{stefano.chiesa,sergio.taraglio}@enea.it

Abstract. The work in progress on a LSTM sequence to sequence variational autoencoder generative model is presented. The architecture is trained on a floating car data dataset in order to grasp the statistical features of the traffic demand in the city of Rome. An analysis of parameters influence is furnished. The generated trajectories are briefly compared with the ones in the dataset. Further work direction is provided.

Keywords: Generative models · Variational auto encoder · Urban mobility · Traffic model

1 Introduction

Transportation forecasting is the attempt of estimating the number of vehicles or people that will use a specific transportation facility in the future. This can be considered as a relevant subject of Urban Computing: the process of tackling the major issues which cities face using big and heterogeneous data collected by a diversity of sources in urban areas [1].

Among the different approaches to the problem of traffic modelling, activity-based models are a group that try to predict for individuals where and when specific activities (e.g. work, leisure, shopping) are carried out [2]. The idea behind activity-based models is that travel demand is derived from activities that people need or wish to perform; travel is then seen as just a sort of by-product of an agenda, as a component of an activity scheduling decision. In this work we are mainly interested in modelling the traffic flows in an urban environment.

Traffic modelling begins with the collection of data. People can be of extreme help in this, since they generate data while travelling e.g. by carrying and using a mobile phone. Data can be collected through many different means, e.g. using CDR (Call Data Records) from the cellular phones, social network posting, vehicle or smartphone navigation data, floating car data (FCD) et cetera. The approach here presented is loosely related to activity-based models since it exploits car trajectory stops in a floating car dataset (FCD) which can be linked to human activities such as working, sleeping, shopping. However the FCD can account for a very partial description of human activities, only the car related ones, but they can help in reconstructing some features of mobility using machine learning.

© Springer Nature Switzerland AG 2021
O. Gervasi et al. (Eds.): ICCSA 2021, LNCS 12950, pp. 598–606, 2021.
https://doi.org/10.1007/978-3-030-86960-1_43

Human mobility can be seen as a highly structured domain composed of mostly regular daily/weekly schedules, showing high predictability across diverse populations [3]. Further analysis also suggests that human mobility shows temporal and spatial regularities [4, 5].

All these mobility data suffer from a major drawback: the issue of anonymization, the travelling information should be gathered namelessly in order to protect the people's right to privacy. Besides this basic fact, there are further concerns linked to the actual availability of data, since they often belong to private enterprises or governmental agencies, and linked to their representativeness, being often biased or with a low sampling rate.

A possible way to overcome these concerns is to employ generative models, as it has recently been proposed in several fields of application [6], and also to tackle the modelling of traffic demand [7, 8].

In a nutshell a generative model can be described as follows. Let us assume a dataset of observations with a given unknown statistical distribution P_{data}. A generative model P_{mod} is designed to mimic P_{data}, thus in order to generate new observations distributed as P_{data} it would be possible to sample P_{mod} [6]. The model is trained with the observation data and it can learn a compact feature space, whose sampling outputs new observation data with the same statistical properties of the original ones. This clearly overcomes on one hand the problem of user privacy and on the other the risk of too small datasets.

A reliable approach for the generation of new patterns on the basis of the knowledge of a given domain is represented by the use of auto encoders. A further refinement of these is represented by the variational auto encoders (VAE) [6], which are able to add a stochastic component that helps in producing both good data modelling and generation at the same time.

In the following a machine learning approach and some preliminary results are presented on a FCD dataset composed of the trajectories of a set of private cars in the town of Rome, in order to reproduce the traffic flows and vehicle geographical density. The architecture used is a sequence to sequence Long Short Term Memory variational auto encoder, the LSTM being a recursive neural network proficient in modelling sequences [9].

In the second section of the paper a brief description of the chosen architecture is provided together with a sketch of standard and variational auto encoders and LSTM neural networks. In the third the dataset is described and some data processing disclosing some interesting features together with the data representation is presented. In the fourth the training results and some preliminary analysis on generated data are provided. Finally in the fifth section some conclusions and future work directions are given.

2 LSTM Networks and the Variational Auto Encoders

Recurrent neural networks (RNNs) have become the state of the art for sequence modelling and generation, the language and translation domains being the more instantiated subjects. Long Short Term Memory (LSTM) networks [9] are one of the most popular variations of RNNs with proven ability to generate sequences in various domains, such as text [10], images [11] and handwriting [10]. The success of RNNs in these domains, which present similarities with the traffic one, has motivated several works applying RNNs for human activity sequence modelling and generation [7, 8, 12].

An auto encoder is an artificial neural network used to learn efficient data encodings in an unsupervised manner. The auto encoder is generally composed of two subsections: an encoder and a decoder, they are linked by an internal latent space. The trick is that the latent space dimension is significantly smaller than the input one, which is the same as output, so the model is forced to discover and efficiently internalize the essence of the data. During the training, the system learns to reproduce in output the same data given in input. Once that it has been trained, it is possible to generate new output data via the sampling of the latent space. A further refinement of these models is represented by the variational auto encoders which are able to add a stochastic component in the latent space that helps in producing a better data modelling and generation, see [13].

In this paper, a variational auto encoder based on a sequence-to-sequence LSTM architecture has been used. A LSTM encoder encodes a sequence of places visited by a driver during a temporal interval of one day to its stochastic latent space and a symmetrical LSTM decodes it to the output trajectory sequence, which is made as matching as possible to the input one during the training phase. Once trained the latent space can be normally sampled in order to produce trajectory sequences similar, in a statistical sense, to the ones of the training dataset.

3 The Dataset

The available Octo-Telematics dataset is a FCD containing one month of records from a set of about 150K cars. Each record consists of time, position, speed, distance from preceding record point, engine status (running, turned off, started) and numerical ID of the car. The records are stored as soon as either a given time interval has passed or a given distance is travelled; also the starting or switching off of the engine are recorded events. In short, the dataset is composed of geographical points along car routes.

The first step has been that of studying the dataset in order to extract some global characteristics. As an example in Fig. 1 is shown an analysis in terms of stops, where a stop has been defined as a turning off of car engine for more than 5 min. The average number of stops per day on the whole dataset is 4.99 and 4.04, respectively in weekdays or weekends/holidays. It is remarkably similar to the results in [7] hinting to a sort of universality of human travel behaviour.

In Fig. 1, the bivariate histogram of stops in terms of the stop starting time and the stop duration is plotted. As it can be seen, three areas are clearly recognizable: very short-term stops, overnight stops and the 9-to-5 working stops.

From this and similar analyses it has been chosen to use a simplified dataset composed of sequences containing a maximum number of eight stops in a day. In the following, we will refer to these sequences as trajectories. Each stop is described by six quantities: longitude, latitude, duration, start time, day of week and trip distance covered to reach the current stop. Thus a trajectory is represented by 48 values; in case the number of daily stops is smaller than eight, the data are padded with zeroes; if a given driver performs more than eight stops, it is discarded (all the trajectories of the driver are discarded, to avoid data distortion).

With the above assumption a fraction of 50% of all the available drivers are considered.

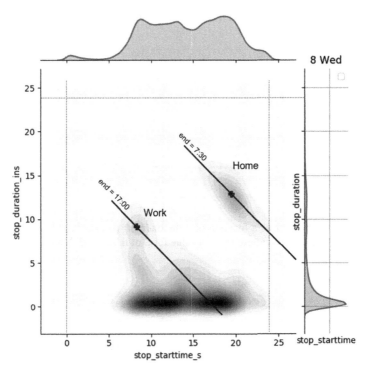

Fig. 1. Histogram of the stops on Wed. 8 May 2013. In x the starting time of the stops and in y the relative duration, both in hours. The red dotted lines delimit the day, the intensity of color increases with the number of stops. The two diagonal lines represent the set of stops ending respectively at 17:00 and 7:30 next day, hinting at the workplace stops and nightly residential ones

4 Training and Generation

4.1 Training

An LSTM sequence-to-sequence VAE has been trained on the subset of the dataset comprising the drivers with the considered number of stops, that means more than 900 K different trajectories. An analysis on the influence on training of the various features of the architecture has been performed and is here presented.

The loss function of a VAE is usually composed of two terms: a reconstruction loss describing how well the output matches the input and a regularisation term helping in furnishing a multivariate normal distribution in the latent space. In this work the two terms are tuned through the:

$$L_{TOT} = \beta L_{KL} + (1 - \beta)L_{MSE} \qquad (1)$$

where L_{MSE} is the reconstruction loss, typically measured by the mean squared error, and L_{KL} is the regularisation loss, typically assessed with the Kullback-Leibler divergence. In Fig. 2 is shown the final loss behaviour versus the β term. It is evident that it is sufficient a very low regularisation term contribution to obtain good results.

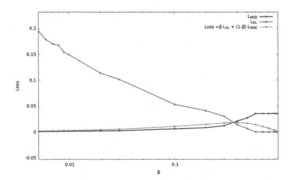

Fig. 2. Final loss behaviour as a function of β the trade-off parameter between reconstruction error (MSE) and regularisation term (KL), a value for β of $0.02 \div 0.03$ assures a low MSE with good regularisation

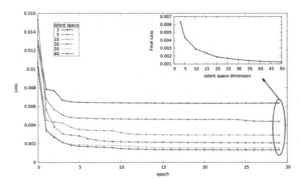

Fig. 3. Training with different latent space dimensions. Larger graph: the total loss as a function of training epochs for different latent space dimensions. Inset: the final loss behaviour as a function of latent space dimension ($\beta = 0.02$)

Several latent space dimensions have been tested and in Fig. 3 is summarized both the training loss as a function of epochs and the final one. It is evident that the larger the latent space dimension the smaller the loss. In the following, for further analyses, the latent space dimension has been chosen at 20.

In Fig. 4 is shown the loss function behaviour versus the LSTM size, i.e. the number of LSTM neurons in both the encoder and decoder parts of the auto encoder. The final loss becomes rapidly insensitive to the number of LSTM cells in the autoencoder, thus a number of 32 LSTM neurons has been chosen.

In Fig. 5 is shown, instead, the final loss behaviour as a function of the number of different drivers considered, i.e. 0.1 K, 1 K, 3 K, 10 K and 30 K. It is noteworthy that

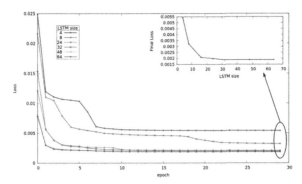

Fig. 4. Training with different LSTM numerosity. Larger graph: the total loss as a function of training epochs for different LSTM sizes. Inset: the final loss behaviour as a function of LSTM size ($\beta = 0.02$)

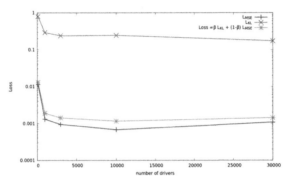

Fig. 5. Final loss behaviour as a function of the number of drivers, with $\beta = 0.02$, and a latent space dimension of 50

the error remains roughly constant while the dataset increases in number in both the re-construction and the regularisation terms, the plot is relative to runs with 100 training epochs, a latent dimensionality of 20, 32 LSTM neurons and $\beta = 0.02$.

The overall reconstruction error is in the order of 0.1%, that can be translated into a geographical position error in the order of a kilometer.

4.2 Generation and Results

Once trained, the system has been used for its true purpose: the generation of new trajectories for synthetic cars and drivers. The here reported data are representative of the generation of 30 K trajectories.

In Fig. 6 the normalized distribution of the number of daily stops in both the training dataset and the generated trajectories are presented. The comparison shows that the statistical distribution of this quantity has been remarkably grasped by the generative model.

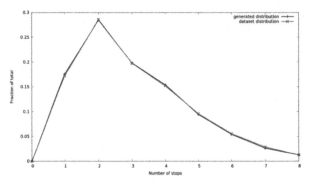

Fig. 6. The normalized distribution of the number of stops in the trajectories: comparison between the one from the training dataset and the one from the generated trajectories

In Fig. 7 the same bivariate histogram of Fig. 1 is presented, i.e. stop start time versus stop duration.

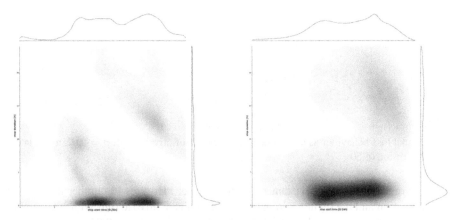

Fig. 7. Histogram of the stop start time vs duration: left the training dataset, right the generated trajectories

On the left of the figure there is the original training dataset, while on the right the generated trajectories. The short-term stops are in the same time slots and with slightly longer durations, the nightly stops in the generated trajectories are there but somewhat longer in duration and later in their beginning, nonetheless there is an evident similarity between the two graphs and the overall quality of the generated trajectories is similar to the original dataset behaviour.

The overall results are quite encouraging hinting to the ability of the generative system to grasp the general statistical distribution of the trajectories in the training dataset.

5 Conclusions and Future Work

It has been presented a generative model based upon a LSTM sequence to sequence variational auto encoder used to model and generate human motorized mobility pat-terns deduced from a FCD dataset.

The here presented preliminary results show that the proposed approach is able to generate trajectories with overall statistical features similar to the ones of the training set. Some tuning of the model has been performed in terms of latent space dimensionality, encoder/decoder size and loss function trade-off between the reconstruction term and the regularization one.

Future work is directed towards two different paths: on one side the experimentation of architectures different from the LSTM to build the encoder and decoder will be considered, specifically both a CNN (convolutional neural network) and a vanilla dense network are going to be tested. On the other side a tiling approach for the geographic localization will be tailored, subdividing the spatial domain into small tiles, e.g. hexagons with a diameter of 700 m, since the goal of the work is the modelling of traffic needs in terms of origin-destination matrices. The validation of the foreseen implementations will be performed also against different and larger traffic datasets.

Acknowledgements. This work has been partly carried out in the framework of the Triennial Plan 2019–2021 of the National Research on the Electrical System (Piano Triennale 2019–2021 della Ricerca di sistema elettrico nazionale), funded by the Italian Ministry of Ecologic Transition.

References

1. Zheng, Y., Capra, L., Wolfson, O., Yang, H.: Urban computing: concepts, methodologies, and applications. ACM Trans. Intell. Syst. Technol. (TIST) **5**(3), 38 (2014)
2. Chu, Z., Cheng, L., Chen, H.: A review of activity-based travel demand modeling. In: Proceedings of the Twelfth COTA International Conference of Transportation Professionals, pp. 48–59 (2012). https://doi.org/10.1061/9780784412442.006
3. Gonzalez, M.C., Hidalgo, C.A., Barabasi, A.L.: Understanding individual human mobility patterns. Nature **453**(7196), 779–782 (2008)
4. Song, C., Qu, Z., Blumm, N., Barabasi, A.L.: Limits of predictability in human mobility. Science **327**(5968), 1018–1021 (2010)
5. Alessandretti, L., Sapiezynski, P., Lehmann, S., Baronchelli, A.: Evidence for a conserved quantity in human mobility. Nat. Hum. Behav. **2**, 485–491 (2018). https://doi.org/10.1038/s41562-018-0364-x
6. Harshvardhan, G.M., Gourisaria, M.K., Pandey, M., Rautaray, S.S.: A comprehensive survey and analysis of generative models in machine learning. Comput. Sci. Rev. **38**, 100285, ISSN 1574–0137 (2020). https://doi.org/10.1016/j.cosrev.2020.100285
7. Yin, M., Sheehan, M., Feygin, S., Paiement, J., Pozdnoukhov, A.: A generative model of urban activities from cellular data. IEEE Trans. Intell. Transp. Syst. **19**(6), 1682–1696 (2018). https://doi.org/10.1109/TITS.2017.2695438
8. Lin, Z., Yin, M., Feygin, S., Transportation, M.S., Paiement, J., Cee, A.P.: Deep generative models of urban mobility. In: Proceedings of KDD 2017, 13–17 August 2017, Halifax, Nova Scotia, Canada (2017)

9. Hochreiter, S., Schmidhuber, J.: Long short-term memory. Neural Comput. **9**(8), 1735–1780 (1997). https://doi.org/10.1162/neco.1997.9.8.1735
10. Graves, A.: Generating sequences with recurrent neural networks. arXiv:1308.0850 (2013)
11. Gregor, K., Danihelka, I., Graves, A., Rezende, D.J., Wierstra, D.: Draw: a recurrent neural network for image generation. arXiv:1502.04623 (2015)
12. Huang, D. et al.: A variational autoencoder based generative model of urban human mobility. In: IEEE Conference on Multimedia Information Processing and Retrieval (MIPR), pp. 425–430, San Jose, CA, USA (2019). https://doi.org/10.1109/MIPR.2019.00086
13. Kingma, D.P., Welling, M.: An introduction to variational autoencoders. Found. Trends Mach. Learn. **12**, 307–392 (2019)

The Influence of Sponsorship on Purchase Intent: Oakley Brand Case Study

António Macedo[1], Manuel Sousa Pereira[1], and Helena Sofia Rodrigues[1,2](✉)

[1] Escola Superior de Ciências Empresariais, Instituto Politécnico de Viana do Castelo, Valença, Portugal
sofiarodrigues@esce.ipvc.pt
[2] CIDMA - Centro de Investigação e Desenvolvimento em Matemática e Aplicações, Universidade de Aveiro, Aveiro, Portugal

Abstract. The income of the sponsorship activities is a theme that requires attentiveness at the time of their implementation. The effectiveness of sponsorship and the perceptible contribution to the organization are significant points to the sponsoring entity. It is expected that contribute to the success of its business performance and the proper fulfillment of its goals.

Oakley, a brand recognized for its optical products and services, uses sports sponsorship as its dominant marketing and communication tool. This work aims to understand the impact of Oakley sponsorship on the purchase intention of Portuguese consumers. Through a survey, it intends to study the brand recognition in sports events, as well as the consumer's attitude and engagement to the brand.

Keywords: Sponsorship · Buy intention · Oakley · Sporting events · Consumer engagement · Purchase intent

1 Introduction

Over time, companies have been discovering the benefits and potentials of sponsorship as a means of communication, using this marketing component as a support in the formulation of strategies and the measurement of communicational and organizational objectives. According to Pope [26], sponsorship can be defined as the provision of resources (e.g., money, people, equipment) by an organization (the sponsor) directly to an individual, authority or body (the sponsee).

Sponsorship can also be seen as a mix of advertising with its capacity for message control and public relations. Companies that want to sponsor have a wide variety of entities to choose from, according to the target or values that wish to transmit, such as charities, teams, tournaments, arts, and sports. According to

Supported by FCT - Fundação para a Ciência e a Tecnologia.

O. Gervasi et al. (Eds.): ICCSA 2021, LNCS 12950, pp. 607–621, 2021.
https://doi.org/10.1007/978-3-030-86960-1_44

Dolphin [5], the most popular is sports sponsorship, reaching 75% of all support giving.

Sports sponsorships have characteristics that differentiate them from other promotional strategies, having as added values the possibility of a relationship, the segmentation of the target audience, the sale of products or services, and the significant exposure in the media [11,15]. Sport is the leading type of sponsorship mainly for the propensity to attract large audiences not only at each event, but also through the media attached to these activities. Besides, it provides a simplistic measure of segmentation and higher visibility opportunities for the sponsor because of the duration of each event.

Meenaghan [17] argues that one of the advantages of sports sponsorship over other communication tools is interaction with the target audience, and positive feelings are triggered in consumers, which are later manifested by purchase behaviors. Sports sponsorship allows influencing consumers' attitudes, generating positive emotional connections with them. Sponsors drive consumers to engage with the product and gain business-friendly conduct [2,4,14].

Oakley, Inc. has long used athlete endorsements to authenticate its products and tell its brand story. It uses sponsorship and athlete endorsements to build relevancy, credibility, and authenticity through third-party validation. The company leverages ties to access content for its social media platforms and those of its retail partners.

The brand's recognition and engagement with the Portuguese market were never studied. Besides, the number of studies that connect brands and sports events in the country is still small. The main aim of this study is to understand the relationship of the Oakley brand with consumers. Analyzing the communication tool adopted, with dimensions as consumer engagement, consumer attitude, and purchase intention, the final intention is to give some guidelines to marketing managers when they consider using this communication technique.

The paper is composed of five sections. Section 2 describes concepts related to sponsorship, brand recognition, and purchase intention. It is also presented the companies focused on the study. Section 3 is regarding the methodology used for this work, such as the survey dimensions and the statistical tools involved. In Sect. 4, the results reached with the survey are analyzed. Finally, the main conclusions are carried out in Sect. 5, and some limitations and future work are pointed.

2   Theoretical Background

Sports marketing aims to increase public recognition, reinforce a corporate image, define the identification of market requirements, overcome competition actions, involve a company with a community and give credibility to products and services [10,11].

According to Cardia [3], sports marketing and sports sponsorship are two concepts that are interconnected and complement each other, being a resource that business entities associated with segmentation. The desired return from the application of these two components comes from the increase in the brand image, the public's appreciation, the return at the institutional level, and the increase in sales.

2.1 Sponsorship

Meenaghan [16] characterizes sponsorship as an investment, which can be applied in cash or goods, in an activity, individual, cause, or event. The trade-off based on sponsorship activity is based on the sponsor's access to the exploitable commercial potential, associated with that activity, person, project, or event so that political, financial, or promotional returns are obtained.

Neto [22] argues that sponsorship derives from a promotional investment strategy, applied at sports, cultural or social levels. The sponsorship activity focuses on obtaining return and the occurrence of interaction and compliance contexts between the brand, the product, the sponsoring entity, and the consumer.

The sponsorship of sporting events has attracted the attention of marketing technicians, who seek to cover specific markets, looking for a profitable situation for attracting consumers [6,28]. Similar to sponsoring sports celebrities, sponsoring sports events indirectly influences the consumer, providing the association of the event's image with the sponsoring entity. Sponsorship relating to an event establishes a synergistic relationship between the sponsor and the sponsor. Event sponsorship is an effective way for companies to promote their brands and products [8,24]. These, within the scope of merchandising, may contain the logo of a company, to facilitate consumer retention and recognition [1,3].

2.2 Consumer Engagement and Purchase Intent

The engagement of the audience is crucial because the perceived associations with the sponsor became stronger. Several studies highlighting the importance of attitudes towards an unconditioned stimulus [29]: a positive attitude toward the event will be associated with a positive response towards the sponsor. Sponsorship can be used to modifying the brand perception, create a new relationship with potential clients, and enhance the organization's image.

Grohs *et al.* [7] states that companies employing sponsorship for the attainment of corporate communications goals tend to have a more positive public image than companies not involved in sponsorship. Therefore, it is fundamental to evaluate the image's transfer. Related to the previous one, building goodwill could be another benefit of sponsorship, by boosting brand image and positioning [18–21].

Finally, the essential purpose of sponsorship is sales increase. To accomplish this goal, companies evaluate their objectives and then decide what specific sponsorship would suitable. Pope and Voges [27] state that consumers who believe that a company's brand is involved in sponsorship tend to have a higher intention to purchase the company's product. Purchase intentions have often been accepted in the literature as a final measure of sponsorship effectiveness. The purchase intent could be seen as the outcome for the sponsor, to have effective impact on consumers' behavior.

One of the brands that use Sponsorship as a part of brand communication is Oakley.

2.3 Oakley Company Charaterization

Oakley, Inc. is a subsidiary of the Italian business group Luxottica Group S.p.A and was founded in 1975 by Jim Jannard. Its founder started the business of the company over forty years ago, has reinvented the concept of solar optics, especially in sports. Oakley's current CEO is Colin Baden, headquartered at Foothill Ranch in the United States [23].

Oakley communicates with its audience in a variety of ways and means, adopting a global communication strategy that is tailored to the brand's reality. The brand establishes multi-channel communication, using forms of communication that encompass media such as sponsorship, patronage, and advertising, but also digital and interactive media such as e-commerce, social networking, and digital platforms.

Throughout its activity, the Oakley brand has used and continues to use sponsorship as one of its key communication components, linking its name to athletes and events. Many athletes have become icons and benchmarks, such as Bubba Watson, Greg Lemond, Lance Armstrong, Olympic athletes Shaun White and Lindsey Vonn, and NBA star Michael Jordan. Wishing to familiarize themselves with recognized athletes and promote the brand at international sporting events, the Oakley brand has taken the initiative to become the official sponsor and supplier of optical products for the US Olympic Committee and the US Team.

This investigation sought to verify the existence of a constant relationship between Oakley brand sponsorship actions and the consumer behavior, to prove a positive effect of these actions ond consumer's engagement and attitude towards them.

3   Methodology

This section presents an overview of the main steps adopted in this research to achieve the goal planned. The work was designed to assess the relationships between Oakley and consumer and was developed as follows:

- the research began with the analysis of the current literature about sports sponsorship;
- based on the literature review, a questionnaire was designed, taking into account scales related to the theme; this research was based on a case study, adopting a quantitative research methodology through a survey;
- questionnaire was applied: the sample consists of two hundred and fourteen inquiries;
- statistical analysis was developed resorting to IBM SPSS version 24, and the main results were drawn.

3.1   Research Design

The survey consisted of three parts; the first contains general questions about the characterization of the respondents, such as gender, age, employment situation, and educational qualifications. The second part is related to consumer behavior concerning optical products. The main idea was to know how often the consumers buy optical products if they recognize the brand Oakley and also if they associated this brand with specific sportsmen and sports. The last part refers to the sponsorship activity. The scales of Speed and Thompson [29] and Martensen *et al.* [13] served as a reference in the design of the survey. The questions were divided into five dimensions present the answer option based on the Likert scale, which ranges from 1 (Strongly Disagree), 2 (Disagree), 3 (Neither Agree nor Disagree), 4 (Agree) to 5 (Strongly Agree). The questions for each dimension are compiled in Table 1.

Speed and Thompson [29] model was a model developed to verify the effect of the consumer's attitude towards the sponsoring entity and the event, as well as the effect of the consumer's perception regarding the sponsoring entity's relationship with the sponsor. The aim was the orientation given to marketing managers, motivated by a consulting market that provides support to sponsoring entities during their sponsorship actions, promoting their effectiveness.

The theoretical model of Martinsen *et al.* [13] was a model developed to verify the effectiveness of sponsorship of a given event. This model analyzes the impact of the event sponsorship, and how it influences the level of purchase intention and the consumer perception of a brand. The authors analyzed a golf sporting event sponsored by the Danish brand Bang & Olufsen and the results confirmed that the event in the question had a positive effect on purchase and consumer perception of the sponsoring entity.

3.2   Data Collection

The questionnaire was developed in the Google Forms platform. It is an easy and intuitive question development system and has a simple and practical interface for the respondent. Another advantage is the fast collection of information and the absence of non-responses.

The survey was subject to a pre-test with a small sample, to assess the language used and the redundancy of the questions. Some minors corrections were implemented.

Speed and Thompson [29] model was a model developed to verify the effect of the consumer's attitude towards the sponsoring entity and the event, as well as the effect of the consumer's perception regarding the sponsoring entity's relationship with the sponsor. The aim was the orientation given to marketing managers, motivated by a consulting market that provides support to sponsoring The sample taken is a convenient one due to time and budget constraints. There were obtained 214 answers from consumers of both genders, belonging to different age groups, educational qualifications and occupations, and with internet access. Despite a convenient sample not could be population representative,

Table 1. Dimensions used for the sponsorship activity and the brand Oakley

Dimension	Item	Description of the item	Reference
A - Actions of sponsorship	A1	Oakley's sponsorship actions add value to the brand	[29]
	A2	Oakley's sponsorship actions promote the memorization of the brand's name	
	A3	Oakley's sponsorship actions promote a favorable attitude towards the brand	
	A4	Oakley's sponsorship actions increase the likelihood of experimentation with Oakley products	
	A5	Consumer purchase intention regarding Oakley brand products/services	
S - Sponsorship in sport events	S6	There is a relationship of agreement between sporting events and Oakley sponsorship actions	[29]
	S7	The sporting events in which Oakley participates as a sponsor convey relevant information about the brand	
	S8	Attending the sporting events in which Oakley participates as a sponsor contributes to increase brand awareness	
	S9	Oakley's image conforms to sporting events in which Oakley participates as a sponsor	
CE - Consumer Engagement	E10	Oakley is an interesting brand	[13]
	E11	Oakley is a significant brand	
	E12	Oakley is an appealing brand	
	E13	Oakley is a fascinating brand	
	E14	Oakley is an essential brand	
CA - Consumer attitude	CA15	I consider Oakley a quality brand	[13]
	CA16	I have a positive attitude towards the Oakley brand	
	CA17	Oakley has benefits compared to brands in the same industry	
	CA18	Oakley has a higher level of quality compared to brands in the same industry	
CP - Consumer Purchase Intent	CP19	Oakley has added relevance to me	[13]
	CP20	Next time I need optics, I'll opt for the Oakley brand	
	CP21	I am willing to pay more for Oakley over other brands	
	CP22	I will recommend to others the Oakley brand	

could also bring some useful information for the brand and its relationship with some potential costumers as an exploratory approach.

3.3   Data Analysis

SPSS 24.0 was used for data analyses (IBM, Chicago). Descriptive and inferential statistics were applied. Descriptive statistics included frequency distributions (range, mean, standard deviations, and percentage).

The scales' reliability was calculated with Cronbach's alpha. Cronbach's alpha is a statistical tool that quantifies the internal reliability of a survey; it ranges from 0 to 1 and the minimum acceptable value is 0.7 [9].

After that, principal component analysis is carried out to show the most influencing determinants in the survey design. Finally, the Mann-Whitney test was performed to analyze if there is a difference of gender factor in the responses of the survey. The statistical significance was set at a p-value <.05.

4   Analysis of the Results

4.1   Sample Characterization

The respondents had a large spectrum of characteristics, as summarized in Table 2. The sample is mostly male (72%). The age group with the largest representativeness of the sample is between 24 to 34 years old (52.3%). Individuals aged 35 to 45 years represent 23.8% of the sample. Regarding the employment situation, the sample is largely composed of individuals who work for others (49.5%). However, 21.5% of the sample is represented by students and 19.6% by self-employed individuals. Most of the sample (44.4%) consisted of individuals with Secondary Education, 39.2% of individuals with a Bachelor degree and 6% of individuals with a Master's degree. Given the theme (sports), young men tend to be more inclined to participate in the research. Our resulting sample is not balanced, but it may well be representative of the distribution of gender groups of the consumers that Oakley could try to achieve.

4.2   Oakley and Its Relation with Consumers and Sports

This subsection is related to data analysis concerning the Oakley brand, regarding its relation to sports events and consumers. For the 22 questions, descriptive statistics and principal components analysis are presented (PCA) to give a panoramic view of the results. In Table 3, all items were answered using the totality of the scale (from 1 to 5). The average value of the answers was higher than three, meaning at least a neutral position to the statements provided.

Regarding the dimension related to actions of sponsorship (Dimension A), the sample also agreed that Oakley's sponsorship actions add value to its brand, promote the memorization of its name, promote a favorable attitude from individuals towards its brand, and increases the likelihood of trying its products/services. It should be noted that the authors Martinsen *et al.* [13], obtained similar results, showing an influence on the consumer's purchase intention, influenced by the characteristics and brand image, as well as its memorization.

Table 2. Sample characterization

Gender	%	Age	%
Male	72	13–23	17.8
Female	28	24–34	52.3
		35–45	23.8
		46–56	5.2
		57–67	0.9
Employment situation	%	Education qualifications	%
Unemployed	8.9	Basic education	2.3
Student	21.5	Secondary education	44.4
Retired	0.5	BSc	39.2
Employee	49.5	Master Degree	11.7
Self employed	19.6	PhD	2.3

Table 3. Descriptive statistics and Cronbach'alpha

Dim.	Item	Min	Max	Mean	St. Dev	Alpha
A	A1	1	5	3.82	0.954	0.921
	A2	1	5	3.93	0.942	
	A3	1	5	3.85	0.843	
	A4	1	5	3.87	0.898	
	A5	1	5	3.66	0.954	
S	S6	1	5	3.69	0.878	0.896
	S7	1	5	3.55	0.864	
	S8	1	5	3.89	0.918	
	S9	1	5	3.79	0.882	
CE	CE10	1	5	3.81	0.886	0.919
	CE11	1	5	3.72	0.891	
	CE12	1	5	3.71	0.946	
	CE13	1	5	3.40	0.962	
	CE14	1	5	3.20	0.998	
CA	CA15	1	5	3.83	0.868	0.935
	CA16	1	5	3.72	0.890	
	CA17	1	5	3.49	0.838	
	CA18	1	5	3.46	0.891	
CP	CP19	1	5	3.16	1.042	0.938
	CP20	1	5	3.26	1.047	
	CP21	1	5	3.19	1.106	
	CP22	1	5	3.34	0.998	

The results obtained from sponsorship in sports (Dimension S) showed that the relationship between the Oakley brand and the sporting events in which it is inserted as a sponsoring entity, is a fruitful relationship that benefits the brand at various levels. The results presented are consistent with the study by Speed and Thompson [29]. The authors have proven that sponsors and agents of sporting events can maximize the value of sponsorship by understanding the attitudes held by their audience.

Regarding the third dimension analyzed (Dimension CE), there is a positive level of engagement between the consumer and the Oakley brand, with the sample considering it an interesting, significant and appealing brand. Regarding adjectives that classify Oakley as a fascinating and essential brand, there was a considerable percentage of respondents who showed a certain indifference. These results are distinct from a study conducted on the essential features of the Oakley brand in 2005 [25]. Respondents, despite having a positive attitude towards the brand, also found that the brand was not fascinating or attractive to the majority of the female audience.

The consumer's attitude towards the Oakley brand (CA Dimension) is positive. Most respondents agreed that Oakley is a quality brand, expressing also a positive attitude towards it, thus meeting the study by Peters [25].

In terms of purchase intention (CP Dimension), it was possible to verify a considerable number of elements that showed a sense of indifference towards the purchase intention of the analyzed brand's products/services. Even so, the number of respondents who said they fully agree or agree with the relevance of the brand and are willing to pay more for it is higher than the discordant responses. Martinsen et al. [13], similarly to the sport of golf, demonstrated that the fact of proving a close relationship between the sporting event and the studied brand, produced a transfer of value for the brand in question, leading to a higher purchase intention relative to the sponsoring brand.

Making a general analysis of all items, the ones with higher means are A2 with the mean 3.93 and item S8 with the value 3.89. Both items are related to the memory of the brand in consumers and sports; the inquiries agree that sponsorship, especially in sport events, increase brand awareness. The lowest means, although in the positive spectrum, are related to the dimension "Consumer Purchase Intent", with the values 3.16 (item CP19) and 3.19 (item CP21). Although consumers believe in brand quality, they are not completely committed to pay more to acquire this product. The standard deviation does not present great discrepancies between items. attitude To evaluate the reliability of the scales, the Cronbach's alpha coefficients were calculated. In this survey, the minimum obtained was 0.896, which shows the reliable of the survey.

Principal Component Analysis is a factorial model in which factors are based on total variance. With this analysis it is necessary to evaluate the Kaiser-Meyer-Olkin (KMO) sampling adequacy measure and the Bertlett sphericity test. Both indicate the adequacy of the data for the accomplishment of the factorial analysis [9]. For each dimension, one component was obtained. Through the results (Table 4) of the factorial weights we conclude that all the items present a positive

Table 4. Principal component analysis

Dimension	Item	Communalities	Factor loadings	KMO measure	Bartlett's test (sig)	Total variance (%)
A - Actions of sponsorship	A1	0.668	0.818	0.888	<0.001	76.4
	A2	0.831	0.912			
	A3	0.816	0.904			
	A4	0.793	0.891			
	A5	0.713	0.844			
S - Sponsorship in sport events	S6	0.803	0.896	0.829	<0.001	76.3
	S7	0.686	0.828			
	S8	0.763	0.873			
	S9	0.801	0.895			
CE - Consumer Engagement	CE10	0.799	0.894	0.826	<0.001	75.4
	CE11	0.790	0.889			
	CE12	0.847	0.920			
	CE13	0.775	0.881			
	CE14	0.557	0.746			
CA - Consumer Attitude	CA15	0.828	0.910	0.791	<0.001	83.9
	CA16	0.831	0.911			
	CA17	0.867	0.931			
	CA18	0.829	0.911			
CP - Consumer Purchase Intent	CP19	0.825	0.908	0.840	<0.001	78.3
	CP20	0.831	0.911			
	CP21	0.638	0.799			
	CP22	0.837	0.915			

correlation. Regarding the suitability measure of the KMO sample, the minimum result obtained of 0.791, which is close to 1, indicates the average degree of adequacy of the PCA. In this case, the calculated chi-squared statistic was significant at the level of <0.001, indicating the presence of correlation between the various items. Regarding the communalities (after extraction), it is possible to see, that the lowest value is in dimensions "Consumer Engagement", which translate inn lower factor loading for the correspondent item in the component.

4.3   Influence of Sponsorship by Gender

A last analysis was done, regarding the different behavior between gender and consumption of optical products, for the dimensions related with sponsorship actions and consumer purchase intention.

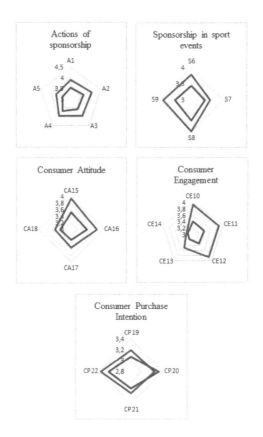

Fig. 1. Average level of agreement, by gender (blue - male; red - female) (Color figure online)

Figure 1 shows the average answers, by gender. It is notorious that the degree of agreement with the sentences of the survey is higher in the male gender. One exception in all surveys, which is related with item CP20, related with the buyer option of Oakley brand which is higher in women. The most visible differences in the answers are in the dimension "Consumer Engagement". These results, meet the study results of Peters [25], where the Oakley brand was considered less attractive to the female target.

In order to prove that these differences are statistically significant, a Mann-Whitney U test was done [12]. It compares the differences between two independent groups (in this case divided by gender), when the dependent variable is either ordinal or continuous, but not normally distributed. It is used to test the null hypothesis that two samples come from the same population, i.e., have the same median. The significance level used was 0.05. The results meet the graphical analysis done in Fig. 1. The answers given by male are statistically different than woman, regarding the importance of sponsorship for the Oakley brand recognition. Except the item CP20, the level of engagement with this brand is higher for men, when compared with women's answers. Curiously, the answers given related to the purchase intention are significantly similar for both genders.

Table 5. Mann-Whitney U test, using gender as grouping variable

Dimension	Item	Asymp. Sig. (2-tailed)	Decision
A	A1	0.002	Reject
	A2	0.002	Reject
	A3	<0.001	Reject
	A4	0.011	Reject
	A5	0.007	Reject
S	S6	0.003	Reject
	S7	0.097	Retain
	S8	0.004	Reject
	S9	0.002	Reject
CE	CE10	<0.001	Reject
	CE11	<0.001	Reject
	CE12	<0.001	Reject
	CE13	0.019	Reject
	CE14	0.839	Retain
CA	CA15	<0.001	Reject
	CA16	0.004	Reject
	CA17	0.066	Retain
	CA18	0.231	Retain
CP	CP19	0.341	Retain
	CP20	0.593	Retain
	CP21	0.541	Retain
	CP22	0.140	Retain

5 Conclusions

As the main conclusion of this study, it was verified that Oakley brand sponsorship actions positively influenced the consumer's attitude towards the brand. In this sense, its target audience establishes a positive global assessment concerning the sponsorship activities.

Concerning sporting events in which the Oakley brand participates as a sponsoring entity, the results are in line with the literature review, which argues that the higher the level of association and identification between the sponsor and the sponsored event, the greater the involvement and the more profitable will be the attitude of individuals towards the event and the brand.

Despite the consumer purchase intention is the dimension where the survey values are smaller, the consumer recognizes the Oakley value and will recommend it to others.

As the main contributions of the investigation, it is considered that all the work carried out during this investigation presents a special contribution for the Luxottica group, more specifically for the Oakley brand, which based on this study, will be able to focus on relevant aspects focused on their brand, on its relationship with sponsorship activity and consumer response to this marketing tool.

The study's limitations are related to the fact that the sample was selected by convenience, which could lead to some bias of the sample. the use of a stratified sample in the next questionnaire application, could contribute to a more reliable conclusions.

As future investigations, this study can be applied in other contexts and realities based on interviews and focus groups that equally address the theme of sponsorship, and that analyze, in a more personal way, its influence on consumer behavior, another approach consists of analysis of the sponsorship effect, in a purely relational perspective, focused on the influence of the sponsorship on the relationship with the consumer, to the detriment of the purchase intention.

Acknowledgements. This work is supported by The Center for Research and Development in Mathematics and Applications (CIDMA) through the Portuguese Foundation for Science and Technology (FCT - Fundação para a Ciência e a Tecnologia), references UIDB/04106/2020 and UIDP/04106/2020 (Rodrigues).

This research was part of a Marketing' Master thesis of the student António Macedo.

References

1. Abiodun, O.: The Significance of Sponsorship as a Marketing Tool in Sport Events. Arcada University of Applied Sciences, Helsínquia (2011)
2. Bai, Y., Yim, B.H., Breedlove, J., Zhang, J.J.: Moving away from category exclusivity deals to sponsorship activation platforms: the case of the Ryder cup. Sustainability **13**(3), 1151 (2021)
3. Cardia, W.: Marketing e Patrocínio Esportivo. Porto Alegre, Bookman (2004)

4. Contursi, E.: Patrocínio. Coleção Marketing de Sucesso. Rio de Janeiro, Sprint (2003)
5. Dolphin, R.R.: Sponsorship: perspectives on its strategic role. Corp. Commun. Int. J. **8**(3), 173–186 (2003)
6. Ge, Q., Humphreys, B.R.: Athlete misconduct and team sponsor stock prices: the role of incident type and media coverage. J. Sport Manage. **35**, 1–12 (2021)
7. Grohs, R., Wagner, U. e Vsetecka, S.: Assessing the effectiveness of sport sponsorship - an empirical examination. Schmalenbach Bus. Rev. **56**, 119–138 (2004)
8. Jobber, D.: Principles and Practice of Marketing, 5th ed. McGraw-Hill (2007)
9. Kline, R.B.: Principles and Practice of Structural Equation Modelling, 3rd edn. Guilford Press, New York (2011)
10. Lee, S., Harris, J., Lyberger, M.: Recreational golfers' attitudes and awareness of sponsorship: a case study of the 2008 Ryder Cup. Manag. Leis. **16**(3), 192–206 (2011)
11. Lois, N.: Estratégias empíricas em patrocínio esportivo à luz da experiência de organizações esportivas e empresas investidoras. Federal University of Santa Catarina, Florianópolis (2013)
12. Maroco, J.: Análise Estatística com utilizaçao do SPSS, 3a edn. Lisboa, Ediçoes Silabo (2007)
13. Martensen, A., Gronholdt, L., Bendtsen, L., Jensen, M.: Application of a model for the effectiveness of event marketing. J. Advert. Res. **47**(3), 283–301 (2007)
14. Mason, K.: How corporate sport sponsorship impacts consumer behavior. J. Am. Acad. Bus. **7**(1), 32–35 (2005)
15. Mattar, M.: Tomada de decisão em ações de patrocínio esportivo: análise descritiva do processo decisório e critérios de seleção em empresas patrocinadoras no Brasil. University of São Paulo, São Paulo (2007)
16. Meenagham, T.: The role of sponsorship in the marketing communications mix. Int. J. Advert. **10**(1), 35–47 (1991)
17. Meenaghan, T.: Ambush marketing - a threat to corporate sponsorship. Sloan Manag. Rev. **38**(1), 103–113 (1996)
18. Meenaghan, T., Osullivan, P.: The passionate embrace - consumer response to sponsorship. Psychol. Market. **18**(2), 87–94 (2001)
19. Meenaghan, T.: Understanding sponsorship effects. Psychol. Market. **18**(2), 95–122 (2001)
20. Meenaghan, T.: From sponsorship to marketing partnership: the Guinness sponsorship of the GAA All - Ireland Hurling Championship. Ir. Mark. Rev. **15**(1), 3–23 (2002)
21. Meenaghan, T.: Current developments and future directions in sponsorship. Int. J. Advert. **17**(1), 3–28 (2008)
22. Neto, F.: Marketing de Patrocínio. Sprint Editora, Rio de Janeiro (2003)
23. Oakley Company. http://www.luxottica.com/en/about-us/company-profile. Accessed 20 Jan 2021
24. Pertzborn, J.: Comunicação no setor desportivo: o caso do basquetebol. Tese de mestrado, Escola Superior de Comunicaçao Social (2020)
25. Peters, W.: A Study of the Brand Characteristics of Oakley. University of Stellenbosch - Department of Business Management, Matieland (2005)
26. Pope, N.: Consumption values, sponsorship awareness, brand and product use. J. Prod. Brand Manag. **7**(2), 124–136 (1998)
27. Pope, N.K.L., Voges, K.E.: The impact of sport sponsorship activities, corporate image, and prior use on consumer purchase intention. Sport Mark. Q. **9**(2), 96–102 (2000)

28. Roy, D. e Cornwell, B.: Brand equity's influence on responses to event sponsorships. J. Prod. Brand Manag. **12**(6), 377–393 (2005)
29. Speed, R., Thompson, P.: Determinants of sports sponsorship response. J. Acad. Mark. Sci. **28**(2), 226–238 (2000)

Asymptotic Quadrature Based Numerical Integration of Stochastic Damped Oscillators

Raffaele D'Ambrosio and Carmela Scalone[✉]

Department of Engineering and Computer Science and Mathematics,
University of L'Aquila, Via Vetoio, Loc. Coppito, 67100 L'Aquila, Italy
{raffaele.dambrosio,carmela.scalone}@univaq.it

Abstract. This paper is focused on the numerical solution of stochastic harmonic damped oscillators, characterized by a nonlinear high-oscillating time-varying forcing term, coupled with a random force driven by an additive Wiener noise. The scheme is based on asymptotic quadratures together with the application of the variation of constants formula. Theoretical issues are provided and the numerical evidence on selected popular related physical models is also given.

Keywords: Stochastic differential equations · Stochastic oscillators · Periodic time varying force · Asymptotic quadrature

1   Introduction

Mathematical modelling of oscillating phenomena plays a crucial role in the scientific literature, both for deterministic and stochastic scenarios. Usually, stochastic oscillators are obtained by introducing noisy components into existing deterministic oscillatory models as described, for instance, in [1–5, 17–20, 27, 33] and references therein.

In this work, we are interested in oscillators described a second order differential equation of the form:

$$\ddot{x} = -\omega^2 x + \eta\dot{x} + g_\sigma(t) + \varepsilon\xi(t), \tag{1}$$

where $\xi(t)$ is a white noise process satisfying $\mathbb{E}|\xi(t)\xi(t')| = \delta(t-t')$, being $\delta(\cdot)$ the Dirac delta function. The parameters ω and η are real constants, representing the frequency and the damping of the oscillations. Equations of the form (1) typically model stochastic harmonic damped oscillators driven by both a deterministic time-dependent force and a random Gaussian forcing term. In particular, the high oscillations in the model are determined by the deterministic forcing term

---

This work is supported by GNCS-INDAM project and by PRIN2017-MIUR project. The authors are member of the INDAM Research group GNCS.

O. Gervasi et al. (Eds.): ICCSA 2021, LNCS 12950, pp. 622–629, 2021.
https://doi.org/10.1007/978-3-030-86960-1_45

$g_\sigma(t)$ via the parameter σ and, in general, not by the high frequencies, as in [25] for the deterministic setting.

Stochastic oscillators (1) are very popular in literature as a relevant physical model (see, for instance, [19,20,27] and references therein), therefore a robust numerics to approximate their solutions is a crucial topic to be properly addressed.

The existing numerical literature offers a focus on several aspects regarding the discretizations of stochastic oscillators; we refer, for instance, to [3,4,6–8,11–13,15,16,28,29,31,32] and references therein. Here we present a numerical strategy suited for problem (1), combining the variation of constants formula and a quadrature rule based on the asympotic expansion [23,24]. The idea consists in proposing an appropriate extension to the approach of [26] and [25], in a stochastic setting, for a problem of the form (1). In particular, in [26], the author introduces methods for the efficient numerical approximation of linear and non-linear systems of highly oscillatory ordinary differential equations and shows how an appropriate choice of quadrature rule improves the accuracy of approximation as the frequencies of the oscillations grow. The variation of constants formula, as a starting point to design numerical methods suited for oscillatory problems, is also very employed in the deterministic setting, see [21,22] and reference therein.

The aforementioned discretization of problem (1) is described in Sect. 2, where we also recall the essential aspects of quadrature formulas based on asymptotic expansions. In Sect. 3, we provide the numerical evidence, confirming the superiority of the proposed method in comparison to a standard solver.

2   Description of the Numerical Scheme

Equation (1) is equivalent to the following first order system of two equations in the variables X_t (the position of the oscillating particle) and V_t (its velocity):

$$\begin{cases} dX_t = V_t dt, \\ dV_t = -\omega^2 X_t dt + \eta V_t dt + g_\sigma(t)dt + \varepsilon dW_t. \end{cases} \tag{2}$$

We are particularly interested in the case

$$g_\sigma(t) = a(t)e^{i\sigma h(t)},$$

where $g'(t) \neq 0$, for $t \geq 0$ (see [25]). By setting the matrix

$$A = \begin{bmatrix} 0 & 1 \\ -\omega^2 & \eta \end{bmatrix},$$

the variation of constants formula applied to the system (2) is given by

$$\begin{pmatrix} X_t \\ V_t \end{pmatrix} = e^{tA} \begin{pmatrix} X_0 \\ V_0 \end{pmatrix} + \int_0^t e^{sA} \mathbf{e} \ g_\sigma(s)ds + \varepsilon \int_0^t e^{sA} \mathbf{e} \ dW_s, \tag{3}$$

where $\mathbf{e} = [0 \ 1]^T$.

Moreover, setting $\mathbf{y}_{n+1} = \begin{pmatrix} X_{n+1} \\ V_{n+1} \end{pmatrix}$, a discretization in the stochastic part reads as

$$\mathbf{y}_{n+1} = e^{hA}\mathbf{y}_n + e^{hA} \int_0^h e^{-sA}\mathbf{e} \; g_\sigma(s)ds + \varepsilon e^{hA} \, \mathbf{e} \, \Delta W_n, \tag{4}$$

where $\Delta W_n = W(t_{n+1}) - W(t_n)$ denotes the Wiener increment. Clearly, in order to obtain a full discretization to (2), we must approximate the integral involving the deterministic force $g_\sigma(t)$.

2.1   Discretization Based on Asymptotic Expansions

Following [25], we recall that an asymptotic expansion of the integral

$$I = \int_0^t e^{-sA}g_\sigma(s)ds \tag{5}$$

is readily available, in fact,

$$I = \int_0^t e^{-sA}a(s)e^{i\omega h(s)}ds \sim -\sum_{r=0}^\infty \frac{1}{(-i\omega)^{r+1}} \left[x_r(t)\frac{e^{ih(t)}}{h'(t)} - x_r(t)\frac{e^{ih(0)}}{h'(0)} \right],$$

where

$$x_0(t) = e^{-tA}a(t) \quad x_{r+1} = \frac{d}{dt}\frac{x_r(t)}{g'(t)}, \quad r \in \mathbb{Z}.$$

Note that the functions x_r are non-oscillatory. By considering [24] (in particular Lemma 2.1), a truncation of the series to the s-th term

$$Q_A^s = -\sum_{r=0}^s \frac{1}{(-i\omega)^{r+1}} \left[x_r(t)\frac{e^{ih(t)}}{h'(t)} - x_r(t)\frac{e^{ih(0)}}{h'(0)} \right] \tag{6}$$

provides an efficient approximation to the integral (5) if ω is a large number.

The numerical strategy we propose is based on the semi-discretization (4), where the integral (5) is approximated by employing formula (6) with $s = 2$.

Finally, the following result (see [24]) analyzes the accuracy of the quadrature formula (6), establishing its asymptotic order of convergence.

Theorem 1. *For every smooth h and a it is true that*

$$I - Q_A^s \sim \mathcal{O}(\omega^{-s-1}) \quad \omega \longrightarrow \infty.$$

3   Numerical Experiments

Motivated by [1, 19, 25], we propose the following system

$$
\begin{cases}
dX_t = V_t dt, \\
dV_t = -\omega^2 X_t dt + \eta V_t dt + e^{i\sigma h(t)} dt + \varepsilon dW_t,
\end{cases}
\tag{7}
$$

where $\varepsilon = 0.3$. We set $\omega^2 = 50$, $\eta = -10$, and the initial conditions $X_0 = 0.5$, $V_0 = 0.5$. We test the method introduced in Sect. 2, considering the cases $h(t) \equiv 1$ and $h(t) = -t^2 + t$.

In Fig. 1, we show the behaviour of the method when $h(t) = -t^2 + t$ and $\sigma = 80$. Moreover, we observe the behaviour of the mean-square error in the position of the numerical solution, choosing $h(t) \equiv 1$, with $\sigma = 70$ (Fig. 2) and $\sigma = 150$ (Fig. 3). Furthermore, we propose a comparison between our approach and the classical Stochastic Exponential Euler method [30].

We can see in all the provided experiments that the Stochastic Exponential Euler method has problem of convergence, as visible in Tables 1, 2 and 3, beacuse it is not able to accurately catch the high oscillations of the solution, as our approach is, evidently, able to do. Indeed, as visibles from Figs. 1, 2 and 3 and the aforementioned tables, the introduced method is clearly convergent, also for high values of the frequencies.

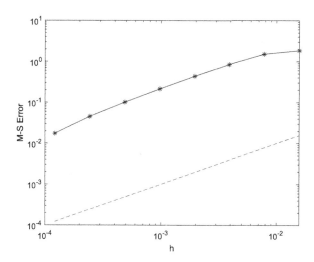

Fig. 1. Mean-square error in the position of the numerical solution to problem (7) with $h(t) = -t^2 + t$ and $\sigma = 80$, computed by the asymptotic quadrature method introduced in Sect. 2. The pattern of the mean-square error is plotted vs the employed stepsize, in log-log scale.

Table 1. Observed mean-square orders of convergence of the stochastic exponential Euler method and the asymptotic quadrature method introduced in Sect. 2 and the stochastic exponential Euler method applied to problem (7) with $h(t) = -t^2 + t$ and $\sigma = 80$.

h	Exponential euler	Asymptotic quadrature
0.0001	–	–
0.0002	0.0107	0.0570
0.0005	0.0088	0.1117
0.0010	0.0103	0.2255
0.0020	0.0108	0.4452
0.0039	0.0103	0.8620
0.0078	0.0105	1.5243
0.0156	0.0102	1.8475

Table 2. Observed mean-square orders of convergence of the stochastic exponential Euler method and the asymptotic quadrature method introduced in Sect. 2 and the applied to problem (7) with $f(t) \equiv 1$ and $\sigma = 70$.

h	Exponential euler	Asymptotic quadrature
0.0005	–	–
0.0010	0.0245	0.0199
0.0020	0.0232	0.0387
0.0039	0.0233	0.0757
0.0078	0.0230	0.1493
0.0156	0.0222	0.2879
0.0313	0.0247	0.4933
0.0625	0.0226	0.4559

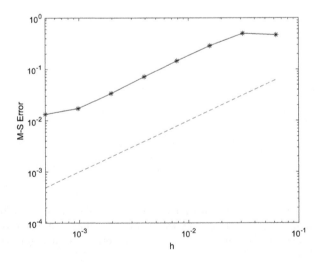

Fig. 2. Mean-square error in the position of the numerical solution to problem (7) with $f(t) \equiv 1$ and $\sigma = 70$, computed by the asymptotic quadrature method introduced in Sect. 2. The pattern of the mean-square error is plotted vs the employed stepsize, in log-log scale.

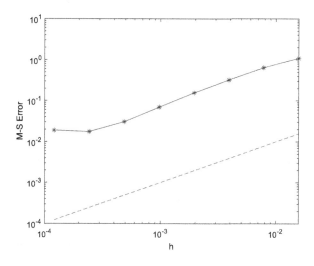

Fig. 3. Mean-square error in the position of the numerical solution to problem (7) with $f(t) \equiv 1$ and $\sigma = 150$, computed by the asymptotic quadrature method introduced in Sect. 2. The pattern of the mean-square error is plotted vs the employed stepsize, in log-log scale.

Table 3. Observed mean-square orders of convergence of the stochastic exponential Euler method and the asymptotic quadrature method introduced in Sect. 2 and the stochastic exponential Euler method applied to problem (7) with $f(t) \equiv 1$ and $\sigma = 150$.

h	Exponential euler	Asymptotic quadrature
0.0001	–	–
0.0002	0.0107	0.0317
0.0005	0.0112	0.0504
0.0010	0.0103	0.0946
0.0020	0.0108	0.0183
0.0039	0.0110	0.3531
0.0078	0.0107	0.6746
0.0156	0.0104	1.1184

4 Conclusions

In this paper, we have combined two typical ingredients of the numerics for deterministic differential problems, i.e., the variation of constants formula and the asymtpotic expansion quadrature, to define a numerical strategy for second order stochastic equation of type (1). The effectiveness of the approach, also briefly described by the provided numerical tests, is also particularly relevant in comparison with existing schemes, such as the classical exponential Euler method for stochastic differential equations. Clearly, such strategy remains tied to the type of system considered. In this work, we enforce the idea to properly extend, when possibile, efficient numerical strategies for deterministic ODEs to the SDEs setting; such an idea inspired other works in different stochastic settings, as [14] for the development of a theory of stochastic two-step Runge-Kutta methods or [9,10] for the assessment of a theory of nonlinear stability for stochastic numerical schemes.

References

1. Bulsara, A.R., Lindenberg, K., Shuler, K.E.: Spectral analysis of a nonlinear oscillator driven by random and periodic forces. I. Linearized theory. J. Stat. Phys. **27**, 787–808 (1982). https://doi.org/10.1007/BF01013448
2. Burrage, K., Cardone, A., D'Ambrosio, R., Paternoster, B.: Numerical solution of time fractional diffusion systems. Appl. Numer. Math. **116**, 82–94 (2017)
3. Burrage, K., Lenane, I., Lythe, G.: Numerical methods for second-order stochastic differential equations. SIAM J. Sci. Comp. **29**, 245–264 (2007)
4. Burrage, K., Lythe, G.: Accurate stationary densities with partitioned numerical methods for stochastic differential equations. SIAM J. Numer. Anal. **47**, 1601–1618 (2009)
5. Burrage, K., Lythe, G.: Accurate stationary densities with partitioned numerical methods for stochastic partial differential equations. Stochast. Partial Diff. Eq. Anal. Comput. **2**(2), 262–280 (2014). https://doi.org/10.1007/s40072-014-0032-8
6. Citro, V., D'Ambrosio, R.: Long-term analysis of stochastic θ-methods for damped stochastic oscillators. Appl. Numer. Math. **150**, 18–26 (2019)
7. Cohen, D.: On the numerical discretisation of stochastic oscillators. Math. Comput. Simul. **82**, 1478–1495 (2012)
8. Cohen, D., Sigg, M.: Convergence analysis of trigonometric methods for stiff second-order stochastic differential equations. Numer. Math. **121**, 1–29 (2012)
9. D'Ambrosio, R., Di Giovacchino, S.: Mean-square contractivity of stochastic theta-methods. Comm. Nonlin. Sci. Numer. Simul. **96**, 105671 (2021)
10. D'Ambrosio, R., Di Giovacchino, S.: Nonlinear stability issues for stochastic Runge-Kutta methods. Comm. Nonlin. Sci. Numer. Simul. **94**, 105549 (2021)
11. D'Ambrosio, R., Moccaldi, M., Paternoster, B.: Numerical preservation of long-term dynamics by stochastic two-step methods. Disc. Cont. Dyn. Sys. Ser. B **23**(7), 2763–2773 (2018)
12. D'Ambrosio, R., Scalone, C.: On the numerical structure preservation of nonlinear damped stochastic oscillators. Num. Algorithms **86**(3), 933–952 (2020). https://doi.org/10.1007/s11075-020-00918-5
13. D'Ambrosio R., Scalone C.: Filon quadrature for stochastic oscillators driven by time-varying forces, to appear in Appl. Numer. Math
14. D'Ambrosio, R., Scalone, C.: Two-step Runge-Kutta methods for stochastic differential equations. Appl. Math. Comput. **403**, 125930 (2021)
15. D'Ambrosio, R., Scalone C.: A Magnus integrator for nonlinear stochastic oscillators with non-constant frequency, submitted
16. de la Cruz, H., Jimenez, J.C., Zubelli, J.P.: Locally linearized methods for the simulation of stochastic oscillators driven by random forces. BIT **57**(1), 123–151 (2017)
17. Failla, G., Pirrotta, A.: On the stochastic response of a fractionally-damped duffing oscillator. Commun. Nonlinear Sci. Numer. Simul. **17**(12), 5131–5142 (2012)

18. Gardiner, C.W.: Handbook of Stochastic Methods, for Physics Chemistry and the Natural Sciences, 3rd edn. Springer, Heidelberg (2004)
19. Gitterman, M.: The Noisy Oscillator, The First Hundred Years From Einstein Until Now. World Scientific (2005)
20. Gitterman, M.: Oscillator subject to periodic and random forces. J. Mod. Phys. **4**, 94–98 (2013)
21. Hairer, E., Lubich, C., Wanner, G.: Geometric Numerical Integration. Springer, Heidelberg (2006). https://doi.org/10.1007/3-540-30666-8
22. Hochbruck, M., Ostermann, A.: Exponential integrators. Acta Numerica **19**, 209–286 (2010)
23. Iserles, A., Nørsett, S.P.: On quadrature methods for highly oscillatory integrals and their implementation. BIT **44**, 755–772 (2004)
24. Iserles, A., Nørsett, S.P.: Efficient quadrature of highly oscillatory integrals using derivatives. Proc. R. Soc. A **461**, 1383–1399 (2006)
25. Condon, M., Iserles, A., Nørsett, S.P.: Differential equations with general highly oscillatory forcing terms. Proc. R. Soc. A **470**, 20130490 (2015)
26. Khanamiryan, M.: Quadrature methods for highly oscillatory linear and nonlinear systems of ordinary differential equations: part I. BIT **48**, 743 (2008)
27. Lingala, N., Namachchivaya, N., Pavlyukevich, I.: Random perturbations of a periodically driven nonlinear oscillator: escape from a resonance zone. Nonlinearity **30**, 1376–1404 (2017)
28. Scalone C.: A numerical scheme for harmonic stochastic oscillators based on asymptotic expansions. submitted
29. Senoisian, M.J., Tocino, A.: On the numerical integration of the undamped harmonic oscillator driven by independent additive gaussian white noises. Appl. Numer. Math. **137**, 49–61 (2019)
30. Shi, C., Xiao, Y., Zhang, C.: The convergence and MS stability of exponential Euler method for semilinear stochastic differential equations, abstract and applied analysis (2012)
31. Strömmen Melbö, A.H., Higham, D.J.: Numerical simulation of a linear stochastic oscillator with additive noise. Appl. Numer. Math. **51**, 89–99 (2004)
32. Tocino, A.: On preserving long-time features of a linear stochastic oscillator. BIT **47**, 189–196 (2007)
33. Yalim, J., Welfert, B.D., Lopez, J.M.: Evaluation of closure strategies for a periodically-forced Duffing oscillator with slowly modulated frequency subject to Gaussian white noise. Commun. Nonlinear Sci. Numer. Simulat. **44**, 144–158 (2017)

Computational Issues in the Application of Functional Data Analysis to Imaging Data

Juan A. Arias-López[1]([⊠])(iD), Carmen Cadarso-Suárez[1](iD), and Pablo Aguiar-Fernández[2,3](iD)

[1] Biostatistics and Biomedical Data Science Unit. Department of Statistics, Mathematical Analysis, and Operational Research, Universidade de Santiago de Compostela, Santiago, Spain
juanantonio.arias.lopez@usc.es
[2] Nuclear Medicine Department and Molecular Imaging Group, University Clinical Hospital (CHUS) and Health Research Institute of Santiago de Compostela (IDIS), Santiago, Spain
[3] Molecular Imaging Group, Department of Psychiatry, Radiology and Public Health, Faculty of Medicine, Universidade de Santiago de Compostela, Santiago, Spain

Abstract. Functional Data Analysis (FDA) is the field of statistics which deals with the analysis of data expressed in the form of functions, which is extensible to data in the form of images. In a recent publication, Wang et al. [1] settled the mathematical groundwork for the application of FDA to the estimation of mean function and simultaneous confidence corridors (SCC) for a group of images and for the difference between two groups of images. This approach presents at least two advantages compared to previous methodologies: it avoids loss of information in complex data structures and also avoids the multiple comparison problem which arises from pixel-to-pixel comparison techniques. However, the computational costs of applying these procedures are yet to be fully explored and could outweigh the benefits resulting from the use of an FDA approach. In the present study, we aim to apply these novel procedures to simulated data and measure computing times both for the estimation of mean function and SCC for a one-group approach, for the comparison between two groups of images, and for the construction of Delaunay triangulations necessary for the implementation of the methodology. We also provide the computational tools to ensure replicability of the results herein presented.

Keywords: Functional data analysis · Image processing · Computational biology

1 Introduction

1.1 Functional Data Analysis

Functional Data Analysis (FDA) can be defined as the field of statistics concerned with the theoretical basis and the development of analytical tools

© Springer Nature Switzerland AG 2021
O. Gervasi et al. (Eds.): ICCSA 2021, LNCS 12950, pp. 630–638, 2021.
https://doi.org/10.1007/978-3-030-86960-1_46

oriented towards the study of data in the form of functions. From this scope of view, a function is the minimum unit of the analysed data and usually every subject or participant in a study has one or more functions associated. In the last decades, FDA has experimented a rise in popularity in several research areas and different publications - including monographs [2] and review articles [3] - are now available to the public explaining its theoretical basis and applications.

However, FDA is still in its infancy and thus a strict definition of the field is difficult and not recommended. Instead, there are a few characteristics generally assumed to be inherent to functional data. First, functional data are continuously defined and single instances of this data are only considered as realizations of the underlying function, a necessary constraint in order to use this data with the computational means available to us. Second, the whole function - rather than individual points - is the basic element of the analytical process carried out in FDA. Additionally, functional data is usually associated with time as a variable and often this data is smoothed or at least some regularity conditions are imposed on it [2]. Functional data usually consist of a sample of independent functions whose values are located in a compact interval (I) and are usually assumed to be in a Hilbert space (L^2):

$$X_1(t), X_2(t), ..., X_n(t); I = [0, T] \in L^2 \tag{1}$$

1.2 FDA for Imaging Data

FDA techniques can easily be extended to work with images, and especial attention has been paid to its use with biomedical imaging data such as brain scanner data and images of tumor tissue [2]. However, smoothing methods proposed to date in the scientific literature for imaging analysis (e.g. kernel smoothing, tensor product smoothing...) suffer from a problem of *leakage* when the data structure is complex, showing difficulties in boundary regions' estimation resulting in inappropriate smoothing.

Besides, when analysing medical imaging data, problems arise not only for the estimation of the mean value for a given point, but also for the estimation of the associated uncertainty (i.e. confidence band). This problem becomes even more complicated when considering that the spatial correlation has to be taken in account. To date, the predominant analytic techniques usually rely on *mass univariate approaches*, which consider pixels in the image as independent units and then perform comparisons between them with classical methods such as a simple *T-test*. This brings up the problem of multiple comparisons, which is usually addressed with popular approaches such as the Bonferroni correction or extensions of random field theory [4], which are *ad hoc* corrections heavily dependent on the appropriate choice of a threshold.

In a recent publication, Wang et al. [1] proposed a way to avoid these problems using FDA techniques. First, the challenge of *leakage* on complex domains is addressed by using bivariate splines over Delaunay triangulations (see Sect. 3.1), thus preserving important features of the imaging data. Second, the proposed methodology treats imaging data as an instance of functional data continuously defined and only observed on a regular grid. If the image is considered as functional, attention moves from the pixel as an individual unit towards the analysis of images as a whole, allowing for calculations such as the simultaneous confidence corridors (SCC; also known as *simultaneous confidence bands*), an approach which has been proven superior to conventional multiple comparison approaches [5].

In their article, Wang and colleagues [1] describe the proposed bivariate spline estimators, test their asymptotic properties, describe the attributes of SCC based on these estimators, and extract coverage probability for the obtained mean function, concluding that the proposed SCC account for the correct probability coverage both in one-group and two-group setups. However, although the proposed methodology accounts for the correct probability coverage, the computational resources necessary for its application are not addressed by the authors and thus its utility for a practical case is yet to be tested, as computing times, which include among others the calculation of Delaunay triangulation parameters, could outweigh the benefits of choosing an FDA approach.

2   Objectives

Given the gap in knowledge mentioned in the previous section, we set the goal of testing the practical utility of this novel methodology by studying the computational efforts necessary to implement it. In order to do that, we evaluate computing times for the calculation of the polygonal domain of Delaunay triangulations for imaging data (necessary for the technique's implementation), evaluate computing times for the calculation of mean function and SCC for a sample of simulated imaging data, and also for the calculation of mean function and SCC for the difference between two samples of simulated imaging data.

3   Methods

3.1   Delaunay Triangulations

In mathematical terms, Delaunay triangulations consist on a series of triangles created by the union of different vertices in which no vertices falls inside the circumcircle of a given triangle. The examined FDA approach relies on applying bivariate splines over these triangulations to preserve imaging data features with the minimum loss of information. These Delaunay triangulations are essential for the construction of SCC and thus its computing times also need to be tested. In order to do this, we will be using the *Triangulation* R package [6] and test its performance for growing values of the triangulation fineness degree. Computing times and examples of the results are analysed in Sect. 4.1.

3.2   One-Group Mean Function and SCC

The methodology herein tested allows for the estimation of the mean function for a series of images and also for the calculation of its associated SCC. With the aim of evaluating performance in terms of computing times we simulate a growing number of data using functions implemented in *ImageSCC* R package [7] with what would be the boundaries of a brain used in a hypothetical neuroimaging study. We then evaluate computing times for the estimation of mean function and SCC for this data using different triangulation parameters. Computing times and examples of the results are analysed in Sect. 4.2.

3.3   Two-Group Mean Function and SCC

This technique is also extensible to a two-sample setup. Following this approach, estimated mean functions are calculated for the two different groups separately and then SCC are estimated for the difference between the two groups' mean functions. The results display areas in the analysed images whose difference fall outside estimated SCC for a given α level, thus suggesting a significant difference in that region in one group compared to another. With the aim of testing this methodology's performance, we simulate a growing number of neuroimaging data for two groups and then compute the estimated mean functions for each group and SCC for the difference between these groups' mean functions, visualizing areas falling outside estimated SCC. Computing times and examples of the results are analysed in Sect. 4.3.

4   Results

In the following sections we summarize results for the analysis proposed in our Objectives (Sect. 2), including computing times and sample graphics of the obtained results.

4.1   Delaunay Triangulations

In Fig. 1 we visually represent three output examples of Delaunay triangulations for an hypothetical brain imaging study. The brain is naturally tri-dimensional, however, we will be working with a two-dimensional axial cut of the brain for practical reasons. These triangulations are the basis for imaging mean function and SCC estimation and will be used in the following sections in order to evaluate computing times of this methodology.

In Fig. 2 we visually assess computing times for these triangulations depending on the selected degree of fineness. We can see a pattern of growing computing times with the increase of n value. According to the authors [1] a value of $n = 8$ is sufficient, although higher values such as $n = 15$ and $n = 25$ still take reasonable amounts of time and will be considered in this study.

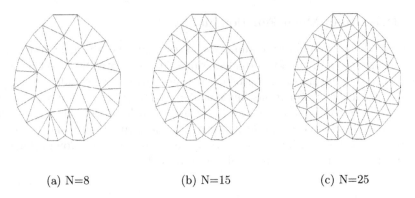

(a) N=8   (b) N=15   (c) N=25

Fig. 1. Delaunay triangulations with increasing N values representing increases in triangulation's degree of fineness.

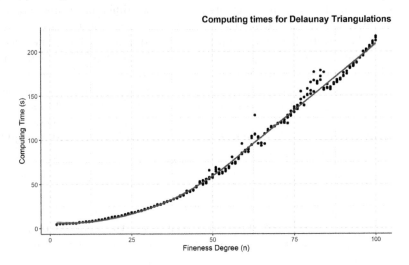

Fig. 2. Computing times for Delaunay triangulations with growing fineness degree values. Curve fitted with LOESS regression (interactive version).

4.2   One-Group Mean Function and SCC Estimation

In Fig. 3 we present an example of the images obtained using this methodology with one-group imaging data, displaying estimated mean function, lower, and upper confidence intervals in the form of images.

Simulation of the data presented in Fig. 3 was carried out with ImageSCC [7] package using a $n = 50$, with cubic function for data generation and eigenvalues to adjust subject-level variation in the simulated data of $\lambda_1 = 0.5$ and $\lambda_2 = 0.2$. Besides for triangulation degree of fineness ($n = 25$), estimated mean function and SCC were calculated using parameters recommended by

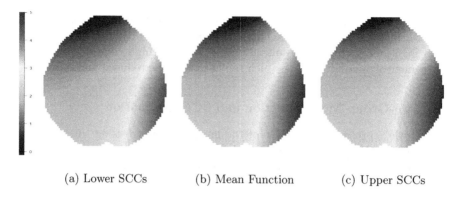

(a) Lower SCCs   (b) Mean Function   (c) Upper SCCs

Fig. 3. Results for one-sample SCC estimation with estimated mean function (center), lower SCC (left), and upper SCC (right) calculated for $\alpha = 0.05$ using triangulation's fineness degree $N = 25$.

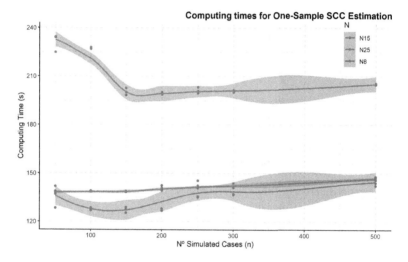

Fig. 4. Computing times for one-group mean function and SCC estimation for imaging data with growing number of simulated cases and three Triangulation fineness degrees (N = 8, N = 15, N = 25). Curves fitted using local (LOESS) regression (interactive version).

Wang et al. [1] including: degree of bivariate spline for mean estimation $d.est = 5$, degree of bivariate spline for construction of SCC $d.band = 2$, smoothness parameter $r = 1$, and a vector of candidates for penalty parameter with values ranging from 10^{-6} to 10^3. In addition, SCC were calculated for three different α levels.

In Fig. 4 we visually assess computing times for mean function and SCC estimation for a one-group setup with growing number of simulated subjects and three different triangulation parameters. We see that these times are similar for fineness degrees of $N = 8$ and $N = 15$, and much larger for $N = 25$. We can also observe that these times are stable and tend to orbit around a certain value with slow increases as the number of simulated subjects grows. It is also worth mentioning that the results for $N = 25$ show high computing times for low number of simulated cases which tends to then be reduced and stabilize.

4.3 Two-Group Mean Function and SCC Estimation

In Fig. 5 we represent an example of the images obtained using this methodology with two-group imaging data, displaying estimated mean function for both groups and also areas whose values fall outside estimated SCC for the difference between groups of images.

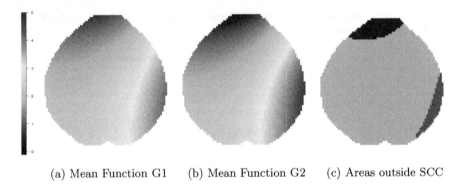

(a) Mean Function G1   (b) Mean Function G2   (c) Areas outside SCC

Fig. 5. Results for a two-sample setup showing estimated mean function for Group 1 (left) and Group 2 (center), together with estimated areas with significant ($\alpha = 0.05$) inter-group differences (blue: hypo-activity; red: hyper-activity). Triangulation's fineness degree $N = 25$. (Color figure online)

Simulation of data for the creation of estimated mean functions and SCC was carried out following the same parameters described in Sect. 4.2. In addition, the additional parameter $\delta = 0.8$ was used, indicating the scale of difference between the two simulated groups' mean functions. In Fig. 6 we visually assess computing times for obtaining the two simulated groups' mean function and SCC for the difference between them. We carry out this calculation for a growing number of simulated subjects and three triangulation setups. The obtained timings are similar for $N = 8$ and $N = 15$, and much larger for $N = 25$. We can also see that computing times are extremely stable around a certain value and present very slow growth with regards to growing number of simulated subjects. Compared to Fig. 4, the variability of computing times is almost null.

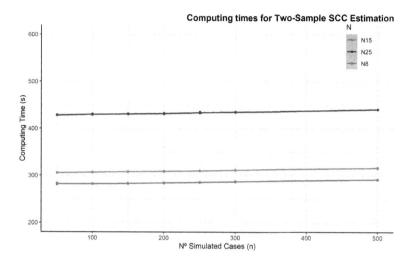

Fig. 6. Computing times for two-group mean function and SCC estimation for the differences between groups with growing number of simulated cases and three Triangulation parameters. Curves fitted using LOESS regression (interactive version).

5 Discussion

In this article we aimed to assess computational costs for Wang and colleagues' [1] proposal of an FDA methodology in order to obtain mean functions and SCC for imaging data. Our appraisal consisted on the calculation of computing times for Delaunay triangulations necessary for this method, simulation of data, and evaluation of time necessary for the estimation of mean function and SCC (one-group approach) and estimation of mean functions and SCC for the difference between groups (two-groups approach).

The obtained results suggest the following. First, computation times for Delaunay triangulations grow with the increase of selected fineness degree, although these computing times are low, especially regarding the fineness degree values recommended by the authors (see Fig. 2). Second, computing times for a one-group approach are much more dependent on the selected triangulation parameters than the number of simulated cases, as shown in Fig. 4; besides, it seems sensible to use higher values of triangulation's degree of fineness than the recommended by the authors, as computing times for $N = 8$ and $N = 15$ are almost identical. Third, computing times for the two-group approach are also heavily dependent on the selected triangulation parameters rather than the variation in number of simulated subjects, for which the results remain stable as shown in Fig. 6.

The strong dependence of computing times on the triangulation parameters, rather than the number of studied cases, goes in line with expected outcomes for functional data methodologies, which are meant to be applied to a high number of cases, whereas increases in the intricacy of the triangulation meshes tend to produce cumulative effects deriving in increased computation times due to the higher complexity of the calculations involved.

In conclusion, due to the utility of this methodology to avoid other recurrent problems in imaging analysis and also due to the relatively low computing times of this methodology, it seems reasonable to recommend further application of this technique to practical cases, especially bearing in mind the great time stability displayed in the two-sample case for big numbers of simulated cases, a property which is valuable in medical research dealing with a great number of patients. Nevertheless, this is a simulation study and there is still a need for testing this methodology and its results in a complete setup, with real data over more realistic image structures. In short, these preliminary results suggest that functional data analysis techniques can be useful for imaging analysis, displaying desirable properties such as stability and low computing times.

Computer Specifications

This study was carried out using a personal computer with the following specifications and R version. Computer model: MSI Modern 14 A10M; operating system: Windows 10 Enterprise LTSC 64-bit; CPU: Intel(R) Core(TM) i5-10210U CPU @ 1.60 GHz (8 CPUs); RAM memory: 16384 Mb; R version: 4.0.3 (2020-10-10).

Complementary Materials

Complementary material are stored in our GitHub code repository.

References

1. Wang, Y., Wang, G., Wang, L., Ogden, R.T.: Simultaneous confidence corridors for mean functions in functional data analysis of imaging data. Biometrics **76**(2), 427–437 (2020). https://doi.org/10.1111/biom.13156
2. Ramsay, J.O.: Functional data analysis. In: Encyclopedia of Statistical Sciences, vol. 4 (2004). https://doi.org/10.1002/0471667196.ess3138
3. Wang, J.-L., Chiou, J.-M., Müller, H.-G. : Functional data analysis. Ann. Rev. Stat. Appl. **3**, 257–295 (2016). https://doi.org/10.1146/annurev-statistics-041715-033624
4. Worsley, K.J., Taylor, J.E., Tomaiuolo, F., Lerch, J.: Unfied univariate and multivariate random field theory. NeuroImage **23**, S189–S195 (2004). https://doi.org/10.1016/j.neuroimage.2004.07.026
5. Degras, D.A.: Simultaneous confidence bands for nonparametric regression with functional data. Statistica Sinica, 1735–1765 (2011). https://doi.org/10.5705/ss.2009.207
6. Lai, M.J., Wang, L.: Triangulation: Triangulation in 2D domain. R package version 0.1.0 (2020)
7. Wang, Y., Wang, G., Wang, L.: ImageSCC: SCC for Mean Function of Imaging Data. R package version 0.1.0 (2020)

Spatial Modelling of Seablite Distribution

Thanapong Chaichana$^{(\boxtimes)}$ (iD) and Yasinee Chakrabandhu$^{(\boxtimes)}$ (iD)

College of Maritime Study and Management, Chiang Mai University, Samut Sakhon, Thailand
{thanapong.c,yasinee.c}@cmu.ac.th

Abstract. Foods demands are increasing today opposed to the change in environmental degradation. In this work, we studied new seablite distribution modelling. We aimed to identify substantial basic factors related to spatial distribution of seablites to make a meaningful explanation or accurate prediction in a coastal region of Samut Sakhon, Thailand. Virtual field survey and field survey data of physical geography were used to form a structure of spatial model and build a predictive model. We found that important underlying factors of spatial distribution of seablites were soil salinity, soil pH, soil moisture, air temperature, height above sea level, distance from seashore, and wind direction. Our predictive model improves understanding upon a distribution of seablites and environments. This preliminary work supported to simulate the environments to establish and thrive seablites for smart agriculture system.

Keywords: Digital agriculture · Parameter estimation · Prediction · Smart farming · Spatial distribution model

1 Introduction

Seablite, generally known as Annual Seablite or Suaeda maritima, originally found in coastline regions, for example, in Thailand, in Samut Sakhon province. Indigenous people, such as, in Australia, Canada, Thailand and the United States, used seablite as food sources [1–4]. One reason purposed by them, for instance, tribes, Mon people in Thailand used seablite as alternative vegetable to cook a main dish, instead of using vegetation came from highland areas. These days, seablite has been used to produce foods, cosmetics, and medicine [5–7], and has shown a great potential to enter the local grocery stores and supermarkets. Thus, seablite itself reveals economic plant capabilities, and abilities to farm it elsewhere. Now, computer simulation has become a useful tool for modeling many systems, such as, agriculture, medicine, physics, geography, and climatology [8–13]. Early physical geography research, Janet Franklin purposed the predictive vegetation mapping to predict the vegetation distribution through the landscape from mapped environment variables [14]. In 1995, Franklin's model revealed that climate is a key parameter to influence soil development, moisture regime, and temperature regime for potential natural vegetation. This model was determined by computerized techniques, and useful for environment planning and basic research upon the role of the biota to the application of the earth system science.

© Springer Nature Switzerland AG 2021
O. Gervasi et al. (Eds.): ICCSA 2021, LNCS 12950, pp. 639–647, 2021.
https://doi.org/10.1007/978-3-030-86960-1_47

In 2012, Jennifer A. Miller described the species distribution models to predict broad-scale of ecological patterns for characterizations of non-stationarity and auto-correlation data [15]. Miller found that species richness had connected with environment factors between climate and normalized difference vegetation index. Later, in 2015, Miller and Holloway investigated the species distribution models incorporated species' ability to access suitable habitat [16]. Movement ability was found that involved with landscape factors (e.g., barriers, disturbance, and patch configuration), migration rates or dispersal constraints [17], and demographic factors (e.g., age at maturity, fecundity, and mortality). In 2018, Kosanic et al. studied the changes in the geographical distribution of plant species impacted by climate change at South West Peninsula, England, the United Kingdom [18]. Ellenberg values and climate indicators were used as the markers to respond the changes of both environments and climates. Kosanic's study suggested that Ellenberg and climate indicator values can contribute directly to plan appropriate environmental strategies. In 2019, Lewis et al. investigated long-term implication of crop planning thru climate change [19]. Precipitation and temperature factors were predicted for the next five decades until 2071, based on optimization models for crop planning. Lewis's study recommended that the crop mixes will have to change due to climate change. Sustainable crop choices will change in the future, because of the large-scale irrigated agriculture may become unavailable. For these reasons, the spatial model of specific plant distribution in certain area will improve our knowledge on parameter estimation for plants framing, however, there were not yet studied on geographical distribution of seablite origin and its environmental factors. This paper developed model for seablite spatial distribution.

This paper is organized as follows. Section 2 describes the proposed method, especially present seablite data, environmental determinants, survey techniques, as well as spatial and predictive model. Section 3 illustrates quantitative quantity of seablite distribution. Section 4 discusses on how our finding will improve future farming. Finally, Sect. 5 concludes seablite distribution model and describes future research.

2 Methods

In this study, we first explored current seablite data across Samut Sakhon, Thailand. Then, virtual field survey and field survey have been conducted to investigate the key factors correlated to spatial distribution of seablites. Subsequently, structure of spatial model and predictive model were built. Environmental factors were determined using coastal environments. Finally, how to investigate the geographical distribution of seablite origin and its environmental factors are introduced.

2.1 Current Seablite Data

Current spatial records of seablite distribution in Samut Sakhon were obtained from the online "Food from Cha Khram (seablite) leaves" of the Samut Sakhon Provincial Cultural Office (SSPCO) [20]. The SSPCO data contains the distribution of seablite situated in a coastal region, characteristics of seablite's leaves, benefits of seablite for cooking food or main dishes, medicinal properties. The SSPCO data records were published on 21st June 2017 and have been accessed since July 2020.

2.2   Virtual Field Survey

Virtual field survey was conducted for three months from July to September 2020 in the following sub-districts of Mueang Samut Sakhon: Na Khok, Kalong, Bang Tho Rat, Ban Bo, Bang Ya Praek, Krok Krak, Khok Kram, and Panthai Norasing, which located in coastal regions started from the left-to-right, as shown in Fig. 1, respectively. We conducted a virtual field survey to identify the seablite sources using Google Maps, Google Earth, and GoodNotes software. Consequently, we spotted the locations of seablite origins as the red dots in digital illustrative maps.

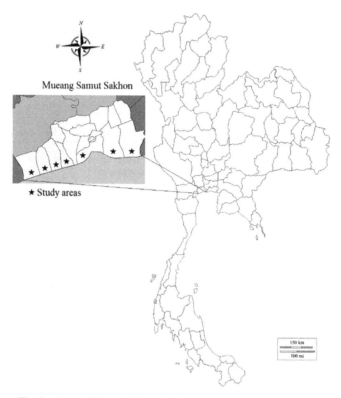

Fig. 1. Map of Thailand showing Mueang Samut Sakhon district.

2.3   Field Survey

Field survey was conducted in September 2020 for an arbitrarily verification of the results obtained from virtual field survey in the following subdistricts: Na Khok, Khok Kram, and Panthai Norasing. In addition, we conducted the second field survey in February 2021 through the study areas for quantification of spatial data of seablite origin including soil salinity, soil pH, soil moisture, air temperature, elevation, distance from seashore, wind direction, and height and width of seablite.

We then collected a sample pack of seablite for each subdistrict is around 100–200 g. Figure 2(a) revealed sample from Bang Ya Praek subdistrict. Hence, there were seven sample packs in total. Sodium chloride was extracted in grams per hundred grams of seablite in room temperature at chemical laboratory (Central Laboratory (Thailand) Co.,Ltd., Samut Sakhon, Thailand) with reference method is "AOAC (2019) 937.09" on 4th March 2021. Finally, sodium chloride was extracted from seablite to predict the condition of soil moisture, and soil salinity.

2.4 Structure of Spatial Model

The structure of spatial model of seablite was constructed from virtual field survey and field survey. Figure 2(b) reveals one spot of spatial model of seablite origin (e.g., at latitude x and longitude y-coordinate (13.49200, 100.16349) in geographic coordinate system.

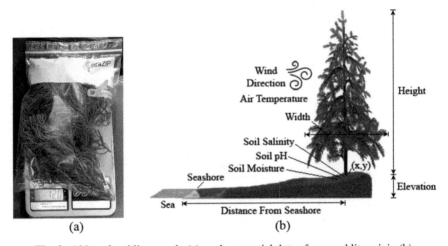

Fig. 2. 120 g of seablite sample (a), and geospatial data of one seablite origin (b).

2.5 Predictive Model

Predictive model was built from geospatial data in previous section. The modelling method was used Boolean Logic incorporated with environmental determinants (soil conditions, wind direction, and maritime climate), and direct gradients (salinity, pH, moisture, and temperature). Predictive model was successfully developed to predict seablite origin, as shown in Fig. 3.

In addition, climate has a direct impact upon soil developments and their environmental factors, clearly demonstrated by the relationship between geology and topology in Fig. 3. We purposed smart agriculture system mapping based on the predictive model. Figure 4 shows digital agriculture mapping for farming seablite with using Logic AND Gates, and digital circuit was designed using CircuitVerse.

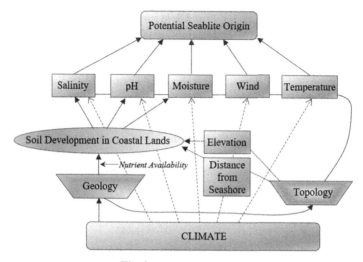

Fig. 3. Conceptual model.

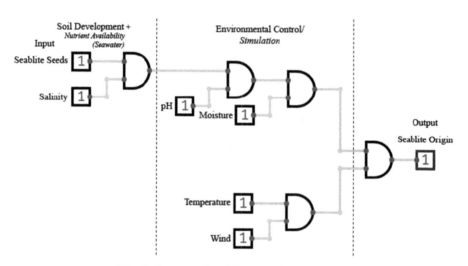

Fig. 4. Conceptual seablite smart farming model.

Seablite origin was predicted using the conceptual seablite smart farming model in Fig. 4. A binary digit was assigned, a value of bit '1' (succeed) and bit '0' (not succeed) was configurated as inputs of environmental determinants and direct gradients.

3   Results

3.1   Geographic Distribution of Seablite Origin

Geographic distribution of seablite origin in Mueang Samut Sakhon district has been analyzed using the spatial model structure and predictive model described in previous sections. The computing machine for virtual field survey operated on Microsoft Windows 10, 64-bit system with 8 GB RAM, and Intel i3-10100F processor for 3.60 GHz CPU. The computation duration of seablite origins distributed along coastal regions in each sub-district was estimated to be 4–5 days. Field survey mainly verified results achieved from virtual field survey. Table 1 demonstrates the number of seablite distribution retrieved from virtual field survey and field survey.

Table 1. Distribution of seablite origins in coastal lands

Mueang Samut Sakhon subdistricts	Seablite origins
Na Khok	490
Kalong	267
Bang Tho Rat	443
Ban Bo	1,365
Bang Ya Praek	104
Krok Krak	0
Khok Kram	701
Panthai Norasing	560

3.2 Prediction of Seablite Origin

Table 2 demonstrates the simulated 2-states of the outputs achieved from conceptual seablite smart farming model in Fig. 4.

Table 2. Seablite forecast

States	Seablite seeds	Salinity	pH	Moisture	Temperature	Wind	Seablite distribution
A	1	1	1	0	1	1	0
B	1	1	1	1	1	1	1

Bit '1' = succeed, and bit '0' = not succeed.

3.3 Virtual Field Survey and Field Survey

Figure 5 shows the geographical analysis of seablite origin of both virtual field survey and field survey. Figure 6 demonstrates an example of digital illustrative maps achieved from virtual field survey in Na Khok subdistrict. Thus, a total figure of 490 seablite origins has been quantified (see Table 1).

Fig. 5. Both spatial data of seablite origins. Red circle spotted matching the same pillar. These data were taken in Bang Tho Rat subdistrict at latitude and longitude (13.49200, 100.16349). (Color figure online)

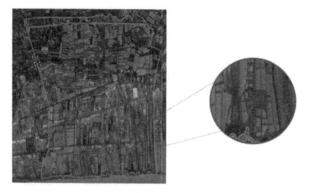

Fig. 6. Seablite origins recorded as red dots. (Color figure online)

3.4 Analysis of Sodium Chloride Extracted from Seablite

Table 3 examined sodium chloride extracted from seablite samples in all study areas.

Table 3. Extraction of sodium chloride (NaCl) from seablite

Mueang Samut Sakhon subdistricts	NaCl (g/100 g)
Na Khok	3.97
Kalong	3.36
Bang Tho Rat	3.23
Ban Bo	3.69
Bang Ya Praek	4.37
Khok Kram	4.89
Panthai Norasing	4.01

4 Discussions

Human being centered activities mainly depend on agriculture to produce foods. The current climate change issues will impact agriculture to produce foods in the next five decades, and farming will have to change due to environmental change [19]. Studies of specific plant distribution will help to improve farming. Our work aimed to explain spatial model and its important factors to support smart farming. We modeled seablite distribution using virtual field survey. This is quick and efficient way to obtain the results. It reduced travel expenses, avoided the road accidents, and was given the same quality as the field survey conducted (see Fig. 5). Ban Bo subdistrict was found 1,365 seablite origins, whilst taking longest time for computation. In contrast, Krok Krak subdistrict was full housings led to unidentified seablite sources (see Table 1). We found that soil development contained salinity, pH, and moisture for seablite origin. These factors depended on climate. Thus, climate was a key controller for agriculture (see Fig. 3). Seablite sample collected from Khok Kram subdistrict (see Table 2) displayed a greatest quantity of NaCl (4.89 g). It proved that Khok Kram had a good condition of soil moisture, and soil salinity. Agricultural forecast of seablite was determined by our conceptual model (see Fig. 4). Output of seablite relied on environmental determinants and direct gradients. Nonetheless, sunlight was uncontrolled variable, and excluded in present study. This study has some limitations. It was mainly conducted using virtual field survey, however seablite outputs were verified by physical field survey, and study area was only in seashore regions.

5 Conclusions

This work conducted simulation of environmental determinants and direct gradients for seablite smart farming. Our results are preliminary intended to accurately describe the substantial basic factors related to seablite origins. All study areas had an average distance of 8.495 km from seashore, computed seablite origins spatially distributed in Mueang Samut Sakhon district. This finding reported sufficiently data and models to simulate key environmental factors for farming seablite. Our work helps to improve understanding on spatial data of seablite distribution in coastal areas. Further studies focusing on the development of seablite smart farming based on our primary results and spatial model of seablite distribution.

Acknowledgments. This research project is supported by TSRI.

References

1. Kuhnlein, H.V., Turner, N.: Traditional Plant Foods of Canadian Indigenous Peoples: Nutrition, Botany and Use. Routledge, London (2020). https://doi.org/10.4324/978100305 4689
2. Chalmers, L.: Native vegetation of estuaries and saline waterways in south Western Australia. Water and Rivers Commission, Perth, Western Australia (1997)

3. Norton, S.: World's first indoor saltwater farm brings sea beans to Charleston menus. https://charlestonmag.com/features/the_worlds_first_indoor_saltwater_farm_brings_sea_beans_to_charleston_menus. Accessed 25 Mar 2021

4. Vorasiha, E.: The travelling route for gastronomic tourism via salt in western region of Thailand. Afr. J. Hospitality Tour. Leisure **7**(3), 1–9 (2018)

5. Jabiyeon Cream. http://www.saenggreen.com/eng/product/product010301.html. Accessed 14 Jan 2021

6. Harikrishnan, R., et al.: Effect of dietary supplementation with Suaeda maritima on blood physiology, innate immune response, and disease resistance in olive flounder against Miamiensis avidus. Exp. Parasitol. **131**(2), 195–203 (2011)

7. Song, J., Wang, B.: Using euhalophytes to understand salt tolerance and to develop saline agriculture: Suaeda salsa as a promising model. Ann. Bot. **115**(3), 541–553 (2015)

8. Chaichana, T., Sun, Z., Jewkes, J.: Computation of hemodynamics in the left coronary artery with variable angulations. J. Biomech. **44**(10), 1869–1878 (2011)

9. Chaichana, T., Wiriyasuttiwong, W., Reepolmaha, S., Pintavirooj, C., Sangworasil, M.: Extraction blood vessels from retinal fundus image based on fuzzy C-median clustering algorithm. In: 4th International Conference on Fuzzy Systems and Knowledge Discovery, pp. 144–148 (2007)

10. Sun, Z., Mwipatayi, B., Chaichana, T., Ng, C.: Hemodynamic effect of calcified plaque on blood flow in carotid artery disease: a preliminary study. In: 3rd International Conference on Bioinformatics and Biomedical Engineering, pp. 1–4 (2009)

11. Chaichana, T., Brennan, C.S., Osiriphun, S., Thongchai, P., Wangtueai, S.: Development of local food growth logistics and economics. AIMS Agric. Food **6**(2), 588–602 (2021)

12. Chaichana, T.: Renewable, sustainable and natural materials on food packaging: primary data for robotically detect packaging shape in logistics. J. Phys. Conf. Ser. **1681**(1), 012024 (2020)

13. Chaichana, T.: Analysis of computing progress in maritime studies. ACM International Conference Proceeding Series, pp. 150–154 (2019)

14. Franklin, J.: Predictive vegetation mapping: geographic modelling of biospatial patterns in relation to environmental gradients. Prog. Phys. Geogr. **19**(4), 474–449 (1995)

15. Miller, J.A.: Species distribution models: Spatial autocorrelation and non-stationarity. Prog. Phys. Geogr. **36**(5), 681–692 (2012)

16. Miller, J.A., Holloway, P.: Incorporating movement in species distribution models. Prog. Phys. Geogr. **39**(6), 837–849 (2015)

17. Vasudev, D., Fletcher, R.J., Goswami, V.R., Krishnadas, M.: From dispersal constraints to landscape connectivity: lessons from species distribution modeling. Ecography **38**, 967–978 (2015)

18. Kosanic, A., Anderson, K., Harrison, S., Turkington, T., Bennie, J.: Changes in the geographical distribution of plant species and climatic variables on the West Cornwall peninsula (South West UK). PLoS One **13**(2), e0191021 (2018)

19. Lewis, A., Randall, M., Elliott, S., Montgomery, J.: Long term implications of climate change on crop planning. In: Rodrigues, J.M.F., et al. (eds.) ICCS 2019. LNCS, vol. 11536, pp. 369–382. Springer, Cham (2019). https://doi.org/10.1007/978-3-030-22734-0_27

20. The Samut Sakhon Provincial Cultural Office (SSPCO): Food from Cha Khram leaves. https://www.m-culture.go.th/samutsakhon/ewt_news.php?nid=459&filename=index. Accessed 25 Mar 2021

A LSTM Recurrent Neural Network Implementation for Classifying Entities on Brazilian Legal Documents

Rafael Mecheseregian Razeira and Ildeberto Aparecido Rodello[✉]

University of São Paulo, Ribeirão Preto, SP 14049-900, Brazil
rodello@usp.br

Abstract. Although the use of Natural Language Processing and Named Entity Recognition methods to deal with classification problems in the most diverse areas is something well-established, applications in Law offer challenges due to the specific terminology, broader vocabulary and the presence of more complex semantic and syntactic structures compared to spoken Portuguese. In this short paper, we present a Recurrent Neural Network implementation using LSTM to classify entities such as class, subject, value, individuals, among others items, in lawsuits. The initial dataset focused on a sample of 100 thousand lawsuits of São Paulo state court, in Brazil. The proposed method achieved an accuracy and F1-score of approximately 90% in the tested data. Such preliminary results indicate that it is possible to create a model capable of generalizing such classifications on a large scale even regarding the specifics of Brazilian legal texts terminology.

Keywords: Natural Language Processing · Named entity recognition · Legal documents

1 Introduction

Currently, information has been recognized as a strategic asset for any organization to be effective in decision-making. The data gathered from internal and external sources constitute this strategic asset and are a particularly big challenge for modern Information and Communication Technology (ICT) systems and organizations [1].

According to this perspective, by harnessing the power of computational resources, lawyers can predict from past experiences more accurately how events will unfold in litigation. Thus, lawyers who adopt data-driven decision-making will have a clear advantage over their colleagues who still cling to their outdated instruments [2].

In many ways, artificial neural networks (ANN) have been successfully applied to diverse problems. Among them, it is possible to mention pattern recognition [3]. Regarding Law, it is possible to find citations of ANN usage to aid the measurement of punishment [4] and in fighting cybercrime [5]. Other attempts to use machine learning models in the Law area come across a lack of data, especially in Named Entity Recognition (NER) problems. Even so, some attempts to create a public database for the application

© Springer Nature Switzerland AG 2021
O. Gervasi et al. (Eds.): ICCSA 2021, LNCS 12950, pp. 648–656, 2021.
https://doi.org/10.1007/978-3-030-86960-1_48

of Law written in Brazilian Portuguese had good results [6]. However, the amount of data made available is a trammel since they are in the hundreds or thousands, which is difficult to train deep learning models with results expressively good to overcome the human performance [6].

In this context, the project "Mediation and Conciliation empirically evaluated: jurimetry for proposing efficient actions" was an empirical research that studied ways to improve the delivery of the judicial provision [7]. It consists of a database with an extensive collection of lawsuits ranging from the years 2015 to 2018 collected from public data courts and the Court of Auditors Justice (TJ) in Brazil. Based on this database, initial analysis was performed on quantitative information about the lawsuits, which are extremely relevant to decision-making and action proposals, such as the demands average duration and the values being negotiated, among others.

In contrast to other areas for NER task applications, legal texts in Brazilian Portuguese have some issues that hinder such application. For instance, non-regular patterns in capitalization, punctuation and the text structure, as well as semantic structure and the jargon of Law. These particularities tend to increase the difficulty in obtaining a good dataset for training the model. In addition, the NER datasets for general purpose, do not have specific tags to Law.

This paper presents an approach for developing a dataset to train a deep leaning model in legal texts and an implementation of a Recurrent Neural Network (RNN), applied in Law, to perform some types of classification of legal texts published in official journals, as interest in specific types of elements belonging to the lawsuit, such as its subject, classes, monetary values involved, among others.

The paper is organized as follows. Section 2 describes the data source as well as the tagging procedure. Section 3 presents the methodology, specifically how the model was implemented. The preliminary results are presented and discussed on Sect. 4. Finally, the Sect. 5 briefly presents our conclusions and the future work.

2   The Database

The data source used by this study is the database generated from the research project "Mediation and Conciliation empirically evaluated: jurimetry to propose efficient actions" [7]. The database contains lawsuits referring to some Brazilian states with each sample built up of the following fields: *id*, the database unique identifier. *Dados Fórum*, the judge and court identification. *Conteúdo,* the content of the lawsuit, i. e., a plain text written by the judge for describing it as well as indicating decisions and following, among other information. *Processo,* the lawsuit id according to the CNJ (National Council of Justice). *Assunto*, the subject of the lawsuit. *Classe*, the class of the lawsuit. *Data*, the date of publication. *Autor,* the plaintiffs involved in the lawsuit. *Reu,* the defendants. *Valores*, the monetary values. Table 1 shows a sample of a record from the database.

Among the mentioned fields, *Conteúdo* refers to the source for training the deep learning model for recognizing words that have the meaning of classification to subjects, classes, legal entities, values, etc. The procedure of building the new data set will be further elaborated in the next section.

Table. 1. Sample of a record of the database (names were suppressed).

Field	Description	Content
id	Database ID	5e6135e5a15735ce218a9960
Dados Fórum	Lawsuit forum information	JUÍZO DE DIREITO DA 4ª VARA CÍVEL JUIZ(A) DE DIREITO ███████████ █████████ ESCRIVÃ(O) JUDICIAL ██████████████
Processo	Lawsuit number	0001543-25.2013.8.26.0344
Conteúdo	The text of the lawsuit	Processo 0001543-25.2013.8.26.0344 (034.42.0130.001543) - Procedimento Comum - Evicção ou Vicio Redibitório - ███ ████████████████████ - Sobre a contestação apresentada pelo Curador Especial nas fls. 211, manifeste-se o requerente. Prazo: 15 (quinze) dias. - ADV: ███████████████ ████████████████████████ ████████████████████████
Assunto	The subject of the lawsuit	Null
Classe	The class of the lawsuit	Null
Data	The date of the lawsuits was published	2017/07/11
Autor	Plaintiffs	'Nome': █████████████████, 'Gênero': 'Masculino', 'Pessoa': 'Física'
Reu	Defendants	'Nome': ███████████████, 'Gênero': 'Masculino', 'Pessoa': 'Física','Nome': █████████████, 'Gênero': 'Masculino', 'Pessoa': 'Física'
Valores	The monetary values in the lawsuit	Null

Note in the sample record of Table 1 that some fields are null or missing. That is one of the reasons to create a model capable of correctly predicting such information, since, among all sample cases in the mentioned before database, which reaches at more than 100 million, only a few of them have this classification. Something which is a hindrance for new studies on this data, for example, if some interested researcher lawyer in study all the cases in which the class is *"Mandato de Segurança"* (Writ of Mandamus) to know what's the best arguments, or the history line, etc. Without this classification the research will be depreciated or not even possible

2.1 The Dataset Tagging Procedure

The database used for creating a dataset is composed of lawsuits from the state of *São Paulo* court. It contains around 90 million records representing around 19.183.228 unique lawsuits. From that, a sample of one hundred thousand was taken for the training base.

Analyzing the *Conteúdo* field, at first glance there is a possible division by the character ' - ' in order of process number, class, subject, parts and content as also shown in Table 1. However, not all documents followed the same pattern and around 12.080 documents from the sample were not used for constructing the database.

After that, the words were separated and tagged, comparing lists of classes and subjects according to CNJ class or subject label pattern (https://www.cnj.jus.br/sgt/con sulta_publica_classes.php). Finally, for the remaining content, it was verified who were the plaintiffs and defendants, as well as their gender, if it were a legal entity, government or legal confidentiality. Such words were searched for in the content and tagged.

3 Methodology

It was used different methods for labeling the tags of each word: 1) IOB (Inside-Outside-Beginning) tagging scheme [8], where "B-" indicates the beginning of the Tag, "I-" indicates that the word is within a Tag and "O" indicates that the word does not belong to any Tag and 2) The joining of IOB tags for only a single classification of the words. The description of each acronym is described in Table 2.

Table 2. The Tags and respective descriptions.

IOB TAG	Description	NOT IOB	Description
O	Does not belong to any tag of interest	O	Does not belong to any tag of interest
B-Cla	Begin of the class	Cla	The word is a class
I-Cla	Inside of the class	–	–
B-Ass	Begin of the subject	Ass	The word is a subject
I-Ass	Inside of the subject	–	–
B-Jur	Begin of legal person	Jur	The word is a legal person
I-Jur	Inside of legal person	–	–
B-FiM	Begin of male person	FiM	The word is a male person
I-FiM	Inside of a male person	–	–
B-FiF	Begin of a female person	FiF	The word is a female person
I-FiF	Inside of a female person	–	–
B-Gov	Begin of a government name	Gov	The word is a government name

(continued)

Table 2. (*continued*)

IOB TAG	Description	NOT IOB	Description
I-Gov	Inside of a government name	–	–
B-Seg	Begin of a legal confidentiality name person	Seg	The word is a legal confidentiality person
I-Seg	Inside of a legal confidentiality name person	–	–
B-Val	Begin of a value in process	Val	The word is a value in the process
I-Val	Inside of a value in process	–	–

3.1 The Implemented Model

The architecture of the implemented model consists of 3 layers: Embedding layer, Bidirectional LSTM layer and Dense layer. Such a network was implemented using Python 3 and deep learning libraries as Tensorflow [9] and Keras [10].

The size tested of the embedding vector was 50 and 100. The Bidirectional layer groups LSTM neurons [11], with a spatial dropout of 0.5, and finally, a last layer called Dense, is a layer of perceptrons containing 17 neurons, with a softmax activation function to calculate the probability of the output for each word. Each neuron is associated with a tag that words can belong to.

3.2 Pipelines

Before training the model, we replace each word in the documents with numbers that represented them. For instance, the word "Processo" was replaced by the number 16 in that position, the word "Procedimento" was replaced by the number 15679, and so on. In addition, the padding-post technique [12] was used in documents with a fixed length of 150 words. The padding technique consists of adding an arbitrary word to the document if it is less than 150 words, or removing words from the end until we are left with 150. Table 3 exemplifies the pre-processing. In addition, two forms of class balancing were applied: Balanced Under Sample (BUS) [13] and Random Under Sample (RUS) [14]. Each balancing method was tested for both forms of tagging.

The BUS algorithm consists of keeping the context of the interesting tags. In contrast to the RUS algorithm, in which each word is removed randomly, the BUS algorithm maintains a rate of words in the proximity of a tag of interest, while removing the words that are in the middle of the two tags. For instance, if the entire raw text is like in Table 3 (column "Original Document") that the interesting word tags go from "Procedimento" to "Santos" the final results of a under the sample is a like the phrase starting in "0130" (2 words before) to "contestação" (2 words after), in the column "Word to Numbers with Padding".

Finally, the dataset was divided into 90% for training and 10% for testing using the metric F1-score and precision to assess the overall performance of the model. During training Adam [15] was used as an optimizer, sparse categorical cross entropy as a loss

Table 3. Example of a padding post method (names were suppressed).

Original Document	Words Extracted	Word to Numbers without Padding	Word to Numbers with Padding
Processo 0001543-25.2013.8.26.0344(034.4 2.0130.001543) - Procedimento Comum - Evicção ou Vicio Redibitório - Erick ███████ ████ s - Sobre a contestação apresentada pelo Curador Especial nas fls. 211, manifeste-se o requerente. Prazo: 15 (quinze) dias. - ADV: ███████	['Processo','0001543' ,'25','2013','8','26','03 44','034','42','0130','0 01543','**Procedimen to**','Comum','Evicçã o','ou','Vicio','Redibit ório',███████ ,'a','contestação' ,'apresentada','pelo',' Curador','Especial','n as','fls','211','manifes te','se','o','requerente' ,'Prazo','15','quinze',' dias','ADV',███████]	[**16**,45585,417,59,1 5,17,21506,8342,53 8,610,45586,44,**231** ,21360,30,21005,21 361,24630,1472,25, 260,120686,1547,1 787,25,260,3210,17 87,25,260,11507,3, 230,638,42,4967,11 1,344,31,4662,478, 8,5,193,880,109,21 50,33,18,15995,109 06,165150,10,7487 0,11,2588,1356,131 6,10,179818,11]	[6,45585,417,59,1 5,17,21506,8342,5 38,610,45586,44,2 31,21360,30,21005 ,21361,24630,1472 ,25,260,120686,15 47,1787,25,260,32 10,1787,25,260,11 507,3,230,638,42,4 967,111,344,31,46 62,478,8,5,193,880 ,109,2150,33,18,15 995,10906,165150, 10,74870,11,2588, 1356,1316,10,1798 18,11,0, 0, 0, ... ,0]

function, with a batch size of 128, during 50 epochs and the test data was separated for validation during training.

In short, given the dataset, all of its words were changed to numbers, a padding was applied, a form of tagging was chosen (IOB or NOT IOB), one of the class balance methods was applied (BUS, RUS or none), a size of embedding was chosen and finally trained the network. Thus, totaling 12 tested methodologies.

4 Preliminary Results

The main metric used to assess the overall performance of the models was F1-score. Also, sensitivity and specificity were calculated for each tag and in the scope of measuring the performance of the methodologies. In addition to the F1-Score, the overall precision of the model was calculated. Such metrics were calculated for both each balancing method tested and only on the test data. Some results are available in Table 4 and Table 5, presenting the performance of tested methodologies.

According to Table 5, the model can predict with an F1-score above 89% and accuracy above 97%. Whatever the methodology tried, there are not significant differences between them. At the most the two undersample methods increase the overall accuracy in NOT IOB methods in 1%.

Table 4. Values of model performance using BUS and IOB Tags with size 50 of embedding.

Tag	Precision	Recall	F1-Score	Samples
O	0.99	0.99	0.99	741782
I-Cla	1.00	1.00	1.00	18093
I-Seg	0.85	0.85	0.85	3539
I-Jur	0.79	0.78	0.79	16520
I-FiM	0.79	0.79	0.79	13174
I-FiF	0.82	0.79	0.80	9960
I-Gov	0.85	0.73	0.78	6246
I-Ass	1.00	1.00	1.00	9746
I-Val	0.95	0.97	0.96	2273
B-Cla	1.00	1.00	1.00	8655
B-Seg	0.92	0.96	0.94	1385
B-Jur	0.83	0.85	0.84	3224
B-FiM	0.81	0.88	0.84	4202
B-FiF	0.85	0.84	0.85	2849
B-Gov	0.83	0.82	0.83	1067
B-Ass	1.00	1.00	1.00	5636
B-Val	0.86	0.84	0.85	1032

Table 5. General performance of the methodologies.

Methodology	Embedding dimensions	F1-Score	Recall	Precision	Accuracy
BUS + IOB Tag	50	0.89	0.89	0.89	0.98
BUS + IOB Tag	100	0.89	0.89	0.89	0.98
BUS + NOT IOB	50	0.89	0.89	0.89	0.98
BUS + NOT IOB	100	0.89	0.90	0.89	0.98
RUS + IOB Tag	50	0.89	0.89	0.88	0.98
RUS + IOB Tag	100	0.89	0.88	0.90	0.98
RUS + NOT IOB	50	0.89	0.89	0.89	0.98
RUS + NOT IOB	100	0.89	0.89	0.89	0.98
NO Undersample + IOB Tag	50	0.89	0.90	0.88	0.98
NO Undersample + IOB Tag	100	0.89	0.89	0.89	0.97

(*continued*)

Table 5. (*continued*)

Methodology	Embedding dimensions	F1-Score	Recall	Precision	Accuracy
NO Undersample + NOT IOB	50	0.89	0.90	0.88	0.97
NO Undersample + NOT IOB	100	0.89	0.90	0.89	0.97

The point to be observed is that two subject and class tags present both an F1-Score and precision stagnant at 1.00 or 0.99, something that would normally be considered a good level. However, in this case, the words of class and subject appear at the beginning of the text, which provide an inconvenience in case of applying the model to other legal texts that contain its class or subject not in the same region of the text. Probably these words will not be classified correctly.

5   Conclusion and Future Work

The short paper presented preliminary results of a LSTM Recurrent Neural Network implementation for classifying entities on Brazilian legal Documents. In the sample of 100 thousand dataset lawsuits from Brazilian courts, the model achieves good results. These results suggest that the applied technique based on Recurrent Neural Network using LSTM can be successful to classify entities for Brazilian legal documents.

In addition, we intend to slice a 5% of train set to fine-tuning other hyperparameters such as the size of embedding vectors, the learning rate and dropout, among others and tests on the test set previous apart.

Finally, implement the methodology with the best results obtained in a data set using the entire database of 90 million of lawsuits.

Acknowledgments. This project has been partially supported by Huawei do Brasil Telecomunicações Ltda (Fundunesp process # 3123/2020) and by University of São Paulo (Portaria PRP Nº 668, DE 17 DE OUTUBRO DE 2018).

References

1. Laudon, K., Laudon, J.: Sistemas de informações gerenciais. 11a. ed. [s.l.] Pearson/Prentice Hall, Upper Saddle River (2015)
2. Lettieri, N., et al.: Ex machina: analytical platforms, law and the challenges of computational legal science. Future Internet **10**(5), 26 (2018)
3. Bishop, C.M.: Neural Networks for Pattern Recognition. Oxford University Press (1995). ISBN: 0198538642
4. Grimm, C.: Dosimetria da pena utilizando redes neurais. Monografia. Curso de Direito. Universidade Federal do Paraná. Curitiba (2006)
5. Lossio, C.J.B.: O anticrime nas redes sociais: os algoritmos e a rede neural artificial (RNA) em face do cybercrime (2017)

6. Luz, P., de Araujo, T., Campos, R., Oliveira, M., Couto, S., Bermejo, P.: Lener-br: a dataset for named entity recognition in brazilian legal text. In: Villavicencio, A., et al. (eds.) PROPOR 2018. LNCS (LNAI), vol. 11122, pp. 313–323. Springer, Cham (2018). https://doi.org/10.1007/978-3-319-99722-3_32

7. Ventura, C.A.A., et al.: Justiça Pesquisa - Mediações e Conciliações Avaliadas Empiricamente. https://www.cnj.jus.br/wp-content/uploads/2011/02/e1d2138e482686bc5b66d18f0b0f4b16.pdf&gt, Acessed 15 Jan 2021

8. Ramshaw, L.A., Marcus, M.P.: Text chunking using transformation-based learning. In: Armstrong, S., Church, K., Isabelle, P., Manzi, S., Tzoukermann, E., Yarowsky, D. (eds.) Natural language processing using very large corpora, pp. 157–176. Springer Netherlands, Dordrecht (1999). https://doi.org/10.1007/978-94-017-2390-9_10

9. Tensorflow. https://www.tensorflow.org/, Accessed 12 Aug 2020

10. Keras. https://keras.io, Accessed 12 Aug 2020

11. Bengio, Y., Simard, P., Frasconi, P.: Learning long-term dependencies with gradient descent is difficult. IEEE Trans. Neural Netw. 5(2), 157–166 (1994)

12. Dwarampudi, M., Reddy, N.V.: Effects of padding on LSTMs and CNNs. arXiv preprint arXiv:1903.07288 (2019)

13. Akkasi, A., Varoğlu, E., Dimililer, N.: Balanced undersampling: a novel sentence-based undersampling method to improve recognition of named entities in chemical and biomedical text. Appl. Intell. 48(8), 1965–1978 (2017)

14. Ganganwar, V.: An overview of classification algorithms for imbalanced datasets. Int. J. Emerg. Technol. Adv. Eng. 2(4), 42–47 (2012)

15. Kingma, D.P., Jimmy, B.A.: Adam: A method for stochastic optimization. arXiv preprint arXiv:1412.6980 (2014)

Covid-19 Impact on Higher Education Institution's Social Media Content Strategy

Tiago Coelho[1,2] and Alvaro Figueira[1,2(✉)]

[1] CRACS - INESC TEC, 4200-465 Porto, Portugal
[2] Faculty of Sciences, University of Porto, 4169-007 Porto, Portugal
`arf@dcc.fc.up.pt`

Abstract. In recent years we have seen a large adherence to social media by various Higher Education Institutions (HEI) with the intent of reaching their target audiences and improve their public image. These institutional publications are guided by a specific editorial strategy, designed to help them better accomplish and fulfill their mission. The current Covid-19 pandemic has had major consequences in many different fields (political, economic, social, educational) beyond the spread of the disease itself. In this paper, we attempt to determine the impact of the pandemic on the HEI content strategies by gauging if these social-economical, cultural and psychological changes that occurred during this global catastrophe are actively reflected in their publications. Furthermore, we identified the topics that emerge from the pandemic situation checking the trend changes and the concept drift that many topics had. We gathered and analyzed more than 18k Twitter publications from 12 of the top HEI according to the 2019 Center for World University Rankings (CWUR). Utilizing machine learning techniques, and topic modeling, we determined the emergent content topics for each institution before, and during, the Covid-19 pandemic to uncover any significant differences in the strategies.

Keywords: Text mining · Covid-19 · Higher education institutions · Twitter · Topic modeling

1 Introduction

As competitiveness and the expectations of current and future students increase, and as the government support and new enrollments decrease, Higher Education Institutions (HEI) need to ensure their subsistence [1]. As such, it becomes vital for a HEI to obtain additional competitive advantages in the form of differentiation, which is achieved through self-promotion and with increased interaction with wider audiences. Most of the time, HEI use social media a tool to do so. As it has been shown in the literature [2], advantageous social media campaigns are one of the most significant drivers of loyalty to an institution.

© Springer Nature Switzerland AG 2021
O. Gervasi et al. (Eds.): ICCSA 2021, LNCS 12950, pp. 657–665, 2021.
https://doi.org/10.1007/978-3-030-86960-1_49

As the Covid-19 pandemic has been unfolding we are seeing several changes in the social, economic, and political landscape at a global scale. While various populations have had to adapt their behaviors to this new context, organizations have also had to adapt their strategies to meet the new needs of the populations, the current policies and regulations.

Therefore, in this project we seek to determine the impact that the Covid-19 pandemic had on Higher Education Institutions' content strategy by: i) creating an automatic system capable of discovering the communication strategy of a given HEI through the analysis of its communication pattern and the use of machine learning techniques, and ii) compare the discovered pre/post Covid-19 editorial strategies to determine if there are significant differences. Due to the recency of the pandemic, there is still a lack of literature that explores it's impact on the publication strategy of different HEI. However, regarding content strategy identification, there are already efforts made to deepen the knowledge about the different publication strategies employed by the institutions.

Our approach differs in several aspects from previous studies. With this project we aim to study HEI communication strategies on Twitter, whereas the studies of [3–5] use Facebook posts as the social network to analyze. The communication strategy of a HEI may vary according to the social network it is employed in, and we expect to find some differences from the previously identified strategies. Unlike the methodology used in the articles [3,4] our approach uses unsupervised machine learning techniques to obtain the topics on which each HEI is focusing their communication, and therefore, does not impose a pre-defined model. We believe our approach creates an editorial model that is more faithful to reality.

The Center for World University Rankings, CWUR, annually publishes a Global University Ranking which measures the quality of education and training of students as well as the prestige of the faculty members and the quality of their research without relying on surveys and university data submissions. By selecting HEI based on the 2019/2020 ranking by CWUR we expect to obtain better defined communication strategies than in [4]. The selected HEI are: Caltech, Cambridge University, ETH Zurich, Harvard University, Imperial College, Johns Hopkins University, MIT, Stanford University, University of Chicago, UC Berkeley, UCL and University of Oxford.

2   Data Set Characterization

Using the Twitter API we gathered all the posts from each of the 12 selected HEI for the period between the 1st of September 2019 to the 31th of October 2020. Even though not all HEI start and end their academic year at these exact dates, we believe that this period is the most fitting in representing an academic year for the majority of them. Taking into account the date on which the World Health Organization (WHO) declared Covid-19 as a Pandemic, March 11 of 2020, the selected period allows us to obtain a somewhat balanced division of the dataset into the pre/post Covid-19 periods, with a time window close to 6 months each.

When observing the posting frequency within the specified period we were able to determine that Imperial College was the HEI that published the most, while Stanford University was the one which published the least. Actually, the numbers are quite different for each organization. If we divide them by publication frequency we have: Stanford, UCL and Caltech in the same group, under 1k publications; Imperial College, Johns Hopkins and UC Berkeley with over 2k and, lastly, the remaining institutions with values in between. These values range between 2 and 8 posts a day, depending on the HEI. The average number of retweets and favorites is also significantly different. In what concerns retweets, it ranges from 10 (ETH Zurich) to 119 (Stanford) and the average number of favorites per tweet ranges from 13 (UCL) to 157 (Harvard). The standard deviations for both these metrics in each HEI are extremely high: this is the result of outlier trending tweets with high retweet and favorite counts that inflate the distribution.

With regards to posting patterns we were able to determine some common preferences by most organizations. All organizations have a more or less restrict set of hours that represent their activity period which results in all of them having a period of the day where little to no publications are made. The same can be said about the weekdays, with every organizations showing less activity during the weekend. To better understand posting frequency of each HEI throughout the year we aggregated their respective posts by month, and no evident similarities were found between the various HEI that could indicate the same trend/event. The same was made for the average number of favorites and retweets (Fig. 1). To facilitate a comparison, we used the same scale for all graphs. The red dashed line represents the date Covid-19 was declared a pandemic.

Analyzing the depicted time series we can find numerous differences between the 12 HEI. However, there are some similarities in regard to seasonality and trend that we cannot overlook. The first evident aspect to point out is that the average number of favorites and retweets usually presents the same behavior due to the high correlation between the two, i.e., it is expected to see that tweets with a high/low number of favorites also have a high/low number of retweets (and vice versa). Although, in Caltech, Cambridge, Stanford and Berkeley that is not always the case. Another noticeable aspect is that in most HEI the peak average number of retweets and favorites occur after the Covid-19 pandemic start, whereas Cambridge and Chicago figure as exceptions. The final interesting similarity is noticed in Harvard, Stanford, UC Berkeley and Oxford, in which taking into account the number of retweets we may say that these HEI did experience very similar events characterized by two major peaks and two major valleys, although not with the same intensity and not always at the same exact periods.

3   Emerging Topics Identification

For the topic modeling task we used the Non-Negative Matrix Factorization (NMF) and Latent Dirichlet Allocation (LDA) techniques because of their better performance in the comparative analysis [6] on datasets which are similar

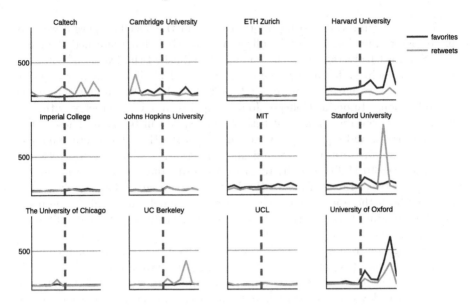

Fig. 1. Monthly average number of favorites and retweets for each HEI.

to ours (short text, sourced from microblogs). NMF is an unsupervised matrix factorization (linear algebraic) method that is able to obtain topics from a collection of documents by performing both dimension reduction and clustering simultaneously. LDA topic modeling discovers topics that are latent in a collection of text documents by using a generative probabilistic model and Dirichlet distributions to infer possible topics based on the words present in the said documents. When choosing the LDA implementation, we opted to use the LDA Mallet wrapper for Genisim because it uses an optimized Gibbs Sampling [7] as the inference algorithm, which is more precise than the faster Variational Bayes algorithm [8] used by Gensim's LdaModel. After data cleaning steps, stop-word removal and tokenization we transformed the text to include bigrams and trigrams with hyperparameters determined trough iterative testing: total collected count above 50 and a score above 100 in Gensim's default bigram scoring function, based on the original [9]. Then, we lemmatized the text, performed POS tagging, keeping only nouns, verbs, adjectives and adverbs and created the two main inputs needed for topic modeling, the dictionary and the corpus. In order to determine the ideal number of topics for LDA and NMF models we decided to create models with an increasing number of topics and plot their measured performance. We evaluate the models in terms of topic coherence: A metric that scores a topic (between 0 and 1) by measuring the degree of semantic similarity between the high scoring words in it. Since the coherence of the models seem to increase with the number of topics (which can lead to an overfitting), we decided to select the models with the highest coherence values before the growth starts flattening out, or it reflects a major decrease. Based on the previously mentioned criteria, the number of topics was selected for both models (k = 5), with the LDA model having a coherence score of 0.37 and the NMF model a coherence score of 0.28. The LDA model also revealed better human interpretability of the derived topics. Observing the inter-topic semantic similarities with the help of the pyLDAvis library, the 5-topic LDA model also proved superior, with greater inter-topic distance, making it easier to differentiate the topics. Therefore, we decided to advance with the LDA model with k = 5 to characterize the publication strategies of the HEI. The topics discovered are represented in the following figure (Fig. 2) using 10 of their most frequent words in the form of word clouds.

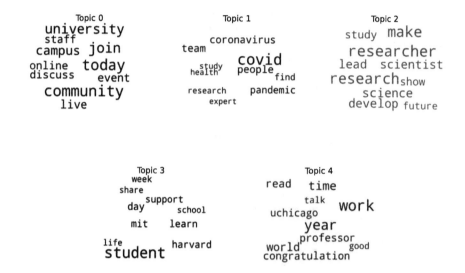

Fig. 2. Word clouds for each LDA generated topic.

Analyzing the word clouds we try to interpret the topics according the context they are inserted in. Topic 0 seems to be about college events: words like "university", "join", "online", "live", "today", "discuss" and "event" emphasize that idea. Topics 1 and Topic 2 are clearly related to public health/Covid-19 pandemic and research/investigation, respectively. Topic 3 seems to relate to education, evidenced by words: "student", "learn", "school", "support". Finally, Topic 4, seems to be the hardest to categorize, but pulling not only from the top 10 most salient words but from the top 30 we can get a clearer picture. We believe this topic regards public image due to words like "congratulation", "work", "good", "great", "woman", "celebrate", "hope", "world" and "history". Having the topics 0 to 4 properly identified, we will refer to them hereafter as Events, Health, Research, Education and Image, respectively. From this basis, to better understand the publication frequency per topic, we assign the dominant topic to each tweet based on the tweet content.

One of the most glaring aspects that is easily observable in every HEI is the sudden and massive increase in the number of Health publications after the

Covid-19 WHO pandemic declaration. Possibly correlated, we can also observe a shared decreasing trend of the Research topic over the academic year, with it's biggest valley coinciding, in many cases, with the Health topic increase. This could be due to the semantic similarities between the Health and Research topics, since their inter-topic distance is the lowest out of any other possible pair, causing them to share some salient terms like "research" or "study". We may say that Imperial College, UCL and University of Oxford possibly shared the same event, since all exhibit similar peaks in the number of Image publications between June 2020 and August 2020. In the University of Oxford's case, this peak coincides with their absolute maximum in average number of retweets and favorites. In the same way, Stanford and MIT also have similar peaks in Research publications in October 2019, as well as Caltech and ETH Zurich, later that academic year, in January 2020.

4   Covid19 Impact Analysis

To determine each HEI strategy before and after the pandemic we displayed their posting behavior, color coded by topic, as horizontal stacked bar charts (Fig. 3).

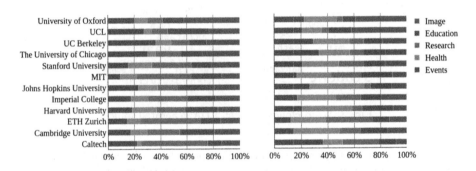

Fig. 3. Pre and Post Covid-19 topic frequencies, respectively

Before the pandemic, We can see that organizations like Stanford (30%), MIT (41%), Imperial College (30%), ETH Zurich (46%) and Caltech (50%) had Research as their primary topic of publications. Only UC Berkeley (37%) and The University of Chicago (31%) had Events as their primary topics, although the latter also had an almost equally matching preference for Image (28%). Focusing on Education oriented tweets we have UCL (31%), Johns Hopkins University (28%), Harvard University (25%) and Cambridge University (26%). However, this topic's degree of dominance is not as high as the previous ones. All the HEI with this primary topic have a much more balanced distributions and a prominent second topic with almost as much focus as the first one. Both UCL and Johns Hopkins University prioritize the Events topic in their editorial model while Harvard University and Cambridge University prioritize Research. Lastly, the University of Oxford has 42% of tweets falling into the Image topic, the only one with this topic preference. Curiously, there are no HEI with a strategy that favors publications regarding the topic Health.

After the Covid-19 Pandemic start, we can clearly see significant changes in every institution's strategy. As expected, the Health topic, who previously wasn't the first topic of choice in any HEI, suddenly becomes the primary focus of UC Berkeley (27%), Stanford University (28%), MIT (25%), Johns Hopkins University (41%), Imperial College (29%), Harvard University (38%) and Cambridge University (34%). UC Berkeley, Imperial College, Harvard University and Cambridge University maintain their previous dominant topic as the second most prevalent, while the remaining do not. The University of Oxford, University of Chicago, and ETH Zurich maintained a similar strategy by keeping the same primary topic of publications. However, its dominance has decreased slightly due to the rise of the Health topic present in every single one of the post Covid-19 strategies. UCL and Caltech have had the most drastic changes, being the only ones with a complete switch of dominant topic, but not a switch to Health dominance. UCL joins Oxford University with 42% Image related content and Caltech switches from Research to Events with 36%. To compare the topics themselves in the 2nd time window, we created the bar charts displayed in Fig. 4, representing the share and favorites/retweets values per topic.

We can see that before Covid, Research was the most common topic of publication, represented by 25% of Tweets. After the pandemic, it became the least common, suffering a decrease of about 11%. Education and Image also suffered a 3% decrease each, making Events and Heath the only topics whose publication share increased after the pandemic. Health went from the least common to most common, with an increase of 17%.

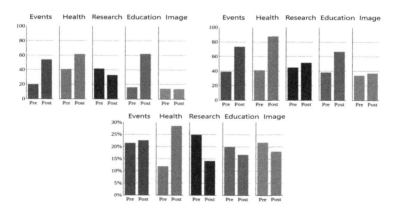

Fig. 4. Average number of retweets (Top-left), Average number of favorites (Top-Right) and Topic share percentage (Bottom), before and after the Covid-19 Pandemic.

In regard to the number of retweets and favorites, Events, Health and Education had similar results, each one increasing their number of interactions during the pandemic. Research and Image had a slight decrease in the average number of retweets but still had an increase in the average favorite count, marking an overall increase in average number of favorite across all topics during the pandemic.

5   Conclusions and Future Work

In this work we used the CWUR ranking to identify 12 of the highest ranking HEI, from which we collected 18727 tweets, for the period of 01/09/2019 to 31/08/2020. We then performed LDA topic modeling to identify topic areas present in each HEI content strategy and compare the resulting strategies before and after the Covid-19 WHO pandemic declaration.

Based on the results presented, we found that the editorial strategies of HEI Twitter can be characterized as revolving around five key content topics (events, health, research, education, and image) or editorial domains, and that there are significant strategic changes in topic distribution in each HEI after the pandemic declaration of Covid-19 WHO, indicating a change in content strategy in social media due to the current pandemic situation. In addition, we can also observe that the Covid-19 pandemic had an overall positive impact on public interaction with HEI social media publications. This could possibly be due to an increased interest in the topics discussed by the HEI according to its new content strategy, or even just the increased online presence of the public resulting from the pandemic restrictions leading to remote work and education.

Throughout the studied period, we can observe opposing trends for the Health and Research topics, with Health taking over the strategy focus that was previously placed on Research. This can be explained by the semantic similarity between the topics, as previously stated, but also could be due to the fact that most of the research efforts during that period were actively being directed towards the health industry. The Events topic was the only one besides Health that managed to increase in all 3 of the studied metrics reflecting a prevalent interest/need for social interaction derived from the isolation caused by the pandemic.

Lastly, it's noteworthy to point out that, despite having the lowest average number of publications per day, Stanford has managed to achieve some of the best interaction numbers, with the highest average of retweets per tweet at 120 and the 3rd highest average of favorites per tweet. On the contrary, despite having the highest average number of publications per day, Imperial College was in 9th place in average retweets and the 7th place in average favorites, making a strong case for publication frequency not being a good predictor of engagement.

Apart from the previous conclusions, our contribution in this paper is also in the form of a proposed general methodology for understanding and measure social media content strategies in Higher Education Institutions. As a future work we will explore how the sentiment (expressed in each tweet) and associated with each topic unfolds during the pandemic and how it affected the postings of Higher Education Institutions.

References

1. Mitchell, M., Leachman, M., Masterson, K.: A Lost Decade in Higher Education Funding State Cuts Have Driven up Tuition and Reduced Quality (2017)
2. Çiçek, M., Erdogmus, I.: The impact of social media marketing on brand loyalty. Procedia. Soc. Behav. Sci. **58**, 1353–1360 (2012). https://doi.org/10.1016/j.sbspro.2012.09.1119
3. Figueira, A.: Uncovering social media content strategies for worldwide top-ranked universities. In: CENTERIS/ProjMAN/HCist 2018, vol. 138, pp. 663–670 (2018). https://doi.org/10.1016/j.procs.2018.10.088
4. Oliveira, L., Figueira, A.: Measuring performance and efficiency on social media: a longitudinal study. In: ECSM 2018 5th European Conference on Social Media, pp. 198–207 (2018)
5. Peruta, A., Shields, A.: Social media in higher education: understanding how colleges and universities use Facebook. J. Market. High. Educ. **27**, 131–143 (2017). https://doi.org/10.1080/08841241.2016.1212451
6. Albalawi, R., Yeap, T., Benyoucef, M.: Using topic modeling methods for short-text data: a comparative analysis. Front. Artif. Intell. **3**, 42 (2020). https://doi.org/10.3389/frai.2020.00042
7. Hoffman, M., Blei, D., Bach, F.: Online learning for latent dirichlet allocation. In: Advances in Neural Information Processing Systems 23, pp. 856–864 (2010)
8. Yao, L., Mimno, D., McCallum, A.: Efficient methods for topic model inference on streaming document collections. In: Proceedings of the 15th ACM SIGKDD International Conference on Knowledge Discovery and Data Mining, pp. 937–946 (2009). https://doi.org/10.1145/1557019.1557121
9. Mikolov, T., Sutskever, I., Chen, K, Corrado, G., Dean, J.: Distributed representations of words and phrases and their compositionality. In: Proceedings of the 26th International Conference on Neural Information Processing Systems, vol. 2, pp. 3111–3119 (2013). https://doi.org/10.5555/2999792.2999959

Towards a Smart Water Distribution Network for Assessing the Effects by Critical Situations in Electric Networks. The Pilot Case of Castel San Giorgio

Valeria Ottobrino[✉], Tony Esposito, and Stefano Locoratolo

Protection of Water Resources, GORI S.p.A., Scafati, Italy
{vottobrino,tesposito,slocoratolo}@goriacqua.com

Abstract. Water and electricity networks are among the main national critical infrastructures. One of their characteristics is mutual functional interdependence. This work presents the technological advancements will affect the pilot water network of Castel San Giorgio within the RAFAEL Project. Predictive modeling capabilities coupled with networked sensors to be installed along the water distribution system will allow the water supplier GORI to control network hydraulics during critical situations due to electrical system failures in order to minimize the effects on customers service.

Keywords: Critical Infrastructures (IC) · Water distribution network · Water supplier

1 Introduction

Critical infrastructures are elements, systems or parts of them which are essential for the maintenance of the vital functions of society, health, safety and the economic and social well-being of citizens and whose damage or destruction would have a significant impact due to the impossibility of maintaining those functions (Directive 2008/114/CE). Critical Infrastructure Systems (CI) include utility networks such as energy, water, telecommunications and transport systems, or discrete critical structures such as hospitals, ports and airports, among others. The objective to be achieved by GORI, as water supplier, within the funded RAFAEL Project [1, 2] will be the development of a smart water network able to effectively manage critical situations due to electrical network failures. In particular, the water distribution network of Castel San Giorgio will be analysed from the perspective of the water risk, the vulnerability and resilience of water networks in relation to the failure of the electrical networks, quantifying their impacts on the provided water service. The RAFAEL Project aims to integrate CIs, with technologies developed ad hoc, within a Decision Support System (DSS) CIPCast, which will become the reference platform to provide services to companies and the Public Administration.

CIPCast is an advanced DSS for predicting the occurrence and impact of natural or anthropogenic events on Critical Infrastructures and Essential Service Operators as well

© Springer Nature Switzerland AG 2021
O. Gervasi et al. (Eds.): ICCSA 2021, LNCS 12950, pp. 666–673, 2021.
https://doi.org/10.1007/978-3-030-86960-1_50

as on the population and the industrial system. Within RAFAEL Project, this system will extend its use to the areas of Southern Italy in order to gradually expand its functionality to the whole national territory. CIPCast System implements tools for assessing the Ics risks due to essentially disastrous natural events tacking into account ICs operative interdependences.

For achieving the project objectives, an optimal sensor placement method will be developed for monitoring physical and chemical parameters along the entire pilot network. Optimal sensor schemes will be defined on the basis of on which a distributed sensor network will be installed by GORI aimed at optimizing the modelling accuracy. In particular, IoT-derived sensory nodes will be integrated for the pervasive sensing of water parameters based on new generation data transmission networks. Predictive modelling capabilities will be developed refereed at the pilot water distribution network of Castel San Giorgio in order to assess the vulnerability and resilience of the water supply network in relation to electricity network failures, quantifying the impacts on the water service.

GORI will be able to quickly analyse alternative operating scenarios and compare them for determining the most appropriate solution to address the critical situations due to eventual electrical service failures.

2   The Water Supplier: GORI SpA

GORI S.p.A. is the water supplier of the Integrated Water Service of the "Sarnese-Vesuviano" District Area in Campania and its main target is to make efficient, effective the management of water resources. To date, GORI manages the Integrated Water Service of 74 municipalities. Total area of the district is about 897 km^2 with a resident population of $1'422'435$ as of 31.12.2019 (source ISTAT _ National Institute of Statistics) and a number of users as of 31.12.2019 equal to $526'775$ of which $467'554$ domestic and $59'221$ non-domestic.

2.1   The Water Supply of the "Sarnese Vesuviano" District

The Water System of the "Sarnese Vesuviano" District is divided into three main subsystems (Fig. 1):

- The Vesuvian System
- The Lattari Mountains System
- The Ausino System

The three water systems draw water resources from:

- Managed endogenous sources (wells and springs)
- Campania Region Supply (Vesuvian Aqueduct System and Monti Lattari Aquedottistic System)
- ABC Supply (Vesuvian Aqueduct System)
- Ausino Supply (Ausino Aqueduct System)

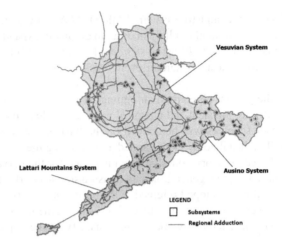

Fig. 1. "Sarnese Vesuviano" District– Regional Adduction.

In the Table 1, geometric data related to the entire water system managed by GORI are listed.

Table. 1 The geometry data of the water distribution network

Infrastructure	Consistency	Measurement unit
Water network distribution	4'226	km
Water network adduction	861	km
Wells	116	n
Springs	13	n
Water lifts	122	n
Tanks	206	n

The table summarizes the data relating to the consistency of the water service infrastructures currently managed by GORI (up to 31/12/2019).

The company guarantees users the supply of drinking water (the quality of which is assured by continuous controls) and a service appropriate to a modern European country, obtained also through investments aiming at the improvement of infrastructures and concerning above all technological innovation. Constant attention is assured to environmental protection. The innovation spirit has addressed the participation of the water supplier GORI in the research project of RAFAEL. As said, the main target of this Project is to optimize and integrate methodologies and technologies developed in the field of Critical Infrastructure Protection and GORI as stakeholder and project partner will give its contribution by the water network pilot.

The network pilot chosen in the Project is the water network of the Castel San Giorgio Municipality.

3 The Water Distribution Network of Castel San Giorgio

3.1 The Territory

The Municipality of Castel San Giorgio is located in the South of Italy, in Campania Region. The city is located at an altitude of 90 m above sea level and covers an area of 13,363 km². Castel San Giorgio is an agricultural, commercial and industrial centre and home to canning industries, which represent one of the main production activities, together with that of building materials. The population based on ISTAT data updated on 31st December 2019 is 13′695 inhabitants (Fig. 2).

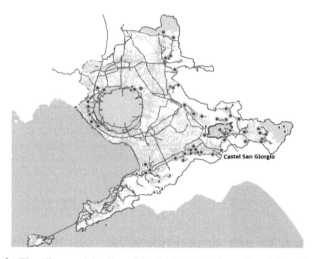

Fig. 2. The pilot municipality of the RAFAEL Project: Castel San Giorgio.

3.2 Consistency Status of the Water Network and Service Data of the Municipality of Castel San Giorgio

GORI, with a complex water supply and distribution network, supplies the entire territory of Castel San Giorgio serving about 5′715 consumers including domestic, commercial, artisan, agricultural and industrial users.

The Municipality of Castel San Giorgio, although it falls within the boundaries of the Ausino Aqueduct System, which is the system of adduction for the municipalities of the "Sarnese–Vesuvian" District Area covering the easternmost outcrops of the territory served, is fed exclusively through the use of the water outlet from three wells located within the municipal territory: Pozzo Traiano 1, Pozzo Traiano 2 and Pozzo Santa Croce. In particular, the water released from the two wells of the Campo Pozzi Traiano, located in the eastern part of the territory at an altitude of about 100 m.a.s.l., is accumulated in the tank having the same name, which has a capacity of about 130 m³, and then is raised by two separate lifting systems towards the tanks of Santa Croce (125 m.a.s.l.) and Torello Cortedomini (160 m.a.s.l.) with a storage capacity of 930 and 1188 m³ respectively.

At the Torello Cortedomini Tank, fed only and exclusively by the water of the Campo Pozzi Traiano, belong two distinct districts: a larger one powered by fall (Distr. Idr. CGG Serb. Torello Cortedomini) and a smaller one placed higher than the tank which is fed by means of a small lifting system (Distr. Idr. CGG Serb. Torello Cortedomini ZA) (Fig. 3). The third district of Castel San Giorgio (Distr. Idr. CGG Serb. Santa Croce) is instead subtended to the Santa Croce Tank, which is fed by the water of the Santa Croce and Traiano wells. There is also a fourth district within the administrative boundaries of the Municipality of Castel San Giorgio which is instead hydraulically fed by the Municipality of Siano and specifically by the Boscoborbone Tank.

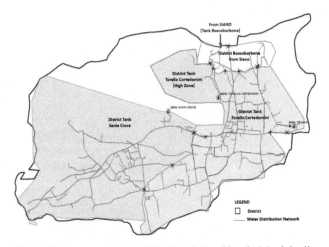

Fig. 3. The water network of the Castel San Giorgio Municipality.

A simplified diagram of the power supply of the Municipality and the individual plants will be shown in the following sections. In the meantime, some summary tables (Table 2, 3, 4 and Table 5) containing important information on both the hydraulic and management level are given in the following.

Table. 2 Hydraulic service data of the municipality of Castel San Giorgio

Hydraulic service data		
Inhabitants		13'695
Territorial area		13,363 km^2
Active users		5'715
Typology users	Domestic	85,34%
	Public	1,14%
	Commercial and craft industries	13,4%

(continued)

Table. 2 (*continued*)

Hydraulic service data		
	Industrial	0,12%
Km of adductors		8,38
Km of water distribution network		63,08
Wells		3
Springs		0
Tanks		3
Water lifts		3

Table. 3 Hydraulic data of wells

Well	Height [m.a.s.l.]	Average flow rate emitted [l/s]
Traiano 1	103	32,9
Traiano 2	105	14,1
Santa Croce	125	11,3

Table. 4 Hydraulic data of tanks

Tank	Height [m.a.s.l.]	Compensation capacity [m^3]
Traiano	100	130
Santa Croce	125	930
Torello Cortedomini	160	1188

Table. 5 Hydraulic data of water lifts

Water lift	Height [m.a.s.l.]	Medium lifted range [l/s]
Traiano 1	100	33,4
Traiano 2	100	12,2
Torello Cortedomini	160	N.D

4   Remote Control Schemes

Using a remote control platform developed in a SCADA environment, the company remotely manages the operating rules of the individual systems and the active alarms, while all the hydraulic/topographic part is recorded in a GIS system and is updated daily.

The water supply diagram of the Municipality of Castel San Giorgio can be schematically represented in a simplified way as shown on the remote control platform (Fig. 4).

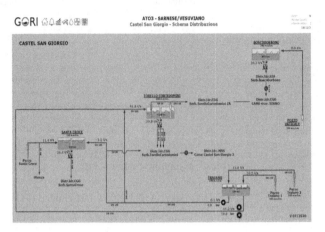

Fig. 4. TLC simplified water scheme of the Municipality of Castel San Giorgio Municipality.

Details of each individual plant related to the power supply scheme of the Municipality can then be displayed on its dedicated page of TLC (Remote Control System).

The remote control of each plant allows a constant monitoring of following values: flow rate (input and output) [l/s], pressure [bar], wells level [m] and tank levels [m] with alarms (set at the minimum and overflow threshold). Operating rules of the individual plants, pumps and automations applied to inverters, metering pumps and automated flow control valves are indicated.

5   Conclusions

The water supplier Gori as partner within the RAFAEL Project aims to analyse several hydraulic behaviour scenarios that can occur in ordinary or critical management situations and evaluate the impact on network and customers.

The participation in the RAFAEL Project will represent for GORI a further positive note in terms of innovation because it will return to the operator a water distribution network equipped with sensors that will allow a continuous monitoring of physical/chemical parameters distributed along the entire network. At the same time this participation will allow GORI to assess the effects due to critical situations in the electrical service on the eater distribution network and the impact on the customers service. All this at the scope to prepare efficient emergency management plans and promptly and properly respond in order to minimize the impacts on the served people. The participation of GORI in the RAFAEL Project shows, once again, how careful the Company is in guaranteeing the supply of drinking water to the users, and in providing an adequate service in compliance with the Italian regulation framework by ARERA, also through the realization of investments aimed at improving infrastructures and concerning above all technological innovation, with constant attention to the safeguarding of the environment.

References

1. Pollino, M., Di Pietro, A., La Porta, L., Fattoruso, G., Giovinazzi, S., Longobardi, A.: Seismic-crisk simulations of a water distribution network in Southern Italy. In: Computational Science and Its Applications – ICCSA 2021. Lecture Notes in Computer Science
2. Longobardi, A., et al.: Water distribution network perspective in RAFAEL Project. A system for critical infrastructure risk analysis and forecast – ICCSA 2021. Lecture Notes in Computer Science
3. TLC GORI Homepage (Wonderware). https://tse01.goriacqua.com
4. GIS Gori Homepage GIS (ESRI). https://gis20web.aceaspa.it/aceaidrico/GOWEB/

Spatial Justice Models: An Exploratory Analysis on Fair Distribution of Opportunities

Fillipe Oliveira Feitosa$^{(\boxtimes)}$ ⓘ, Jan-Hendrik Wolf ⓘ, and João Lourenço Marques ⓘ

Universidade de Aveiro, 3810-193 Aveiro, Portugal
fillipefeitosa@ua.pt

Abstract. Equity, fairness, and justice are related concepts widely discussed in several areas of study but remain an open field in terms of spatial justice and support decision systems application. Uneven spatial development have shown a tendency to amplify social inequalities alongside territories. To better understand the spatial configuration and spatial distribution of resources for different social groups, multiple objective criteria can be used to formulate optimal resource allocation. This work discusses spatial justice by utilitarianism and Rawlsian difference principle perspectives to formulate two models based on facility location problem (FLP) framework. Assuming the proximity to a desired opportunity (service or resource) as a measure of wellbeing and satisfaction, we weight the distances to the nearest facility by a social factor based on exponential function. Optimization results tend to favor outliers for weighted FLP, while the regular distances FLP formulation tend to favor heavy urban areas. We found that results are heavy context based, as the distribution of social groups are determinant in optimization process.

Keywords: Spatial justice · Social justice · Optimization · Facility location problem

1 Introduction

Cities can be considered the centres of modern societies. Although urbanization is an ancient phenomenon, city growth accelerated after the industrial revolution in the 19th century, when concerns regarding infrastructures such as transportation, security, and health led to public and private investments that reshaped the urban environment [1]. But resources flow unevenly into urban development. Business and private resources tend to be allocated to zones where they can generate the highest returns on revenue, considering that location is related with business success [2].

Public investments, while being committed to a certain level of equity, also tend to concentrate in more central locations. This led to a focus on spatial inequalities and their consequences to society. Recent studies have, for example, shown that territorial

This work has been supported by Portuguese national and EU funds through the Foundation for Science and Technology, FCT, I.P., in the context of the JUST_PLAN project (PTDC/GES-OUT/2662/20) and the DRIVIT-UP project (POCI-01-0145-FEDER-031905) and the research unit on Governance, Competitiveness and Public Policy (UIDB/04058/2020)+(UIDP/04058/2020).

O. Gervasi et al. (Eds.): ICCSA 2021, LNCS 12950, pp. 674–683, 2021.
https://doi.org/10.1007/978-3-030-86960-1_51

inequalities may lead to poor accessibility to services of general interest [3], low economic potential [4], or precarious living conditions [5], which lay the foundation to demographic decline by increasing unemployment and social exclusion [6].

The way the spatial allocation of resources and services relates to this policy goal is a subject open to debate. From a business perspective, it is important to achieve an optimal allocation based on profit-maximizing goals. From a public policy perspective, justice in the distribution of spatial amenities, must also be addressed. But it is not always easy to separate these two perspectives, since, as observed by Stiglitz, politics tend to shape the market in such a way that the structure became favourable to wealthier groups at the expense of the rest [7]. While the political aspects of resource distribution are not a specific focus of this work, this argument stresses the need to consider the complex dynamics of spatial resource allocation and the, often conflicting, goals it maximizes.

Previous work in this field discussed the balance between efficiency and equity, concluding that even efficiency-focused scenarios can be considered 'fair', depending on the understanding of equity [8]. Therefore, the challenge is to define the desired goals that need to be satisfied and to translate what is a just distribution of resources into objective. From a planning perspective, the spatial distribution of services or facilities is often approached through location analysis and, namely, by solving facility location problems [9]. This kind of problem represents an established tool to determine the optimal locations for the installation of facilities to provide for existing consumer demand [10], using models that range in complexity from simple linear, uncapacitated to non-linear probabilistic models [11].

In this work, this modelling framework is used to analyse equity in the spatial distribution of public services, considering that there are many different definitions of spatial equity and, also, many understandings of what they mean in practice. More specifically, it aims to show how different justice principles can be incorporated in the formulation of objective functions that can be optimized to provide different spatial distributions of facilities (opportunities). This approach will be illustrated in the case of primary school locations in a peripheral municipality of Lisbon, Portugal (Loures), attributing weights to the accessibility of different socioeconomic groups and analysing the outcomes according to different indicators.

The solutions are, of course, not deterministic, in the sense that contextual problems may have distinct optimal solutions, specific restrictions, and even special conditions that cannot be optimized at all. Nonetheless, we believe that this analysis con-tributes to location analysis, providing operational insights and practical discussions of different ways to incorporate equity principles in spatial planning. The methodology and the analysis details are described in the data and methods section, following a brief review of principles of justice for location analysis. Afterwards, the results are presented and discussed. The final section includes conclusions and main findings.

1.1 Principles of Justice for Location Analysis

Assuming that the spatial location of services of general interest should follow principles of spatial equity does not answer the question of what these principles should be. Two major lines of reasoning can be identified in this regard. The first is a utilitarian perspective, where individuals are treated as equals and the benefits of resource distribution are, accordingly, weighted equally. As noted by Mandal et al., (2016), utilitarian approaches, aiming to maximize aggregate welfare, are concerned with providing the maximum benefit to the greatest number of people, disregarding the differences in individuals initial position of the inequalities that might arise from this process (for a more thorough discussion of utilitarian principles see, for example, [13]).

Regarding spatial accessibility Pereira et al. [10] consider that, from an utilitarian perspective, promoting the spatial accessibility of higher income classes at the expense of lower income classes can be justified, as long as it brings net benefits. A schematical representation of optimum locations under utilitarianism assumptions is shown in Fig. 1(left).

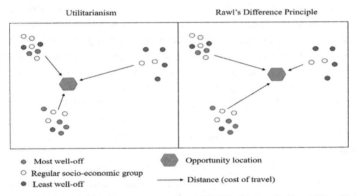

Fig. 1. Illustration of spatial justice representation of Utilitarianism (left) and Rawl's Difference Principle (right).

A different perspective can be found in authors which assume justice as the main goal which should guide public policies, such as Fainstein (2009), Sen (1992) or Rawls (1971, 1981). Rawlsian principles, in particular, are frequently evoked to justify setting limits to the inequalities generated by the spatial distribution of resources. According to Rawls, inequalities should be structured in a way that confers the most benefit to the least well-off. This definition assumes that a certain level of inequality will always exist, but that public policies should aim to redistribute resources and promote justice, to guarantee a maximum degree of freedom to different social groups and individuals, taking into consideration the different initial conditions with which they enter society.

In location analysis this principle can broadly be understood as meaning that accessibility levels should be highest for those whose initial position is worse. The solutions that best fit with this perspective are those that locate spatial amenities closest to the less privileged (see Fig. 1 right).

One major question that is not answered by this discussion is exactly who are the worse of in a given state-of-affairs. Or, in other words, which are the dimensions in which inequalities should be evaluated. Regarding spatial accessibility there are two major approaches. The first one only considers inequalities in accessibility levels. This approach is concerned with guaranteeing that no-one is above certain thresholds or that the spatial distribution of amenities minimizes the inequalities between different territories (examples of this approach can be found in [18, 19]). The second one considers the levels of accessibility that are provided to different social groups. It has, in this regard, been noted that lower socioeconomic groups tend to have worse accessibility to spatial resources such as public parks [20] or, indeed, schools [3, 21].

This work is in line with the second approach, considering the way spatial accessibility to public facilities (opportunities) is related with pre-existing inequalities. This means it will maximize accessibility goals but weight the distances of different socio-economic groups differently. We intend to create scenarios of fairness that can be compared to the existing opportunities distribution in terms of accessibility metrics.

2 Data and Methods

Accessibility can be defined as the ease of reaching relevant activities, individuals, or opportunities [22], and its meaning is closed related to the spatial distribution of destinations. Our work assumes that proximity to opportunities plays an important factor in individual development and well-being. This is not a novel concept and has already been stressed by Sen [23], who argues that access to opportunities is vital for justice and individual freedom, and was revisited by the planning theory [14] regarding the justice in the spatial distribution of services.

In this work, we go a step further by addressing the "easiness" of access in accessibility. As previously argued, Ralw's Difference Principle states that inequalities may occur if they promote the least well-off, but it is not clear how the least well-off are to be identified and, if these social differences do exist, how this reflects on the access to opportunities. In other words, how hard is for the least privileged in a given context to reach their relevant destination? This subject is addressed by adjusting the distance to the nearest opportunity to a "social cost" factor and using facility location problem (FLP) as a framework to create optimal services locations to analyze how the real distribution relates to them. By doing this weighed FLP formulation, we intend to model Ralw's justice principles tending optimization to the least well-off, in contrast with the regular FLP model, that will maximize the overall welfare by satisfying the populated areas demand without taking socio-economic factor into account.

Facility location problems optimize objective functions which express different goals in the spatial distribution of resources. A detailed formulation can be found on [24], and several use cases can be found on [25]. Applications on justice principles often requires clever approaches to the objection function and its restrictions. For instance, Hooker [26] uses objectives functions to find just resource distribution from both utilitarian and distributive justice perspectives.

Here, these two perspectives are applied to analyze the accessibility to primary schools in the municipality of Loures, assessing how the attribution of different weights to different social groups affects the spatial layout of the school network. For this, we consider the student population of the 628 census tracts, which leads to a total demand of 2407 school aged children. To classify the region socio-economic situation a qualification index (QI) was formulated based on the years of schooling (Table 1), discussed in Marques et al. (2020). Three different groups were considered in this index:

group 0 includes those with less than 6 years of schooling, group 1 those from 6 to 11 years, and group 2 those with more than 11 years.

We chose to limit the capacity of school facilities to 80 students, which is the number considered necessary for their good functioning according to Portugal national planning guidelines (DGOTDU, 2002).

The objective function (Eq. 1) will minimize the sum of the distances (or "costs") c_{ij} from the location i to the potential opportunity j (in this case a primary school), times the amount of demand that is correspondent times the fraction of demand y_{ij} allocated to that opportunity. The restrictions (Eq. 2) indicate that the demand of every census tract must be satisfied, respecting every capacity.

$$minimize: \sum_{i=1}^{n} \sum_{j=1}^{m} c_{ij}y_{ij} \tag{1}$$

$$subject\ to: \sum y_{ij} = d_i \quad for\ i, = 1, \ldots.n \tag{2}$$

$$\sum y_{ij} \leq M_j x_j,\ for\ j = 1, \ldots, m;\ for\ i = 1, \ldots, n$$

$$y_{ij} \leq d_i x_j, for\ i = 1, \ldots, n;\ j = 1, \ldots, m$$

$$x_j \in \{0, 1\} for\ j = 1, \ldots, m$$

All 628 census tracts are considered potential location for a school. The model ran with two configurations. The first one considered the real distances to the nearest facility (to maximize overall utility) and the second one weighted the real distance by an exponential cost function (our Rawlzian distributive justice modeling proposal), giving more weight to lower socio-economic groups (group 0) than privileged ones (groups 1 and 2). There are many different cost functions (impedance functions) which can be used for this effect. There, we chose an exponential function which, according to Vale [27] is the most frequently used in the scientific literature. While these authors tested several combinations of factors to verify their influence on accessibility, we chose to use different B for each social group (see Table 1) to create the aggregate socioeconomic difficulty to reach the destination.

Finally, we compared the results using exploratory data analysis. We considered the mean, the standard deviation, the minimum and maximum values to understand how the different models used for optimization differ from each other, and how the real distribution of opportunities relates to them.

Table 1. Data and measures overview.

Indicator/measure	Description	Initial data	Source
Distance to the nearest opportunity (school)	Distance calculated for different nodes of the urban network and then matched with parishes geometry by intersecting them with these nodes	i) schools by coordinates for the council (Loures) ii) Street Network	i) DGEEC ii) OpenStreetMap
Qualification index	The qualification index is calculated for every parish, based on the following formula: $QI = \frac{\sum y_i x_i}{\sum y_i}$ Where x are the years in each cycle of studies, y is the number of people in that cycle, and i is the parish	i) Spatially delimited statistical units (parishes) ii) data for population by educational level for the Census 2011	i) BGRI ii) INE
Cost function	Impedance factors that weight the distance to reflect how hard is for a social group reach a relevant opportunity $f(C_{ij}) = e^{(\beta C_{ij})}$ Where C_{ij} is the cost (or distance) of moving from i to j, and β is a factor that adjust the function	i) $\beta = 0.007$ if Group 0 ii) $\beta = 0.005$ if Group 1 iii) $\beta = 0.003$ if Group 2	[27]

3   Results

The weight imposed by the cost function upon the distances induces gaps between the social groups. As the outliers can get quite distant from the urban center, their cost gets too high, so the model adjusts opportunities in a sparse fashion that generates high dispersion. The zero in minimum distances on Table 2 means that the one opportunity (school) got located in the exact place as the one of the group census tracts during the optimization process.

Benefited by both their proximity to the denser urban center and the middle cost factor that is still quite forgiving to the distance, group 1 was the most benefited by our cost function choice. The occupation pattern plays an important part, as this groups tend to live close to populated areas, where the difference between the weighted costs can get quite low, even among different social groups. This pattern can also be observed on the higher minimum lower mean distances for the weighted FLP.

Group 2 accessibility improved with weighted FLP, even if the cost does not benefit them directly. This behavior was not expected, as their cost of travel is lower and optimal opportunity location would tend to avoid being too close of this group, in favor off better accessibility for group 0 and group 1. Closer analysis reveals that occupation pattern of group 2 tends to be around denser areas, even more than group 1, in such a way that they remain close to local schools. Therefore, the group 2 accessibility metrics improvement are directly related to the occupation pattern, and probably would differ if the illustration case selected were any different.

Table 2. Accessibility metrics aggregated by social group.

Scenarios	Group	Minimum distance	Maximum distance	Mean distance	Standard deviation
Real scenario	Group 0	122	5221	1832	1220.731
	Group 1	0	6489	1451	1246.203
	Group 2	0	3219	1315	812.8781
FLP - real distances	Group 0	43	4836	1330	1051.289
	Group 1	0	5432	1049	1006.182
	Group 2	0	2797	518	629.064
FLP weighted distances	Group 0	197	3770	1998	1091.904
	Group 1	0	2291	893	573.3682
	Group 2	0	1748	564	422.3245

To support our argument about Rawlzian difference principle modeling as a weighed version of FLP formulation, we compare the different metrics for the least well-off group on Fig. 2. Although minimum and mean distances of the regular FLP (considered a utilitarian approach) benefits group 0 compared to the weighed FLP, max distance is greatly reduced. This indicator is important because it represents the outliers that live far away from the desired service. Sustained by a low standard deviation, meaning that the distance is evenly distributed among group 0, applying a cost factor to the distance tends to create a more suitable scenario for lower socio-economic groups.

4 Summary and Outlook

In this work we discussed how justice principles can relate to spatial justice and proposed two models based on facility location problem optimization. The first model was a capacitated facility location that aimed to minimize the sum of the distances pondered by the location demand. By doing that, we aimed to create a scenario that maximizes the overall welfare as the opportunities tend to be close to denser areas. The second model is our proposal for a Rawl's difference principle. It is a variation of the first one,

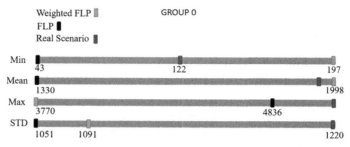

Fig. 2. Accessibility metrics comparison for Group 0 (least well-off).

weighting the distance by an exponential cost function, or "social cost", that will force the optimization to favor lesser well-off regions.

The results have shown that the occupation pattern and, therefore, the spatial context of analyzed area, plays an important role in opportunity location. The fact that the most well-off group was also benefited when our cost function was deliberately chosen to favor the least well-off, does not imply that Rawlzian model will improve overall wellbeing. It seems that spatial distribution of social groups displays complex dynamics during the optimization. This behavior is directly related to the B chosen in the exponential factor, as the cost of denser and mixed areas on urban centers can get not that different, leaving the cost function effectiveness to outliers. An experimental approach to tune the B to impact relevant distances can be done, similar as conducted by [22]. But the relevant distances and affected groups can be context dependent.

The weighed FLP model was able to turn the configuration in favor of the least well-off group, even though not by all accessibility metrics. By greatly reducing the maximum distance, this group tend to stay closer to desired opportunities. In future works, the standard deviation can be also optimized to create even distribution of services. Also, the fact that our utilitarianism model (FLP) created more even opportunity distribution than the Rawlsian exemplifies that fair and justice can be achieved by different perspectives.

This discussion illustrates how spatial justice modeling is not a trivial task. Nevertheless, we argue that "social cost" functions can act as a proxy to understand the relation between relevant opportunities and the socio-economic distribution in a given territory. By weighting distances with social cost function, it can be possible to propose a support decision system to make interventions on the desired services configuration. For instance, it can be possible to suggest a new location for a service that minimizes the overall mean distance by the two different justice perspectives, in a similar way that Fig. 2 illustrates.

References

1. Hall, P.: Cities of Tomorrow: An Intellectual History of Urban Planning and Design Since 1880 (2014)
2. Porter, M.E.: Clusters and the new economics of competition. Harv. Bus. Rev. **76**, 77–90 (1998)
3. Marques, J., Wolf, J., Feitosa, F.: Accessibility to primary schools in Portugal: a case of spatial inequity? Reg. Sci. Policy Pract. n/a (2020). https://doi.org/10.1111/rsp3.12303

4. De Toni, A., Di Martino, P., Dax, T.: Location matters. Are science and policy arenas facing the Inner Peripheries challenges in EU? Land Use Policy **100**, 105111 (2021). https://doi.org/10.1016/j.landusepol.2020.105111

5. Scandurra, R., Cefalo, R., Kazepov, Y.: Drivers of youth labour market integration across European regions. Soc. Indic. Res. **154**(3), 835–856 (2021). https://doi.org/10.1007/s11205-020-02549-8

6. Copus, A., Mantino, F., Noguera, J.: Inner peripheries: an oxymoron or a real challenge for territorial cohesion? Ital. J. Plan. Pract. **7**, 24–49 (2017)

7. Stiglitz, J.: The price of inequality. New Perspect. Q. **30**, 52–53 (2013). https://doi.org/10.1111/npqu.11358

8. Wolf, J., Feitosa, F., Marques, J.L.: Efficiency and equity in the spatial planning of primary schools. Int. J. E-Planning Res. **10**, 21–38 (2021). https://doi.org/10.4018/IJEPR.2021010102

9. Drezner, Z., Hamacher, H.W.: Facility Location: Applications and Theory. Springer, Heidelberg (2002)

10. Pereira, R.H.M., Schwanen, T., Banister, D.: Distributive justice and equity in transportation. Transp. Rev. **37**, 170–191 (2017). https://doi.org/10.1080/01441647.2016.1257660

11. Klose, A., Drexl, A.: Facility location models for distribution system design. Eur. J. Oper. Res. **162**, 4–29 (2005)

12. Mandal, J., Ponnambath, D.K., Parija, S.C.: Utilitarian and deontological ethics in medicine. Trop. Parasitol. **6**, 5–7 (2016). https://doi.org/10.4103/2229-5070.175024

13. Kymlicka, W.: Contemporary Political Philosophy: An Introduction. Oxford University Press, Oxford (2002)

14. Fainstein, S.S.: Spatial justice and planning. In: Readings in Planning Theory: Fourth Edition, pp. 258–272. Wiley, Chichester (2016)

15. Sen, A.: Inequality Reexamined. Harvard University Press, New York (1992)

16. Rawls, J.: A Theory of Justice. Harvard University Press, Cambridge (1971)

17. Rawls, J.: The Basic Liberties and Their Priority. University of Utah Press, Salt Lake City (1981)

18. Tsou, K.W., Hung, Y.T., Chang, Y.L.: An accessibility-based integrated measure of relative spatial equity in urban public facilities. Cities **22**, 424–435 (2005). https://doi.org/10.1016/j.cities.2005.07.004

19. Taleai, M., Sliuzas, R., Flacke, J.: An integrated framework to evaluate the equity of urban public facilities using spatial multi-criteria analysis. Cities **40**, 56–69 (2014). https://doi.org/10.1016/j.cities.2014.04.006

20. Talen, E.: The social equity of urban service distribution: an exploration of park access in Pueblo, Colorado, and Macon, Georgia. Urban Geogr. **18**, 521–541 (1997). https://doi.org/10.2747/0272-3638.18.6.521

21. Lee, J., Lubienski, C.: The impact of school closures on equity of access in Chicago. Educ. Urban Soc. **49**, 53–80 (2017). https://doi.org/10.1177/0013124516630601

22. Vale, D.S., Saraiva, M., Pereira, M.: Active accessibility: a review of operational measures of walking and cycling accessibility. J. Transp. Land Use **9**, 209–235 (2016)

23. Sen, A.: Human rights and capabilities. J. Hum. Dev. **6**, 151–166 (2005). https://doi.org/10.1080/14649880500120491

24. Raghavan, S., Sahin, M., Salman, F.S.: The capacitated mobile facility location problem. Eur. J. Oper. Res. **277**, 507–520 (2019). https://doi.org/10.1016/j.ejor.2019.02.055

25. Farahani, R.Z., Asgari, N., Heidari, N., et al.: Covering problems in facility location: a review. Comput. Ind. Eng. **62**, 368–407 (2012)

26. Hooker, J.N.: Optimality conditions for distributive justice. Int. Trans. Oper. Res. **17**, 485–505 (2010). https://doi.org/10.1111/j.1475-3995.2009.00742.x

27. Vale, D.S., Pereira, M.: The influence of the impedance function on gravity-based pedestrian accessibility measures: a comparative analysis. Environ. Plan. B Urban Anal. City Sci. **44**, 740–763 (2017). https://doi.org/10.1177/0265813516641685

Hybrid Urban Model (CA + Agents) for the Simulation of Real Estate Market Dynamics and Sea-Level Rise Impacts

Guilherme Kruger Dalcin[(⊠)] and Romulo Krafta

Federal University of Rio Grande do Sul (UFRGS), Porto Alegre, Brazil
{guilherme.dalcin,krafta}@ufrgs.br

Abstract. The paper presents a proposal for a hybrid model - based on cellular automata and agents - that simulates the spatial distribution of population and built area according to real estate market's dynamics and the risk of flooding due to sea-level rise. Its main differential is the integration of network analysis metrics to the functioning of the cellular automata. This proposal was motivated by the interest in analysing future development scenarios for the coast of Rio Grande do Sul, a state located in southern Brazil. Its demographic dynamics have been generating pressure for urban growth to the detriment of the surrounding natural environment, making cities in the region more susceptible to natural phenomena such as the sea-level rise. The proposed model is presented through the ODD + D description protocol and the results of simulations executed for Imbé and Tramandaí, two municipalities located on the coast of Rio Grande do Sul. The results show that the model represents the effect of current planning policies on long-term urban development. However, some urban dynamics are not yet precisely represented by the proposal at its current stage of development.

Keywords: Urban modelling · Cellular automata · Agents

1 Introduction

The north coast of Rio Grande do Sul - a state located in southern Brazil - has been presenting the highest demographic growth in the state, contrasting with the population decrease that occurs in its neighbouring regions [1]. In addition, during the summer, the region's population increase up to 250% in relation to the winter period due to the touristic attractiveness of local beaches [2]. These dynamics cause an intense real estate market that generates pressure for more urban growth [3], creating challenges for urban management and affecting the surrounding natural environment [1]. In this sense, the suppression of dunes and vegetation makes the region's urban areas more susceptible to natural disasters such as storms or even the sea level rise due to climate changes, expected to impact the region by the end of the century [4].

In the literature, no analyses are found on how these issues will affect cities on the north coast of Rio Grande do Sul, although studies that apply modelling and simulation techniques to speculate on the development of cities have become commonplace in

© Springer Nature Switzerland AG 2021
O. Gervasi et al. (Eds.): ICCSA 2021, LNCS 12950, pp. 684–692, 2021.
https://doi.org/10.1007/978-3-030-86960-1_52

other regions of the world in order to anticipate possible urban trends according to different scenarios [5–8]. Aiming to contribute to this set of studies by adapting existing modelling techniques to the specific context of the coast of Rio Grande do Sul, the authors' ongoing research seeks to analyse development scenarios of municipalities in the mentioned region using cellular automata and agents.

This work presents a hybrid model - based on cellular automata and agents - to assist the analysis of the study area and whose main objective is the representation of: i) locational preferences of the population groups; ii) land value variation; and iii) the impact of sea-level rise on the region's urban configuration. The differential of this proposal is the use of network analysis metrics as attributes of the cellular automata, enabling the direct communication of non-adjacent cells. Experimental simulations show that the model is able to represent the effect of planning policies on long-term urban development.

2 Research Background

Complex systems are systems whose components are other complex systems and in which the interaction between such components originates endogenous and decentralized processes of organization that influence the system's behaviour at all scales [9]. Cities are classified as complex self-organizing systems since they consist of numerous individuals with complex behaviour interacting with each other and the environment on the micro-level and influencing the patterns that emerge in the city's macro-level [10]. Because cities are complex self-organizing systems, their operation presents the specific characteristics of such systems: emergence [11], non-linearity and far-from-equilibrium functioning [12], path dependence [9] and robustness [13].

To model cities as complex self-organizing systems, it is necessary to treat them as processes, incorporating time in the model and assuming an algorithmic approach, in which the behaviour of the system's components is disaggregated into actions performed iteratively step by step [14]. Two specific model types are commonly applied to implement such an approach: cellular automata (CA) and multi-agent models (MAM). CA consists of the division into cells of the represented territory, each representing a fraction of it. These cells present: i) a state, which assumes a value from a pre-defined list of possibilities; ii) a set of transition rules indicating how states change over time; and iii) attributes which describe the characteristics of the territory and are also used to define transition rules [15]. States and attributes generally provide a geographical description of the territory - land cover, occupation, use - while transition rules tend to incorporate theoretical spatial or economic statements referring to dynamics of cities [16]. In this sense, CA's general objective in urban modelling is to capture the rules that determine how the state of cells changes according to what happens in its neighbourhood and how these changes produce patterns representing future possibilities for the system [17].

The agents of a MAM are able to move within the limits of the model, and each also has its own attributes and states, as well as transition rules; however, due to their mobility, it is not required for them to have an immutable neighbourhood [16], and interactions between agents usually occur between those who are close enough during simulation or between those connected by means of communication, such as social networks [18, 19].

Agents enable the simulation of a heterogeneous system in which the specificities of different population groups are considered [18]. They also can interact with the territory when CAs and MAMs are used in a hybrid model, enabling the territory's cells to be modified according to the interaction with agents and also to modify them [16].

However, for such models of complex systems, the validation process presents dilemmas regarding the difficulty of testing all the assumptions that are present in the model [20] and the lack of consensus on how to demonstrate the reliability of future predictions, considering that past or present empirical observations do not necessarily indicate how the system will develop into the future [14]. It is commonly accepted that such models serve to inform the debate about specific urban issues and not to provide completely accurate forecasts [21]. Given that cities are path-dependent, one way to demonstrate the model's reliability is to show that simulated results correspond to possible behaviours of cities and that, therefore, they could be one of the many future possibilities for the analysed urban system [14]. One can accomplish this by using: i) map comparison metrics that define the goodness-of-fit between reality and simulation, showing if simulations represent specific aspects of the study area [22]; and ii) fractal metrics for which the literature has already defined values that typically represent cities, showing that simulation represents general behaviours of cities [14].

3   Model Description

The proposal is composed of a CA model representing the analysed environment and a MAM representing the population's locational preferences. It is described in the following subsections with the use of the ODD + D protocol for ABMs [23].

3.1   Overview

Purpose. The proposed model aims to provide a methodological platform to integrate land value variation dynamics and the impact of natural phenomena into the analysis of urban environments. The main question that the model seeks to answer is: how the referred factors influence urban growth and the population's spatial distribution? In this paper, the only represented natural phenomenon is the sea level rise, but other factors could also be included in future studies.

Entities, State Variables, and Scales. The model is composed of agents, cells, and real estate developers (REDs). Cells correspond to hexagonal fractions of the territory with 50-m-long edges. Their state represents its urbanization capacity and reachability, dividing them into: i) cells that can be urbanized; ii) those that cannot be urbanized but are reachable, such as beaches; and iii) cells that cannot be urbanized and are unreachable, such as lakes or the sea. The attributes that influence the cells' state are presented in Table 1. Empirical GIS data is uploaded at initialization to define the cell's initial state and attributes.

Agents represent individuals who settle in cells to perform residential or commercial activities. Their state variables are: i) their income; ii) the cell's minimum attractiveness for the agent to settle in it; and iii) the maximum number of iterations that the agent may remain in the same cell.

REDs are implemented as a computational function that looks for cells where they could build more units to increase their capacity and profit from such operation. They do not have state variables and are not physically represented in simulations.

Table 1. Attributes of CA's cells.

Attribute	Description
Occupancy	The number of agents inhabiting the cell
Capacity	The maximum number of agents that can simultaneously inhabit the cell
Load	An estimate of the agent attractors in the cell. It is divided into natural (such as beaches) and urban load (active residences and commerce). It weighs the results of the spatial opportunity and convergence metrics
Spatial opportunity	A network analysis metric which estimates the accessibility to commercial use and services from the point of view of residences [24]. In this study, it represents the cells' attractiveness to residential agents
Spatial convergence	A network analysis metric which measures the locational privileges of commercial activities [25]. In this study, it represents the cell's attractiveness to commercial agents
Land value	An estimate of the cell's sale value

Process Overview and Scheduling. The simulation process works according to the following steps: i) agents are inserted in the model and seek, among the cells compatible with their income, the most attractive ones; ii) as agents settle in or leave, cells update their attributes; iii) agents check if their current cells still meet their attractiveness and cost requirements and, if this does not happen, they search for another one; and iv) the most sought cells become more expensive, attracting REDs who expand these cells' capacity to profit with the sale of the created units. During this process, some cells are affected by sea-level rise.

Since spatial opportunity and convergence metrics are based on the analysis of the territory as a network, the shortest paths between all pairs of cells are computed using a graph representation as described in Sect. 3.3.

3.2 Design Concepts

Theoretical and Empirical Background. Real estate market dynamics follow the modelling of microeconomic supply and demand as proposed by Filatova [26], where the land value is defined endogenously in the model as agents decide which cells they find attractive. Agents' locational preferences and the behaviour of REDs are based on Krafta [24, 25], who proposed metrics to characterize the territory according to the preferences of residences or commerce and models of urban development based on such metrics. The logic of allocating different land uses according to their own preferences was loosely based on the proposals of White, Engelen and Uljee [14].

Individual Decision-Making and Learning. Agents decide where to settle in by analysing, for a randomly selected set of cells, if its land value is lower than the agent's income and if it is more attractive than the agent's requirement. After settling in, an agent decides to leave the cell if: i) its land value becomes higher than the agent's income; ii) it becomes less attractive than the agent's requirement; and iii) the agent's maximum number of iterations in the same cell is reached.

RED's decision to expand a cell is based on the developer's equation as proposed by Krafta [25]: if the cost of buying a cell and constructing in it is lower than the expected gains for selling the additional units, which is computed based on the neighbouring cells' land value, than REDs decide to perform such operation.

Individual Sensing and Predictions, Heterogeneity, and Collectives. Agents can perceive a cell's attractiveness and land value, and they use such information when settling in a cell or deciding whether to leave it. REDs also perceive a cell's land value, and they make predictions about its future price by analysing the neighbours' land value and considering their average as the value that the cell will probably have in the future.

Agents are distinct from one another because they have different incomes and attractiveness requirements. However, although such attributes are the same for some agents, they do not form any kind of collectives.

Stochasticity. When agents and REDs analyse cells, their search is performed considering a randomly selected set of cells with random size. Such randomness represents the fact that one cannot gather information about the whole system when deciding for a place to live in or when considering where to insert a building.

Observation. The main objective of simulations is to observe the spatial distribution of population and buildings, identifying the main factors that caused the appearance of the densest urban areas and analysing how the restriction of cells due to sea-level rise affect the population locational choices. Besides, the cells' land value and the agents' income may also indicate the appearance of economic spatial segregation.

3.3 Details

Implementation Details, Input Data and Initialization. The model is being developed in Gama Platform [27]. As input, it uses data on territory and roads from OpenStreetMap [28] and sea-level rise from Climate Central's Coastal Risk Screening Tool [4]. Currently, the model is initialized with an unoccupied territory.

Submodels. To enable the use of network analysis metrics as attributes of CA's cells, a graph representation was added as an auxiliary layer of the CA. In such representation, cells' centroids are nodes - weighted according to their load attribute - and their connections are represented as edges whose impedances are defined according to the hierarchy of the roads they intersect - it increases as the road's hierarchy increases. Based on these, the shortest paths (i.e., the ones with the lowest sum of impedances) between all nodes are computed, and they are used for computing the spatial opportunity and convergence metrics as described by Krafta [24, 25].

Validation. The authors expect to validate such results using the comparison between simulated results for the period 2010 to 2020 and the existing situation in 2020 through the measure of the goodness-of-fit for multiple spatial resolutions [22], and, for future scenarios simulations, metrics whose typical values for cities are described in the literature, such as the radial dimension and the cluster-size frequency distribution [14].

4 Results

To demonstrate the feasibility of the proposed model, an experimental simulation was performed for the municipalities of Imbé and Tramandaí, which compose a conurbation on the coast of Rio Grande do Sul, using a CA containing 16,000 cells. Since it was an experiment to test the model in its initial development stage, the territory was considered as initially unoccupied, and a fixed number of random agents was added per iteration of the simulation. Besides, the sea-level rise was updated in the 100th, 150th and 200th iterations with forecasts' data of, respectively, 2040, 2060 and 2080.

Figure 1 shows how the spatial distribution of agents evolved during simulation. High agent concentrations can be seen in areas close to the seashore, especially those close to several important roads, with areas further inland only being occupied when free cells next to the beaches become scarce. Figure 2 shows the variation of cells' land value and how the most sought areas become more expensive, making lower-income agents look for cheaper cells elsewhere.

Such observations indicate that current planning legislation – represented as an attribute of the CA - creates a tendency for spatial concentration in areas whose building restrictions are permissible, and the increase of sea-level rise reinforces such concentration. Therefore, one could consider modifying such building restrictions, aiming at a more homogeneous distribution of the population that needs to relocate due to sea-level rise, making the urban changes caused by it less drastic.

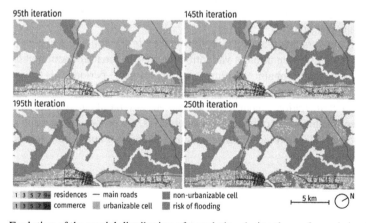

Fig. 1. Evolution of the spatial distribution of population during the performed simulation.

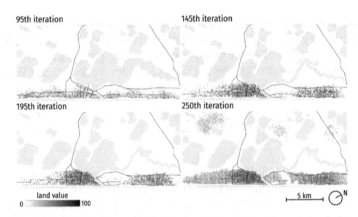

Fig. 2. Evolution of the cells' land value during the performed simulation.

The simulations also indicate that the model tends to overestimate forces of urban agglomeration to the detriment of those responsible for urban dispersion. This can be seen in the 145th and 195th iterations, where there is a concentration of agents in reduced areas of the territory which expands elsewhere only when available cells in central areas become scarce. This is probably due to the absence of factors that contribute to dense areas becoming less attractive, such as congestion.

5   Final Remarks

This study aimed to propose a predictive model of the urban configuration of municipalities on the north coast of Rio Grande do Sul. The presented simulation seems to be coherent with the theoretical bases used to support the study and demonstrate the relevance of simulating the impacts of sea-level rise and land market dynamics to inform discussions about planning policies. However, the results also show that the model overemphasizes agglomeration forces since it does not consider some of the sprawl forces existing in actual urban development.

Other improvements planned to be introduced in the following stages of development are: i) the implementation of validation metrics; ii) the introduction of the seasonal agents that repeatedly inhabit the municipalities for short periods; iii) the introduction of institutional equipment - such as parks, schools and hospitals - that affect the attractiveness of specific regions in particular ways since commonly their location is the result of top-down planning. Despite these considerations, the proposed approach still has the potential to provide more accurate results and, therefore, could be a relevant tool for urban modelling.

References

1. Bertê, A., Lemos, B., Testa, G., Zanella, M., Oliveira, S.: Perfil Socioeconômico COREDE Litoral. Department of Planning, Mobility and Regional Development of the State Government of Rio Grande do Sul, Porto Alegre, Brazil (2015). https://planejamento.rs.gov.

br/upload/arquivos/201512/15134132-20151117102724perfis-regionais-2015-litoral.pdf. Accessed 18 Mar 2021

2. Zuanazzi, P., Bartels, M.: Estimativas para a População Flutuante do Litoral Norte do RS. Fundação de Economia e Estatística Siegfried Emanuel Heuser, Porto Alegre, Brazil (2016)

3. Kluge, I.: A Articulação entre Urbanização, Economia e Mercado Imobiliário em Cidades Litorâneas e a Relação com o Ambiente Construído: o estudo de caso do município de Capão da Canoa – RS. Federal University of Rio Grande do Sul - UFRGS, Porto Alegre, Brazil (2015). https://lume.ufrgs.br/handle/10183/130719. Accessed 18 Mar 2021

4. Climate Central's Coastal Risk Screening Tool. https://coastal.climatecentral.org/map. Accessed 18 Mar 2021

5. Casali, Y., Heinimann, H.R.: A topological characterization of flooding impacts on the Zurich road network. PLOS ONE **14**(7), e0220338 (2019). https://doi.org/10.1371/journal.pone.022 0338

6. Kim, Y., Newman, G.: Advancing scenario planning through integrating urban growth prediction with future flood risk models. Comput. Environ. Urban Syst. **82**, 101498 (2020). https://doi.org/10.1016/j.compenvurbsys.2020.101498

7. Li, S., Li, X., Liu, X., Wu, Z., Ai, B., Wang, F.: Simulation of spatial population dynamics based on labor economics and multi-agent systems: a case study on a rapidly developing manufacturing metropolis. Int. J. Geogr. Inf. Sci. **27**, 2410–2435 (2013). https://doi.org/10.1080/13658816.2013.826360

8. Taberna, A., Filatova, T., Roy, D., Noll, B.: Tracing resilience, social dynamics and behavioral change: a review of agent-based flood risk models. Socio-Environ. Syst. Model. **2**, 17938 (2020). https://doi.org/10.18174/sesmo.2020a17938

9. Batty, M. Complexity in city systems: understanding, evolution and design. Centre for Advanced Spatial Analysis (CASA), London (2007). https://discovery.ucl.ac.uk/id/eprint/3473/. Accessed 18 Mar 2021

10. Portugali, J.: What makes cities complex? In: Portugali, J., Stolk, E. (eds.) Complexity, Cognition, Urban Planning and Design. SPC, pp. 3–19. Springer, Cham (2016). https://doi.org/10.1007/978-3-319-32653-5_1

11. Moroni, S., Cozzolino, S.: Action and the city: emergence, complexity, planning. Cities **90**, 42–51 (2019). https://doi.org/10.1016/j.cities.2019.01.039

12. De Roo, G.: Spatial planning, complexity and a world 'out of equilibrium': outline of a nonlinear approach to planning. In: de Roo, G., Hillier, J., van Wezemael, J. (eds.) Complexity and Spatial Planning: Systems, Assemblages and Simulations. Ashgate Publishing, Farnham (2012)

13. Zellner, M., Campbell, S.D.: Planning for deep-rooted problems: what can we learn from aligning complex systems and wicked problems? Plan. Theor. Pract. **16**, 457–478 (2015). https://doi.org/10.1080/14649357.2015.1084360

14. White, R., Engelen, G., Uljee, I.: Modeling Cities and Regions as Complex Systems: From Theory to Planning Applications. The MIT Press, Cambridge, Massachusetts (2015).978-0-262-02956-8

15. Torrens, P.M.: How Cellular Models of Urban Systems Work (1. Theory). Centre for Advanced Spatial Analysis (CASA), London (2000). https://discovery.ucl.ac.uk/id/eprint/1371/. Accessed 18 Mar 2021

16. Torrens, P.M.: Automata-based models of urban systems. In: Longley, P., Batty, M. (eds.) Advanced Spatial Analysis, pp. 61–79. ESRI Press, Redlands (2003)

17. Liu, Y., Batty, M., Wang, S., Corcoran, J.: Modelling urban changes with cellular automata: contemporary issues and future research directions. Prog. Hum. Geogr. **45**, 3–24 (2021). https://doi.org/10.1177/0309132519895305

18. Crooks, A.T., Patel, A., Wise, S.: Multi-agent systems for urban planning. In: Pinto, N., Tenedório, J.A., Antunes, A.P., Caldera, J.R. (eds.) Technologies for Urban and Spatial Planning: Virtual Cities and Territories, pp. 29–56. IGI Global (2014)

19. Dahal, K.R., Chow, T.E.: An agent-integrated irregular automata model of urban land-use dynamics. Int. J. Geogr. Inf. Sci. **28**, 2281–2303 (2014). https://doi.org/10.1080/13658816.2014.917646

20. Batty, M., Torrens, P.M.: Modelling and prediction in a complex world. Futures **37**, 745–766 (2005). https://doi.org/10.1016/j.futures.2004.11.003

21. Batty, M.: Cities and Complexity: Understanding Cities with Cellular Automata, Agent-Based Models, and Fractals. 1. Paperback edn. MIT Press, Cambridge, Massachusetts (2007). ISBN 978-0-262-52479-7

22. Ngo, T.A., See, L.: Calibration and validation of agent-based models of land cover change. In: Heppenstall, A.J., Crooks, A.T., See, L.M., Batty, M. (eds.) Agent-Based Models of Geographical Systems, pp. 181–197. Springer Netherlands, Dordrecht (2012). https://doi.org/10.1007/978-90-481-8927-4_10

23. Müller, B., et al.: Describing human decisions in agent-based models – ODD + D, an extension of the ODD protocol. Environ. Model. Softw. **48**, 37–48 (2013). https://doi.org/10.1016/j.envsoft.2013.06.003

24. Krafta, R.: Spatial self-organization and the production of the city. Cybergeo: Eur. J. Geogr. (1999). Document 350. https://doi.org/10.4000/cybergeo.4985

25. Krafta, R.: Urban convergence: morphology and attraction. Environ. Plan. B Plan. Des. **23**, 37–48 (1996). https://doi.org/10.1068/b230037

26. Filatova, T.: Empirical agent-based land market: integrating adaptive economic behavior in urban land-use models. Comput. Environ. Urban Syst. **54**, 397–413 (2015). https://doi.org/10.1016/j.compenvurbsys.2014.06.007

27. Taillandier, P., et al.: Building, composing and experimenting complex spatial models with the GAMA platform. GeoInformatica **23**(2), 299–322 (2018). https://doi.org/10.1007/s10707-018-00339-6

28. OpenStreetMap. https://www.openstreetmap.org/search?query=imb%C3%A9#map=12/-29.9355/-50.1156. Accessed 19 Mar 2021

Automatic Sown Field Detection Using Machine Vision and Contour Analysis

Mikhail Shirobokov$^{(\boxtimes)}$ ⓘ, Valeriy Grishkin ⓘ, and Diana Kayumova ⓘ

Saint-Petersburg State University, University Embankment 7/9,
199034 Saint-Petersburg, Russia
v.grishkin@spbu.ru, st026914@student.spbu.ru

Abstract. The paper proposes a prototype of an algorithm based on the use of machine vision methods, which allows automatic identification and selection of fields sown with agricultural crops on images. The algorithm works with satellite images and consists of two stages. At the first stage, the image undergoes initial processing, after which edge detection and contour finding algorithms are applied to it. At the second stage, the obtained image areas enclosed within the contours are represented as a set of numerical and logical parameters which are used for filtering and classification of the areas.

Keywords: Image processing · Machine vision · Contour analysis

1 Introduction

The need to analyze and control the state of arable land in the north-west of the Caspian Sea has existed for several decades, since the degradation of lands suitable for use as pastures is observed due to the retreat of the sea [1,5]. One of the tasks is to track the use and condition of the fields sown with agricultural crops, instantaneously and in dynamics. Currently, the processing of data from satellite images at facilities in this region is carried out in manual mode. In this paper, we propose an algorithm that allows to automatically detect the areas corresponding to the sown fields on satellite images, as well as to determine their properties for the purpose of further processing, monitoring (e.g. sowing date detection [12] or yield analysis [9]) and assessing their condition.

2 Region Boundaries Locating

2.1 Preprocessing

The image preprocessing consists of the following steps:

1. Bilateral filtering [15] is applied to reduce the amount of noise while preserving the boundaries of objects.
2. The image is converted from the RGB color space to grayscale.

© Springer Nature Switzerland AG 2021
O. Gervasi et al. (Eds.): ICCSA 2021, LNCS 12950, pp. 693–701, 2021.
https://doi.org/10.1007/978-3-030-86960-1_53

Bilateral filtering is performed before converting the image to grayscale, since the areas of interest may differ significantly from the neighboring ones in color while being quite similar in brightness. This processing allows more efficient use of edge detection algorithms based on gradient values.

2.2   Contour Detection

For the contour detection, the image undergoes the following transformations:

1. Canny edge detection algorithm is applied [3].
2. The resulting image goes through the erode and dilate stages with a rectangular kernel.
3. Artificial edges are added along the boundaries of the image.
4. The Suzuki contour detection algorithm [14] is applied.

Erode and dilate stages are necessary to close small gaps in contours that may have emerged at the first stage due to insufficient difference in brightness and color between, for example, a sown field and a neighboring natural meadow. The rectangular core of erode and dilate is used because the sown fields that we are looking for in the image have pronounced corners that can be lost when using, for example, an elliptical core.

The edge that runs along the boundaries of the image is necessary in order to create closed contours for fields that have been separated in the middle by the image border, since we only consider the field as a closed region. Figure 1 shows the result of the described process.

3   Region Filtering

The resulting set of contours contains both the sown fields and unseeded fields, along with various natural objects, parts of urban infrastructure and various noise. To distinguish the sown fields from other elements of the set, we need to determine a vector of parameters for each contour, so we can make a decision on the object classification depending on its values.

3.1   Region Properties Selection

We select the following properties of the contour to compose the vector of its parameters:

Closedness. A contour is considered to be closed if its area exceeds its perimeter. The presence of an artificial edge at the borders of the image allows us to close all the fields near it. Any other contour that touches the border of the image in two places (but does not coincide with it) will also be closed.

Fig. 1. The obtained raster edges drawn on the original image. (Color figure online)

Area. The area of the region within a contour is an important parameter, since the required fields are industrial and take up a fairly large space. Knowing the scale of the image and having an expert estimate of the minimum size of the industrial field, we can cut off areas that are too small. In addition, due to the presence of edges at the borders of the image, we can get very large closed regions, which area will be comparable to the area of the entire image. They are also easily identified using this parameter.

Parent Region (if exists). If one region is entirely inside another, the larger region is considered the parent of the smaller region.

Minkowski-Bouligand Dimension. We estimate the Minkowski-Bouligand dimension of a region through counting the boxes of different sizes that completely cover the area. The sizes of the boxes are chosen as 2^k for all k from 2 to $p - 1$, where $p = \lfloor \log_2 m \rfloor$ and m is the size in pixels of the smaller side of the bounding rectangle of the region.

Median Brightness. We use the median brightness and not the average, since in some cases crops on the field can be partially collected or the field can be partially eroded, which affects the accuracy of the results.

Standard Deviation of Brightness Distribution. Since we locate the boundaries after bilateral filtering, a significant spread in brightness may indicate the presence of a color gradient within the area.

Number of Corners of the Approximating Polygon. We use the Ramer-Douglas-Peucker algorithm [6,11] to approximate the region with a polygon. In addition to counting the number of corners in it, this approximation also allows an efficient storage of the detected regions.

3.2   Initial Filtering

We use the following rules to mark some regions as ineligible for being considered a sown field:

– The region is open.
– The area of the region does not exceed a predetermined minimum threshold.
– The area of the region exceeds a predetermined maximum threshold.
– Minkowski-Bouligand of the region dimension exceeds the specified threshold.

The result of applying these basic filters is shown in Fig. 2.

Fig. 2. Regions left in the set after applying the initial filters. (Color figure online)

3.3   Region Clustering

Applying the Clustering Algorithm. The remaining regions undergo the clustering procedure. Since we do not know the number of clusters, and their sizes can vary greatly, we use the DBSCAN [7,13] algorithm to determine them. Of the remaining three parameters, not used in the previous steps, we use the median value and standard deviation of the area brightness to build the clusters. The clustering result is shown in Fig. 3.

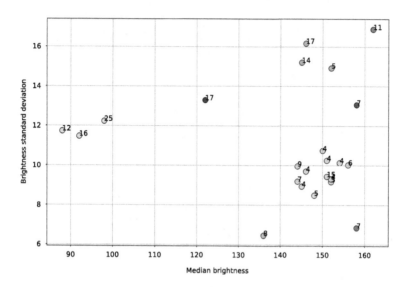

Fig. 3. The result of clustering the regions by the median value and standard deviation of brightness. Areas belonging to the same cluster are highlighted with the same color. Each point is annotated with a number of corners of the Ramer-Douglas-Peucker approximating polygon. (Color figure online)

Clustering Results Processing. For each of the resulting clusters, we calculate the average value of the standard deviation of brightness and the median number of corners of the approximating polygon. Among all the clusters, those for which the value of the median number of corners does not exceed 6 are selected (the fields tend to have a rectangular shape, but this is not always the case), and the cluster with the maximum number of elements in it is selected and marked as the "main cluster". This cluster is considered to contain the desired elements - the areas corresponding to the seeded fields. Among all the remaining areas, ones that meet the following conditions are selected:

- The median brightness of the region differs from the median brightness of the main cluster by no more than two standard deviations.
- The number of corners of the approximating polygon does not exceed 8.
- The parent area, if it exists, does not belong to the main cluster.

These areas are also marked as sown fields. For example, partially harvested or eroded fields fall into this category. The result can be seen in Fig. 4.

Fig. 4. Regions marked as "sown fields" after going through all the steps of the algorithm. Regions that belong to the main cluster are drawn in green. Regions that were subsequently selected are drawn in yellow. Ramer-Douglas-Peucker polygons are drawn in blue. (Color figure online)

4   Limitations and Disadvantages of the Method

4.1   Large Number of Parameters

The main disadvantage of the method is the large number of parameters, the values of which must be somehow determined before using it. According to the stages of processing the original image, they can be distributed as follows:

Preprocessing

– Image scale.
– Color and space restrictions of bilateral filtering.

Region Boundaries Locating

- Parameters of the Canny algorithm (in particular, upper and lower bounds for the hysteresis procedure).
- Number of iterations and kernel size for erode and dilate stages.

Region Filtering

- Minimum area size.
- Maximum area size.
- Maximum Minkowski-Bouligand dimension.

Region Clustering

- Parameters of the DBSCAN clustering algorithm (in particular, the maximum distance between points).

Clustering Results Processing

- Accuracy of polygon approximation by the Ramer-Douglas-Peucker algorithm.
- Maximum allowed median number of corners of an approximating polygons in the main cluster.
- Maximum number of corners of the approximating polygon for the appended regions.

While for some of these parameters (for example, the image scale, the minimum and maximum allowed sizes of the regions, the maximum allowed Minkowski-Bouligand dimension, the accuracy of the polygon approximation using the Ramer-Douglas-Peucker algorithm) sufficiently adequate values can be obtained using an expert judgment, or based on limitations of the initial data (for example, available image scales), for other parameters it is much more convenient to search for them by training the algorithm on a sample of known and pre-labeled data. We do not currently have such a sample.

4.2 Region Distorting

Image filtering allows to get rid of noise, but in the case of a fuzzy border, it can become more blurry even when using filtering algorithms that preserve borders. The erode and dilate stages allow to close open contours, but at the same time distort the boundaries. Both approaches represent a trade-off between false-negative and false-positive results, and the choice of their parameters greatly affects the quality of the algorithm.

In some cases roads are taken as a part of the region, both outside and inside of the actual boundaries.

4.3   Overlapping Regions

As can be seen from the central part of the Fig. 4, sometimes drawn regions are overlapping. In this case, the reason is that the region from the main cluster is inside of one of the appended regions. At the moment, we are not sure how to deal with such areas (leave one of them, leave both, calculate the difference between areas, or add a new heuristic to dynamically determine this behavior) and are awaiting for an expert opinion of the agricultural workers. A similar problem is observed in the case of one of the appended areas being the parent of another (in Fig. 4 this case can be seen in the lower-right corner).

5   Conclusion

The proposed algorithm has shown its fundamental applicability to the available source data. With its help, fields sown with agricultural crops can be automatically located and highlighted on satellite images with sufficient accuracy, and their boundaries are preserved. Its application to a single image makes it possible to estimate the use of arable lands and their parameters, while, by using the algorithm on the same region over the several years period, it is possible to monitor the state of the same agricultural land, and also estimate the number of fields subject to erosion.

Its disadvantages include a significant number of parameters, as well as some distortion of areas. One of the ways to reduce their impact on the quality of the algorithm is training and validation of the method on a training set.

6   Further Work

6.1   Obtaining a Training Set

The presence of a training set will allow finding the optimal parameters of the algorithm, objectively validating the quality of its work, and applying more complex algorithms for filtering the region set based on decision trees, clustering or classifying the objects by metrics which are now used in the rules based on the maximum and minimum thresholds.

6.2   Extended Expert Assessment

Additional involvement of experts in the field of agriculture can help with the selection of new metrics, improvement of existing ones, as well as with the selection or construction of a training set.

Acknowledgements. The proposed algorithm is implemented in Python using the OpenCV [2] image processing library and the Scikit-learn [10] machine learning library. The Matplotlib [8] library was used for graph visualization. We use Copernicus Open Access Hub [4] for satellite images.

References

1. Balinova, T.A., Muchkaeva, G.M., Dzharinov, O.V.: Socio-economic and environmental aspects of the development of the caspian region. In: Materials of the International scientific-practical conference, pp. 597–602. Kalmyk State University, Elista (2019)
2. Bradski, G.: The OpenCV library. Dr. Dobb's J. Softw. Tools **25**, 120–125 (2000)
3. Canny, J.: A computational approach to edge detection. IEEE Trans. Pattern Anal. Mach. Intell. PAMI **8**, 679–698 (1986)
4. Copernicus open access hub homepage (2021). https://scihub.copernicus.eu. Accessed 24 April 2021
5. Dedova, E.B., Goldvarg, B.A., Tsagan-Mandzhiev, N.L.: Land degradation of the republic of kalmykia: problems and ways of their restoration. Arid Ecosyst. **26**(2), 63–71 (2020). https://doi.org/10.24411/1993-3916-2020-10097
6. Douglas, D.H., Peucker, T.K.: Algorithms for the reduction of the number of points required to represent a digitized line or its caricature. Cartographica: Int. J. Geogr. Inf. Geovisualization **10**(2), 112–122 (1973)
7. Ester, M., Kriegel, H.P., Sander, J., Xu, X., et al.: A density-based algorithm for discovering clusters in large spatial databases with noise. Kdd **96**, 226–231 (1996)
8. Hunter, J.D.: Matplotlib: a 2d graphics environment. Comput. Sci. Eng. **9**(3), 90–95 (2007)
9. Lobell, D.B.: The use of satellite data for crop yield gap analysis. Field Crops Res. **143**, 56–64 (2013). https://doi.org/10.1016/j.fcr.2012.08.008, https://www.sciencedirect.com/science/article/pii/S0378429012002754, crop Yield Gap Analysis-Rationale, Methods and Applications
10. Pedregosa, F., et al.: Scikit-learn: machine learning in python. J. Mach. Learn. Res. **12**, 2825–2830 (2011)
11. Ramer, U.: An iterative procedure for the polygonal approximation of plane curves. Comput. Graph. Image Process. **1**(3), 244–256 (1972)
12. Sadeh, Y., Zhu, X., Chenu, K., Dunkerley, D.: Sowing date detection at the field scale using cubesats remote sensing. Comput. Electron. Agric. **157**, 568–580 (2019)
13. Schubert, E., Sander, J., Ester, M., Kriegel, H.P., Xu, X.: Dbscan revisited, revisited: why and how you should (still) use dbscan. ACM Trans. Database Syst. (TODS) **42**(3), 1–21 (2017)
14. Suzuki, S., et al.: Topological structural analysis of digitized binary images by border following. Comput. Vis. Graph. Image Process. **30**(1), 32–46 (1985)
15. Tomasi, C., Manduchi, R.: Bilateral filtering for gray and color images. In: Sixth International Conference on Computer Vision (IEEE Cat. No. 98CH36271), pp. 839–846. IEEE (1998)

BIM Interoperability and Data Exchange Support for As-Built Facility Management

Andrea di Filippo[1]([⊠]) [iD], Victoria Andrea Cotella[1] [iD], Caterina Gabriella Guida[1] [iD],
Victoria Molina[2] [iD], and Lorena Centarti[3]

[1] Dipartimento di Ingegneria Civile, Università degli Studi di Salerno, Fisciano, Italy
anddifilippo1@unisa.it
[2] Voyansi, Córdoba, Argentina
[3] Corbis Studio, Córdoba, Argentina

Abstract. Nowadays, the increasing complexity of buildings highlights the need for the architecture, engineering, construction, and operation (AECO) sector to manage a large amount of data. In this scenario, the building digitisation process offers the opportunity to create virtual databases to collect data from different disciplines efficiently. At this point, Building Information Modelling represents the innovative methodology able to manage a building in its entire life cycle.

Its application has been studied and documented in the design phase to construction, but it is important to underline its relevance in the Facility Management (FM) sector. While the literature highlights the many potentials, it also draws attention to the criticalities of using BIM for FM, such as inconsistent and organic asset data acquisition guidelines, inadequate knowledge integration due to different schemas, syntaxes and semantics, and lack of systematic cataloguing of different sources.

The contribution of this study is the identification of a generic set of information requirements for FM systems, which should be included in BIM As-Built models for efficient facility operations and maintenance using an open standard format such as Construction-Operations Building Information Exchange (COBie). To this aim, the process is structured in three steps starting with Data Acquisition with digital survey techniques to obtain a 3D point cloud. Then, Data Integration in a Revit platform, and finally, defining what information from the BIM model could be useful, Data Management to transfer the identified specific information requirements into FM systems.

Keywords: Digital survey · Scan-to-BIM · COBie · Information hierarchy

1 Introduction

1.1 State of the Art in Facility Management for Cultural Heritage

The spread of the BIM methodology is contributing substantially to revolutionising the way in which data relating to structures is managed and exchanged between all the actors involved in the processes of the Architecture, Engineering, Construction, and

O. Gervasi et al. (Eds.): ICCSA 2021, LNCS 12950, pp. 702–711, 2021.
https://doi.org/10.1007/978-3-030-86960-1_54

Operations (AECO) sector. In addition to gaining greater consistency between all the deliverables produced by the central model, professionals have been able to benefit from numerous functionalities such as object clash detection, cost analysis, quality take-off, energy analysis, optimised construction sequencing, all of which can be traced back to the so-called seven 'dimensions' of BIM [1–3]. The fast pace at which the methodology is being disseminated, especially for the design and construction phases, is stimulating interest in its use also during the stages following the commissioning of the building, such as facility management (FM), renovation, control of installations [4, 5].

In this way, it is possible to ensure the extension of BIM to the entire building lifecycle, resulting in reduced operating costs, access to structured information, the presence of a central database and less dependence on the workforce for data extraction [6].

While the literature highlights the many potentials, it also draws attention to the criticalities of using BIM for FM, such as inconsistent and organic asset data acquisition guidelines, inadequate knowledge integration due to different schemas, syntaxes and semantics, and lack of systematic cataloguing of different sources.

Most of these problems could be mitigated, if not completely solved, by using a standard for information management and exchange. Industry Foundation Class (IFC), a vendor-independent data model, is developed as a standardised digital description of assets for collaboration and information sharing throughout the project lifecycle.

Being designed to accommodate many different configurations and Levels of Development, the IFC is often too generic and complex [7]. The solution comes from the Model View Definition (MVD), a purpose-built subset of the IFC itself that filters the information required for a particular workflow. Some examples of such configurations contain the data required for clash detection, costing, fire risk management. The Construction Operations Building Information Exchange (COBie) is an MVD designed for FM and asset management [8] and collecting the minimum data required for such operations. This emphasises the need to collect and organise data already at the design stage to be used at later stages.

COBie is gaining the attention not only of the academic world and software houses but also of government bodies, as technicians have understood that the real innovation of BIM lies in the information component, which makes the final model something more than a simple three-dimensional visualization of the project. Such a change in perception should ensure awareness that there is no point in pushing for virtualisations that are overly rich in geometric detail (Level of Geometry - LOG) but instead, seek a balance between this and information content (Level of Information - LOI). To facilitate data acquisition for COBie, the guide and responsibility matrix help identify what information is needed, at what stages and from which stakeholders, highlighting a significant reliance on practitioners to ascertain that the data collected is in the right format and verified for consistency.

Some contributions point out that data management for FM is unfortunately disconnected from the design phases. This results in the loss of much information that could be useful in the operational phases, as it is not systematised properly [9]. An immediate consequence is that while the Level of Geometry increases along the life cycle (at least for new constructions requiring a direct engineering methodology), the LOI remains in the background and undergoes a strong growth only before the delivery of the building. A possible solution, proposed in some writings, foresees that the COBie datasheet is drafted in stages preceding the operational one, using the implementation of the guidelines and the responsibility matrix. The implementation of this solution is however costly and requires considerable effort in terms of human resources.

Implementing new information along the supply chain is understood as an evolution of the information and not as simple addition and therefore requires a consistency check. This operation is made even more complex by the fragmentation of data and their correlation [10].

1.2 Aims of the Contribution and Case Study

The present work aims to highlight the potentialities of BIM implementation for FM and operations, starting from the identification of a series of operative protocols that enrich the BIM content through the development of appropriate guidelines compatible with different case studies with the same model uses [11].

The results achieved may contribute to the creation of an operational background, composed of examples able to overcome the actual lack of standards and interoperability, making the proposed method more effective since the field of facility management and BIM are beginning to formulate just what one means to the other [12].

To investigate the proposed standardisation, the definition of a data pipeline of the BIM model used connected to the FM field becomes essential.

The main steps are three: the 3D point cloud obtained from laser scanner survey; the implementation of an As-Built Model for FM [13]; the integration with a COBie platform. These could be reached through the operational declination of the corresponding objectives of Data Acquisition, Data Integration and Data Management. In these final steps, it will be essential to filter the information content to optimise the exchange process and identify the data that would be useful to maintain the data structure.

The research case study, represented by *La Usina del Arte*, is a multidisciplinary cultural centre located in *La Boca* neighbourhood, in the city of Buenos Aires (Fig. 1). Built in 1916, the building was used as a power plant by the Italian-Argentine Electricity Company for many years. It was opened as a cultural centre in 2012 following an ambitious regeneration project, taking inspiration from international projects such as Park Avenue Armory in New York and Radial System in Berlin. The centre has a 1,200-capacity symphony hall with state-of-the-art acoustics and multiple spaces dedicated to dance and art exhibitions.

As already stated, the developed BIM guidelines for FM contain the investigated methodology standardisation defined as a set of activities related to the acquisition, integration, and management of models. The subsequent sections of the document are specifically aimed at proving the adopted procedure. Section 2 presents data acquisition methods with a Scan to BIM approach and management information in a FM environment, defining a procedural pipeline. Section 3 reports the results obtained and, finally, Sect. 4 closes this paper with the necessary reflections and conclusions.

2   Methods

2.1   Our Scan-to-BIM Approach

The creation of a BIM model for existing buildings is a reverse engineering methodology. Unlike new construction processes, where starting from a high level of abstraction and logical design rules, we arrive at the production of a physical object, the documentation of existing structures aims to analyse their functioning in detail, reconstructing their underlying logic, and deriving useful information for subsequent management phases. Regarding the reproduction of geometries, the literature offers interesting applications [14], but a holistic treatment is still lacking.

The technique opted for is Scan-to-BIM, a reverse modelling technique that uses digital sensing technologies to obtain point clouds that become the basis for BIM modelling. In detail, a four-stage workflow is applied: purpose identification, data acquisition, analysis, redesign.

Starting from the decision-making aspect, it is important to remember that the development of this BIM model is linked to a precise purpose, i.e., its connection to an information database that can correctly systematise the data and exchange it between the figures involved in the operational phase of the life cycle. Therefore, the goal is the production of a BIM model that can become the access tool to all the heterogeneous information gathered so far.

At this point, a Level of Development (LOD) for the virtual objects should be defined [15]. However, the first difficulties arise. Standard levels are conceived with reference to a forward engineering methodology, where geometric and informative contents increase as one moves from the idea to the real system.

Based on these observations, referring to an existing building surveyed and then modelled, one could be led to attribute the product to a LOD 500 (built structure), where the digital objects express the verified virtualisation of the single building elements (As-Built situation), containing the trace of the management, maintenance, repairs, replacements carried out during the whole life cycle.

If this direct correspondence can be valid for the geometrical aspect (we will see that the problem is more complex than it seems), the same is not true for the information content, which depends on the cognitive process. A solution to the problem could be to decouple the two aspects, identified as Level of Geometry (LOG) and Level of Information (LOI) which, however, cannot be treated separately. A major problem concerns geometry. Just think of the detailed knowledge of the stratigraphy that must be achieved to ensure a LOD 500. This is certainly a simple but revealing example.

Returning to the case study, based on the above considerations, we could indicatively attribute it a LOD 350, while reiterating the need to specifically define the contents necessary to describe the heritage and differentiating the geometric and material characterisation of the surfaces, attributable to a LOD 500.

Turning to the analysis, automatic data segmentation is not employed in the proposed application. The structure investigated required manual cataloguing of information, relying on the skills of qualified personnel capable of differentiating the elements studied by interweaving numerous parameters, which are difficult to manage simultaneously by an algorithm.

Fig. 1. Main entrance of *La Usina del Arte* in the city of Buenos Aires.

The final stage of the workflow involves redesign (Fig. 2). Whenever measured data undergo transformation into a digital object, a method of representation must be chosen. In our case study, we do not opt for a well-defined one, but we prefer to start by assuming a level of Represented Accuracy to be respected, corresponding to USIBD LOA 20 (1.5 cm–5 cm). In the first instance, the elements are modelled with a simple fit. Where the deviations between model and measured data exceeded this LOA, a detailed fit is applied (Fig. 3).

Fig. 2. Detail of the main entrance model with cloud overlay (Voyansi).

2.2   Information Management in a BIM Environment

One of the key issues for the proposed methodology is to define what information from the BIM model could be useful for FM [12]. In this regard, this procedure was considered. In the first instance, we proceeded studying the literature, standards, and best practice guidelines to define a set of information content. The second phase involved the use of a survey questionnaire based on the results of the first phase to prioritize the information content of BIM models to support FM systems specifically. The questionnaire was submitted to the client, asking them to rank the importance of the information requirements provided, with opportunities to add more.

As a result of the prioritisation phase, the following requirements needed to be registered to identifying objects: Manufacturer, Product model name and Product model number, included in Revit as type parameters, Unique equipment tag and an Uniformat code, which were instead included as instance parameters.

These information requirements were identified and ranked according to their potential to support FM systems, as suggested by the client. Based on these results, the next phase involved the development of a generic list of BIM model information contents to boost processes for identifying and exchanging data between BIM models and FM systems using an open data format. From a practical point of view, a plugin of Autodesk Revit, the platform used for parametric geometric modelling of the structure, was employed to export the predetermined information content in the form of COBie spreadsheetML file [16, 17].

Using COBie categories, the information provided in the Revit Parameters was linked to the common language of the COBie Spreadsheet using a properly written COBie.Type tab. The choice in the file format was tied up also to the fact that, in practice, the modelling teams do not coincide with those of information management, and these last ones often do not possess abilities of programming. Formatting as a spreadsheet allows this potential barrier to be overcome, while preserving the ability to interface with the IFC model. After creating the COBie spreadsheet based on the linked information from the BIM model, this was then imported into YouBIM, the FM management platform.

3   Results

For this project, the end goal is leveraging technology with the aim of creating an As-Built model and drawings that could be used for future reference in remodels or maintenance [18]. As for the acquisition campaign, a FARO Focus^{3D} X330 scanner is employed. To ensure complete coverage of the site, avoiding shaded areas, 300 stations are planned, with scans acquired at a resolution of about 6 mm at 10 m. In total it took three days of work for a team of 2 operators.

The single clouds are characterized by a high degree of overlap and for this reason are registered employing a global bundle adjustment procedure, accomplished after a top-view-based pre-registration. Given the set of scans, the first step involves the selection of a reference scan, those that will be considered fixed and those that can be moved. The algorithm searches for all the possible connections between the pairs of point clouds with overlap. For each connection, a pairwise ICP is performed defining a maximum search distance, the dimension of the sample on the cloud to register (number of control points)

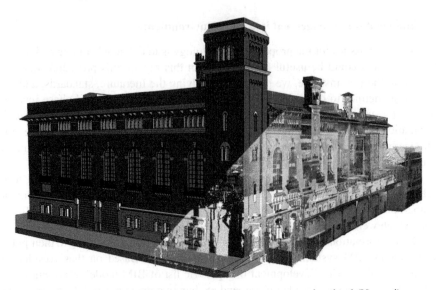

Fig. 3. Overlay of the BIM model with the laser scanner point cloud (Voyansi).

and the number of iterations. The best matching control point pairs between the two scans are saved. A final non-linear minimisation is run only among these matching point pairs of all the connections with the least square method. The global registration error of these point pairs is minimized, having as unknown variables the scan poses. The maximum value of the RMSE on the individual registration pairs is about 1.26 cm. Even the original versions of the scans are preserved to conduct more checks. The overall cloud consists of about 500 million points. For this reason, it has been divided into multiple regions to be imported smoothly into the software used for modelling. The team spent approximately 200 h to create an As-built BIM model using tools like Trimble and Edgewise to improve accuracy and speed. Based on the model, the team also created As-Built 2D drawings, including information related to room names, dimensions, gridlines, etc.

After this stage, the team had the challenge of starting from scratch with the asset data entry and development. All existing information was received in PDFs or images, so the team had to load them into the models through specific parameters, export the data as COBIe (Fig. 4) and finally upload contents to a FM Management software called YouBIM to manage the data.

As previously mentioned, one of our key tasks was to filter this content to streamline the interchange process and identify the data that would be useful when maintaining the structure. From the results of the analysis, it emerged that the most relevant data concerns the asset location, description and specification, room tags, system/equipment performances, maintenance history and other general information.

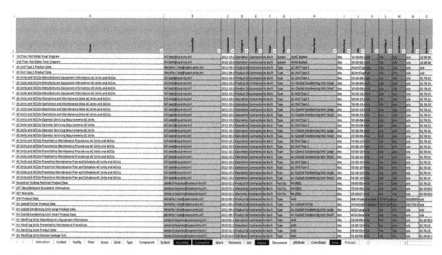

Fig. 4. Example of COBie spreadsheetML file exported from the BIM model (Voyansi).

4 Conclusions

This study aims to propose a combined BIM-FM methodology to manage the information content of a building during its life cycle. One of the goals is to identify a general set of BIM information requirements to support FM systems, which could be used as a basis for developing an automated information exchange process. Information is gathered and identified through a literature review of similar studies, existing standards, and best practices. These results are then organised into a survey questionnaire that includes information items, according to feedback from the client interviewed. The results identify some relevant information: asset content, room labels, system performance, and maintenance history.

Most of them are related to maintenance and asset management, all factors perceived as a priority by customers to ensure maximum utilisation of both facility and equipment assets. The investigation also developed the proposed process of data identification and exchange using an open standard format such as COBie to develop facility information requirements at an early stage, to check for compliance to these requirements and, finally, to transfer the identified specific information requirements into FM systems.

The distinctive element of the methodology is that the information is filtered already at an early stage, streamlining the output file from the model. This will reduce the redundant efforts of sorting and organising the COBie spreadsheet to extract the required data.

Another contribution is the identification of a generic set of information requirements for FM systems, which should be included in BIM As-Built models for efficient facility operations and maintenance. The study findings provide the basis of the integration of BIM with FM systems. Such integration will enable automated data exchange, which will support efficient operations and reduce costs.

Some future developments will be addressed to expand the catalogue of transferred information content, building for them reference categories useful for a better systematisation of data. A further step forward will be the automation of data export process from the model and the consequent correlation to these categories.

References

1. Choi, J., Kim, H., Kim, I.: Open BIM-based quantity take-off system for schematic estimation of building frame in early design stage. J. Comput. Des. Eng. **2**(1), 16–25 (2015). https://doi.org/10.1016/j.jcde.2014.11.002
2. Pärn, E.A., Edwards, D.J., Sing, M.C.P.: The building information modelling trajectory in facilities management: a review. Autom. Constr. **75**, 45–55 (2017)
3. Caetano, I., Leitão, A.: Integration of an algorithmic BIM approach in a traditional architecture studio. J. Comput. Des. Eng. **6**(3), 327–336 (2019)
4. Volk, R., Stengel, J., Schultmann, F.: Building Information Modeling (BIM) for existing buildings - literature review and future needs. Autom. Constr. **38**, 109–127 (2014)
5. Kumar, V., Teo, E.A.L.: Conceptualizing "COBieEvaluator": an application for data mining COBie datasets to track asset changes throughout project lifecycle. In: Proceedings of the International Congress and Conferences on Computational Design and Engineering 2019 (2019)
6. Becerik-Gerber, B., Jazizadeh, F., Li, N., Calis, G.: Application areas and data requirements for BIM-enabled facilities management. J. Constr. Eng. Manag. **138**(3), 431–442 (2012). https://doi.org/10.1061/(ASCE)CO.1943-7862.0000433
7. Koo, B., Shin, B.: Applying novelty detection to identify model element to IFC class misclassifications on architectural and infrastructure building information models. J. Comput. Des. Eng. **5**(4), 391–400 (2018). https://doi.org/10.1016/j.jcde.2018.03.002
8. William East, E., Nisbet, N., Liebich, T.: Facility management handover model view. J. Comput. Civ. Eng. **27**(1), 61–67 (2013). https://doi.org/10.1061/(ASCE)CP.1943-5487.0000196
9. Lavy, S., Liu, R., Issa, R.R.J.F.: Design for maintenance accessibility using BIM tools. Facil. **32**(3/4), 153–159 (2014)
10. Yana, A.A.G.A., Rusdhi, H.A., Wibowo, M.A.: Analysis of factors affecting design changes in construction project with Partial Least Square (PLS). Procedia Eng. **125**, 40–45 (2015)
11. Barba, S., Barbato, D., di Filippo, A., Napoletano, R., Ribera, F.: BIM-oriented modelling and management of structured information for cultural heritage. In: Agustín-Hernández, L., Vallespín Muniesa, A., Fernández-Morales, A. (eds.) Graphical Heritage, EGA 2020. Springer Series in Design and Innovation, vol. 5. Springer, Cham (2020). https://doi.org/10.1007/978-3-030-47979-4_54
12. Hardin, B., McCool, D.: BIM and Construction Management: Proven Tools, Methods, and Workflows. Wiley, Hoboken (2015)
13. Ugliotti, F.M., Osello, A., Rizzo, C., Muratore, L.: BIM-based structural survey design. Procedia Struct. Integrity **18**, 809–815 (2019)
14. Camagni, F., Colaceci, S., Russo, M.: Reverse modeling of cultural heritage: pipeline and bottlenecks. Int. Arch. Photogram. Remote Sens. Spat. Inf. Sci. **XLII-2/W9**, 197–204 (2019). https://doi.org/10.5194/isprs-archives-XLII-2-W9-197-2019
15. Alferieff, D., Bomba, M.J.B.: Level of Development (LOD) Specification Part I & Commentary for Building Information Models and Data, pp. 13–16 (2019)
16. Gouda Mohamed, A., Abdallah, M.R., Marzouk, M.: BIM and semantic web-based maintenance information for existing buildings. Autom. Constr. **116**, 103209 (2020)

17. Alnaggar, A., Pitt, M.: Towards a conceptual framework to manage BIM/COBie asset data using a standard project management methodology. J. Facil. Manage. **17**(2), 175–187 (2019). https://doi.org/10.1108/JFM-03-2018-0015

18. Eastman, C.M., Eastman, C., Teicholz, P., Sacks, R., Liston, K.: BIM Handbook: A Guide to Building Information Modeling for Owners, Managers, Designers, Engineers and Contractors. Wiley, Hoboken (2011)

Assessment and Management of Pharmaceutical Substances in the Environment

Reda Beigaite, Inga Baranauskaite-Fedorova, and Jolanta Dvarioniene$^{(\boxtimes)}$

Institute of Environmental Engineering, Kaunas University of Technology, Gedimino Street 50, 44239 Kaunas, Lithuania
{reda.beigaite,inga.baranauskaite}@ktu.edu,
jolanta.dvarioniene@ktu.lt

Abstract. The occurrence of pharmaceutical substances and its migration in aquatic ecosystems has recently raised a serious public concern as well as relevant environmental issues all over the world. The main objective of the research was to perform the flow analysis of the investigated substances and to provide recommendations on how to reduce pollution of the selected pharmaceuticals in Lithuania.

Keywords: Pharmaceuticals · Pollution · Environmental risk assessment · Sustainability · Management

1 Introduction

The 2030 Agenda for Sustainable Development, adopted by all United Nations Member States in 2015, provides a shared blueprint for peace and prosperity for people and the planet, now and into the future. The ambitious aims by 2030, substantially reduce the number of deaths and illnesses from hazardous chemicals and air, water and soil pollution and contamination [1]. In its communication introducing The European Green Deal (EGD) [2], the European Commission stressed the importance of transforming the EU's economy to a sustainable future, and to mainstream sustainability in all EU policies, linking it also to the United Nation's 2030 Agenda and the Sustainable Development Goals (SDGs). While an unprecedented worldwide effort is under way to fight the COVID-19 pandemic, the persistent threats to the health of our planet call for urgent remedies too [3].

In 2018, 5th of June decision of European Commission established a *Watch list* of substances for Union-wide monitoring in the field of water policy [4]. This list was for supporting future prioritization exercises was established in all EU countries. Pharmaceutical substances, such as *antibiotics - erythromycin, azithromycin and clarithromycin* were included in this list, in order to gather Union-wide monitoring data for facilitating the determination of appropriate measures to address the risk posed by those substances. As for water and soil pollution from pharmaceuticals, in addition to the EU target on antimicrobial sale reduction, the Commission will encourage international cooperation to address the environmental risks in other countries where pharmaceutical emissions

© Springer Nature Switzerland AG 2021
O. Gervasi et al. (Eds.): ICCSA 2021, LNCS 12950, pp. 712–717, 2021.
https://doi.org/10.1007/978-3-030-86960-1_55

from manufacturing and other sources may contribute, among other things, to the spread of antimicrobial resistance (AMR) [3].

Article 8c of the Priority Substances Directive (2008/105/EC1 as amended by Directive 2013/39/EU2) requires the European Commission to propose a strategic approach to the pollution of water by pharmaceutical substances. With the present Communication, the Commission is delivering on that legal obligation, as well as on a call referred to in the pharmacovigilance legislation to examine the scale of the problem of pollution of water and soils with pharmaceutical residues [6]. In particular to achieving SDG 6 on clean water and sanitation.

The overall quality of water in Lithuania's rivers has improved significantly over the past decade thanks to large investments in wastewater management system. However, the Curonian Lagoon, the Baltic Sea shore, about two-thirds of Lithuania's rivers and about one-third of lakes do not meet yet the requirements for a good status of water (share of water bodies of good water status accounted for 54%). The occurrence of pharmaceutical substances and its migration in aquatic ecosystems has recently raised a serious public concern as well as relevant environmental issues not only all over the world, but in Lithuania as well.

A holistic approach taking into consideration the sources of these substances, discharges and related risks for the environment, it is therefore essential to anticipate and manage those compounds in the environment. Until today, no many investigations were made regarding these pharmaceutical substances investigations in Lithuania. Therefore, made research is important for both the actual substances identification in the aquatic environment and the management with the aim to reduce the toxicological risk posed to the environment and human health.

2 Sources and Pathways

Pharmaceuticals are a widely recognized, effective treatment for many human and animal diseases. The 2019 EU Strategy on Medicinal Products for the Environment states that the EU currently allows about 3,000 active pharmaceutical ingredients. The availability of medicines for human and veterinary use has doubled in 6 years (2005–2014) [7]. The EU sells about 25% of medicines to humans and 31% to veterinary medicine. With the increase in the consumption of pharmaceutical active substances in society, the problem of increased pollution is observed. There is growing evidence of the release of active substances into the environment. Even low concentrations of pharmaceuticals can be toxic to humans and ecosystems.

Various pharmaceuticals, such as antibiotic residues, enter the aquatic and terrestrial environment through a variety of routes, such as municipal wastewater discharged from urban treatment plants. Due to the composition and properties of antibiotics, up to 90% of pharmaceutical residues enter the environment in their original form [7]. In this way, some pharmaceutical substances end up due to improper disposal of unused medicines (discharges into sinks or toilets). Another way is by spreading animal manure. It is one of the untreated sources of livestock pollutants. Pharmaceutical substances also reach the environment in other ways: directly from with industrial effluents, through treated or untreated wastewater discharged and sewage sludge, from veterinary services for pets and from improper disposal of drug-contaminated waste in landfills, animal feed and aquaculture ponds (Fig. 1) [8].

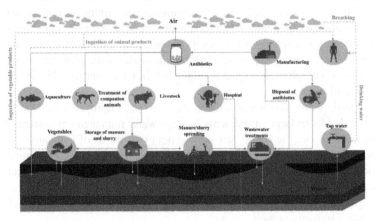

Fig. 1. Human exposure to antibiotic resistance associated with antibiotic residues in the environment [8]

Of all the other antibiotics in use in the world, the *macrolide antibiotics* are *erythromincin, clarithromycin and azithromycin*. The organic composition does not allow them to degrade naturally, so these antibiotics are also included in the group of persistent contaminants [9]. According to the Eurobarometer 478 survey published by the EC, about 31% of the Lithuanian population in 2017 used antibiotics in the form of tablets, powder or treacle. As many as 92% of respondents received antibiotics from a health care professionals. While 27% had a prescription, 65% received antibiotics directly from a doctor, 4% of respondents used available antibiotic residues, and as many as 3% purchased over-the-counter medications [10] (Table 1).

Table 1. Consumption of research materials in Lithuania in 2015–2018

Researched substance	Year	DDD/1000 inhabitant/per day
Erythromycin	2015	0,0167
	2016	0,0144
	2017	0,0114
	2018	0,0103
Clarithromycin	2015	1,42
	2016	1,55
	2017	1,63

(continued)

Table 1. (*continued*)

Researched substance	Year	DDD/1000 inhabitant/per day
	2018	1,68
Azithromycin	2015	0,506
	2016	0,433
	2017	0,470
	2018	0,459

3 Management of the Risks

According the results of the substance flow analyses (SFA) the highest consumption of pharmaceuticals occurs in households, where it is quite difficult to control wastewater (Table 2).

Table 2. Quantities of pharmaceutical substances in effluents in 2017 [11]

Pharmaceuticals	Concentrations before treatment, ng/l/Concentrations after treatment, ng/l				
	City 1	City 2	City 3	City 4	
Azithromycin	37,0/13,4	76,4/19,9	182,4/12,1	11,9/11,0	2017
Clarithromycin	126,5/229,3	474,8/150,2	1326,7/73,7	243,6/62,0	
Erythromycin	95,5/75,2	49,9/29,3	359,2/33,0	2,5/3,9	

A possible solution would be to separate domestic effluent streams from other streams in order to allow for analysis that is more extensive and monitoring in the environment to assess which streams generate the highest pollutant loads after domestic effluents. It would be possible to monitor individual flows more actively and to find new solutions to control them. For example, separating flows from hospitals and the outpatient sector could treat wastewater before handing it over to urban wastewater treatment plants (WWTPs). Assessing the efficiency of wastewater facilities could also lead to more efficient and easy-to-implement solutions for further treatment.

The scientific literature indicates that 50% to 67% of erythromycin is excreted in the environment with wastewater, with other antibiotics entering the environment in an amount of 30% to 90%, depending on their chemical composition [12, 13]. Because macrolide antibiotics have similar chemical composition and properties, it is estimated that approximately 60% of erythromycin, clarithromycin, and azithromycin enter the environment through the effluent.

The effects of the pharmaceutical substances on human health and the natural environment are not widely known. Monitoring of pharmaceuticals is limited and requires more accurate analytical solutions and methods. More information on the effects and concentrations of relevant substances in the aquatic environment could lead to more research.

It is important to educate the population and medical staff about the proper use of medicines, the consequences of improper unused pharmaceuticals disposal. It would be possible to draw the attention of the population to informative surveys about the risks of pharmaceuticals to the environment and their use. Encourage citizens to use antibiotics responsibly, to buy only prescription medicines from reliable sources, to stop using obsolete and unsuitable medicines, and not to share medicines with relatives or engage in self-medication. The government should organise the campaign to inform the public about the correct disposal of pharmaceuticals by delivering them to the pharmacy and about unauthorized disposal methods such as disposal in containers or disposal directly into sewer.

It is likely that new legal requirements regarding the generation of wastewater polluted by pharmaceutical substances will come very soon in country, as there are many different initiatives regarding the micropollutants, as well pharmaceuticals, in the waster cycles. A lot of attention should be paid to the qualitative parameters at the points of discharge wastewater to the sewage systems as well to the natural environment, as well use of the pharmaceuticals and discharge from the health care institutions.

Particular attention must be paid to informing consumers about not only the benefits of antibiotics, but also the damage they cause to the environment and to humans.

Acknowledgement. This research was developed in the framework of the INTERREG EUROPE programme project "Improving the European Rivers Water Quality through Smart Water Management Policies - BIGDATA4RIVERS" funded by the EU's European Regional Development fund through the INTERREG EUROPE, 2014–2020.

References

1. Transforming our world: the 2030 Agenda for Sustainable Development. Resolution adopted by the General Assembly on 25 September 2015
2. The European Green Deal. Communication from the Commission to the European Parliament, the European Council, the Council, the European Economic and Social committee and the Committee of the Regions. Brussels, 11.12.2019 COM(2019) 640 final
3. Communication from the Commission to the European Parliament, the Council, the European economic and social committee and the committee of the regions. Pathway to a Healthy Planet for All EU Action Plan: 'Towards Zero Pollution for Air, Water and Soil'. Brussels, 12.5.2021 COM(2021) 400 final
4. EC: Commission Implementing Decision (EU) 2018/840 of 5 June 2018 establishing a watch list of substances for Union-wide monitoring in the field of water policy pursuant to Directive 2008/105/EC of the European Parliament and of the Council and repealing Comm. Off. J. Eur. Union **L 141**(October 2000), 9–12 (2018)
5. Communication from the Commission to the European Parliament, the Council and the European economic and social committee. European Union Strategic Approach to Pharmaceuticals in the Environment. Brussels, 3 March, 2019. COM 2019 (128)
6. Regulation (EU) No 1235/2010 of the European Parliament and of the Council of 15 December 2010 amending, as regards pharmacovigilance of medicinal products for human use,

Regulation (EC) No 726/2004 laying down Community procedures for the authorisation and supervision of medicinal products for human and veterinary use and establishing a European Medicines Agency, and Regulation (EC) No 1394/2007 on advanced therapy medicinal products, OJ L 348, 31.12.2010, p. 1; Directive 2010/84/EU of the European Parliament and of the Council of 15 December 2010 amending, as regards pharmacovigilance, Directive 2001/83/EC on the Community code relating to medicinal products for human use, OJ L 348, 31.12.2010, p. 74

7. Communication from the Commission to the European Parliament, the Council and the European economic and social committee. European Union Strategic Approach to Pharmaceuticals in the Environment. Brussels, 11.3.2019 COM(2019) 128 final

8. Ben, Y., Caixia, F., Min, H., Liu, L., Wong, M.H., Zheng, C.: Human health risk assessment of antibiotic resistance associated with antibiotic residues in the environment: a review. Environ. Res. **169**, 483–493 (2019)

9. Islas-Espinoza, M., Aydin, S., de las Heras, S., Ceron, C.A., Martínez, S.G., Vázquez-Chagoyán, J.C.:Sustainable bioremediation of antibacterials, metals and pathogenic DNA in water. J. Cleaner Prod. **183**, 112–120 (2018)

10. Eurobarometer, Special Eurobarometer 478: Antimicrobial Resistance, November 2016

11. Determination of the Regional Pharmaceutical Burden in 15 Selected WWTPs and Associated Water Bodies using Chemical Analysis. Status in four coastal regions of the South Baltic Sea Germany, Lithuania, Poland and Sweden. Project MORPHEUS 2017–2019, p. 126 (2020)

12. Jessick, A.M.: Detection, fate, and bioavailability of erythromycin in environmental matrices, pp. 1–99 (2010)

13. Aydin, S., Aydin, M.E., Ulvi, A., Kilic, H.: Antibiotics in hospital effluents: occurrence, contribution to urban wastewater, removal in a wastewater treatment plant, and environmental risk assessment. Environ. Sci. Pollut. Res. **26**(1), 544–558 (2018). https://doi.org/10.1007/s11356-018-3563-0

The Impact of Corrosion on the Seismic Assessment of Reinforced Concrete Bridge Piers

Alessandro Rasulo[1]([⊠]) ⓘ, Angelo Pelle[2] ⓘ, Giuseppe Quaranta[3] ⓘ,
Davide Lavorato[2] ⓘ, Gabriele Fiorentino[2] ⓘ, Camillo Nuti[2] ⓘ,
and Bruno Briseghella[4] ⓘ

[1] Department of Civil and Mechanical Engineering, University of Cassino and Southern Lazio,
03043 Cassino, FR, Italy
alessandro.rasulo@unicas.it
[2] Department of Architecture, Roma Tre University, 00153 Roma, Italy
{angelo.pelle,davide.lavorato,gabriele.fiorentino,
camillo.nuti}@uniroma3.it
[3] Department of Structural and Geotechnical Engineering, Sapienza University of Rome, 00184
Rome, Italy
giuseppe.quaranta@uniroma1.it
[4] College of Civil Engineering, Fuzhou University, Fuzhou 350108, China
bruno@fzu.edu.cn

Abstract. Reinforced Concrete (RC) bridges are a key elements for any transportation network, so that their safety and serviceability is of crucial importance especially after the occurrence of a seismic event. Frequently, RC bridges are exposed to the action of aggressive chemicals, such as in marine or in cold climate environments, for the presence of chloride ions, either due to the salty water or the use of deicing salt. Generally those aggressive environmental actions affect the steel rebars inducing corrosion phenomena with a degradation the mechanical properties of reinforcement. In this paper some modeling issues related to corroded RC piers are presented and discussed, with particular emphasis to their seismic assessment.

Keywords: Earthquake engineering · Bridges · Seismic assessment · Reinforced Concrete · Rebar corrosion

1 Introduction

In most of the earthquake-prone European countries like Italy, the seismic vulnerability of road infrastructures is a topic of serious economic and social concern [16, 17, 24]. The most vulnerable elements within a road network are indubitably the bridges.

Due to the fact that the infrastructures have been built in the economic boom after the Second World War (from late 1950s to early 1960s), when most of the territory was not yet recognized to be subject to earthquakes, in the design process the seismic action was completely neglected. Even when the sites were classified as seismically prone, the

© Springer Nature Switzerland AG 2021
O. Gervasi et al. (Eds.): ICCSA 2021, LNCS 12950, pp. 718–725, 2021.
https://doi.org/10.1007/978-3-030-86960-1_56

level of the seismic demand was much lower than the one currently adopted in design codes [2, 5, 6, 22, 23, 25, 26].

Furthermore, details should be considered substandard according to modern design practice. The stirrup spacing in the piers of existing bridges exceeds the maximum stirrup spacing limit in current seismic design codes. According to the reconnaissance surveys on structures damaged during recent severe earthquake, the longitudinal steel rebars have been found to have buckled due to the insufficient confinement offered by cover concrete and transversal stirrups.

Finally, since existing bridges have been functional along the infrastructures for an extended period of time, their performances can be affected by specific aging factors that may have induced degradation phenomena in the materials. In the case of reinforced concrete bridge piers, corrosion of steel rebars is undoubtedly the most important degradation phenomenon for RC bridges piers, due to the chloride ion attack from marine, airborne and deicing sources.

The corrosion phenomena influence directly the mechanical properties of steel rebars in terms of strength and ductility, negatively affecting the serviceability and capacity of RC bridges and increasing their vulnerability in earthquakes.

The most common corrosion morphologies over steel rebars may be either uniform or localized. The former is mainly due to carbonation and reduces equally the diameter of the steel bars over its full length, while the latter, mainly due to the action of chloride ions penetrated in concrete, is characterized by localized cavities and notches, known as pits, over the reinforcing surface. Pitting corrosion may appear in a heterogeneous way along the length of the reinforcement resulting in a more severe attack than a uniform one. Localized cavities, in fact, not only reduce the strength of the reinforcing bars but also their ductility. Indeed, if the steel bar is stretched by a tensile force, the deformation will be concentrated in deeper notches. Thus, the overall elongation will be developed in small zones and, consequently, the average strain will be less at failure compared to an uncorroded bar.

When compressed, longitudinal bars, if not properly retained by transversal reinforcement, may buckle and corrosion is expected to increase the likelihood of premature bar buckling, by reducing the average bar diameter and increasing the buckling length of reinforcement. Even though Pelle et al. [19] showed that the pattern of corrosion may or may not lead bar buckling depending on the pit configuration. In addition, the transversal reinforcement is less protected by the concrete cover, so that is subjected to corrosion more than longitudinal one, therefore corroded stirrups have a reduced ability to retain the longitudinal reinforcement, making easier local and global bar buckling.

In certain circumstances, a strong reduction of cross-section of the transversal rebars may facilitate shear failure.

The present study aims at contributing to the assessment of existing reinforced concrete (RC) bridge piers under seismic loading, when affected by corrosion problems. A fiber beam element implemented in OpenSEES [11] is proposed as a numerical tool able to provide both computational efficiency and appropriate accuracy.

2 Numerical Investigations

The RC pier used for the present numerical investigation, even if not referring to any specific structure, is representative of many real bridge columns existing along Italian roads. It has been placed in Reggio Calabria (Italy), where expected PGA (peak ground acceleration) is 0.35 g.

The pier has a diameter of 2.5 m and a height of 7.0 m. The steel rebars have yield strength equal to 450 MPa and the ultimate strain for uncorroded rebars is assumed to be equal to 0.075.

Two longitudinal reinforcement configurations have been considered, as shown in Fig. 1 and Table 1 (marked as Type A and B); in all of them the concrete cover was assumed to be 50 mm measured from the external surface to the centroid of spiral reinforcement. The longitudinal reinforcement ratio ρl for the two reinforcement configurations is about $\rho l = 3.0\%$. The axial load ratio is $v = 15.0\%$ (P = 13936.7 kN).

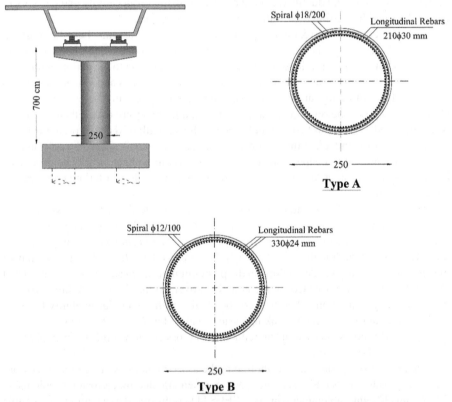

Fig. 1. RC pier bridge geometry and reinforcements: For description of reinforcement arrangements see Table 1.

Table 1. Pier Geometry and reinforcement arrangements.

Type	Longitudinal reinforcement	Transverse reinforcement
A	210ϕ30 – on 2 layers	2 ϕ18 mm spiral/200
B	330ϕ24 – on 3 layers (2nd and 3rd in adherence)	2 ϕ12 mm spiral/100

3 Numerical Model

The finite element model adopted in this research has been developed using the computation framework OpenSEES (Open System for Earthquake Engineering Simulation). The model, as schematically depicted in Fig. 2, has been developed according to Kashani et al. [8]: it is constituted by two nonlinear force-based beam-column elements, used to model the flexural response and a zero-length element used to model the bonding response.

The first element, in the proximity of the plastic hinge zone, has three Gauss-Lobatto integration points and its length is 6 times the longitudinal steel buckling length L_b (herein assumed equal to the spiral pitch). The second element has five integration points (Fig. 2a). The behavior of the beam-column is obtained through the integration of the responses obtained at the section level. The element cross-sections are discretized in fibres (Fig. 2b).

Since it is expected that bar buckling will occur on the first critical section of the column, the modified Monti and Nuti with corrosion stress-strain law (SteelMN implemented in OpenSEES by the authors), which is able to account for bar buckling, is employed to represent the response of the steel bars in the element 1. Instead, the constitutive model by Menegotto and Pinto (Steel02) is adopted for the bars in the 2nd element. Variations of steel bar mechanical proprieties due to corrosion are considered only for the outermost layer.

The corrosion level at different time of exposure is evaluated by a multiphysics approach as explained in [19, 21]. In Table 2 is reported the loss cross-section area for the considered cases. To account for the effect of strain concentration due to pitting corrosion which reduces the reinforcements ductility, the method proposed by Imperatore et al. [7] is employed. The greater slenderness ratio ($\lambda = L/D$, where L is the spiral pitch and D is the diameter of the longitudinal reinforcement) of the corroded longitudinal rebars under compression is accounted for as suggests in [3, 9, 14, 15, 27].

The concrete is modelled by the Concrete04 uniaxial material, the concrete on the section cover is considered unconfined, whilst the concrete in the section core is considered as confined, using the Mander et al. model [10, 20]. The effect of concrete cover cracking and spalling due to rust expansion is also taken into account reducing the strength of the cover concrete as suggest by Coronelli and Gamabrova [1], no variations of mechanical proprieties of the confined concrete core are considered since internal spiral confinement is expected to be uncorroded.

Table 2. Pier geometry and reinforcement arrangements.

Case	Time [years]	Reinforcement area reduction A(t)/A0 [%]	
		Longitudinal reinf.	Transversal reinf.
A.0, B.0	0.00	100.0%	100.0%
A.68	68.0	89.3%	60.5%
B.68	68.0	81.5%	27.2%

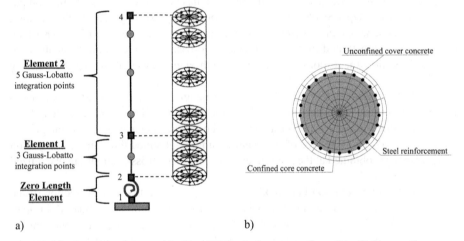

a)   b)

Fig. 2. Model implemented in OpenSEES: a) element configuration, b) fibre section.

The bonding is responsible for the extra displacement due to the slippage of the longitudinal reinforcing bars anchored in the concrete foundation. A linear moment-rotation relationship has been used for the rotational slip spring placed at the bottom of the column.

Table 2 reports a general overview of the numerical investigations considered in the present paper: the cases are designed by a two character designation, the first character (the letter A, B) specifies the section typology (as indicated in Table 1), the second one the time of exposure to the ageing agents (0 representing the uncorroded case). The cyclic history displacement applied to the top of the column is plotted in Fig. 3.

4 Results

In Fig. 4 the pier shear-drift behaviors for the uncorroded and corroded cases are compared. For Typer A (Fig. 4a) there are slight differences between the uncorroded and

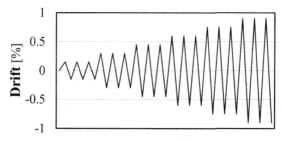

Fig. 3. Cyclic displacements applied to the fiber-element model.

corroded responses. This may be attributed to the modest loss cross-section area of the longitudinal reinforcement due to pitting corrosion. In fact, there is a loss of about 10% at 68 years of exposure in the external circumference while the internal remains uncorroded. Similar conclusions can be drawn for Type B (Fig. 4b), despite a greater loss cross-section area (about 20% at 68 years of exposure for the external layer) only one-third of longitudinal rebars are expected to be corroded.

It is important to underline that the external spiral reinforcements are heavily affected by larger corrosion with respect to longitudinal reinforcement. Therefore, the pier may undergo a shear failure in the elastic branch or immediately after the formation of the plastic hinge [18]. The FE model developed herein is not able to directly account for shear behavior. However, the shear strengths for the corroded pier evaluated with the model of Priestley [18] and with a truss model ($\theta = 45°$) are also reported in Fig. 4.

The latter model is quite conservative than the former one. It is evident that in Type B the failure may be in shear, while in Type A the shear strength is still higher than shear demand. This is due to the different transverse arrangements; the greater is the spiral reinforcement diameter (18 Vs 12), the lower is the damage due to corrosion.

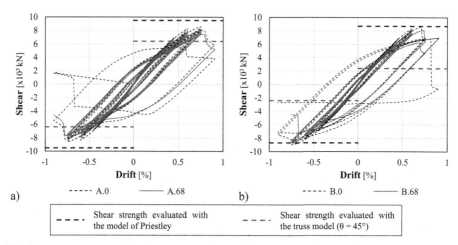

Fig. 4. Comparison between uncorroded and corroded behaviours. a) pier shear-drift responses for Type A, b) pier shear-drift responses for Type B. Shear strength is evaluated only for corroded cases.

5 Conclusions

The present paper aimed to investigate the effect of corrosion on different steel arrangement. The cases of bridge columns with code complying for design for corrosion are investigated at time 0 (zero) and after 68 years. The numerical model of the RC pier proposed needs of further validation against experimental data.

The time-dependent evolution of steel corrosion was evaluated by a Multiphysics approach which require a large computational effort, but some simplification can be introduced to reduce the elaboration time.

The following conclusions can be drawn:

- transversal reinforcement are more affected by corrosion than longitudinal ones due to minor concrete cover depth and a lower diameter. Moreover large longitudinal diameter adoption reduces the amount of corrosion;
- shear failure may arise in the steel arrangement B due to a smaller diameter of the transversal reinforcement than arrangement A;
- longitudinal reinforcement arranged on multi-layer may mitigate the effect of the steel corrosion.

References

1. CEB-FIB. Model Code 2010-Final Draft. The International Federation for Structural Concrete (FIB CEB-FIP), Lausanne, Switzerland (2010)
2. Decanini, L., De Sortis, A., Goretti, A., Langenbach, R., Mollaioli, F., Rasulo, A.: Performance of masonry buildings during the 2002 Molise, Italy, earthquake. Earthq. Spectra **20**(S1), 191–220 (2004)
3. Dhakal, R.P., Maekawa, K.: Modeling for postyield buckling of reinforcement. J. Struct. Eng. **128**(9), 1139–1147 (2002)
4. Eligehausen, R., Bertero, V.V., Popov, E.P.: Local bond stress-slip relationships of deformed bars under generalized excitations: Tests and analytical model. Report No EERC., pp. 83–23, Earthquake Engineering Research Center, University of California: Berkeley, CA, USA (1983)
5. Grande, E., Rasulo, A.: A simple approach for seismic retrofit of low-rise concentric X-braced steel frames. J. Constr. Steel Res. **107**, 162–172 (2015)
6. Grande, E., Rasulo, A.: Seismic assessment of concentric X-braced steel frames. Eng. Struct. **49**, 983–995 (2013)
7. Imperatore, S., Rinaldi, Z.: Experimental behavior and analytical modeling of corroded steel rebars under compression. Constr. Build. Mater. **226**, 126–138 (2019)
8. Kashani, M.M.: Size effect on inelastic buckling behavior of accelerated pitted corroded bars in porous media. J. Mater. Civ. Eng. **29**(7), 04017022 (2017)
9. Lavorato, D., et al.: A corrosion model for the interpretation of cyclic behavior of reinforced concrete sections. Struct. Concr. **21**(5), 1732–1746 (2020). https://doi.org/10.1002/suco.201900232
10. Mander, J.B., Priestley, M.J., Park, R.: Theoretical stress-strain model for confined concrete. J. Struct. Eng. **114**(8), 1804–1826 (1988)
11. McKenna, F., Fenves, G.L., Scott, M.H., Jeremic, B.: Open System for Earthquake Engineering Simulation (OpenSEES). University of California, Berkeley, CA, USA (2000)

12. Meda, A., Mostosi, S., Rinaldi, Z., Riva, P.: Experimental evaluation of the corrosion influence on the cyclic behaviour of RC columns. Eng. Struct. **76**, 112–123 (2014)
13. Menegotto, M., Pinto, P.E.: Method of analysis of cyclically loaded RC plane frames including changes in geometry and non-elastic behavior of elements under normal force and bending. International Association for Bridge and Structural Engineering, vol. 11, pp. 15–22 (1973)
14. Monti, G., Nuti, C.: Nonlinear cyclic behavior of reinforcing bars including buckling. J. Struct. Eng. **118**(12), 3268–3284 (1992)
15. Monti, G., Nuti, C.: Un modello analitico del comportamento ciclico di barre in acciaio con svergolamento post-elastico. Dipartimento di ingegneria strutturale e geotecnica, Università degli studi di Roma "La Sapienza", Report No. 6 (1991)
16. Nuti, C., Rasulo, A., Vanzi, I.: Seismic safety evaluation of electric power supply at urban level. Earthq. Eng. Struct. Dyn. **36**(2), 245–263 (2007)
17. Nuti, C., Rasulo, A., Vanzi, I.: Seismic safety of network structures and infrastructures. Struct. Infrastruct. Eng. Maintenance Manag. Life-Cycle **6**(1–2), 95–110 (2010)
18. Paulay, T., Priestley, M.N.: Seismic Design of Reinforced Concrete and Masonry Buildings, 1st edn. Wiley, Hoboken (1992)
19. Pelle, A., et al.: Time-dependent cyclic behavior of RC bridge columns under chlorides-induced corrosion and rebars buckling. submitted to Structural Concrete (2021)
20. Popovics, S.: A numerical approach to the complete stress-strain curve of concrete. Cem. Concr. Res. **3**(5), 583–599 (1973)
21. Quaranta, G., et al.: Multiphysics finite element analysis and capacity assessment of reinforced concrete bridge piers exposed to chlorides. In: Di Menegotto, P. (eds.) Proceedings of the Italian Concrete Days (ICD) 2020, Naples, 14–16 April 2021 (2021)
22. Rasulo, A., Testa, C., Borzi, B.: Seismic risk analysis at urban scale in Italy. In: Gervasi, O., et al. (eds.) ICCSA 2015. LNCS, vol. 9157, pp. 403–414. Springer, Cham (2015). https://doi.org/10.1007/978-3-319-21470-2_29
23. Rasulo, A., Fortuna, M.A., Borzi, B.: A seismic risk model for Italy. In: Gervasi, O., et al. (eds.) ICCSA 2016. LNCS, vol. 9788, pp. 198–213. Springer, Cham (2016). https://doi.org/10.1007/978-3-319-42111-7_16
24. Rasulo, A., Goretti, A., Nuti, C.: Performance of lifelines during the 2002 Molise, Italy, earthquake. Earthq. Spectra **20**(S1), 301–314 (2004)
25. Rasulo, A., Pelle, A., Lavorato, D., Fiorentino, G., Nuti, C., Briseghella, B.: Finite element analysis of reinforced concrete bridge piers including a flexure-shear interaction model. Appl. Sci. **10**(7), 2209 (2020). https://doi.org/10.3390/app10072209
26. Rasulo, A., Pelle, A., Lavorato, D., Fiorentino, G., Nuti, C., Briseghella, B.: Seismic assessment of reinforced concrete frames: influence of shear-flexure interaction and rebar corrosion. In: Gervasi, O., et al. (eds.) ICCSA 2020. LNCS, vol. 12252, pp. 463–478. Springer, Cham (2020). https://doi.org/10.1007/978-3-030-58811-3_34
27. Zhou, Z., Nuti, C., Lavorato, D.: Modeling of the mechanical behavior of stainless reinforcing steel. In: Proceedings of the 10th Fib International PhD Symposium in Civil Engineering, Québec, QC, Canada, pp. 21–23 (2014)

Author Index